SEVENTH EDITION

STATISTICAL METHODS

Statistical Methods

SEVENTH EDITION

GEORGE W. SNEDECOR
Professor Emeritus of Statistics
and Former Director, Statistical Laboratory
Iowa State University

WILLIAM G. COCHRAN
Professor Emeritus of Statistics
Harvard University

THE IOWA STATE UNIVERSITY PRESS
Ames, Iowa, U.S.A.

William G. Cochran was professor emeritus of statistics, Harvard University, at the time of his death in 1980. He served formerly on the faculties of Johns Hopkins University, North Carolina State University, and Iowa State University. He was a member of the National Academy of Sciences and of the American Academy of Arts and Sciences; he held honorary LLD degrees from Glasgow University and Johns Hopkins University. He was past president of the International Statistical Institute, the American Statistical Association, the Institute of Mathematical Statistics, and the Biometric Society. His writings include many research papers in professional journals and the books *Sampling Techniques,* 3rd ed., 1977, and *Experimental Designs* (with Gertrude M. Cox), 2nd ed., 1957.

Before his death in 1974, **George W. Snedecor** was professor emeritus of statistics, Iowa State University, where he taught from 1913 to 1958 and where he was for fourteen years director of the statistical laboratory. His writings include a body of scientific journal articles, research bulletins, and books, including *Correlation and Machine Calculation* (with H. A. Wallace) and *Calculation and Interpretation of Analysis of Variance and Covariance.* He held a master of science degree from the University of Michigan and honorary doctor of science degrees from North Carolina State University and Iowa State University. He was a member of the International Statistical Institute, past president of the American Statistical Association, and an honorary Fellow of the British Royal Statistical Society. He served also as consultant, Human Factors Division, U.S. Navy Electronics Laboratory, San Diego, California.

The publisher gratefully acknowledges the kind cooperation and invaluable help of **David F. Cox,** professor of statistics, Iowa State University, in completing the remaining authorial work on the seventh edition after the untimely death of William Cochran.

Composed by Science Press, Division of Monroe Printing Co., Inc.

Sixth edition by George W. Snedecor and William G. Cochran, © 1967 The Iowa State University Press; first through fifth editions by George W. Snedecor, © 1937, 1938, 1940, 1946, 1956 The Iowa State University Press
Seventh edition, 1980
Second printing, 1982

Library of Congress Cataloging in Publication Data

Snedecor, George Waddel, 1881-
Statistical methods.
Includes bibliographies and index.
1. Statistics. I. Cochran, William Gemmell, 1909- joint author. II. Title.
QA276.12.S59 1980 519.5 80-14582
ISBN 0-8138-1560-6

CONTENTS

Chapter 7. The binomial distribution

Chapter 8. Shortcut and nonparametric methods

Chapter 9. Regression

Chapter 10. Correlation

A SHORT COURSE IN THE ELEMENTS OF STATISTICAL METHODS

PREFACE

EARLIER EDITIONS of this book have been used extensively both as texts for introductory courses in statistics and as reference sources for statistical techniques needed by those who collect and analyze data. As a text, the book contains ample material for a course extending throughout the academic year. A suggested list of topics for a one-term course precedes the preface. In most of the book the mathematical level required is little more than elementary algebra.

In this edition some rewriting and rearranging has been done in the hope of making the book easier to use and to make the chapters somewhat shorter. Chapter 1 now introduces the types of data with which the book mainly deals—sample surveys, controlled experiments, and comparative observational studies—plus the concept of random sampling. Then follow chapters on frequency distributions, the mean and standard deviation, the normal distribution, and tests of hypotheses.

Later chapters follow the same general course as in the sixth edition except that the material on the random effects model in the analysis of variance is now in a separate chapter, as is the material on failures in the assumptions in the analysis of variance.

The calculations for fitting a multiple linear regression have been changed to the use of the sweep operator. This is a routine method that is easily learned and easily programmed on an electronic computer. Also, by the use of dummy X-variables to represent classifications, the calculations for analyses of variance and covariance can be done by a computer program for fitting multiple linear regressions. Teachers are not agreed on whether a standard programming language should be learned and used for an introductory course in statistics. Perhaps the most common technique at present is to construct simple programs that can be used by the students without learning any programming language and that will do the calculations for the numerical examples in the introductory course. No programs are presented in this edition, but comments are made from time to time on how computers are likely to present results and on calculations that may be available on demand in computer outputs.

Topics that appear for the first time in this edition are:

Probability paper (section 2.6)
The probability of at least one success in a series of trials (section 7.5)
Levene's robust test for the equality of a set of estimated variances (section 13.11)
Balancing the order in which treatments are given in a repeated measurements experiment (section 14.9)
Simultaneous study of the different effects of a transformation in the analysis of variance (section 15.15)
Yates' algorithm in factorial experiments (section 16.12)
Experiments with repeated measurements (section 16.16)
Mallows C_p criterion for choosing a subset of predictor variables in multiple regression (section 17.14)
Nonsampling errors in sample surveys (section 21.13)

A number of tables for the detection of outliers have been added: Table A 16 of the 5% and 1% levels maximum normed residual $\max|X - \bar{X}| / \sqrt{\Sigma (X - \bar{X})^2}$ for a normal sample and of the related quantity $\max|d| / \sqrt{\Sigma d^2}$ for independent normal ds; table A 18 of the significance levels of the maximum normed residual in a two-way classification; table A 17 of Dixon's criteria for quick tests of outliers. In addition, some previous tables have been expanded.

A list of symbols frequently used in this book follows the preface.

For their generous permission to reprint tables we are indebted to the late Sir Ronald Fisher and his publisher, Oliver & Boyd; to Maxine Merrington, Catherine M. Thompson, Joyce M. May, E. Lord, and E. S. Pearson, whose work was published in *Biometrika;* to the late C. I. Bliss. E. L. Crow, W. J. Dixon, S. Katti, W. von Türk (formerly Stefansky), C. White, R. Wilcox, and the late F. Wilcoxon; to Bernard Ostle and his publisher, the Iowa State University Press; and to the American Mathematical Society. Thanks are due also to N. R. Draper and W. G. Hunter for permission to reproduce one of their figures, to the many investigators who made data available for this book as illustrative examples, and to teachers who gave advice from their experience in using prior editions as texts. Finally, we wish to thank Holly Grano for her fine work in the typing and preparation of this edition.

William G. Cochran

Symbols frequently used

General

Greek letters $\alpha, \beta, \ldots$: population parameters
Latin letters $a, b, \ldots$ or symbols $\hat{\alpha}, \hat{\beta}, \ldots$: sample estimates
X, Y: observations
μ_X, μ_Y: population means; $\bar{X}, \bar{Y}$: sample means
$x = X - \bar{X}$: deviations from sample means
n: size of sample

a	$\Sigma\lvert x\rvert/(ns)$ = mean deviation/standard deviation in test for kurtosis
α_i, A_i	effect of ith class with fixed and random effects
$[A]$	factorial effect total for factor A
β_i, b_i	regression coefficient of Y on X_i in multiple regression
$\sqrt{\beta_1} = \gamma_1$	$E(X - \mu)^3/\sigma^3$, measure of skewness
$\sqrt{b_1} = g_1$	$m_3/m_2^{3/2}$, sample estimate of measure of skewness
c_{ij}	multipliers used in multiple regression: $c_{ij}\sigma_\epsilon^2 = \text{Cov}(b_i b_j)$
C_p	Mallows' criterion for selecting a set of variates for prediction
χ^2	chi-square
cv	coefficient of variation: σ/μ or $s/\bar{X}$
D_i	$X_{1i} - X_{2i}$ in a paired sample ($i = 1, 2, \ldots, n$)
$d_{y.x}$	deviation of Y from its sample regression on X
$d_{y.12\ldots p}$	deviation of Y from its sample regression on $X_1, X_2, \ldots, X_p$
E	expected value: $E(x)$ = expected value (mean value) of X
ϵ	random residual from a model
e	error of measurement in X; also e = 2.71828, base of natural logs
f	observed frequency in a class
$f(X)$	function of X; e.g., X^3, e^X, log X
F	expected frequency
F	variance ratio, s_1^2/s_2^2
γ_2, g_2	measures of kurtosis: $\gamma_2 = E(X - \mu)^4/\sigma^4 - 3$; $g_2 = m_4/m_2^2 - 3$
I	class interval width
L	linear function: $\Sigma\lambda_i X_i$ or $\Sigma L_i X_i$, where λ_i, L_i are numbers
$m_1, \ldots, m_4$	first four sample moments: $m_i = \Sigma(X - \bar{X})^i/n$
n_h	harmonic mean

N	size of finite population
ν	number of degrees of freedom
P	probability
p	probability of success in a binomial population: $q = 1 - p$
Q	Studentized range; (sample range)$/s$
ϕ	n/N, sampling fraction
ρ, r	correlation coefficient between Y and X
ρ_I, r_I	intraclass correlation coefficient
r_s	Spearman's rank correlation coefficient
R^2	square of the multiple correlation coefficient
R_p	reduction in Σy^2 due to regression on $X_1, X_2, \ldots, X_p$
σ, s	standard deviation
$\sigma_{y.x}$, $s_{y.x}$	standard deviation of residuals from regression of Y on X
$\sigma_{\bar{X}}$, $s_{\bar{X}}$	standard deviation of sample mean
Σ	sum, $\Sigma_{i=1}^{n} X_i = X_1 + X_2 + \ldots + X_n$
t	Student's $t = (\bar{X} - \mu)/s$
U	coded value of X: $(X - X_0)/I$
V	variance
w	sample range; also, weight in weighted sums of squares
W_h	weight for stratum h in stratified sampling: $W_h = N_h/N$
X_0	value of X at zero on coded (U) scale
Z	$(X - \mu)/\sigma$; the observation X expressed in standard form

Contractions

AH	alternative hypothesis
ANOVA	analysis of variance
Cov(XY)	covariance of X and $Y = E(X - \mu_X)(Y - \mu_Y)$
CR	completely randomized
deff	design effect
df	degrees of freedom
ESD	extreme Studentized deviate
H_0, H_A	null hypothesis, alternative hypothesis
ln	natural logarithm
LSD	least significant difference
MLR	multiple linear regression
MNR	maximum normed residual
MS	mean square
MSE	mean square error
NH	null hypothesis
RB	randomized blocks
SD	standard deviation
SE	standard error
SS	sum of squares

SEVENT

STATISTICAL METHODS

1

Introduction

1.1—Introduction. The subject matter of the field of statistics has been described in various ways. According to one definition, statistics deals with techniques for collecting, analyzing, and drawing conclusions from data. This description helps to explain why an introduction to statistical methods is useful to students who are preparing for a career in one of the sciences and to persons working in any branch of knowledge in which much quantitative research is carried out. Such research is largely concerned with gathering and summarizing observations or measurements made by planned experiments, by questionnaire surveys, by the records of a sample of cases of a particular kind, or by combing past published work on some problem. From these summaries, the investigator draws conclusions hoped to have broad validity.

The same intellectual activity is involved in much other work of importance. Samples are extensively used in keeping a continuous watch on the output of production lines in industry, in obtaining national and regional estimates of crop yields and of business and employment conditions, in auditing financial statements, in checking for the possible adulteration of foods, in gauging public opinion and voter preferences, in learning how well the public is informed on current issues, and so on.

Acquaintance with the main ideas in statistical methodology is also an appropriate part of a general education. In newspapers, books, television, radio, and speeches we are all continuously exposed to statements that draw general conclusions; for instance, that the cost of living rose by 0.3% in the last month, that the smoking of cigarettes is injurious to health, that users of "Blank's" toothpaste have 23% fewer cavities, that a television program had 18.6 million viewers. When an inference of this kind is of interest to us, it is helpful to be able to form our own judgment about the truth of the statement. Statistics has no magic formula for doing this in all situations, for much remains to be learned about the problem of making sound inferences. But the basic ideas in statistics assist us in thinking clearly about the problem, provide some guidance to the conditions that must be satisfied if sound inferences are to be made, and enable us to detect many inferences that have no good logical foundation.

1.2—Purpose of this chapter. To introduce students to applications of statistical methods in many different fields, a joint committee of the American

Statistical Association and the National Council of Teachers of Mathematics published a book in 1972 titled: *Statistics: A Guide to the Unknown,* (2nd ed., 1978) (1).* The book contains 45 articles describing a wide range of uses of statistical methods. In this chapter we give examples from these articles of two important types of applications: (*i*) sample surveys, in which information about a large aggregate or population is obtained by selecting and measuring a sample drawn from the population, and (*ii*) comparative studies of the effects of different agents (for instance, different foods, different medications, different methods of teaching).

Our objectives are: (*i*) to indicate some of the problems we meet when trying to draw sound conclusions from the results of a survey or comparative study and (*ii*) to introduce a technique in the collection of data—randomization—that helps greatly in tackling these problems.

1.3—Examples of sample surveys. 1. About 50,000 families in the United States are visited every month by interviewers from the Census Bureau (2). The interviewer asks questions to determine the work status of every family member aged 14 and over during the previous week. Specifically, the interviewer wants to know whether the member was working at paid employment, has a job but was temporarily laid off, was not employed but was seeking paid work, or was not employed and is not currently seeking paid work. From this survey, monthly estimates are published of the number of employed and the number of unemployed in the United States. Although the sample is large, only a very small proportion of those over 14 in the United States are included in a monthly sample. For cost reasons the method of selecting the sample is complex, but at present it gives everyone aged 14 or over a chance of about 1 in 1240 of being in the sample in a specific month.

2. When freight travels from place A to place B by two different railroads, the total freight charge is divided between them. A document called a waybill states the amount due each railroad for any shipment. Reviewing all the waybills during a 6-month period to determine how much money is due each railroad is time-consuming. The Chesapeake & Ohio Railroad conducted a number of experiments to determine whether this allocation could be estimated accurately enough from a sample of the waybills (3). They used past time periods in which the correct allocation was already known so that the error in any sampling method could be determined.

In one experiment, there were 22,984 waybills in the population during a 6-month period. They chose a sample of 2072 waybills, about 9%. The results were:

Amount due Chesapeake & Ohio from complete population	$64,651
Estimated amount from sample	64,568
Difference	$ 83

*References (1), (2), etc., are given at the end of each chapter.

Since the sample saved over $4000 in clerical costs, sampling is clearly advantageous economically. For the same reason, sampling is used in allocating airline passenger ticket receipts for journeys involving several airlines.

3. We are all familiar with election and public opinion polls (4). Samples of around 1500 persons are quite common in nationwide polls. They can provide estimates of the attitude of the public to recent events, types of behavior, and new proposals and laws and are of considerable interest to social scientists and representatives of government as well as to the general public.

1.4—Problems of sampling. A *sample* is a set of items or individuals selected from a larger aggregate or *population* about which we wish quantitative information. As the preceding examples show, substantial savings in resources, time, and money can be made through sampling. Much valuable information could not have been collected without sampling. We must remember, however, that the results of a sample are of interest only in what they tell us about the population from which the sample was drawn—which implies also what they tell us about the part of the population that was not measured.

Sampling would not be a problem were it not for ever present variation. If all individuals in a population were alike, a sample of one would give complete information about the population. But variability is a basic feature of the populations to which statistical methods are applied; successive samples from the same population give different results. Clearly, the facts observed in a sample cannot be taken as facts about the population.

Major problems of sampling are to learn how to select samples that enable the investigator to make reliable statements about the population and to learn what kinds of statements can be made.

In many applications these statements about the population have two parts. The first is an *estimate* of some characteristic of the population, sometimes called a *point estimate*. It answers questions like: How many people in the country were employed last month? What percentage of the people in Boston, Massachusetts, support this proposal? As a simple example (data furnished by T. A. Brindley), a sample of 100 farmers from the population of 2300 farmers in Boone County, Iowa, were asked in 1950 whether they had sprayed their cornfields for control of the European corn borer. In the sample, 23 farmers (23%) said they had applied the treatment. A sample estimate of the number of farmers in Boone County who sprayed is made by assuming that 23% of *all* farmers sprayed. This assumption gives the estimate of $23 \times 2300/100 = 529$ farmers in Boone County who sprayed.

Owing to variation we cannot trust the sample point estimate to be fully correct. Consequently, we add to it a statement indicating how much the point estimate is likely to be in error. One way is to supplement the point estimate by *interval estimate*. We can say, for example, that from the sample evidence we are confident that the number of farmers in Boone County who sprayed was between 345 and 736. By "confident" we mean that the probability is 95 chances in 100 that the interval from 345 to 736 contains the true but unknown number of farmers in Boone County who sprayed. The interval is called a 95% *confidence interval*. Note that the interval in this example is wide. Evidently a sample of only 100 farmers cannot be expected to estimate closely the number in the county who sprayed.

The process of making statements about the population from the results of samples is called *statistical inference* and is one of the basic areas in our subject. Discussion of statistical inference is postponed until some background material has been covered. Now we consider the prior problem of selection of the sample.

1.5—Biased sampling. It is easy to select a sample that gives a distorted picture of the population. Suppose an interviewer in an opinion poll picks sample families mainly among friends who are pleasant to visit or collects data only from those who are at home and willing to answer on the first visit. The sample may not at all represent the opinions of the rest of the population. A sample may give a distorted picture of a population even when selected with no conscious attempt to choose one kind of individual rather than another and with no refusals. The embryologist Corner (5) describes how publication of a method for the extraction of the hormone progesterone was held up for months because of biased sampling. When Corner selected a rabbit from the supply batch to test the extract of progesterone, he tended (unknown to himself) to pick a small, young rabbit that was too young to respond to progesterone. Thus a succession of successful extracts of progesterone were pronounced failures for reasons that were baffling until Corner's biased sample selection was discovered.

Human judgment may result in biased sampling even when the population can be inspected carefully before the sample is drawn. In a class exercise, a population of rocks of varying sizes was spread out on a large table so that all the rocks could be seen. After inspecting the rocks and lifting any that they wished, students were asked to select samples of five rocks such that their average weight would estimate as accurately as possible the average weight of the whole population of rocks. The average weights of the students' samples consistently overestimated the average weight for the population. A larger and heavier rock is more easily noticed than a smaller one.

This is not to say that the average weights of the five rocks chosen by the students were poor estimates. Consdering the great variability in the weights of individual rocks, the students did a good job. Biased samples can give good estimates if the biases are small. The trouble with biased sampling of this type is that we almost never know the size of the bias. Consequently, in developing reliable methods of sampling, investigators looked for methods that would give protection against bias.

1.6—Random sampling. A sampling method widely used is *random sampling*. Suppose the individuals in the population can be numbered from 1 to N, the total size of the population, and members of the sample are drawn one by one until the desired sample size n has been obtained. In one form of random sampling, each member of the population has an equal chance of being selected at every draw. With a population that is not too large, we can select a random sample by having discs numbered from 1 to N representing the members of the population. The discs are put in a receptacle, mixed thoroughly, and one sample member is drawn out blindfold. After its number has been noted we replace it, mix again, draw a second member, and so on. This method is called random sampling *with replacement*. The replacement is necessary to ensure that at each draw *every* member in the population has an equal chance of being drawn.

For example, suppose that the population has four members, numbered 1, 2, 3, 4, and we are drawing samples of size 2. There are 16 possible samples: (1, 1), (1, 2), (1, 3), (1, 4), (2, 1), (2, 2), (2, 3), (2, 4), (3, 1), (3, 2), (3, 3), (3, 4), (4, 1), (4, 2), (4, 3), (4, 4). Every population member appears once in six samples and twice in one; the sampling shows no favoritism between one member and another. Each of the 16 samples has an equal chance of being the sample that is drawn.

A variant of this method of sampling is to allow any member of the population to appear only once in the sample. This restriction is intuitively appealing—since the purpose of the sample is to provide information about the population, it makes sense to have as many *distinct* members of the population as possible in the sample.

To select this kind of sample we do not replace a disc in the receptacle after it has been drawn. If we ignore the order of drawing there are six distinct equiprobable samples of size 2—(1, 2), (1, 3), (1, 4), (2, 3), (2, 4), (3, 4)—again showing no leaning towards one part of the population or another. In sample surveys, sampling is nearly always done *without replacement,* as this method is called. If the sample is a small fraction of the population (say less than 2%), samplings with and without replacement give practically identical results, since an individual is very unlikely to appear more than once in a sample.

The intuitive appeal of random sampling lies in its fairness—it gives every item in the population an equal chance of being selected and measured and thus should protect us against distortion or misrepresentation of the population. On the other hand, random sampling does not use any available advance knowledge about the structure of the population. For this reason, or to cope with operational problems, more complex types of random sampling have been developed and are commonly used.

The samples of railroad waybills and Boone County farmers both employed *stratified random sampling.* In the waybills sample, the waybills were first classified (or stratified) according to the total size of the bill. *All* waybills with total bills over \$40 were selected because we need an accurate sample of large bills over \$40. A random sample of 50% of the waybills with totals between \$20 and \$40 was chosen, and so on. Only 1% of waybills with totals under \$5 was selected; we do not need an accurate sample of bills under \$5, since little money is at stake. The monthly sample of the employed, which covers the whole country, is more complex still but employs randomization in the actual selections.

In selecting random samples, numbered discs in a box or bag that are thoroughly mixed and drawn out blindfold have often been used with relatively small populations. In class exercises, an alternative for selecting repeated samples of the same size is to have numbered balls mixed in a box, with scoops having 5, 10, or 20 semicircular indentations that draw out samples of these sizes quickly. In studying two-class populations (male-female, yes-no, dead-alive) we need only discs or balls of two colors.

With large populations such methods are slow and unwieldy. Instead, random digits are used to select random samples.

1.7—Tables of random digits. These tables are created by a process designed to give each digit from 0 to 9 an equal chance of appearing at every draw. Before publication, such tables are checked in numerous ways to ensure that the series of digits do not depart materially from randomness in a manner that would vitiate the commonest uses of the tables. Table A 1 contains 10,000 such digits, arranged in 5 × 5 blocks, with 100 rows and 100 columns each numbered from 00 to 99. Table 1.7.1 shows the first 100 numbers from this table.

The chaotic appearance of the digits is evident. To illustrate their use, suppose we want a random sample of size $n = 10$ from a population of size $N = 82$. Draw two-digit numbers from columns 00–01, 02–03, and so on in table 1.7.1, ignoring 00 and numbers greater than 82 if they are drawn. The sample consists of the population members numbered 54, 15, 61, 5, 46, 38, 14, 21, 32, 14. Note that 14 is drawn twice—this sampling is with replacement. (If we are sampling without replacement we ignore the second 14 and continue, finding the last sample member to be 26 in columns 06–07.)

If the first digit in N is 1, 2, or 3, this method requires skipping many numbers in the table because they are too large. With $N = 270$, for example, one alternative is to use the three-digit numbers 001–270, 301–570, 601–870 from three columns of digits in table 1.7.1. Mentally subtract 300 from any number drawn in the range 301–570 and 600 from any in the range 601–870. This method uses 3 × 270 = 810 of the 1000 three-digit numbers. If we start with columns 00–02 in table 1.7.1 and want $n = 5$, the five sample members are 244, 153, 259, 11, 52.

TABLE 1.7.1
ONE HUNDRED RANDOM DIGITS FROM TABLE A 1

	00–04	05–09	10–14	15–19
00	54463	22662	65905	70639
01	15389	85205	18850	39226
02	85941	40756	82414	02015
03	61149	69440	11286	88218
04	05219	81619	10651	67079

1.8—Sampling distributions of estimates. Successive samples from a variable population differ from one another and lead to different estimates of a population characteristic. However, a valuable property of random sampling is that with random digits we can draw repeated random samples of a specific size from a known population, estimate the population characteristic from each sample, and learn how close these estimates are to the correct value for the population. Such investigations have taught us the kinds of objective statements (statistical inferences) that we can make about populations from the results of random samples. They also lead to rules that enable us to estimate the size of random sample needed for estimates of prescribed accuracy.

Consider, for example, a proposal that 50% of a large population approve and 50% disapprove. To learn how well the percent approving in the population would be estimated by a random sample of size 10, one group of students drew 200 random samples each of size 10. In a table of random digits an odd number represented approval, an even number disapproval. In any sample the number of odd digits is of course one of the numbers 0, 1, 2, . . . , 10. Table 1.8.1 shows the frequency out of 200 with which each number of odd digits (and hence each estimated percent approval) appeared. Such tables are called *frequency distributions* or *sampling distributions*.

The sample estimates of the percentage who approve vary from 0% to 90%. If we regard a sample estimate of 40%, 50%, or 60% approval as satisfactorily close to the population 50%, we see that 120 samples or 60% met this standard. But occasionally we get very bad estimates. You may verify that 30 samples, or 15%, are in error by 30% or more. In practical sampling we do not know the correct population value and cannot tell whether any specific sample is one of the better or one of the poorer ones. But from studies like that in table 1.8.1 in which the population value is known, we can find out how frequently our estimate is in error by less than a given amount, determining the frequency distribution of the error of estimate. The technique is called the *Monte Carlo* method. The name is appropriate, since the luck of the draw creates the sampling distribution. Monte Carlo studies have become such an important tool that electronic computers have programs designed to create sets of digits that can be regarded as random for sampling purposes. Some small pocket calculators also have keys that produce such digits when pressed.

In many problems, the frequency distributions of estimates made from random samples can also be worked out by probability theory. This avoids the need for Monte Carlo studies, which of course give only approximations to sampling distributions, being themselves subject to sampling error. The distribution in table 1.8.1 is an example; you will later work out the probability that a random sample of size 10 contains 0, 1, 2, . . . , 10 odd digits.

Some exercises will familiarize the reader with random digits as used in drawing random samples and studying sampling distributions. If a class using table A 1 intends to combine results, each student should use a different part of the table. Further random digits may be available from computers or from reference 10, which contains 1 million digits.

TABLE 1.8.1

NUMBERS OF ODD DIGITS AND CORRESPONDING % APPROVAL IN 200 SAMPLES WITH $n = 10$

Number	0	1	2	3	4	5	6	7	8	9	10
Estimated %	0	10	20	30	40	50	60	70	80	90	100
Frequency	1	1	8	25	39	45	36	25	16	4	0

EXAMPLE 1.8.1—Draw a random sample of size $n = 10$ without replacement from a numbered population of size $N = 100$, recording the numbers drawn.

EXAMPLE 1.8.2—Draw a random sample of size $n = 15$ without replacement from a numbered population of size $N = 123$. How many of the 1000 three-digit random numbers does the subtraction method in the text use? Can you think of a method (with more troublesome subtractions) that uses 984 of the digits?

EXAMPLE 1.8.3—Select a random sample of 20 pages of this book, regarding the population as consisting of pages 1–100. Record the number of pages in your sample on which at least one new section begins. (Do not count References as a section.) The population proportion is $56/100 = 0.56$.

EXAMPLE 1.8.4—When the doors of a clinic are opened, 12 patients enter simultaneously. Each patient wishes to be seen first. Use random digits to arrange the patients in a random order.

EXAMPLE 1.8.5—(For a class.) For a large population in which 30% favor some proposal, draw a set of around 200 random samples of size 10. Appearance of any of the digits 1, 2, 3 can represent someone favoring the proposal. Construct the combined frequency distribution. What proportion of your sample estimates are correct to within $\pm 10\%$? (That is, give estimates of 20%, 30%, or 40%.) The theoretical proportion is 0.7.

EXAMPLE 1.8.6—Draw 100 random samples of 5 digits. For each sample write down the smallest and largest digit in the sample. Construct the frequency distribution of (*i*) the largest digit, (*ii*) the average of the largest and the smallest digits, (*iii*) the sample range—the largest minus the smallest digits. These cases illustrate three different shapes of frequency distributions: distribution (*i*) has the highest frequencies at 9 and 8 and is called *negatively skew;* (*ii*) is symmetrical and gives good estimates of the average 4.5 of all the digits; (*iii*) is also skew negative but less so than (*i*).

EXAMPLE 1.8.7—Proceeding along a series of random digits, stop whenever you reach a 0 and record the number of digits used (including the 0). Continue 50 times, constructing the frequency distribution of the number of digits needed to reach a 0. Going along the rows of table 1.7.1, for instance, the numbers of digits needed in the first five trials are 14, 3, 12, 6, 12. The average number needed is 10, but the distribution has a wide spread and is slightly skew positive.

1.9—Studies of the comparative effects of different agents. We turn now to the second type of application described in reference (1). As a first illustration (6), General Foods Corporation developed an easy-to-prepare product *H* for a meal, available either in liquid or solid form. They wished to compare the palatability of liquid and solid *H* with that of a product *C* already on the market. As tasters to rate the palatability the company recruited 75 males and 75 females, who were paid for their work. In each sex, 25 persons rated *C*, 25 rated liquid *H*, and 25 rated solid *H*. Palatability was scored by each person on a seven-point scale, the lowest rating being -3 and the highest $+3$. The average score for the 25 persons who had rated a product was the criterion for comparing the three products.

Random digits were used in dividing the 75 males and 75 females into three groups—25 to taste *C*, 25 to taste liquid *H*, 25 to taste solid *H*. Thus, to protect against any systematic bias, conscious or unconscious, the investigators deliberately had no say in who tasted what. This choice was determined by random selection that gave everyone an equal chance of being assigned to a given food.

Incidentally, one way of carrying out the random assignment of 25 tasters is first to number the tasters from 1 to 75. Go down two columns of random digits

TABLE 1.9.1
RANDOM ALLOTMENT OF 75 PERSONS TO THREE FOODS, IN STEM AND LEAF FORM

	C		Liquid H		Solid H
0	5	0	2, 9, 8, 3, 7	0	1, 4, 6
1	0, 1, 6, 5, 2	1		1	3, 4, 7, 8, 9
2	2, 5, 0, 6, 7	2	1, 9, 3	2	4, 8
3	7	3	6, 5, 4, 0, 3	3	1, 2, 8, 9
4	0	4	2, 4, 7, 8, 5, 9	4	1, 3, 6
5	9, 4, 7, 6	5	2, 1, 5	5	0, 3, 8
6	0, 4, 5	6	2, 9	6	1, 3, 6, 7, 8
7	1, 5, 4, 3, 2	7	0	7	

until 25 *distinct* numbers between 01 and 75 have been selected. These are the persons who taste C. Then find a further 25 for liquid H, the remainder getting solid H. A method that helps to detect quickly numbers already drawn that cannot be used is to record the draws in a form known as *stem and leaf,* due to Tukey (7). The stem is the first digit, 0, 1, 2, . . . , 7, of the two-digit random number drawn, the leaf is the second digit. The results of one random drawing are shown in table 1.9.1.

The first four distinct random numbers drawn were 59, 10, 11, 54—allotted to C. In the stem and leaf arrangement it is easy to check whether a number has already appeared and to spot the numbers remaining for solid H when 50 have been drawn.

1.10—Problems in drawing conclusions from comparative studies. For a brief look at the results of the palatability comparisons, table 1.10.1 shows the average scores for males and females.

It looks from table 1.10.1 as if both males and females give substantially higher palatability ratings to H than to C. Liquid H was rated more highly than solid H by both sexes but only by very small amounts. In reporting on the average difference in results between two treatments we must remember, however, that because of person-to-person variability, two groups of 25 would have different average scores even if given exactly the same food.

Consequently, we often begin with the preliminary question: Are we convinced that there is a real difference between the effects of different treatments? This question can be tackled by considering the hypothesis (called the *null hypothesis*) that there is no difference between the effects of the two treatments in the population from which these tasters were drawn. We then ask:

TABLE 1.10.1
AVERAGE PALATABILITY SCORES

	C	Liquid H	Solid H
Males (25 persons)	0.20	1.24	1.08
Females (25 persons)	0.24	1.22	1.04
Both sexes (50 persons)	0.22	1.18	1.06

Do our sample data agree or disagree with this hypothesis? To answer this we calculate the probability that an average difference between the treatments as great as that observed in our sample could arise solely from person-to-person variability if the null hypothesis were true. If this probability is small, say 1 in 20 or 1 in 100, we reject the hypothesis and claim that there was a real difference between the effects of the two treatments. This technique, sometimes called a *test of statistical significance,* is a much used tool in statistical inference. It is discussed more fully in chapter 5 and subsequent chapters.

The palatability comparisons in table 1.10.1 show real differences between the average ratings given by either sex to H and to C. The small differences between the ratings for liquid H and solid H, on the other hand, could easily be accounted for by person-to-person variability.

In comparing two agents or treatments that appear to have different effects, the question of primary interest is: How large is the difference between their effects? For instance, in table 1.10.1 the tasters rated H as superior to C by an overall average amount, $1.12 - 0.22 = 0.90$ units. This result is a point estimate of the population difference in palatability ratings. A further question is: How accurate is this estimate? In these two questions, we are facing the same problems of estimation encountered in sample surveys (section 1.4) except that we are now dealing with differences instead of means.

1.11—Effectiveness of the Salk polio vaccine. After the Salk polio vaccine was developed, a study was carried out in 1954 to assess its effectiveness in protecting young children against paralysis or death from polio (8). Two different study plans were used, each participating state deciding which plan it would follow. Two groups of children took part in the plan described here. One group received the prescribed three inoculations of the Salk vaccine. The other group received no vaccine. The presence of this group was necessary for comparison in judging the effectiveness of the vaccine. The most important criteria in the comparison were the numbers of paralytic polio cases and the resulting numbers of deaths that developed in the two groups. Since polio was (fortunately) rare, large numbers of children were needed in the two groups so that sufficient paralytic cases would develop to provide a basis for comparison. Meier's article (8) refers to the study as "the biggest public health experiment ever." Just over 200,000 children were in each of the two groups.

As in the palatability study, randomization was used in deciding which 200,000 children received inoculations and which did not. In the polio study, however, a restricted or stratified randomization was employed, as follows. Many schools all over the country took part. Randomization was carried out separately within each participating school so that every school had about equal numbers of inoculated and control children. Thus, schools in relatively high risk areas and in relatively low risk areas each had about equal numbers of vaccinated and control children, chosen at random. This method was an improvement over simple randomization, which would merely guarantee that *overall* the two groups would have about equal numbers.

Each uninoculated child received three shots of a saline solution believed to have no effectiveness against polio. These shots were given at the same times and in the same manner as the polio vaccine. In medical terminology this treatment

TABLE 1.11.1
PARALYTIC CASES FOR VACCINATED AND PLACEBO GROUPS

Group	Number of Children	Paralytic Cases	Case Rate/ 100,000
Vaccinated	200,745	33	16
Placebo	201,229	115	57

is called a *placebo,* a Latin word meaning, literally, "I shall please." A dictionary definition of a placebo is: "An inactive medicine given merely in order to satisfy a patient."

The purpose of the placebo in this trial was to conceal the knowledge of whether a child had received vaccine or saline from the child, from the parents, from those giving the inoculations, and from physicians making a later diagnosis when a child became ill. Otherwise, in case of doubt with a sick child, a physician might be more likely to diagnose polio if the child was known not to have received vaccine and thus bias the result in favor of the polio vaccine. When feasible, placebo controls are often used for similar reasons in comparative trials in medicine if the handling of the patient, the patient's report of progress, and the doctor's diagnosis might be affected by knowledge of the particular medication received.

The decision to subject more than 200,000 children to three unpleasant shots of unprotective saline was a drastic one. This decision was made by those states participating and by the parents who gave permission because a placebo control was considered essential to ensure a fully objective comparison in this important trial. The term "placebo," however, seems misused here—the saline shots were certainly not given to placate or satisfy the children.

Table 1.11.1 shows the numbers of children in the vaccinated and placebo groups, the numbers of paralytic cases that developed, and the case rates per 100,000 children.

In considering these results we encounter the same two questions as in the palatability study. Was there a real difference between the case rates for vaccinated and placebo children? What are the point estimates and the interval estimates of the size of the difference? Methods of answering questions of this type form an important part of statistical inference.

1.12—Death rates of different smoking groups. Seven large comparative studies of the death rates of males with different smoking habits were started between 1951 and 1959 (9). One study was conducted in Britain, one in Canada, and five in the United States. Except for minor variations the study plans were the same. First, a questionnaire was sent to the selected population group asking about current and past smoking habits and a few other facts, such as age. A system was set up to ensure that if any of those who answered the questionnaire died, this fact was reported and recorded and a diagnosis of the cause of death obtained (by death certificate and by autopsy where available). The number of men in a study ranged from 34,000 to 448,000.

These studies contained numerous sampling groups whose death rates and causes can be compared: (1) different types of smokers—nonsmokers, cigarette, cigar, pipe, mixed; (2) different amounts smoked by a given type; (3) different ages at which smoking was started, by a given type and amount; and (4) classification of ex-smokers by how long ago they stopped and by how much they smoked before they stopped.

1.13—Observational studies. When we try to draw conclusions from comparisons among these groups, we find a major logical difference between the smoking studies and the palatability or polio studies. In the two latter studies, the investigators could decide which group of subjects received a given treatment. In making the assignment of treatments to subjects they used randomization or stratified randomization. The objective was to try to ensure no systematic differences between the groups being compared except the difference in treatment.

In smoking studies, on the other hand, the investigators could not assign subjects to groups. Subjects assign themselves to groups by their smoking habits. Thus smokers of two different types may differ systematically in many ways, apart from smoking, that affect their death rates. For example, cigar and pipe smokers were usually found to be substantially older than nonsmokers at the start of the smoking studies. Cigarette smokers tended to be a little younger. It is well known that after middle age the death rate increases steadily with age. Consequently, comparisons of the overall death rates of nonsmokers, cigarette smokers, and cigar or pipe smokers would be biased in favor of cigarette smoking and heavily biased against cigar or pipe smoking. Furthermore, smokers of different types may differ systematically in their eating and drinking habits, amount of exercise, and numerous other factors that might affect their death rates.

To avoid such biases the investigators must try to adjust the death rates so that they apply to groups in which these extraneous factors are similar. The adjustments make the statistical analysis more complicated than in randomized studies. Moreover, conclusions are usually less confident, because we can seldom be sure that we measured and adjusted correctly for all important variables in which our groups differed systematically. Studies like the smoking studies are sometimes called *observational,* as a reminder that the investigator lacks the power to create the groups to be compared but is restricted to his or her choice of the observations or data that are collected and analyzed.

The planning and analysis of observational studies is an area of great importance because such studies often provide the only methods feasible for comparative investigations.

1.14—Summary. The purpose of this introductory chapter is to introduce two broad classes of statistical applications—sample surveys and comparative studies of the effects of different agents or treatments.

Statistics deals with techniques for collecting, analyzing, and drawing conclusions from data.

A *sample* is a collection of items from some larger aggregate (the *population*) about which we wish information. Collection of information by sample surveys saves time and money.

Variability from item to item is characteristic of the populations encountered in statistical studies. For this reason, problems in conducting sample surveys deal with (*i*) how to select the sample so that it does not give a distorted view of the population and (*ii*) how to make statements about the population from the results of the sample. Methods for attacking these problems are an important part of the subject matter of statistics.

A widely used method of sample selection is some variant of *random sampling*. If the sample is drawn item by item, *random sampling with replacement* gives every item in the population an equal chance of being chosen at any draw. In *random sampling without replacement* the chance of selection at any draw is the same for all items in the population not previously drawn. The properties of the two methods are essentially the same if the sample is a small fraction (e.g., less than 2%) of the population. These methods give protection against biased sample selection but do not utilize information that we may possess about the structure of the population. For this purpose restrictions on simple random sampling have been developed.

Tables of random digits are tables in which the digits 0, 1, 2, . . . , 9 have been drawn by some process that gives each digit an equal chance of being selected at any draw. Many electronic computers also produce random digits for use in sample selection.

An important use of tables of random digits and of computers is to draw repeated random samples of a given size from a population. By estimating a desired population characteristic from each sample we obtain the *sampling distribution* of the estimates. From studies of this kind, sometimes called *Monte Carlo* studies, we can learn how well estimates from random samples perform.

From Monte Carlo studies and from mathematical investigations of the same type, two kinds of statements about the population can be made from the results of the sample. One is a *point estimate* of some characteristic (e.g., number of unemployed, proportion of farmers who sprayed against European corn borer). Since point estimates are inevitably subject to error, the second is an *interval estimate*—a statement that the population characteristic has a value lying between two specified limits (e.g., the population contains between 56,900 and 60,200 children under five years). Owing to variability, even interval estimates cannot be made with certainty. But we can attach to the interval a probability (95%)—meaning that if such interval statements are made repeatedly in statistical work, the interval will include the true population value in 95% of the cases.

The second broad class of applications of statistics introduced in this chapter is comparative studies of the effects of different agents, often called *treatments*. In some studies the investigator can decide which subjects will receive any given treatment. These studies, illustrated by the palatability trial (3 treatments) and the Salk polio vaccine trial (2 treatments) are known as *controlled experiments*.

In both studies the investigators assigned each treatment to about an equal number of subjects. The specific choices were made by random numbers so that every subject had the same chance of receiving any treatment. The objective was to guard against any consistent differences in the subjects given different treatments that might bias the comparisons. Further control was exercised by

TABLE 2.2.1
FREQUENCY DISTRIBUTION OF DISEASED PLANTS PER AREA OF 9 PLANTS

Number of Diseased Plants	Frequency
0	36
1	48
2	38
3	23
4	10
5	3
6	1
7	1
8	0
	$n = 160$

The frequency distribution in table 2.2.1 does not change or distort the original data in any way. It presents as complete a picture of the nature of the variation from area to area as the data allow. We note that 1 is the most frequent number of diseased plants and that relatively few areas have over 50% diseased plants.

2.3—Grouped frequency distributions. With discrete data in which the observations include a large number of distinct values and with continuous data, the observations are grouped into classes when forming frequency distributions. Between 10 and 20 classes are usually sufficient to give a general idea of the shape of the distribution and to calculate the principal descriptive summary statistics that we meet later.

Table 2.3.1 shows two grouped frequency distributions. The first comes from a sample by K. Pearson and A. Lee (2) of the heights of 1052 mothers, grouped into 1-inch classes. The second is the frequency distribution (3) of the 1970 census of inhabitants of the 150 U.S. cities whose sizes ranged between 100,000 and 1 million. (We use the term *city sizes* rather than *city populations*. In statistics the word *population* has a technical meaning, as noted in section 1.4.)

Figure 2.3.1 shows the distribution of mothers' heights in graphic form. When all classes are of equal width, as in this example, each class is represented by a rectangle whose height is proportional to the class frequency. This diagrammatic form, known as a *histogram,* gives a good picture of the general shape of the distribution. We notice that the distribution of mothers' heights is nearly symmetrical, with the highest frequency, 183, occurring in the middle. This histogram is often described as bell shaped. On the other hand, the highest frequency in city sizes occurs in the lowest class, the frequencies falling off fairly steadily towards zero. A distribution of this shape is called *skew positive.*

With highly variable data that have small frequencies over a considerable part of the range, using classes of unequal widths in different parts of the distribution is advisable. This situation exists with city sizes in table 2.3.1. The

TABLE 2.3.1
TWO GROUPED FREQUENCY DISTRIBUTIONS: (1) HEIGHTS OF 1052 MOTHERS; (2) 1970 SIZES OF 150 U.S. CITIES

Heights of Mothers		1970 City Sizes		
Class Limits (in.)	Frequency	Class Limits (1000s)	Frequency	Frequency/ 25,000 Width
52–53	0.5	100–125	38	38
53–54	1.5	125–150	27	27
54–55	1	150–175	15	15
55–56	2	175–200	11	11
56–57	6.5	200–300	16	4
57–58	18	300–400	16	4
58–59	34.5	400–500	7	1.8
59–60	79.5	500–600	8	2
60–61	135.5	600–800	10	1.2
61–62	163	800–1000	2	0.2
62–63	183	Total	150	
63–64	163			
64–65	114.5			
65–66	78.5			
66–67	41			
67–68	16			
68–69	7.5			
69–70	4.5			
70–71	2			
Total	1052			

range extends from 100,000 to 1 million, but 60% of the cities have sizes between 100,000 and 200,000. We need at least 4 classes of width 25,000 to represent this part of the range, preferably more. But with classes of width 25,000, an additional 36 classes are needed to reach 1 million and these classes would have an average of 1.6 cities per class. Consequently, we used classes of width 100,000

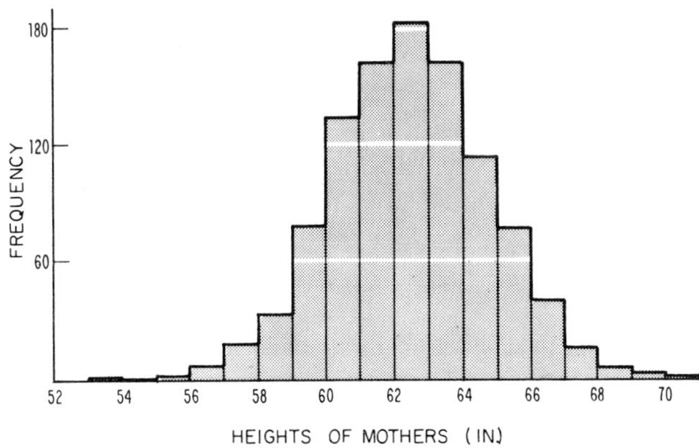

Fig. 2.3.1—Histogram of heights of mothers (in.).

from 200,000 to 600,000 and of width 200,000 from 600,000 to 1 million, as shown in table 2.3.1.

A disadvantage of classes of unequal width is that one does not get a picture of the shape of the distribution by glancing at the frequencies because the frequencies are not comparable—they apply to classes with different widths. This distortion is removed in part by calculating the frequencies per class of 25,000 width (or of any convenient constant width) as shown in the right-hand column of table 2.3.1. A better picture is given by drawing a histogram from the classes of unequal widths. Make the heights of the rectangles proportional to the frequencies per class of width 25,000, but give each rectangle its proper width: 25,000, 100,000, or 200,000. The histogram will then give an undistorted picture of the shape of the distribution.

2.4—Class limits. In forming classes some thought must be given to the choice of class limits. In both examples of table 2.3.1 the upper limit of any class coincides with the lower limit of the class above. This raises the question: Where do we put an observation that falls on a boundary? For a mother whose height fell on a boundary between two classes, Pearson and Lee put 1/2 person in the lower class and 1/2 person in the upper class—note the 10 classes whose frequencies are not integers. This rule handles the problem without serious distortion, but the appearance of halves and the fact that the frequencies in table 2.3.1 are not actual frequencies are awkward.

This problem did not occur with the city sizes. In the original data the sizes were recorded as numbers of persons without any rounding. No city had a size of exactly 150,000 or fell exactly on any boundary. Use of boundaries that overlap in this way is convenient when the chance of a boundary value is very small, as may happen when the original data are recorded to several significant figures.

The more common practice is to form clearly distinct classes. If mothers' heights were recorded originally to the nearest inch, Pearson and Lee might have used as class limits 52–53 in., 54–55 in., 56–57 in., and so on. When listing class limits in a table, it is important that the reader of the table know exactly which values have gone into any class.

With conceptually continuous variables, it is necessary for some purposes to distinguish between recorded class limits and what are called *true class limits*. Consider the class with heights 54–55 in. If heights were recorded to the nearest inch, any mother with true height between 53.5 and 54.5 in. would have been recorded as 54 in. Similarly, the recorded 55-in. heights would represent mothers whose true heights were between 54.5 and 55.5 in. The true limits for this class are therefore 53.5 in. and 55.5 in.

2.5—Cumulative frequency distributions. Sometimes the investigator is interested in estimating the proportion of the members of a population whose measured values exceed some stated level or fall short of the level. What proportion of families have annual incomes less than $8000? What proportion of couples who have been married ten years have 2 or more children? What

TABLE 2.5.1
CUMULATIVE DISTRIBUTION OF CITY SIZES FROM TABLE 2.3.1

Upper Limit of Class (1000s)	Cumulative Value up to Class Upper Limits	
	Frequency	Percent
125	38	25.3
150	65	43.3
175	80	53.3
200	91	60.7
300	107	71.3
400	123	82.0
500	130	86.7
600	138	92.0
800	148	98.7
1000	150	100.0

proportion of mothers have heights between 5 ft and 5 ft 6 in.? For questions like these, estimates are obtained from *cumulative frequency distributions.*

To form a cumulative distribution, add the class frequencies from one end. Then divide each cumulated frequency by the total frequency, expressing the cumulated frequencies as proportions or percents. Table 2.5.1 shows the cumulative distribution for the city sizes in table 2.3.1, starting with the class containing the smallest cities.

We note, for example, that 86.7% of the cities in this population are less than half a million in size and only 13.3% exceed half a million. Estimated answers to various questions about a distribution can be obtained from the cumulative table.

EXAMPLE 1—What percent of the cities are less than a quarter of a million in size? No class in table 2.5.1 has an upper limit of 250,000. Our problem can be indicated as follows:

City size	200,000	250,000	300,000
% less	60.7	?	71.3

since 250,000 is halfway between 200,000 and 300,000, *linear interpolation* suggests the estimate: $(60.7 + 71.3)/2 = 66.0\%$.

EXAMPLE 2—What size is exceeded by half the cities? This problem can be laid out as follows from the information in table 2.5.1.

City size	150,000	?	175,000
% less	43.3	50.0	53.3

Now, $(50 - 43.3)/(53.3 - 43.3) = 0.67$. The desired city size should therefore exceed 150,000 by 0.67 times the distance to 175,000; that is, by $(0.67)(25{,}000) = 17{,}000$. Our answer is 167,000. A note on interpolation is given at the beginning of the appendix tables.

Answers to such questions can be obtained more simply by a graphic method. Plot the cumulative frequencies as ordinates against the class *upper limits* as abscissas. Join the points by straight lines to form a *cumulative frequency polygon,* shown in figure 2.5.1. To estimate the percent of cities with sizes under 250,000, erect a vertical line at 250,000. At the point *A* where this

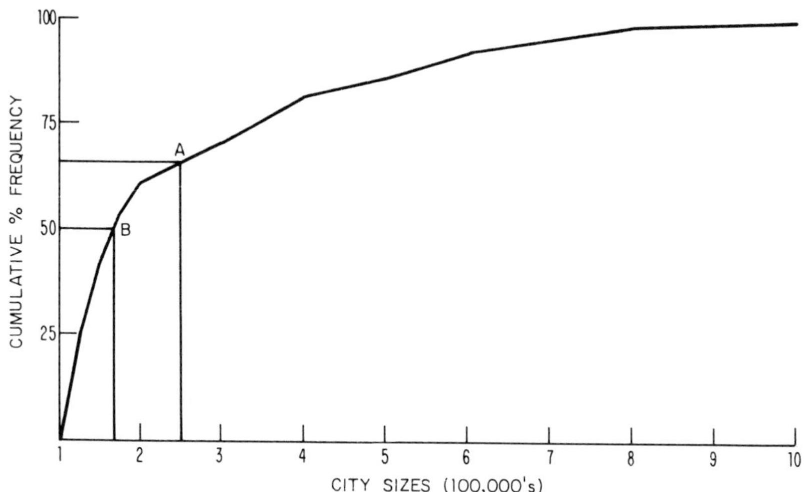

Fig. 2.5.1—Cumulative frequency polygon of city sizes.

line meets the polygon, draw a horizontal line. This meets the ordinate scale at 66%, the desired estimate. For example 2, draw a horizontal line at 50%. It meets the polygon at *B*—about 168,000 on the abscissa. The straight lines in the polygon perform the linear interpolation automatically.

If most of our questions concern the percentages of population members that *exceed* certain values, we start the cumulation with the class containing the largest values. The cumulative percents are then plotted against the *lower* limits of the classes. Furthermore, we nearly always use true class limits in tables like table 2.5.1, since we usually want to know what percentage of the population has true rather than recorded values above or below some limit.

Some examples to be worked by the reader follow.

EXAMPLE 2.5.1—A random sample of 30 households from an urban district gave the following numbers of persons per household: 5, 6, 3, 3, 2, 3, 3, 3, 4, 4, 3, 2, 7, 4, 3, 5, 4, 3, 3, 4, 3, 3, 1, 2, 4, 3, 4, 2, 4, 4. Construct a frequency distribution. Would you say that its shape is closest to that of lengths of run (table 2.6.3), mothers' heights, city sizes, or numbers of diseased tomato plants?

EXAMPLE 2.5.2—The data show the frequency distribution of reported taxable incomes in Canada in 1972 (4).

Class limits (1000s)	Frequency, f
0–2	250
2–4	1589
4–6	1768
6–8	1473
8–10	1172
10–15	1298
15–20	306
20–50	200
50–100	21
100–200	3
over 200	0
	8080

Draw a histogram of the distribution of reported taxable income up to $50,000. Note that widths of class intervals are unequal.

EXAMPLE 2.5.3—Cumulate the frequencies in example 2.5.2 from the smallest incomes upwards and find the percent cumulative distribution. Draw the cumulative polygon up to incomes of $20,000. From this, estimate (*i*) the percent of incomes less than $5000, (*ii*) the percent greater than $12,000, (*iii*) the *median* income (income exceeded by 50% of the population), (*iv*) the income exceeded in only 10% of the returns. Ans. (*i*) 33.7%, (*ii*) 16.2%, (*iii*) $6600, (*iv*) $13,900.

EXAMPLE 2.5.4—Check your answers in the preceding example by linear interpolation.

EXAMPLE 2.5.5—Estimate from the polygon the percent of returns reporting taxable incomes between $7500 and $17,500. (The polygon is very convenient for this type of question because it avoids two interpolations.) Ans. Our polygon gave 37.2%, while interpolation gave 37.0%.

2.6—Probability distributions. The frequency distributions in the preceding examples were obtained from data that had been collected. In some problems, however, an investigator can construct from theoretical considerations the probability or relative frequency distribution that a variable is expected to follow in a population. A simple example is the distribution followed by random digits. A random digit can take any of the values 0, 1, 2, . . . , 9. At any draw, each of these digits has an equal probability, 1/10, of being drawn. The *probability distribution* of random digits is therefore as shown in table 2.6.1.

With a large sample of truly random digits, the observed relative frequencies of the digits should all be very close to 0.1. In a moderate or small sample the agreement will not be as close. For example, a sample of 300 digits was produced by a pocket calculator that purports to give pairs of random digits when a key is pressed. Table 2.6.2 compares the observed sample frequencies with the *expected frequencies* derived from the probability distribution by multiplying the probabilities by the sample size, 300. The largest discrepancy between observed and expected numbers is the excess of 1s—we drew 40 as against an expected 30. Do the results in table 2.6.2 support or contradict an assumption that the calculator gives each digit equal probability? A method of answering this question is given in section 5.12.

A second test often applied with a table of presumably random digits is to compare the observed and expected distributions of lengths of run of odd and

TABLE 2.6.1
PROBABILITY DISTRIBUTION OF RANDOM DIGITS

Digit, X	0	1	2	3	4	5	6	7	8	9
Probability, P	0.1	0.1	0.1	0.1	0.1	0.1	0.1	0.1	0.1	0.1

TABLE 2.6.2
COMPARISON OF OBSERVED AND EXPECTED FREQUENCIES

Digit	0	1	2	3	4	5	6	7	8	9	Total
Obs. freq.	26	40	32	28	30	27	31	27	29	30	300
Exp. freq.	30	30	30	30	30	30	30	30	30	30	300

even digits. Suppose that we start along row 00 of the second page of table A 1. The first 15 digits are 593915803052098. The first run, which is of odd digits, is of length 6, given by the digits 593915. The run stops at 6 because the next digit, 8, is even. Successive runs are of lengths 2 (80), 1 (3), 1 (0), 1 (5), 2 (20), 1 (9), with the parentheses showing the digits in the runs. The run that starts with the final 8 is of length 3 (882). To this point we have 8 runs: 4 of length 1, 2 of length 2, 1 of length 3, and 1 of length 6.

It is not difficult to work out the probability distribution of lengths of runs of truly random digits. Suppose we first draw an odd digit. (The same argument applies if we start with an even digit.) This run will be of length 1 if the next digit is even and of length greater than 1 if the next digit is odd. Thus, half of the runs should be of length 1, half of the runs of length greater than 1.

But half the odd digit runs of length greater than 1 should be of length 2, since they will be of length 2 if the third digit is even. Hence, the probability of a run of length 1 is 1/2, and the probability of a run of length 2 is $(1/2) \times (1/2) = (1/2)^2$. By continuing this argument we show that the probability of a run of length j is $(1/2)^j$.

A sample of 200 lengths of run was drawn from table A 1. Table 2.6.3 compares the observed frequencies with the frequencies $(200/2)^j$ expected from truly random digits. This comparison tests a different type of departure from independent random drawing than does table 2.6.2 and is a useful supplement to that comparison. For instance, a process in which an odd digit was more likely to be followed by an even digit than by an odd digit could still pass the test in table 2.6.2 but would produce too few longer runs.

There is a slight deficiency of runs of length 5 or more (8 observed against 12.5 expected), but as we will see, this is an amount that occurs moderately often with truly random digits.

In the preceding examples the theoretical probability distributions were obtained by knowledge of or assumptions about the nature of the mechanism that produced our data. Theoretical frequency distributions obtained in this way are widely employed in statistical work. When they apply to our data, they enable us to predict many useful results about variable populations, saving the time and expense of obtaining approximate answers by repeated sampling.

NOTATION. With a discrete probability distribution, P denotes a probability, X an observed or measured value, and k the number of distinct values of X. Individual values of X are denoted by a subscript.

The only restrictions on the P_j values are that every $P_j \geq 0$ and that their sum is unity, that is,

TABLE 2.6.3
OBSERVED AND EXPECTED FREQUENCIES OF LENGTHS OF RUN

Length	1	2	3	4	5	6	7	8	≥9	Total
Obs. freq.	110	40	26	16	2	4	1	1	0	200
Exp. freq.	100	50	25	12.5	6.2	3.1	1.6	0.8	0.8	200

$$P_1 + P_2 + P_3 + \ . \ . \ . + P_k = 1 \qquad (2.6.1)$$

(The three dots are read "and so on.")

This sum can be written in a shorter form,

$$\sum_{j=1}^{k} P_j = 1 \qquad (2.6.2)$$

by use of the summation sign Σ. The expression $\Sigma_{j=1}^{k}$ denotes the sum of the terms immediately following the Σ from the term with $j = 1$ to the term with $j = k$.

The sample size is denoted by n. Expected frequencies are therefore nP_j.

EXAMPLE 2.6.1—The probability distribution of lengths of run is a discrete distribution in which the number of distinct lengths—values of X_j—has no limit. Given an endless supply of random digits, you might sometimes find a run of length 20, or 53, or any stated length. If you add the P_j values, 1/2, 1/4, 1/8, and so on, you will find that the sum gets nearer and nearer to 1. Hence the result $\Sigma P_j = 1$ in (2.6.2) and holds for such distributions. By the time you add runs of length 9 you will have reached a total slightly greater than 0.998.

EXAMPLE 2.6.2—(*i*) Form a cumulative frequency distribution from the sample frequency distribution of 200 lengths of run in table 2.6.3. Cumulate from the longest run downwards; i.e., find the number of runs ≥ 8, ≥ 7, and so on. Find the cumulative percent frequencies. What percent of runs are of length (*ii*) greater than 2; (*iii*) less than 2; (*iv*) between 3 and 6, inclusive, i.e., of length 3, 4, 5, and 6? Ans. (*ii*) 25%, (*iii*) 55%, (*iv*) 24%.

EXAMPLE 2.6.3—How well do the estimates from example 2.6.2 compare with the correct percentages obtained from the probability distribution? Ans. Quite well; the correct percentages are (*ii*) 25%, (*iii*) 50%, (*iv*) 22.4%.

TECHNICAL TERMS

class limits
continuous data
cumulative frequency distribution
cumulative frequency polygon
discrete data
expected frequencies
histogram
linear interpolation
probability distribution
sample size
skew positive
true class limits

REFERENCES

1. Cochran, W. G. *Suppl. J. R. Stat. Soc.* 3 (1936):53.
2. Pearson, K., and Lee, A. *Biometrika* 2 (1902):364.
3. *World Almanac* 210 (1976).
4. *World Almanac* 54 (1975).

3

The mean and standard deviation

3.1—Arithmetic mean. The frequency distribution provides the most complete description of the nature of the variation in a population or in a sample drawn from it. For many purposes for which data are collected, however, the chief interest lies in estimating the average of the values of X in a population—the average number of hours worked per week in an industry, the average weekly expenditure on food of a family of four, the average corn acres planted per farm. The average palatability scores of the three foods were the criteria used in comparing the foods in the comparative study in section 1.9.

Given a random sample of size n from a population, the most commonly used estimate of the population mean is the *arithmetic mean* $\overline{X}$ of the n sample values. Algebraically, $\overline{X}$, called *X bar*, is expressed as

$$\overline{X} = (X_1 + X_2 + \ldots + X_n)/n = (1/n)\sum_{i=1}^{n} X_i \tag{3.1.1}$$

Also, investigators often want to estimate the *population total* of a variable (total number employed, total state yield of wheat). If the population contains N individuals, $N\overline{X}$ is the estimate of the population total derived from the sample mean.

As an example of the sample mean, the weights of a sample of 11 forty-year-old men were 148, 154, 158, 160, 161, 162, 166, 170, 182, 195, and 236 lb. The arithmetic mean is

$$\overline{X} = (148 + 154 + \ldots + 236)/11 = 1892/11 = 172 \text{ lb}$$

Note that only 3 of the 11 weights exceed the sample mean. With a skew distribution, $\overline{X}$ is not near the middle observation (in this case 162 lb) when they are arranged in increasing order. The 11 deviations of the weights from their mean are, in pounds,

$$-24, -18, -14, -12, -11, -10, -6, -2, +10, +23, +64$$

The sum of these deviations is zero. This result, which always holds, is a consequence of the way in which $\overline{X}$ was defined. The result is easily shown as follows.

In this book, deviations from the sample mean are represented by lowercase letters. That is,

$$x_1 = X_1 - \overline{X}$$
$$x_2 = X_2 - \overline{X}$$
$$\cdot \quad \cdot \quad \cdot$$
$$x_n = X_n - \overline{X}$$

Add the above equations:

$$\sum_{i=1}^{n} x_i = \sum_{i=1}^{n} X_i - n\overline{X} = 0 \qquad (3.1.2)$$

since $\overline{X} = \Sigma X_i/n$ as defined in (3.1.1).

With a large sample, time may be saved when computing $\overline{X}$ by first forming a frequency distribution of the individual Xs. If a particular value X_j occurs in the sample f_j times, the sum of these X_j values is f_jX_j. Hence,

$$\overline{X} = (1/n)(f_1X_1 + f_2X_2 + \ldots + f_kX_k) = (1/n)\sum_{j=1}^{k} f_jX_j \qquad (3.1.3)$$

As before, k in (3.1.3) is the number of distinct values of X.

For example, in the ratings of the palatability of the food product C by a sample of $n = 25$ males, 1 male gave a score of $+3$, 2 gave $+2$, 7 gave $+1$, 8 gave 0, 5 gave -1, and 2 gave -2. Hence the average sample rating is

$$\overline{X} = (1/25)[(1)(3) + (2)(2) + (7)(1) + (5)(-1) + (2)(-2)] = 5/25 = 0.2$$

Since every sample observation is in some frequency class, it follows that

$$\sum_{j=1}^{k} f_j = n \qquad \sum_{j=1}^{k} (f_j/n) = 1 \qquad (3.1.4)$$

The sample relative frequency f_j/n is an estimate of the probability P_j that X has the value X_j in this population.

3.2—Population mean. The Greek symbol μ is used to denote the population mean of a variable X. With a discrete variable, consider a population that contains N members. As the sample size n is increased with $n = N$, the relative frequency f_j/n of class j becomes the probability P_j that X has the value X_j in this population. Hence, when applied to the complete population, (3.1.3) becomes

$$\mu = \sum_{j=1}^{k} P_jX_j \qquad (3.2.1)$$

In discrete populations, (3.2.1) is used to define μ even if N, the number of units in the population, is infinite. For example, the population of random digits is conceptually infinite if we envisage a process that produces a never ending supply of random digits. But there are only 10 distinct values of X, that is, 0, 1, 2, . . . , 9, with $P_j = 0.1$ for each distinct digit. Then, by (3.2.1),

$$\mu = (0.1)(1 + 2 + 3 + 4 + 5 + 6 + 7 + 8 + 9) = (0.1)(45) = 4.5$$

One result that follows from the definition of μ is

$$\sum_{j=1}^{k} P_j(X_j - \mu) = \sum_{j=1}^{k} P_jX_j - \sum_{j=1}^{k} P_j\mu = \mu - \mu \sum_{j=1}^{k} P_j = 0 \qquad (3.2.2)$$

since $\Sigma_{j=1}^{k} P_j = 1$ from (2.6.2). Result (3.2.2) is the population analog of result (3.1.2) that $\Sigma x_j = 0$.

EXAMPLE 3.2.1—The following are the lengths of 20 runs of odd or even random digits drawn from table A 1: 1, 4, 1, 1, 3, 6, 2, 2, 3, 1, 1, 5, 2, 2, 1, 1, 1, 2, 1, 1. Find the sample arithmetic mean (*i*) by adding the numbers, (*ii*) by forming a frequency distribution and using (3.1.3). Ans. Both methods give 41/20 = 2.05.

EXAMPLE 3.2.2—The weights of 12 staminate hemp plants in early April at College Station, Texas, were approximately (1): 13, 11, 16, 5, 3, 18, 9, 9, 8, 6, 27, and 7 g. Calculate the sample mean, 11 g, and the deviations from it. Verify that $\Sigma x = 0$.

EXAMPLE 3.2.3—Ten patients troubled with sleeplessness each received a nightly dose of a sedative for a 2-week period. In another 2-week period they received no sedative (2). The average hours of sleep for each patient during each period were as follows:

Patient	1	2	3	4	5	6	7	8	9	10
Sedative	1.3	1.1	6.2	3.6	4.9	1.4	6.6	4.5	4.3	6.1
None	0.6	1.1	2.5	2.8	2.9	3.0	3.2	4.7	5.5	6.2

(*i*) Calculate the mean hours of sleep $\bar{X}_S$ under sedative and $\bar{X}_N$ under none. (*ii*) Find the ten differences D = (sedative − none). Show that their mean $\bar{X}_D$ equals $\bar{X}_S - \bar{X}_N$. (*iii*) Prove algebraically that this always holds.

EXAMPLE 3.2.4—In repeated throws of a six-sided die, appearance of a 1 is scored 1, appearance of a 2, 3, 4, is scored 2, and appearance of 5 or 6 is scored 3. If X_j denotes the score (j = 1, 2, 3), (*i*) write down the probability distribution of the X_j, (*ii*) find the population mean μ, (*iii*) verify that $\Sigma P_j(X_j - \mu) = 0$. Ans. (*i*) For $X_j = 1, 2, 3$, $P_j = 1/6, 3/6, 2/6$; (*ii*) $\mu = 13/6$.

EXAMPLE 3.2.5—In a table of truly random digits the probability P_j of a run of length j of even or odd digits is known to be $1/2^j$. With an infinite supply of random digits we could conceivably get runs of any length, so the number of distinct lengths of run in this population is infinite. By adding the terms in $\Sigma P_jX_j = \Sigma j/2^j$ for $j = 1, 2, 3, \ldots$, satisfy yourself that the population mean μ is 2. By $j = 10$ you should have reached a total equal to 1.9883 and by $j = 14$, 1.9990. The sum steadily approaches but never exceeds 2 as j increases.

EXAMPLE 3.2.6—In a card game each card from 2 to 10 is scored at its face value, while the ace, king, queen, and jack are each scored 10. (*i*) Write down the probability distribution of the scores in repeated sampling with replacement from a well-shuffled pack. (*ii*) Show that the population mean score $\mu = 94/13$. (*iii*) Write down the deviations $X_j - \mu$. (It may help to express the successive scores as 26/13, 39/13, and so on.) Verify that $\Sigma P_j(X_j - \mu) = 0$.

EXAMPLE 3.2.7—J. U. McGuire (3) counted the numbers X of European corn borers in 1296 corn plants. The following is the sample frequency distribution:

X	0	1	2	3	4	5	6	7	8	9	≥10
f	423	414	253	117	53	22	4	5	3	2	0

Calculate the sample mean $\bar{X} = 1.31$. This is another skew distribution, with 65% of the observations less than the mean and only 35% greater.

3.3—Population standard deviation. Roughly speaking, the population standard deviation is a measure of the amount of variation among the values of X in a population. Along with μ, the standard deviation is important as a descriptive statistic. In describing a variable population, often the first two things that we want to know are: What is the average level of X? How variable is X? Moreover, the standard deviation plays a dominant role in determining how accurately we can estimate the population mean from a sample of a given size. To illustrate, suppose we are planning a sample and would like to be almost certain to estimate μ from $\bar{X}$ correct to within one unit. If all the values of X in the population lie between 98 and 102, it looks as if a small sample (say $n < 10$) ought to do the job. If the values of X are spread over the range from 0 to 5000, we face a much tougher problem.

It is not obvious how to construct a measure of the amount of variation in a population. Presumably, this measure will depend on the sizes of the deviations $X - \mu$ of the individual values of X from the population mean. The larger these deviations, the more variable the population will be. The average of these deviations, $\Sigma P_j(X_j - \mu)$, is not a possible measure, since by (3.2.2) it is always zero. The average of the absolute deviations, $\Sigma P_j|X_j - \mu|$, could be used, but for several reasons a different measure was adopted around 1800—the population *standard deviation*. It is denoted by the Greek letter σ and defined as

$$\sigma = \sqrt{\sum_{j=1}^{k} P_j(X_j - \mu)^2} \qquad (3.3.1)$$

We square each deviation, multiply by the probability P_j of its occurrence, add, and finally take the square root.

For the population of random digits, $\mu = 4.5$. The successive deviations $X_j - \mu$ are $0 - 4.5 = -4.5, -3.5, -2.5, -1.5, -0.5, +0.5, +1.5, +2.5, +3.5, +4.5$. Each deviation has probability $P_j = 0.1$. Combining the two deviations that have the same size but opposite signs (e.g., -4.5 and $+4.5$), we get

$$\sigma^2 = (0.2)(4.5^2 + 3.5^2 + 2.5^2 + 1.5^2 + 0.5^2) = 8.25$$

This gives the standard deviation $\sigma = \sqrt{8.25} = 2.872$. The quantity

$$\sigma^2 = \sum_{j=1}^{k} P_j(X_j - \mu)^2 \qquad (3.3.2)$$

is called the population *variance*. The following result enables us to calculate σ^2

and hence σ without finding any deviations $X_j - \mu$. Expand the quadratic terms $(X_j - \mu)^2$ in (3.3.2). This step gives

$$\sigma^2 = \sum_{j=1}^{k} P_j X_j^2 - 2\mu \sum_{j=1}^{k} P_j X_j + \mu^2 \sum_{j=1}^{k} P_j \quad (3.3.3)$$

Since $\Sigma P_j X_j = \mu$ and $\Sigma P_j = 1$, (3.3.3) may be rewritten

$$\sigma^2 = \sum_{j=1}^{k} P_j X_j^2 - 2\mu^2 + \mu^2 = \sum_{j=1}^{k} P_j X_j^2 - \mu^2 \quad (3.3.4)$$

The student may ask: In what sense does σ measure the amount of variation in a population? If we know the value of σ for a population, what simple statements can we make about the extent of its variation? We can make only rather qualified statements, since the specific shape of the distribution also plays a role. However, for most distributions (both real and theoretical) met in statistical work, more than 94% of all observations in the population are less than a distance $\pm 2\sigma$ from the mean μ, that is, are *inside* the interval from $\mu - 2\sigma$ to $\mu + 2\sigma$. This interval is often written as $(\mu \pm 2\sigma)$.

For example, in the probability distribution of random digits, *all* the observations in the population lie inside this interval. With the random digits we found $\mu = 4.5$, $\sigma = 2.872$. The interval $(\mu \pm 2\sigma)$ therefore extends from -1.244 to 10.244 and contains all the digits 0, 1, 2, . . . , 9. At the other extreme is a simple probability distribution (see example 3.5.3) in which only 3/4 (or 75%) of the distribution lies inside the interval $(\mu \pm 2\sigma)$.

3.4—Two-class populations. Populations consisting of two classes (success-failure, yes-no, etc.) are very common. The purpose in collecting such data is usually to estimate the proportion of units in the population that belongs to one class—say the first class. By a simple device the formulas developed in preceding sections can be applied to this problem.

Construct a variate X that takes the value 1 for every unit in the population that belongs to class 1 and the value 0 for every unit that belongs to class 2. What does the sample mean $\bar{X}$ represent? We find this mean by adding all the sample Xs, which amounts to counting all the sample members of class 1, and dividing by n. Thus $\bar{X}$ is the proportion in the sample belonging to class 1. Similarly μ is the proportion in the population belonging to class 1. In two-class problems, it is customary to let p represent this proportion and to let $q = 1 - p$.

Thus in the population, X follows a very simple probability distribution as follows:

Values of X	Probability, P	$X - \mu$
0	q	$-p$
1	p	$1 - p = q$

The population mean for this distribution is, of course,

$$\mu = \Sigma PX = q(0) + p(1) = p \tag{3.4.1}$$

For the variance σ^2, we have from (3.3.2),

$$\sigma^2 = \Sigma P(X - \mu)^2 = qp^2 + pq^2 = pq(p + q) = pq$$

Alternatively, (3.3.4) also gives

$$\sigma^2 = \Sigma PX^2 - \mu^2 = p - p^2 = pq \tag{3.4.2}$$

The standard deviation $\sigma = \sqrt{pq}$. With a two-class population you may verify that the complete population lies inside the limits $(\mu \pm 2\sigma)$ if $0.2 < p < 0.8$. At $p = 0.21$, for instance, $\sigma = 0.407$. The limits $(\mu \pm 2\sigma)$ are $(-0.604, 1.024)$, which contain both $X_j = 0$ and $X_j = 1$. At $p = 0.2$, however, the limits are $(-0.6, 1)$; and $X_j = 1$, which has probability 0.2, lies on the boundary of the interval, not inside it. Therefore, for a two-class population the percent of the population lying inside $(\mu \pm 2\sigma)$ may fall to 80%.

3.5—Sample standard deviation. In seeking a sample estimator of σ, it is natural to start with the deviations $x_i = x_i - \overline{X}$ of the sample observations from the sample mean.

In statistical work the most widely used estimator of σ is the sample standard deviation, denoted by s. If the sample observations have not been arranged in a frequency distribution, the formula defining s is

$$s = \sqrt{\sum_{i=1}^{n} (X_i - \overline{X})^2/(n - 1)} = \sqrt{\sum_{i=1}^{n} x_i^2/(n - 1)} \tag{3.5.1}$$

First, each deviation is squared. Next, the sum of squares Σx^2 is divided by $(n - 1)$, one less than the sample size. The result is the *sample variance* s^2. Finally, extraction of the square root recovers the original scale of measurement.

As an example, a random sample of 10 mothers' heights was drawn from the data leading to table 2.4.1. The calculation of s is shown in table 3.5.1. Note that s is in the same units (in.) as the original observations.

Before further discussion of s, its calculation should be fixed in mind by working a couple of examples. Table A 22 is a table of square roots. Hints on

TABLE 3.5.1
CALCULATION OF SAMPLE STANDARD DEVIATION

Sample number	1	2	3	4	5	6	7	8	9	10	Total	$\overline{X}$
Heights (in.)	59	58	65	68	66	63	66	61	65	59	630	63
Deviations, x_i	−4	−5	+2	+5	+3	0	+3	−2	+2	−4	0	
	$\Sigma x_i^2 = 112$		$s^2 = 112/9 = 12.44$				$s = \sqrt{12.44} = 3.53$ in.					

finding square roots from this table are given on p. 462. Many pocket calculators have a key that gives square roots when pressed.

EXAMPLE 3.5.1—Five patients with pneumonia are treated with sodium penicillin G. The numbers of days required to bring the temperature down to normal were 1, 4, 7, 5, 3. Compute s for this sample. Ans. $s = 2.24$ days.

EXAMPLE 3.5.2—Calculate s for the hemp plant weights in example 3.2.1. Ans. $s = 6.7$ g.

The appearance of the divisor $(n - 1)$ instead of n in computing s^2 and s is puzzling at first sight. The quantity $(n - 1)$ is called the number of *degrees of freedom* in s. Later we shall meet situations in which the number of degrees of freedom is not $(n - 1)$ but some other quantity. If the practice of using the degrees of freedom as divisor is followed, the same statistical tables that are needed in important applications serve for a wide variety of purposes—a considerable advantage.

Division by $(n - 1)$ leads to a property that is often cited. If random samples are drawn with replacement from any population, *the average value of s^2 taken over all random samples is exactly equal to σ^2*. An estimate whose average value over all possible samples is equal to the population parameter being estimated is called *unbiased*. Thus, s^2 is an unbiased estimator of σ^2 in random sampling with replacement. This property, which says that *on the average* the estimator gives the correct answer, seems desirable for an estimator to possess, but it is not essential. For example, the standard deviation s is slightly biased as an estimate of σ.

More important are the questions: Is s^2 an accurate estimate of σ^2? Is s an accurate estimate of σ? By ordinary standards, the answer is no: s^2 and s are accurate only with large samples. For illustration, suppose that we are drawing a random sample and want to be reasonably sure that the sample estimate s will be correct to within $\pm 10\%$. That is, s lies between 0.9σ and 1.1σ. With a bell-shaped distribution like that of the mothers' heights in table 2.3.1, it can be shown that a random sample of size around $n = 200$ is required to make the probability 0.95 that s will lie within $\pm 10\%$ of σ. The trouble comes mainly from the extreme observations in the population, which have large deviations that are squared when calculating s^2. If the sample happens to have fewer of these than expected, s may be much too small, if more than expected, much too large.

By a well-known formula, the sum of squares of deviations Σx^2 in the numerator of s^2 can be calculated without finding any deviations. Expand each of the quadratic terms $(X_i - \overline{X})^2$ in Σx^2. Then

$$\sum_{i=1}^{n} (X_i - \overline{X})^2 = \sum_{i=1}^{n} X_i^2 - 2\sum_{i=1}^{n} X_i\overline{X} + \sum_{i=1}^{n} \overline{X}^2 = \sum_{i=1}^{n} X_i^2 - 2\overline{X}\sum_{i=1}^{n} X_i + n\overline{X}^2$$

since the term $\Sigma\overline{X}^2$ amounts to adding $\overline{X}^2$ to itself n times. Further, if we write $\Sigma X_i = n\overline{X}$ in the second term above, the second and third terms combine to give $-n\overline{X}^2$. Hence

$$\sum_{i=1}^{n} x_i^2 = \sum_{i=1}^{n} (X_i - \overline{X})^2 = \sum_{i=1}^{n} X_i^2 - n\overline{X}^2 \tag{3.5.2}$$

Since $\overline{X} = \Sigma X_i/n$, (3.5.2) can be rewritten as

$$\sum_{i=1}^{n} x_i^2 = \sum_{i=1}^{n} X_i^2 - \left(\sum_{i=1}^{n} X_i\right)^2 \Big/ n \tag{3.5.3}$$

We take the sum of the squares of the original observations and subtract from this sum the square of their total divided by n.

WARNING. If the observations X_i are large and we round $\overline{X}$ when using (3.5.2), it can give results that are very wrong. Equation (3.5.3) works well if no rounding is done until we have seen the values of its two terms.

For example, consider a sample with $n = 3$ and $X_i = 63, 60, 59$. Equation (3.5.3) gives, correctly,

$$\Sigma x_i^2 = 63^2 + 60^2 + 59^2 - 182^2/3 = 11{,}050 - 11{,}041.33 = 8.67$$

or fully correct, 8 2/3. The sample mean $\overline{X} = 182/3 = 60\ 2/3$. If this is rounded to 61 when using (3.5.2), we get

$$\Sigma x_i^2 = 63^2 + 60^2 + 59^2 - 3(61^2) = 11{,}050 - 11{,}163 = -113$$

a nonsensical value.

Incidentally, rounding $\overline{X}$ to the same degree as the original X_i gives a roughly correct answer if Σx_i^2 is computed from the deviations. Taking $\overline{X} = 61$, the deviations are $+2, -1, -2$, giving $\Sigma x_i^2 = 9$—the correct answer (8 2/3) rounded to an integer.

EXAMPLE 3.5.3—In a population a variable X takes only three distinct values: -1 with probability $P_1 = 1/8$; 0 with $P_2 = 3/4$; and $+1$ with $P_3 = 1/8$. Calculate μ and σ and show that the probability is 3/4 that X lies inside the limits $(\mu \pm 2\sigma)$. Ans. $\mu = 0$; $\sigma = 1/2$. The limits are $(-1, 1)$. In a famous result, Chebyshev (4) proved that there is no population with finite μ and σ for which this probability is less than 3/4. More generally, he proved that the probability that X lies inside the interval from $(\mu - r\sigma)$ to $(\mu + r\sigma)$ is greater than or equal to $(1 - 1/r^2)$.

EXAMPLE 3.5.4—Here are some simple data from which to verify that $\Sigma x^2 = \Sigma X^2 - \Sigma X^2/n$: 5, 7, 9, 4, 3, 2. Show by using the deviations and the formula above that $\Sigma x^2 = 34$, $s^2 = 6.8$.

EXAMPLE 3.5.5—For the 10 patients in example 3.2.3, the average differences in hours of sleep per night between sedative and no sedative were: 0.7, 0.0, 3.7, 0.8, 2.0, -1.6, 3.4, -0.2, -1.2, -0.1. Calculate s. Ans. $s = 1.79$ hr.

EXAMPLE 3.5.6—Subtraction of any constant amount X_0 from all sample values of X reduces $\overline{X}$ by X_0 but does not change s. These results are sometimes useful when no calculator is available. The constant X_0 is sometimes called a *working mean*. The values in a sample with $n = 5$ are 478, 482, 484, 469, 475. (i) Let $U = X - 480$. Write down the 5 values of U. Show that $\overline{X} = 477.6$ and $\overline{U} =$

$-2.4 = \overline{X} - 480$. (*ii*) By use of the formula $\Sigma u_i^2 = \Sigma U_i^2 - (\Sigma U_i)^2/n$, show that $s_u = 5.94$. Hence $s_x = s_u = 5.94$. (*iii*) Repeat the process with $x_0 = 477$ and check that you get the same values of $\overline{X}$ and s.

EXAMPLE 3.5.7—Without finding deviations from $\overline{X}$ or using a calculating machine, compute Σx^2 for the following measurements: 961, 953, 970, 958, 950, 951, 957. Ans. 286.9.

3.6—A note re computers. Availability of computers and computer programs saves students from hours of calculations. Instructors are not yet fully agreed on how best to introduce students to computers in introductory courses that are not intended to include a standard programming language. At present the most common method is for the instructor to construct a set of programs to carry out the principal calculations used in the course. These programs are given names or contracted names, such as ANOVA (for analysis of variance) or COMPARE (for comparisons among group means, as in the palatability study). The student learns how to enter *(input)* the data into the computer, how to instruct the computer to perform the main calculation, and how to request that certain intermediate or subsidiary results be printed. The results of a computation are usually called the *output*. The program may have some optional calculations related to the interpretation of the main calculation, and the computer may ask the student which, if any, of these options are wanted.

With computers available, the student need not spend time learning the speediest methods for doing the calculations without a computer. The student does need to be able to answer the following questions. Exactly what is the main calculation? What is its purpose? To what kinds of data can it be applied and to what kinds should it not be applied? Are any important assumptions made in the calculation? Do these assumptions seem to hold reasonably well in the calculations to be performed? Which options are necessary or may help in the interpretation of the results of the main analysis? The programs themselves are usually checked beforehand by running calculations already done on a desk machine.

Any such set of computer programs will include one that calculates $\overline{X}$ and s. Also, many pocket calculators have a built-in program that computes $\overline{X}$ and s by entering the data with a special key and then pressing two keys marked MEAN and S.D. or S. DEV. However, the next sections give an example of the calculation of $\overline{X}$ and s from data in the form of a frequency distribution, as is commonly done with large samples. This calculation may not be given by pocket calculators or in small sets of programs constructed for introductory courses.

3.7—Use of frequency distributions to calculate $\overline{X}$ and s. The purpose in first constructing a frequency distribution when calculating $\overline{X}$ and s is to save time. Two cases occur. In the first, the distribution of X is discrete, with X taking only a limited number k of distinct values X_j ($j = 1, 2, \ldots, k$). The frequency f_j with which any X_j appears in the sample is counted. The method for calculating $\overline{X}$ in this case was given in (3.1.3):

$$\overline{X} = \sum_{j=1}^{k} f_j X_j / n \tag{3.7.1}$$

The sample total sum of squares of the Xs is

$$\sum_{i=1}^{n} X_i^2 = \sum_{j=1}^{k} f_j X_j^2 \qquad (3.7.2)$$

Hence, using (3.5.3) for Σx_i^2,

$$s^2 = [1/(n-1)] \sum_{i=1}^{n} x_i^2 = [1/(n-1)] \left[\sum_{j=1}^{k} f_j X_j^2 - \left(\sum_{j=1}^{k} f_j X_j\right)^2 \Big/ n\right] \qquad (3.7.3)$$

In the second case, we have grouped the Xs into k classes with known class limits. This grouping may be done either for a continuous variable or for a discrete variable that takes many distinct values, as with the mothers' heights and city sizes in table 2.3.1. In using (3.7.1) and (3.7.3) we know f_j, the class frequencies. For X_j we use as an approximation the class midpoints—the average of the two class limits. The approximation assumes that the values of X in a class are evenly distributed throughout the class. Caution is necessary if unusual groupings are present in the scale of measurement. A change in the class limits may bring the mean of the Xs into a class nearer the class midpoint.

The same problem with class midpoints can occur in the extreme classes of a frequency distribution, where the frequencies are tailing off. With class intervals 0–9, 10–19, etc., we might notice that the first class contained six 0s, one 2, and one 6, giving $\bar{X}_1 = 1$ against a class midpoint of 4.5. For more accurate work, use $\bar{X}$ instead of the class midpoint in such extreme classes.

3.8—Numerical example. The data in table 3.8.1 come from a sample of 533 weights of swine, grouped into 11 classes. Class midpoints are given in the

TABLE 3.8.1
CALCULATION OF MEAN AND STANDARD DEVIATION FROM THE FREQUENCY DISTRIBUTION OF WEIGHTS OF 533 SWINE (X_0 = 185 lb, I = 20 lb)

Class Midpoint (lb)	Frequency f	Coded Numbers U	fU	fU^2	Calculations
85	1	−5	−5	25	$\Sigma fU = -120$
					$I\bar{U} = 20\,(-120)/533$
					$= -4.5$ lb
105	7	−4	−28	112	$\bar{X} = X_0 + I\bar{U} = 185 - 4.5$
125	39	−3	−117	351	$= 180.5$ lb
145	66	−2	−132	264	
165	143	−1	−143	143	$s_U^2 = \dfrac{\Sigma fU^2 - (\Sigma fU)^2/n}{n-1}$
185	112	0	0	0	
205	80	1	80	80	$= \dfrac{(1650) - 120^2/533}{532}$
225	53	2	106	212	
245	16	3	48	144	$s_U^2 = 3.051,\ s_U = 1.747$
265	9	4	36	144	
285	7	5	35	175	
Total	533		−120	1650	$s_X = Is_U = 20s_U = 34.9$ lb

first column. The steps in the calculation of $\overline{X}$ and s are given to the right in the table.

A further simplification comes from *coding* the class midpoints. The symbol U in the third column denotes the coded values. On the coded U scale, let $U = 0$ correspond to a class midpoint on the X scale for a class near the middle. In table 3.8.1, $U = 0$ was chosen at $X = 185$ lb. Call this value X_0. With class intervals of equal widths I, the class midpoints above this class are coded 1, 2, 3, etc.; those below are -1, -2, -3, etc.

The relation between the X scale and the U scale is important:

$$X = X_0 + IU \tag{3.8.1}$$

Two results that follow are

$$\overline{X} = X_0 + I\overline{U} \qquad s_X = Is_U \tag{3.8.2}$$

Equations (3.8.2) use two general results worth remembering. If every observation in a sample is multiplied by a constant I, the sample standard deviation (SD) is multiplied by I. Hence $s_{IU} = Is_U$. Second, addition of a constant to every member of a sample does not change the value of s, since the constant cancels out in every deviation $x_i = X_i - \overline{X}$. Hence $s_X = s_{IU} = Is_U$.

The steps in finding $\overline{X}$ and s in table 3.8.1 should now be easy to follow. We first find $\overline{U}$ and s_U. With a calculator that finds a sum of products, the values fU^2 need not be written down, since $\Sigma fU^2 = \Sigma(fU)U$. From $\overline{U}$ and s_U, $\overline{X}$ and s_x are found by (3.8.2).

EXAMPLE 3.8.1—Compute s for the frequency distribution of numbers of corn borers per plant in example 3.2.7. Ans. $s = 1.36$.

EXAMPLE 3.8.2—The data show the frequency distribution of the heights of 8585 men, arranged in 2-in. classes. Compute $\overline{X}$ and s, using a convenient coding. Ans. $\overline{X} = 67.53$ in., $s = 2.62$ in.

Class Midpoint	Frequency	Class Midpoint	Frequency
58	6	68	2559
60	55	70	1709
62	252	72	594
64	1063	74	111
66	2213	76	23

EXAMPLE 3.8.3—The yields in grams of 1499 rows of wheat are recorded by Wiebe (5). In a tabulation, the class midpoints (grams) and frequencies f are as follows:

Midpoint	400	475	550	625	700	775	850	925	1000	Total
f	57	314	420	389	207	77	24	9	2	1499

Compute $\overline{X} = 588.0$ g and $s = 101.6$ g, using a coded scale.

EXAMPLE 3.8.4—In the preceding example, estimate the percentage of this large sample that lies within the limits $\overline{X} - 2s$ and $\overline{X} + 2s$. This estimation can be done by constructing the cumulative distribution and using either a frequency polygon or interpolation. Regard the true class limits as (362.5, 437.5), (437.5, 512.5), and so on. Ans. Our polygon gave 95.0%; interpolation gave 95.2%.

EXAMPLE 3.8.5—This example illustrates how the accuracy of the grouped frequency method of calculating $\overline{X}$ and s improves when the class midpoints are replaced by the actual means of the observations in the classes. The data ($n = 14$) are as follows: 0, 0, 10, 12, 14, 16, 20, 22, 24, 25, 29, 32, 34, 49. (*i*) Compute $\overline{X}$ and s directly from these data. (*ii*) Form a grouped frequency distribution with classes 0–9, 10–19, etc. Compute $\overline{X}$ and s using the usual class midpoints of 4.5, 14.5, etc. (*iii*) Find the actual means of the observations in each class and use these means instead of the midpoints in finding $\overline{X}$ and s. (Coding does not help here.) Ans. (*i*) $\overline{X} = 20.5$, $s = 13.4$; (*ii*) $\overline{X} = 21.6$, $s = 11.4$ (both quite inaccurate); (*iii*) $\overline{X} = 20.5$ (correct, of course), $s = 13.2$.

3.9—Coefficient of variation. A measure often used in describing the amount of variation in a population is the *coefficient of variation:* $cv = \sigma/\mu$. The sample estimate is $s/\overline{X}$. The standard deviation is expressed as a fraction, or sometimes as a percentage, of the mean. For data from different populations or sources, the mean and standard deviation often tend to change together so that the coefficient of variation is relatively stable or constant. Furthermore, being dimensionless, *cv* is easy to remember.

For example, the following data come from the frequency distributions of the heights of 1052 mothers in table 2.3.1 and of 8585 men in example 3.8.2:

	$\overline{X}$ (in.)	s (in.)	*cv*
Mothers	62.5	2.3	3.7%
Men	67.5	2.6	3.8%

In other studies (6), the coefficient of variation of stature for both males and females appeared to be close to 3.75% from age 18 onwards.

As a second example (7), the question was raised as to what method would give the most consistent results in experiments involving chlorophyll determinations in pineapple plants. Three bases of measurement were tried, each involving 12-leaf samples, resulting in the following data:

Statistic	100 g Wet Basis	100 g Dry Basis	10 cm^2 Basis
Sample mean, $\overline{X}$ (mg)	61.4	33.7	13.71
Sample standard deviations (mg)	5.22	31.2	1.20
Coefficient of variation, *cv* (%)	8.5	9.3	8.8

In appraising consistency, we obviously do not compare the s values, since the population means will apparently be widely different for the different bases. Instead, the *cv*s are compared. It was judged that the methods were about equally reliable so that the most convenient basis could be chosen.

TECHNICAL TERMS

arithmetic mean
class midpoints
coded scale
coding
coefficient of variation, *cv*
degrees of freedom

deviations
input
output
population mean
population standard deviation
population total
population variance
sample mean
sample standard deviation
sample variance
unbiased
working mean

REFERENCES

1. Talley, P. J. *Plant Physiol.* 9 (1934):737.
2. Cushny, A. R., and Peebles, A. R. *Am. J. Physiol.* 32 (1905):501.
3. McGuire, J. U. Mimeogr. Ser. 2, Stat. Lab., Iowa State Univ., Ames.
4. Feller, W. 1957. *An Introduction to Probability Theory and its Applications,* vol. 1, 2nd ed. Wiley, New York.
5. Wiebe, G. A. *J. Agric. Res.* 50 (1935):331.
6. Council on Foods. *J. Am. Med. Assoc.* 110 (1938):651.
7. Tam, R. K., and Magistad, O. C. *Plant Physiol.* 10 (1935):161.

4

The normal distribution

4.1—Normally distributed populations. In section 2.6 you were introduced to probability distributions for discrete variates that take on a limited number of distinct values. Now we turn to variables that are essentially continuous, such as heights, weights, or yields. The variable flows without a break from one value to the next with no limit to the number of distinct values. Consider the histogram of mothers' heights in figure 2.3.1. Imagine that the size of the sample is increased without limit and the class intervals on the horizontal axis are decreased correspondingly. Figure 2.3.1 would gradually become a continuous curve. Continuous variables are distributed in many ways; we first consider the *normal distribution.*

As you know, a theoretical frequency distribution for a discrete variable is described by stating each distinct value X_j of X and the probability P_j that X equals X_j, with these probabilities adding to 1. With a continuous variable X, we give instead a function $f(X)$ that states the height or ordinate of the continuous curve at the abscissa value X. The normal distribution is completely determined by its mean μ and its standard deviation σ. The height $f(X)$ at the value X is

$$Y = f(X) = [1/(\sigma\sqrt{2\pi})]e^{-(X-\mu)^2/(2\sigma^2)} \tag{4.1.1}$$

The quantity $e = 2.7183$ is the base for natural logarithms and π is of course 3.1416. Figure 4.1.1 shows a normal curve fitted to the 1052 mothers' heights given in figure 2.3.1. Visually, the agreement between the histogram and the curve looks good, as it usually does with human heights.

To illustrate the role of the standard deviation σ in determining the shape of the curve, figure 4.1.2 shows two curves. The solid curve has $\mu = 0, \sigma = 1$, while the dotted curve has $\mu = 0, \sigma = 1.5$. The curve with larger σ is lower at the mean and spread wider. Values of X that are far from the mean are more frequent when $\sigma = 1.5$ than when $\sigma = 1$. A curve with $\sigma = 0.5$ would have a height of nearly 0.8 at the mean and would have frequency nearly zero beyond $X = 1.5$. In any normal curve, over 2/3 of the measurements lie in the interval $\mu \pm \sigma$, while some 95% are in the interval $\mu \pm 2\sigma$. Only 0.26% of the total frequency lies beyond $\pm 3\sigma$.

To indicate the effect of a change in the mean μ, the curve with $\mu = 2, \sigma = 1$ is obtained by lifting the solid curve bodily and centering it at $X = 2$ without changing its shape in any other way. This property explains why μ is called the

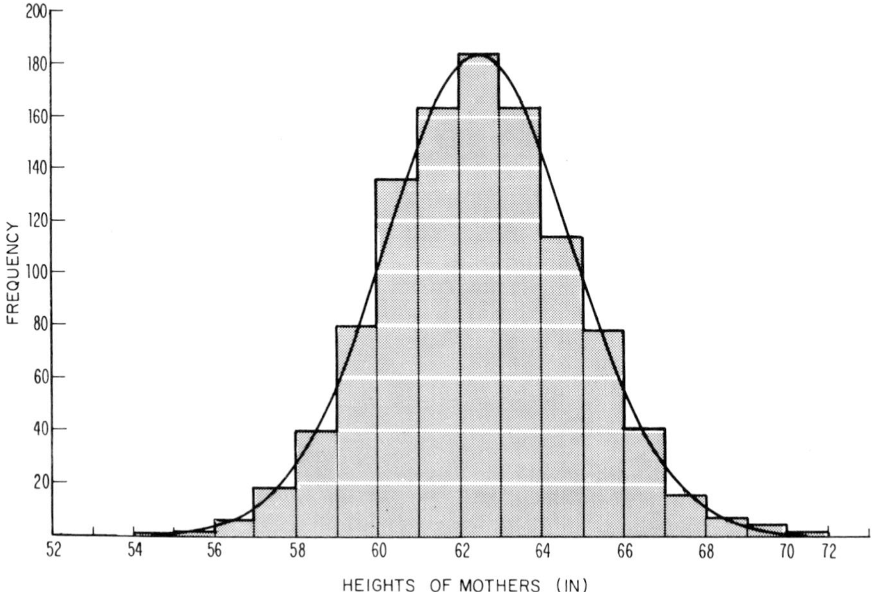

Fig. 4.1.1—Normal distribution fitted to the heights of 1052 mothers.

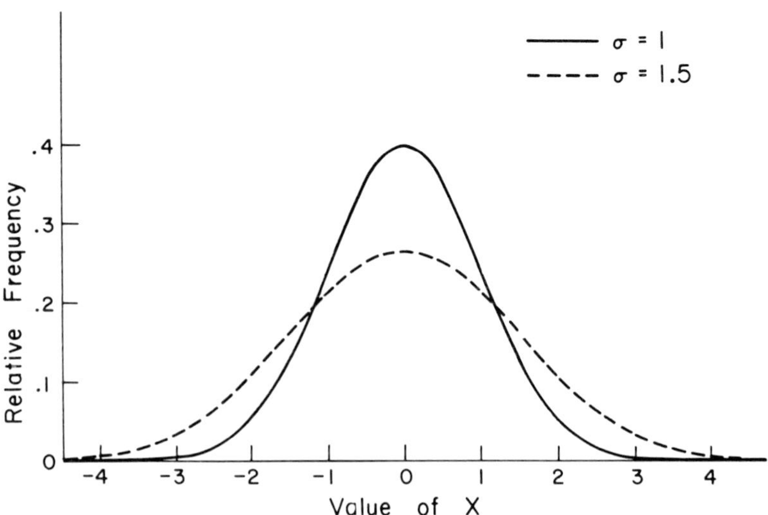

Fig. 4.1.2—Solid curve: normal distribution with $\mu = 0$ and $\sigma = 1$. Dotted curve: normal distribution with $\mu = 0$ and $\sigma = 1.5$.

parameter of location, the word *parameter* denoting constants like μ or σ that enter into the height $f(X)$ of the curve.

4.2—Reasons for use of the normal distribution. The normal distribution was one of the earliest to be developed. De Moivre published its equation in 1733. It is astonishing that the normal distribution has dominated statistical practice as well as theory. Briefly, the main reasons are as follows:

1. The distributions of some variables are approximately normal, such as heights of people; lengths of ears of corn; and more generally, many linear dimensions—for instance, those of numerous manufactured articles.
2. For measurements whose distributions are not normal, a simple transformation of the scale of measurement may induce approximate normality. The square root, $\sqrt{X}$, and the logarithm, $\ln X$, are often used as such transformations. The scores made by students in examinations are frequently rescaled so that they appear to follow a normal curve.
3. The normal distribution is relatively easy to work with mathematically. Many results useful in statistical work can be derived mathematically when the population is normal. Such results may hold well enough for rough-and-ready use when samples come from nonnormal populations, including populations with discrete variables. When presenting such results we shall try to indicate how well the results stand up under nonnormality.
4. Even if the distribution in the original population is far from normal, the distribution of sample means $\bar{X}$ tends to become normal under random sampling as the size of sample increases. This result, given in more detail in sections 4.5 and 4.6, is the single most important reason for the use of the normal curve.

4.3—Tables of the normal distribution. Since the normal curve depends on the two parameters μ and σ, there are a great many different normal curves. All standard tables of this distribution are for the distribution with $\mu = 0$ and $\sigma = 1$. Consequently if you have a measurement X with mean μ and standard deviation σ and wish to use a table of the normal distribution, you must rescale X so that the mean becomes 0 and the standard deviation becomes 1. The rescaled measurement is given by relation

$$Z = (X - \mu)/\sigma$$

The quantity Z goes by various names—*standard normal variate; standard normal deviate; normal variate in standard measure;* or, in education and psychology, *standard score* (although this term sometimes has a slightly different meaning). To transform from the Z scale back to the X scale, the formula is

$$X = \mu + \sigma Z$$

There are two principal tables.

TABLE OF ORDINATES. Table A 2 gives the ordinates or heights of points on the standard normal distribution curve. The formula for the ordinate is

$$y = (1/\sqrt{2\pi})e^{-Z^2/2}$$

These ordinates are used when graphing the normal curve. Since the curve is symmetrical about the origin, the heights are presented only for positive values of Z. Here is an example.

EXAMPLE 1—Suppose that we wish to sketch the normal curve for a variate X for which $\mu = 3$ and $\sigma = 1.6$. What is the height of this curve at $X = 2$?

Step 1. Find $Z = (2 - 3)/1.6 = -0.625$.

Step 2. Read the ordinate in table A 2 for $Z = 0.625$. In the table, the Z entries are given to two decimal places only. For $Z = 0.62$ the ordinate is 0.3292 and for $Z = 0.63$ the ordinate is 0.3271. Hence we take 0.328 for $Z = 0.625$.

Step 3. Finally, divide the ordinate 0.328 by σ, getting $0.328/1.6 = 0.205$ as the answer. This step is needed because if you look back at (4.1.1) for the ordinate of the general normal curve, you will see a σ in the denominator that does not appear in the tabulated curve.

TABLE OF THE CUMULATIVE DISTRIBUTION. Table A 3 is much more frequently used than table A 2. It gives the area under the curve from 0 to Z. The table has positive values of Z but by the symmetry of the normal curve, the same tabulated values hold for negative Z. Thus the probability that a variate drawn at random from the standard normal distribution has a value lying between -1.2 and 0, or between 0 and $+1.2$, are each equal to 0.3849 as shown in table A 3.

Let us take a quick look at table A 3. If we want the probability for $Z = 1.64$, we look along the row beginning 1.6 until we come to the column headed 0.04 (the second decimal place in Z). The entry at this point is 0.4495—the probability that a normal deviate lies between 0 and 1.64, or between -1.64 and 0.

At $Z = 3.9$, or any larger value, the area is 0.5000 to four decimal places. It follows that the probability of a value of Z lying between -3.9 and $+3.9$ is 1.0000 to four decimals, since the curve is symmetrical about the origin. This means that any value drawn from a standard normal distribution is practically certain to lie between -3.9 and $+3.9$. At $Z = 1.0$, the area is 0.3413. Thus the probability of a value lying between -1 and $+1$ is 0.6826. This verifies a previous remark (section 4.1) that over 2/3 of the observations in a general normal distribution lie in the interval $\mu \pm \sigma$. Similarly, for $Z = 2$ the area is 0.4772, corresponding to the result that about 95% of the observations (more accurately 95.44%) will lie between $\mu - 2\sigma$ and $\mu + 2\sigma$.

When using table A 3, you will often want probabilities represented by areas different from those tabulated. If A is in the area in table A 3, table 4.3.1 shows how to obtain the probabilities most commonly needed.

Verification of these formulas is left as an exercise. A few more complex examples follow.

EXAMPLE 2—What is the probability that a normal deviate lies between -1.62 and $+0.28$? We have to split the interval into two parts: from -1.62 to 0, and from 0 to 0.28. From table A 3, the areas for the two parts are, respectively, 0.4474 and 0.1103, giving 0.5577 as the answer.

TABLE 4.3.1

FORMULAS FOR FINDING PROBABILITIES RELATED TO THE NORMAL DISTRIBUTION

Probability of a Value	Formula
(1) Lying between 0 and Z	A
(2) Lying between $-Z$ and Z	$2A$
(3) Lying outside the interval $(-Z, Z)$	$1 - 2A$
(4) Less than Z (Z positive)	$0.5 + A$
(5) Less than Z (Z negative)	$0.5 - A$
(6) Greater than Z (Z positive)	$0.5 - A$
(7) Greater than Z (Z negative)	$0.5 + A$

EXAMPLE 3—What is the probability that a normal deviate lies between -2.67 and -0.59? In this case we take the area from -2.67 to 0, namely 0.4962, and subtract from it the area from -0.59 to 0, namely 0.2224, giving 0.2738.

EXAMPLE 4—The heights of a large sample of men were found to be approximately normally distributed with mean 67.56 in. and standard deviation 2.57 in. What proportion of the men have heights less than 5 ft 2 in.? We must first find Z.

$$Z = (X - \mu)/\sigma = (62 - 67.56)/2.57 = -2.163$$

The probability wanted is the probability of a value less than Z, where Z is negative. We use (5) in table 4.3.1. Reading table A 3 at $Z = 2.163$, we get $A = 0.4847$, interpolating mentally between $Z = 2.16$ and $Z = 2.17$. From formula (5), the answer is $0.5 - A$, or 0.0153. About 1 1/2% of the men have heights less than 5 ft 2 in.

EXAMPLE 5—What height is exceeded by 5% of the men? The first step is to find Z. We use (6) in table 4.3.1, writing $0.5 - A = 0.05$, so that $A = 0.45$. We now look at table A 3 for the value of Z such that $A = 0.45$. The value is $Z = 1.645$. Hence the actual height is

$$X = \mu + \sigma Z = 67.56 + (2.57)(1.645) = 71.79 \text{ in.}$$

just under 6 ft.

Some examples to be worked by the reader follow:

EXAMPLE 4.3.1—Using table A 2, (*i*) what is the height of a normal curve at the mean, with $\sigma = 2$? (*ii*) For any normal curve, at what value of X is the height of the curve one-tenth of the highest height? Ans. (*i*) 0.1994, (*ii*) at the value $X = \mu + 2.15\sigma$.

EXAMPLE 4.3.2—Using table A 3, show that 92.16% of the items in a normally distributed population lie between -1.76σ and $+1.76\sigma$.

EXAMPLE 4.3.3—Show that 65.24% of the items in a normal population lie between $\mu - 1.1\sigma$ and $\mu + 0.8\sigma$.

EXAMPLE 4.3.4—Show that 13.59% of the items lie between $Z = 1$ and $Z = 2$.

EXAMPLE 4.3.5—Show that half the population lies in the interval from $\mu - 0.6745\sigma$ to $\mu + 0.6745\sigma$. The deviation 0.6745σ, formerly often used, is called the *probable error* of X. Ans. You will have to use interpolation. You are seeking a value of Z such that the area from 0 to Z is 0.2500; $Z = 0.67$ gives 0.2486 and $Z = 0.68$ gives 0.2517. Since $0.2500 - 0.2486 = 0.0014$, and $0.2517 - 0.2486 = 0.0031$, we need to go 14/31 of the distance from 0.67 to 0.68. Since $14/31 = 0.45$, the desired result is $Z = 0.6745$.

EXAMPLE 4.3.6—Show that 1% of the population lies outside the limits $Z = +2.575$.

EXAMPLE 4.3.7—If $\mu = 67.56$ in. and $\sigma = 2.57$ in. for the heights of men, what percentage of the population has heights between 5 ft 5 in. and 5 ft 10 in.? Compute your Zs to two decimals only. Ans. 67%.

EXAMPLE 4.3.8—The specification for a manufactured component is that the pressure at a certain point must not exceed 30 lb. A manufacturer who would like to enter this market can make components with a mean pressure $\mu = 28$ lb, but the pressure varies from one specimen to another with a standard deviation $\sigma = 1.6$ lb. What proportion of specimens will fail to meet the specification? Ans. 10.6%.

EXAMPLE 4.3.9—By quality control methods it may be possible to reduce σ in the previous example while keeping μ at 28 lb. If the manufacturer wishes only 2% of the specimens to be rejected, what must σ be? Ans. 0.98 lb.

4.4—Standard deviation of sample means. With measurement data, as mentioned in section 3.1, the purpose of an investigation is often to estimate an average or total over a population (average selling price of houses in a town, total wheat crop in a region). If the data are a random sample from a population, the sample mean $\bar{X}$ is often used to estimate the corresponding average over the population. This brings up the question: How accurate is a sample mean as an estimator of the population mean?

In addition to Monte Carlo experiments as described in section 1.8, much mathematical work has been done on this problem. It has produced two of the most exciting and useful results in the whole of statistical theory. These results, which are part of every statistician's stock in trade, are stated first; some experimental verification will then be presented for illustration. The first result gives the mean and standard deviation of $\bar{X}$ in repeated sampling; the second gives the shape of the frequency distribution of $\bar{X}$.

MEAN AND STANDARD DEVIATION OF $\bar{X}$. If repeated random samples of size n are drawn from any population (not necessarily normal) with mean μ and standard deviation σ, the frequency distribution of the sample means $\bar{X}$ in these repeated samples has mean μ and standard deviation $\sigma/\sqrt{n}$.

This result says that under random sampling the sample mean $\bar{X}$ is an unbiased estimator of μ; on the average, in repeated sampling, it will be neither too high nor too low. Further, the sample means have less variation about μ than the original observations. The larger the sample size, the smaller this variation becomes.

Students sometimes find it difficult to reach the point at which *standard deviation of $\bar{X}$* has a concrete meaning for them. Having been introduced to the idea of a standard deviation, it is not too hard to feel at home with a phrase like "the standard deviation of a man's height," because every day we see tall men and short men and realize that this standard deviation is a measure of the extent to which heights vary from one man to another. But usually when we have a sample, we calculate a *single* mean. Where does the variation come from? It is the variation that would arise if we drew repeated samples from the population that we are studying and computed the mean of each sample. The experimental samplings presented in this chapter may make this concept more realistic.

The standard deviation of $\bar{X}$, $\sigma/\sqrt{n}$, is often called, alternatively, the *standard error of* $\bar{X}$. The terms *standard deviation* and *standard error* are synonymous. When we are studying the frequency distribution of an estimator like $\bar{X}$, its standard deviation supplies information about the amount of error in $\bar{X}$ when it is used to estimate μ. Hence, the term "standard error" is rather natural when applied to a mean. Normally, we would not speak of the standard error of a man's height because height does not imply error.

The quantity $N\bar{X}$, often used to estimate a total over the population, is also an unbiased estimator under random sampling. Since N is simply a fixed number, the mean of $N\bar{X}$ in repeated sampling is $N\mu$, which, by the definition of μ, is the correct population total. The standard error of $N\bar{X}$ is $N\sigma/\sqrt{n}$. Another frequently used result is that the sample total, $\Sigma X = n\bar{X}$, has a standard deviation $n\sigma/\sqrt{n}$, or $\sigma\sqrt{n}$.

Strictly, the result $\sigma/\sqrt{n}$ for the standard deviation of $\bar{X}$ requires random sampling with replacement. But if the ratio n/N of sample size to population size is small, the result can be used also for sampling without replacement.

4.5—Frequency distribution of sample means. The *second major result* from statistical theory is that, whatever the shape of the frequency distribution of the original population of Xs, the frequency distribution of $\bar{X}$ in repeated random samples of size n tends to become normal as n increases. To put the result more specifically, recall that if we wish to express a variable X in *standard measure* so that its mean is 0 and its standard deviation is 1, we change the variable from X to $(X - \mu)/\sigma$. For $\bar{X}$, the corresponding expression in standard measure (sm) is

$$\bar{X}_{sm} = \frac{(\bar{X} - \mu)}{\sigma/\sqrt{n}}$$

As n increases, the probability that $\bar{X}_{sm}$ lies between any two limits L_1 and L_2 approaches the probability that the standard normal deviate Z lies between L_1 and L_2. This result, known as the *central limit theorem* (1), explains why the normal distribution and results derived from it are so commonly used with sample means, even when the original population is not normal. Apart from the condition of random sampling, the theorem requires very few assumptions; it is sufficient that σ is finite and that the sample is a random sample from the population.

To the practical worker, a key question is: How large must n be to use the normal distribution for $\bar{X}$? No simple general answer is available. With variates like the heights of men, the original distribution is near enough normal so that normality may be assumed for most purposes. In this case a sample with $n = 1$ is large enough. There are also populations, at first sight quite different from the normal, in which $n = 4$ or 5 will do. At the other extreme, some populations require sample sizes well over 100 before the distribution of $\bar{X}$ becomes at all near the normal distribution.

4.6—Three illustrations. To illustrate the central limit theorem, the results of three sampling experiments are presented. In the first, the population is the population of random digits 0, 1, 2, . . . , 9, which we met in section 2.6. This population is discrete. The variable X has ten possible values 0, 1, 2, . . . , 9 and has an equal probability 0.1 of taking any of these values. The frequency distribution of X is represented in the upper part of figure 4.6.1. Clearly, the distribution does not look like a normal distribution. Distributions of this type are sometimes called *uniform,* since every value is equally likely. In chapter 3 we found that this distribution has mean $\mu = 4.5$ and standard deviation $\sigma = 2.872$.

Four hundred random samples of size 5 were drawn from table A 1 of random digits, each sample being a group of five consecutive numbers in a column. The frequency distribution of the sample means appears in the lower half of figure 4.6.1. A normal distribution with mean μ and standard deviation $\sigma/\sqrt{5}$ is also shown. The agreement is surprisingly good, considering that the samples are only of size 5.

Drawing samples of size 5 from the random digit tables is recommended as an easy way of seeing the central limit theorem at work. The total of each sample is quickly obtained mentally. To avoid divisions by 5, work with sample totals instead of means. The sample total $5\bar{X}$ has mean $(5)(4.5) = 22.5$ and standard deviation $(5)(1.284) = 6.420$ in repeated sampling. In forming the frequency distribution, put the totals 20, 21, 22, 23 in the central class, each class containing four consecutive totals. Although rather broad, this grouping is adequate unless, say, 500 samples have been drawn.

The second sampling experiment illustrates the case in which a large sample size must be drawn if $\bar{X}$ is to be nearly normal. This happens with populations that are markedly skew, particularly if a few values are very far from the mean. The population chosen consisted of the sizes (number of inhabitants) of U.S. cities having over 50,000 inhabitants in 1950 (2), excluding the four largest cities. All except one had sizes ranging between 50,000 and 1,000,000. The exception contained 1,850,000 inhabitants. The frequency distribution is shown at the top of figure 4.6.2. Note how asymmetrical the distribution is, the smallest class having much the highest frequency. The city with 1,850,000 inhabitants is not shown on this histogram; it would appear about 3 in. to the right of the largest class.

A set of 500 random samples with $n = 25$ and another set with $n = 100$ were drawn. The frequency distributions of the sample means appear in the middle and lower parts of figure 4.6.2. With $n = 25$, the distribution has moved towards the normal shape but is still noticeably asymmetrical. Some further improvement towards symmetry occurs with $n = 100$, but a normal curve would still be a poor fit. Evidently samples of 400–500 would be necessary to use the normal approximation with any assurance. Part of the trouble is caused by the 1,850,000 city; the means for $n = 100$ would be more nearly normal if this city had been

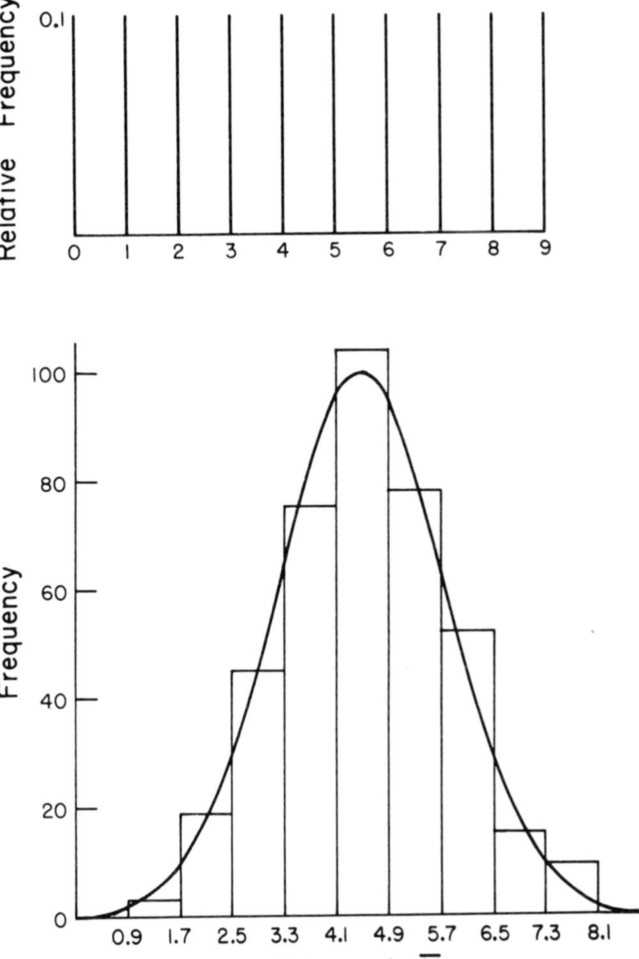

Fig. 4.6.1—Above: theoretical probability distribution of the random digits from 0 to 9. Below: histogram showing the distribution of 400 means of samples of size 5 drawn from the random digits. The curve is the normal distribution with mean $\mu = 4.5$ and standard deviation $\sigma/\sqrt{n} = 2.872/\sqrt{5} = 1.284$.

excluded from the population. On the other hand, the situation would be worse if the four largest cities had been included.

Combining the theorems in this and the previous sections, we now have the very useful result that in samples of reasonable size, $\overline{X}$ is approximately normally distributed about μ, with standard deviation or standard error $\sigma/\sqrt{n}$.

In the two preceding examples we sample nonnormal populations to illustrate the working of the central limit theorem. If we have random samples

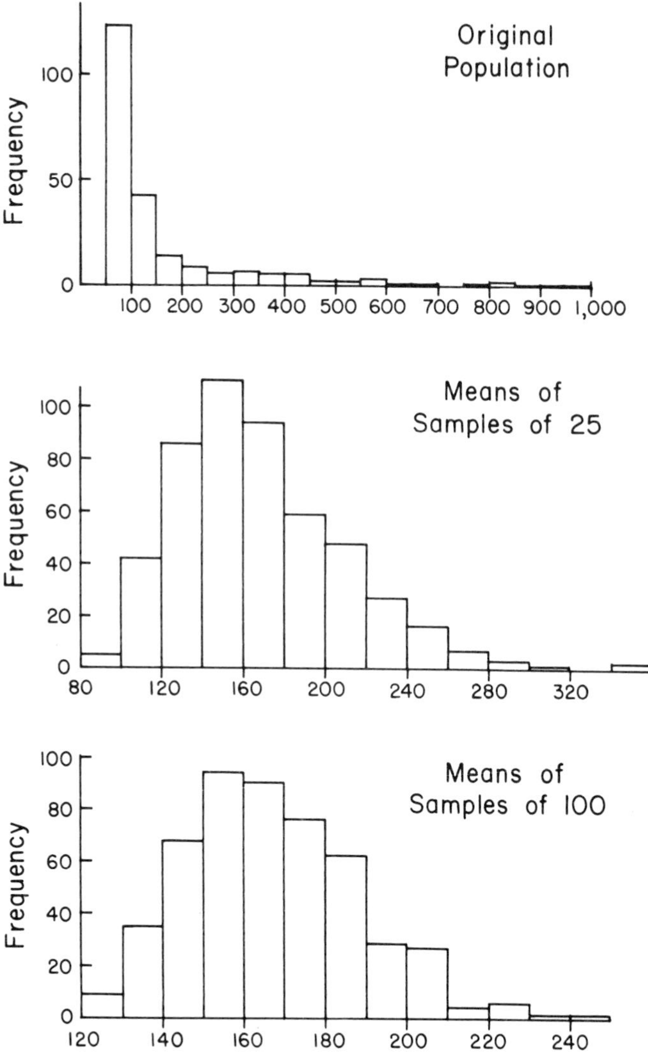

Fig. 4.6.2—Top: frequency distribution of the populations of 228 U.S. cities having populations over 50,000 in 1950. Middle: frequency distribution of the means of 500 random samples of size 25. Bottom: frequency distribution of the means of 500 random samples of size 100. The four largest cities are excluded.

from a normal distribution itself, the sample means $\bar{X}$ are normally distributed for *any* size of sample, as can be proved mathematically. An experimental verification of this result can be obtained by having students in a class draw random samples of size $n = 5$ or $n = 10$ from a normal population and pool their results. As we shall see, the same samples can be used to verify other properties, useful in statistical work, that are derived from the normal distribution. For such student or class exercises, table 4.14.1 at the end of this chapter gives the gains in weights of 100 swine during 20 days. The data were slightly modified to simulate a distribution that is approximately normal with $\mu = 30$ lb, $\sigma = 10$ lb. The 100 gains are numbered from 00 to 99 so that random samples can easily be drawn by the use of pairs of random digits from a table or computer. Sampling with replacement should be used, since this population is quite small.

As a third illustration, one class drew 511 random samples of size $n = 10$. By theory, the sample means should follow a normal distribution with $\mu = 30$ lb, $\sigma = 10/\sqrt{10} = 3.162$ lb. The class limits in the frequency distribution of the means were taken as $-\infty$–20.4, 20.5–22.4, and so on. The second and third columns of table 4.6.1 show the observed class frequencies and the theoretical frequencies expected in these classes in sampling from a normal population with $\mu = 30$, $\sigma = 3.162$. Visually, the agreement between observed and expected frequencies in table 4.6.1 looks good. In section 5.12 an objective test of the degree of agreement is given. The three right-hand columns of table 4.6.1 are used in fitting the normal distribution with $\mu = 30$, $\sigma = 3.162$.

FITTING A NORMAL DISTRIBUTION. The steps in fitting a normal distribution to a grouped frequency distribution are:

1. Calculate the Z values in standard measure corresponding to the *true* upper class limits. The successive true upper class limits are 20.45, 22.45, and so on. The Z values (third column from the right in table 4.6.1) are $(20.45 - 30)/3.162 = -3.020$, $(22.45 - 30)/3.162 = -2.388$, and so on. The true upper limit of the highest class is $+\infty$, since the abscissa of the normal curve extends from $-\infty$ to $+\infty$.

2. From table A 3, read the probability of a value less than Z (second column from the right in table 4.6.1). For Z negative, use $0.5 - A$; for Z positive, $0.5 + A$, as noted in table 4.3.1. Thus for $Z = -3.020$ we have $0.5 - 0.4987 = 0.0013$.

TABLE 4.6.1

FREQUENCY DISTRIBUTION OF 511 SAMPLE MEANS WITH $n = 10$, DRAWN FROM SWINE GAINS IN TABLE 4.14.1

Class Limits (lb)	Observed Frequency	Expected Frequency	Z for True Upper Class Limit	Cumulative $P(<Z)$	Class Probabilities
$-\infty$–20.4	2	0.66	−3.020	0.0013	0.0013
20.5–22.4	7	3.68	−2.388	0.0085	.0072
22.5–24.4	15	15.89	−1.756	0.0396	.0311
24.5–26.4	49	46.55	−1.123	0.1307	.0911
26.5–28.4	89	92.70	−0.490	0.3121	.1814
28.5–30.4	138	124.89	+0.142	0.5565	.2444
30.5–32.4	102	114.62	+0.775	0.7808	.2243
32.5–34.4	67	71.23	+1.407	0.9202	.1394
34.5–36.4	36	30.20	+2.040	0.9793	.0591
36.5–38.4	5	8.64	+2.672	0.9962	.0169
38.5–∞	1	1.94	$+\infty$	1.0000	0.0038
Total	511	511.00			

For most values of Z, linear interpolation in table A 3 is required. Thus for $Z = +0.142$, we have $0.5 - (0.2)(0.0596) + (0.8)(0.0557) = 0.5565$.

3. The probabilities of values lying in the successive classes are now given by subtracting successive cumulative probabilities. For the class $-\infty$ to 20.45 we have 0.0013; for the class 20.45 to 22.45 we have $0.0085 - 0.0013 = 0.0072$, and so on (right-hand column of table 4.6.1).

4. Finally, to obtain the expected frequencies, multiply the class probabilities by $n = 511$.

EXAMPLE 4.6.1—A population of heights of men has a standard deviation $\sigma = 2.6$ in. What is the standard error of the mean of a random sample of (*i*) 25 men, (*ii*) 100 men? Ans. (*i*) 0.52 in., (*ii*) 0.26 in.

EXAMPLE 4.6.2—In order to estimate the total weight of a batch of 720 bags that are to be shipped, each of a random sample of 36 bags is weighed, giving $\overline{X} = 40$ lb. Assuming $\sigma = 3$ lb, estimate the total weight of the 720 bags and give the standard error of your estimate. Ans. 28,800 lb; standard error, 80.5 lb.

EXAMPLE 4.6.3—In estimating the mean height of a large group of boys with $\sigma = 1.5$ in., how large a sample must be taken if the standard error of the mean height is to be 0.2 in.? Ans. 56 boys.

EXAMPLE 4.6.4—If perfect dice are thrown repeatedly, the probability is 1/6 that each of the faces 1, 2, 3, 4, 5, 6 turns up. Compute μ and σ for this population. Ans. $\mu = 3.5$, $\sigma = 1.71$.

EXAMPLE 4.6.5—If males and females are equally likely to be born, the probabilities that a family of size 2 contains 0, 1, 2 boys are, respectively, 1/4, 1/2, and 1/4. Find μ and σ for this population. Ans. $\mu = 1$, $\sigma = 1/\sqrt{2} = 0.71$.

EXAMPLE 4.6.6—The following sampling experiment shows how the central-limit theorem performs with a population simulating what is called a U-shaped distribution. In the random digits table, score 0, 1, 2, 3 as 0; 4, 5 as 1; and 6, 7, 8, 9 as 2. In this population, the probabilities of scores of 0, 1, 2 are 0.4, 0.2, and 0.4, respectively. This distribution is discrete and the central ordinate, 0.2, is lower than the two outside ordinates, 0.4. Draw a number of samples of size 5, using the random digits table. Record the total score for each sample. The distribution of total scores will be found to be fairly similar to the bell-shaped normal curve. The theoretical distribution of the total scores is as follows:

Score	0 or 10	1 or 9	2 or 8	3 or 7	4 or 6	5
Probability	0.010	0.026	0.077	0.115	0.182	0.179

That is, the probability of a 0 and that of a 10 are both 0.010.

TABLE 4.6.2
MOTHERS' HEIGHTS (2-IN. CLASSES)

True Class Limits (in.)	Frequencies		True Class Limits (in.)	Frequencies	
	Obs.	Exp.		Obs.	Exp.
$-\infty$–55	3	1.0	63–65	277.5	279.3
55–57	8.5	11.6	65–67	119.5	125.5
57–59	52.5	67.3	67–69	23.5	29.7
59–61	215.0	204.3	69–∞	6.5	4.0
61–63	346.0	329.3	Total	1052.0	1052.0

EXAMPLE 4.6.7—To provide an example of the fitting of a normal curve, table 4.6.2 shows the grouped frequency distribution of 1052 mothers' heights in 2-in. classes. Take $\mu = 62.49$ in., $\sigma = 2.435$ in. (μ and σ are actually the sample $\bar{X}$ and s). The true class limits for fitting the normal curve are given, as are the observed class frequencies. The expected class frequencies that we obtained in fitting are shown as a check, though you may find minor discrepancies.

4.7—Confidence limits for μ when σ is known. As we noted in chapter 1, it is important when making an estimate of a population mean to have some idea of the accuracy of the estimate. With a random sample of size n from a population, where n is large enough that $\bar{X}$ can be assumed to be normally distributed, we can calculate the probability that $\bar{X}$ is in error by less than plus or minus any given amount, provided that σ is known for this population. Although in most applications the value of σ is not known, it is sometimes known from past data in repeated surveys, from surveys on similar populations, or from theoretical considerations. The following example shows how the probability of an error of a given size in $\bar{X}$ is calculated.

EXAMPLE 6—In a large city the distribution of incomes per family had $\sigma = \$3250$. For a random sample of 400 families from this population, what is the probability that the sample mean $\bar{X}$ is correct to within (*i*) $\pm\$100$ and (*ii*) $\pm\$500$? Ans. The standard deviation of $\bar{X}$ is $\sigma/\sqrt{n} = \$3250/20 = \$162.5$. The probability that the error $\bar{X} - \mu$ is less than \$100 equals the probability that $Z = (\bar{X} - \mu)/162.5$ is less than $100/162.5 = 0.615$ in absolute value. From table A 3, the probability that Z lies between 0 and 0.615 is $(0.2291 + 0.2324)/2 = 0.23075$. We multiply this by 2 to include the probability that Z lies between -0.615 and 0, giving 0.46 as the answer. In (*ii*) the limit on $|Z|$ is 3.077, and the probability desired is $2 \times 0.4990 = 0.998$—almost certainty.

The preceding argument can be used to answer the question: Now that we have the sample results, what have we learned about the population mean μ that is being estimated? We are assuming that our sample estimate $\bar{X}$ is normally distributed in repeated sampling, with mean μ and standard deviation $\sigma/\sqrt{n}$. Thus, unless an unlucky 5% chance has come off, $\bar{X}$ will lie between $\mu - 1.96\sigma/\sqrt{n}$ and $\mu + 1.96\sigma/\sqrt{n}$. Expressing this as a pair of inequalities, we write

$$\mu - 1.96\sigma/\sqrt{n} \le \bar{X} \le \mu + 1.96\sigma/\sqrt{n} \qquad (4.7.1)$$

apart from a 5% chance. These inequalities can be rewritten so that they provide limits for μ when we know $\bar{X}$. The left-hand inequality is equivalent to the statement that

$$\mu \le \bar{X} + 1.96\sigma/\sqrt{n}$$

In the same way, the right-hand inequality implies that

$$\mu \ge \bar{X} - 1.96\sigma/\sqrt{n}$$

Putting the two together, we reach the conclusion that unless an unlucky 5% chance occurred in drawing the sample,

$$\bar{X} - 1.96\sigma/\sqrt{n} \le \mu \le \bar{X} + 1.96\sigma/\sqrt{n} \qquad (4.7.2)$$

This interval is called the 95% *confidence interval* for μ.

Similarly, the 99% confidence interval for μ is

$$\bar{X} - 2.58\sigma/\sqrt{n} \le \mu \le \bar{X} + 2.58\sigma/\sqrt{n}$$

because the probability is 0.99 that a normal deviate Z lies between the limits -2.58 and $+2.58$.

When we make a 95% confidence interval statement, the population mean μ has a fixed but unknown value. The uncertainty attached to a confidence interval for μ comes from the variability in the sampling process. We do not know whether any individual confidence interval includes μ or not. We do know that in the long run 95% of the intervals will include μ, and similarly, 95% of the statements that "μ lies between the lower and the upper limit" will be correct. Incidentally, the phrase *confidence probability* is sometimes used as a shortcut for "probability attached to a confidence interval."

To find the confidence interval corresponding to any confidence probability P, read from the cumulative normal table (table A 3) a value Z_P, say, such that the area given in the table is $P/2$. Then the probability that a normal deviate lies between $-Z_P$ and $+Z_P$ is P. The confidence interval is

$$\bar{X} - Z_P\sigma/\sqrt{n} \le \mu \le \bar{X} + Z_P\sigma/\sqrt{n} \qquad (4.7.3)$$

ONE-SIDED CONFIDENCE LIMITS. Sometimes we want to find only an upper limit or a lower limit for μ but not both. A company making large batches of a chemical product might have, as part of its quality control program, a regulation that each batch must be tested to ensure that it does not contain more than 25 parts per million of a certain impurity, apart from a 1-in-100 chance. The test consists of drawing out n amounts of the product from the batch and determining the concentration of impurity in each amount. If the batch is to pass the test, the 99% upper confidence limit for μ must be not more than 25 parts per million. Similarly, certain roots of tropical trees are a source of a potent insecticide whose concentration varies considerably from root to root. The buyer of a large shipment of these roots wants a guarantee that the concentration of the active ingredient in the shipment exceeds some stated value. It may be agreed between buyer and seller that the shipment is acceptable if, say, the 95% lower confidence limit for the average concentration μ exceeds the desired minimum.

To find a *one-sided* or *one-tailed* limit with confidence probability 95%, we want a normal deviate Z such that the area beyond Z in one tail is 0.05. In table A 3, the area from 0 to Z will be 0.45, and the value of Z is 1.645. Apart from a 5% chance in drawing the sample,

$$\bar{X} \le \mu + 1.645\sigma/\sqrt{n}$$

This gives, as the *lower* 95% confidence limit for μ,

$$\mu \ge \bar{X} - 1.645\sigma/\sqrt{n}$$

The *upper* limit is $\bar{X} + 1.645\,\sigma/\sqrt{n}$. For a 99% limit the value of Z is 2.326. For a one-sided limit with confidence probability P (expressed as a proportion), read table A 3 to find the Z that corresponds to probability $P - 0.5$.

4.8—Size of sample. The question, How large a sample must I take? is frequently asked by investigators. The question is not easy to answer. But if the purpose of the investigation is to estimate the mean of a population from the results of a sample, the methods in the preceding sections are helpful.

First, the investigator must state how accurate the sample estimate should be. Should it be correct to within 1 unit, 5 units, or 10 units on the scale on which it is measured? In answering this question, the investigator thinks of the purposes to which the estimate will be put and tries to envisage the consequences of having errors of different amounts in the estimate. If the estimate is to be made to guide a specific business or financial decision, calculations may indicate the level of accuracy necessary to make the estimate useful. In scientific research it is often hard to do this, and an element of arbitrariness may be in the answer finally given.

By one means or another, the investigator states that the estimate is desired to be correct to within some limit $\pm L$, say. Since the normal curve extends from minus infinity to plus infinity, we cannot guarantee that $\bar{X}$ is certain to lie between the limits $\mu - L$ and $\mu + L$. We can, however, make the probability that $\bar{X}$ lies between these limits as large as we please. In practice, this probability is usually set at 95% or 99%. For the 95% probability, we know that there is a 95% chance that $\bar{X}$ lies between the limits $\mu - 1.96\sigma/\sqrt{n}$ and $\mu + 1.96\sigma/\sqrt{n}$. This gives the equation

$$1.96\sigma/\sqrt{n} = L$$

which is solved for n.

The equation requires a knowledge of σ, although the sample has not yet been drawn. From previous work on this or similar populations, the investigator guesses a value of σ. Since this guess is likely to be somewhat in error, we might as well replace 1.96 by 2 for simplicity. This gives the formula

$$n = 4\sigma^2/L^2$$

The formula for 99% probability is $n = 6.6\sigma^2/L^2$.

To summarize, the investigator must supply (*i*) an upper limit L to the amount of error that can be tolerated in the estimate, (*ii*) the desired probability that the estimate will lie within this limit of error, and (*iii*) an advance guess at the population standard deviation σ. The formula for n is then very simple.

EXAMPLE 4.8.1—When sample evidence is submitted to justify deductions from taxable profits for expenses incurred in a long series of business transactions, it may be stipulated that the sample estimate of the average expense per transaction must be correct to within ±$500, apart from a 1-in-20 chance. From past experience, a firm knows that the distribution of expenses per transaction has σ close to \$1800. It decides to take a random sample of $n = 40$ transactions to estimate μ, the average expense per transaction. (*i*) What is the probability that the sample estimate $\overline{X}$ is correct to within $\pm$\$500? (*ii*) Will $n = 40$ satisfy the stipulation? Ans. (*i*) Probability = 0.921; (*ii*) no, you may verify that about $n = 50$ is needed.

EXAMPLE 4.8.2—In estimating μ from random samples of size 10 drawn from the approximately normal population of swine gains in table 4.14.1, confidence intervals of width 11 lb for μ are desired; that is, the confidence interval is to extend from $\overline{X} - 5.5$ to $\overline{X} + 5.5$ (*i*) From normal theory, what is the probability that these intervals contain μ? (In the original population, $\mu = 30$ lb, $\sigma = 10$ lb.) (*ii*) The set of 511 random samples from this population with $n = 10$ had 24 values of $\overline{X}$ below 24.5 and 17 values above 35.5. Does the sample estimate of the confidence probability agree well with the probability in (*i*)? Ans. (*i*) 91.8%, (*ii*) 92.0%. The agreement is almost too good to be true.

EXAMPLE 4.8.3—In table 4.6.1 the mean of the 511 sample values of $\overline{X}$ is 29.823, which of course is the mean of a random sample of size $n = 5110$. (*i*) Calculate a 95% confidence interval for μ. (*ii*) Does the interval cover μ? Ans. (*i*) $29.549 \le \mu \le 30.097$, (*ii*) yes.

EXAMPLE 4.8.4—Find (*i*) the 80% and (*ii*) the 90% confidence limits for μ, given $\overline{X}$ and σ. Ans. (*i*) $\overline{X} \pm 1.28\sigma/\sqrt{n}$, (*ii*) $\overline{X} \pm 1.64\sigma/\sqrt{n}$.

EXAMPLE 4.8.5—The heights of a random sample of 16 men from a population with $\sigma = 2.6$ in. are measured. What is the probability that $\overline{X}$ does not differ from μ by more than 1 in.? Ans. $P = 0.876$.

EXAMPLE 4.8.6—The buyer of roots that contain insecticide wants assurance that the average content of the active ingredient is at least 8 lb per 100 lb, apart from a 1-in-100 chance. A sample of 9 bundles of roots drawn from the batch gives, on analysis, $\overline{X} = 10.2$ lb active ingredient per 100 lb. If $\sigma = 3.3$ lb per 100 lb, find the lower 99% confidence limit for μ. Does the batch meet the specification? Ans. Lower limit = 7.6 lb per 100 lb. No.

EXAMPLE 4.8.7—In auditing a firm's accounts receivable, 100 entries were checked in a ledger containing 1000 entries. For these 100 entries, the auditor's check showed that the stated total amount receivable exceeded the correct amount receivable by \$214. Calculate an upper 95% limit for the amount by which the reported total receivable in the whole ledger exceeds the correct amount. Assume σ = \$1.30 in the population of bookkeeping errors. Ans. \$2354. Note: For an estimated population *total*, the formula for a one-sided upper limit for $N\mu$ is $N\overline{X} + NZ\sigma/\sqrt{n}$, where $Z = 1.645$, $N = 1000$, $n = 100$.

EXAMPLE 4.8.8—When measurements are rounded to the nearest whole number, it can often be assumed that the error due to rounding is equally likely to lie anywhere between -0.5 and $+0.5$. That is, rounding errors follow a *uniform* distribution between the limits -0.5 and $+0.5$. From theory, this distribution has $\mu = 0$, $\sigma = 1/\sqrt{12} = 0.29$. If 100 independent, rounded measurements are added, what is the probability that the error in the total due to rounding does not exceed 5 in absolute value? Ans. $P = 0.916$.

EXAMPLE 4.8.9—In an area of a large city in which houses are rented, an economist wishes to estimate the average monthly rent correct to within $\pm$\$20, apart from a 1-in-20 chance. If the economist guesses that σ is about \$60, how many houses must be included in the sample? Ans. $n = 36$.

EXAMPLE 4.8.10—Show that if we wish to cut the limit of error from L to $L/2$, the sample size must be quadrupled. With the same L, if we wish 99% probability of being within the limit rather than 95% probability, what percentage increase in sample size is required? Ans. About 73% increase.

4.9—Student's *t* distribution. In most applications in which sample means are used to estimate population means, the value of σ is not known. We can, however, obtain an estimate s of σ from the sample data that give us the value of $\overline{X}$. If the sample is of size n, the estimate s is based on $(n - 1)$ degrees of freedom. We require a distribution that will enable us to compute confidence limits for μ, knowing s but not σ. Known as *Student's t distribution,* this result was published by W. S. Gossett in 1908 (3) and perfected by R. A. Fisher in 1926 (4). This distribution (under the pseudonym "Student") has revolutionized the statistics of small samples.

The quantity t is given by the equation

$$t = \frac{\overline{X} - \mu}{s/\sqrt{n}} \tag{4.9.1}$$

That is, t is the deviation of the estimate mean from the population mean, measured in terms of the unit $s/\sqrt{n}$. The quantity $s/\sqrt{n}$ is the sample estimate of $\sigma/\sqrt{n}$, the *standard error* of $\overline{X}$.

The distribution of t is laid out in table A 4.

The cumulative normal table, A 3, and the t table, A 4, show different quantities. Table A 3 gives, for a value Z, the probability that a normal deviate lies between 0 and Z. Table A 4 gives, for each probability at the top of a column, the value of t, (say t^*), such that $P[|t| > t^*]$ has the stated probability for the number of degrees of freedom (df) in the row. For instance, consider the column in table A 4 headed 0.050. With this probability table A 4 shows that for ∞ df, $t^* = 1.960$ (as we know); for 30 df, $t^* = 2.042$ (only about a 4% increase); and for 4 df, $t^* = 2.776$, as shown also in figure 4.9.1. Like the normal, the t distribution is symmetrical about the mean, so the area in each tail in figure 4.9.1 is 0.025.

When t is calculated from a single random sample of size n, remember that the degrees of freedom are $(n - 1)$. In more complex statistical analysis, rules are given for determining the number of degrees of freedom in t, so table A 4 serves many applications.

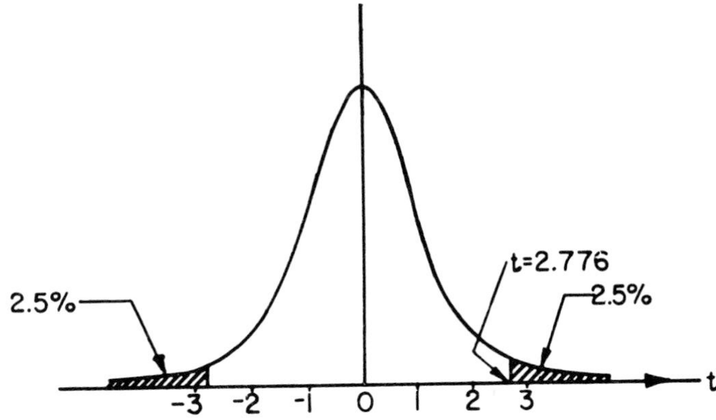

Fig. 4.9.1—Distribution of t with 4 df. The shaded areas comprise 5% of the total area. The distribution is more peaked in the center and has higher tails than the normal.

EXAMPLE 4.9.1—If samples of size $n = 17$ are randomly drawn from a normal population and t is calculated for each sample, what is the probability that t will fall between -1.75 and $+1.75$? Ans. Nearly 0.90.

EXAMPLE 4.9.2—In the preceding samples, what is the probability that t is less than -0.69? Ans. 0.25.

EXAMPLE 4.9.3—What size of sample has $|t| > 2$ in 5% of all random samples from normal populations? Ans. $n = 61$.

EXAMPLE 4.9.4—In random samples of size 16 drawn from a normal population with $\mu = 10$, one sample gave $\bar{X} = 12.1$, $s = 5.0$. For this size of sample, is this $\bar{X}$ an unusually poor estimate of μ? One way to answer this question is to calculate the sample t from (4.9.1) and find from table A 4 how rarely values of $|t|$ greater than this occur. Ans. Sample $t = (12.1 - 10)/1.25 = 1.68$, df $= 15$. From table A 4, values of $|t| > 1.753$ occur 10% of the time and values of $|t| > 1.341$ occur 20% of the time. The deviation in this sample is of a magnitude that occurs slightly more than 10% of the time—not too rarely.

4.10—Confidence limits for μ based on the t distribution. With σ known, the 95% limits for μ are given by the relations

$$\bar{X} - 1.96\sigma/\sqrt{n} \le \mu \le \bar{X} + 1.96\sigma/\sqrt{n}$$

When σ is replaced by s, the only change needed is to replace the number 1.96 by a quantity that we call $t_{0.05}$. To find $t_{0.05}$, read table A 4 in the column headed 0.050 and find the value of t for the *number of degrees of freedom in s*. When the degrees of freedom are infinite, $t_{0.05} = 1.960$. With 40 df, $t_{0.05}$ has increased to 2.021; with 20 df it has become 2.086; and it continues to increase steadily as the number of degrees of freedom decline.

The inequalities giving the 95% confidence limits become

$$\bar{X} - t_{0.05}s/\sqrt{n} \le \mu \le \bar{X} + t_{0.05}s/\sqrt{n} \tag{4.10.1}$$

The proof of this result is similar to that given when σ is known. Although μ is unknown, the drawing of a random sample creates a value of

$$t = \frac{\bar{X} - \mu}{s/\sqrt{n}}$$

that follows Student's t distribution with $(n - 1)$ df. Now the quantity $t_{0.05}$ in table A 4 was computed so that the probability is 0.95 that a value of t drawn at random lies between $-t_{0.05}$ and $+t_{0.05}$. Thus, there is a 95% chance that

$$-t_{0.05} \le \frac{\bar{X} - \mu}{s/\sqrt{n}} \le +t_{0.05}$$

Multiply the inequality throughout by $s/\sqrt{n}$ and then add μ to each term. This gives, with 95% probability,

$$\mu - t_{0.05}s/\sqrt{n} \le \bar{X} \le \mu + t_{0.05}s/\sqrt{n}$$

The remainder of the proof is exactly the same as when σ is known. The limits may be expressed more compactly as $\bar{X} \pm t_{0.05} s_{\bar{X}}$. For a one-sided 95% limit, use $t_{0.10}$ in place of $t_{0.05}$.

As an illustration, consider the sample of the weights of 11 forty-year-old men in section 3.1. The weights were 148, 154, 158, 160, 161, 162, 166, 170, 182, 195, and 236 lb. They give $n = 11$, $\bar{X} = 172$ lb, $s = 24.95$ lb, $s_{\bar{X}} = 24.95/\sqrt{11} = 7.52$ lb. To get the 95% confidence interval for the population mean:

1. Enter table A 4 with $11 - 1 = 10$ df. In the column headed 0.050 and the row for 10 df, take the entry $t_{0.05} = 2.228$.
2. Calculate the quantity $t_{0.5} s_{\bar{X}} = (2.228)(7.52) = 16.75$ lb.
3. The 95% confidence interval extends from $172 - 16.75 = 155.25$ lb to $172 + 16.75 = 188.75$ lb.

A claim that μ lies between 155.25 lb and 188.75 lb will be correct unless a 1-in-20 chance has occurred in the sampling. Since $16.75/172 = 0.097$, we can say, alternatively, that the estimate $\bar{X} = 172$ of μ is highly likely to be correct to within $\pm 10\%$.

EXAMPLE 4.10.1—The yields of alfalfa from 10 plots were 0.8, 1.3, 1.5, 1.7, 1.7, 1.8, 2.0, 2.0, 2.0, and 2.2 T/acre. Set 95% limits on the mean of the population of which this is a random sample. Ans. 1.40 and 2.00 T/acre.

EXAMPLE 4.10.2—In an investigation of growth in school children in private schools, the sample mean height of 265 boys of age 13 1/2–14 1/2 years was 63.84 in. with standard deviation $s = 3.08$ in. What is the 95% confidence interval for μ? Ans. 63.5 to 64.2 in.

EXAMPLE 4.10.3—In a check of a day's work for each of a sample of 16 women engaged in tedious, repetitive work, the average number of minor errors per day was 5.6, with a sample standard deviation of 3.6. Find (*i*) a 90% confidence interval for the population mean number of errors and (*ii*) a one-sided upper 90% limit to the population number of errors. Ans. (*i*) 4.0 to 7.2, (*ii*) 6.8.

EXAMPLE 4.10.4—(*i*) What are the 90% confidence limits for μ given by the sample of the weights of 11 men? (*ii*) What is the 95% one-sided lower limit? Ans. (*i*) (158.37, 185.63) lb, (*ii*) 158.37 lb.

EXAMPLE 4.10.5—Suppose, for simplicity, that we use the rule $\bar{X} \pm 2s_{\bar{X}}$ in calculating 95% confidence limits for any size sample. (*i*) For what sample sizes are the probabilities attached to our limits actually less than 90%? With the rule $\bar{X} \pm 2.6s_X$ for calculating 99% limits, for what sample sizes are the confidence probabilities (*ii*) less than 97.5% and (*iii*) less than 95%? Ans. (*i*) $n \le 6$, (*ii*) $n \le 11$, (*iii*) $n \le 5$. With small samples, s is a poor estimator of σ and the confidence limits must widen to allow for the chance that the sample s is far removed from σ.

EXAMPLE 4.10.6—During the fall of 1943, a random sample of 300 families in Iowa towns was visited to learn the number of quarts of food canned. The sample gave an average of 165 qt per family with a standard deviation $s = 153$ qt. Calculate the 95% confidence limits for the population mean. Ans. 165 ± 17.3 qt.

EXAMPLE 4.10.7—Approximately 312,000 families lived in Iowa towns in 1943. Estimate the number of quarts of food canned in Iowa towns in 1943, with 95% confidence limits. Ans. 51.5 million qt, with 95% confidence limits 46.1 million to 56.9 million qt.

4.11—Experimental sampling of the *t* distribution. A set of 511 samples of size 10 drawn from the normally distributed gains in weight of swine are

TABLE 4.11.1
FOUR SAMPLES OF 10 ITEMS DRAWN AT RANDOM FROM THE SWINE GAINS OF TABLE 4.13.1

Item Number and Formulas	Sample Number 1	2	3	4
1	33	32	39	17
2	53	31	34	22
3	34	11	33	20
4	29	30	33	19
5	39	19	33	3
6	57	24	39	21
7	12	53	36	3
8	24	44	32	25
9	39	19	32	40
10	36	30	30	21
$\bar{X}$	35.6	29.3	34.1	19.1
s^2	169.8	151.6	9.0	112.3
s	13.0	12.3	3.0	10.6
$s_{\bar{X}}$	4.11	3.89	0.95	3.35
t	1.36	−0.18	4.32	−3.25
$t_{0.05}s_{\bar{X}}$	9.3	8.8	2.2	7.6
$\bar{X} \pm t_{0.05}s_{\bar{X}}$	26.3–44.9	20.5–38.1	31.9–36.3	11.5–26.7

described and used at the end of section 4.6. These samples also give an experimental verification of the t distribution with 9 df and of confidence interval statements based on this distribution. Table 4.11.1 shows the results of four of the samples with some of the statistics of interest calculated for each sample: $\bar{X}$, s^2, s, $s_{\bar{X}} = s/\sqrt{10}$, $t = (\bar{X} - 30)/s_{\bar{X}}$, and the 95% confidence interval based on t with 9 df. These samples are discussed in section 4.12.

The 511 sample values of t are shown in a grouped frequency distribution in

TABLE 4.11.2
SAMPLE AND THEORETICAL DISTRIBUTION OF t SAMPLES OF 10; DEGREES OF FREEDOM, 9

Class Intervals		Class Frequencies		Percentage Frequencies	
From	To	Observed	Expected	Sample	Theoretical
−∞	−3.250	3	2.6	0.6	0.5
−3.250	−2.821	4	2.6	0.8	0.5
−2.821	−2.262	5	7.7	1.0	1.5
−2.262	−1.833	16	12.8	3.1	2.5
−1.833	−1.383	31	25.5	6.1	5.0
−1.383	−0.703	77	76.6	15.1	15.0
−0.703	0.0	132	127.7	25.8	25.0
0.0	0.703	126	127.7	24.6	25.0
0.703	1.383	73	76.6	14.3	15.0
1.383	1.833	18	25.5	3.5	5.0
1.833	2.262	13	12.8	2.5	2.5
2.262	2.821	8	7.7	1.6	1.5
2.821	3.250	2	2.6	0.4	0.5
3.250	∞	3	2.6	0.6	0.5
Total		511	511.0		

table 4.11.2. The class intervals in table 4.11.2 are of unequal width. The class limits were chosen from the t table, A 4, so as to make it easy to calculate the sample estimates of confidence probabilities. Thus in table A 4, the two-tailed 1% value of t with 9 df is 3.250. By the symmetry of the t distribution, 0.5% of the values lie below -3.250, the upper limit of the first class in table 4.11.2.

The agreement between the observed and expected class frequencies in table 4.11.2 is excellent—probably better than is found in 90% of such comparisons between a theoretical distribution and random samples drawn from it. Note also the near symmetry of the observed frequencies about the origin.

4.12—Sample check on confidence interval statements. For each sample a 95% confidence interval for μ was calculated as shown in table 4.11.1. If you say for any particular sample that the interval includes μ, you will either be right or wrong. Which it is may be determined readily because you know that $\mu = 30$ lb. The theory will be verified if about 95% of your statements are right and about 5% wrong.

The 95% intervals for μ given by the four samples in table 4.11.1 are, respectively, 26.3 to 44.9, 20.5 to 38.1, 31.9 to 36.3, and 11.5 to 26.7.

The intervals from samples 1 and 2 both contain $\mu = 30$. On the contrary, samples 3 and 4 illustrate cases leading to false statements, sample 3 because of an unusually small standard deviation $s = 3.0$ as against $\sigma = 10$ and sample 4 because of an unusually divergent mean $\overline{X} = 19.1$. Sample 3 is particularly misleading; not only does it miss the mark, but the narrow confidence interval suggests wrongly that we have an unusually accurate estimate.

From table 4.11.2, it is easy to calculate how many of the 95% confidence interval statements were wrong. A statement will be wrong if a sample t is less than -2.262 or is greater than $+2.262$.

Table 4.11.2 has 12 samples with $t < -2.262$ and 13 samples with $t > 2.262$. Hence the proportion of correct sample interval statements is 486/511 or 95.1%, very close to the theoretical 95%. You may also verify from table 4.11.2 that 457/511 or 89.4% of the statements for the 90% confidence interval are correct.

4.13—Probability paper. Since many of the common statistical techniques require the assumption of normality, methods for checking the normality of a set of data before deciding on a method of analysis are helpful. A graphic check is provided by a special type of graph paper called *arithmetic probability paper*; see figure 4.13.1. The abscissa scale is of the usual type. The ordinates are percentages, descending on the left scale from 99.99 to 0.01 and rising on the right scale from 0.01 to 99.99. The ordinate scale is stretched at both ends; the distance from 50 to 60 is smaller than that from 60 to 70, which is smaller than that from 70 to 80, and so on. The distortion is such that if we plot for any X the percentage of the standard normal population that exceeds X, the points lie on a straight line. Figure 4.13.1 shows five points on this line (the values of X for which the probabilities are 99%, 80%, 50%, 20% and 1%). The crosses on this figure give the corresponding plot for a t distribution with 3 df, a distribution that is symmetrical but has much longer tails than the normal. Note that with t

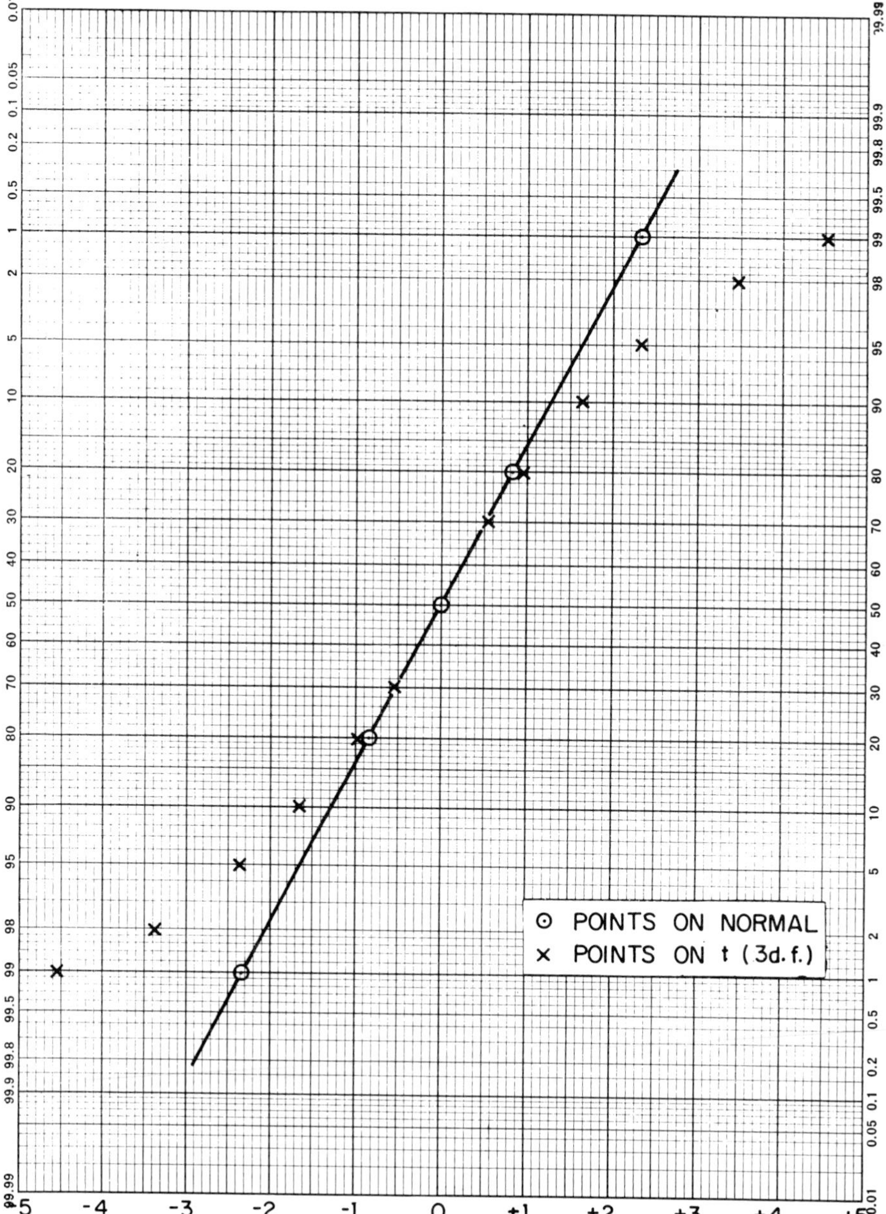

Fig. 4.13.1—Points on the cumulative normal distribution and on the t distribution with 3 df, plotted on probability paper.

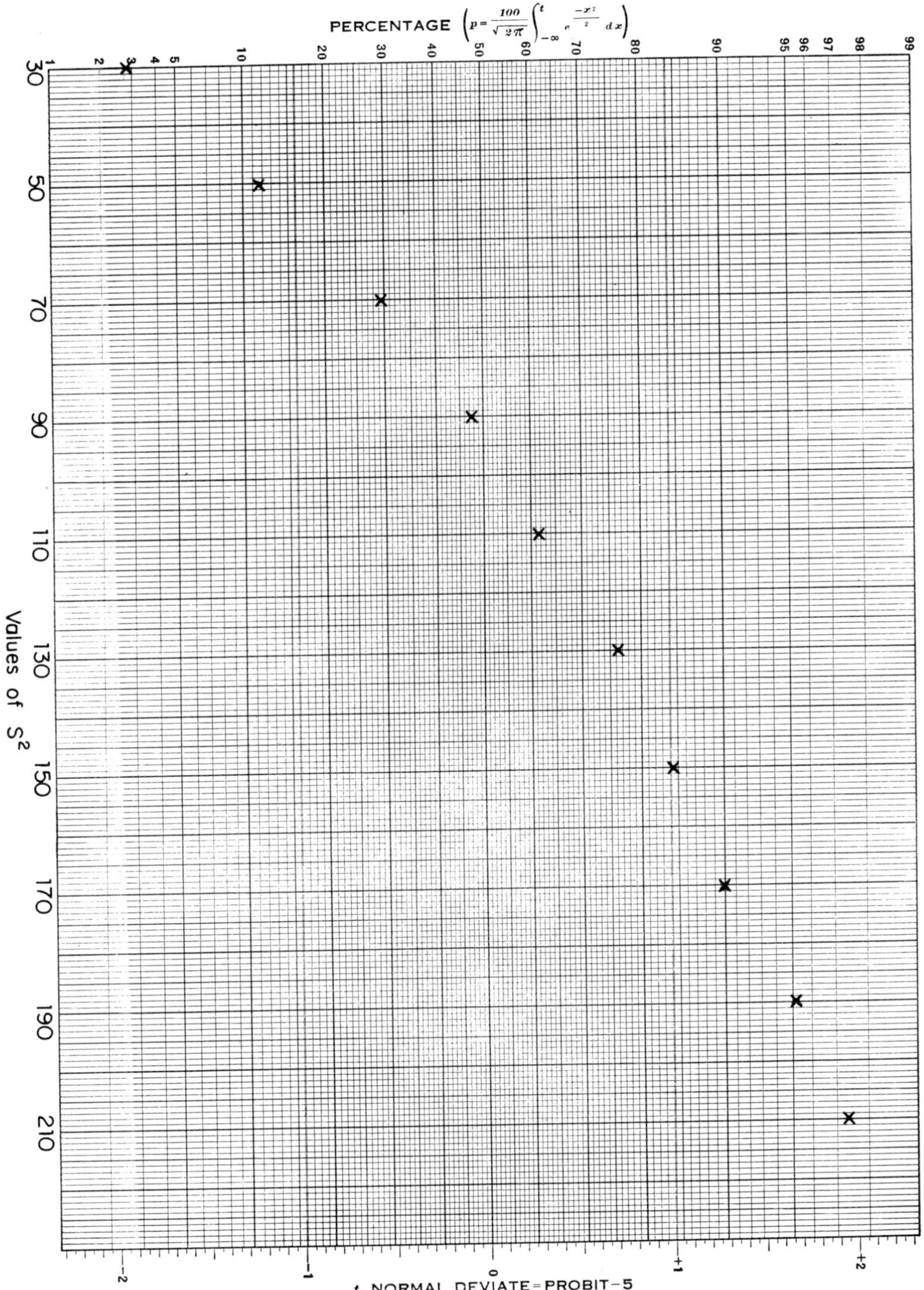

Fig. 4.13.2—Points on the cumulative distribution of s^2 with 9 df, plotted on probability paper.

the crosses do not lie on a straight line but on something more like part of an S curve.

In checking normality with a large sample of data, plot for each of a set of values of X the percentage of the sample that exceeds X. If more convenient, plot for each X the percentage of the sample that is less than X, using the right-hand ordinate scale. Owing to sampling variations, the points will depart somewhat from the line even with normal data. We look for systematic deviations of the type indicating a long-tailed distribution as in figure 4.13.1 or those indicating a skew distribution as in figure 4.13.2.

Figure 4.13.2 is a plot from a sample of 511 values of s^2, each with 9 df, on a type of probability paper. The values of s^2 were computed from samples of size 10 drawn from the simulated normal population in table 4.13.1. Note that as s^2 increases in figure 4.13.2, the slope of the line joining two consecutive points steadily decreases. This is the typical sign of positive skewness.

With relatively small samples we may want to plot the probability corresponding to the lowest value of X. For this value, the sample percentage less than X is zero, which cannot be plotted. One way around this trouble is to argue that the lowest value of a sample of size 10 ought to lie, roughly, in the middle of the lowest tenth of the frequency distribution and therefore should be plotted at

TABLE 4.14.1

ARRAY OF GAINS IN WEIGHT (POUNDS) OF 100 SWINE DURING A PERIOD OF 20 DAYS (gains approximate a normal distribution with μ = 30 lb and σ = 10 lb)

Item Number	Gain	Item Number	Gain	Item Number	Gain	Item Number	Gain
00	3	25	24	50	30	75	37
01	7	26	24	51	30	76	37
02	11	27	24	52	30	77	38
03	12	28	25	53	30	78	38
04	13	29	25	54	30	79	39
05	14	30	25	55	31	80	39
06	15	31	26	56	31	81	39
07	16	32	26	57	31	82	40
08	17	33	26	58	31	83	40
09	17	34	26	59	32	84	41
10	18	35	27	60	32	85	41
11	18	36	27	61	33	86	41
12	18	37	27	62	33	87	42
13	19	38	28	63	33	88	42
14	19	39	28	64	33	89	42
15	19	40	28	65	33	90	43
16	20	41	29	66	34	91	43
17	20	42	29	67	34	92	44
18	21	43	29	68	34	93	45
19	21	44	29	69	35	94	46
20	21	45	30	70	35	95	47
21	22	46	30	71	35	96	48
22	22	47	30	72	36	97	49
23	23	48	30	73	36	98	53
24	23	49	30	74	36	99	57

TABLE 4.14.2
FREQUENCY DISTRIBUTION OF GAINS IN WEIGHT OF 100 SWINE
(a finite population approximating the normal)

Class midpoint (lb)	5	10	15	20	25	30	35	40	45	50	55
Frequency	2	2	6	13	15	23	16	13	6	2	2

5% probability less than X. Similarly, the second lowest X should be plotted at 15%, and so on. This rule plots the ith X from the bottom in a sample of size n against $100(2i - 1)/(2n)$% less than X. Another rule that can be shown to give a slightly better straight-line plot with normal data is: Plot the ith X from the bottom against $100(3i - 1)/(3n + 1)$% less than X.

In sections 5.13 and 5.14, numerical tests for *skewness* and *kurtosis* (presence of long or short tails) are described.

4.14—A finite population simulating the normal. Table 4.14.1 gives the gains in weight of 100 swine over 20 days to provide a simple means of sampling an approximately normal distribution.

Table 4.14.2 gives the frequency distribution of the data in table 4.14.1. The class intervals are from 2.5 to 7.5 lb, etc.

EXAMPLE 4.14.1—Verify the theoretical expected frequency 76.6 for the class from 0.703 to 1.383 in table 4.11.2.

EXAMPLE 4.14.2—A random sample of 10 swine gains gave the following values: 18, 24, 33, 34, 30, 35, 39, 12, 18, 30. Calculate (*i*) the 80% two-tailed confidence limits for μ and (*ii*) the 80% upper one-tailed limit, using the t table. Ans. (*i*) 23.4 to 31.2, (*ii*) 29.8—wrong.

EXAMPLE 4.14.3—In the preceding example, what answers would be given to (*i*) and (*ii*) if we knew that $\sigma = 10$ for the swine gains and used the normal table instead of the t table? Ans. (*i*) 23.2 to 31.4. Note that normal limits can be wider than t limits if the sample s is less than σ, as in this case. (*ii*) Rounding to one decimal—30.0, acceptable. If we report two decimals, the upper limit is 29.96—a trifle too small.

TECHNICAL TERMS

central limit theorem
confidence limits
normal distribution
parameter
probability paper
standard normal deviate
Student
t distribution

REFERENCES

1. Mood, A. M., and Graybill, F. A. 1963. *Introduction to the Theory of Statistics*, 2nd ed. McGraw-Hill, New York.
2. Statistical Abstract of the United States. 1959. U.S. Government Printing Office. Washington, D.C.
3. Student. *Biometrika* 6 (1908):1.
4. Fisher, R. A. *Metron* 5 (1926):90.

5

Tests of hypotheses

5.1—Introduction. A tool widely used in statistical analysis is a *test of a hypothesis*, or alternatively, a *test of significance*. For instance, the palatability experiment described in section 1.9 was conducted to compare the average palatability scores of protein food C and of food H in liquid (LH) or solid (SH) forms. Before reporting the average differences found in the data, however, we might ask: Do the results show *real* differences among the average palatability scores in three populations, or can the differences observed in the samples be explained by subject-to-subject variation? Thus we might set up the hypothesis $\mu_C = \mu_H$ and the hypothesis $\mu_{LH} = \mu_{SH}$. For each hypothesis we would examine whether the sample results support the hypothesis. Unless our analyses reject these hypotheses, we are not in a position to claim that there are real differences among the C, LH, and SH populations in average palatability scores.

In such applications, the hypothesis under test is called the *null·hypothesis* and is denoted by H_0. The adjective *null* seems to have been chosen to indicate that a null hypothesis is merely a hypothesis put up for consideration. Both *Webster's* and the *Oxford English Dictionary* give the first meaning of *null* as "of no legal or binding force; of no efficacy." Nevertheless, the word *null* is used quite generally in statistics to denote the hypothesis being tested.

5.2—A test of the mean of a normal population (σ known). To illustrate the steps in a test of significance, we will test the null hypothesis that the mean of a normal population has a known value μ_0. For simplicity, the population standard deviation σ is assumed known. The data are a random sample of size n drawn from the population. Construction of a test involves the following steps.

Given the null hypothesis H_0, the first step is to note the *alternative hypothesis* or hypotheses. If our data seems inconsistent with the null hypothesis and we reject it, what hypothesis do we put in its place? In the test of a mean, the obvious alternative hypothesis H_A is that μ has some value μ_A different from μ_0.

In statistical analyses with specific objectives, the investigator usually knows the null and alternative hypotheses in advance (though frequently only the null hypothesis is stated explicitly). This type of analysis differs from exploratory data analysis, in which the objective is often to find hypotheses that are consistent with the data.

Let $\overline{X}$ be the sample mean. With these null and alternative hypotheses, it is natural to look at the size of the deviation $(\overline{X} - \mu_0)$ as a possible criterion by which to judge whether the data favor H_0 or H_A. If H_0 holds, $(\bar{X} - \mu_0)$ is normally distributed with mean 0, standard deviation $\sigma/\sqrt{n}$. If H_A holds, $(\bar{X} - \mu_0)$ has mean $(\mu_A - \mu_0)$, different from zero. Consequently, if H_A holds, $(\bar{X} - \mu_0)$ will tend to become large, positively or negatively. How large will depend on the size of $(\mu_A - \mu_0)$ and on the value of $\sigma/\sqrt{n}$. The quantity $(\bar{X} - \mu_0)$ is called the *test criterion.*

With $(\bar{X} - \mu_0)$ as the test criterion, large values, positive or negative, cause us to reject H_0 in favor of H_A. How large? There is no single answer. The larger we make the value of $(\bar{X} - \mu_0)$ required for rejection of H_0, the smaller is the probability of rejecting H_0 when it is true, but the smaller is the probability of rejecting H_0 when H_A is true, that is, when H_0 *should* be rejected. The practice most commonly followed in statistics is to choose a critical value of $(\bar{X} - \mu_0)$ such that the probability of rejecting H_0 when it is true is 0.05, or 5%. Such a test is called a *test at the 5% level.* A level of 0.01, or 1%, is sometimes employed when incorrect rejection of the null hypothesis is regarded as a serious mistake. Other levels—10% or 20%—may be used when considered appropriate.

In reporting results, an asterisk is often added to a critical value to denote that it is significant at the 5% level. Two asterisks denote significance at the 1% level.

The method of testing the hypothesis $\mu = \mu_0$ at the 5% level is as follows:

1. Convert $(\bar{X} - \mu_0)$ to a standard normal deviate $Z = \sqrt{n}(\bar{X} - \mu_0)/\sigma$. Since this alternative hypothesis makes no reference to the sign of $(\bar{X} - \mu_0)$, only absolute values $|Z|$ are relevant. A test of this type is called a *two-sided* or *two-tailed* test.

2. Reject the null hypothesis if $|Z| > 1.960$. By the normal tables, this rule ensures a probability 0.05 of rejection when H_0 is true. In terms of $\bar{X}$ the rule is:

$$\text{reject } H_0{:}\mu = \mu_0 \text{ if } \bar{X} \text{ is outside the limits } \mu_0 \pm 1.96\sigma/\sqrt{n} \qquad (5.2.1)$$

The set of values of $\bar{X}$ that cause rejection is called the *critical region* or the *region of rejection.* For two-tailed tests at the 1% level, replace 1.96 by 2.576.

Another way of describing rejection is to say that the difference $\bar{X} - \mu_0$ is *statistically significant,* or more succinctly, *significant,* at the stated level.

As a numerical example, let us take the population of swine gains in table 4.14.1, assumed normal with $\mu = 30$, $\sigma = 10$. We test the null hypothesis $\mu = 30$ from the large random sample of size $n = 5110$ drawn from this population, with $\bar{X} = 29.823$. We do not expect to reject this null hypothesis unless something went wrong in the sampling or recording.

In this case, $Z = \sqrt{5110}(29.823 - 30)/10 = -1.265$. Since $|Z| < 1.960$, we do not reject H_0 at the 5% level. There is something to be said for reporting also the actual probability of a deviation greater than the sample $|Z|$ if H_0 is true. This probability shows whether the sample deviation is close to the 5% level or far short of it and may be useful to others. By interpolation, table A 3 gives 0.3971 for the area from $Z = 0$ to $Z = 1.265$. Hence, the total probability in the two tails is $2(0.5 - 0.3971) = 0.206$. We might report, "The null hypothesis $\mu = 30$ was not rejected ($P = 0.21$)." Reports of tests in computer programs normally give this P value.

In summary, the choice of a significance level determines the relative frequency of two kinds of mistakes that we can make—rejecting H_0 when H_0 is correct and failing to detect the truth of H_A when it is correct. The two mistakes are called, respectively, errors of *Type I* and *Type II*.

5.3—Tests of significance and confidence intervals. A simple relation exists between a 5% two-tailed test of the null hypothesis $\mu = \mu_0$ and the 95% confidence interval for μ. If the test fails to reject H_0 at the 5% level, then μ_0 lies inside the 95% confidence interval for μ, and vice versa. If the test rejects H_0, then μ_0 lies outside the 95% confidence interval for μ, and vice versa.

For if μ_0 lies in the 95% confidence interval for μ,

$$\bar{X} - 1.96\sigma/\sqrt{n} < \mu_0 < \bar{X} + 1.96\sigma/\sqrt{n} \tag{5.3.1}$$

That is, $\bar{X} - \mu_0 < 1.96\sigma/\sqrt{n}$, giving $Z < 1.96$; and $\bar{X} - \mu_0 > -1.96\sigma/\sqrt{n}$, giving $Z > -1.96$. In both cases, $|Z| < 1.96$ and we do not reject H_0 at the 5% level. The same algebra shows that if $|Z| > 1.96$ and we reject H_0, then μ_0 lies outside the 95% interval for μ.

The two techniques have different purposes. The confidence interval gives an answer to the question, How accurately do we know μ? while the test relates to the question, Could μ have the value μ_0?

5.4—Practical uses of tests of significance. One use of tests of significance has become very common. If investigators claim that method A is superior to method B for some purposes, the claim is unlikely to be accepted by their peers or in scientific journals unless they can present results of a comparative study that show A superior to B in a test of significance at the 5% or 1% level. This practice has been beneficial in preventing false claims from spreading and in saving the time and experience needed to show later that the claims were false, though the practice presumably has discouraged some work that would have been fruitful.

However, it is not clear just what we should conclude from a nonsignificant result. A test of significance is most easily taught as a rule for deciding whether to *accept* or *reject* the null hypothesis. But the meaning of the word *accept* requires careful thought. A nonsignificant result does not prove that the null hypothesis is correct—merely that it is tenable—our data do not give adequate grounds for rejecting it.

Nevertheless, given a nonsignificant result, we often have to decide, both in applications of our results and in subsequent statistical analyses, whether to act as if the null hypothesis *is* correct. Should we assume that varieties A and B give the same yield per acre? Should we regard a variable Y as normally distributed? Before such decisions are made, we should consider any additional knowledge available to us that may guide a decision. The consequences of wrongly assuming the null hypothesis to be correct should also be weighed. An assumption that medications C and D have the same frequency of side effects may not be critical if the side effects are minor temporary annoyances, but it is a different matter if these effects can be fatal. Issues of this type are taken into account in the branch of the subject known as *statistical decision theory*.

The size of the sample on which the test of significance is made is also important. With a small sample, a test is likely to produce a significant result only if the null hypothesis is very wrong. With a large sample, small departures from the null hypothesis that might be quite unimportant in practice can be detected as statistically significant. After seeing the results from a large sample, we may decide to act as if the null hypothesis is correct even if the test has rejected it.

In this connection, a look at the 95% confidence limits for μ is helpful. For our sample with $n = 5110$, these limits are $29.823 \pm (1.96)(0.140) = 29.549$ and 30.097. Seeing how close both limits are to $\mu = 30$, we may decide that the assumption $\mu = 30$ can safely be made for future work or action. Suppose, however, that our evidence is a sample of size $n = 10$ with $\overline{X} = 29.8$. We find that $\mu_0 = 30$ is accepted by the test of significance, but the 95% confidence limits are $29.8 \pm (1.96)(3.162) = 23.6$ and 36.0. We may decide that we need more data to narrow these limits before we assume $\mu = 30$ in a future action.

5.5—One-sided or one-tailed tests. In one-sided or one-tailed tests of a mean or of a difference between two means, the investigator pays attention only to deviations from the null hypothesis in one direction, ignoring deviations in the other direction. This happens primarily in two situations. Occasionally (situation I) the investigator knows enough about the circumstances of the test to be certain that if μ is not equal to μ_0, then μ is greater than μ_0; or if the difference δ between two population means is not 0, then δ is greater than 0. For instance, corn borers decrease corn yields by attacking the corn. With a spray that was being tried as protection against the corn borer, it was known that at the concentration used the spray had no direct effect on corn yield. It was reasonable to assume that in fields exposed to possible corn borer attack, the spray would either have no effect on corn yields (the null hypothesis) or would increase them. Furthermore, it is sometimes clear from the nature of the test criterion that when the null hypothesis is false the criterion tends to increase, that is, to change in one direction only. An example is the χ^2 test of goodness of fit in section 5.12.

As an example of situation II, an investigator is comparing a new treatment A with a standard treatment S, of mean μ_0, that has proved satisfactory. The investigator chooses $\mu_A \leq \mu_0$ as the null hypothesis and $\mu_A > \mu_0$ as the alternative hypothesis, since the new treatment is of no interest unless it is superior to the standard treatment.

In making a one-tailed test when the alternative hypothesis H_A is $\mu_A > \mu_0$, accept H_0 if $(\overline{X} - \mu_0)$ is negative or zero. If $(\overline{X} - \mu_0)$ is positive, calculate

$$Z = \sqrt{n}(\overline{X} - \mu_0)/\sigma$$

and read the probability that a normal deviate in table A 3 exceeds Z. The 5% significance level of Z in a one-tailed test is 1.645 and the 1% level is 2.326, these being, of course, the 10% and 2% levels in a two-tailed test.

EXAMPLE 5.5.1—In a normal population with $\sigma = 20$, a random sample of size n gave $\overline{X} = 56.53$. At the 5% level, make a two-tailed test of the null hypothesis $\mu = 50$ when (*i*) $n = 25$, (*ii*) $n = 36$, (*iii*) $n = 64$. In each case, report also the significance probability, that is, the probability P of getting a larger deviation that than observed if H_0 is true. Ans. (*i*) Not significant, $P = 0.103$; (*ii*) borderline, $P = 0.05$; (*iii*) significant, $P = 0.009$.

EXAMPLE 5.5.2—For each of the three sample sizes in example 5.5.1, calculate 95% confidence limits for μ from $\overline{X} = 56.53$. Verify that the confidence interval is consistent in each case with the verdict of the test of significance. (*i*) Limits (48.69, 64.37), (*ii*) limits (50.00, 63.06), (*iii*) limits (51.63, 61.43).

EXAMPLE 5.5.3—In a two-tailed test of $H_0{:}\mu = 80$ in a normal population with $\sigma = 15$, an investigator reported, "Since $\overline{X} = 71.91$, the null hypothesis was rejected at the 1% level." What can be said about the sample size used? Ans. n was at least 23.

EXAMPLE 5.5.4—National student exam scores are sometimes scaled so that the population of scores is normal with $\mu = 500$, $\sigma = 100$. In a particular class of 25 students, the mean score was 472. Wondering if this is a class of below average exam performance, the instructor makes a 5% one-tailed test of $H_0{:}\mu \geq 500$ against $H_A{:}\mu < 500$. What is its verdict? Ans. $Z = -1.4$; not significant, one-tailed $P = 0.081$.

EXAMPLE 5.5.5—In mass production a certain component will fit into its proper place only if its length is between 6.2 cm and 6.3 cm. The lengths of individual components vary slightly with $\sigma = 0.021$ cm and are approximately normal. The average length μ of a batch of components can be set by a dial at an approximate chosen value. Verify the value of μ that minimizes the percentage of nonfitting components: $\mu = 6.25$ cm. Give the resulting percent that will not fit at this setting. Ans. $\mu = 6.25$ cm gives 1.74% failures to fit; $\mu = 6.24$ or 6.26, for example, gives 3.05% failures.

EXAMPLE 5.5.6—In example 5.5.5 the operator sets the batch average length at approximately $\mu_0 = 6.25$ cm. As a check that it is near the mark, the operator measures 100 lengths, getting $\overline{X} = 6.2516$ cm. (*i*) Test $H_0{:}\mu_0 = 6.25$ at the 10% level and (*ii*) calculate 95% confidence limits for μ. Ans. (*i*) Not significant, $Z = 0.762$; (*ii*) limits (6.2475, 6.2557).

5.6—Power of a test of significance. Ideally, a test of significance should reject the null hypothesis when it is false. The probability that the test does this is called the *power* of the test. For the test of a normal mean, the power can be calculated from table A 3. We begin with an example to illustrate the method.

Consider the large sample of size $n = 5110$ from the swine gains, in which $\sigma = 10$. In section 5.2, using the sample mean $\overline{X}$, we performed a two-tailed test at the 5% level of the null hypothesis $\mu_0 = 30$. Suppose that unknown to us the population mean μ had been 30.5. What is the power of this test against this alternative? That is, what is the probability that the hypothesis $\mu_0 = 30$ would have been rejected if actually $\mu = 30.5$?

To answer this question, we first find the values of $\overline{X}$ that cause H_0 to be rejected. The standard error of $\overline{X}$ is $\sigma/\sqrt{n} = 10/\sqrt{5110} = 0.140$. At the 5% level, we reject H_0 if $\overline{X}$ lies outside the limits $30 \pm (1.96)(0.140) = 30 \pm 0.274$. That is, we reject if either

$$\overline{X} < 29.726 \quad \text{or} \quad \overline{X} > 30.274 \tag{5.6.1}$$

We now calculate the probability that $\overline{X}$ lies in either of these regions if μ is actually 30.5. For $\overline{X} = 29.726$, the normal deviate Z is $(29.726 - 30.5)/$

0.140 = −5.529. For $\overline{X}$ = 30.274, $Z = (30.274 - 30.5)/0.140 = -1.614$. From table A 3, $P(Z) < -5.529$ is practically zero, while $P(Z) > -1.614 = 0.947$. Thus the power of the test, the sum of these two probabilities, is 0.947. In summary, the test is almost certain to detect this small departure from the null hypothesis because the sample size, 5110, is so large.

Figure 5.6.1 illustrates this argument. The curve on the left is the normal distribution of $\overline{X}$ on the null hypothesis. The two regions of rejection are $\overline{X} <$ 29.726 and $\overline{X} >$ 30.274. From the distribution of $\overline{X}$ on the alternative hypothesis, centered about $\mu = 30.5$, we see that practically all of this distribution lies to the right of 30.274—hence the power value P of 0.947.

The preceding argument may be carried through algebraically, H_0 being $\mu = \mu_0$ and the alternative H_A being $\mu = \mu_A$. In a two-tailed test of H_0 at significance level α, the power of the test against H_A is the probability that a normal deviate Z lies in one of the two regions $Z < Z_1$, $Z > Z_2$, where

$$Z_1 = (-\sqrt{n}/\sigma)(\mu_A - \mu_0) - Z_\alpha \quad \text{and} \quad Z_2 = (-\sqrt{n}/\sigma)(\mu_A - \mu_0) + Z_\alpha \tag{5.6.2}$$

Here, α denotes the two-tailed significance level, with $Z_{0.05} = 1.960$ and $Z_{0.01} =$ 2.576.

In a one-tailed test at level α, the alternative being $\mu_A > \mu_0$, the power is the probability that the normal deviate

$$Z_1 > (-\sqrt{n}/\sigma)(\mu_A - \mu_0) + Z_{2\alpha} \tag{5.6.3}$$

where, of course, $Z_{2\alpha} = 1.645$ at the 5% level and $Z_{2\alpha} = 2.326$ at the 1% level.

The power of a test is a fairly complex quantity. It depends not only on the size $(\mu_A - \mu_0)$ of the departure from H_0 but also on n, σ, and the significance level and type of test (one-sided or two-sided). Table 5.6.1 shows some power values for tests at different levels. The key quantity on which the power depends is $\phi = \sqrt{n}(\mu_A - \mu_0)/\sigma$.

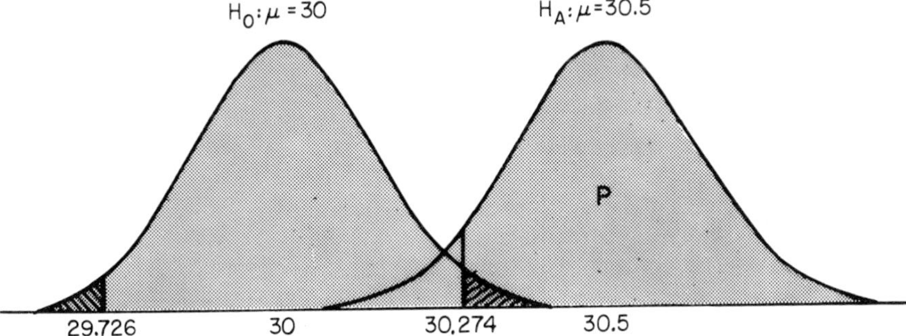

Fig. 5.6.1—Left curve: distribution of $\overline{X}$ for H_0: $\mu = 30$, showing the regions of rejection, $\overline{X} < 29.726$ and $\overline{X} > 30.274$. Right curve: distribution of $\overline{X}$ for H_A: $\mu = 30.5$, the power P being the area to the right of 30.274.

TABLE 5.6.1
POWER OF A TEST OF THE NULL HYPOTHESIS $\mu = \mu_0$ MADE FROM THE MEAN $\overline{X}$ OF A NORMAL SAMPLE WITH KNOWN σ

Level of Test	One- or Two-Tailed	$\phi = \sqrt{n}(\mu_A - \mu_0)/\sigma =$				
		1.5	2	2.5	3	3.5
0.05	one	0.44	0.64	0.80	0.91	0.97
.05	two	.32	.52	.71	.85	.94
.01	one	.20	.37	.57	.75	.88
0.01	two	0.14	0.22	0.47	0.61	0.82

The power does not become high until $\phi = \sqrt{n}(\mu_A - \mu_0)/\sigma$ reaches 2.5 or 3 for tests at the 5% level and 3.5 for tests at the 1% level. At the same level, one-tailed tests naturally have consistently more power than two-tailed tests, and of course, 5% tests have more power than 1% tests. With $\sqrt{n}(\mu_A - \mu_0)/\sigma = 1.5$ we see that a two-tailed 1% test has only about 1 chance in 7 of revealing that H_0 is false. In applications, power calculations are used frequently to decide on sample size when planning comparative tests (section 6.14).

EXAMPLE 5.6.1—Verify the powers given in the following entries in table 5.6.1: (*i*) 5% two-tailed test, $\phi = \sqrt{n}(\mu_A - \mu_0)/\sigma = 2$; (*ii*) 1% two-tailed test, $\phi = 3.5$; (*iii*) 5% one-tailed test, $\phi = 3$; (*iv*) 1% one-tailed test, $\phi = 1.5$.

EXAMPLE 5.6.2—In a swine gains population with $\mu = 30.5$, $\sigma = 10$, what is the probability of rejecting H_0:$\mu_0 = 30$ in a two-tailed 5% test if (*i*) $n = 10$, (*ii*) $n = 100$, (*iii*) $n = 1000$. Ans. (*i*) 0.053, (*ii*) 0.079, (*iii*) 0.353. Only very large samples give a good chance of detecting this small departure from the null hypothesis.

EXAMPLE 5.6.3—It has been suggested that tests at the 5% level give too much protection against Type I errors and too little against Type II errors, i.e., too little power in detecting failure of H_0. In table 5.6.1, how much would the power of a two-tailed test improve (*i*) for $\phi = 1.5$ and (*ii*) for $\phi = 2$ if we test at the 20% level, for which $Z_{0.20} = 1.282$, instead of at the 5% level? Ans. (*i*) For $\phi = 1.5$, the power increases from 0.32 to 0.59; (*ii*) for $\phi = 2$, from 0.52 to 0.76.

EXAMPLE 5.6.4—In calculating the power of a two-tailed test at the 5% or a lower significance level, verify from (5.6.2) that the probability of Z being less than Z_1 can be regarded as zero with little loss of accuracy if $\phi = \sqrt{n}(\mu_A - \mu_0)/\sigma \geq 0.5$. For this value of ϕ the power is only 0.079. Thus we rarely need two areas from table A 3 when calculating the power.

5.7—Testing a mean when σ is not known. In most applications in which a null hypothesis about a population mean is to be tested, the population standard deviation σ is not known. In such cases we use the sample standard deviation s as an estimate of σ and replace the normal deviate by

$$t = \sqrt{n}(\overline{X} - \mu_0)/s \quad \text{with } (n - 1) \text{ df}$$

For two-tailed tests the significance levels of t with different degrees of freedom are read directly from table A 4. In terms of $\overline{X}$, H_0 is rejected at level α if $\overline{X}$ lies outside the limits $\mu_0 \pm t_\alpha s/\sqrt{n}$.

For a one-tailed test at level α, read the t value for level 2α from table A 4.

The following is an example of historical interest. In an early comparative experiment (conducted in 1764), Arthur Young (1) chose an area (1 acre or 1/2 acre) on each of 7 fields on his farm. He divided each area into halves, "the soil exactly the same in both." On one half he broadcast the wheat seed (the old husbandry) and on the other half he drilled the wheat in rows (the new husbandry). All expenses on each half and the returns at harvest were recorded. Table 5.7.1 shows the differences D_i in profit per acre (pounds sterling) for drilling minus broadcasting on the seven fields. On eye inspection, drilling does not look promising. The natural test is a two-sided test of the null hypothesis $\mu_D = 0$. The sample data in table 5.7.1 gives $s_D = 1.607$. Hence

$$t = \sqrt{n}\bar{D}/s_D = \sqrt{7}(-1.01)/1.607 = -1.663 \quad (6 \text{ df})$$

The null hypothesis that the two methods give equal profit in the long run cannot be rejected. From table A 4 the significance probability (the probability that $|t| > 1.663$ given H_0) is about 0.15.

Following is an example of a one-tailed t test of a nonzero μ_D. On 14 farms in Boone County, Iowa, in 1950, the effect of spraying against the corn borer was evaluated by measuring the corn yields on both sprayed and unsprayed strips in a field on each farm. The 14 differences (sprayed minus unsprayed) in yields (bushels per acre) were as follows: -5.7, 3.7, 6.4, 1.5, 4.3, 4.8, 3.3, 3.6, 0.5, 5.0, 24.0, 8.8, 4.5, 1.1.

The sample mean difference was $\bar{D} = 4.70$ bu per acre with $s_D = 6.48$ bu/acre and $s_{\bar{D}} = 6.48/\sqrt{14} = 1.73$ bu/acre. The cost of commercial spraying was \$3/acre and the 1950 crop sold at about \$1.50/acre. So a gain of 2 bu/acre would pay for the spraying.

Does the sample furnish strong evidence that spraying is profitable? To examine this question, we make a one-sided test with $H_0{:}\mu_D \leq 2$, $H_A{:}\mu_D > 2$. The test criterion is $t = (4.7 - 2.0)/1.3 = 1.56$ with $= 13$ df. The one-tailed 5% level in table A 4 is 1.771, the significance probability for $t = 1.56$ being about 0.075—not quite at the 5% level. The lower 95% limit in a one-sided confidence interval statement is $4.70 - (1.771)(1.73) = 1.64$ bu/acre. As expected, the confidence interval agrees with the verdict of the test in indicating that there might be a small loss (equivalent to $2 - 1.64 = 0.36$ bu/acre) from spraying. In the 14 differences, note that the eleventh difference, 24.0, is much larger than any of the others and looks out of line. Tests described in sections 15.5 and 15.6 confirm that this value comes from a different population than the other 13 differences, although we know of no explanation. This discrepancy casts doubt on the test of significance of the mean difference.

TABLE 5.7.1

DIFFERENCES (DRILLED MINUS BROADCASTING) IN PROFIT PER ACRE (POUNDS STERLING)

Field	1	2	3	4	5	6	7	Total	Mean
D_i	−0.3	−0.3	−2.3	−0.7	+1.7	−2.2	−3.0	−7.1	−1.01

EXAMPLE 5.7.1—Samples of blood were taken from each of 8 patients. In each sample, the serum albumen content of the blood was determined by each of two laboratory methods A and B. The objective was to discover whether there was a consistent difference in the amount of serum albumen found by the two methods. The 8 differences ($A - B$) were as follows: 0.6, 0.7, 0.8, 0.9, 0.3, 0.5, −0.5, 1.3, the units being gm/100 ml. Compute t to test the null hypothesis (H_0) that the population mean of these differences is zero and report the approximate value of your significance probability. What is the conclusion? Ans. $t = 3.09$ with 7 df, P between 0.025 and 0.01. Method A has a systematic tendency to give higher values.

EXAMPLE 5.7.2—In an investigation of the effect of feeding 10 μg of vitamin B_{12} per pound of ration to growing swine (5), 8 lots (each with 6 pigs) were fed in pairs. The pairs were distinguished by being fed different levels of Aureomycin, an antibiotic that did not interact with the vitamin; that is, the differences were not affected by the Aureomycin. The average daily gains (to about 200 lb liveweight) are summarized as follows:

	Pairs of Lots							
Ration	1	2	3	4	5	6	7	8
With B_{12}	1.60	1.68	1.75	1.64	1.75	1.79	1.78	1.77
Without B_{12}	1.56	1.52	1.52	1.49	1.59	1.56	1.60	1.56
Difference, D	0.04	0.16	0.23	0.15	0.16	0.23	0.18	0.21

For the differences, calculate the statistics $\overline{D} = 0.170$ lb/day and $s_D = 0.0217$ lb/day. Clearly, the effect of B_{12} is statistically significant.

EXAMPLE 5.7.3—The effect of B_{12} seems to be a stimulation of the metabolic processes including appetite. The pigs eat more and grow faster. In the experiment above, the cost of the additional amount of feed eaten, including that of the vitamin, corresponded to about 0.130 lb/day of gain. Make a one-sided test of $H_0{:}\mu_D \le 0.130$ lb/day, $H_A{:}\mu_D > 0.130$ lb/day. Ans. $t = 1.843$, just short of the 5% level.

EXAMPLE 5.7.4—In example 5.7.3, what is the lower 95% one-sided confidence limit for μ_D? Ans. 0.129 lb/day of gain, indicating (apart from a 5% chance) only a trivial money loss from B_{12}.

5.8—Other tests of significance. As we have noted, the method of making a significance test is determined by the null and the alternative hypotheses. Suppose we have data that are thought to be a random sample from a normal distribution. In preceding sections we showed how to make one-tailed and two-tailed tests when the null hypothesis is $\mu = \mu_0$ and the alternative is $\mu = \mu_A$.

In statistical analysis, there are many different null and alternative hypotheses that are on occasion useful to consider and therefore many different tests of significance. Four are listed:

1. With the same data, our interest might be in the amount of variability in the population, not in the population mean. The null hypothesis H_0 might be that the population variance has a known value σ_0^2, while H_A is that it has some other value. As might be anticipated, this test (section 5.11) is made from the sample value of s^2.

2. In other circumstances we might have no hypothesis about the values of μ or σ^2 and no reason to doubt the randomness of the sample. Our question might be, Is the population normal? For problems in which the alternative says no more than that the sample is drawn from *some kind* of nonnormal population, a test of goodness of fit is given (section 5.12).

3. With the same null hypothesis, the alternative might be slightly more specific, namely that the data come from a skew, asymmetrical distribution. For this alternative, a test for skewness (section 5.13) is used.

4. In other applications, the alternative might be that the distribution differs from the normal primarily in having longer tails. A test criterion whose aim is to detect this type of departure from normality (kurtosis) appears in section 5.14.

5.9—Frequency distribution of s^2. In studying the variability of populations and the precision of measuring instruments and of mass production processes, we face the problem of estimating σ^2—in particular, of constructing confidence intervals for σ^2 from the sample variances s^2 and of testing null hypotheses about the value of σ^2. As before, $s^2 = \Sigma_{i=1}^{n} (X_i - \bar{X})^2/(n - 1)$.

If the X_i are randomly drawn from a normal distribution, the frequency distribution of the quantity $(n - 1)s^2/\sigma^2$ is another widely used distribution, the chi-square distribution with $(n - 1)$ df. *Chi-square* (χ^2) *with* ν *degrees of freedom* is defined as the distribution of the sum of squares of ν independent normal deviates. Thus, if $Z_1, Z_2, \ldots, Z_\nu$ are independent normal deviates, the quantity

$$\chi^2 = Z_1^2 + Z_2^2 + \ldots + Z_\nu^2 \tag{5.9.1}$$

follows the chi-square distribution with ν df. The form of the distribution was worked out mathematically. Table A 5 presents the percentage points of the distribution—the values of χ^2 that are *exceeded* with the stated probabilities 0.995, 0.990, and so on.

The relation between s^2 and χ^2 is made a little clearer by the following algebra. Remember that $(n - 1)s^2$ is the sum of squares of deviations, $\Sigma(X_i - \bar{X})^2$, which may be written $\Sigma[(X_i - \mu) - (\bar{X} - \mu)]^2$, where μ is the population mean. Hence,

$$\frac{(n - 1)s^2}{\sigma^2} = \frac{(X_1 - \mu)^2}{\sigma^2} + \frac{(X_2 - \mu)^2}{\sigma^2} + \ldots + \frac{(X_n - \mu)^2}{\sigma^2} - \frac{n(\bar{X} - \mu)^2}{\sigma^2} \tag{5.9.2}$$

Now the quantities $(X_i - \mu)/\sigma$ are all in standard measure—they are random normal deviates Z_i. And $\sqrt{n}(\bar{X} - \mu)/\sigma$ is another normal deviate, since the standard error of $\bar{X}$ is $\sigma/\sqrt{n}$. Hence we may write

$$(n - 1)s^2/\sigma^2 = Z_1^2 + Z_2^2 + \ldots + Z_n^2 - Z_{n+1}^2 \tag{5.9.3}$$

Thus $(n - 1)s^2/\sigma^2$ is a χ^2 with n df *minus* a χ^2 with 1 df. This difference can be shown to be distributed as χ^2 with $(n - 1)$ df.

The 511 sample values of s^2 calculated from the swine gains in section 4.14 provide an experimental verification of the theoretical distribution of $s^2 = \chi^2\sigma^2/(n - 1)$. The comparison of observed and expected frequencies is shown in

TABLE 5.9.1
DISTRIBUTION OF 511 VALUES OF s^2 WITH 9 DF FROM NORMAL SAMPLES WITH $\sigma^2 = 100$

Class Limits	Frequencies Obs. f_i	Exp. F_i	$(f_i - F_i)^2/F_i$	Class Limits	Frequencies Obs. f_i	Exp. F_i	$(f_i - F_i)^2/F_i$
0–30	12	12.8	0.05	150–170	29	29.6	0.12
30–50	47	50.8	0.28	170–190	26	18.4	3.14
50–70	92	84.8	0.61	190–210	11	10.8	0.00
70–90	93	94.7	0.03	210–230	8	6.1	0.59
90–110	72	84.5	1.85	230–250	2	3.4	0.58
110–130	73	65.2	0.93	over 250	4	3.9	0.00
130–150	42	45.9	0.33	Total	511	510.9	8.51

table 5.9.1. The expected frequencies were obtained from the table in reference (2), which gives the probabilities of exceeding specified values of χ^2 and is more convenient than table A 5. We note in table 5.9.1. that (*i*) the agreement between the observed and expected frequencies seems close; (*ii*) the distribution of s^2 is skew, with a longer tail of high values; (*iii*) as mentioned previously, s^2 is not an accurate estimate of σ^2—only 72 or 14% of the 511 sample values come within $\pm 10\%$ of $\sigma^2 = 100$. The columns headed $(f_i - F_i)^2/F_i$ are discussed in section 5.12.

Like the normal distribution, the theoretical distribution of χ^2 is continuous. Unlike the normal, χ^2, being a sum of squares, cannot take negative values. The distribution extends from 0 to ∞. A result from theory is that the mean value of χ^2 with ν df is exactly ν.

5.10—Interval estimates of σ^2. Given an estimated variance s^2 with ν df computed from normal data, the χ^2 table (A 5) can be used to obtain a confidence interval for σ^2. For a 95% interval the relevant quantities are $\chi^2_{0.975}$ and $\chi^2_{0.025}$, the values of chi-square exceeded with probabilities 0.975 and 0.025, respectively. The probability that a value of χ^2 drawn at random lies between these limits is 0.95. Since $\chi^2 = \nu s^2/\sigma^2$, the probability is 95% that when our sample was drawn,

$$\chi^2_{0.975} \le \nu s^2/\sigma^2 \le \chi^2_{0.025} \tag{5.10.1}$$

These inequalities are equivalent to

$$\nu s^2/\chi^2_{0.025} \le \sigma^2 \le \nu s^2/\chi^2_{0.975} \tag{5.10.2}$$

Expression (5.10.2) is the general formula for two-sided 95% confidence limits for σ^2. With s^2 computed from a sample of size n, $\nu = n - 1$.

As an example, we show how to compute the 95% limits for σ^2, given one of the s^2 values obtained from $n = 10$ swine gains. For 9 df, table A 5 gives 19.02 for $\chi^2_{0.025}$ and 2.70 for $\chi^2_{0.975}$. Hence, from (5.10.2) the 95% interval for σ^2 is

$$19s^2/10.02 < \sigma^2 < 9s^2/2.70 \quad \text{or} \quad 0.47s^2 < \sigma^2 < 3.33s^2$$

Note that s^2 is not in the middle of the interval and that the confidence interval is wide, extending from about $s^2/2$ to over $3s^2$.

5.11—Test of a null hypothesis value of σ^2. This test is not frequently required. It arises when a quality standard about the amount of variability allowable in some process specifies a value of σ^2 or when σ^2 is predicted from a theory that is to be tested or is known for a population with which sample data are being compared.

Usually, one-tailed tests are wanted. Let the null hypothesis be $\sigma^2 \leq \sigma_0^2$. When the alternative hypothesis is $\sigma^2 > \sigma_0^2$, compute

$$\chi^2 = \nu s^2/\sigma_0^2 = \Sigma x^2/\sigma_0^2 \tag{5.11.1}$$

This value is significant at the 5% level if $\chi^2 > \chi^2_{0.050}$ with ν df. Suppose an investigator has used for years a stock of inbred rats whose weights have $\sigma_0 = 26$ g. The investigator considers switching to a cheaper source of supply although the new rats may show greater variability, which would be undesirable. A sample of 20 new rats gave $s^2 = 23{,}297/19$, $s = 35.0$, in line with suspicions. As a check, the investigator tests $H_0{:}\sigma \leq 26$ g against $H_A{:}\sigma > 26$ g.

$$\chi^2 = 23{,}297/26^2 = 34.46 \quad (\text{df} = 19)$$

In table A 5, $\chi^2_{0.050} = 30.14$, so the null hypothesis is rejected.

To test $H_0{:}\sigma^2 \geq \sigma_0^2$ against the alternative $\sigma^2 < \sigma_0^2$, reject at the 5% level if $\chi^2 = \Sigma x^2/\sigma_0^2 < \chi^2_{0.950}$. For a two-sided test, reject if either $\chi^2 < \chi^2_{0.975}$ or $\chi^2 > \chi^2_{0.025}$. For example, the 511 samples of means $\bar{X}$ of 10 swine gains in weight in table 4.6.1 gave $s^2 = 10{,}201$ with 510 df. Since the original population (table 4.14.1) had $\sigma^2 = 100$, the theoretical $\sigma^2_{\bar{X}}$ is $\sigma_0^2 = 100/10 = 10$. We will make a two-tailed test of this null hypothesis.

Table A 5 stops at $\nu = 100$. For values of $\nu > 100$, an approximation is that the quantity

$$Z = \sqrt{2\chi^2} - \sqrt{2\nu - 1} \tag{5.11.2}$$

is $\mathcal{N}(0, 1)$, where $\mathcal{N}(\mu, \sigma^2)$ denotes a normal variate with mean μ, variance σ^2. In our example, $\chi^2 = (510)(10.201)/10 = 520.25$; $\nu = 510$. Hence

$$Z = \sqrt{1040.5} - \sqrt{1019} = 0.335$$

This value lies far short of the two-tailed 5% level, 1.96. The null hypothesis $\sigma^2 \doteq 100$ is accepted.

5.12—The χ^2 test of goodness of fit. In table 5.9.1 we compare the frequency distribution of 511 sample estimates s^2 (9 df) grouped into 13 classes, with the theoretical distribution that s^2 should follow if the original distribution of swine gains is normal. The comparison is made by looking at the frequency f_i observed in any class and the frequency F_i expected from the theoretical distribution. We note that the agreement between the observed and the expected frequencies appears good.

Another important use of the χ^2 distribution is to provide a quantitative test of the discrepancies between the observed (f_i) and the expected (F_i) frequencies in such a comparison. This gives a test of the null hypothesis: Can the observations be regarded as randomly drawn from the theoretical distribution? The test can be used with either continuous or discrete theoretical distributions. This test is the χ^2 *test of goodness of fit*.

To apply the test, calculate for each class the quantity

$$(f_i - F_i)^2/F_i = (\text{observed} - \text{expected})^2/\text{expected}$$

The test criterion is

$$\chi^2 = \sum_{i=1}^{k} (f_i - F_i)^2/F_i \tag{5.12.1}$$

summed over the k classes. If the null hypothesis holds, K. Pearson (3) showed in 1899 that the test criterion follows the theoretical χ^2 distribution in large samples. If the observations come from some other distribution so that the null hypothesis is false, the observed f_i tend to agree poorly with the expected F_i and the computed χ^2 becomes large. Consequently, a value of χ^2 greater than $\chi^2_{0.050}$ in table A 5 causes rejection at the 5% level. This is a one-sided test.

NUMBER OF DEGREES OF FREEDOM IN χ^2. The theoretical frequency distribution of s^2 under normality depends on the population variance σ^2 and on the degrees of freedom ν in s^2. Since both of these were known when the theoretical distribution in table 5.9.1 was fitted, we say that the theoretical distribution is completely specified. In this case, the number of degrees of freedom in χ^2 is $(k - 1)$, where k is the number of classes whose contributions are summed in finding χ^2. Note that the degrees of freedom in χ^2 do not directly involve the sample size n.

If we want to test whether this theoretical distribution fits but do not know the value of σ^2, we can use the average of the 511 sample variances s^2 as an estimate of σ^2 when fitting. The number of degrees of freedom in χ^2 is then $(k - 2)$. The subtracted number 2 can be thought of as the number of ways in which we have forced the observed and expected frequencies to agree (in their totals n and in the variances σ^2). Similarly, in fitting a normal distribution, the degrees of freedom in χ^2 are $(k - 1)$ if μ and σ are both known for the population when fitting, as in table 4.6.1, but the degrees of freedom are $(k - 3)$ if $\bar{X}$ and s from the sample are used as estimates of μ and σ when fitting. The rule is:

$$\text{df} = (k - 1 - \text{number of fitted parameters})$$

The individual values $(f_i - F_i)^2/F_i$ in testing the theoretical distribution of s^2 under normality are given in table 5.9.1. We find $\chi^2 = 8.51$ with $13 - 1 = 12$

df, far short of the 5% level of 21.03. Actually, 8.51 is close to $\chi^2_{0.750}$ for 12 df. When H_0 is true, a χ^2 value as large as 8.51 turns up about three times out of four.

Since Pearson's result is a large sample result, the expected numbers must not be too small. Small expectations are likely to occur only in extreme classes. A working rule (4) is that no class expectation should be below 1; two extreme expectations may be close to 1 provided that most of the other expected values exceed 5. Classes with expectations below 1 should be combined to meet this rule. When counting degrees of freedom, note that k is the number of classes *after such combinations have been made.*

The χ^2 test of goodness of fit is a nonspecific test in that the test criterion is not directed against any particular pattern that the deviations $(f_i - F_i)$ may follow. When we have reason to suspect a particular type of departure, the χ^2 goodness of fit test may be supplemented or replaced by a more specific test.

EXAMPLE 5.12.1—In a sample of 61 patients, the amount of anesthetic required to produce anesthesia suitable for surgery was found to have a standard deviation (from patient to patient) of $s = 10.2$ mg. Compute 90% confidence limits for σ. Ans. 8.9 and 12.0 mg. Use $\chi^2_{0.950}$ and $\chi^2_{0.050}$.

EXAMPLE 5.12.2—For routine equipment (such as light bulbs, which wear out after a time), the standard deviation of the length of life is an important factor in determining whether it is cheaper to replace all the pieces at fixed intervals or to replace each piece individually when it breaks down. An industrial statistician has calculated that it will pay to replace a certain gadget at fixed intervals if $\sigma < 6$ days. A sample of 71 pieces gives $s = 4.2$ days. Examine this question (*i*) by finding the upper 95% limit for σ from s, (*ii*) by testing the null hypothesis $\sigma \geq 6$ days against the alternative $\sigma < 6$ days. Ans. (*i*) The upper 95% limit is 5.0, (*ii*) H_0 is rejected at the 5% level. Notice that the two procedures are equivalent; if the upper confidence limit had been 6.0 days, the chi-square value would be at the 5% significance level.

EXAMPLE 5.12.3—For df > 100, which are not shown in table A 5, it is suggested (section 5.11) that $\sqrt{2\chi^2}$ is approximately normal with mean $\sqrt{2\nu - 1}$ and standard deviation 1, where ν is the number of degrees of freedom in χ^2. Check this approximation by finding the value that it gives for $\chi^2_{0.025}$ when $\nu = 100$, the correct value being 129.56. Ans. 129.1

EXAMPLE 5.12.4—Table 4.6.1 shows a normal distribution with $\mu = 30$, $\sigma = \sqrt{10}$ fitted to the means of 511 samples of size 10 drawn from the swine gains. Make a χ^2 test of goodness of fit. (Combine the first two classes.) Ans. $\chi^2 = 11.45$ with 9 df, since μ and σ are given; H_0 accepted.

EXAMPLE 5.12.5—A normal distribution was fitted to the heights of 1052 mothers arranged in 2-in. classes, using $\overline{X} = 62.49$ and $s = 2.435$ as the estimates of μ and σ. Results were as follows:

Class Upper Limit (in.)	f_i	F_i
55	3	1.1
57	8.5	11.6
59	52.5	67.2
61	215.0	204.5
63	346.0	329.3
65	277.5	279.3
67	119.5	125.5
69	23.5	29.7
∞	6.5	4.0

Apply the χ^2 goodness of fit test. Ans. $\chi^2 = 11.85$ with 6 df; it approaches the 5% value, 12.59, but is not quite significant. The small expectation 1.1 in the first class gives the largest contribution to χ^2.

EXAMPLE 5.12.6—In 100 samples with $n = 5$, the observed frequencies of the largest digit in the sample and the frequencies expected from the known probability distribution are as follows:

Frequency	≤4	5	6	7	8	9	Total
f_i	2	3	12	11	34	38	100
F_i	3.1	4.7	9.0	16.0	26.3	41.0	100.1

Make the χ^2 test of goodness of fit. Ans. $\chi^2 = 6.04$, df = 5; H_0 accepted.

EXAMPLE 5.12.7—With a random sample of 5 random digits, it was suggested (example 1.8.6) that the average $\hat{\mu}$ of the largest and the smallest digit is a good estimate of the population mean $\mu = 4.5$. In exploring whether $\hat{\mu}$ is more accurate than the sample mean $\overline{X}$, an investigator drew 60 random samples of 5 digits, calculated $\hat{\mu}$ for each sample, and used $s_{\hat{\mu}}^2 = \Sigma(\hat{\mu} - 4.5)^2/60$ as an estimate of the error variance of $\hat{\mu}$. (The divisor and the degrees of freedom in $s_{\hat{\mu}}^2$ are $n = 60$, since the deviations are taken from the population mean, 4.5.) The investigator found $s_{\hat{\mu}}^2 = 1.212$ (60 df), whereas $\sigma_{\overline{X}}^2 = 8.25/5 = 1.65$. (*i*) Make a one-sided test of $H_0{:}\sigma_{\hat{\mu}}^2 \geq 1.65$ against $H_A{:}\sigma_{\hat{\mu}}^2 < 1.65$. (*ii*) Find 90% two-sided confidence limits for $\sigma_{\hat{\mu}}^2$. Ans. (*i*) $\chi^2 = 44.07$ (60 df), the lower tail P is just over 0.05; (*ii*) 90% limits, (0.98, 1.68). The estimate $\hat{\mu}$ looks promising for this population, but a larger sample is needed to establish its superiority to $\overline{X}$. In using the χ^2 distribution to test $s_{\hat{\mu}}^2$, we support the assumption that $s_{\hat{\mu}}^2$ is distributed as a multiple of χ^2 by claiming that $\hat{\mu}$ is nearly normally distributed. When random digits are given to many decimals so that *any* value between 0 and 10 is equally likely, it is known from theory (9) that the variances of $\hat{\mu}$ and $\overline{X}$ are 1.19 and 1.67.

5.13—Test of skewness. A measure of the amount of skewness in a population is given by the average value of $(X - \mu)^3$ taken over the population. This quantity is called the *third moment about the mean.* If low values of X are bunched close to the mean μ but high values extend far above the mean, this measure will be positive, since the large positive contributions $(X - \mu)^3$ obtained when X exceeds μ will predominate over the smaller negative contributions $(X - \mu)^3$ obtained when X is less than μ. Populations with negative skewness, in which the lower tail is the extended one, are also encountered. To render this measure independent of the scale on which the data are recorded, it is divided by σ^3. The resulting *coefficient of skewness* is denoted sometimes by $\sqrt{\beta_1}$ and sometimes by γ_1.

The sample estimate of this coefficient is denoted by $\sqrt{b_1}$ or g_1. We compute

$$m_3 = \Sigma(X - \overline{X})^3/n \tag{5.13.1}$$

$$m_2 = \Sigma(X - \overline{X})^2/n \tag{5.13.2}$$

and take

$$\sqrt{b_1} = g_1 = m_3/(m_2\sqrt{m_2}) \tag{5.13.3}$$

Note that in computing m_2, the sample variance, we have divided by n instead of the customary $(n - 1)$. This makes subsequent calculations slightly easier.

The calculations are illustrated for the means of city sizes in table 5.13.1. Coding is worthwhile. Since $\sqrt{b_1}$ is dimensionless, the whole calculation can be done in the coded scale with no need to decode. Having chosen coded values U,

TABLE 5.13.1
COMPUTATIONS FOR TESTS OF SKEWNESS AND KURTOSIS

Lower Class Limit (1000s)	f	U	U^2	U^3	U^4
120–	9	−4	16	−64	256
130–	35	−3	9	−27	81
140–	68	−2	4	−8	16
150–	94	−1	1	−1	1
160–	90	0	0	0	0
170–	76	1	1	1	1
180–	62	2	4	8	16
190–	28	3	9	27	81
200–	27	4	16	64	256
210–	4	5	25	125	625
220–	5	6	36	216	1296
230–	1	7	49	343	2401
240–	1	8	64	512	4096

Test of Skewness

$n = 500 \qquad h_1 = \Sigma f U/n = +0.172$

$\Sigma f U = +86 \qquad h_2 = \Sigma f U^2/n = 4.452$

$\Sigma f U^2 = 2226 \qquad h_3 = \Sigma f U^3/n = +6.664$

$\Sigma f U^3 = +3332$

$$m_2 = h_2 - h_1^2 = 4.4224$$

$$m_3 = h_3 - 3h_1h_2 + 2h_1^3 = 4.3770$$

$$\sqrt{b_1} = m_3/(m_2\sqrt{m_2}) = 4.3770/(4.4224\sqrt{4.4224}) = 0.4707$$

Test of Kurtosis

$\Sigma f U^4 = 32{,}046 \qquad h_4 = \Sigma f U^4/n = 64.092$

$$m_4 = h_4 - 4h_1h_3 + 6h_1^2h_2 - 3H_1^4 = 60.2948$$

$$b_2 = m_4/m_2^2 = 60.2948/(4.4224)^2 = 3.083$$

write down their squares and cubes (paying attention to signs). Form the sums of products with the fs as indicated, and divide each sum by n to give the quantities h_1, h_2, h_3. Carry two extra decimal places in the hs. The moments m_2 and m_3 are then obtained from the algebraic identities given in the table. Finally, we obtain $\sqrt{b_1} = 0.4707$.

If the sample comes from a normal population, $\sqrt{b_1}$ is approximately normally distributed with mean 0 and standard deviation $\sqrt{6/n}$, or in this case $\sqrt{6/500} = 0.110$. Since $\sqrt{b_1}$ is over four times its standard deviation, the positive skewness is confirmed. The assumption that $\sqrt{b_1}$ is normally distributed is accurate enough for this test if n exceeds 150. For sample sizes between 25 and 200, the *one-tailed* 5% and 1% significance levels of $\sqrt{b_1}$, computed from a more accurate approximation, are given in table A 20(i).

5.14—Tests for kurtosis. A further type of departure from normality is called *kurtosis*. A measure of kurtosis in a population is the average value of $(X - \mu)^4$ divided by σ^4. For the normal distribution, this ratio has the value 3. If the ratio, denoted by β_2, exceeds 3, the distribution has longer tails than a normal distribution with the same σ. This is the manner in which the t distribution departs from the normal.

A sample estimate of the amount of kurtosis is given by

$$g_2 = b_2 - 3 = m_4/m_2^2 - 3 \quad (5.14.1)$$

where $m_4 = \Sigma(X - \overline{X})^4/n$ is the fourth moment of the sample about its mean. Notice that the normal distribution value 3 has been subtracted, with the result that long-tailed distributions show positive kurtosis and flat-topped distributions show negative kurtosis.

The manual computation of m_4 and b_2 from the coded values U is shown in table 5.13.1. For this sample, $g_2 = b_2 - 3$ has the value $+0.083$. In very large samples from the normal distribution, g_2 is normally distributed with mean 0 and standard deviation $\sqrt{24/n} = 0.219$, since n is 500. The sample value of g_2 is much smaller than its standard error, so the amount of kurtosis in the population appears to be trivial.

Unfortunately, the distribution of g_2 does not approach the normal closely until the sample size is over 1000. For sample sizes between 50 and 1000 table A 20(ii) contains better approximations to the 5% and 1% significance levels. Since the distribution of g_2 is skew, the two tails are shown separately. For $n = 500$, the upper 5% value of g_2 is $+0.37$, much greater than the value 0.083 found in this sample.

R. C. Geary (6) developed an alternative test criterion for kurtosis,

$$a = (\text{mean deviation})/(\text{standard deviation}) = \Sigma|X - \overline{X}|/(n\sqrt{m_2})$$

and tabulated its significance levels for sample sizes down to $n = 11$. If X is a normal deviate, the value of a when computed for the whole population is 0.7979. Positive kurtosis produces lower values and negative kurtosis higher values of a. When applied to the same data, a and g_2 usually agree well in their verdicts. The advantages of a are that tables are available for smaller sample sizes and a is easier to compute. Some writers (7, 8) recommend a in preference to g_2.

An identity simplifies the calculation of the numerator of a. This will be illustrated for the coded scale in table 4.14.1. Let

$$\Sigma' = \text{sum of all observations that exceed } \overline{U}$$

$$n' = \text{number of observations that exceed } \overline{U}$$

$$\Sigma|U - \overline{U}| = 2(\Sigma' - n'\overline{U})$$

Since $\overline{U} = 0.172$, all observations in the classes with $U = 1$ or more exceed $\overline{U}$, giving $\Sigma' = 457$, $n' = 204$. Hence,

$$\Sigma|U - \overline{U}| = 2[457 - (204)(0.172)] = 843.82$$

Since $m_2 = 4.4224$, we have

$$a = 843.82/(500\sqrt{4.4224}) = 0.802$$

which is little greater than the value 0.7979 for the normal distribution, in agreement with the result given by g_2. For $n = 500$, the upper 5% level of a is about 0.814.

The quantities γ_1 and $\gamma_2 = \beta_2 - 3$ have been useful in studying the distributions of $\overline{X}$ and s^2 when the original population is nonnormal. Two results are quoted. For the distribution of $\overline{X}$ in random samples of size n,

$$\gamma_1(\overline{X}) = \gamma_1/\sqrt{n} \qquad \gamma_2(\overline{X}) = \gamma_2/n \tag{5.14.2}$$

Thus, in the distribution of $\overline{X}$, the measures of skewness and kurtosis both approach zero when the sample size increases, as would be expected from the central limit theorem. Since the kurtosis is damped much faster than the skewness, it is not surprising that in our sample means g_1 is substantial but g_2 small.

Second, the exact variance of s^2 with ν df is known to be

$$V(s^2) = \frac{2\sigma^4}{\nu}\left[1 + \frac{\nu}{(\nu + 1)}\frac{\gamma_2}{2}\right] \tag{5.14.3}$$

The factor outside the brackets is the variance of s^2 in samples from a normal population. The term inside the brackets is the factor by which the normal variance is multiplied when the population is nonnormal. For example, if the measure of kurtosis γ_2 is 1, the variance of s^2 is about 1.5 times as large as it is in a normal population. With $\gamma_2 = 2$, the variance of s^2 is about twice as large as in a normal population. These results show that the distribution of s_2 is sensitive to amounts of kurtosis that may pass unnoticed in handling the data.

EXAMPLE 5.14.1—Table 4.14.2 gives the frequency distribution of a sample with $n = 100$, thought to be approximately normal. Compute $g_1 = -0.0139$ and $g_2 = 0.0460$, showing that the distribution is practically normal in these respects.

TECHNICAL TERMS

accept
alternative hypothesis
chi square
critical region
errors of Type I
errors of Type II
kurtosis
mean deviation
one-tailed test
power of a test
region of rejection
skewness
statistical decision theory
statistically significant
test criterion
test of goodness of fit
test of significance
two-tailed test

REFERENCES

1. Young, A. 1771. *A Course of Experimental Agriculture*, vol. 1. Exshaw, Dublin.
2. Pearson, E. S., and Hartley, H. O. 1954. *Biometrika Tables for Statisticians*, vol. 1. Cambridge Univ. Press.

3. Pearson, K. *Phil. Mag.*, ser. 5, 50 (1899):157.
4. Cochran, W. G. *Biometrics* 14 (1954):480.
5. Crampton, E. W. *J. Nutr.* 7 (1934):305.
6. Geary, R. C. *Biometrika* 25 (1936):295.
7. D'Agostino, R. B. *Psychol. Bull.* 74 (1970):138.
8. Gastwirth, J. L., and Owens, M. B. *Biometrika* 64 (1977):135.
9. Hastings, C., et al. *Ann. Math. Stat.* 18 (1947):413.

6

The comparison of two samples

6.1—Estimates and tests of differences. Comparative studies are designed to discover and evaluate *differences* between effects rather than the effects themselves. It is the difference between the amounts learned under two methods of teaching, the difference between the lengths of life of two types of glassware, or the difference between the degrees of relief reported from two pain-relieving drugs that is wanted. In this chapter we consider the simplest investigation of this type, in which two groups or two procedures are compared. In experimentation, these procedures are often called *treatments*. Such a study may be conducted in two ways.

PAIRED SAMPLES. Pairs of *similar* individuals or things are selected. In an experiment, one treatment is applied to one member of each pair, the other treatment to the second member. The members of a pair may be two students of similar ability, two patients of the same age and sex who have just undergone the same type of operation, or two male mice from the same litter. A common application occurs in *self-pairing* in which a single individual is measured on two occasions. For example, the blood pressure of a subject might be measured before and after heavy exercise. For any pair, the difference between the measurements given by the two members is an estimate of the difference in the effects of the two treatments or procedures.

The aim of the pairing is, of course, to make the comparison more accurate by having the members of any pair as alike as possible except in the treatment difference that the investigator deliberately introduces.

INDEPENDENT SAMPLES. This case arises whenever we wish to compare the means of two populations and have drawn a sample from each quite independently. We might have a sample of men aged 50–55 and one of men aged 30–35 and compare the amounts spent on life insurance. Or we might have a sample of high school seniors from rural schools and one from urban schools and compare their knowledge of current affairs as judged by a special examination on this subject. Independent samples are widely used in experimentation when no suitable basis for pairing exists, as, for example, in comparing the lengths of life of two types of drinking glass under the ordinary conditions of restaurant use.

Although independent samples are more common in practice, we begin with the analysis of paired experiments, which is easier. In fact, you have already encountered the analysis of Young's experiment (section 5.7), which was paired.

6.2—A simulated paired experiment. Eight pairs of random normal deviates were drawn from a table of random normal deviates. The first member of each pair represents the result produced by a standard procedure, while the second member is the result produced by a new procedure that is being compared with the standard. The eight differences D_i, new − st., are shown in the column headed Case I in table 6.2.1.

Since the results for the new and standard procedures were drawn from the same normal population, Case I stimulates a situation in which there is no difference in effect between the two procedures. The observed differences represent the natural variability that is always present in experiments. It is obvious on inspection that the eight differences do not indicate any superiority of the new procedure. Four of the differences are + and 4 are −, and the mean difference is small.

The results in Case II were obtained from those in Case I by adding +10 to every figure to represent a situation in which the new procedure is actually 10 units better than the standard. On looking at the data, most investigators would reach the judgment that the superiority of the new procedure is definitely established and would probably conclude that the average advantage in favor of it is not far from 10 units.

Case III is more puzzling. We added +1 to every figure in Case I, so the new procedure gives a small gain over the standard. The new procedure wins 6 times out of the 8 trials, and some workers might conclude that the results confirm the superiority of the new procedure. Others might disagree. They might point out that it is not too unusual for a fair coin to show heads in 6 tosses out of 8 and that the individual results range from an advantage of 0.7 units for the standard procedure to an advantage of 4.2 units for the new procedure. They

TABLE 6.2.1
A SIMULATED PAIRED EXPERIMENT

Pair	Case 1 New − st., D_i	Case II New − st., D_i	Case III New − st., D_i
1	+3.2	+13.2	+4.2
2	−1.7	+ 8.3	−0.7
3	+0.8	+10.8	+1.8
4	−0.3	+ 9.7	+0.7
5	+0.5	+10.5	+1.5
6	+1.2	+11.2	+2.2
7	−1.1	+ 8.9	−0.1
8	−0.4	+ 9.6	+0.6
Mean, ($\bar{D}$)	+0.28	+10.28	+1.28
s_D	1.527	1.527	1.527
$s_{\bar{D}}$	0.540	0.540	0.540

would argue that the results are inconclusive. We shall see what verdicts are suggested by the statistical analyses in these three cases.

The data also illustrate the assumptions made in the analysis of a paired trial. The differences D_i in the individual pairs are assumed to be distributed about a mean μ_D, which represents the average difference in the effects of the two treatments over the population of which these pairs are a random sample. The deviations $D_i - \mu_D$ may be due to various causes, in particular to inherent differences between the members of the pair and to any errors of measurement of the measuring instruments. Another source of variation is the different effects a treatment may actually have on different members of the population. A lotion for the relief of muscular pains may be more successful with some types of pain than with others. The adage, "One man's meat is another man's poison," expresses this variability in extreme form. For many applications, the extent to which the effect of a treatment varies from one member of the population to another is important. This study requires a more elaborate analysis, and usually a more complex experiment, than we are discussing at present. In the simple paired trial we compare only the *average* effects of the two treatments or procedures over the population.

In the analysis, the deviations $D_i - \mu_D$ are assumed to be normally and independently distributed with population mean zero. Consequences of failures in these assumptions are discussed in chapter 15.

When these assumptions hold, the sample mean difference $\bar{D}$ is normally distributed about μ_D with standard deviation or standard error $\sigma_D/\sqrt{n}$, where σ_D is the standard deviation of the population of differences. The value of σ_D is seldom known, but the sample furnishes an estimate:

$$s_D = \sqrt{\frac{\Sigma(D_i - \bar{D})^2}{n-1}} = \sqrt{\frac{\Sigma D_i^2 - (\Sigma D_i)^2/n}{n-1}}$$

Hence, $s_{\bar{D}} = s_D/\sqrt{n}$ is an estimate of $\sigma_{\bar{D}}$, based on $(n-1)$ df.

The important consequence of these results is that the quantity

$$t = (\bar{D} - \mu_D)/s_{\bar{D}}$$

follows Student's t distribution with $(n-1)$ df, where n is the number of pairs. The t distribution may be used to test the null hypothesis that $\mu_D = 0$ or to compute a confidence interval for μ_D.

TEST OF SIGNIFICANCE. The test is applied first to the doubtful Case III. The values of s_D and $s_{\bar{D}}$ are shown at the foot of table 6.2.1. Note that these values are identical in all three cases, since the addition of a constant μ_D to all the D_i does not affect the deviations $D_i - \bar{D}$. For Case III we have

$$t = \bar{D}/s_{\bar{D}} = 1.28/0.540 = 2.370$$

With 7 df, table A 4 shows that the 5% level of t in a two-tailed test is 2.365. The observed mean difference just reaches the 5% level, so the data point to a superiority of the new treatment.

In Case II, $t = 10.28/0.540 = 19.04$. This value lies far beyond even the 0.1% level (5.405) in table A 4. We might report "$P < 0.001$."

In Case I, $t = 0.28/0.540 = 0.519$. From table A 4, an absolute value of $t = 0.711$ is exceeded 50% of the time in sampling from a population with $\mu_D = 0$. The test provides no evidence on which to reject the null hypothesis in Case I. To sum up, the tests confirm the judgment of the preliminary inspection in all three cases.

CONFIDENCE INTERVAL. From inequalities (4.10.1), the 95% confidence interval for μ_D is

$$\overline{D} \pm t_{0.05} s_{\overline{D}} = \overline{D} \pm (2.365)(0.540) = \overline{D} \pm 1.28$$

In the simulated example the limits are

Case I :	−1.00 to 1.56
Case II :	9.00 to 11.56
Case III:	0.00 to 2.56

As always happens, the 95% confidence limits agree with the verdict given by the 5% tests of significance. Either technique may be used.

6.3—Example of a paired experiment. The preceding examples illustrate the assumptions and formulas used in the analysis of a paired set of data but do not bring out the purpose of the pairing. Youden and Beale (1) wished to find out if two preparations of a virus would produce different effects on tobacco plants. Half a leaf of a tobacco plant was rubbed with cheesecloth soaked in one preparation of the virus extract and the second half was rubbed similarly with the second extract. The measurement of potency was the number of local lesions appearing on the half leaf; these lesions appear as small dark rings that are easily counted. The data in table 6.3.1 are taken from the second leaf on each of eight plants. The steps in the analysis are exactly the same as in the preceding. We have, however, presented the deviations of the differences from their mean, $d_i = D_i - \overline{D}$, and obtained the sum of squares of deviations directly instead of by the shortcut formula.

For a test of the null hypothesis that the two preparations produce on the average the same number of lesions, we compute

$$t = \overline{D}/s_{\overline{D}} = 4/1.52 = 2.63 \qquad \text{df} = n - 1 = 7$$

From table A 4, the significance probability is about 0.04, and the null hypothesis is rejected. We conclude that in the population the second preparation produces fewer lesions than the first. From this result we would expect that both 95% confidence limits for μ_D will be positive. Since $t_{0.05} s_{\overline{D}} = (2.365)(1.52) = 3.59$, the 95% limits are $+0.4$ and $+7.6$ lesions per leaf.

TABLE 6.3.1
NUMBER OF LESIONS ON HALVES OF EIGHT TOBACCO LEAVES†

Pair Number	Preparation 1 X_1	Preparation 2 X_2	Difference $D = X_1 - X_2$	Deviation $d = D - \overline{D}$	Squared Deviation d^2
1	31	18	13	9	81
2	20	17	3	−1	1
3	18	14	4	0	0
4	17	11	6	2	4
5	9	10	−1	−5	25
6	8	7	1	−3	9
7	10	5	5	1	1
8	7	6	1	−3	9
Total	120	88	32	0	130
Mean	15	11	$\overline{D} = 4$		$s_D^2 = 18.57$
		$s_{\overline{D}}^2 = 18.57/8 = 2.32$	$s_{\overline{D}} = 1.52$ lesions		

†Slightly changed to make calculation easier.

In this experiment the leaf constitutes the pair. This choice was made as a result of earlier studies in which a single preparation was rubbed on a large number of leaves, the lesions found on each half leaf being counted. In a new type of work, a preliminary study of this kind can be highly useful. Since every half leaf was treated in the same way, the variations found in the numbers of lesions per half leaf represent the natural variability of the experimental material. From the data, the investigator can estimate the population standard deviation and in turn estimate the size of sample needed to ensure a specified degree of precision in the sample averages. The researcher can also look for a good method of forming pairs.

Youden and Beale found that the two halves of the same leaf were good partners, since they tended to give similar numbers of lesions. An indication of this fact is evident in table 6.3.1, where the pairs are arranged in descending order of total numbers of lesion per leaf. Note that with two minor exceptions, this descending order shows up in each preparation. If one member of a pair is high, so is the other; if one is low, so is the other. The numbers on the two halves of a leaf are said to be *positively correlated.* Because of this correlation, the differences between the two halves tend to be mostly small and therefore less likely to mask or conceal an imposed difference due to a difference in treatments.

EXAMPLE 6.3.1—L. C. Grove (2) determined the sample mean numbers of florets produced by seven pairs of plots of Excellence gladiolus, one plot of each pair planted with high (first-year) corms, the other with low (second-year or older) corms. (A corm is an underground propagating stem.) The plot means were as follows:

	Florets						
High corm	11.2	13.3	12.8	13.7	12.2	11.9	12.1
Low corm	14.6	12.6	15.0	15.6	12.7	12.0	13.1

Calculate the sample mean difference. Ans. 1.2 florets. In the population of such differences, test H_0: $\mu_D = 0$. Ans. $P = 0.06$, approximately.

EXAMPLE 6.3.2—Mitchell, Burroughs, and Beadles (3) computed the biological values of proteins from raw peanuts P and roasted peanuts R as determined in an experiment with 10 pairs of rats. The pairs of data P, R are as follows: 61, 55; 60, 54; 56, 47; 63, 59; 56, 51; 63, 61; 59, 57; 56, 54; 44, 63; 61, 58. Compute the sample mean difference, 2.0, and the sample standard deviation of the differences, 7.72 units. Since $t = 0.82$, over 40% of similar samples from a population with $\mu_D = 0$ would be expected to have larger t values.

Note: Of the 10 differences, $P - R$, 9 are positive. One would like some information about the next to last pair, 44, 63. The first member seems abnormal. While unusual individuals like this do occur in the most carefully conducted trials, their appearance demands immediate investigation. Doubtless an error in recording or computation was searched for but not found. What to do about such aberrant observations is a moot question; their occurrence detracts from one's confidence in the experiment.

EXAMPLE 6.3.3—A man starting work in a new town has two routes A and B by which he may drive home. He conducts an experiment to find out which route is quicker. Since traffic is unusually heavy on Mondays and Fridays but does not seem to vary much from week to week, he selects the day of the week as the basis for pairing. The test last four weeks. On the first Monday, he tosses a coin to decide whether to drive by route A or B. On the second Monday, he drives by the other route. On the third Monday, he again tosses a coin, using the other route on the fourth Monday, and similarly for the other days of the week. The times taken, in minutes, are as follows:

	M1	M2	Tu1	Tu2	W1	W2	Th1	Th2	F1	F2
A	28.7	26.2	24.8	25.3	25.1	23.9	26.1	25.8	30.3	31.4
B	25.4	25.8	24.9	25.0	23.9	23.3	26.6	24.8	28.8	30.3

(*i*) Treating the data as consisting of 10 pairs, test whether any real difference in average driving times appears between A and B. (*ii*) Compute 95% confidence limits for the population mean difference. What would you regard as the population in this trial? (*iii*) By eye inspection of the results, does the pairing look effective? (*iv*) Suppose that on the last Friday (F2) there had been a fire on route B, so the time taken to get home was 48 minutes. Would you recommend rejecting this pair from the analysis? Give your reason. Ans. (*i*) $t = 2.651$, with 9 df; P about 0.03. Method B seems definitely quicker. (*ii*) 0.12 to 1.63 min; there really is not much difference. (*iii*) Highly effective.

6.4—Conditions for pairing. The objective of pairing is to increase the accuracy of the comparison of the two procedures. Identical twins are natural pairs. Littermates of the same sex are often paired successfully because they usually behave more nearly alike than do animals less closely related. If the measurement at the end of the experiment is the subject's ability to perform some task (e.g., to do well in an exam), subjects similar in natural ability and previous training for this task should be paired. Often the subjects are tested at the beginning of the trial to provide information for forming pairs. Similarly, in experiments that compare two methods of treating sick persons, patients whose prognoses appear about the same at the beginning of the trial should be paired if feasible.

The variable on which we pair should predict accurately the performance of the subjects *on the measurement by which the effects of the treatments are to be judged*. Little will be gained by pairing students on their IQs if IQ is not closely related to ability to perform the particular task being measured in the experiment.

Self-pairing is highly effective when an individual's performance is consistent on different occasions but exhibits wide variation when comparisons are made from one individual to another. If two methods of conducting a chemical extraction are being compared, the pair is likely to be a sample of the original raw material that is thoroughly mixed and divided into two parts.

Environmental variation often calls for pairing. Two treatments should be laid down side by side in the field or on the greenhouse bench to avoid the effects of unnecessary differences in soil, moisture, temperature, etc. Two plots or pots next to each other usually respond more nearly alike than do two pots far apart. As a final illustration, sometimes the measuring process is lengthy and at least partly subjective, as in certain psychiatric studies. If several judges must be used to make the measurements for comparing two treatments A and B, each scoring a different group of patients, an obvious precaution is to ensure that each judge scores as many A patients as B patients. Even if the patients were not originally paired, they could be paired for assignment to judges.

6.5—Comparison of the means of two independent samples. When no pairing has been employed, we have two independent samples with means $\bar{X}_1, \bar{X}_2$ that are estimates of their respective population means μ_1, μ_2. Tests of significance and confidence intervals concerning the population difference, $\mu_1 - \mu_2$, are again based on the t distribution, where t now has the value

$$t = [(\bar{X}_1 - \bar{X}_2) - (\mu_1 - \mu_2)]/s_{\bar{X}_1 - \bar{X}_2}$$

It is assumed that $\bar{X}_1$ and $\bar{X}_2$ are normally distributed and independent. By theory, their difference is also normally distributed so that the numerator of t is normal with mean zero.

The denominator of t is a sample estimate of the standard error of $\bar{X}_1 - \bar{X}_2$. The background for this estimate is given in the next two sections. First, we need an important new result for the population variance of a difference between any two variables X_1 and X_2:

$$\sigma^2_{X_1 - X_2} = \sigma^2_{X_1} + \sigma^2_{X_2}$$

The variance of a difference is the *sum* of the variances. This result holds for any two variables, whether normal or not, provided they are *independently* distributed.

6.6—Variance of a difference. A population variance is defined (section 3.3) as the average, over the population, of the squared deviations from the population mean. The symbol *avg* will denote "the average of." Thus we may write

$$\sigma^2_{X_1 - X_2} = [(X_1 - X_2) - (\mu_1 - \mu_2)]^2_{\text{avg}}$$

But, $(X_1 - X_2) - (\mu_1 - \mu_2) = (X_1 - \mu_1) - (X_2 - \mu_2)$. Hence, on squaring and expanding,

$$[(X_1 - X_2) - (\mu_1 - \mu_2)]^2 = (X_1 - \mu_1)^2 + (X_2 - \mu_2)^2 - 2(X_1 - \mu_1)(X_2 - \mu_2)$$

Now average over all pairs of values X_1, X_2 that can be drawn from their respective populations. By the definition of a population variance,

$$(X_1 - \mu_1)^2_{\text{avg}} = \sigma^2_{X_1} \qquad (X_2 - \mu_2)^2_{\text{avg}} = \sigma^2_{X_2}$$

This leads to the general result

$$\sigma^2_{X_1-X_2} = \sigma^2_{X_1} + \sigma^2_{X_2} - 2[(X_1 - \mu_1)(X_2 - \mu_2)]_{\text{avg}} \qquad (6.6.1)$$

At this point we use the fact that X_1 and X_2 are independently drawn. Because of this independence, any specific value of X_1 will appear with all the values of X_2 that can be drawn from its population. Hence, for this specific value of X_1,

$$[(X_1 - \mu_1)(X_2 - \mu_2)]_{\text{avg}} = (X_1 - \mu_1)(X_2 - \mu_2)_{\text{avg}} = 0$$

since μ_2 is the mean or average of all the values of X_2. It follows that the overall average of the cross-product term $(X_1 - \mu_1)(X_2 + \mu_2)$ is zero, so

$$\sigma^2_{X_1-X_2} = \sigma^2_{X_1} + \sigma^2_{X_2} \qquad (6.6.2)$$

Apply this result to two means $\bar{X}_1$, $\bar{X}_2$ drawn from populations with the same variance σ^2. With samples of size n each mean has variance σ^2/n, which gives

$$\sigma^2_{\bar{X}_1-\bar{X}_2} = 2\sigma^2/n$$

In this case the variance of a difference is twice the variance of an individual mean.

If σ is known, the preceding results provide the material for tests and confidence intervals concerning $\mu_1 - \mu_2$. To illustrate, from the table of swine gains (table 4.14.1) that we used to simulate a normal distribution with $\sigma = 10$ lb, the first two samples drawn gave $\bar{X}_1 = 35.6$ and $\bar{X}_2 = 29.3$ lb, with $n = 10$. Since the standard error of $\bar{X}_1 - \bar{X}_2$ is $\sqrt{2}\sigma/\sqrt{n}$, the quantity

$$Z = \sqrt{n}\,[(\bar{X}_1 - \bar{X}_2) - (\mu_1 - \mu_2)]/(\sqrt{2}\sigma)$$

is a normal deviate. To test the null hypothesis that $\mu_1 = \mu_2$, we compute

$$Z = \sqrt{n}(\bar{X}_1 - \bar{X}_2)/(\sqrt{2}\sigma) = \sqrt{10}(6.3)/[\sqrt{2}(10)] = 19.92/14.14 = 1.41$$

From table A 3 a larger value of Z, ignoring sign, occurs in about 16% of the trials. As we would expect, the difference is not significant. The 95% confidence limits for $\mu_1 - \mu_2$ are

$$(\bar{X}_1 - \bar{X}_2) \pm (1.96)\sqrt{2}\sigma/\sqrt{n}$$

6.7—A pooled estimate of variance. In most applications the value of σ^2 is not known. However, each sample furnishes an estimate of σ^2; call these estimates s_1^2 and s_2^2. With samples of the same size n, the best combined estimate is their pooled average $s^2 = (s_1^2 + s_2^2)/2$.

Since $s_1^2 = \Sigma x_1^2/(n - 1)$ and $s_2^2 = \Sigma x_2^2/(n - 1)$, where, as usual, $x_1 = X_1 - \overline{X}_1$ and $x_2 = X_2 - \overline{X}_2$, we may write

$$s^2 = (\Sigma x_1^2 + \Sigma x_2^2)/[2(n - 1)]$$

This formula is recommended for routine computing, since it is quicker and extends easily to samples of unequal sizes.

The number of degrees of freedom in the pooled s^2 is $2(n - 1)$, the sum of the degrees of freedom in s_1^2 and s_2^2. This leads to the result that

$$t = \sqrt{n}[(\overline{X}_1 - \overline{X}_2) - (\mu_1 - \mu_2)]/(\sqrt{2}s)$$

follows Student's t distribution with $2(n - 1)$df.

Remember the assumption required, namely that σ is the same in the two populations. The situations in which this assumption is suspect and the comparison of $\overline{X}_1$ and $\overline{X}_2$ when the assumption does not hold are discussed in section 6.11.

It is now time to apply these methods to a real experiment.

6.8—An experiment comparing two samples of equal size. Breneman (4) compared the 15-day mean comb weights of two lots of male chicks, one receiving sex hormone A (testosterone), the other C (dehydroandrosterone). Day-old chicks, 11 in number, were assigned at random to each of the treatments. To distinguish between the two lots, which were caged together, the heads of the chicks were stained red and purple, respectively. The individual comb weights are recorded in table 6.8.1.

The calculations for the test of significance are given at the foot of the table. Note the method recommended for computing the pooled s^2. With 20 df, the value of t is significant at the 1% level. Hormone A gives higher average comb weights than hormone C. The two sums of squares of deviations, 8472 and 7748, make the assumption of equal σ^2 appear reasonable.

The 95% confidence limits for $\mu_1 - \mu_2$ are

$$\overline{X}_1 - \overline{X}_2 \pm t_{0.05} s_{\overline{X}_1 - \overline{X}_2}$$

or, in this example,

$$41 - (2.086)(12.1) = 16 \text{ mg} \qquad 41 + (2.086)(12.1) = 66 \text{ mg}$$

EXAMPLE 6.8.1—Lots of 10 bees were fed two concentrations of syrup, 20% and 65%, at a feeder half a mile from the hive (5). Upon arrival at the hive their honey sacs were removed and the concentration of the fluid measured. In every case there was a decrease from the feeder concentration. The decreases were: from the 20% syrup, 0.7, 0.5, 0.4, 0.7, 0.5, 0.4, 0.7, 0.4, 0.2, and 0.5%; from the 65% syrup, 1.7, 2.8, 2.2, 1.4, 1.3, 2.1, 0.8, 3.4, 1.9, and 1.4%. Every observation in the second

TABLE 6.8.1
TESTING THE DIFFERENCE BETWEEN THE MEANS OF TWO INDEPENDENT SAMPLES

	Weight of Comb (mg)	
	Hormone A	Hormone C
	57	89
	120	30
	101	82
	137	50
	119	39
	117	22
	104	57
	73	32
	53	96
	68	31
	118	88
Total	1,067	616
n	11	11
Mean $\overline{X}$	97	56
ΣX^2	111,971	42,244
$(\Sigma X)^2/n$	103,499	34,496
Σx^2	8,472	7,748
df	10	10

$$\text{Pooled } s^2 = \frac{8472 + 7748}{10 + 10} = 811 \qquad \text{df} = 20$$

$$s_{\overline{X}_1 - \overline{X}_2} = \sqrt{2s^2/n} = \sqrt{2(811)/11} = 12.14 \text{ mg}$$

$$t = (\overline{X}_1 - \overline{X}_2)/s_{\overline{X}_1 - \overline{X}_2} = 41/12.14 = 3.38$$

sample is larger than any in the first, so, rather obviously, $\mu_1 < \mu_2$. Show that $t = 5.6$ if $\mu_1 - \mu_2 = 0$. There is little doubt that under the experimental conditions imposed the concentration during flight decreases more with the 65% syrup. But how about equality of variances? See sections 6.11 and 6.12 for further discussion.

EXAMPLE 6.8.2—Four determinations of the pH of Shelby loam were made with each of two types of glass electrodes (6). With a modified quinhydrone electrode, the readings were 5.78, 5.74, 5.84, and 5.80, while with a modified Ag/AgCl electrode they were 5.82, 5.87, 5.96, and 5.89. With the hypothesis that $\mu_1 - \mu_2 = 0$, calculate $t = 2.66$.

EXAMPLE 6.8.3—In experiments to measure the effectiveness of carbon tetrachloride as a worm-killer, each of 10 rats received an injection of 500 larvae of the worm, *Nippostrongylus muris*. Eight days later 5 of the rats, chosen at random, each received 0.126 cm^3 of a solution of carbon tetrachloride; two days later the rats were killed and the numbers of adult worms counted. These numbers were 378, 275, 412, 265, and 286 for the control rats and 123, 143, 192, 40, and 259 for the rats treated with CCl_4. Find the significance probability for the difference in mean numbers of worms, and compute 95% confidence limits for this difference. Ans. $t = 3.64$ with 8 df, P close to 0.01. Confidence limits are 63 and 280.

EXAMPLE 6.8.4—Fifteen kernels of mature Iodent corn were tested for crushing resistance. Measured in pounds, the resistances were 50, 36, 34, 45, 56, 42, 53, 25, 65, 33, 40, 42, 39, 43, 42. Another batch of 15 kernels was tested after being harvested in the dough stage: 43, 44, 51, 40, 29,

49, 39, 59, 43, 48, 67, 44, 46, 54, 64. Test the significance of the difference between the two means. Ans. $t = 1.38$.

EXAMPLE 6.8.5—In research reports it is sometimes desirable to supply a test of significance that was not considered necessary by the author. As an example, Smith (7) gave the sample mean yields and their standard errors for two crosses of maize as 8.84 ± 0.39 and 7.00 ± 0.18 g. Each mean was the average of five replications. Determine if the mean difference is significant. Ans. $t = 4.29$, df $= 8$, $P < 0.5\%$. For the quickest answer, satisfy yourself that the estimate of the variance of the difference between the two means is the sum of the squares of 0.39 and 0.18, namely 0.1845.

6.9—Samples of unequal sizes. Unequal numbers are common in comparisons made from survey data as, for example, comparing the mean incomes of men of similar ages who have master's and bachelor's degrees or the severity of injury suffered in auto accidents by drivers wearing seat belts and drivers not wearing seat belts. In planned experiments equal numbers, being simpler to analyze and more efficient, are preferable, but equality is sometimes impossible or inconvenient to attain. Two lots of chicks from two batches of eggs treated differently nearly always differ in the number of birds hatched. Occasionally, when a new treatment is in short supply or costly, an experiment with unequal numbers is set up deliberately.

Unequal numbers in experiments occur also because of accidents and losses during the course of the trial. In such cases the investigator should always consider whether any loss represents a failure of the treatment rather than an accident that cannot be blamed on the treatment. Such situations require careful judgment.

The statistical analysis for groups of unequal sizes follows almost exactly the pattern for groups of equal sizes. We assume at present that the variance is the same in both populations. The means $\bar{X}_1$ and $\bar{X}_2$ of samples of sizes n_1, n_2 have variances σ^2/n_1 and σ^2/n_2. The variance of the difference is then

$$\sigma^2/n_1 + \sigma^2/n_2 = \sigma^2(1/n_1 + 1/n_2) = \sigma^2[(n_1 + n_2)/(n_1 n_2)]$$

To form a pooled estimate of σ^2, we follow the rule given for equal-sized samples.

Add the sums of squares of deviations in the numerators of s_1^2 and s_2^2 and divide by the sum of their degrees of freedom.

These degrees of freedom are $(n_1 - 1)$ and $(n_2 - 1)$, so the denominator of the pooled s^2 is $(n_1 + n_2 - 2)$. This quantity is also the number of degrees of freedom in the pooled s^2. The procedure will be clear from the example in table 6.9.1. Note how closely the calculations follow those given in table 6.8.1 for samples of equal sizes.

The high protein diet showed a slightly greater mean gain. Since P is about 0.08, however, a difference as large as the observed one would occur about 1 in 12 times by chance and the observed difference cannot be regarded as established by the usual standards in tests of significance.

TABLE 6.9.1
ANALYSIS FOR TWO SAMPLES OF UNEQUAL SIZES. GAINS IN WEIGHTS OF TWO LOTS OF FEMALE RATS (28–84 DAYS OLD) UNDER TWO DIETS

	Gains (g)	
	High protein	Low protein
	134	70
	146	118
	104	101
	119	85
	124	107
	161	132
	107	94
	83	
	113	
	129	
	97	
	123	
Total	1440	707
n	12	7
Mean $\overline{X}$	120	101
ΣX^2	177,832	73,959
$(\Sigma X)^2/n$	172,800	71,407
Σx^2	5,032	2,552
df	11	6

$$\text{Pooled } s^2 = \frac{5032 + 2552}{11 + 6} = 446.12 \qquad \text{df} = 17$$

$$s_{\overline{X}_1-\overline{X}_2} = \sqrt{s^2[(n_1 + n_2)/(n_1 n_2)]} = \sqrt{(446.12)(19)/84} = 10.04 \text{ g}$$

$$t = 19/10.04 = 1.89 \qquad P \text{ about } 0.08$$

For evidence about homogeneity of variance in the two populations, observe that $s_1^2 = 5.032/11 = 457$ and $s_2^2 = 2552/6 = 425$.

The investigator who is more interested in estimates than in tests may prefer the confidence interval; an observed difference of 19 g in favor of the high protein diet is reported, with 95% confidence limits -2.2 and 40.2 g.

EXAMPLE 6.9.1—The following are the rates of diffusion of carbon dioxide through two soils of different porosity (8). Through a fine soil (f), the diffusion rates are: 20, 31, 18, 23, 23, 28, 23, 26, 27, 26, 12, 17, 25; through a coarse soil (c), 19, 30, 32, 28, 15, 26, 35, 18, 25, 27, 35, 34. Show that pooled $s^2 = 35.83$, $s_{\overline{X}_1-\overline{X}_2} = 2.40$, df $= 23$, and $t = 1.67$. The difference, therefore, is not significant.

EXAMPLE 6.9.2—The total nitrogen content of the blood plasma of normal albino rats was measured at 37 and 180 days of age (9). The results are expressed as g/100 cc of plasma. At age 37 days, 9 rats had 0.98, 0.83, 0.99, 0.86, 0.90, 0.81, 0.94, 0.92, and 0.87; at age 180 days, 8 rats had 1.20, 1.18, 1.33, 1.21, 1.20, 1.07, 1.13, and 1.12 g/100 cc. Since significance is obvious, set a 95% confidence interval on the population mean difference. Ans. 0.21 to 0.35 g/100 cc.

EXAMPLE 6.9.3—Sometimes, especially in comparisons made from surveys, the two samples are large. Time is saved by forming frequency distributions and computing the means and variances as in section 3.7. The following data from an experiment serve as an illustration. The objective was to compare the effectiveness of two antibiotics A and B for treating patients with lobar pneumonia. The numbers of patients were 59 and 43. The data are the numbers of days needed to bring the patients' temperatures down to normal.

Day		1	2	3	4	5	6	7	8	9	10	Total
Number of	A	17	8	5	9	7	1	2	1	2	7	59
patients	B	15	8	8	5	3	1	0	0	0	3	43

What are your conclusions about the relative effectiveness of the two antibiotics in bringing down the fever? Ans. The difference of about 1 day in favor of B has a P value between 0.05 and 0.025. Note that although these are frequency distributions, the only real grouping is in the 10-day groups, which actually represented at least 10 and were arbitrarily rounded to 10. Since the distributions are very skew, the analysis leans heavily on the central limit theorem. Do the variances given by the two drugs appear to differ?

EXAMPLE 6.9.4—Show that if the two samples are of sizes 6 and 12, the standard deviation of the difference in means is the same as when the samples are both of size 8. Are the degrees of freedom in the pooled s^2 the same? In this example, we assume that $\sigma_1^2 = \sigma_2^2$.

EXAMPLE 6.9.5—Show that the pooled s^2 is a weighted mean of s_1^2 and s_2^2, in which each is weighted by its number of degrees of freedom.

6.10—Precautions against bias: randomization. With either independent or paired samples, the analysis assumes that the difference $\overline{X}_1 - \overline{X}_2$ is an unbiased estimate of the population mean difference between the two treatments. Unless precautions are taken when conducting an experiment, $\overline{X}_1 - \overline{X}_2$ may be subject to a bias of unknown amount that makes this assumption and the resulting conclusion false. As noted in chapter 1, one helpful device is *randomization.* When pairs have been formed, the decision as to which member of a pair receives treatment A is made by tossing a coin or by using a table of random numbers. If the random number drawn is odd, the first member of the pair will receive treatment A. With 10 pairs, we draw 10 random digits from table A 1, say 9, 8, 0, 1, 8, 3, 6, 8, 0, 3. In pairs 1, 4, 6, and 10, treatment A is given to the first member of the pair and B to the second member. In the remaining pairs, the first member receives B.

With independent samples, random numbers are used to divide the $2n$ subjects into two groups of n. Number the subjects in any order from 1 to $2n$. Proceed down a column of random numbers, allotting the subject to A if the number is odd, to B if even, continuing until either n As or n Bs have been allotted. The remaining subjects are allotted to the other treatment.

Randomization gives each treatment an equal chance of being allotted to any subject that happens to give an unusually good or unusually poor response, exactly as assumed in the theory of probability on which the statistical analysis, tests of significance, and confidence intervals are based. Randomization does not guarantee to balance out the natural differences between the members of a pair *exactly*. With n pairs, there is a small probability $1/2^{n-1}$ that one treatment will be assigned to the superior member in every pair. With 10 pairs this probability is about 0.002. If the experimenter can predict which member in each pair is likely to be superior, a more sophisticated design (chapter 14) that utilizes this information more effectively than randomization should be used. A great virtue of randomization is that it protects against *unsuspected* sources of bias. Randomization can be used not merely in the allocation of treatments to subjects but at any later stage in which it may be a safeguard against bias, as discussed in (11, 13).

Both independent and paired samples are often much in comparisons made from surveys or observational studies. The problem of avoiding misleading conclusions is formidable in such cases. Suppose we tried to learn something about the value of completing a high school education by comparing some years later the incomes, job satisfaction, and general well-being of a group of boys who completed high school with a group from the same schools who started but did not finish. Obviously, significant differences found between the sample means may be due to factors other than the completion of high school in itself—differences in the natural ability and personal characteristics of the boys, their parents' economic levels, number of useful contacts, and so on. Pairing the subjects on their school performances and parents' economic levels may help, but no randomization within pairs is possible and a significant mean difference may still be due to extraneous factors whose influence has been overlooked.

Remember also that a significant t value is evidence only that the population *means* differ. Popular accounts are sometimes written as if a significant t implies that every member of population 1 is superior to every member of population 2: "The oldest child in the family achieves more in science or in business." In fact, the two populations may largely overlap even though t is significant.

6.11—Analysis of independent samples when $\sigma_1 \neq \sigma_2$. The ordinary method of finding confidence limits and making tests of significance for the difference between the means of two independent samples assumes that the two population variances are the same. Common situations in which the assumption is suspect are as follows:

1. When the samples come from populations of different types, as in comparisons made from survey data. In comparing the average values of some characteristic of boys from public and private schools, we might expect from our knowledge of the differences in the two kinds of schools that the variances will not be the same.
2. When computing confidence limits in cases in which the population means are obviously very different. The frequently found result that σ tends to change, although slowly, when μ changes, would make us hesitant to assume $\sigma_1 = \sigma_2$.
3. When one treatment is erratic in its performance, sometimes giving a high and sometimes a low response. In populations that are markedly skew, the relation between σ and μ is often relatively strong.

When $\sigma_1 \neq \sigma_2$, the formula for the variance of $\bar{X}_1 - \bar{X}_2$ in independent samples still holds, namely,

$$\sigma^2_{\bar{X}_1 - \bar{X}_2} = \sigma_1^2/n_1 + \sigma_2^2/n_2$$

The two samples furnish unbiased estimates s_1^2 of σ_1^2 and s_2^2 of σ_2^2. Consequently, the ordinary t is replaced by the quantity

$$t' = (\overline{X}_1 - \overline{X}_2)/\sqrt{s_1^2/n_1 + s_2^2/n_2}$$

This quantity does not follow Student's t distribution when $\mu_1 = \mu_2$.

Two different forms of the distribution of t', arising from different theoretical backgrounds, have been worked out, one by Behrens (14) and Fisher (15), the other by Welch and Aspin (16, 19). Both require special tables, given in the references. The tables differ relatively little. The following approximation, due to Satterthwaite (10), assigns an approximate number of degrees of freedom to t' so that the ordinary t table may be used. It is sufficiently accurate for tests and confidence intervals.

Let $v_1 = s_1^2/n_1$ with ν_1 df, while $v_2 = s_2^2/n_2$ with ν_2 df. In comparing the means of two independent samples, $\nu_1 = n_1 - 1$ and $\nu_2 = n_2 - 1$. Then ν', the approximate number of degrees of freedom in t', is given by

$$\nu' = (v_1 + v_2)^2/(v_1^2/\nu_1 + v_2^2/\nu_2) \qquad (6.11.1)$$

Round ν' down to the nearest integer when using the t table.

The following artificial example illustrates the calculations. A quick but imprecise method of estimating the concentration of a chemical in a vat has been developed. Eight samples from the vat are analyzed by this method, as well as four samples by the standard method, which is precise but slow. In comparing the means we are examining whether the quick method gives a systematic overestimate or underestimate. Table 6.11.1 gives the computations. Since

TABLE 6.11.1

A TEST OF $\overline{X}_1 - \overline{X}_2$ WHEN $\sigma_1 \neq \sigma_2$. CONCENTRATION OF A CHEMICAL BY TWO METHODS

Standard	Quick
25	23
24	18
25	22
26	28
	17
	25
	19
	16
$\overline{X}_1 = 25$	$\overline{X}_2 = 21$
$n_1 = 4, \nu_1 = 3$	$n_2 = 8, \nu_2 = 7$
$s_1^2 = 0.67$	$s_2^2 = 17.71$
$s_1^2/n_1 = 0.17$	$s_2^2/n_2 = 2.21$

$$t' = 4/\sqrt{2.38} = 2.60$$

$$\nu' = \frac{2.38^2}{0.17^2/3 + 2.21^2/7} = 8.0$$

$t'_{0.05}$ = 5% level of t' = 2.31 (8 df)

$2.60 > 2.31$, the difference is significant at the 5% level; the quick method appears to underestimate.

Approximate 95% confidence limits for $\mu_1 - \mu_2$ are

$$\bar{X}_1 - \bar{X}_2 \pm t'_{0.05} s_{\bar{X}_1 - \bar{X}_2}$$

or in this example, $4 \pm (2.31)(1.54) = 4 \pm 3.6$.

The ordinary t test with a pooled s^2 gives $t = 1.84$, to which we would erroneously attribute 10 df. The t test tends to give too few significant results when the larger sample has the larger variance, as in this example, and too many when the larger sample has the smaller variance.

Sometimes, when it seemed reasonable to assume that $\sigma_1 = \sigma_2$ or when failing to think about the question in advance, the investigator notices that s_1^2 and s_2^2 are distinctly different. A test of the null hypothesis that $\sigma_1 = \sigma_2$, given in the next section, is useful. If the null hypothesis is rejected, the origin of the data should be reexamined. This may reveal some cause for expecting the standard deviations to be different. In case of doubt it is better to avoid the assumption that $\sigma_1 = \sigma_2$.

6.12—A test of the equality of two variances. The null hypothesis is: s_1^2 and s_2^2 are calculated from independent random samples from normal populations with the same variance σ^2. The distribution of $F = s_1^2/s_2^2$ was worked out by Fisher (17) early in the 1920s. Like χ^2 and t, it is one of the basic distributions in modern statistical methods. Table A 14, part I, gives the upper one-tailed 5% and 1% levels of F, while table A 14, part II, gives the corresponding 25%, 10%, 2.5%, and 0.5% levels in the rows labeled $P = 0.250, 0.100, 0.025$, and 0.005.

In situations without prior reason to anticipate inequality of variance, the alternative to the null hypothesis is a two-sided one: $\sigma_1 \neq \sigma_2$. To make this test at the 5% level, put the *larger* of s_1^2, s_2^2 in the numerator of F and read the 2.5% level of F from table A 14, part II. If the observed $F > F_{0.025}$, the null hypothesis is rejected at the 5% level. As an illustration we use the bee data in example 6.8.1. Bees fed a 65% concentration of syrup showed a mean decrease in concentration in their fluid of $\bar{X}_1 = 1.9\%$, with $s_1^2 = 0.589$, while bees fed a 20% concentration gave $\bar{X}_2 = 0.5\%$, with $s_2^2 = 0.027$. Each mean square has 9 df. Hence

$$F = 0.589/0.027 = 21.8$$

In table A 14, part II, the 2.5% level of F (5% two-tailed value) is 4.03. The null hypothesis is rejected. No clear explanation of the inequality in variances was found except that it may reflect the association of a larger variance with a larger mean. The difference between the means is strongly significant whether the variances are assumed the same or not. The reader may verify that Satterthwaite's approximation assigns 9 df to t'—the same as the degrees of freedom in s_1^2, whose sampling error dominates that of the much smaller s_2^2.

Sometimes a one-tailed test is wanted. In the artificial example in table 6.11.1, we might want to examine whether the sample data support the alternative $\sigma_2^2 > \sigma_1^2$ as we anticipated. To make this test, calculate $F = s_2^2/s_1^2 =$

$17.71/0.67 = 26.4$ with $\nu_1 = 7, \nu_2 = 3$ df. The 5% level from table A 14, part I, is $F_{0.05} = 8.88$, so the data are in line with our anticipation. If the alternative H_A is $\sigma_1^2 > \sigma_2^2$, we put s_1^2 in the numerator of F.

EXAMPLE 6.12.1—Young examined the basal metabolism of 26 college women in two groups of $n_1 = 15$ and $n_2 = 11$; $\bar{X}_1 = 34.45$ and $\bar{X}_2 = 33.57$ cal/m²/hr; $\Sigma x_1^2 = 69.36$, $\Sigma x_2^2 = 13.66$. Test H_0: $\sigma_1 = \sigma_2$. Ans. $F = 3.63$ to be compared with $F_{0.05} = 3.55$. (Data from Ph.D. dissertation, Iowa State University, 1940.)

BASAL METABOLISM OF 26 COLLEGE WOMEN
(cal/m²/hr)

7 or More Hours of Sleep				6 or Less Hours of Sleep			
1:	35.3	9:	33.3	1:	32.5	7:	34.6
2:	35.9	10:	33.6	2:	34.0	8:	33.5
3:	37.2	11:	37.9	3:	34.4	9:	33.6
4:	33.0	12:	35.6	4:	31.8	10:	31.5
5:	31.9	13:	29.0	5:	35.0	11:	33.8
6:	33.7	14:	33.7	6:	34.6	$\Sigma X_2 =$	369.3
7:	36.0	15:	35.7				
8:	35.0	$\Sigma X_1 =$	516.8				
$\bar{X}_1 = 34.45$ cal/m²/hr				$\bar{X}_2 = 33.57$ cal/m²/hr			

EXAMPLE 6.12.2—In the metabolism data there is little difference between the group means, and the difference in variances can hardly reflect a correlation between variance and mean. It might arise from nonrandom sampling, since the subjects are volunteers, or it could be due to chance, since F is scarcely beyond the 5% level. As an exercise, test the difference between the means (*i*) without assuming $\sigma_1 = \sigma_2$, (*ii*) making this assumption. Ans. (*i*) $t' = 1.29$, $t'_{0.05} = 2.08$; (*ii*) $t = 1.19$, $t_{0.05} = 2.064$. There is no difference in the conclusions.

EXAMPLE 6.12.3—In the preceding example, show that 95% confidence limits for $\mu_1 - \mu_2$ are -0.52 and 2.28 if we do not assume $\sigma_1 = \sigma_2$ and -0.63 and 2.39 if this assumption is made.

6.13—Paired versus independent samples. The formula for the variance of a difference throws more light on the circumstances in which pairing is effective. Quoting (6.6.1),

$$\sigma^2_{X_1 - X_2} = \sigma^2_{X_1} + \sigma^2_{X_2} - 2[(X_1 - \mu_1)(X_2 - \mu_2)]_{\text{avg}}$$

When pairing, we try to choose pairs such that if X_1 is high, so is X_2. Thus, if $X_1 - \mu_1$ is positive, so is $X_2 - \mu_2$, and their product $(X_1 - \mu_1)(X_2 - \mu_2)$ is positive. Similarly, in successful pairing, when $X_1 - \mu_1$ is negative, $X_2 - \mu_2$ will usually also be negative. Their product $(X_1 - \mu_1)(X_2 - \mu_2)$ is again *positive.* For paired samples, then, the average of this product is positive. This helps, because it makes the variance of $X_1 - X_2$ *less* than the sum of their variances, sometimes very much less. The average value of the product over the population is called the *covariance* of X_1 and X_2. The result for the variance of a difference may now be written

$$\sigma^2_{X_1 - X_2} = \sigma^2_{X_1} + \sigma^2_{X_2} - 2\,\text{Cov}\,(X_1, X_2) \qquad (6.13.1)$$

Pairing is not always effective because X_1 and X_2 may be poorly correlated. Fortunately, it is possible from the results of a paired experiment to estimate what the standard error of $\bar{X}_1 - \bar{X}_2$ would have been had the experiment been conducted without deliberate pairing, with the experimenter arranging the $2n$ sample units in random order and assigning treatment A to the first n units.

With paired samples of size n, the standard error of the mean difference $\bar{D} = \bar{X}_1 - \bar{X}_2$ is $\sigma_D/\sqrt{n}$, where σ_D is the standard deviation of the population of paired differences (section 6.2). With two unpaired groups, the standard error of $\bar{X}_1 - \bar{X}_2$ is of the form $\sqrt{2}\sigma_u/\sqrt{n}$ (section 6.6). Omitting the $\sqrt{n}$, the quantities to be compared are σ_D and $\sqrt{2}\sigma_u$. Usually the comparison is made in terms of variances; we compare σ_D^2 with $2\sigma_u^2$.

From the statistical analysis of the paired experiment, we have an unbiased estimate s_D^2 of σ_D^2. Estimation of σ_u^2 is more difficult. One possibility is to analyze the results of the paired experiment by the method of section 6.8 for two independent samples, using the pooled s^2 as an estimate of σ_u^2. For a comparison applying to the $2n$ units in our experiment, however, this procedure is not quite correct. Random assignment of treatment A to n of the units will usually be found to give A and B together in some pairs so that a partial pairing occurs without deliberate planning. An unbiased estimate of $2\sigma_u^2$ may be shown to be not $2s^2$ but the slightly smaller quantity

$$2\hat{\sigma}_u^2 = 2s^2 - (2s^2 - s_D^2)/(2n - 1)$$

The hat [̂] above a population parameter is often used to denote an estimate of the parameter.)

Let us apply this method to the paired experiment on virus lesions (table 6.3.1), which gave $s_D^2 = 18.57$. You may verify that the pooled s^2 is 45.714, giving $2s^2 = 91.43$. Hence, an unbiased estimate of $2\sigma^2$ is

$$2\sigma^2 = 91.43 - (91.43 - 18.57)/15 = 86.57$$

The pairing has given a much smaller variance of the mean difference—$18.57/n$ versus $86.57/n$. What does this imply in practical terms? With independent samples, the sample size would have to be increased from 8 pairs to 8(86.57)/18.57 or about 37 pairs to give the same variance of the mean difference as does the paired experiment. The saving in amount of work due to pairing is large in this case.

The computation overlooks one point. In the paired experiment, s_D^2 has 7 df, whereas the pooled s^2 would have 14 df for error. The t value used in tests of significance or in computing confidence limits would be slightly smaller with independent samples than with paired samples. Several writers (10, 11, 12) have discussed the allowance that should be made for this difference in number of degrees of freedom. We suggest a rule given by Fisher (11). Multiply the estimated variance by $(\nu + 3)/(\nu + 1)$, where ν is the degrees of freedom that the experimental plan provides. Thus we compare

$$(18.57)(10)/8 = 23.2 \qquad (86.57)(17)/(15) = 98.1$$

D. R. Cox (13) suggests the multiplier $(\nu + 1)^2/\nu^2$. This gives almost the same results, imposing a slightly higher penalty when ν is small.

A comparison like the above from a single experiment is not very precise, particularly if n is small. The results of several paired experiments using the same criterion for pairing give a more accurate picture of the success of the pairing. If the criterion has no correlation with the response variable, a small loss in accuracy results from pairing due to the adjustment for degrees of freedom. A substantial loss in accuracy may even occur if the criterion is badly chosen so that members of a pair are negatively correlated.

When analyzing the results of a comparison of two procedures, the investigator must know whether samples are paired or independent and must use the appropriate analysis. Sometimes a worker with paired data forgets they are paired when it comes to analysis and carries out the statistical analysis as if the two samples were independent. This mistake is serious if the pairing has been effective. In the virus lesions example, the worker would be using $2s^2/n = 91.43/8 = 11.44$ as the variance of $\bar{D}$ instead of $18.57/8 = 2.32$. The mistake throws away all the advantage of the pairing. Differences that are actually significant may be found nonsignificant, and confidence intervals will be too wide.

Analysis of independent samples as if they were paired seems to be rare in practice. If the members of each sample are in essentially random order so that the pairs are a random selection, the computed s_D^2 may be shown to be an unbiased estimate of $2\sigma^2$. Thus the analysis still provides an unbiased estimate of the variance of $\bar{X}_1 - \bar{X}_2$ and a valid t test. There is a slight loss in sensitivity, since t tests are based on $(n - 1)$ df instead of $2(n - 1)$ df.

As for assumptions, pairing has the advantage that its t test does *not* require $\sigma_1 = \sigma_2$. "Random" pairing of independent samples has been suggested as a means of obtaining tests and confidence limits when the investigator knows that σ_1 and σ_2 are unequal.

Artificial pairing of the results by arranging each sample in descending order and pairing the top two, the next two, and so on, produces a great underestimation of the true variance of $\bar{D}$. This effect may be illustrated by the first two random samples of swine gains from table 4.11.1. The population variance σ^2 is 100, giving $2\sigma^2 = 200$. In table 6.13.1 this method of artificial pairing has been employed.

Instead of the correct value of 200 for $2\sigma^2$, we get an estimate s_D^2 of only 8.0. Since $s_{\bar{D}} = \sqrt{8.0/10} = 0.894$, the t value for testing $\bar{D}$ is $t = 6.3/0.894 = 7.04$ with 9 df. This gives a P value of much less than 0.1%, although the two samples were drawn from the same population.

EXAMPLE 6.13.1—In testing the effects of two painkillers on the ability of young men to tolerate pain from a narrow beam of light directed at the arm, each subject was first rated several times as to the amount of heat energy he bore without complaining of discomfort. The subjects were then paired according to these initial scores. In a later experiment the amounts of energy received that caused the subject to complain were as follows (A and B denoting the treatments).

Pair	1	2	3	4	5	6	7	8	9	Sums
A	15	2	4	1	5	7	1	0	−3	32
B	6	7	3	4	3	2	3	0	−6	22

TABLE 6.13.1
TWO SAMPLES OF 10 SWINE GAINS ARRANGED IN DESCENDING ORDER, TO ILLUSTRATE ERRONEOUS CONCLUSIONS FROM ARTIFICIAL PAIRING

Sample 1	57	53	39	39	36	34	33	29	24	12	Mean = 35.6
Sample 2	53	44	32	31	30	30	24	19	19	11	Mean = 29.3
Differences	4	9	7	8	6	4	9	10	5	1	Mean = 6.3
		$\Sigma d^2 = 469 - 63^2/10 = 72.1$					$s_D^2 = 72.1/9 = 8.0$				

To simplify calculations, 30 was subtracted from each original score. Show that for appraising the effectiveness of the pairing, comparable variances are 22.5 for the paired experiment and 44.6 for independent groups (after allowing for the difference in degrees of freedom). The preliminary work in rating the subjects reduced the number of subjects needed by almost one-half.

EXAMPLE 6.13.2—In a previous experiment comparing two routes A and B for driving home from an office (example 6.3.3), pairing was by days of the week. The times taken (−23 min) for the 10 pairs were as follows:

A	5.7	3.2	1.8	2.3	2.1	0.9	3.1	2.8	7.3	8.4
B	2.4	2.8	1.9	2.0	0.9	0.3	3.6	1.8	5.8	7.3
$A - B$	3.3	0.4	−0.1	0.3	1.2	0.6	−0.5	1.0	1.5	1.1

Show that if the 10 nights on which route A was used had been drawn at random from the 20 nights available, the variance of the mean difference would have been about 8 times as high as with this pairing.

EXAMPLE 6.13.3—If pairing has not reduced the variance so that $s_D^2 = 2\sigma^2$, show that allowance for the error degrees of freedom by Fisher's rule makes pairing 15% less effective than independent groups when $n = 5$ and 9% less effective when $n = 10$. In small experiments, pairing is inadvisable unless a sizable reduction in variance is expected.

6.14—Sample size in comparative experiments. In planning an experiment to compare two treatments, the following method is often used to estimate the size of sample needed. The investigator first decides on a value δ that represents the size of difference between the true effects of the treatments regarded as important. If the true difference is as large as δ, the investigator would like the experiment to have a high probability of showing a statistically significant difference between the treatment means. In the terminology of section 5.5, the experimenter would like the test to have high power when the time difference is δ. Probabilities or powers of 0.80 and 0.90 are common. A higher probability, say 0.95 or 0.99, can be set, but the sample size required to meet these severe specifications is often too expensive.

This way of stating the aims in planning the sample size is particularly appropriate when (*i*) the treatments are a standard treatment and a new treatment hoped to be better than the standard, and (*ii*) the experimenter intends to discard the new treatment if the experiment does not show it to be significantly superior to the standard. The experimenter does not mind dropping the new treatment if it is at most only slightly better than the standard but does not want to drop it on the evidence of the experiment if it is substantially superior. The value of δ measures the idea of a substantial true difference.

In order to make the calculation the experimenter supplies:

1. The value of δ.
2. The desired probability P' of obtaining a significant result if the true difference is δ.
3. The significance level α of the test, which may be either one-tailed or two-tailed.

In addition, we assume at first that the population standard deviation σ_D for paired samples or σ for independent samples is known. We assume $\sigma_1 = \sigma_2 = \sigma$.

Given the mean $\bar{X}$ of a normal sample, (5.6.2) showed that the power of a two-tailed test of $H_0{:}\mu = \mu_0$ against $H_A{:}\mu = \mu_A$ at significance level α is the probability that a normal deviate Z lies in one of the regions

$$Z < (-\sqrt{n}/\sigma)(\mu_A - \mu_0) - Z_\alpha \quad \text{or} \quad Z > (-\sqrt{n}/\sigma)(\mu_A - \mu_0) + Z_\alpha$$

where Z_α denotes the *two-tailed* significance level, with $Z_{0.10} = 1.645$, $Z_{0.05} = 1.960$, and $Z_{0.01} = 2.576$.

With a paired experiment, apply this result to the population of differences between the members of a pair. In terms of differences, H_0 becomes $\mu_0 = 0$ and H_A is $\mu_A = \delta$, which we assume > 0. The test criterion is $\sqrt{n}\,\bar{D}/\sigma_D$. Hence, (5.6.2) is written

$$Z < -\sqrt{n}\delta/\sigma_D - Z_\alpha \quad \text{or} \quad Z > -\sqrt{n}\delta/\sigma_D + Z_\alpha \tag{6.14.1}$$

For power values of any substantial size, the probability that Z lies in the first region will be found to be negligible and we need consider only the second. As an example, suppose we want the power to be 0.80 in a *two-tailed* test at the 5% level so that $Z_{0.05} = 1.960$. From table A 3, the normal deviate exceeded by 80% of the distribution is $Z = -0.842$. Hence, in (6.14.1) we want $-0.842 = -\sqrt{n}\delta/\sigma_D + 1.960$ or

$$\sqrt{n}\delta/\sigma = 2.802 \qquad n = 7.9\sigma_D^2/\delta^2$$

Note that with $\sqrt{n}\delta/\sigma_D = 2.802$, the first region in (6.14.1) is $Z < -4.762$, whose probability is negligible as predicted.

To obtain a general formula, verify that if the power of the test is to be P', the lower boundary of the second region in (6.14.1) is $-Z_{2(1-P')}$. If we write $\beta = 2(1 - P')$, the formula for the needed sample size n is

$$n = (Z_\alpha + Z_\beta)^2\sigma_D^2/\delta^2 \tag{6.14.2}$$

In a one-tailed test at level α, (5.6.3) shows that in terms of differences the power is the probability that a normal deviate Z lies in the region

$$Z > -\sqrt{n}\delta/\sigma_D + Z_{2\alpha} \tag{6.14.3}$$

Hence, with a one-tailed test,

$$n = (Z_{2\alpha} + Z_\beta)^2 \sigma_D^2/\delta^2 \tag{6.14.4}$$

For two independent samples of size n, the test criterion is $\sqrt{n}(\bar{X}_1 - \bar{X}_2)/(\sqrt{2}\sigma)$, instead of $\sqrt{n}\bar{D}/\sigma_D$ for paired samples. Hence, the only change in (6.14.2) and (6.14.3) for n is that $2\sigma^2$ replaces σ_D^2 and $\mu_A - \mu_0$ replaces δ.

Using (6.14.2) and (6.14.4), table 6.14.1 presents the multipliers of σ_D^2/δ^2 or $2\sigma^2/\delta^2$ needed to give n in the cases most frequently used. Note that with either paired or independent samples, the total number of observations is $2n$.

When σ_D and σ are estimated from the results of the experiments, t tests replace the normal deviate tests. The logical basis of the argument, due to Neyman, Iwaskiewicz, and Kolodziejczyk (18), remains the same, but the formula for n becomes an equation that must be solved by successive approximations.

For practical use, the following approximation agrees well enough with Neyman's solution:

1. Find n_1 to one decimal place by table 6.14.1 and round upwards.
2. For tests of significance at the 5% or 10% levels, increase the number of pairs n by 2 with paired samples and the size n of each sample by 1 with independent samples. For tests at the 1% level, change these increases to 3 and 2, respectively.

To illustrate, suppose that a 10% difference δ is regarded as important and that $P' = 0.80$ in a two-tailed 5% test of significance. The samples are to be independent, and past experience has shown that σ is about 6%. The multiplier for $P' = 0.80$ and a 5% two-tailed test in table 6.14.1 is 7.9. Since $2\sigma^2/\delta^2 = 72/100 = 0.72$, $n_1 = (7.9)(0.72) = 5.7$. We round up to 6 and add 1 by step 2, taking $n = 7$ in each sample.

Note that the experimenter must still guess a value of σ_D or σ. Usually it is easier to guess σ. If pairing is to be used but is expected to be only moderately effective, take $\sigma_D = \sqrt{2}\sigma$, reducing this value if something more definite is known about the effectiveness of pairing.

When planning some experiments, an upper limit to n is imposed in advance by cost or other considerations. Inequalities (6.14.1) and (6.14.3) may be used to estimate the power that this value of n will give for specified δ, as a guide to the decision of whether to conduct the experiment.

The preceding method is designed to protect the investigator against

TABLE 6.14.1
MULTIPLIERS OF σ_D^2/δ^2 IN PAIRED SAMPLES, AND OF $2\sigma^2/\delta^2$ IN INDEPENDENT SAMPLES, NEEDED TO DETERMINE THE SIZE OF EACH SAMPLE

Desired Power	Two-Tailed Tests Level			One-Tailed Tests Level		
P'	0.01	0.05	0.10	0.01	0.05	0.10
0.80	11.7	7.9	6.2	10.0	6.2	4.5
.90	14.9	10.5	8.6	13.0	8.6	6.6
0.95	17.8	13.0	10.8	15.8	10.8	8.6

finding a nonsignificant result and consequently dropping a new treatment that is actually effective, because the experiment was too small. The method is therefore most useful in the early stages of a line of work. At later stages, when something has been learned about the sizes of differences produced by new treatments, we may wish to specify the size of the standard error or the half-width of the confidence interval that will be attached to an estimated difference.

For example, previous small experiments have indicated that a new treatment gives an increase of around 20% and σ is around 7%. The investigator would like to estimate this increase, in the next experiment, with a standard error of $\pm 2\%$ and thus sets $\sqrt{2}(7)/\sqrt{n} = 2$, giving $n = 25$ in each group. This type of rough calculation is often helpful in later work.

EXAMPLE 6.14.1—In table 6.14.1, verify the multipliers given for a one-tailed test at the 1% level with $P' = 0.90$ and for a two-tailed test at the 10% level with $P' = 0.80$.

EXAMPLE 6.14.2—In planning a paired experiment, the investigator proposes to use a one-tailed test of significance at the 5% level and wants the probability of finding a significant difference to be 0.90 if (*i*) $\delta = 10\%$, (*ii*) $\delta = 5\%$. How many pairs are needed? In each case, give the answer if (*a*) σ_D is known to be 12%, (*b*) σ_D is guessed as 12% but a t test will be used in the experiment. Ans. (*ia*) 13, (*ib*) 15, (*iia*) 50, (*iib*) 52.

EXAMPLE 6.14.3—In the previous example, how many pairs would you guess to be necessary if $\delta = 2.5\%$? The answer brings out the difficulty of detecting small differences in comparative experiments with variable data.

EXAMPLE 6.14.4—If $\sigma_D = 5$, how many pairs are needed to make the half-width of the 90% confidence interval for the difference between the two population means equal 2? Ans. $n = 17$.

EXAMPLE 6.14.5—In an experiment with two independent samples, the investigator, using a two-tailed test at the 5% level and wanting a 90% chance of getting a significant difference if $\delta = \mu_A - \mu_0 = 10\%$, guesses $\sigma = 10\%$. (*i*) Find the size n of each sample by table 6.14.1. What will be the power of the test if actually $\sigma =$ (*ii*) 8%, (*iii*) 12%, (*iv*) 14%? Ans. (*i*) $n = 21$, (*ii*) $P' = 0.99$, (*iii*) $P' = 0.81$, (*iv*) $P' = 0.68$.

TECHNICAL TERMS

independent samples
paired samples
pooled variance
randomization
self-pairing
variance ratio, F

REFERENCES

1. Youden, W. J., and Beale, H. P. 1934. *Contrib. Boyce Thompson Inst.* 6:437.
2. Grove, L. C. 1939. Iowa Agric. Exp. Stn. Res. Bull. 253.
3. Mitchell, H. H.; Burroughs, W.; and Beadles, J. R. *J. Nutr.* 11 (1936):257.
4. Breneman, W. R. Personal communication.
5. Park, O. W. 1932. Iowa Agric. Exp. Stn. Res. Bull. 151.
6. Dean, H. L., and Walker, R. H. *J. Am. Soc. Agron.* 27 (1935):433.
7. Smith, S. N. *J. Am. Soc. Agron.* 26 (1934):792.
8. Pearson, P. B., and Catchpole, H. R. *Am. J. Physiol.* 115 (1936):90.
9. Swanson, P. P., and Smith, A. H. *J. Biol. Chem.* 97 (1932):745.
10. Satterthwaite, F. E. *Biom. Bull.* 2 (1946):110.
11. Cochran, W. G., and Cox, G. M. 1957. *Experimental Designs*, 2nd ed. Wiley, New York.

12. Fisher, R. A. 1960. *The Design of Experiments,* 7th ed. Oliver & Boyd, Edinburgh.
13. Cox, D. R. 1958. *Planning of Experiments.* Wiley, New York.
14. Behrens, W. V. *Landwirtsch. Jahrb.* 68 (1929):807.
15. Fisher, R. A., and Yates, F. 1957. *Statistical Tables,* 5th ed., tables VI, VI_1 and VI_2. Oliver & Boyd, Edinburgh.
16. Aspin, A. A. *Biometrika* 36 (1949):290.
17. Fisher, R. A. 1924. *Proc. Int. Math. Conf.* Toronto, p. 805.
18. Neyman, J; Iwaskiewicz, K.; and Kolodziejczyk, St. *J. R. Stat. Soc.* 2 (1935):114.
19. Trickett, W. H.; Welch, B. L.; and James, G. S. *Biometrika* 43 (1956):203.

7

The binomial distribution

7.1—Introduction. As mentioned previously, many populations consist of only two kinds of elements: odd or even, pass or fail, male or female, infested or free, dead or alive. The investigator's interest is in the proportion, the percentage, or the number of individuals in one of the two classes.

A two-class population has a very simple structure; it can be described by giving the proportion p of the members of the population that fall in one class. When we sample, we shall call an observation falling in the class of primary interest a success (S). In a random sample of size n, the probability of getting 0, 1, 2, . . . , n successes is easily worked out. This distribution was first discovered by Bernoulli and was published posthumously in 1713. It is known as the binomial or Bernoulli distribution.

In this chapter the binomial distribution and its relation to the normal distribution is derived and examined. Further, just as we compare the means of two normal samples, independent or paired, we compare proportions from independent and from paired samples.

The binomial distribution is derived from some rules in the theory of probability.

7.2—Some simple rules of probability. The study of probability began around three hundred years ago. At that time, gambling and games of chance had become a fashionable pastime, and there was much interest in questions about the chance that a certain type of card would be drawn from a pack or that a die would fall in a certain way.

In a problem in probability, we are dealing with a trial, about to be made, that can have a number of different outcomes. A six-sided die when thrown may show any of the numbers 1, 2, 3, 4, 5, 6 face up—these are the possible outcomes. Simpler problems in probability can often be solved by writing down all the different possible outcomes of the trial and recognizing that these are *equally likely*. Suppose that the letters a, b, c, d, e, f, g are written on identical balls that are placed in a bag and mixed thoroughly. One ball is drawn out blindly. Most people would say without hesitation that the probability that an a is drawn is 1/7, because there are 7 balls, one of them is certain to be drawn, and all are equally likely. In general terms, this result may be stated as follows.

Rule 1. If a trial has k equally likely outcomes, of which one and only one will happen, the probability of any individual outcome is $1/k$.

The claim that the outcomes are equally likely must be justified by knowledge of the exact nature of the trial. For instance, dice to be used in gambling for stakes are manufactured with care to ensure that they are cubes of even density. They are discarded by gambling establishments after a period of use in case the wear, though not detectable by the naked eye, has made the six outcomes no longer equally likely. The statement that the probability is 1/52 of drawing the ace of spades from an ordinary pack of cards assumes a thorough shuffling that is difficult to attain, particularly when the cards are at all worn.

In some problems the event in which we are interested will happen if any one of a specific group of outcomes turns up when the trial is made. With the letters a, b, c, d, e, f, g, suppose we ask, What is the probability of drawing a vowel? The event is now, A vowel is drawn. This will happen if either an a or an e is the outcome. Most people would say that the probability is 2/7, because 2 vowels are present out of 7 competing letters, and each letter is equally likely. Similarly, the probability that the letter drawn is one of the first 4 letters is 4/7. These results are an application of a second rule of probability.

Rule 2 (Addition Rule). If an event is satisfied by any one of a group of mutually exclusive outcomes, the probability of the event is the sum of the probabilities of the outcomes in the group.

In mathematical terminology, this rule is sometimes stated as

$$P(E) = P(O_1) + P(O_2) + \ldots + P(O_m)$$

where $P(O_i)$ denotes the probability of the ith outcome.

Rule 2 contains one condition: The outcomes in the group must be *mutually exclusive.* This phrase means that if any one of the outcomes happens, all the others fail to happen. The outcomes "a is drawn" and "e is drawn" are mutually exclusive. But the outcomes "a vowel is drawn" and "one of the first four letters is drawn" are not mutually exclusive, because if a vowel is drawn, it might be an a, in which case the event "one of the first four letters is drawn" has also happened.

The condition of mutual exclusiveness is essential. If it does not hold, Rule 2 gives the wrong answer. To illustrate, consider the probability that the letter drawn is either one of the first four letters or a vowel. Of the seven original outcomes a, b, c, d, e, f, g, five satisfy the event in question, namely a, b, c, d, e. The probability is given correctly by Rule 2 as 5/7, because these five outcomes are mutually exclusive. But we might try to shortcut the solution by saying, The probability that one of the first four letters is drawn is 4/7 and the probability that a vowel is drawn is 2/7; therefore, by Rule 2, the probability that one or the other of these happens is 6/7. This answer is wrong.

In leading up to the binomial distribution we must consider the results of repeated drawings from a population. The successive trials or drawings are assumed *independent* of one another. This term means that the outcome of a trial does not depend in any way on what happens in the other trials.

With a series of trials the easier problems can again be solved by Rules 1 and 2. For example, a bag contains the letters a, b, c. In trial 1 a ball is drawn after thorough mixing. The ball is replaced, and in trial 2 a ball is again drawn after thorough mixing. What is the probability that both balls are a? First, we list all possible outcomes of the two trials. These are (a, a), (a, b), (a, c), (b, a), (b, b), (b, c), (c, a), (c, b), (c, c), where the first letter in a pair is the result of trial 1 and the second that of trial 2. Then we claim that these nine outcomes of the pair of trials are equally likely. Challenged to support this claim, we might say: (*i*) a, b, and c are equally likely at the first draw because of the thorough mixing, and (*ii*) at the second draw the conditions of thorough mixing and of independence make all nine outcomes equally likely. The probability of (a, a) is therefore 1/9.

Similarly, suppose we are asked the probability that the two drawings contain no cs. This event is satisfied by four mutually exclusive outcomes: (a, a), (a, b), (b, a), and (b, b). Consequently, the probability (by Rule 2) is 4/9.

Both the previous results can be obtained more quickly by noticing that the probability of the combined event is the *product* of the probabilities of the desired events in the individual trials. In the first problem the probability of an a is 1/3 in the first trial and also 1/3 in the second trial. The probability that both events happen is $1/3 \times 1/3 = 1/9$. In the second problem, the probability of not drawing a c is 2/3 in each individual trial. The probability of the combined event (no c at either trial) is $2/3 \times 2/3 = 4/9$. A little reflection shows that the numerator of this product (1 or 4) is the number of equally likely outcomes of the two drawings that satisfy the desired combined event. The denominator (9) is the total number of equally likely outcomes in the combined trials. The probabilities need not be equal at the two drawings. For example, the probability of getting an a at the first trial but not at the second is $1/3 \times 2/3 = 2/9$, the outcomes that produce this event being (a, b) and (a, c).

Rule 3 (Multiplication Rule). In a series of independent trials, the probability that each of a specified series of events happens is the product of the probabilities of the individual events. In mathematical terms,

$$P(E_1 \text{ and } E_2 \text{ and } \ldots \text{ and } E_m) = P(E_1)P(E_2) \ldots P(E_m)$$

In practice, the assumption that trials are independent, like the assumption that outcomes are equally likely, must be justified by knowledge of the circumstances of the trials. In complex probability problems there have been disputes about the validity of these assumptions in particular applications, and some interesting historical errors have occurred.

This account of probability provides only the minimum background needed for working out the binomial distribution. Reference (1) is recommended as a more thorough introduction to this important subject at an elementary mathematical level.

EXAMPLE 7.2.1—A bag contains the letters A, b, c, D, e, f, G, h, I. If each letter is equally likely to be drawn, what is the probability of drawing: (*i*) a capital letter, (*ii*) a vowel, (*iii*) either a capital or a vowel. Ans. (*i*) 4/9, (*ii*) 1/3, (*iii*) 5/9. Does Rule 2 apply to the two events mentioned in (*iii*)? No.

EXAMPLE 7.2.2—Three bags contain, respectively, the letters a, b; c, d, e; f, g, h, i. A letter is drawn independently from each bag. Write down all 24 equally likely outcomes of the three drawings. Show that six of them give a consonant from each bag. Verify that Rule 3 gives the correct probability of drawing a consonant from each bag (1/4).

EXAMPLE 7.2.3—Two six-sided dice are thrown independently. Find the probability that: (*i*) the first die gives a 6 and the second at least a 3, (*ii*) one die gives a 6 and the other at least a 3, (*iii*) both give at least a 3, (*iv*) the sum of the two scores is not more than 5. Ans. (*i*) 1/9, (*ii*) 2/9, (*iii*) 4/9, (*iv*) 5/18.

EXAMPLE 7.2.4—From a bag with the letters a, b, c, d, e, a letter is drawn and laid aside; then a second is drawn. By writing down all equally likely pairs of outcomes, show that the probability that both letters are vowels is 1/10. Rule 3 does not apply to this problem. Why not?

EXAMPLE 7.2.5—If two trials are not independent, the probability that event E_1 happens at the first trial and E_2 at the second is obtained (1) by a generalization of Rule 3: $P(E_1 \text{ and } E_2) = P(E_1)P(E_2$, given that E_1 has happened). This last factor is called the *conditional probability* of E_2, given E_1, and is usually written $P(E_2 | E_1)$. Show that this rule gives the answer, 1/10, in example 7.2.4, where E_1, E_2 are the probabilities of drawing a vowel at the first and second trials, respectively. See reference (1).

EAMPLE 7.2.6—In some statistical applications we are interested in the probability distribution of the largest or the smallest member of a sample. Find the probability that the largest member of a sample of five random digits is 9. Hint: It is easier to do this by first finding the probability that the largest member is less than 9. This implies that all five selected digits lie between 0 and 8, for which the probability is 0.9^5 by Rule 3. Ans. $0.4095 = 1 - 0.9^5$.

EXAMPLE 7.2.7—In example 7.2.6, what is the probability that the largest digit in the sample is 8? This task is harder, but from example 7.2.6 you know how to find the probability that the largest digit is less than 9. If it is less than 9 it must either be 8 or less than 8. But the probability that it is less than 8 is 0.8^5. Ans. $0.9^5 - 0.8^5 = 0.2628$. In this way you can find the complete probability distribution of the largest digit in the sample, which is given in table 5.12.6.

EXAMPLE 7.2.8—In each of three independent sets of data an investigator tests the same hypothesis H_0 at the 5% level. If H_0 is correct, what is the probability that the investigator finds at least one result significant? Ans. 0.1426, nearly three times the individual significance level.

In many applications the probability of a particular outcome must be determined by a statistical study. For instance, insurance companies are interested in the probability that a man aged sixty will live for the next ten years. This quantity, calculated from national statistics of the age distribution of males and of the age distribution of deaths of males, is published in actuarial tables. Provided that the conditions of independence and of mutually exclusive outcomes hold where necessary, Rules 2 and 3 are applied to probabilities of this type also. Thus, the probability that three men aged sixty, selected at random from a population, will all survive for ten years would be taken as p^3, where p is the probability that an individual sixty-year-old man will survive for ten years.

7.3—The binomial distribution. A proportion p of the members of a population possesses some attribute. A sample of size $n = 2$ is drawn. The result

TABLE 7.3.1

BINOMIAL DISTRIBUTION FOR $n = 2$

(1) Outcomes		(2)	(3)	(4)
1	2	Probability	Successes	Probability
F	F	qq	0	q^2
F	S	qp	1	$2pq$
S	F	pq		
S	S	pp	2	p^2
Total		1		1

of a trial is denoted by S (success) if the member drawn has the attribute and by F (failure) if it does not. In a single drawing, p is the probability of obtaining an S, while $q = 1 - p$ is the probability of obtaining an F. Table 7.3.1 shows the four mutually exclusive outcomes of the two drawings in terms of successes and failures.

The probabilities in column (2) are obtained by applying Rule 3 to the two trials. For example, the probability of two successive Fs is qq, or q^2. This assumes, of course, that the two trials are independent, as is necessary if the binomial distribution is to hold. The third column notes the number of successes. Since the two middle outcomes, FS and SF, each give 1 success, the probability of 1 success is $2pq$ by Rule 2. The third and fourth columns present the binomial distribution for $n = 2$. As a check, the probabilities in columns 2 and 4 each add to unity, since

$$q^2 + 2pq + p^2 = (q + p)^2 = 1^2 = 1$$

In the same way, table 7.3.2 lists the eight relevant outcomes for $n = 3$. The probabilities in the second and fourth columns are obtained by Rules 3 and 2 as before. Three outcomes provide 1 success with total probability $3pq^2$, while three provide 2 successes with total probability $3p^2q$. Check that the eight outcomes are mutually exclusive.

TABLE 7.3.2

BINOMIAL DISTRIBUTION FOR $n = 3$

(1) Outcomes			(2)	(3)	(4)
1	2	3	Probability	Successes	Probability
F	F	F	qqq	0	q^3
F	F	S	qqp		
F	S	F	qpq	1	$3pq^2$
S	F	F	pqq		
F	S	S	qpp		
S	F	S	pqp	2	$3p^2q$
S	S	F	ppq		
S	S	S	ppp	3	p^3

TABLE 7.3.3
BINOMIAL COEFFICIENTS GIVEN BY PASCAL'S TRIANGLE

Sample Size	Binomial Coefficients																
n									1								
1								1		1							
2							1		2		1						
3						1		3		3		1					
4					1		4		6		4		1				
5				1		5		10		10		5		1			
6			1		6		15		20		15		6		1		
7		1		7		21		35		35		21		7		1	
8	1		8		28		56		70		56		28		8		1
etc.									etc.								

The general structure of the binomial formula is now apparent. The formula for the probability of r successes in n trials has two parts. One part is the term $p^r q^{n-r}$. This part follows from Rule 3, since any outcome of this type must have r Ss and $(n - r)$ Fs in the set of n draws. The other part is the number of mutually exclusive ways in which the r Ss and the $(n - r)$ Fs can be arranged. In algebra this term is called the *number of combinations* of r letters out of n letters. It is denoted by the symbol $\binom{n}{r}$. The formula is

$$\binom{n}{r} = \frac{n(n-1)(n-2)\ldots(n-r+1)}{r(r-1)(r-2)\ldots(2)(1)}$$

For small samples these quantities, the *binomial coefficients,* can be written down by an old device known as *Pascal's triangle,* shown in table 7.3.3. Each coefficient is the sum of the two just above it to the right and the left. Thus, for $n = 8$, the number $56 = 21 + 35$. Note that for any n the coefficients are symmetrical, rising to a peak in the middle.

Putting the two parts together, the probability of r successes in a sample of size n is

$$\binom{n}{r} p^r q^{n-r} = \frac{n(n-1)(n-2)\ldots(n-r+1)}{r(r-1)(r-2)\ldots(2)(1)} p^r q^{n-r}$$

These probabilities are the successive terms in the expansion of the binomial expression $(q + p)^n$. This fact explains why the distribution is called binomial and also verifies that the sum of the probabilities is 1, since $(q + p)^n = 1^n = 1$.

For $n = 8$, figure 7.3.1 shows these distributions for $p = 0.2$, 0.5, and 0.9. The distribution is positively skew for $p < 0.5$ and negatively skew for $p > 0.5$. For $p = 0.5$ the general shape, despite the discreteness, bears some resemblance to a normal distribution.

Reference (16) contains extensive tables of individual and cumulative terms of the binomial distribution for n up to 49; reference (17) has cumulative terms up to $n = 1000$.

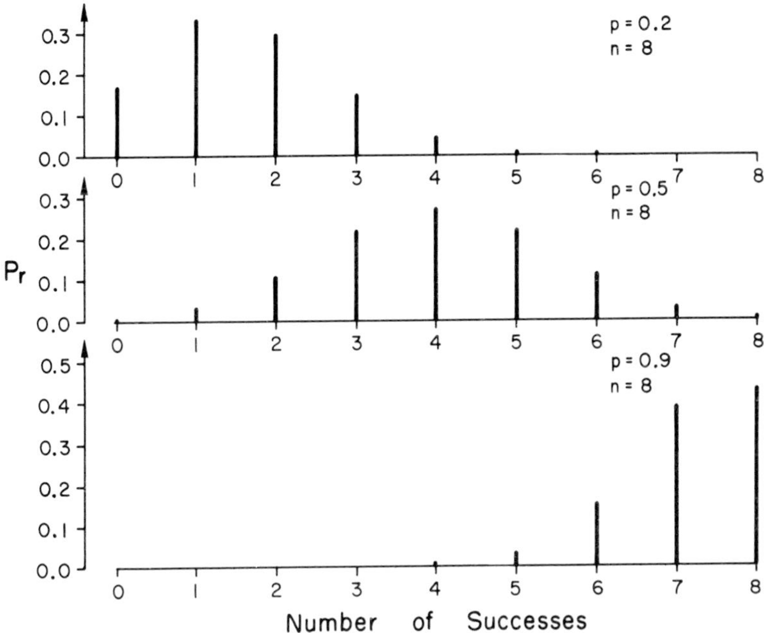

Fig. 7.3.1—Binomial distributions for $n = 8$. Top: $p = 0.2$; middle: $p = 0.5$; bottom: $p = 0.9$.

7.4—Sampling the binomial distribution. As usual, you will find it instructive to verify the preceding theory by sampling. The table of random digits (table A 1) is convenient for drawing samples from the binomial with $n = 5$, since the digits in a row or column are arranged in groups of 5. For instance, to sample the binomial with $p = 0.2$, let the digits 0 and 1 represent a success and each other digit a failure. By recording the total number of 0s and 1s in each group of 5, many samples from $n = 5, p = 0.2$ can be drawn quickly. Table 7.4.1 shows the results of 100 drawings of this type.

To fit the corresponding theoretical distribution, first calculate the terms $p^r q^{n-r}$. For $r = 0$ (no successes), it is $q^n = 0.8^5 = 0.32768$. For $r = 1$, it is $pq^{n-1} =$

TABLE 7.4.1

Tallying of 100 Drawings from the Binomial with $n = 5, p = 0.2$

Successes		Total
0	卌 卌 卌 卌 卌 卌 ||	32
1	卌 卌 卌 卌 卌 卌 卌 卌 ||||	44
2	卌 卌 卌 ||	17
3	卌 |	6
4	|	1
5		0
		100

TABLE 7.4.2

FITTING THE THEORETICAL BINOMIAL FOR $n = 5$, $p = 0.2$

Successes (r)	$p^r q^{n-r}$	Binomial Coefficient	$\binom{n}{r} p^r q^{n-r}$	Expected Frequency	Observed Frequency
0	0.32768	1	0.32768	32.77	32
1	.08192	5	.40960	40.96	44
2	.02048	10	.20480	20.48	17
3	.00512	10	.05120	5.12	6
4	.00128	5	.00640	0.64	1
5	0.00032	1	0.00032	0.03	0
Total			1.00000	100.00	100

$(0.2)(0.8)^4$. Note that this term can be written $(q^n)(p/q)$. It is computed from the previous term by multiplying by $p/q = 0.2/0.8 = 1/4$. Thus for $r = 1$ the term is $0.32768/4 = 0.08192$. Similarly, the term for $r = 2$, p^2q^{n-2} is found by multiplying the term for $r = 1$ by p/q, and so on for each successive term.

The details appear in table 7.4.2. The binomial coefficients are read from Pascal's triangle. These coefficients and the terms in p^rq^{n-r} are multiplied to give the theoretical probabilities of 0, 1, 2, . . . , 5 successes. Finally, since $N = 100$ samples were drawn, we multiply each probability by 100 to give the expected frequencies of 0, 1, 2, . . . , 5 successes.

In applying the χ^2 test of goodness of fit between the observed and expected frequencies, we combine classes 3, 4, and 5 to make the smallest expectation greater than 1. We find $\chi^2 = 1.09$, df $= 4 - 1 = 3$, P about 0.75.

EXAMPLE 7.4.1—With $n = 2$, $p = 1/2$, show that the probability of one success is 1/2. If p differs from 1/2, does the probability of one success increase or decrease?

EXAMPLE 7.4.2—A railway company claims that 95% of its trains arrive on time. If a man travels on three of these trains, what is the probability (assuming the claim is correct) that: (*i*) all three arrive on time, (*ii*) one of the three is late. Ans. (*i*) 0.857, (*ii*) 0.135.

EXAMPLE 7.4.3—Assuming the probability that a child is male is 1/2, find the probability that in a family of 6 children there are: (*i*) no boys, (*ii*) exactly 3 boys, (*iii*) at least 2 girls, (*iv*) at least 1 girl and 1 boy. Ans. (*i*) 1/64, (*ii*) 5/16, (*iii*) 57/64, (*iv*) 31/32.

EXAMPLE 7.4.4—Work out the terms of the binomial distribution for $n = 4$, $p = 0.4$. Verify that: (*i*) the sum of the terms is unity, (*ii*) 1 and 2 successes are equally probable, (*iii*) 0 success is about five times as probable as 4 successes.

EXAMPLE 7.4.5—By extending Pascal's triangle, obtain the binomial coefficients for $n = 10$. Hence compute and graph the binomial distribution for $n = 10$, $p = 1/2$. Does the shape appear similar to a normal distribution? Hint: When $p = 1/2$, the term $p^rq^{n-r} = 1/2^n$ for any r. Since $2^{10} = 1024 \doteq 1000$, the distribution is given accurately enough for graphing by simply dividing the binomial coefficients by 1000.

EXAMPLE 7.4.6—In a binomial sample with $n = 5$, $p = 0.6$, the expected number of successes is $(5)(0.6) = 3$. Which is larger, $P(X > 3)$ or $P(X < 3)$? Ans. $P(X < 3) = 0.317$, $P(X > 3) = 0.337$ (slightly larger).

EXAMPLE 7.4.7—A baseball batter whose hitting probability is 0.3 throughout a season would be considered a good hitter. Show that this batter, appearing three times at bat in a game, has a chance of about 2/3 of getting at least one hit. Note: Some evidence indicates that this probability does not remain constant from game to game, depending on the pitcher and the wind and light conditions.

7.5—Probability of at least one success. For some applications we need the probability that at least one of a series of events happens or its complement, the probability that none of the events happens. If the n events are independent and each has probability p, these probabilities are given by the binomial distribution. The probability that none happens is $q^n = (1 - p)^n$, while the probability that at least one happens is $1 - (1 - p)^n$. A result in algebra shows that

$$P(\text{at least one success}) = 1 - (1 - p)^n < np \qquad (7.5.1)$$

With rare events such that $np < 0.1$, this upper limit np is quite close to the correct answer. For instance, with $p = 0.025$ and $n = 4$, $np = 0.1$, while $(1 - 0.975^4) = 0.0963$, just slightly less.

If the events are independent but have different probabilities of success p_1, $p_2, \ldots, p_n$, the probability of at least one success can be calculated from Rule 3, (the multiplication rule). Since the events are independent, the probability that none happens is $(1 - p_1)(1 - p_2)\ldots(1 - p_n)$. Hence the probability of at least one success is

$$P(\text{at least one success}) = 1 - (1 - p_1)(1 - p_2)\ldots(1 - p_n) \qquad (7.5.2)$$

From (7.5.2), it may be shown by algebra that if every $p_i > 0$,

$$P(\text{at least one success}) < p_1 + p_2 + \ldots + p_n \qquad (7.5.3)$$

Inequality (7.5.3) is the extension of (7.5.1) to this case. Furthermore, we can show that (7.5.3) holds when the events are not independent if we replace the sign $<$ by the sign $\leq$.

With two events, let the symbol $+ +$ indicate that both events happen, the symbol $+ -$ that the first event happens but the second does not, and so on. The four outcomes, $+ +$, $+ -$, $- +$, $- -$, are clearly mutually exclusive. Let p_{++}, p_{+-}, p_{-+}, p_{--} be the respective probabilities of the four outcomes, while (as before) p_1 and p_2 are the probabilities that event 1 and event 2 happen. By Rule 2,

$$p_1 = p_{++} + p_{+-} \qquad p_2 = p_{++} + p_{-+} \qquad (7.5.4)$$

Hence,

$$p_1 + p_2 = 2p_{++} + p_{+-} + p_{-+} \qquad (7.5.5)$$

Now at least one of the events 1 and 2 happens if we get any of the three outcomes $+\ -$, $-\ +$, or $+\ +$. Hence, by Rule 2,

$$P(\text{at least one success}) = p_{+-} + p_{-+} + p_{++} \tag{7.5.6}$$

$$= p_1 + p_2 - p_{++} \le p_1 + p_2 \tag{7.5.7}$$

by (7.5.5). Equality occurs in (7.5.7) only if $p_{++} = 0$, that is, if the events are mutually exclusive.

With three events, we get at least one success with either of two mutually exclusive outcomes: (*i*) we get at least one success in events 1, 2; (*ii*) the outcome $-\ -\ +$ occurs, i.e., 1 and 2 fail but 3 succeeds. By Rule 2,

$$P(\text{at least one success}) = P(\text{at least one success in } 1, 2) + p_{--+}$$

But by (7.5.7),

$$P(\text{at least one success in } 1, 2) \le p_1 + p_2 \quad \text{and} \quad p_{--+} \le p_3$$

Hence, with three events,

$$P(\text{at least one success}) \le p_1 + p_2 + p_3 \tag{7.5.8}$$

By repeated extension of the preceding argument, it follows that with n events (independent or not),

$$P(\text{at least one success}) \le p_1 + p_2 + \ldots + p_n \tag{7.5.9}$$

This inequality is the first of a series of useful inequalities known as *Bonferroni's inequalities*. In its complementary form, (7.5.9) can be written

$$P(\text{no success}) = P(\text{all } n \text{ events fail}) \ge 1 - (p_1 + p_2 + \ldots + p_n) \tag{7.5.10}$$

Hence, if $q_i = 1 - p_i$ is the probability that the ith event fails to happen, (7.5.10) gives P(no failure) if we replace p_i by q_i; that is,

$$P(\text{all } n \text{ events happen}) \ge 1 - (q_1 + q_2 + \ldots + q_n) \tag{7.5.11}$$

This form of the inequality is used later. For instance, suppose we want the probability to be at least 0.95 that each of n events happens. If we can control the probabilities p_i that the individual events happen, we can achieve our goal by making every $p_i = 0.995$. For with $q_i = 0.005$, $10q_i = 0.05$; and (7.5.11) gives:

$$P(\text{all 10 events happen}) \ge 1 - 10q_i = 1 - 0.05 = 0.95$$

More generally, if we want the probability to be $\ge (1 - \alpha)$ that all n events happen, we take $p_i = (1 - \alpha/n)$ so that $q_i = \alpha/n$.

EXAMPLE 7.5.1—In the +, − notation, the probabilities of the eight mutually exclusive outcomes for three events are as follows:

+ + +	+ + −	+ − +	+ − −	− + +	− + −	− − +	− − −	Total
0.11	0.03	0.04	0.04	0.08	0.10	0.20	0.40	1.00

Calculate the probability that (*i*) both events 1 and 3 happen, (*ii*) neither event 1 nor 3 happens, (*iii*) at least one of events 1 and 2 happens, (*iv*) at least one of events 1, 2, and 3 happens. (*v*) Verify that in (*iv*) the probability is less than $p_1 + p_2 + p_3$. Ans. (*i*) 0.15, (*ii*) 0.50, (*iii*) 0.40, (*iv*) 0.60.

EXAMPLE 7.5.2—(*i*) What choice of p guarantees a probability of at least 0.6 that each of four events happens? (*ii*) With this value of p, what is the probability that all four happen if the events are independent? Ans. (*i*) $p = 0.9$, (*ii*) $P = 0.6561$.

7.6—The normal approximation and the correction for continuity. The formulas $\mu = p$ and $\sigma^2 = pq$ for the mean and variance of a single observation from a binomial population are given in section 3.4. Hence, the following results hold for estimates made from a random binomial sample of size n. Number of successes, r:

$$\mu = np \qquad \sigma^2 = npq \qquad \sigma = \sqrt{npq} \tag{7.6.1}$$

Proportion of successes, $\hat{p} = r/n$:

$$\mu = p \qquad \sigma^2 = pq/n \qquad \sigma = \sqrt{pq/n} \tag{7.6.2}$$

Formulas (7.6.2) also apply to the estimated *percentage* of successes if p in the formulas now stands for the population percentage and $q = 100 - p$.

For a sample of fixed size n, the standard deviations $\sqrt{npq}$ for the number of successes and $\sqrt{pq/n}$ for the proportion of successes are greatest when $p = 1/2$. As p moves toward either 0 or 1, the standard deviation declines, though quite slowly at first, as the following table shows.

p	0.5	0.4 or 0.6	0.3 or 0.7	0.2 or 0.8	0.1 or 0.9
$\sqrt{pq}$	0.500	0.490	0.458	0.400	0.300

EXAMPLE 7.6.1—For the binomial distribution of the number of successes with $n = 2$ (given in table 7.3.1), verify from 7.5.1 that $\mu = 2p$, $\sigma^2 = 2pq$.

EXAMPLE 7.6.2—For the binomial distribution with $n = 5$, $p = 0.2$, given in table 7.4.2, compute $\Sigma\, rf_r$ and $\Sigma\, (r - \mu)^2 f_r$ and verify that the results are $\mu = 1$ and $\sigma^2 = 0.80$.

EXAMPLE 7.6.3—For $n = 96$, $p = 0.4$, calculate the standard deviations of (*i*) the number and (*ii*) the percentage of successes. Ans. (*i*) 4.8, (*ii*) 5.

EXAMPLE 7.6.4—An investigator intends to estimate, by random sampling from a large file of house records, the percentage of houses sold in a town in the last year. The investigator thinks that p is about 10% and would like the standard deviation of the estimated percentage to be about 1%. How large should n be? Ans. 900 houses.

Further, since $X = r/n$ is the mean of a sample from a population that has a finite variance pq, we can quote the central limit theorem (section 4.5). It states that the mean $\overline{X}$ of a random sample from any population with finite variance tends to normality. Hence, as n increases, the binomial distribution of r/n or of r approaches the normal distribution.

For $p = 0.5$, the normal is a good approximation when n is as low as 10, as figure 7.6.1 indicates. As p approaches 0 or 1, some skewness remains in the binomial distribution until n is large.

The *solid* vertical lines in figure 7.6.1 show the binomial distribution or r for $n = 10$, $p = 0.5$. Also shown is the approximating normal curve, with mean $np = 5$ and $SD = \sqrt{npq} = 1.581$.

One difference between the binomial and the normal is that the binomial is discrete, having probability greater than zero only at $r = 0, 1, 2, \ldots, 10$, while the normal has probability greater than zero in any interval from $-\infty$ to ∞. This raises a problem: In estimating the binomial probability of say 4 successes, what part of the normal curve do we use as an approximation? We need to set up a correspondence between the set of binomial ordinates and the areas under the normal curve.

The simplest way of doing this is to regard the binomial as a grouping of the normal into unit class intervals. Under this rule the binomial ordinate at 4 corresponds to the area under the normal curve from 3.5 to 4.5. The ordinate at 5 corresponds to the area from 4.5 to 5.5, and so on. The ordinate at 10 corresponds to the normal area from 9.5 to ∞. These class boundaries are the dotted lines in figure 7.6.1.

In the commonest binomial problems we wish to calculate the probabilities at the ends of the distribution, for instance, the probability of 8 or more successes. The exact result, found by adding the binomial probabilities for $r = 8$, 9, 10 is $56/1024 = 0.0547$. By our rule, the corresponding area under the normal curve is the area from 7.5 to ∞, *not* the area from 8 to ∞. The normal deviate is therefore $z = (7.5 - 5)/1.581$, which by coincidence is also 1.581. The

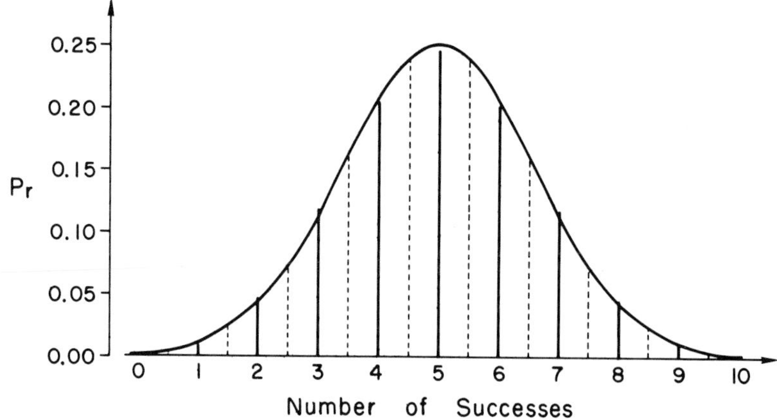

Fig. 7.6.1—Solid vertical lines show the binomial distribution of the number of successes for $n = 10, p = 0.5$. The curve is the normal approximation to this distribution, which has mean $np = 5$ and standard deviation $\sqrt{npq} = 1.581$.

approximate probability from the normal table is $P = 0.0570$, close enough to 0.0547. Use of $z = (8 - 5)/1.581$ gives $P = 0.0288$, a poor result.

Similarly, the probability of 4 or fewer successes is approximated by the area of the normal curve from $-\infty$ to 4.5. When adding probabilities of values of r on a *single* tail of the distribution, as in a one-tailed test, the procedure is to decrease the *absolute* value of $r - np$ by 0.5. Thus,

$$z_c = (|r - np| - 0.5)/\sqrt{npq} \qquad (7.6.3)$$

The subtraction of 0.5 is called a *correction for continuity*. It usually improves the accuracy of the normal approximation, though with some tendency to overcorrect.

With a two-tailed test, the preceding method of correction can be improved, as shown in (20). The general rule that figure 7.6.1 suggests is: Find the next largest deviation that the data can provide. In calculating z_c, use a deviation halfway between the observed and the next largest deviation. In a two-tailed test, the next largest deviation often occurs on the other tail of the distribution of deviations and is not one less than the observed deviation as in a one-tailed test.

To illustrate, suppose that $n = 41$, $r = 5$, and we are making a two-tailed test of the null hypothesis $p = 0.3$. Here, $np = 12.3$ and the observed deviation is -7.3. For a two-tailed test, the next largest absolute deviation that the data provide is given by $r = 19$ on the other tail and is $|6.7|$. Hence, in finding z_c, we use the corrected deviation $(7.3 + 6.7)/2 = 7.0$. Since $\sqrt{npq} = 2.9343$, $z_c = 7.0/2.9343 = 2.386$. The two-tailed normal approximation gives $P = 0.017$; the exact value from the binomial is $P = 0.0156$.

For two-tailed tests, this improved correction is easily made, as follows: In correcting, delete the fractional part of $|r - np|$ if this part is greater than 0 and less than or equal to 0.5. Thus, 7.3 is corrected to 7 when finding z_c. If the fractional part lies between 0.5 and 1.0 inclusive, replace it by 0.5 when correcting. Thus $|r - np| = 8.73$ is corrected to 8.5, $|r - np| = 5.0$ is corrected to 4.5, and $|r - np| = 3.5$ is corrected to 3.0. Note that in two-tailed tests, this rule agrees with the 0.5 rule in (7.6.3) only when $|r - np|$ is a whole number or ends in 0.5.

EXAMPLE 7.6.5—For $n = 10$, $p = 1/2$, calculate: (*i*) the exact probability of 4 or fewer successes; (*ii*) the normal approximation, corrected for continuity; and (*iii*) the normal approximation, uncorrected. Ans. (*i*) 0.377, (*ii*) 0.376, (*iii*) 0.263.

EXAMPLE 7.6.6—In a sample of size 49 with $p = 0.2$, the expected number of successes is 9.8. An investigator is interested in the probability that the observed number of successes will be (*i*) 15 or more or (*ii*) 5 or less. Estimate these two probabilities by the 0.5 normal approximation. Ans. (*i*) 0.0466, (*ii*) 0.0623. The exact answers obtained by summing the binomial are (*i*) 0.0517, (*ii*) 0.0547. Because of the skewness ($p = 0.2$), the normal curve underestimates in the long tail and overestimates in the short tail. For the sum of the two tails the normal curve does better, giving 0.1089 as against the exact 0.1064. Try the new method. Ans. 0.0407.

EXAMPLE 7.6.7—With $n = 16$, $p = 0.9$, estimate by the normal curve the probability that 16 successes are obtained. The exact result is, of course, $0.9^{16} = 0.185$. Ans. 0.180.

EXAMPLE 7.6.8—With $n = 49$, $r = 15$ successes were observed. Make a two-tailed test of the null hypothesis $p = 0.2$, using the correction for continuity. Ans. Since $r - np = 15 - 9.8 = 5.2$, we take $z_c = 5/\sqrt{npq} = 5/2.8 = 1.786$, $P = 0.074$, not significant at the 5% level. Note: In finding the exact P from the binomial tables, this problem differs from example 7.6.6. The critical region in a two-tailed test is the region giving deviations at least as great as $|5.2|$. This region is $r = 4$ or less and $r = 15$ or more. The combined exact P is 0.0732.

EXAMPLE 7.6.9—In a two-tailed test of the null hypothesis $p = 0.4$, $r = 9$ successes were observed in a random sample of size $n = 12$. Calculate the probability of a deviation as large as that observed or larger (*i*) by the normal approximation, corrected for continuity, and (*ii*) exactly. Note: $0.6^{12} = 0.0021768$, $0.4^{12} = 0.00001678$. Ans. (*i*) $P = 0.0184$, (*ii*) $P = 0.01745$. The exact critical region consists of $r = 0$ and $r = 9, 10, 11, 12$.

7.7—Test of significance of a binomial proportion. In section 7.6 we made one- and two-tailed tests of significance of a binomial proportion by means of the normal approximation with a correction for continuity. The tests can be conveniently made by comparing the observed number of successes r with the number np expected on H_0. For a one-tailed test we take

$$z_c = (|r - np| - 0.5)/\sqrt{npq}$$

and read one tail of the normal distribution, provided, of course, that $r - np$ is in the direction consistent with H_A. Note, however, that if $|r - np| \leq 0.5$, the observed value of r gives the closest possible agreement with the null hypothesis. In this event the correction reduces $|r - np|$ to 0.

To correct for continuity in a two-tailed test, look at the fractional part g of $r - np$. If $0 < g \leq 0.5$, delete the fractional part. If $0.5 < g \leq 1$, reduce g to 0.5.

The same test can be made as an application of the χ^2 test of goodness of fit (section 5.12). In the notation of the χ^2 test our results are as follows:

	Successes	Failures
Observed, f	r	$n - r$
Expected, F	np	$nq = n - np$
Observed − expected, $f - F$	$r - np$	$-(r - np)$

Hence,

$$\chi^2 = \sum \frac{(f-F)^2}{F} = \frac{(r-np)^2}{np} + \frac{(r-np)^2}{nq}$$

$$= \frac{(r-np)^2}{npq}(q+p) = \frac{(r-np)^2}{npq} = z^2$$

since the normal deviate $z = (r - np)/\sqrt{npq}$ if no correction for continuity is used. Further, the χ^2 distribution with 1 df is the distribution of the square of a normal deviate; the 5% significance level of χ^2, 3.84, is simply the square of 1.96. Thus the two tests are identical.

To correct χ^2 for continuity, we use the square of z corrected for continuity. The z method is more informative, since we can note the size and direction of $\hat{p} - p$.

7.8—Confidence limits for a binomial proportion. If r members of a sample of size n are found to possess some attribute, the sample estimate of the proportion in the population possessing this attribute is $\hat{p} = r/n$. In large samples, as we have seen, the binomial estimate $\hat{p}$ is approximately normally distributed about the population proportion p with standard deviation $\sqrt{pq/n}$. For the true but unknown standard deviation $\sqrt{pq/n}$, we substitute the sample estimate $\sqrt{\hat{p}\hat{q}/n}$. Hence, the probability is approximately 0.95 that $\hat{p}$ lies between the limits

$$p - 1.96\sqrt{\hat{p}\hat{q}/n} \quad \text{and} \quad p + 1.96\sqrt{\hat{p}\hat{q}/n}$$

But this statement is equivalent to saying that p lies between

$$\hat{p} - 1.96\sqrt{\hat{p}\hat{q}/n} \quad \text{and} \quad \hat{p} + 1.96\sqrt{\hat{p}\hat{q}/n} \qquad (7.8.1)$$

unless we were unfortunate in drawing one of the extreme samples that turns up once in twenty times. The limits (7.8.1) are therefore the approximate 95% confidence limits for p.

For example, suppose that 200 individuals in a sample of 1000 possess the attribute. The 95% confidence limits for p are

$$0.2 \pm 1.96\sqrt{(0.2)(0.8)/1000} = 0.2 \pm 0.025$$

The confidence interval extends from 0.175 to 0.225, that is, from 17.5% to 22.5%. For 99% limits, we replace 1.96 by 2.576.

If the above reasoning is repeated with the 0.5 correction for continuity included, the 95% limits for p become

$$\hat{p} \pm [1.96\sqrt{\hat{p}\hat{q}/n} + 1/(2n)]$$

The correction amounts to widening the limits a little. We recommend that the correction be used as a standard practice, although it makes little difference when n is large. To illustrate the correction in a smaller sample, suppose that 10 families out of 50 report ownership of more than one car, giving $\hat{p} = 0.2$. The 95% confidence limits for p are

$$0.2 \pm (1.96\sqrt{0.16/50} + 0.01) = 0.2 \pm 0.12$$

or 0.08 and 0.32. More exact limits for this problem, computed from the binomial distribution itself, are 0.10 and 0.34. The normal approximation gives the correct width of the interval, 0.24, but the normal limits are symmetrical about $\hat{p}$, whereas the correct limits are displaced upwards because an appreciable amount of skewness still remains in the binomial when $n = 50$ and p is not near 0.5.

7.9—Comparison of proportions in paired samples. A comparison of two sample proportions may arise either in paired or in independent samples. To illustrate paired samples, suppose that a lecture method is being compared with a method that uses a machine for programmed learning, but no lecture, to teach workers how to perform a rather complicated operation. The workers are first grouped into pairs by an initial estimate of their aptitudes for this kind of task. One member of each pair is assigned at random to each method. At the end, each student is tested for success or failure on the operation.

With 100 pairs, the results might be presented as follows:

A	*B*	Number of Pairs
S	*S*	52
S	*F*	21
F	*S*	9
F	*F*	18

In 52 pairs, both workers succeeded in the test; in 21 pairs, the worker taught by method *A* succeeded but the partner taught by method *B* failed; and so on.

As a second illustration (2), different media for growing diphtheria bacilli were compared. Swabs were taken from the throats of a large number of patients with symptoms suggestive of the presence of diphtheria bacilli. From each swab, a sample was grown on each medium. After allowing time for growth, each culture was examined for presence or absence of the bacilli. A successful medium is one favorable to the growth of the bacilli so that they are detected. This example illustrates self-pairing, since each medium is tested on every patient. It is also an example in which a large number of *FF*s would be expected because diphtheria is now rare and many patients would actually have no diphtheria bacilli in their throats.

Usually our primary interest is in a comparison of $\hat{p}_A$ and $\hat{p}_B$, the overall proportions of successes by methods *A* and *B* in the n trials. There is a very simple method of making the test of significance of the null hypothesis that the proportion of successes is the same for the two methods. The *SS* and *FF* pairs are ignored in the test of significance, since they give no indication in favor of either *A* or *B*. We concentrate on the *SF* and *FS* pairs. If the null hypothesis is true, the population must contain as many *SF* as *FS* pairs. The numerical example contains $21 + 9 = 30$ pairs of the *SF* or *FS* types. Under the null hypothesis we expect 15 of each type as against 21 and 9 observed.

Hence, the null hypothesis is tested by either the χ^2 or the z test of section 7.7. (In the z test we take $n = 30$, $r = 21$, $p = 1/2$.) When $p = 1/2$, χ^2 takes the particularly simple form,

$$\chi_c^2 = \frac{(|n_{SF} - n_{FS}| - 1)^2}{n_{SF} + n_{FS}} = \frac{(|21 - 9| - 1)^2}{30} = \frac{121}{30} = 4.03 \tag{7.9.1}$$

with 1 df. The null hypothesis is rejected at the 5% level (3.84). Method *A* has given a significantly higher proportion of successes. Remember that in this test the denominator of χ^2 is always the total number of *SF* and *FS* pairs. This test is the same as the sign test described in section 8.4.

Some users are uncomfortable with this test in a paired experiment. They note that the sample size n (total number of pairs) does not appear explicitly in z or χ^2 and that at first sight the test does not seem to be a comparison of $\hat{p}_A$ and $\hat{p}_B$. Also, cannot we somehow use the n_{SS} and n_{FF} pairs?

These features are just a consequence of the pairing. We get the same test if we compare $\hat{p}_A$ and $\hat{p}_B$. For this test we divide $\hat{p}_A - \hat{p}_B$ by its estimated standard error (SE), and treat the quotient as an approximate normal deviate. In our data, $\hat{p}_A = 73/100 = 73\%$, $\hat{p}_B = 61/100 = 61\%$. Since, algebraically, $\hat{p}_A = (n_{SS} + n_{SF})/n$, while $\hat{p}_B = (n_{SS} + n_{FS})/n$, we have

$$\hat{p}_A - \hat{p}_B = (n_{SF} - n_{FS})/n = \hat{p}_{SF} - \hat{p}_{FS} \tag{7.9.2}$$

In finding $SE(\hat{p}_A - \hat{p}_B)$ we must remember that because of the pairing, $\hat{p}_A$ and $\hat{p}_B$ are not independent. If H_0 holds, the estimated variance of $\hat{p}_A - \hat{p}_B$ is simply $(\hat{p}_{SF} + \hat{p}_{FS})/n$ (see example 7.9.2). Hence, before correction for continuity, the normal deviate z used in testing H_0 is

$$z = \sqrt{n}(\hat{p}_{SF} - \hat{p}_{FS})/\sqrt{\hat{p}_{SF} + \hat{p}_{FS}} = (n_{SF} - n_{FS})/\sqrt{n_{SF} + n_{FS}} \tag{7.9.3}$$

When we substitute $\hat{p}_{SF} = n_{SF}/n$ and $\hat{p}_{FS} = n_{FS}/n$ in z, note that the total sample size n cancels out. For our data, $z = 12/\sqrt{30}$, so $z^2 = \chi^2 = 4.8$ (before correction for continuity). To correct z, take

$$z_c = (|n_{SF} + n_{FS}| - 1)/\sqrt{n_{SF} + n_{FS}} \tag{7.9.4}$$

Once again, $z_c^2 = \chi_c^2$. Hence, this χ_c^2 is exactly the same as χ_c^2 in (7.9.1), which ignores the n_{SS} and n_{FF} pairs.

If we do not assume the null hypothesis $p_A = p_B$, the formula for the estimated $SE(\hat{p}_A - \hat{p}_B)$ is slightly more complicated, namely

$$SE(\hat{p}_A - \hat{p}_B) = \sqrt{\hat{p}_{SF} + \hat{p}_{FS} - (\hat{p}_{SF} - \hat{p}_{FS})^2/n} \tag{7.9.5}$$

With our data this gives $\sqrt{[0.21 + 0.09 - (0.21 - 0.09)^2]/10} = 0.0534$. Example 7.9.2 gives a method of proving (7.9.5).

The 95% confidence limits for $(p_A - p_B)$ are

$$(\hat{p}_A - \hat{p}_B) \pm [1.96\, SE(\hat{p}_A - \hat{p}_B) + 1/(2n)] \tag{7.9.6}$$

For our data they are $0.120 \pm 0.110 = 0.01$ and 0.23.

EXAMPLE 7.9.1—If pairing is ineffective, i.e., essentially random, $\hat{p}_A$ will be independent of $\hat{p}_B$. With $\hat{p}_A = 0.73$, $\hat{p}_B = 0.61$, show by (6.6.2) that the estimated $SE(\hat{p}_A - \hat{p}_B)$ would then be 0.0660. The smaller value 0.0534 obtained in our data indicates that this pairing was effective to some extent. Of course, $n = 100$ in each sample.

EXAMPLE 7.9.2—With continuous data we analyzed paired samples by forming a column of the differences d_i in results between members of the ith pair (section 6.3) and calculating $t = \sqrt{n}\,\bar{D}/s_{\bar{D}}$. If 1 denotes a success and 0 denotes a failure, application of the same method to a two-class population gives the following table:

A	B	d_i	Probability
1	1	0	p_{SS}
1	0	+1	p_{SF}
0	1	−1	p_{FS}
0	0	0	p_{FF}
			1

(*i*) For the population of differences, show that the mean $\mu_D = p_{SF} - p_{FS} = p_A - p_B$, while the variance $\sigma_D^2 = (p_{SF} + p_{FS}) - (p_{SF} - p_{FS})^2$. (*ii*) Hence show that for a random sample of n pairs, the variance V of the sample mean difference is

$$V(\hat{p}_A - \hat{p}_B) = [(p_{SF} + p_{FS}) - (p_{SF} - p_{FS})^2]/n$$

If H_0 holds, $p_{SF} = p_{FS}$ and $V(\hat{p}_A - \hat{p}_B)$ reduces to $(p_{SF} + p_{FS})/n$. For sample estimates we substitute $\hat{p}_{SF} = n_{SF}/n, \hat{p}_{FS} = n_{FS}/n$, getting the formulas given in the text, i.e., $\hat{V}(\hat{p}_A - \hat{p}_B) = (\hat{p}_{SF} + p_{SF})/n$.

7.10—Comparison of proportions in independent samples: the 2 × 2 table. This problem occurs very often in investigative work. Many controlled experiments that compare two procedures or treatments are carried out with independent samples because no effective way of pairing the subjects or animals is known to the investigator. Comparison of proportions in independent groups is also common in observational studies. A manufacturer compares the proportions of defective articles in two separate sources of supply for these articles, or a safety engineer compares the proportions of head injuries sustained in automobile accidents by passengers with seat belts and those without seat belts.

The data used for illustration come from a large Canadian study (3) of the relation between smoking and mortality. By an initial questionnaire in 1956, male recipients of war pensions were classified according to their smoking habits. We shall consider two classes: (*i*) nonsmokers and (*ii*) those who reported that they smoked pipes only. A report of the death of any pensioner during the succeeding six years was obtained. Thus, the pensioners were classified also according to their status (dead or alive) at the end of six years. Since the probability of dying depends greatly on age, the comparison given here is confined to men aged 60–64 at the beginning of the study. The numbers of men falling in the four classes are given in table 7.10.1, called a *2 × 2 contingency table.*

Note that 11.0% of the nonsmokers had died as against 13.4% of the pipe smokers. Can this difference be attributed to sampling error, or does it indicate a

TABLE 7.10.1

MEN CLASSIFIED BY SMOKING HABITS AND MORTALITY IN SIX YEARS

	Sample (1), Nonsmokers	Sample (2), Pipe Smokers	Total
Dead	117	54	171
Alive	950	348	1298
Total	$n_1 = 1067$	$n_2 = 402$	1469
Proportion dead	$\hat{p}_1 = 0.1097$	$\hat{p}_2 = 0.1343$	$\hat{p} = 0.1164$

real difference in the death rates in the two groups? The null hypothesis is that the proportions dead, 117/1067 and 54/402, are estimates of the same quantity.

Since $\hat{p}_1 = 0.1097$ and $\hat{p}_2 = 0.1343$ are approximately normally distributed, their difference $\hat{p}_1 - \hat{p}_2$ is also approximately normally distributed. The variance of this difference is the sum of the two variances; see (6.6.2).

$$V(\hat{p}_1 - \hat{p}_2) = \sigma^2_{\hat{p}_1} + \sigma^2_{\hat{p}_2} = p_1q_1/n_1 + p_2q_2/n_2 \qquad (7.10.1)$$

Under the null hypothesis, $p_1 = p_2 = p$, so $\hat{p}_1 - \hat{p}_2$ is approximately normally distributed with mean 0 and standard error

$$SE = \sqrt{pq/n_1 + pq/n_2} \qquad (7.10.2)$$

The null hypothesis does not specify the value of p. As an estimate, we suggest $\hat{p} = 0.1164$ as given by the combined samples. Hence, the normal deviate z is

$$z = \frac{\hat{p}_1 - \hat{p}_2}{\sqrt{\hat{p}\hat{q}(1/n_1 + 1/n_2)}} \qquad (7.10.3)$$

$$= \frac{0.1097 - 0.1343}{\sqrt{(0.1164)(0.8836)(1/1067 + 1/402)}} = \frac{-0.0246}{0.01877} = -1.311$$

In the normal table, ignoring the sign of z, we find $P = 0.19$ in a two-tailed test. We conclude that the observed difference between the death rates of pipe smokers and nonsmokers in this population may well be due to sampling errors.

In finding confidence limits for the population difference $p_1 - p_2$, the standard error of $p_1 - p_2$ should be estimated from (7.10.1) as

$$\sqrt{\hat{p}_1\hat{q}_1/n_1 + \hat{p}_2\hat{q}_2/n_2} \qquad (7.10.4)$$

The standard error given by the null hypothesis formula (7.10.2) is no longer valid. Often the change is small, but it can be material if n_1 and n_2 are very unequal. Incidentally, (7.10.4) could possibly be used as the denominator of z in the test of H_0: $p_1 = p_2$. Robbins (18) has suggested that (7.10.4) might give the test criterion more power than (7.10.2). There is some evidence (19) that this is so in large samples. But z based on (7.10.2) is suggested here because in small samples it should better approximate the distribution of $\hat{p}_1 - \hat{p}_2$ when H_0 holds.

If $n_1 = n_2$, z is corrected for continuity by subtracting 0.5 from the numerator of the *larger* proportion and adding 0.5 to the numerator of the smaller proportion. The denominator of z remains unchanged. If $n_1 \neq n_2$, a correction for continuity is more conveniently made by showing that the z test can be expressed as a χ^2 test of goodness of fit, as in the next section.

7.11—The χ^2 test in a 2 × 2 contingency table. As usual,

$$\chi^2 = \Sigma\,(f - F)^2/F$$

where the fs are the observed numbers 117, 950, 54, 348 in the four cells. The Fs are the numbers that would be expected in the four cells if the null hypothesis $p_1 = p_2$ were true.

The Fs are computed as follows. If the proportions dead are the same for the two smoking classes, our best estimate of this proportion is the proportion 171/1469, found in the combined sample. Since there are 1067 nonsmokers, the expected number dead, by the null hypothesis, is

$$(1067)(171)/1469 = 124.2$$

To find the expected number in any cell, the rule is: Multiply the corresponding column and row totals and divide by the grand total. The expected number of nonsmokers who are alive is

$$(1067)(1298)/1469 = 942.8$$

and so on. Alternatively, having calculated 124.2 as the expected number of nonsmokers who are dead, we find the expected number alive more easily as $1067 - 124.2 = 942.8$. Similarly, the expected number of pipe smokers who are dead is $171 - 124.2 = 46.8$. Finally, the expected number of pipe smokers who are alive is $402 - 46.8 = 355.2$. Thus, only *one* expected number need be calculated; the others are found by subtraction.

The observed numbers, expected numbers, and the differences $f - F$ appear in table 7.11.1. Except for their signs, all four deviations $f - F$ are equal. This result holds in any 2×2 table.

Since $(f - F)^2$ is the same in all cells, χ^2 may be written

$$\chi^2 = (f - F)^2 \sum_{i=1}^{4} 1/F_i \qquad (7.11.1)$$

$$= 7.2^2\,(1/124.2 + 1/46.8 + 1/942.8 + 1/355.2)$$

$$= (51.85)(0.0333) = 1.73$$

How many degrees of freedom has χ^2? The fact that all four deviations are the same apart from sign suggests that χ^2 has only 1 df, as was proved by Fisher. In fact, it can be shown algebraically that $\chi^2 = z^2$, where z is calculated as in the preceding section. Note that from (7.10.3), $z^2 = 1.311^2 = 1.73 = \chi^2$ in this example.

To correct χ^2 for continuity, we use the same method as in testing a single

TABLE 7.11.1

VALUES OF f (OBSERVED), F (EXPECTED), AND $f - F$ IN THE FOUR CELLS

f		F		$f - F$	
117	54	124.2	46.8	−7.2	+7.2
950	348	942.8	355.2	+7.2	−7.2

binomial p (section 7.7). Look at the fractional part g of the deviation $f - F$. If $0 < g \leq 0.5$, delete the fractional part when correcting for continuity. If $0.5 < g \leq 1.0$, reduce the fractional part to 0.5. In this example, $f - F = 7.2$, so $\chi_c^2 = 7^2(0.0333) = 1.632$, giving $z_c = 1.277$, $P_c = 0.20$. This value is very little changed from the uncorrected $P = 0.19$. In small samples the correction makes a substantial difference.

Note: While the χ^2 test is essentially two-tailed, it can be used in this problem to make a one-tailed test for the case in which the only alternative of interest is $p_1 > p_2$. To make a one-tailed test, we find z and use only a single tail of the normal table. For a continuity correction in such a one-tailed test, subtract 0.5 from $|f - F|$.

In interpreting the results of these χ^2 tests in observational studies, caution is necessary, as noted in chapter 1. The two groups being compared may differ in numerous ways, some of which may be wholly or partly responsible for an observed significant difference. For instance, pipe smokers and nonsmokers may differ to some extent in their economic levels, residence (urban or rural), and eating and drinking habits, and these variables may be related to the risk of dying. Before claiming that a significant difference is caused by the variable under study, it is the investigator's responsibility to produce evidence that disturbing variables of this type could not have produced the difference. Of course, the same responsibility rests with the investigator who has done a controlled experiment. But the device of randomization and the greater flexibility that usually prevails in controlled experimentation make it easier to ensure against misleading conclusions from disturbing influences.

The preceding χ^2 and z methods are approximate, the approximation becoming poorer as the sample size decreases. Fisher (14) has shown how to compute an exact test of significance. For accurate work the exact test should be used if (*i*) the total sample size N is less than 20 or (*ii*) if N lies between 20 and 40 and the smallest expected number is less than 5. For those who encounter these conditions frequently, reference (15), which gives tables of the exact tests covering these cases, is recommended.

EXAMPLE 7.11.1—In a study as to whether cancer of the breast tends to run in families, Murphy and Abbey (4) investigated the frequency of breast cancer found in relatives of (*i*) women with breast cancer and (*ii*) a comparison group of women without breast cancer. The data below, slightly altered for easy calculation, refer to the mothers of the subjects.

		Breast Cancer in Subject		
		Yes	No	Total
Breast Cancer in Mother	Yes	7	3	10
	No	193	197	390
Total		200	200	400

Calculate χ^2 and P (*i*) without correction and (*ii*) with correction for continuity for testing the null hypothesis that the frequency of cancer in mothers is the same in the two classes of subjects. Ans. (*i*) $\chi^2 = 1.64$, $P = 0.20$; (*ii*) $\chi_c^2 = 0.92$, $P_c = 0.34$. Note that the correction for continuity always increases P, that is, makes the difference less significant.

With these data, a case can be made for either a two-tailed test or a one-tailed test on the supposition that it is very unlikely that breast cancer in mothers would decrease the risk of breast cancer in the daughters. With a one-tailed test, $P_c = 0.17$, neither test giving evidence of a relation between breast cancer in subject and mother.

EXAMPLE 7.11.2—C. H. Richardson has furnished the following numbers of aphids (*Aphis rumicis* L.) dead and alive after spraying with two concentrations of solutions of sodium oleate:

	Concentration of Sodium Oleate (%)		
	0.65	1.10	Total
Dead	55	62	117
Alive	13	3	16
Total	68	65	133
Percent dead	80.9	95.4	

Has the higher concentration given a significantly different percent kill? Ans. $\chi_c^2 = 5.76$, $P < 0.025$.

EXAMPLE 7.11.3—In 1943 a sample of about 1 in 1000 families in Iowa was asked about the canning of fruits or vegetables during the preceding season. Of the 392 rural families, 378 had done canning, while of the 300 urban families, 274 had canned. Calculate 95% confidence limits for the difference in the percentages of rural and urban families who had canned. Ans. 1.42% and 8.78% (without correction for continuity).

EXAMPLE 7.11.4—In examining the effects of sprays in the control of codling moth injury to apples, Hansberry and Richardson (5) counted the wormy apples on each of 48 trees. Two trees sprayed with the same amount of lead arsenate yielded

A: 2130 apples, 1299 or 61% of which were injured
B: 2190 apples, 1183 or 45% of which were injured

The value $\chi^2 = 21.16$ is conclusive evidence that the chance of injury was different in these two trees. This result is characteristic of spray experiments. For unknown reasons, injuries under identical experimental treatments differ significantly. Hence it is undesirable to compare sprays on single trees because a difference in percentage of injured apples might be due to these unknown sources rather than to the treatments. A statistical determination of the homogeneity or heterogeneity of experimental material under identical conditions, sometimes called a *test of technique*, is often worthwhile, particularly in new fields of research.

7.12—Test of the independence of two attributes. The preceding test is sometimes described as a test of the independence of two attributes. A sample of people of a particular ethnic type might be classified into two classes according to hair color and also into two classes according to color of eyes. We might ask, Are color of hair and color of eyes independent? Similarly, the numerical example in the previous section might be referred to as a test of the question, Is the risk of dying independent of smoking pipes?

In this way of speaking, the word *independent* carries the same meaning as it does in Rule 3 in the theory of probability. Let p_A be the probability that a member of a population possesses attribute A and p_B the probability that the member possesses attribute B. If the attributes are independent, the probability of possessing both attributes in $p_A p_B$. Thus, on the null hypothesis of independence, the probabilities in the four cells of the 2×2 contingency table are as follows:

	Attribute A		
Attribute B	(1) Present	(2) Absent	Total
(1) Present	$p_A p_B$	$q_A p_B$	p_B
(2) Absent	$p_A q_B$	$q_A q_B$	q_B
Total	p_A	q_A	1

Two points emerge from this table. The null hypothesis can be tested either by comparing the proportions of cases in which B is present in columns (1) and (2) or by comparing the proportions of cases in which A is present in rows (1) and (2). These two χ^2 tests are exactly the same as is easily verified.

Second, the table provides a check on the rule given for calculating the expected number in any cell. In a single sample of size N, we expect to find $Np_A p_B$ members possessing both A and B. The sample total in column (1) is our best estimate of Np_A, while that in row (1) similarly estimates Np_B. Thus (column total)(row total)/(grand total) is $(N\hat{p}_A)(N\hat{p}_B)/N = N\hat{p}_A\hat{p}_B$ as required.

7.13—Sample size for comparing two proportions. The question, How large a sample do I need? is naturally of great interest to investigators. For comparing two means, an approach that is often helpful is given in section 6.14. It should be reviewed carefully, since the same principle applies to the comparison of two proportions. The approach assumes that a test of significance of the difference between the two proportions is planned and that future actions will depend on whether the test shows a significant difference or not. Consequently, if the true difference $p_2 - p_1$ is as large as some amount δ chosen by the investigator, the test is desired to have a high probability P' of declaring a significant result.

Formula (6.14.2) for n, the size of each sample, can be applied. For two independent samples, use $\delta = p_2 - p_1$ and $\sigma_D^2 = p_1q_1 + p_2q_2$, which gives

$$n = (Z_\alpha + Z_\beta)^2(p_1q_1 + p_2q_2)/(p_2 - p_1)^2 \qquad (7.13.1)$$

where Z_α is the normal deviate corresponding to the significance level to be used in the test, $\beta = 2(1 - P')$, and Z_β is the normal deviate corresponding to the *two-tailed* probability β. Table 6.14.1 gives $(Z_\alpha + Z_\beta)^2$ for the commonest values of α and β. In using this formula, we substitute the best advance estimate of $p_1q_1 + p_2q_2$ in the numerator.

For instance, suppose that a standard antibiotic has been found to protect about 50% of experimental animals against a certain disease. Some new antibiotics become available that seem likely to be superior. In comparing a new antibiotic with the standard, we would like a probability $P' = 0.9$ of finding a significant difference in a one-tailed test at the 5% level if the new antibiotic will protect 80% of the animals in the population. For these conditions, table 6.14.1 gives $(Z_\alpha + Z_\beta)^2$ as 8.6. Hence

$$n = (8.6)[(50)(50) + (80)(20)]/30^2 = 39.2$$

Thus, 40 animals should be used for each antibiotic.

Some calculations of this type will soon convince you of the sad fact that large samples are necessary to detect small differences between two percentages. When resources are limited, it is sometimes wise, before beginning the experiment, to calculate the probability that a significant result will be found. Suppose that an experimenter is interested in the values $p_1 = 0.8$, $p_2 = 0.9$ but cannot make $n > 100$. If formula (7.13.1) is solved for Z_β, we find

$$Z_\beta = \frac{(p_2 - p_1)\sqrt{n}}{\sqrt{p_1 q_1 + p_2 q_2}} - Z_\alpha = \frac{(0.1)(10)}{0.5} - Z_\alpha = 2 - Z_\alpha$$

If the experimenter intends a two-tailed 5% test, $Z_\alpha \doteq 2$ so that $Z_\beta \doteq 0$. This gives $\beta = 1$ and $P' = 1 - \beta/2 = 0.5$. The proposed experiment has only a 50–50 chance of finding a significant difference in this situation.

Formula (7.13.1), although a large sample approximation, should be accurate enough for practical use, even though there is usually some uncertainty about the values of $\hat{p}_1$ and $\hat{p}_2$ to insert in the formula. Reference (6) gives tables of n based on a more accurate approximation.

EXAMPLE 7.13.1—One difficulty in estimating sample size in biological work is that the proportions given by a standard treatment may vary over time. An experimenter has found that the standard treatment has a failure rate lying between $p_1 = 30\%$ and $p_1 = 40\%$. With a new treatment whose failure rate is 20% lower than the standard, what sample sizes are needed to make $P' = 0.9$ in a two-tailed 5% test? Ans. $n = 79$ when $p_1 = 30\%$ and $n = 105$ when $p_1 = 40\%$.

EXAMPLE 7.13.2—The question of sample size was critical in planning the 1954 trial of the Salk poliomyelitis vaccine (7), since it was unlikely that the trial could be repeated and an extremely large sample of children would obviously be necessary. Various estimates of sample size were therefore made. One estimate assumed that the probability an unprotected child would contract paralytic polio was 0.0003, or 0.03%. If the vaccine was 50% effective (that is, decreased this probability to 0.00015, or 0.015%), it was desired to have a 90% chance of finding a 5% significance difference in a two-tailed test. How many children are required? Ans. 210,000 in each group (vaccinated and unprotected).

EXAMPLE 7.13.3—An investigator has $p_1 = 0.4$ and usually conducts experiments with $n = 25$. In a one-tailed test at the 5% level, what is the chance of obtaining a significant result if (*i*) $p_2 = 0.5$, (*ii*) $p_2 = 0.6$? Ans. (*i*) 0.18, (*ii*) 0.42.

7.14—The Poisson distribution. As we have seen, the binomial distribution tends to the normal distribution as n increases for any fixed value of p. The value of n needed to make the normal approximation a good one depends on the value of p; this value of n is smallest when $p = 0.5$. For $p < 0.5$, a general rule, usually conservative, is that the normal approximation is adequate if the mean $\mu = np$ is greater than 15.

In many applications, however, we are studying rare events, so that even if n is large, the mean np is much less than 15. The binomial distribution then remains noticeably skew and the normal approximation is unsatisfactory. A different approximation for such cases was developed by S. D. Poisson (8). He worked out the limiting form of the binomial distribution when n tends to

TABLE 7.14.1

POISSON DISTRIBUTION FOR $\mu = 1$ COMPARED WITH THE BINOMIAL DISTRIBUTIONS FOR $n = 100$, $p = 0.01$ AND $n = 25$, $p = 0.04$

	Relative Frequencies		
r	(1) Poisson 1	(2) Binomial $n = 100, p = 0.01$	(3) Binomial $n = 25, p = 0.04$
0	0.3679	0.3660	0.3604
1	.3679	.3697	.3754
2	.1839	.1849	.1877
3	.0613	.0610	.0600
4	.0153	.0149	.0137
5	.0031	.0029	.0024
6	.0005	.0005	.0003
≥7	0.0001	0.0001	0.0000
Total	1.0000	1.0000	0.9999

infinity and p tends to zero at the same time in such a way that $\mu = np$ is constant. The binomial expression for the probability of r successes tends to the simpler form,

$$P(r) = \frac{\mu^r}{r!} e^{-\mu} \qquad r = 0, 1, 2, \ldots \tag{7.14.1}$$

where $e = 2.71828$ is the base of natural logarithms. The initial terms in the Poisson distribution are

$$P(0) = e^{-\mu} \qquad P(1) = \mu e^{-\mu} \qquad P(2) = \frac{\mu^2}{2} e^{-\mu} \qquad P(3) = \frac{\mu^3}{(2)(3)} e^{-\mu} \tag{7.14.2}$$

Table 7.14.1 shows in column (1) the Poisson distribution for $\mu = 1$. The distribution is markedly skew. The mode (highest frequency) is at either 0 or 1, these two having the same probability when $\mu = 1$. To give an idea of the way in which the binomial tends to approach the Poisson, column (2) shows the binomial distribution for $n = 100$, $p = 0.01$ and column (3) the binomial for $n = 25$, $p = 0.04$; for both of these, $np = 1$. The agreement with the Poisson is very close for $n = 100$ and quite close for $n = 25$. Tables of individual and cumulative terms of the Poisson are given in (9) and of individual terms up to $\mu = 15$ in (10).

The fitting of a Poisson distribution to a sample is illustrated by the data (11) in table 7.14.2. These show the number of noxious weed seeds in 98 subsamples of *Phleum praetense* (meadow grass). Each subsample weighed 1/4 oz and of course contained many seeds, of which only a small percentage were noxious. The first step is to compute the sample mean.

$$\hat{\mu} = \Sigma fr/\Sigma f = 296/98 = 3.0204 \text{ noxious seeds per subsample}$$

TABLE 7.14.2
NUMBERS OF NOXIOUS WEED SEEDS IN $n = 98$ SUBSAMPLES, WITH FITTED POISSON DISTRIBUTION

Number of Noxious Seeds	Observed Frequency, f	Expected Frequency, F	Observed − Expected, $f - F$	Contribution to χ^2, $(f - F)^2/F$
0	3	4.78	−1.78	0.66
1	17	14.44	+2.56	0.45
2	26	21.81	+4.19	0.80
3	16	21.96	−5.96	1.62
4	18	16.58	+1.42	0.12
5	9	10.03	−1.02	0.10
6	3	5.04	−2.04	0.83
7	5	2.18	+2.82	3.65
8	0 } 1	0.82 } 1.20	−0.20	0.03
9	1	0.27		
10	0	0.08		
11 or more	0	0.03		
Total	98	98.00	−0.01	8.26

From the Poisson distribution, the expected number of subsamples with 0 noxious seeds is

$$Ne^{-\hat{\mu}} = (98)(e^{-3.0204}) = 4.7806$$

From (7.14.2), the expected numbers of subsamples with 1, 2, 3, . . . noxious seeds are found by multiplying 4.7806 successively by 3.0204, 3.0204/2, 3.0204/3, and so on, easily done on a pocket calculator.

The agreement between observed and expected frequencies seems good except perhaps for $r = 2$ and $r = 3$, which have almost equal expected numbers but have observed numbers 26 and 16. In making the χ^2 test of goodness of fit, shown in table 7.14.2, we combine the last four classes to give a minimum expectation of at least 1. This makes number of classes $k = 9$. Since one parameter μ was estimated in fitting the distribution, χ^2 has $9 - 1 - 1 = 7$ df by the rule in section 5.10. With $\chi^2 = 8.26$, the P value is about 0.4, so the discrepancies $f_i - F_i$ are not unusually large.

Two important properties hold for a Poisson variate. The variance of the distribution is equal to its mean μ. This would be expected, since the binomial variance npq tends to np when q tends to 1. Second, if a series of *independent* variates $X_1, X_2, X_3, \ldots$ each follow Poisson distributions with means $\mu_1, \mu_2, \mu_3, \ldots$, their sum follows a Poisson distribution with mean $(\mu_1 + \mu_2 + \mu_3 + \ldots)$.

In the inspection and quality control of manufactured goods, the proportion of defective articles in a large lot should be small. Consequently, the number of defectives in the lot might be expected to follow a Poisson distribution. For this reason, the Poisson distribution plays an important role in the development of plans for inspection and quality control. Further, the Poisson is often found to serve remarkably well as an approximation when μ is small, even if the value of n is ill-defined and if both n and p presumably vary from one sample to another. A much-quoted example of a good fit of a Poisson distribution, due to Bortkewitch, is the number of men in a Prussian army corps who were killed during a year by the kick of a horse. He had $N = 200$ observations, one for each of 10 corps for each of 20 years. On any given day, some men were exposed to a small probability of being kicked, but it is not clear what value n has nor that p would be constant.

The Poisson distribution can also be applied to events in time and space. Suppose signals are being transmitted and the probability that a signal reaches a given point in a very small time interval τ is $\lambda\tau$, irrespective of whether previous signals have arrived recently. The number of signals arriving in a finite time interval of length T may be shown to follow a Poisson distribution with mean λT (example 7.14.4). Similarly, if particles are distributed at random in a liquid with density λ per unit volume, the number found in a sample of volume V is a Poisson variable with mean λV. From these illustrations it is not surprising that the Poisson distribution has found applications in many fields, including communications theory and the estimation of bacterial densities.

EXAMPLE 7.14.1—If $n = 1000$ independent trials are made of an event with probability 0.001 at each trial, give approximate results for the chances that (*i*) the event does not happen, (*ii*) the event happens twice, (*iii*) the event happens at least five times. Ans. (*i*) 0.368, (*ii*) 0.184, (*iii*) 0.0037.

EXAMPLE 7.14.2—A. G. Arbous and J. E. Kerrich (12) report the numbers of accidents sustained during their first year by 155 engine shunters aged 31–35 as follows:

Number of accidents	0	1	2	3	4 or more
Number of men	80	61	13	1	0

Fit a Poisson distribution to these data and test the fit by χ^2. Note: The data were obtained as part of a study of accident-proneness. If some men are particularly liable to accidents, the Poisson would not be a good fit, since p would vary from man to man. Ans. $\chi^2 = 4.54$, df $= 2$, P about 0.1. Since 3.29 cases with 3 or more accidents are expected as against 1 observed, the data do not support an accident-proneness hypothesis.

EXAMPLE 7.14.3—Student (13) counted the number of yeast cells on each of 400 squares of a hemacytometer. In two independent samples, each of which gave a satisfactory fit to a Poisson distribution, the *total* numbers of cells were 529 and 720. (*i*) Test whether these totals are estimates of the same quantity, that is, whether the density of yeast cells per square is the same in the two populations. (*ii*) Compute 95% limits for the difference in density *per square*. Ans (*i*) $z = 5.41$, P very small; (*ii*) 0.30 to 0.65. Note: The normal approximation to the Poisson distribution, or to the difference between two independent Poisson variates, may be used when the observed numbers >15.

EXAMPLE 7.14.4—The *Poisson process* formula for the number of signals arriving in a finite time interval T can be justified by the following rough argument. Let the time interval T be divided into a large number $n = T/\tau$ of small intervals of length τ. In a single small interval, the probability that a signal arrives is $\lambda\tau = p$. These probabilities are assumed independent from one small interval to another. Thus the number of signals in time T is a binomial number of successes in $n = T/\tau$ trials with $p = \lambda\tau$. As τ becomes very small, the binomial tends to become Poisson with $\mu = np = \lambda T$.

TECHNICAL TERMS

Bernoulli distribution
binomial coefficients
binomial distribution
Bonferroni's inequalities
contingency table
correction for continuity
independent
mutually exclusive
number of combinations
Pascal's triangle
Poisson distribution
probability

REFERENCE

1. Mosteller, F.; Rourke, R. E. K.; and Thomas, G. B., Jr. 1970. *Probability with Statistical Applications,* 2nd ed. Addison-Wesley, Reading, Mass.
2. Data made available by Martin Frobisher.
3. Best, E. W. R., et al. 1966. A Canadian Study on Smoking and Health (Final Report). Dept. Natl. Health and Welfare, Canada.
4. Murphy, D. P., and Abbey, H. 1959. *Cancer in Families.* Harvard Univ. Press, Cambridge.
5. Hansberry, T. R., and Richardson, C. H. *Iowa State Coll. J. Sci.* 10 (1935):27.
6. Cochran, W. G., and Cox, G. M. 1957. *Experimental Designs,* 2nd ed. Wiley, New York, p. 17.
7. Francis, T. J., et al. 1957. *Evaluation of the 1954 Field Trial of Poliomyelitis Vaccine.* Edwards Bros., Ann Arbor.
8. Poisson, S. D. 1837. *Recherches sur la probabilité des jugements.* Paris.
9. Molina, E. C. 1942. *Poisson's Exponential Binomial Limit.* Van Nostrand, New York.
10. Pearson, E. S., and Hartley, H. O. 1966. *Biometrika Tables for Statisticians,* vol. I, 2nd ed. Cambridge Univ. Press.
11. Leggatt, C. W. *Comptes rendus de l'association internationale d'essais de semences* 5 (1935):27.
12. Arbous, A. G., and Kerrich, J. E. *Biometrics* 7 (1951):340.
13. Student. *Biometrika* 5 (1907):351.
14. Fisher, R. A. 1941. *Statistical Methods for Research Workers.* Oliver & Boyd, Edinburgh, §21.02.
15. Finney, D. J.; Latscha, R.; Bennett, B. M.; and Hsu, P. 1963. *Tables for Testing Significance in a 2 × 2 Contingency Table.* Cambridge Univ. Press, New York.
16. National Bureau of Standards. 1950. *Tables of the Binomial Probability Distribution.* Appl. Math. Ser. 6.
17. Annals of the Computation Laboratory. 1955. *Tables of the Cumulative Binomial Probability Distribution,* vol. 35. Harvard Univ.
18. Robbins, H. *Am. Stat.* 31 (1977):97.
19. Eberhardt, K. R., and Fligner, M. A. *Am. Stat.* 31 (1977):151.
20. Conover, W. J. *J. Am. Stat. Assoc.* 69 (1974):374.

8

Shortcut and nonparametric methods

8.1—Introduction. The two preceding chapters deal with the comparison of the means of two samples, paired or independent, from normal or binomial populations. This chapter presents a number of alternative methods related to the same problem. In one group—the shortcut methods—the primary aims are speed and simplicity in calculation at the expense of some loss of efficiency. Since electronic computers are now widely available, gains in speed have become small and less important. But the shortcut methods are still useful for a quick initial look at a body of data; in places where aids to calculation are not available; and in repetitive operations, for example, in industry, where hundreds of samples must be processed daily.

The second group of methods has a different objective. Since frequency distributions of many different types appear in statistical investigations, a strong effort has been made during the past 15 years or so to develop techniques that can be described as *robust,* in that they work well for a wide variety of population types from which the samples are drawn. Furthermore, since gross mistakes occur occasionally in measuring, recording, or copying data for statistical analysis, another useful property of statistical techniques is that they should be little affected by gross errors in a few observations. This property is sometimes called *resistance to outliers.*

8.2—Sample median. The median of a population has the property that half the values in the population exceed it and half fall short of it. If the sample size is odd, the sample median is the middle term when the sample values are arranged in increasing order. The sample median of a small sample may be found quickly by inspection. For example, the median of the observations 5, 1, 8, 3, 4 is 4. With n even there is no middle term, and the median is defined as the average of the term just below the middle and the term just above the middle. The median of the observations 1, 3, 4, 5, 7, 8 is $(4 + 5)/2 = 4.5$.

In a larger sample it is harder to spot the median without first arranging the observations in increasing order. When the sample values are arranged in this way, they are often called the 1st, 2nd, 3rd, . . . *order statistics.* In general, if n is odd the median is the order statistic whose number is $(n + 1)/2$. If n is even, the median is the average of the order statistics whose numbers are $n/2$ and $(n + 2)/2$. With a large sample that has been grouped into classes, the sample median may be found accurately enough by the cumulative frequency polygon, as described in section 2.5.

Like the mean, the median is a measure of the middle of a distribution. If the population distribution is symmetrical about its mean, μ, its mean and its median coincide. On the other hand, with highly skewed populations like that of income per family, the median is often reported as a descriptive statistic because it seems to represent people's concept of an average better than the mean, μ. In the distribution of family income, it is not unusual to find that about 65% of families have incomes below the population mean, and only 35% above it. Thus the mean does not seem a good indicator of middle income.

Two advantages of the sample median over the mean are: (1) the median is more quickly found in small samples; and (2) the median is unaffected by erratic extreme values, whereas the mean can be greatly affected. For example, the median of the observations 1, 3, 4, 5, 8 is 4, while the mean is 4.2. Suppose the sample values are 1, 3, 4, 5, 24, where the 24 simulates the presence of a very wealthy family or a gross error in recording the value for this observation. The sample median is still 4 but the mean has changed to 7.4, a value higher than all but one of the sample values. Thus the median is resistant to outliers. In fact, the sample median is still 4 if we do not know the values of the two highest observations but merely that they both exceed 4. In this situation, which occurs in certain practical applications, the sample mean cannot be calculated at all.

In large samples of size n from a normal distribution (14), the sample median becomes normally distributed about the population median with standard deviation $1.253\sigma/\sqrt{n}$, as against $\sigma/\sqrt{n}$ for the mean. For this distribution, in which the sample mean and median are both unbiased estimates of μ, the median is clearly less accurate than the mean. If $\hat{\theta}_1$, $\hat{\theta}_2$ are two unbiased estimates of a parameter θ, the *relative efficiency* of $\hat{\theta}_1$ to $\hat{\theta}_2$ is usually measured as $V(\hat{\theta}_2)/V(\hat{\theta}_1)$—the efficiency of an estimate being inversely proportional to its variance. Thus in large normal samples the relative efficiency of the median to $\overline{X}$ is $1/1.253^2 = 0.637$. In small normal samples the relative efficiency is somewhat higher: 0.669, 0.679, 0.697, and 0.743 for $n = 9, 7, 5$, and 3, respectively.

One use of the relative efficiency is in calculating the relative sample sizes needed so that two estimates have the same variance and can be regarded as equally accurate. Since the large sample variance of the median is $1.253^2\sigma^2/n = 1.57\sigma^2/n$, we need a sample size $1.57n$ to equal the variance σ^2/n of the mean of a sample of size n. For the median, the sample must be 57% larger. Note the relative sample size 1.57 is just the *reciprocal* of the relative efficiency, 0.637.

In nonnormal distributions with long tails, the relative efficiency of the median to the mean rises and may become larger than 1. We can now define more specifically what is meant by a robust estimate—namely, one whose efficiency relative to competitors is high (i.e., seldom much less than 1) over a wide range of parent populations. The median is more robust than the mean as well as more resistant to erratic extreme observations, although it is inferior to the mean with symmetrical distributions not far from normal.

There is a simple method (1) of calculating conservative confidence limits for the population median that is valid for any continuous distribution. Two of the order statistics serve as the upper and lower confidence limits. These are those statistics whose numbers are

$$(n + 1)/2 \pm z\sqrt{n}/2 \tag{8.2.1}$$

where z is the normal deviate corresponding to the desired confidence probability. Taking $z = 2$ for 95% limits, these numbers become simply

$$(n + 1)/2 \pm \sqrt{n} \tag{8.2.2}$$

Round the lower limit down and the upper limit up to the nearest integers. As stated, these limits are *conservative,* the exact confidence probability varying with n. For $n \leq 15$ with 95% limits, the confidence probabilities actually average 97.5%; for n between 15 and 50, they average 96.8%. As an example of the method, the systolic blood pressures (mm) of 11 male subjects aged between 60 and 70 were as follows when arranged in order: 110, 112, 116, 120, 126, 132, 142, 147, 154, 156, 179. The median is 132; the mean is slightly higher, 135.8. The conservative 95% limits for the population median are the order statistics whose numbers are $6 \pm \sqrt{11}$, namely 2 and 10. The corresponding blood pressures are 112 mm and 156 mm.

EXAMPLE 8.2.1—From a sample whose values are 8, 9, 2, 7, 3, 12, 15, 16, 23, estimate (*i*) the median; (*ii*) 95% limits for the population median; (*iii*) for comparison, 95% limits for the population mean. Ans. (*i*) 9; (*ii*) (3, 16); (*iii*) (5.4, 15.7).

EXAMPLE 8.2.2—The conservative confidence limits for the median are obtained by a simple use of the binomial with $p = 1/2$. If the population median is at or below the 1st order statistic in example 8.2.1, our sample is a rare one in which all 9 observations lie in the upper half of the population distribution. The probability that this will happen is $1/2^9 = 1/512$.

Similarly, if the population median lies between the 2nd and the 1st order statistic, our sample is one in which 8 out of 9 lie in the upper half of the distribution, with binomial probability $9/2^9 = 9/512$. Verify by symmetry that the probability is $2 \times 10/512 = 0.039$ that the interval from the 2nd to the 8th order statistic does not cover the population median. The confidence probability for this interval is therefore 96.1%. Find the confidence probability for the example worked in the text of section 8.2 with $n = 11$. Ans. 98.8%—much too conservative.

8.3—Estimation of σ from the sample range. The sample range, usually denoted by w, is the difference between the largest and the smallest measurements. It gives a quick estimate of σ in small samples from distributions thought to be normal or nearly so. In normal samples up to size 20, table A 6 shows (*i*) the number by which the range must be multiplied to give an unbiased estimate of σ; (*ii*) the variance of this estimate, divided by σ^2; and (*iii*) the efficiency of this estimate relative to s.

The relative efficiency of the range to s remains high in normal samples up to $n = 8$ or 10, decreasing slowly but steadily as n increases. Neither the range nor s is a robust estimate; both perform poorly for nonnormal distributions with long tails. Further, the range, being calculated from extreme values, is very sensitive to erratic extreme values (outliers). In short, the role of the range is to give a quick estimate of σ for distributions thought to be close to normal.

The range is extensively used in this situation, particularly in industry. In many operations, σ is estimated in practice by combining the estimates from a substantial number of small samples. In controlling the quality of an industrial process, small samples of the product are taken out frequently, say every 15 minutes. Samples of 5 are often used, the range estimator being computed on each sample and plotted on a time chart. The efficiency of the range estimate in a sample of 5 is 0.955, and the average of a series of ranges has the same efficiency. Often the purpose of the chart is to detect situations in which the industrial process becomes out of control, as revealed by the appearance of increasing or sudden high values of the estimated σ.

The estimate from the range is also an easy rough check on the computation of s. Programs on electronic computers are commonly used for routine computations. The programmer may not fully understand what is wanted or the instructions given to the computer may contain errors. There is therefore a need for quick checks on all standard statistical computations, which the investigator can apply when the output is received.

In the next three sections we present three alternatives to the t test for paired samples.

EXAMPLE 8.3.1—In a sample of size 2, with measurements X_1 and X_2, show that s is $|X_1 - X_2|/\sqrt{2} = 0.707|X_1 - X_2|$ and the range estimator is $0.886|X_1 - X_2|$. The reason for different multipliers is that the range estimator is constructed to be an unbiased estimator of σ, while s is not, as already mentioned.

EXAMPLE 8.3.2—The birth weights of 20 guinea pigs were 30, 30, 26, 32, 30, 23, 29, 31, 36, 30, 25, 34, 32, 24, 28, 27, 38, 31, 34, 30 g. Estimate σ (*i*) by use of the fraction 0.268 in table A 6 and (*ii*) by calculating s. Observe the time required to calculate s. Ans. (*i*) 4.0 g, (*ii*) 3.85 g.

EXAMPLE 8.3.3—In the preceding example, how many birth weights would be required to yield the same precision if the range were used instead of s? Ans. About 29 weights.

EXAMPLE 8.3.4—In a sample of 3 the values are, in increasing order, X_1, X_2, and X_3. The range estimate of σ is $0.591(X_3 - X_1)$. If you are ingenious at algebra, show that s always lies between $(X_3 - X_1)/2 = 0.5(X_3 - X_1)$, and $(X_3 - X_1)/\sqrt{3} = 0.577(X_3 - X_1)$. Verify the two extreme cases from the samples 0, 3, 6, in which $s = 0.5(X_3 - X_1)$ and 0, 0, 6, in which $s = 0.577(X_3 - X_1)$. Hint: Prove the identity:

$$2s^2 = \frac{(X_3 - X_1)^2}{2} + \frac{(X_3 + X_1 - 2X_2)^2}{6}$$

8.4—Sign test. This test applies to paired samples and is used in three situations:

1. Without a quantitative scale of measurement, the investigator can distinguish whether method A or method B is superior in any pair with respect to the objective of the study. The null hypothesis is that in the population of pairs giving a verdict, each method wins in half the pairs. This hypothesis might be chosen because it is plausible that A and B are equal in merit, any recorded superiority of one or the other being an error of measurement or judgment. A second hypothesis leading to the same test is that A and B have different effects, half the population preferring or performing better with A and half with B. We have already encountered this use of the test in section 7.8 when comparing proportions in paired samples—method A is superior in the SF pairs and method B in the FS pairs.

2. The sign test is sometimes used with a quantitative scale of measurement when the investigator doubts that the distribution of the differences between members of a pair is at all close to normal and hesitates to use the t test. The differences are replaced by their signs (+ or −); the sizes of the differences are ignored. Hence the name *sign test.*

3. A third use is as a quick substitute for the t test, particularly in a first look over the data.

As an example, 8 judges ranked the flavor of ground beef patties stored for eight months in two home freezers. One freezer was set at 0°F; the second freezer's temperature fluctuated between 0°F and 15°F. The results (*2*) are shown in table 8.4.1.

TABLE 8.4.1
RANKINGS OF FLAVOR OF GROUND BEEF PATTIES BY 8 JUDGES
(rank 1 is high; rank 2, low)

Judge	A	B	C	D	E	F	G	H
0°F	1	1	2	1	1	1	1	1
0°F to 15°F	2	2	1	2	2	2	2	2
+ or −	+	+	−	+	+	+	+	+

In the sample, 7 of 8 judges preferred the 0°F patty. If r of n pairs show one sign, then in testing the hypothesis H_0 that half the population pairs have this sign, the approximate normal deviate z_c (corrected for continuity) is as follows:

$$z_c = (|2r - n| - 1)/\sqrt{n} = 5/\sqrt{8} = 1.77 \tag{8.4.1}$$

Similarly,

$$\chi_c^2 = z_c^2 = (|2r - n| - 1)^2/n = 3.13 \tag{8.4.2}$$

The two-tailed significance level is $P = 0.078$, indicating nonsignificance. This type of result is common with small samples. From the sample, it looks at first sight as if the great majority of judges in the population would prefer the 0°F patties. However, the test result warns that because of the small sample the data do not provide convincing evidence that *any* majority of judges in favor of 0°F patties holds in the population.

When n is small, the exact significance probability is easily calculated from the terms of the binomial $(1/2 + 1/2)^n$. In the present example, the two-tailed probability is $2 \times 9/2^8 = 0.070$.

If a variate X has a continuous or discrete distribution, a null hypothesis leading to the sign test is that X has the same distribution under the two treatments. This null hypothesis does not need to specify the shape of the distribution. On the other hand, the t test assumes normality of the differences and specifies that the parameter δ (the population mean difference) is zero. For this reason, tests like the t test are sometimes called *parametric,* while the sign test is called *nonparametric.* Similarly, the median and other order statistics are nonparametric estimates, since they estimate percentiles of any continuous distribution without the shape of the distributions specifically defined by a formula and parameters.

Table A 8 shows the *smaller* number of like signs required for significance at the 1%, 2%, 5%, and 10% levels in two-tailed tests and the corresponding exact significance probabilities, which are usually well below the nominal levels.

In section 8.2 we introduced the useful concept of the relative efficiency of two estimators of a population parameter, applying it to the mean and the median as estimates of μ and to s and a multiple of the range as estimators of σ in normal samples. When the sign test is being considered as a quick substitute for the t test in normal samples, we can similarly speak of the relative efficiency of the two tests (sometimes called the relative *power efficiency*). In large normal samples, the efficiency of the sign test relative to the t test is about 64%. This means that if the null hypothesis is false, the sign test applied to a sample of 50 pairs has about the same power as a t test applied to (50)(0.64) = 32 pairs in detecting a given difference δ in the means of the two populations. In small samples the relative efficiency of the sign test increases (as well as depends to some extent on the significance level and the power). Near the 5% level, the relative efficiency is around 70% for 20 pairs, 80% for 10 pairs, and 90% for 5 pairs.

8.5—Substitute *t* test based on range. Lord (3) has developed an alternative to the t test in which the range replaces $s_{\bar{X}}$ in the denominator of t. This test is used in the same way as t for testing a hypothesis or making interval estimates. Pillai (4) has shown that for interval estimates the efficiency of this procedure relative to t stays above 95% in samples up to $n = 20$. Like t, the range test assumes a normal distribution. It has become popular, particularly in industrial work, though it is not resistant to outliers.

Table A 7(i) applies to single samples or to a set of differences obtained from paired samples. The entries are the values of $(\bar{X} - \mu)/w$, where w denotes the range of the sample. This ratio is called t_w, since it plays the role of t.

The test of a null hypothesis by means of t_w is illustrated by the paired samples in chapter 6 showing the numbers of lesions on the two halves of tobacco leaves under two preparations of virus. The 8 differences between the halves were 13, 3, 4, 6, −1, 1, 5, 1. Here the mean difference $\bar{D} = 4$, while $w = 14$ and $n = 8$. For the null hypothesis that the two preparations produce on the average equal numbers of lesions,

$$t_w = \bar{D}/w = 4/14 = 0.286$$

which is practically at the 5% level (0.288). The ordinary t test gave a significance probability of about 4%.

To obtain 95% confidence limits for the population mean difference δ by Lord's method, we have

$$4 \pm (14)(0.288) = 4 \pm 4.03$$

The limits are (−0.03, 8.03) as against somewhat narrower limits (0.41, 7.59) by the t distribution.

EXAMPLE 8.5.1—On being presented with a choice between two sweets, differing in color but otherwise identical, 15 out of 20 children chose color *B*. Test whether this is evidence of a general preference for *B*(*i*) by χ^2, (*ii*) by reference to table A 8. Do the results agree? Ans. Yes. *P* by $\chi^2 =$ 0.044. Exact $P = 0.041$.

EXAMPLE 8.5.2—Two ice creams were made with different flavors but otherwise similar. A panel of 6 expert dairy industry men all ranked flavor *A* as preferred. Is this statistical evidence that the consuming public will prefer *A?* Ans. $P = 0.0156$.

EXAMPLE 8.5.3—To illustrate the difference between the sign test and the *t* test in extreme situations, consider the two samples, each of 9 pairs, in which the actual differences are as follows. Sample I: −1, 1, 2, 3, 4, 4, 6, 7, 10. Sample II: 1, 1, 2, 3, 4, 4, 6, 7, −10. In both samples the sign test indicates significance at the 5% level, with $P = 0.039$ from table A 8. In sample I, in which the negative sign occurs for the smallest difference, we find $t = 3.618$ with 8 df, the significance probability being 0.007. In sample II, where the largest difference is the one with the negative sign, $t = 1.212$ with $P = 0.294$. Verify that Lord's test shows $t_w = 0.364$ for sample I and 0.118 for sample II and gives verdicts in good agreement with the *t* test. When the aberrant signs represent extreme observations, the sign test and the *t* test do not agree well. This does not necessarily mean that the sign test is at fault; if the extreme observation were caused by an undetected gross error, the verdict of the *t* test might be misleading.

EXAMPLE 8.5.4—The differences in serum albumen found by two methods *A* and *B* in 8 blood samples were 0.6, 0.7, 0.8, 0.9, 0.3, 0.5, −0.5, 1.3 g/100 ml. Apply the range method to test the null hypothesis that there is no consistent difference in the amount of serum albumen found by the two methods. Ans. $t_w = 0.32$, $P < 0.05$.

8.6—Signed-rank test. The *signed-rank* test, due to Wilcoxon, is another substitute for the *t* test in paired samples. First, the absolute values (ignore signs) of the differences are ranked, the *smallest* difference being assigned rank 1. Then the signs are restored to the rankings. The method is illustrated from an experiment by Collins et al. (6). One member of a pair of corn seedlings was treated by a small electric current; the other was untreated. After a period of growth, the differences in elongation (treated − untreated) are shown for each of 10 pairs.

In table 8.6.1 the ranks with the negative signs total 15 and those with positive signs total 40. The test criterion is the *smaller* of these totals, in this case, 15. The ranks with the less frequent sign will usually, though not always,

TABLE 8.6.1
EXAMPLE OF WILCOXON'S SIGNED-RANK TEST
(differences in elongation: treated − untreated seedlings)

Pair	Difference (mm)	Signed Rank
1	6.0	5
2	1.3	1
3	10.2	7
4	23.9	10
5	3.1	3
6	6.8	6
7	− 1.5	−2
8	−14.7	−9
9	− 3.3	−4
10	11.1	8

give the smaller rank total. This number, sign ignored, is referred to table A 9. For 10 pairs a rank sum ≤ 8 is required for rejection at the 5% level. Since $15 > 8$, the data support the null hypothesis that elongation was unaffected by the electric current treatment.

In a signed-rank test, when will the sum of the ranks with negative signs be small and cause rejection? First, when there are none or very few negative signs, that is, when the treated seedling wins in nearly all pairs. Second, when the negative ranks that are present are small, implying that the untreated seedling wins by only small amounts in such pairs. We interpret the rejection of H_0 as meaning that on the whole the treatment has stimulated growth.

The null hypothesis in this test is that the frequency distribution of the original measurements is the same for the treated and untreated members of a pair, but as in the sign test the shape of this frequency distribution need not be specified. A consequence of this null hypothesis is that each rank is equally likely to be + or −. On this assumption, the frequency distribution of the smaller rank sum was worked out by Wilcoxon. Since this distribution is discontinuous, the significance probabilities for the entries in table A 9 are not exactly at the stated levels but are close enough for practical purposes.

If the two or more differences are equal, it is often sufficiently accurate to assign to each of the ties the average of the ranks that would be assigned to the group. Thus, if two differences are tied in the fifth and sixth positions, assign rank 5 1/2 to each of them.

If the number of pairs n exceeds 16, calculate the approximate normal deviate, corrected for continuity,

$$Z_c = (\mu - T - 0.5)/\sigma \qquad (8.6.1)$$

where T is the smaller rank sum, and

$$\mu = n(n + 1)/4 \qquad \sigma = \sqrt{(2n + 1)\mu/6} \qquad (8.6.2)$$

The number -0.5 is a correction for continuity. As usual, $Z_c > 1.96$ signifies rejection at the 5% level in a two-tailed test.

EXAMPLE 8.6.1—From skew distributions, two samples of $n = 10$ were drawn and paired at random:

Sample 1	1.98	3.30	5.91	1.05	1.01	1.44	3.42	2.17	1.37	1.13
Sample 2	0.33	0.11	0.04	0.24	1.56	0.42	0.00	0.22	0.82	2.54
Difference	1.65	3.19	5.87	0.81	−0.55	1.02	3.42	1.95	0.55	−1.41
Rank	6	8	10	3	−1.5	4	9	7	1.5	−5

The difference $\delta = \mu_1 - \mu_2$ between the population means is 1. Apply the signed-rank test. Ans. The smallest two absolute differences are tied, so each is assigned the rank $(1 + 2)/2 = 1.5$. The sum of the negative ranks is 6.5. This lies between the critical sums 5 and 8 in table A 9; H_0 is rejected with $P = 0.04$, approximately.

EXAMPLE 8.6.2—If you had not known that the differences in the foregoing example were from a nonnormal population, you might have applied the t test. Would you have drawn any different conclusions? Ans. $t = 2.48$, $P = 0.04$.

EXAMPLE 8.6.3—Apply the signed-rank test to samples I and II of example 8.5.3. Verify that the results agree with those given by the t test and not with those given by the sign test. Is this what you would expect?

EXAMPLE 8.6.4—For 16 pairs, table A 9 states that the 5% level of the smaller rank sum is 30, the exact probability being 0.051. Check the normal approximation in this case by showing that in (8.6.2), $\mu = 68$, $\sigma = 19.34$, so for $T = 30$ the value of Z_c is 1.94, corresponding to a significance probability of 0.052.

We come now to two alternatives to the t test in comparisons between the means of two independent samples.

8.7—Lord's test for two independent samples. In this test the sample ranges w_1, w_2 replace s_1, s_2. This is a quick test, no more robust under nonnormality than the t test, and even more vulnerable to erroneous sample extreme values.

Table A 7(*ii*) applies to two independent samples of *equal* size. The mean of the two ranges, $\overline{w} = (w_1 + w_2)/2$, replaces the w of the paired test and $\overline{X}_1 - \overline{X}_2$ takes the place of $\overline{D}$.

The test of significance is applied to the numbers of worms found in two samples of 5 rats, one sample treated previously by a worm-killer. See table 8.7.1. We have $\overline{X}_2 - \overline{X}_1 = 171.8$ and $\overline{w} = (219 + 147)/2 = 183$. From this, $t_w = (\overline{X}_2 - \overline{X}_1)/\overline{w} = 171.8/183 = 0.939$, which is beyond the 1% point, 0.896, shown in table A 7(*ii*) for $n = 5$.

TABLE 8.7.1
NUMBER OF WORMS PER RAT

	Treated	Untreated
	123	378
	143	275
	192	412
	40	265
	259	286
Means, $\overline{X}$	151.4	323.2
Ranges, w	219	147

To find 95% confidence limits for the reduction in number of worms per rat due to the treatment, we use the formula

$$(\overline{X}_2 - \overline{X}_1) - t_w\overline{w} \le \mu_2 - \mu_1 \le (\overline{X}_2 - \overline{X}_1) + t_w\overline{w}$$
$$171.8 - (0.613)(183) \le \mu_2 - \mu_1 \le 171.8 + (0.613)(183)$$
$$60 \le \mu_2 - \mu_1 \le 284$$

The confidence interval is wide, owing both to the small sample sizes and the high variability from rat to rat. Student's t, used in example 6.8.3 for these data, gave closely similar results for the significance level and the confidence limits.

For two independent samples of unequal sizes, Moore (7) has given tables for the 10%, 5%, 2%, and 1% levels of Lord's test to cover all cases in which the sample sizes n_1 and n_2 are both 20 or less.

The range method can also be used when the sample size exceeds 20. With two samples each of size 24, for example, each sample may be divided at random into two groups of size 12. The range is found for each group, and the average of the four ranges is taken. Lord (3) gives the necessary tables. This device keeps the efficiency of the range test high for samples greater than 20, though the calculation takes a little longer.

To summarize, the range test is convenient for normal samples if a 5% to 10% loss in information can be tolerated. It is often used when many routine tests of significance or calculations of confidence limits have to be made.

EXAMPLE 8.7.1—In this example, given by Lord (3), the data are the times taken for an aqueous solution of glycerol to fall between two fixed marks. In five independent determinations in a viscometer, these times were 103.5, 104.1, 102.7, 103.2, and 102.6 s. For satisfactory calibration of the viscometer, the mean time should be accurate to within $\pm 1/2$ s, apart from a 1-in-20 chance. By finding the half-width of the 95% confidence interval for μ by (*i*) the t_w method and (*ii*) the t method, verify whether this requirement is satisfied. Ans. No. Both methods give $\pm$ 0.76 for the half-width.

EXAMPLE 8.7.2—The crushing resistance of 15 kernels of corn ranged from 25 to 65 lb with a mean of 43.0. Another sample of 15 kernels, harvested at a different stage, ranged from 29 to 67 with a mean of 48.0. Test whether the difference between the means is significant. Ans. No, $t_w = 0.128$. Note that since the ranges of the two samples indicate much overlap, one can guess that the test will not show a significant difference.

8.8—The rank sum test for two independent samples. This test was also developed by Wilcoxon (5), though it is often called the Mann-Whitney test (8). A table due to White (9) gives two-tailed 5% and 1% significance levels for samples of sizes n_1, n_2, where $n_1 + n_2 \leq 30$. All observations in both samples are put into a single array in increasing order, care being taken to tag the numbers of each sample so that they can be distinguished. Ranks are then assigned to the combined array, the smallest observation being given rank 1. Finally, the smaller sum of ranks, T, is referred to table A 10 to determine significance. Note that *small* values of T cause rejection.

An example is drawn from the corn borer project in Boone County, Iowa. It is well established that in an attacked field more eggs are deposited on tall plants than on short ones. For illustration we took records of numbers of eggs found in 20 plants in a rather uniform field. The plants were in two randomly selected sites, 10 plants each. Table 8.8.1 contains the egg counts.

TABLE 8.8.1
NUMBER OF CORN BORER EGGS ON CORN PLANTS, BOONE COUNTY, IOWA, 1950

Height of Plant	Number of Eggs									
Less than 23 in.	0	14	18	0	31	0	0	0	11	0
More than 23 in.	37	42	12	32	105	84	15	47	51	65

In years such as 1950 the frequency distribution of number of eggs tends to be U- or J-shaped rather than normal. At the low end, many plants have no eggs; but there is also a group of heavily infested plants. Normal theory cannot be relied on to yield correct inferences from small samples.

TABLE 8.8.2

EGG COUNTS ARRANGED IN INCREASING ORDER, WITH RANKS
(boldface type indicates counts on plants 23 in. or more)

Count	0	0	0	0	0	0	11	**12**	14	**15**	18	31
Rank	3½	3½	3½	3½	3½	3½	7	**8**	9	**10**	11	12

For convenience in assigning ranks, the counts were rearranged in increasing order (table 8.8.2). The counts for the tall plants are in boldface type. The eight highest counts are omitted from table 8.8.2, since they are all on tall plants and it is clear that the small plants give the smaller rank sum.

By the rule suggested for tied ranks, the six ties are given the rank 3 1/2, this being the average of the numbers 1 to 6. In this instance the averaging is not necessary, since all the tied ranks belong to one group; the sum of the six ranks, 21, is all that we need. But if the tied counts were in both groups, averaging would be required.

With $n_1 = n_2$, the next step is to add the n_1 rank numbers in the group (plants less than 23 in.) that has the smaller sum. We find

$$T = 21 + 7 + 9 + 11 + 12 = 60$$

This sum is referred to table A 10 with $n_1 = n_2 = 10$. Since T is less than $T_{0.01} = 71$, the null hypothesis is rejected with $P \leq 0.01$. The anticipated conclusion is that plant height affects the number of eggs deposited.

When the samples are of unequal sizes n_1, n_2, an extra step is required. First, find the total T_1 of the ranks for the sample that has the *smaller* size, say n_1. Compute $T_2 = n_1(n_1 + n_2 + 1) - T_1$. Then T, which is referred to table A 10, is the smaller of T_1 and T_2. To illustrate, White quotes Wright's data (12) on the survival times, under anoxic conditions, of the peroneal nerves of 4 cats and 14 rabbits. For the cats, the times were 25, 33, 43, and 45 min; for the rabbits, 15, 16, 16, 17, 20, 22, 22, 23, 28, 28, 30, 30, 35, and 35 min. The ranks for the cats are 9, 14, 17, and 18, giving $T_1 = 58$. Hence, $T_2 = 4(19) - 58 = 18$ and is smaller than T_1, so $T = 18$. For $n_1 = 4$, $n_2 = 14$, the 5% level of T is 19. The mean survival time of the nerves is significantly higher for the cats than for the rabbits.

For values of n_1 and n_2 outside the limits of the table, calculate

$$Z = (|\mu - T| - 0.5)/\sigma \tag{8.8.1}$$

where

$$\mu = n_1(n_1 + n_2 + 1)/2 \qquad \sigma = \sqrt{n_2\mu/6} \tag{8.8.2}$$

The approximate normal deviate Z is referred to the tables of the normal distribution to give the significance probability P.

Table A 10 was calculated from the assumption that if the null hypothesis is true, the n_1 ranks in the smaller sample are a random selection from the $n_1 + n_2$ ranks in the combined samples. In rank sum tests, the specific nature of the alternative hypothesis has not been stated. We usually interpret rejection of H_0 to imply that one population has a higher median.

8.9—Comparison of rank and normal tests. When the t test is used on nonnormal data, two things happen. The significance probabilities are changed; the probability that t exceeds $t_{0.05}$ when the null hypothesis is true is no longer 0.05, but may be, say, 0.041 or 0.097. Second, the sensitivity or power of the test in finding a significant result when the null hypothesis is false is altered. Much of the work on nonparametric tests is motivated by a desire to find robust tests whose significance probabilities do not change and whose power relative to competing tests remains high when the data are nonnormal.

The significance levels in rank tests remain the same for any continuous distribution except that they are affected to some extent by ties and by zeros in the signed-rank test. In large normal samples, the rank tests have an efficiency of about 95% relative to the t test (10); and in small normal samples, the signed-rank test has been shown to have an efficiency slightly higher than this (11). With nonnormal data from a continuous distribution, the efficiency of the rank tests relative to t never falls below 86% in large samples and may be much greater than 100% for distributions that have long tails (10). Since they are relatively quickly made, the rank tests are highly useful for the investigator who is doubtful that the data can be regarded as normal.

The beginner may wish to compute both the rank tests and the t test for some data to see how they compare. Needless to say, the practice of quoting the test that agrees with one's predilections vitiates the whole technique.

As has been stated previously, after the preliminary stages most investigations are designed to estimate the sizes of differences in means rather than simply to test null hypotheses. The rank methods can furnish estimates and confidence limits for the difference between two treatments (see example 8.9.3). The calculations require no assumption of normality but are more tedious than with normal data. Some work has also been done in extending rank methods to the more complex types of data that we meet in following chapters, though the available techniques still fall short of the flexibility of the standard methods based on normality.

EXAMPLE 8.9.1—The normal approximation (8.8.1) for the rank sum test works quite well even within the limits of table A 10. In the example with cats and rabbits in section 8.8, show that the normal approximation gives $z_c = 2.071$, with $P = 0.038$ (two-tailed). The exact probability is found to be 0.035.

EXAMPLE 8.9.2—Apply the rank sum test to the data in table 6.9.1 used to illustrate the t test for independent samples of unequal sizes. Ans. With $T = 49.5$, table A 10 gives $P > 0.05$. The normal approximation to table A 10 gives $P = 0.091$, while the t test gave $P = 0.076$.

EXAMPLE 8.9.3—In Wright's data, section 8.8, show that if the survival time for each cat is reduced by 2 min, the value of T in the signed-rank test becomes 18 1/2, while if the cat survival times are reduced by 3 min, $T = 21$. Show further that if 23 min are subtracted from each cat, we find $T = 20$ 1/2, while for 24 min, $T = 19$. Since $T_{0.05} = 19$, any hypothesis stating that the average survival time of cats exceeds that of rabbits by a figure between 3 and 23 min is accepted in a 5% test. The limits 3 and 23 min are conservative 95% confidence limits as found from the rank sum test.

8.10—Discrete scales with limited values: randomization test. In some lines of work the scales of measurement are restricted to a small number of values, perhaps to $-1, 0, 1$ or 1, 2, 3, 4, 5. Persons who favor a proposal might be recorded as 1, those who oppose it as -1, and those who are neutral as 0—an extension of the binomial to a trinomial scale. Investigators are sometimes puzzled as to how to test the mean differences between two groups in this case, because the data do not look normal and rank tests may involve numerous tied ranks. We suggest that the t test, with the inclusion of a correction for continuity, be used as an approximation. To illustrate, consider a paired comparison in which the original data are on a 0, 1, 2 scale. The differences between the members of a pair can then assume only the values 2, 1, 0, -1, and -2.

With $n = 12$ pairs, suppose that the differences D_i between the responses to two treatments A and B are 2, 2, 2, 1, 1, 1, 0, 0, 0, 0, -1, -1. The $\Sigma D_i = 7$. On the null hypothesis that A and B have the same effect, the signs of the D_i are equally likely to be + or $-$. To apply the correction for continuity, find the next highest distinct value of ΣD_i that alteration of any signs could produce. This value is $\Sigma D_i = 5$, which would be obtained if three 1's instead of two had a minus sign. Since 6 is halfway between 5 and 7, we compute t_c as

$$t_c = (|\Sigma D| - 1)/(\sqrt{n}S_D) = 6/[(\sqrt{12})(1.084)] = 1.598 \tag{8.10.1}$$

where $S_D = 1.084$ is computed in the usual way. With 11 df, P is 0.138, not significant.

In applying the correction for continuity, the general rule for paired samples is to find the next highest distinct value of $|\Sigma D|$ that the $\pm$ signs provide. The numerator of t_c is halfway between this value and the observed $|\Sigma D|$. The amount subtracted from $|\Sigma D|$ is often 1 as in (8.10.1), but not always, so care is necessary when making the correction.

For such problems, an alternative robust test is Fisher's *randomization test* (13). This test requires no assumption about the nature of the distribution of the differences D_i. The test uses the argument already mentioned—if there is no difference in effect between treatments A and B, each of the 12 differences is equally likely to be + or $-$. Thus, under H_0 there are $2^{12} = 4096$ possible sets of sample results in our example. Since, however, +0 and -0 are the same, only $2^8 = 256$ cases need be examined. For a two-tailed test, we count the samples in the set with $|\Sigma D| \geq 7$, the observed $|\Sigma D|$. It is not hard to verify that a two-tailed test has 38 samples of this kind. The significance probability for a two-tailed test is therefore $38/256 = 0.148$, in good agreement with the approximate t_c test.

With two independent samples, the randomization test assumes that on H_0 the $n_1 + n_2$ observations have been divided at random into two samples of sizes n_1 and n_2. There are $(n_1 + n_2)!/(n_1!n_2!)$ cases in the randomization set of possible results. As a simple example with $n_1 = n_2 = 4$, let one sample have the values $X_{1j} = 2, 3, 3, 3$ and the other $X_{2j} = 0, 0, 0, 3$. Then the difference between the sample totals $X_{1.} - X_{2.} = 11 - 3 = 8$. (The subscript dot denotes summation over the values of the subscript.)

In the randomization test, the only arrangement giving a larger difference between the sample totals is $X_{1j} = 3, 3, 3, 3$; $X_{2j} = 0, 0, 0, 2$, and, of course, $X_{1j} = 0, 0, 0, 2$; $X_{2j} = 3, 3, 3, 3$ if the test is two-tailed. Also, there are eight two-tailed arrangements that give the same absolute total difference 8, as was observed, since any of the four 3s can be in the sample with the smaller total. Hence, the two-tailed randomization test gives $P = 10/70 = 0.143$.

To check the t_c test against this value, the arrangement giving the next highest value of $X_{1.} - X_{2.}$ is 0, 3, 3, 3 and 0, 0, 2, 3, with $X_{1.} - X_{2.} = 4$. Hence

$$t_c = (|X_{1.} - X_{2.}| - c)/\sqrt{ns_1^2 + ns_2^2} = 6/(2\sqrt{2.50}) = 1.897 \qquad \text{(6df)}$$

The significance probability in a two-tailed test is $P = 0.107$.

With small samples that show little overlap, the randomization test is easily made and is recommended because t_c tends to give too low P values and therefore too many apparently significant results. For instance, with sample values 2, 3, 3, 3 and 0, 0, 0, 2, the two-tailed randomization test gives $P = 4/70 = 0.057$, while $t_c = 3.58$ with 6df, and $P = 0.012$.

EXAMPLE 8.10.1—In a paired two-sample test the 10 values of the differences D_i were 3, 3, 2, 1, 1, 1, 1, 0, 0, −1. Show that the randomization test gives $P = 6/256 = 0.023$ in a one-tailed test, while $t_c = 10/(1.2867\sqrt{10}) = 2.4577$, giving $P = 0.036$.

EXAMPLE 8.10.2—As an example of independent samples of unequal sizes 8, 4, consider the data: $X_{1j} = 3, 3, 3, 3, 3, 2, 2, 1$; $X_{2j} = 2, 1, 1, 1$. The sample means are $\overline{X}_1 = 2.5$, $\overline{X}_2 = 1.25$. To apply the randomization test, select the sample with the higher mean (sample 1) and count the proportion of samples in the randomization set that give a mean as large or larger. Verify that in this example the proportion is $13/495 = 0.026$ in a one-tailed test. With t_c it is not obvious which quantity to correct for continuity. We might consider t itself, but it is simpler and perhaps closer to the randomization test to correct $\overline{X}_1 - \overline{X}_2$. Show that the next highest value of $\overline{X}_1 - \overline{X}_2$ is 0.875 and $t_c = 1.0625/0.4220 = 2.5175$ with 10 df, giving $P = 0.030$; $t_c = (\overline{X}_1 - \overline{X}_2 - c)/(s\sqrt{1/8 + 1/4})$.

TECHNICAL TERMS

median
nonparametric
order statistics
outliers
parametric
power efficiency
randomization test
range
rank sum (Mann-Whitney) test
relative efficiency
resistant
robust
signed-rank test
sign test
substitute t test

REFERENCES

1. Mood, A. M., and Graybill, F. A. 1963. *Introduction to the Theory of Statistics,* 2nd ed. McGraw-Hill, New York, p. 408.
2. Ehrenkrantz, F., and Roberts, H. *J. Home Econ.* 44 (1952):441.
3. Lord, E. *Biometrika* 34 (1947):56.
4. Pillai, K. C. S. *Ann. Math. Stat.* 22 (1951):469.
5. Wilcoxon, F. *Biom. Bull.* 1 (1945):80.
6. Collins, G. N., et al. *J. Agric. Res.* 38 (1929):585.
7. Moore, P. G. *Biometrika* 44 (1957):482.
8. Mann, H. B., and Whitney, D. R. *Ann. Math. Stat.* 18 (1947):50.
9. White, C. *Biometrics* 8 (1952):33.
10. Hodges, J. L., and Lehmann, E. L. *Ann. Math. Stat.* 27 (1946):324.
11. Klotz, J. *Ann. Math. Stat.* 34 (1963):624.
12. Wright, E. B. *Am. J. Physiol.* 147 (1946):78.
13. Fisher, R. A. 1966. *The Design of Experiments,* 8th ed. Oliver & Boyd, Edinburgh, p. 21.
14. Kendall, M. G., and Stuart, A. 1969. *The Advanced Theory of Statistics,* vol I, 3rd ed. Charles Griffin, London, p. 327.

9

Regression

9.1—Introduction. In preceding chapters the problems considered involve only a single measurement on each individual. In this chapter we begin the study of the relationship of one variable Y to another variable X. In mathematics Y is called a *function* of X, but in statistics the term *the regression of Y on X* is generally used to describe the relationship. The growth curve of height is spoken of as the regression of height on age; in toxicology the lethal effects of a drug are described by the regression of the percent kill on the amount of the drug. The historical origins of the term regression are explained in section 9.13. To distinguish the two variables in regression studies, Y is often called the *dependent* and X the *independent* variable. These terms are fairly appropriate in the toxicology example, in which we think of the percent kill Y as being caused by the amount of drug X, the amount itself being variable at the will of the investigator. Though still widely used, they are not suitable and can even confuse when, for example, Y is the weight of a man and X his maximum girth.

Regression has many uses. The objectives may be to determine if there is a relationship between Y and X, to study the shape of the curve of relationship, and to think about the reasons for the relationship. Prediction of Y from X may be the goal. Often the purpose is to estimate the change in Y to be expected from a given increase in X in cases where we are convinced that changes in Y are in the main caused by changes in X. Sometimes we use the relationship to adjust the values of Y before comparison of two groups that differ in their values of X. To provide for these various uses an extensive account of regression methods is necessary.

In the next section the calculations used in fitting a regression are introduced by a numerical example. The theoretical basis of these calculations and useful applications of regression are discussed in subsequent sections.

9.2—Calculations for fitting a linear regression. The data come from a controlled experiment (1) on the effects of increasing amounts of a mixture of the standard crop fertilizers on the yields of potatoes that was carried out in 1933 at the Midland Agricultural College in England. The mixture contained 1 part of sulphate of ammonia, 3 parts of superphosphate, and 1 part of sulphate of potash. The amounts were 0, 4, 8, and 12 cwt per acre, the cwt unit, called hundredweight, being actually 112 lb. Owing to natural variability, the yields of

TABLE 9.2.1
YIELDS OF POTATOES RECEIVING DIFFERENT AMOUNTS OF FERTILIZERS

	Amount X_i	Yield Y_i	Deviations from Means x_i	y_i	Squares x_i^2	y_i^2	Product $x_i y_i$
	0	8.34	−6	−0.63	36	0.3969	+3.78
	4	8.89	−2	− .08	4	.0064	+0.16
	8	9.16	+2	+ .19	4	.0361	+0.38
	12	9.50	+6	+0.53	36	0.2809	+3.18
Total	24	35.89	0	+0.01	80	0.7203	+7.50
Mean	6	8.97					
		$\Sigma x^2 = 80$	$\Sigma y^2 = 0.7203$		$\Sigma xy = 7.50$		

potatoes under a given amount of the mixture vary from plot to plot. The yield figures shown for each amount in table 9.2.1 are the means of random samples of four plots.

In this type of experiment, any consistent changes in the mean yields are regarded as due primarily to the changes in the amounts of fertilizer given. The objectives of the statistical analysis are to estimate the shape of the response curve of yield on amount of fertilizer and, from this and similar experiments, to recommend the most profitable amount of fertilizer to supply to the crop in practice. The symbol X_i denotes the amount of fertilizer (cwt per acre) and Y_i the corresponding mean yield of potatoes (tons per acre). As usual, $x = X - \overline{X}$ and $y = Y - \overline{Y}$ denote deviations from the sample means.

The first thing to do is to graph Y_i against X_i, figure 9.2.1. The independent variable X_i is plotted along the horizontal axis. Each mean yield is indicated by a black dot above the corresponding X_i. Clearly, the trend of yield is upward and

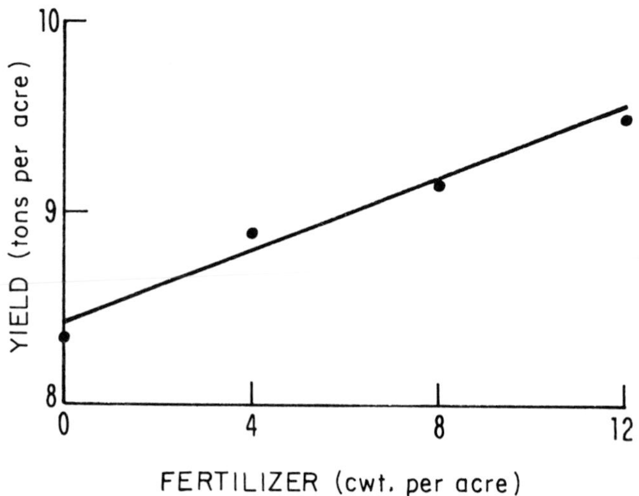

Fig. 9.2.1—Plot of yields of potatoes against the amount of fertilizer applied.

roughly linear. The straight line drawn in the figure is the *sample regression of Y on X*, sometimes called the fitted line. It is computed as follows.

Let the equation of the sample regression be written

$$\hat{Y} = a + b(X - \overline{X}) = a + bx \qquad (9.2.1)$$

where $\hat{Y}$ is the estimated yield of potatoes given by an amount X of the fertilizer. Then the sample estimates a and b are

$$\mathrm{a} = \overline{Y} = 8.97 \qquad (9.2.2)$$

$$b = \frac{\Sigma x_i y_i}{\Sigma x_i^2} = \frac{\Sigma X_i Y_i - (\Sigma X_i)(\Sigma Y_i)/n}{\Sigma X_i^2 - (\Sigma X_i)^2/n} = \frac{+7.50}{80} = +0.09375 \qquad (9.2.3)$$

As will be seen, a and b are often described as *least squares* estimates.

The numerator of b in (9.2.3) is a new quantity—the sum of products of the deviations x and y. In table 9.2.1 the individual values of x_i^2 and $x_i y_i$ have been obtained and summed. As (9.2.3) shows, the numerator and denominator of b can be calculated without finding deviations by the usual device of subtracting the contributions due to the means. In these data, $b = +0.09375$, meaning that yield increases on the average by 0.09375 ton per acre for each cwt per acre increase in the amount of fertilizer. When thinking about or presenting regression coefficients, note that two scales must be reported—the scale in which Y is measured and that in which X is measured.

From (9.2.1), with $a = \overline{Y}$, the fitted line may be written, alternatively,

$$\hat{Y} = (\overline{Y} - b\overline{X}) + bX \qquad (9.2.4)$$

In (9.2.4), $(\overline{Y} - b\overline{X})$ is the estimated intercept of the sample regression line on the Y-axis. Some pocket calculators have programs that fit linear regression in the form (9.2.4) by pressing keys marked "slope" and "intercept" (or a contraction) after feeding in the data. They also give $\hat{Y}$ for any desired X. For the potato data, this form of the sample line is

$$\hat{Y} = 8.97 - (0.09375)(6) + 0.09375X = 8.41 + 0.09375X$$

The line can be drawn in figure 9.2.1 by noting that when $X = 0$, $\hat{Y} = 8.41$ and when $X = 12$, $\hat{Y} = 8.41 + (0.09375)(12) = 9.54$. Plot the points (0, 8.41), (12, 9.54) and join them with a straight line.

The expression for the least squares b, $\Sigma xy/\Sigma x^2$, is unfamiliar at first sight. You have been told that b estimates the average change in Y associated with unit increase in X. But b is not obviously the estimate of this quantity that common sense would suggest to someone unfamiliar with regression theory. In the present example, the following way of looking at the problem may make the rationale of the least squares b more familiar. Table 9.2.2, derived from table 9.2.1, shows the increase in yield of potatoes that accompanies each additional 4 cwt of fertilizers.

Table 9.2.2 provides three direct estimates (0.55, 0.27, and 0.34) of the

TABLE 9.2.2
YIELDS AND INCREASES IN YIELDS OF POTATOES

Amount, X (cwt/acre)	0	4	8	12
Yield, Y (T/acre)	8.34	8.89	9.16	9.50
Increase in yield		+0.55	+0.27	+0.34

increase in yield that accompanies each additional 4 cwt of fertilizers. In estimating the average increase in yield per cwt of fertilizer, a commonsense method is to take the average 0.387 of these three figures and divide by 4, which gives $0.387/4 = 0.0967$. This figure is very close but not quite equal to $b = 0.0938$. Why the difference?

Suppose that we label the yields Y_0, Y_4, Y_8, Y_{12}. The preceding commonsense estimate is

$$[(Y_4 - Y_0) + (Y_8 - Y_4) + (Y_{12} - Y_8)]/12 = (Y_{12} - Y_0)/12$$

Thus, contrary to first appearances, this estimate does not use the data from Y_4 and Y_8 at all. These two values provide an estimate $(Y_8 - Y_4)/4$ that is independent of the commonsense estimate $(Y_{12} - Y_0)/12$ and supplies some additional information.

The least squares b uses both of these estimates. To see this, first verify that $b = \Sigma xy/\Sigma x^2$ may be written $b = \Sigma xY/\Sigma x^2$, since the term $\Sigma x\bar{Y}$ vanishes because $\Sigma x = 0$. In this problem the values of x_i are $-6, -2, +2, +6$, from table 9.2.1, and $\Sigma x^2 = 80$. Hence, b may be written

$$b = \frac{-6Y_0 - 2Y_4 + 2Y_8 + 6Y_{12}}{80} = \frac{9}{10}\left(\frac{Y_{12} - Y_0}{12}\right) + \frac{1}{10}\left(\frac{Y_8 - Y_4}{4}\right)$$

Thus, the least squares b is a weighted mean of the two simple and independent estimates of the increase in yield per unit increase in fertilizer. The estimate $(Y_{12} - Y_0)/12$ receives 9 times the weight of the other estimate because it is much more precise (see example 9.2.4).

EXAMPLE 9.2.1—The following are measurements of heights of soybean plants in a field—a different random selection each week (2).

Age, X (wk)	1	2	3	4	5	6	7
Height, Y (cm)	5	13	16	23	33	38	40

Plot the points. The increase in height per week, while irregular, shows no trend with age. Fit and plot the sample regression and note that the points lie about equally above and below the line. Ans. $\hat{Y} = 6.143X - 0.572$ cm.

EXAMPLE 9.2.2—The winning speeds (mph) in the Indianapolis 500 auto races from 1962 to 1971 were as follows. To save space, the years X have been coded with 1962 = 1 and 1971 = 10, while 100 mph has been subtracted from each speed Y.

Year, X	1	2	3	4	5	6	7	8	9	10
Speed, Y	40.3	43.1	47.4	51.4	44.3	51.2	52.9	56.9	55.7	57.7

(*i*) Satisfy yourself that the regression appears linear rather than curved and fits the least squares line. While extrapolation is risky, what speeds does the regression line estimate for (*ii*) 1958, (*iii*) 1974? Ans. (*i*) $\hat{Y} = 139.97 + 1.841x$ (with *Y*s decoded); (*ii*) $\hat{Y} = 134.4$, actual speed 135.9 mph; (*iii*) $\hat{Y} = 163.9$, actual speed 158.6 mph.

EXAMPLE 9.2.3—What estimate would you give for the average increase in winning speed from (*i*) the results for years 1 and 10 only, (*ii*) the results for years 2 and 9 only? Ans. (*i*) 1.93 mph/yr, (*ii*) 1.80 mph/yr; both are quite close to the least squares estimate.

EXAMPLE 9.2.4—If the yields in table 9.2.1 are independent, with variance σ^2, show that the estimate $(Y_8 - Y_4)/4$ of the increase in yield per added cwt of fertilizers has variance $\sigma^2/8$, while the estimate $(Y_{12} - Y_0)/12$ has variance $\sigma^2/72$, only 1/9 as large.

EXAMPLE 9.2.5—In a controlled experiment on the effects of increasing amounts of mixed fertilizers on sugar beets conducted at Redbourne, Lincs, England, in 1933, the mean yields of sugar beet roots and tops ($n = 5$) for each amount *X* are as follows:

X (cwt/acre)	0	4	8	12	16
Roots (T/acre)	14.42	15.31	15.62	15.94	15.76
Tops (T/acre)	7.48	8.65	9.74	11.00	11.65

Verify, either by plotting or by examining the successive increases to each higher dose, that the response in roots appears curved (higher amounts of fertilizers produce smaller increases in yield) but that the response in tops appears linear. Show that in roots the sample linear regression $\hat{Y}$ overestimates *Y* at the lowest and highest amounts of fertilizer and underestimates *Y* at the three middle amounts. This indication is typical for fitting a curved line (discussed later). For tops, estimate the average increase in yield per additional cwt per acre of fertilizer. Ans. 0.267 T/acre.

9.3—The mathematical model in linear regression.

In standard linear regression, three assumptions are made about the relation between *Y* and *X*:

1. For each specific *X* there is a normal distribution of *Y* from which sample values of *Y* are assumed drawn at random.
2. The normal distribution of *Y* corresponding to a specific *X* has a mean $\mu_{Y.X}$ that lies on the straight line

$$\mu_{Y.X} = \alpha + \beta(X - \overline{X}) = \alpha + \beta x \qquad (9.3.1)$$

This line is called the population regression line. The parameter α is the mean of the population of *Y* values that corresponds to $X = \overline{X}$. The parameter β is the slope of the line, the average change in *Y* per unit increase in *X*.

3. The normal distributions of *Y* for specific *X* are independent and all have the same variance $\sigma^2_{y.x}$.

This mathematical model is described concisely by the equation

$$Y = \alpha + \beta x + \epsilon \qquad (9.3.2)$$

where ϵ is a random variable drawn from $\mathcal{N}(0, \sigma^2_{y.x})$. Thus, as in the potato example (section 9.2), individual observations *Y* do not in general lie on the line. The model assumes that their deviations ϵ from the line are normally distributed with means 0 and constant variance $\sigma^2_{y.x}$.

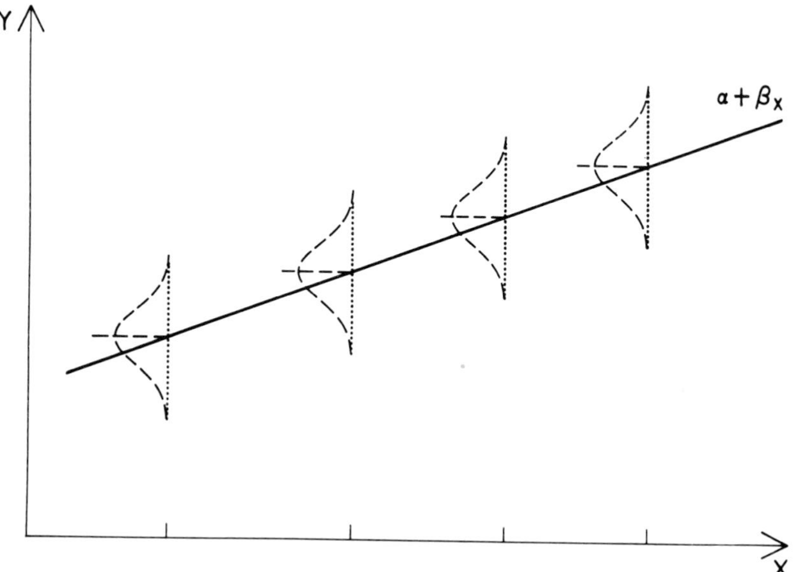

Fig. 9.3.1—Representation of the linear regression model. The normal distribution of Y about the regression line $\alpha + \beta x$ is shown for four selected values of X.

Figure 9.3.1 gives a schematic representation of these populations. For each of four selected values of X the normal distribution of Y about its mean $\mu = \alpha + \beta x$ is sketched. (Strictly, the ordinates of the normal curves should be at right angles to the page, representing a third dimension—frequencies.) These normal distributions would all coincide if their means were superimposed.

With this model we can think of the sample regression coefficient b as an estimate of the population parameter β. We know from the end of section 9.2 that b can be written $\Sigma x_i Y_i / \Sigma x_i^2$. Hence, the relation between b and β is

$$b = \frac{\Sigma x_i(\alpha + \beta x_i + \epsilon_i)}{\Sigma x_i^2} = \beta \frac{\Sigma x_i^2}{\Sigma x_i^2} + \frac{\Sigma x_i \epsilon_i}{\Sigma x_i^2} = \beta + \frac{\Sigma x_i \epsilon_i}{\Sigma x_i^2} \tag{9.3.3}$$

the α term vanishing because $\Sigma x_i = 0$. Thus, b differs from β by a linear expression in ϵ. If we regard the known values of the x_i as fixed, it follows that b is an unbiased estimate of β, since the ϵ_i have zero means in the population.

We now present some consequences of the model that are often used in applications.

1. The deviations or residuals of the points Y from the fitted line are

$$d_{y.x} = Y - \hat{Y} = Y - \overline{Y} - bx = y - bx \tag{9.3.4}$$

These deviations have two properties: When we sum over the sample values,

$$\Sigma d_{y.x} = 0 \qquad \Sigma x d_{y.x} = \Sigma xy - b\Sigma x^2 = 0 \tag{9.3.5}$$

because $b = \Sigma xy/\Sigma x^2$.

2. The variance of b as an estimate of β is

$$V(b) = \sigma^2_{y.x}/\Sigma x^2 \tag{9.3.6}$$

This result follows from (10.7.5), p. 189, for the variance of a linear function, since $b - \beta$ is a linear function of the ϵ_i. Note that the variance of b does not depend explicitly on the number of points n used in fitting the line but rather on the amount of spread Σx^2 among the sample Xs. On reflection, this is not surprising: The denominator of our commonsense estimate of b was $X_n - X_1$, the spread between the extreme Xs.

3. An unbiased sample estimate of $\sigma^2_{y.x}$ is

$$s^2_{y.x} = \sum_{i=1}^{n} (Y_i - \hat{Y}_i)^2/(n - 2) = \Sigma d^2_{y.x}/(n - 2) \tag{9.3.7}$$

Note the divisor, $n - 2$, where n is the number of points used in fitting the line. The use of $n - 2$ instead of n is a consequence of the least squares method of fitting (as indicated in section 9.5).

4. The sample estimate of the standard error of b is $s_b = s_{y.x}/\sqrt{\Sigma x^2}$. Furthermore,

$$t = (b - \beta)/s_b = (b - \beta)\sqrt{\Sigma x^2}/s_{y.x} \tag{9.3.8}$$

is distributed as Student's t with $(n - 2)$ df.

5. The result (9.3.8) is used in two ways. We test the null hypothesis $\beta = 0$ that the means of the Y values are unrelated to X by referring $t = b/s_b$ to the t table with $(n - 2)$ df. In addition, confidence limits for β with confidence probability $(1 - \alpha)$ are $(b \pm t_\alpha s_b)$, where t_α is the two-sided α level significant value of t in table A 4.

We shall apply these results to the numerical example in section 9.2. Table 9.3.1 shows the mean potato yields Y_i, the estimates $\hat{Y}_i = 8.41 + 0.09375X_i$, and the deviations $Y_i - \hat{Y}_i$ of the observed points from the fitted regression.

TABLE 9.3.1

DEVIATIONS FROM REGRESSION AND CALCULATION OF $s^2_{y.x}$

Fertilizer Amount X_i (cwt/acre)	Yield Y_i (T/acre)	Estimated Yield, $\hat{Y}_i$	Deviations $Y_i - \hat{Y}_i = d_{y.x}$
0	8.34	8.410	−0.070
4	8.89	8.785	.105
8	9.16	9.160	.000
12	9.50	9.535	−0.035
$\Sigma d_{y.x} = 0$	$\Sigma d^2_{y.x} = 0.01715$	$s^2_{y.x} = 0.01715/2 = 0.00858$	

With only 2 df, $s_{y.x}$ cannot be expected to be an accurate estimate of $\sigma_{y.x}$. (In this experiment we actually have more information about $\sigma_{y.x}$ from the four plot yields that were measured for each amount of fertilizer.) We will, however, go ahead with the estimate $s_{y.x}$, which is the one used in the common situation with only one measured Y at each level of X.

In these data, $s_b = s_{y.x}/\sqrt{\Sigma x^2} = 0.09263/\sqrt{80} = 0.01036$. To test $H_0{:}\beta = 0$,

$$t = b/s_b = 0.09375/0.01036 = 9.049$$

With 2 df, this value is significant at the 2.5% level but not quite at the 1% level.

The 95% confidence limits for β, the average increase in yield per additional cwt of fertilizer, are

$$b \pm t_{0.05}s_b = 0.09375 \pm (4.303)(0.0104) = (0.049, 0.138)$$

6. Finally, by an algebraic identity, $\Sigma d_{y.x}^2$ can be calculated without finding the individual deviations by the formula

$$\Sigma d_{y.x}^2 = \Sigma y^2 - (\Sigma xy)^2/\Sigma x^2 \qquad (9.3.9)$$

Substituting the potato data from table 9.2.1, $\Sigma d_{y.x}^2 = 0.7203 - 7.50^2/80 = 0.0172$, in agreement with table 9.3.1.

EXAMPLE 9.3.1—Samples of soil were prepared with varying amounts X of inorganic phosphorus. Corn plants grown in each soil were harvested after 38 days and analyzed for phosphorous content. The objective was to see to what extent additional phosphorus in the soil increased the concentration of phosphorus in the plants. The data for 9 samples are as follows:

Inorganic soil phosphorus, X (ppm)	1	4	5	9	13	11	23	23	28
Plant-available phosphorus, Y (ppm)	64	71	54	81	93	76	77	95	109

Plot the points. Observe that although Y increases roughly linearly with X, the variation about the line is substantial. Calculate b, test its significance, and find 90% confidence limits for β. Ans. $b = 1.417$, $s_b = 0.395$, $t = 3.59$ ($P < 0.01$). Confidence limits (0.67, 2.17) for b are rather wide because the points do not fit the line closely. Keep your calculations; they are used in example 9.4.1.

EXAMPLE 9.3.2—An experiment (6) measured the yields of wheat under 7 different levels of nitrogen (N), with the following results:

Units (N/acre)	40	60	80	100	120	140	160
Yield (cwt/acre)	15.9	18.8	21.6	25.2	28.7	30.4	30.7

Plot the yield against the units of N, noting that the response looks linear, though with a hint of a falling-off at the highest level. Estimate the increase in yield per additional unit of N and give 90% confidence limits for this quantity. Ans. 0.133 cwt/acre, 90% limits (0.111, 0.155).

EXAMPLE 9.3.3—The data show the initial weights and gains in weight (grams) of 15 female rats on a high protein diet from 24 to 84 days of age. The point of interest in these data is whether the gain in weight is related to initial weight. If so, feeding experiments on female rats can be made more precise by taking account of the initial weights of the rats, either by pairing on initial weight or by adjusting for differences in initial weight in the analysis. Calculate b and test its significance. Ans.

$b = 1.0641$, $t = b/s_b = 2.02$ with 13 df, not quite significant at the 5% level. Keep your calculations; they are used in example 9.4.2.

Rat number	1	2	3	4	5	6	7	8	9	10	11	12	13	14	15
Initial weight, X	50	64	76	64	74	60	69	68	56	48	57	59	46	45	65
Gain, Y	128	159	158	119	133	112	96	126	132	118	107	106	82	103	104

9.4—Analysis of variance for linear regression. In a linear regression the total variation among the Y_i, as measured by $\Sigma(Y_i - \overline{Y})^2$, may be split into two parts: the variation $\Sigma(\hat{Y}_i - \overline{Y})^2$ among the regression estimates of the Y_i and the residual variation $\Sigma(Y_i - \hat{Y}_i)^2$ not accounted for by the regression. The result may be verified numerically for the potato yield data in table 9.3.1. There we find $\Sigma(Y_i - \overline{Y})^2 = 0.72027$, $\Sigma(\hat{Y}_i - \overline{Y})^2 = 0.70312$, and $\Sigma d_{y.x}^2 = 0.01715$.

To prove the result, note that if Y is any member of the sample

$$(Y - \overline{Y}) = (\hat{Y} - \overline{Y}) + (Y - \hat{Y}) \tag{9.4.1}$$

$$= bx + d_{y.x} \tag{9.4.2}$$

since $\hat{Y} = \overline{Y} + bx$. But from (9.3.5), $\Sigma x d_{y.x} = 0$. Hence, from (9.4.2),

$$\Sigma(Y - \overline{Y})^2 = b^2 \Sigma x^2 + \Sigma d_{y.x}^2 \tag{9.4.3}$$

$$= \Sigma(\hat{Y} - \overline{Y})^2 + \Sigma(Y - \hat{Y})^2 \tag{9.4.4}$$

Thus, as stated, the total variation among the Ys is the sum of the variations among the regression estimates of the Ys and the residual variation. Since $b = \Sigma xy/\Sigma x^2$, you may verify that (9.4.3) is identical with formula (9.3.9), the method of finding $\Sigma d_{y.x}^2$ without calculating any residuals. Incidentally, (9.4.4) supplies a proof of (9.3.9).

Corresponding to the partition of $\Sigma(Y - \overline{Y})^2$ is a partition of its degrees of freedom $(n - 1)$ into 1 for the term $(\Sigma xy)^2/\Sigma x^2$, representing the contribution of the regression estimate, and $(n - 2)$ for the residual variation. These results are presented in an *analysis of variance*, table 9.4.1.

Note the new column of mean squares—the sums of squares each divided by their degrees of freedom. The mean square in the residual line is, of course, $s_{y.x}^2$. The mean square in the total line is s_y^2, the sample variance among the original Ys.

TABLE 9.4.1
ANALYSIS OF VARIANCE FOR LINEAR REGRESSION

Source of Variation	Degrees of Freedom	Sum of Squares	Mean Square
Regression	1	$(\Sigma xy)^2/\Sigma x^2$	$(\Sigma xy)^2/\Sigma x^2$
Residual	$n - 2$	$\Sigma d_{y.x}^2$	$s_{y.x}^2$
Total	$n - 1$	Σy^2	s_y^2

TABLE 9.4.2
ANALYSIS OF VARIANCE OF POTATO YIELDS

Source of Variation	Degrees of Freedom	Sum of Squares	Mean Square
Regression	1	0.70312	0.70312
Residual	2	0.01715	0.00858
Total	3	0.72027	0.24009

The analysis for the potato yield data is in table 9.4.2.

This table gives a neat summary of the contributions to total variability and is very useful when we study regressions with more than one X variate and comparisons among more than two sample means. It is one of the major contributions of R. A. Fisher (5). The standard output of the regression analysis by a computer program always contains an analysis of variance table like 9.4.2.

The analysis of variance table also provides a two-tailed test of significance of b. In table 9.4.1 the ratio (regression mean square)/(residual mean square) = $b^2\Sigma x^2/s_{y.x}^2 = b^2/s_b^2 = t^2$ for testing H_0:$\beta = 0$. If $\beta = 0$, the ratio of these two mean squares is distributed as F with $(1, n - 2)$ df. But these significance levels are just the squares of the corresponding two-tailed levels of t, since t^2 with ν df is distributed as F with 1 and ν df.

As an example, for the potato data in table 9.4.2, $F = 0.70312/0.00858 = 81.9$ with 1 and 2 df. In section 9.3 we found $t = 9.049$ with 2 df, giving $t^2 = 81.9 = F$. A weakness of the F test is that it does not inform the investigator of the sign of b—naturally of major importance.

EXAMPLE 9.4.1—For the phosphorous regression ($n = 9$) in example 9.3.1, verify that the analysis of variance is as follows:

Source of Variation	df	Sum of Squares	Mean Square	F
Regression on inorganic P	1	1473.6	1473.6	12.89
Residual	7	800.4	114.3	
Total	8	2274.0	284.2	

Verify also that $F = t^2$.

EXAMPLE 9.4.2—Show that the analysis of variance for the regression of rat gains in weight on initial weight is as follows:

Source of Variation	df	Sum of Squares	Mean Square	F
Regression		1522.85	1522.85	4.09
Residual	13	4834.88	371.9	
Total	14	6357.73	454.1	

9.5—The method of least squares. The choice of $\overline{Y}$ and b to estimate the parameters α and β is an application of a principle widely used in problems of statistical estimation and known as *the method of least squares*. To explain this method, let $\hat{\alpha}$ and $\hat{\beta}$ denote any two estimators of α and β that we might consider. For the pair of observations (Y, X), the quantity

$$Y - \hat{\alpha} - \hat{\beta}x$$

measures the amount by which the fitted regression is in error in estimating Y. In the method of least squares, $\hat{\alpha}$ and $\hat{\beta}$ are chosen so as to minimize the sum of the squares of these errors, taken over the sample. That is, we minimize

$$\Sigma(Y - \hat{\alpha} - \hat{\beta}x)^2 \tag{9.5.1}$$

The principle seems reasonable to anyone who wants a line that fits the points well. Also, if the model (9.3.1) holds, the scientist Gauss showed around 1800 that estimators obtained in this way are (*i*) unbiased and (*ii*) have the smallest standard errors of any unbiased estimators that are linear expressions in the Ys. Gauss's proof does not require the Ys to be normally distributed but merely that they be independent with means 0 and variances $\sigma^2_{y.x}$.

This result—that (9.5.1) is minimized by taking $\hat{\alpha} = \bar{Y}$, $\hat{\beta} = b$—can be proved by some algebra that also throws light on the relation between $s^2_{y.x}$ and $\sigma^2_{y.x}$. Consider first the deviations $d_{y.x} = y - bx$ from the least squares line. Two properties of the $d_{y.x}$ that we know are

$$\Sigma d_{y.x} = 0 \qquad \Sigma x d_{y.x} = 0 \tag{9.5.2}$$

Now let $\hat{\alpha}$, $\hat{\beta}$ be *any* two estimates of α, β. The error at point X in the resultant fitted line is $Y - \hat{\alpha} - \hat{\beta}x$. It may be written

$$Y - \hat{\alpha} - \hat{\beta}x = Y - \bar{Y} - bx + (\bar{Y} - \hat{\alpha}) + (b - \hat{\beta})x \tag{9.5.3}$$

$$= d_{y.x} + (\bar{Y} - \hat{a}) + (b - \hat{\beta})x \tag{9.5.4}$$

Square both sides and sum over the n values in the sample. Three squared terms and three product terms are on the right side. The squared terms give

$$\Sigma d^2_{y.x} + n(\bar{Y} - \hat{\alpha})^2 + (b - \hat{\beta})^2\Sigma x^2 \tag{9.5.5}$$

Remarkably, the three cross-product terms all vanish when summed over the sample. From (9.5.2),

$$\Sigma d_{y.x}(\bar{Y} - \hat{\alpha}) = (\bar{Y} - \hat{\alpha})\Sigma d_{y.x} = 0$$
$$\Sigma d_{y.x}(b - \hat{\beta})x = (b - \hat{\beta})\Sigma x d_{y.x} = 0$$
$$\Sigma(\bar{Y} - \hat{\alpha})(b - \hat{\beta})x = (\bar{Y} - \hat{\alpha})(b - \hat{\beta})\Sigma x = 0$$

Thus, using (9.5.5) we have shown that for any $\hat{\alpha}$, $\hat{\beta}$,

$$\Sigma(Y - \hat{\alpha} - \hat{\beta}x)^2 = \Sigma d^2_{y.x} + n(\bar{Y} - \hat{\alpha})^2 + (b - \hat{\beta})^2\Sigma x^2 \tag{9.5.6}$$

Our objective is to find the values $\hat{\alpha}$, $\hat{\beta}$ that minimize the sum of squares of the deviations $(Y - \hat{\alpha} - \hat{\beta}x)$ of the points from the fitted line. Now $\Sigma d^2_{y.x}$ is the sum of squares of deviations from the least squares choice $\hat{\alpha} = \bar{Y}$, $\hat{\beta} = b$. But (9.5.6) shows that with any choice of $\hat{\alpha}$ and $\hat{\beta}$ the residual sum of squares is $\Sigma d^2_{y.x}$

plus two terms that are positive unless $\hat{\alpha} = \overline{Y}$ and $\hat{\beta} = b$. This proves that $\overline{Y}$ and b give the smallest possible residual sum of squares.

Equation (9.5.6) can also be used to throw some light on the relation between $s_{y.x}^2 = \Sigma d_{y.x}^2/(n - 2)$ and $\sigma_{y.x}^2$. In (9.5.6), put $\hat{\alpha} = \alpha$, $\hat{\beta} = \beta$, the population parameters. Then (9.5.6) can be rearranged as

$$\Sigma d_{y.x}^2 = \Sigma(Y_i - \alpha - \beta x)^2 - n(\overline{Y} - \alpha)^2 - (b - \beta)^2\Sigma x_i^2$$
$$= \Sigma\epsilon_i^2 - n(\overline{Y} - \alpha)^2 - (b - \beta)^2\Sigma x_i^2 \qquad (9.5.7)$$

In (9.5.7) each ϵ_i has mean 0, variance $\sigma_{y.x}^2$. Thus the term $\Sigma\epsilon_i^2$ is an unbiased estimate of $n\sigma_{y.x}^2$. The two subtracted terms on the right can each be shown to be unbiased estimates of $\sigma_{y.x}^2$. It follows that $\Sigma d_{y.x}^2$ is an unbiased estimate of $(n - 2)\sigma_{y.x}^2$ and hence that $s_{y.x}^2$ is an unbiased estimate of $\sigma_{y.x}^2$. This result does not require the ϵ_i to be normally distributed, but normality is required for the standard tests of significance and confidence interval calculations.

9.6—Regression in observational studies. Often the investigator does not select and apply the values of X but, instead, draws a sample from some population and then measures two characters Y and X for each member of the sample. In our illustration, section 9.7, the sample is a sample of apple trees in which the relation between the percentage of wormy fruits Y on a tree and the size X of its fruit crop is being investigated. In such applications the investigator realizes that if a second sample were drawn, the values of X in that sample would differ from those in the first sample. In the results presented in preceding sections, we regarded the values of X as essentially fixed. The question is sometimes asked, Can these results be used when it is known that the X values will change from sample to sample?

Fortunately, the answer is yes, provided that for any value of X the corresponding Y satisfies the three assumptions stated at the beginning of section 9.3. For each X, the sample value of Y must be drawn from a normal population that has mean $\mu_{y.x} = \alpha + \beta x$ and constant variance $\sigma_{y.x}^2$. Under these conditions the calculations for fitting the line, the t test of b, and the methods given to construct confidence limits for β and for the position of the true line all apply without change.

However, the interpretation of the regression coefficient is usually much clearer in a well-conducted experiment than in an observational study. In the potato experiment the investigator selected and applied specific amounts X of fertilizer to different plots, used randomization in deciding which plots received which specific amount, and took precautions to ensure that the plots were otherwise treated alike. The investigation can therefore reasonably claim that the increase b in yield was caused by the increased fertilizer.

In an observational study we can describe b as an estimate of the average change in Y that accompanies unit increase in X. But many different reasons may contribute to create this change, and the regression calculations give no clue to the explanation. We have to seek it elsewhere. Sometimes X itself is not causally related to Y but is correlated with an unmeasured variable Z that is causally related to Y, so b really represents the causal effect of Z on Y.

Sometimes our explanation has to be put in rather general terms. In both adult males and females there is a positive regression coefficient of weight on height. Tall persons tend to weigh more. Why? We might say that it is a property of the human frame and give arguments to the effect that it is a reasonable property for the human frame to have. Yule (11) has noted that in annual data in England and Wales between 1866 and 1911, the death rate has a linear regression with positive slope on the proportion of marriages performed in the Church of England. Why? In any society, many variables with no direct rational connection move together during a consecutive time period. This particular span of years had improving medical care and decreasing participation in organized religion.

It is easy in a regression calculation to form the habit of referring to b as the average change in Y *due to* unit increase in X. Always ask yourself, What evidence do I have that this is so? Often there is no evidence or only vague indications, and evidence may be hard to find.

What kind of additional evidence is relevant? For instance, among men of a given age it is well known from regression studies that the death rate from heart disease increases with increases in blood pressure, in the number of cigarettes smoked, and with increased obesity. Are any of these effects causal in some sense of this term? Medication usually reduces blood pressure. A man can decrease or give up cigarette smoking. Obesity may be reduced by diet. If several men who make these changes in the three variables have reduced death rates from heart disease as compared with men initially similar who make no changes, some kind of causal hypothesis is supported (although we seldom have the degree of control in the comparison that is fully desirable). Furthermore, such evidence can be practically beneficial whether we believe in a causal hypothesis or not. For additional evidence we can look to medical research on changes in the body that follow cigarette smoking and on differences in blood flow to the heart between persons with high, moderate, and low blood pressure and with obese and average body build.

In short, a causal hypothesis suggested by a regression coefficient can be explored further (*i*) if we are able to plan a comparative study in which the X variable is deliberately changed and note whether any change in Y is consistent with the hypothesis and (*ii*) if we know enough to explain *why* a change in X should be expected to be followed by the predicted change in Y.

There is another distinction between regression in observational studies and in controlled experiments. Suppose that in a study of families, the heights of pairs of adult brothers (X) and sisters (Y) are measured. An investigator might be interested either in the regression of a sister's height on a brother's height,

$$\hat{Y} = \bar{Y} + b_{y.x}(X - \bar{X})$$

or in the regression of a brother's height on a sister's height,

$$\hat{X} = \bar{X} + b_{x.y}(Y - \bar{Y})$$

These two regression lines are *different*. For a sample of 11 pairs of brothers and sisters, they are shown in figure 10.1.1 (p. 176). The line AB in this figure is the regression of Y on X, while the line CD is the regression of X on Y. Since $b_{y.x} = \Sigma xy/\Sigma x^2$ and $b_{x.y} = \Sigma xy/\Sigma y^2$, it follows that $b_{x.y}$ is not in general equal to $1/b_{y.x}$, as it would have to be to make the slopes AB and CD identical.

If the sample of pairs (X, Y) is random, the investigator may use whichever regression is relevant for the purpose. In predicting brothers' heights from sisters' heights, for instance, one uses the regression of X on Y. If, however, the investigator has deliberately selected the sample of values of one of the variates, say X, then only the regression of Y on X has meaning and stability.

9.7—Numerical example. It is generally thought that the percentage of fruits attacked by codling moth larvae is greater on apple trees bearing a small crop. Apparently the density of the flying moths is unrelated to the size of the crop on a tree, so the chance of attack for any particular fruit is augmented if few fruits are on the tree. The data in table 9.7.1 are adapted from the results of

TABLE 9.7.1

REGRESSION OF PERCENTAGE OF WORMY FRUIT ON SIZE OF APPLE CROP

Tree	Size of Crop (hundreds of fruits) X	Percentage of Wormy Fruits Y	Estimate of μ $\hat{Y}$	Deviation from Regression $Y - \hat{Y} = d_{y \cdot x}$
1	8	59	56.14	2.86
2	6	58	58.17	−0.17
3	11	56	53.10	2.90
4	22	53	41.96	11.04
5	14	50	50.06	−0.06
6	17	45	47.03	−2.03
7	18	43	46.01	−3.01
8	24	42	39.94	2.06
9	19	39	45.00	−6.00
10	23	38	40.95	−2.95
11	26	30	37.91	−7.91
12	40	27	23.73	3.27
	$\Sigma X = 228$	$\Sigma Y = 540$		
	$\overline{X} = 19$	$\overline{Y} = 45$		
	$\Sigma X^2 = 5256$	$\Sigma Y^2 = 25{,}522$	$\Sigma XY = 9324$	
	$(\Sigma X)^2/n = 4332$	$(\Sigma Y)^2/n = 24{,}300$	$(\Sigma X)(\Sigma Y)/n = 10{,}260$	

$\Sigma x^2 = 924$ $\quad \Sigma y^2 = 1222$ $\quad \Sigma xy = -936$

$b = \Sigma xy/\Sigma x^2 = -936/924 = -1.013\%/100$ wormy fruits

$\hat{Y} = \overline{Y} + b(X - \overline{X}) = 45 - 1.013\,(X - 19) = 64.247 - 1.013X$

$\Sigma d_{y \cdot x}^2 = 1222 - (-936)^2/924 = 273.84$

$s_{y \cdot x}^2 = \Sigma d_{y \cdot x}^2/(n - 2) = 273.84/10 = 27.384$

an experiment (3) containing evidence about this phenomenon. The 12 trees were all given a calyx spray of lead arsenate followed by five cover sprays made up of 3 lb of manganese arsenate and 1 qt fish oil/100 gal. There is a decided tendency, emphasized in figure 9.7.1, for the percentage of wormy fruits to decrease as the number of apples on the tree increases.

The new feature in the calculations is the majority of negative products xy

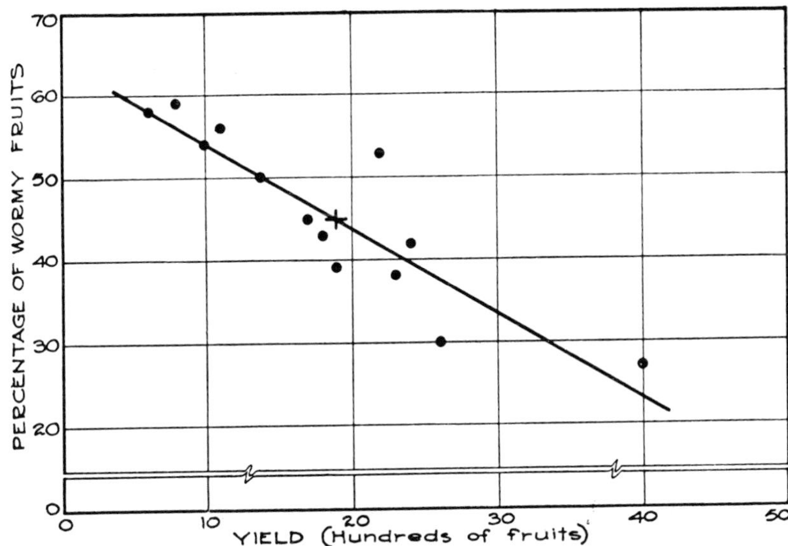

Fig. 9.7.1—Sample regression of percentage of wormy fruits on size of crop in apple trees. The cross indicates the origin for deviations, $0'(\bar{X}, \bar{Y})$.

caused by the tendency of small values of Y to be associated with large values of X. The sample regression coefficient shows that the estimated percentage of wormy apples *decreases* by 1.013% with each increase of 100 fruits in the crop.

Table 9.7.1 shows also the deviations from the fitted line, which measure the failure of crop size to account for variation in the intensity of infestation. Trees 4, 9, and 11 had notably discrepant percentages of injured fruits, while 2 and 5 performed as expected. According to the model these are random deviations from the average (regression) values, but close observation of the trees during the flight of the moths might reveal some characteristics of this phenomenon. Tree 4 might have been on the side from which the flight originated or perhaps its shape or situation caused poor applications of the spray. Trees 9 and 11 might have had some peculiarities of conformation of foliage that protected them. Careful study of trees 2 and 5 might throw light on the kind of tree or location that receives normal infestation. Examination of residuals in this way may add to the investigator's knowledge of the experimental material and may afford clues to the improvement of future experiments.

EXAMPLE 9.7.1—The following weights of body and comb of ten 15-day-old white leghorn male chicks are adapted from Snedecor and Breneman (4):

Body weight, X (g)	83	72	69	90	90	95	95	91	75	70
Comb weight, Y (mg)	56	42	18	84	56	107	90	68	31	48

Graph comb weight against body weight. Note that the relation appears linear, but deviations from the line are substantial. Calculate the sample regression and test $H_0{:}\beta = 0$. Ans. $\hat{Y} = 2.302X - 131.1$, $t = b/s_b = 5.22$ (8 df), $P < 0.01$.

9.8—Prediction of Y from X. In many applications of regression, the purpose is to predict Y from knowledge of the corresponding X. Three related prediction problems are:

1. Prediction of the value $\mu_{y.x} = \alpha + \beta x$ of the population regression line at point X.
2. Prediction of Y for an individual new member of the population for which X has been measured—probably the most frequent application, e.g., in weather prediction.
3. Prediction of the mean of a random sample of m values of Y drawn at a specified value of X—less common, but includes problems 1 and 2 as special cases. If $m = 1$ we have problem 2, prediction of a single Y. As m approaches infinity, we predict $\mu_{y.x}$ as in problem 1, since $\mu_{y.x}$ is the mean of an infinitely large sample of Y values drawn at value X.

To distinguish among the problems, we call our predicted values $\hat{\mu}_{y.x}$, $\hat{Y}_x$, and $\hat{\bar{Y}}_x$. However, in all three problems the predicted value is exactly the same, namely,

$$(\hat{\mu}_{y.x}, \hat{Y}_x, \hat{\bar{Y}}_x) = \bar{Y} + bx = \bar{Y} + b(X - \bar{X}) \tag{9.8.1}$$

Without additional information we have no grounds for making different predictions in the three problems.

On the other hand, the standard errors of our predictions will differ because different quantities are being estimated. The formula for the standard error is presented for problem 3, from which we immediately get the formulas for problems 1 and 2.

From the model $Y = \alpha + \beta x + \epsilon$ it follows that since $\bar{x} = 0$, $\bar{Y} = \alpha + \bar{\epsilon}_n$, where $\bar{\epsilon}_n$ denotes the mean of a random sample of n residuals. Hence, the predicted value $\hat{\bar{Y}}_x$ in problem 3 is

$$\hat{\bar{Y}}_x = \bar{Y} + bx = \alpha + bx + \bar{\epsilon}_n \tag{9.8.2}$$

But from the model the *predictand,* the value being predicted in problem 3, is

$$\bar{Y}_x = \alpha + \beta x + \bar{\epsilon}_m \tag{9.8.3}$$

By subtraction, the error in the prediction is

$$\hat{\bar{Y}}_x - \bar{Y}_x = (\bar{\epsilon}_n - \bar{\epsilon}_m) + (b - \beta)x \tag{9.8.4}$$

The term $(\bar{\epsilon}_n - \bar{\epsilon}_m)$ in this error is the difference between the means of two independent samples of size n, m. Its variance is $\sigma^2_{y.x}(1/n + 1/m)$. In the other term, b is distributed about β with variance $\sigma^2_{y.x}/\Sigma x^2$. Finally, the independence of the ϵs guarantees that the two terms are uncorrelated, so the variance of their sum is the sum of their variances. This gives

$$V(\hat{\bar{Y}}_x) = \sigma^2_{y.x}(1/n + 1/m + x^2/\Sigma x^2) \tag{9.8.5}$$

The estimated standard error of $\hat{Y}_x$ is therefore

$$s(\hat{Y}_x) = s_{y.x}(1/n + 1/m + x^2/\Sigma x^2)^{1/2} \quad (9.8.6)$$

9.9—Prediction of the population regression line. For problem 1 we put $1/m = 0$, giving

$$s(\hat{\mu}_{y.x}) = s_{y.x}\sqrt{1/\text{n} + x^2/\Sigma x^2} \quad (9.9.1)$$

For the apple trees in table 9.7.1, $s_{y.x} = \sqrt{27.38}$, $n = 12$, and $\Sigma x^2 = 924$. Hence

$$s(\hat{\mu}_{y.x}) = \sqrt{27.38}\sqrt{1/12 + x^2/924} = \sqrt{2.282 + 0.0296x^2} \quad (9.9.2)$$

For tree 12 with a large crop, $X = 40$, $x = 21$, and $s(\hat{\mu}_{y.x}) = 3.92\%$, notably greater than $s(\hat{\mu}_{y.x}) = 1.51\%$ at $X = 19$, $x = 0$. This increase in the sampling error as x increases is due to the term $(b - \beta)x$ in the error of $\hat{Y}_x$ in (9.8.4).

The 95% confidence limits for $\mu_{y.x}$ at a specific point X are $\mu_{y.x} \pm t_{0.05}s(\hat{\mu}_{y.x})$, where t has 10 df. However, we are much more likely to be interested in confidence limits that apply to the whole regression line, not just to the height of the line at a single value of X. For these, the multiplier $\pm t_{0.05}$ must be increased. The appropriate multiplier was given by Working and Hotelling (13) in 1929. The multiplier $\pm t_{0.05}$ is replaced by a multiplier $\pm\sqrt{2F_{0.05}}$. The quantity F, which

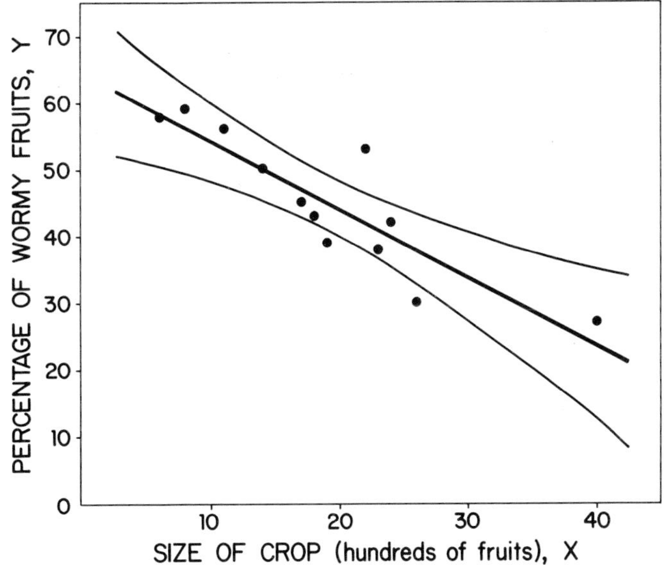

Fig. 9.9.1—Confidence bands (95%) for the population regression line in the regression of percentage of wormy fruits on size of apple crop.

is used frequently in following chapters, is called a *variance ratio*. Its significance levels are given in tables A 14, parts I and II. To use these tables you need to know ν_1, the degrees of freedom in the numerator of F, and ν_2, the degrees of freedom in its denominator. For the regression line problem, $\nu_1 = 2, \nu_2 = (n - 2)$, in our example, 10.

In table A 14, part I, $F_{0.05}$ with 2 and 10 df is 4.10, so the multiplier is $\sqrt{(2)(4.10)} = 2.86$, about 30% larger than $t_{0.05} = 2.23$ with 10 df.

As X covers the range from 5 to 40, the points $Y = \hat{\mu}_{y.x} \pm \sqrt{2F}s(\hat{\mu}_{y.x})$ lie on smooth curves (confidence bands) that are the two branches of a hyperbola. Figure 9.9.1 shows the points (Y, X), the sample regression line, and the two confidence bands. To illustrate, the heights of the confidence bands at $X = 30$ are calculated. In table 9.7.1, $\overline{X} = 19$, $\overline{Y} = 45$, $b = -1.013$. Hence, at $X = 30$,

$$\hat{\mu}_{y.x} = \overline{Y} + bx = 45 - (1.013)(11) = 33.9\%$$

$$\sqrt{2F_{0.05}}s(\hat{\mu}_{y.x}) = 2.86\sqrt{2.282 + 0.0296(11^2)} = \pm 6.9$$

These give 27.0 and 40.8 as the heights of the two branches in figure 9.9.1. The two confidence bands define a region within which the population regression line lies, apart from a 1-in-20 chance.

9.10—Prediction of new individual *Y*s. For the standard error of the prediction $\hat{Y}_x$ for a new value of Y at point X, we put $m = 1$ in (9.8.6), giving

$$s(\hat{Y}_x) = s_{y.x}\sqrt{1 + 1/n + x^2/\Sigma x^2} \tag{9.10.1}$$

For a tree with crop $X = 25$,

$$\hat{Y}_x = \overline{Y} + bx = 45 - (1.013)(6) = 38.9\%$$

$$s(\hat{Y}_x) = \sqrt{27.38}\ \sqrt{1 + 1/12 + 36/924} = \pm 5.54\%$$

The 95% confidence limits are $\hat{Y}_x \pm t_{0.05}s(\hat{Y}_x)$, in this case $38.9 \pm (2.23) \cdot (5.54) = (26.5, 51.3)$.

These limits apply to a *single* prediction made from the sample line. However, if the line has been calculated for the purpose of prediction, it is likely to be used to predict a number of new specimens with known values of X. But if we construct confidence intervals by the preceding formulas for each of say 10 new specimens, the probability that all 10 confidence intervals include the correct values of the new Ys is not 0.95 but substantially less.

If predictions are to be made for n' new specimens, we can control the probability that all n' confidence intervals cover the values of the new Y_xs. The method is given in section 7.5. It uses one of Bonferroni's inequalities (12). To obtain an overall confidence probability $1 - \alpha$, make the individual confidence probabilities $1 - \alpha/n'$; that is, calculate each confidence interval as

$$\hat{Y}_x \pm t_{\alpha/n'}s(\hat{Y}_x) \tag{9.10.2}$$

using $t_{\alpha/n'}$ instead of t_α. For instance, if we want 95% overall confidence probability with $n' = 10$ predictions, we use $t_{0.05/10} = t_{0.005} = 3.58$ (with $n - 2 = 10$ df) instead of $t_{0.05} = 2.23$. With this method the probability that all 10 intervals cover the unknown Y values is greater than or equal to 0.95, or more generally to $1 - \alpha$.

This method often requires interpolation in the t table, A 4. For example, suppose that with $n' = 8$ we want $t_{0.00625}$, which is not tabulated but lies between the values given for $t_{0.010}$ and $t_{0.005}$. Linear interpolation is described at the beginning of the appendix tables. If you have an access to lns in a pocket calculator, linear interpolation against $\ln(1 - \alpha)$ is more accurate than linear interpolation against $(1 - \alpha)$; see example 9.10.2.

EXAMPLE 9.10.1—From the F table (A 14, part II), calculate the two points Y_x on the 90% confidence bands for the regression line in the apple tree table, 9.7.1, at $X = 19$, the point at which the interval is smallest. Ans. 41.35, 48.65.

EXAMPLE 9.10.2—From the t table (A 4) for 10 df, calculate the values of $X = \ln[1000(1 - \alpha)]$ for the points at which $1000(1 - \alpha) = 100, 50, 25, 10, 5$, and 1, i.e., for the 10%, 5%, etc., significance levels of the two-tailed t. Plot the values of t in the table against X. Notice how close the points lie to a straight line. This illustrates why interpolation on $\ln(1 - \alpha)$ works well for finding tail values of t. Draw a straight line as nearly as possible through the points. Use it to calculate the multipliers $t_{0.05/n}$ for finding 95% overall confidence limits when predicting (*i*) 3, (*ii*) 6, (*iii*) 20 new values of Y_x. Ans. By this method we found $t_{0.0167} = 2.87$, $t_{0.0083} = 3.29$, $t_{0.0025} = 4.02$; the correct values are 2.87, 3.28, and 4.00.

EXAMPLE 9.10.3—In work on the relation between annual crop yields and temperatures during the growing season at Rothamsted, England, the crop yields were available from 1852 onwards but Rothamsted temperatures were not recorded until 1878. A study (14) examined how well temperatures at Rothamsted can be predicted from Oxford (38 mi away), which had temperature records from 1852. The regression was constructed from the 30-year period 1878–79 to 1907–8. Following are the seasonal mean temperatures (°F − 40) for Oxford (X) and Rothamsted (Y) for every third year in this period, giving $n = 10$ seasons.

X	6.18	9.94	8.69	5.87	6.65	8.91	8.71	9.00	9.24	8.81
Y	5.59	8.52	7.55	5.32	6.05	7.99	7.65	7.90	8.07	7.76

Compute the sample regression and find $s_{y.x}$. Ans. $\hat{Y} = 0.657 + 0.8028X$, $s_{y.x} = 0.0869$°F.

EXAMPLE 9.10.4—Predict the Rothamsted temperature and 95% confidence limits for a season (not in the sample data) for which the Oxford temperature was 50.24°F. Ans. $\hat{Y} = 48.88$°F, 95% limits (48.65, 49.11). The actual Rothamsted temperature in this year was 49.04°F, comfortably inside the limits.

EXAMPLE 9.10.5—From the Indianapolis speed data (example 9.2.2) calculate 95% confidence limits for the winning speed in 1974. Ans. (156.0, 171.8) mph. The limits are wide, since extrapolation is involved, making $x = X - \bar{X}$ large. The actual speed, 158.6 mph, was near the lower limit.

EXAMPLE 9.10.6—The regression of Rothamsted temperature on Oxford temperatures was calculated in (14) from 30 observations, so t had 28 df. What value of t would be used as multiplier to give overall 90% confidence limits for each of the 26 years 1852–77 inclusive? Ans. $t_{0.0038} = 3.15$ with 28 df.

9.11—Testing a deviation that looks suspiciously large. When Y is plotted against X, one or two points sometimes look as if they lie far from the regression line. When the line has been computed, we can examine this question further by drawing the line and looking at the deviations for these points or by calculating the values of $d_{y.x}$ for them.

In this process one needs some guidance with respect to the question, When is a deviation large enough to excite suspicion? An approximate test of significance can be carried out as follows. Recompute the regression with the suspected point omitted. Since the suspect was not used in computing this line, we can regard it as a new member of the population and test whether its deviation from the recomputed line is within sampling error. For the standard error of this deviation we can use (9.10.1), where $s'_{y.x}$, x', $\Sigma x'^2$ refer to the regression with the suspect omitted and $1/(n-1)$ replaces $1/n$. That is, if $d'_{y.x}$ denotes the deviation of the suspect point from the recomputed line,

$$s^2(d'_{y.x}) = s'^2_{y.x}[1 + 1/(n-1) + x'^2/\Sigma x'^2] \tag{9.11.1}$$

Furthermore, if the suspect actually follows the regression model, the ratio $d'_{y.x}/[s(d'_{y.x})]$ is distributed as t with $(n-3)$ df.

Fortunately, by some algebraic results, this t value can be calculated without going to the trouble of recomputing the line. It can be shown that

$$(n-3)s'^2_{y.x} = \sum^{n} d^2_{y.x} - (Y_i - \hat{Y}_i)^2/(1 - 1/n - x_i^2/\Sigma x^2) \tag{9.11.2}$$

where Y_i, $\hat{Y}_i$, x_i refer to the suspect. Also, the test criterion

$$|t| = |Y_i - \hat{Y}_i|/(s'_{y.x}\sqrt{1 - 1/n - x_i^2/\Sigma x^2}) \tag{9.11.3}$$

To illustrate, this test will be applied to tree 4 in the regression of wormy fruit on size of apple crop, table 9.7.1 and figure 9.7.1. We have already commented that for tree 4, with $X_i = 22$, $Y_i = 53$, the deviation 11.04 looks large. From the data in table 9.7.1,

$$9s'^2_{y.x} = 273.8 - 11.04^2/(1 - 1/12 - 9/924) = 139.4$$

giving $s'_{y.x} = 3.936$. Hence, from (9.11.3),

$$t = 11.04/(3.936\sqrt{11/12 - 9/924}) = 2.94 \qquad (9 \text{ df})$$

If this Y_i value had been picked at random from the sample, the significance probability of t from table A 4 would be about 0.016. However, this probability does not apply, since we selected the deviation 11.04 for testing because it is the largest deviation. If P is the probability that a random t exceeds some value t_0, the probability that the largest of n independent ts exceeds t_0 is $1 - (1 - P)^n$, in this case $1 - 0.984^{12} = 0.18$. (With nP small, this value is roughly nP, which is 0.192 in this example.) The result 0.18 is not quite correct for our problem, since our 12 ts are correlated, but the result should be close enough for guidance. We conclude that the deviation 11.04 is not highly unusual, with P around 0.18.

Note, incidentally, that the largest deviation does not always give the largest value of t, because the standard error of a deviation depends on the value of X_i, as (9.11.1) shows. Check to see that you are testing the most unusual deviation.

9.12—Prediction of X from Y: linear calibration. In some applications the sample regression line is constructed by measuring Y at a series of selected values of X but is used to predict X from Y. This method is employed in problems in which the amount of X in a specimen is difficult or expensive to measure directly but is linearly related to a variable Y that is more easily measured. Examples are photometric methods of estimating the iron or sodium concentration in a specimen or methods for estimating the amounts of the active ingredient in drugs or insecticides from their effects on animals or insects. In one approach, the investigator makes up a series of specimens with *known* amounts of X and measures Y for each specimen. From these data the linear regression of Y on X is computed. This line is the calibration line, from which we will predict X by measuring Y.

In using the line, suppose that Y is measured for a new specimen with an unknown amount X. From the calibration line, the estimate of X is

$$\hat{X} = \bar{X} + (Y - \bar{Y})/b \tag{9.12.1}$$

where $\bar{X}$, $\bar{Y}$ and b refer to the sample calibration line. The estimate $\hat{X}$ involves the ratio of two random variables $(Y - \bar{Y})$ and b. It is biased and no exact expression for its standard error is known.

However, confidence limits for x and X can be obtained from the method in section 9.8 by which we obtained confidence limits for Y, given x. As an illustration we cite the example in table 9.7.1 in which Y = percentage of wormy fruits, X = size of crop (though with these data we would in practice use the regression of X on Y, since both regressions are meaningful).

We shall find 95% confidence limits for the size of crop in a new tree with 40% wormy fruit. For a numerical solution, the fitted line is $\bar{Y} + bx$, where $\bar{Y} = 45$, $b = -1.013$. Hence the value of x when $Y = 40$ is estimated as

$$\hat{x} = (Y - \bar{Y})/b = -(40 - 45)/1.013 = 4.936 \qquad \hat{X} = 23.9 \text{ hundreds}$$

To find the 95% confidence limits for x we start with the confidence limits of Y, given x:

$$Y = \bar{Y} + bx \pm ts_{y.x}\sqrt{1 + 1/n + x^2/\Sigma x^2} \tag{9.12.2}$$

where t is the 5% level for $(n - 2)$ df. Expression (9.12.2) is solved as a quadratic equation in x for given Y. After some manipulation the two roots can be expressed in the following form (which appears the easiest for numerical work):

$$x = \frac{\hat{x} \pm (ts_{y.x}/b)\sqrt{[(n + 1)/n](1 - c^2) + \hat{x}^2/\Sigma x^2}}{1 - c^2} \tag{9.12.3}$$

where

$$c^2 = t^2 s_b^2/b^2 = (1/\Sigma x^2)\left(\frac{ts_{y.x}}{b}\right)^2$$

In this example $n = 12$, $t = 2.228$ (10 df), $s_{y.x} = 5.233$, $\Sigma x^2 = 924$, $b = -1.013$, $\hat{x} = 4.936$. These give

$$\frac{ts_{y.x}}{b} = \frac{(2.228)(5.233)}{-1.013} = -11.509 \qquad c^2 = \frac{11.509^2}{924} = 0.1434$$

From (9.12.3) the limits for x are

$$x = 4.936 \pm (11.509)\sqrt{(1.0833)(0.8566) + 0.0264}$$

This gives -7.4 and $+18.9$ for x or 11.6 and 37.9 for X.

The quantity $c = ts_b/b$ is related to the test of significance of b. If b is significant at the 5% level, $b/s_b > t$, so $c < 1$ and hence $c^2 < 1$. If b is not significant, the denominator (9.12.3) becomes negative, and finite confidence limits cannot be found by this approach. If c is small (b highly significant), c^2 is negligible and the limits become

$$\hat{x} \pm (ts_{y.x}/b)\sqrt{1 + 1/n + \hat{x}^2/\Sigma x^2} \qquad (9.12.4)$$

They are of the form $\hat{x} \pm ts_{\hat{x}}$ where $s_{\hat{x}}$ denotes the factor that multiplies t. In large samples, $s_{\hat{x}}$ can be shown to be approximately the estimated standard error of $\hat{x}$, as this result suggests.

In practice, Y is sometimes the average of m independent measurements on the new specimen. The term $(n + 1)/n$ under the square root sign in (9.12.3) then becomes $(m + n)/(mn)$.

Although statistical theory supports the use of the preceding method, several writers, as noted in (15), have examined the properties of the direct estimate $\hat{X}_2$ made from the sample regression of X on Y:

$$\hat{X}_2 = \overline{X} + b_{x.y}(Y - \overline{Y}) \qquad (9.12.5)$$

The estimate $\hat{X}_2$ has been found to be more accurate than $\hat{X}$ in certain cases with small n when X for the new specimen is near $\overline{X}$, but $\hat{X}_2$ is not recommended for general use when the sample data are obtained by measuring Y at chosen values of X.

EXAMPLE 9.12.1—The following artificial data refer to the spectrophotometric determination of the concentration of iron in plant fibers. A calibration curve was constructed from specimens prepared with known amounts X of iron, the data being as follows:

X	10	20	30	40	50	60	70	80	90	100
Y	4.21	5.92	7.43	9.53	11.50	13.32	14.99	16.74	18.61	20.30

The fitted line is $\hat{Y} = 2.309 + 0.1808X$, while $\Sigma x^2 = 8250$ and $s_{y.x} = 0.1424$. Five independent measurements of Y for a new specimen gave a mean $\bar{Y}_5 = 14.3$. Estimate the iron concentration and 95% confidence limits. Ans. $\hat{X} = 66.3$, limits (65.3, 67.3).

9.13—Galton's use of the term "regression." Galton developed the idea of regression in his studies of inheritance. Of the "law of universal regression" (7) he said, "Each peculiarity in a man is shared by his kinsman, but *on the average* in a less degree." His friend, Karl Pearson, and A. Lee (8) collected more than a thousand records of heights of members of family groups. Figure 9.13.1 shows

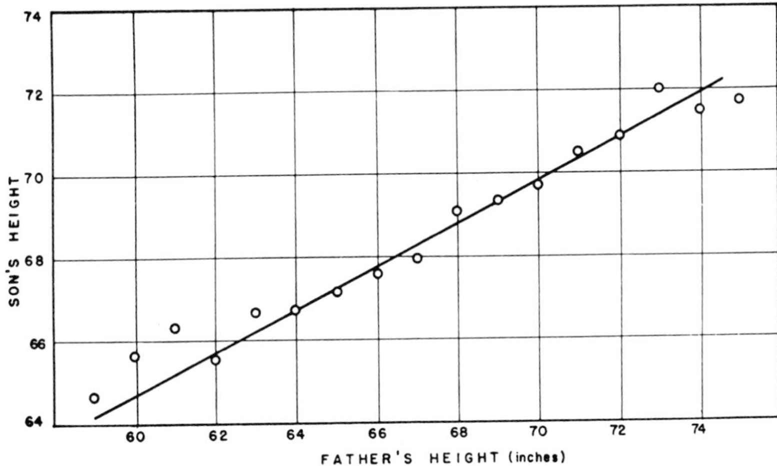

Fig. 9.13.1—Regression of son's stature on father's (8); $\hat{Y} = 0.516X + 33.73$; 1078 families.

their regression of son's height on father's. Though tall fathers do tend to have tall sons, fathers who are 6 in. taller than average have sons who are only 3 in. taller than the average for sons, since $b \doteq 0.5$. There is a regression, or going back, of a son's height toward the average height for sons. The name *regression,* describing the result of Galton's early application, came to be used to describe the statistical technique.

9.14—Regression when X is subject to error. Thus far we have assumed that the X-variable in regression is measured without error. Since no measuring instrument is perfect, this assumption is often unrealistic. A more realistic model assumes $Y = \alpha + \beta(X - \bar{X}) + \epsilon$ as before but regards X as an *unknown* true value. Our measurement of X is $X' = X + e$, where e is the error of measurement. For any specimen we know (Y, X') but not X.

If the measurement is unbiased, e (like ϵ) is a random variable following a distribution with mean zero. The errors e may arise from several sources. For instance, if X is the average price of a commodity or the average family income in a region of a country, it is usually estimated from a sample of stores or of families, so X' is subject to a sampling error. Some concepts like "educational level" or "economic status" may have no fully satisfactory measuring method, so e may represent in part measurement of the wrong concept.

If ϵ, e, and the true X are all normally and independently distributed, it is known that Y and X' follow a bivariate normal distribution (section 10.3). The regression of Y on X' is linear, with regression coefficient

$$\beta' = \beta/(1 + \lambda) \tag{9.14.1}$$

where $\lambda = \sigma_e^2/\sigma_X^2$. (If X is not normal, this result holds in large samples and approximately in small samples if λ is small.) Thus, with errors in X, the sample regression coefficient b' of Y on X' no longer provides an unbiased estimate of β but of $\beta/(1 + \lambda)$.

If the principal objective is to estimate β (often called the *structural* regression coefficient), the extent of this distortion downwards is determined by the ratio $\lambda = \sigma_e^2/\sigma_X^2$. Sometimes it is possible to obtain an estimate s_e^2 of σ_e^2. Since $\sigma_{X'}^2 = \sigma_X^2 + \sigma_e^2$, an estimate of λ is $\hat{\lambda} = s_e^2/(s_{X'}^2 - s_e^2)$. From $\hat{\lambda}$ we can judge whether the downward bias is negligible or not. If it is not negligible, the revised estimate $b'(1 + \hat{\lambda})$ should remove most of the bias.

If the error of measurement e and X are correlated, the effect of the error on β is more complicated (see section 18.7). See also the Berkson case below.

If the objective is to predict the population regression line or the value of an individual Y from the sample of values (Y, X'), the methods of section 9.8 may still be used with X' in place of X. The presence of errors in X decreases the accuracy of the predications, however, because the residual variance is increased (though to a minor extent if λ is small). The relation between $\sigma_{Y.X'}^2$ and $\sigma_{Y.X}^2$ is

$$\sigma_{Y.X'}^2 = \sigma_{Y.X}^2 + [\lambda/(1 + \lambda)](\sigma_Y^2 - \sigma_{Y.X}^2) \tag{9.14.2}$$

Berkson (10) has pointed out an exception to the above analysis. Many laboratory experiments are conducted by setting X' at a series of *fixed* values. For instance, a voltage may be set at a series of predetermined levels $X_1', X_2', \ldots$ on a voltmeter. Owing to errors in the voltmeter or other defects in the apparatus, the true voltages $X_1, X_2, \ldots$ differ from the set voltages.

In this situation we still have $Y = \alpha + \beta X + \epsilon$, $X' = X + e$. In both our original case (X normal) and in Berkson's case (X' fixed) it follows that

$$Y = \alpha + \beta X' + (\epsilon - \beta e) \tag{9.14.3}$$

With X' set by the experimenter, however, e is correlated with X but not with X' and model (9.14.3) satisfies the assumptions for a linear regression. The important practical conclusion is that b', the regression coefficient of Y on X', remains an unbiased estimate of β.

9.15—Fitting a straight line through the origin. From some data the nature of the variables Y and X makes it clear that when $X = 0$, Y must be 0. If a straight-line regression appears to be a satisfactory fit, we have the relation

$$Y = \beta X + \epsilon \tag{9.15.1}$$

where, in the simplest situations, the residual ϵ follows $\mathcal{N}(0, \sigma^2)$. The least squares estimate of β is $b = \Sigma XY/\Sigma X^2$. The residual mean square is

$$s_{y.x}^2 = [\Sigma Y^2 - (\Sigma XY)^2/\Sigma X^2]/(n - 1) \tag{9.15.2}$$

with $(n - 1)$ df. Confidence limits for β are $b \pm ts_b$, where t is read from the t table, with $(n - 1)$ df and the appropriate probability.

This model should not be adopted without first plotting the data. If the sample values of X are all some distance from zero, plotting may show that a straight line through the origin is a poor fit, although a straight line that is not forced to go through the origin seems adequate. The explanation may be that the population relation between Y and X is curved, the curvature being marked near zero but slight in the range within which X has been measured. A straight line of the form $a + bx$ will then be a good approximation within the sample range, though untrustworthy for extrapolation. If the mathematical form of the curved relation is known, it may be fitted by methods outlined in chapter 19.

It is sometimes useful to test the null hypothesis that the line, assumed straight, goes through the origin. The first step is to fit the usual two-parameter line $\alpha + \beta x$, i.e., $\alpha + \beta(X - \overline{X})$, by the methods given earlier in this chapter. The condition that the population line goes through the origin is $\alpha - \beta\overline{X} = 0$. The sample estimate of this quantity is $\overline{Y} - b\overline{X}$, with estimated variance

$$s_{y.x}^2(1/n + \overline{X}^2/\Sigma x^2)$$

Hence, the value of t for the test of significance is

$$t = \frac{\overline{Y} - b\overline{X}}{s_{y.x}\sqrt{1/n + \overline{X}^2/\Sigma x^2}} \tag{9.15.3}$$

with $(n - 2)$ df.

The following example comes from a study (9) of the forces or drafts necessary to draw plows at the speeds commonly attained by tractors. The results of the regression calculations that are needed are shown in table 9.15.1.

One might suggest that logically the line should go through the origin, since when the plow is not moving there is no draft. However, a plot of the points makes it clear that when the line is extrapolated to $X = 0$, the predicted Y is well above 0, as would be expected, since inertia must be overcome to get the plow

TABLE 9.15.1
Draft and Speed of Plows Drawn by Tractors

Draft, Y (lb)	425	420	480	495	540	530	590	610	690	680
Speed, X (mph)	0.9	1.3	2.0	2.7	3.4	3.4	4.1	5.2	5.5	6.0

$\overline{X} = 3.45$ mph $\overline{Y} = 546$ lb $n = 10$

$\Sigma x^2 = 27.985$ $\Sigma y^2 = 82{,}490$ $\Sigma xy = 1492.0$

$b = 53.31$ lb/mi

$s_{y \cdot x}^2 = 368.1$ (8 df)

moving. From (9.15.3) we have

$$t = \frac{546 - (53.31)(3.45)}{\sqrt{(368.1)(1/10 + 3.45^2/27.985)}} = \frac{362.1}{13.90} = 26.0 \qquad (8 \text{ df})$$

confirming that the line does not go through the origin.

When the line is straight and passes through (0, 0), the variance of the residual ϵ is sometimes not constant but increases as X moves away from zero. On plotting, the points lie close to the line when X is small but diverge from it as X increases. The extension of the method of least squares to this case gives the estimate $b = \Sigma w_x XY/\Sigma w_x X^2$, where w_x is the reciprocal of the variance of ϵ at the value of X in question. The method is called *weighted linear regression.*

If numerous observations of Y have been made at each selected X, the variance of ϵ can be estimated directly for each X and the form of the functions w_x determined empirically. If there are not enough data to use this method, simple functions that seem reasonable are employed. A common one when all Xs are positive is to assume that the variance of ϵ is proportional to X, so $w_x = k/X$ (where k is a constant). This gives the simple estimate $b = \Sigma Y/\Sigma X = \bar{Y}/\bar{X}$. The weighted mean square of the residuals from the fitted line is

$$s_{y \cdot x}^2 = [\Sigma(Y^2/X) - (\Sigma Y)^2/\Sigma X]/(n - 1) \qquad (9.15.4)$$

and the estimated standard error of b is $s_{y \cdot x}/\sqrt{\Sigma X}$ with $(n - 1)$ df.

TECHNICAL TERMS

analysis of variance
confidence band
dependent variable
deviations from regression
function
independent variable
least squares
linear calibration
predictand
regression
variance ratio
weighted linear regression

REFERENCES

1. Rothamsted Experimental Station Report. 1933.
2. Wentz, J. B., and Stewart, R. T. *J. Am. Soc. Agron.* 16 (1924):534.
3. Hansberry, T. R., and Richardson, C. H. *Iowa State Coll. J. Sci.* 10 (1935):27.
4. Snedecor, G. W., and Breneman, W. R. *Iowa State Coll. J. Sci.* 19 (1945): 33.
5. Fisher, R. A. 1925. *Statistical Methods for Research Workers.* Oliver & Boyd, Edinburgh.
6. Boyd, D. A. Exp. Husb. Farms and Exp. Hortic. Stn. Rep. 38 (1970).
7. Galton, F. 1889. *Natural Inheritance.* Macmillan, London.
8. Pearson, K, and Lee, A. *Biometrika* 2 (1903):357.
9. Collins, E. V. *Trans. Am. Soc. Agric. Eng.* 14 (1920):164.
10. Berkson, J. *J. Am. Stat. Assoc.* 45 (1950):164.
11. Yule, G. U. *J. R. Stat. Soc.* 89 (1926):1.
12. Feller, W. 1957. *An Introduction to Probability Theory and Its Applications,* 2nd ed. Wiley, New York.
13. Working, H., and Hotelling, H. *J. Am. Stat. Assoc.* 24 (1929):73.
14. Boyd, D. A. *Ann. Eugen.* 9 (1939):341.
15. Shukla, G. K. *Technometrics* 14 (1972):547.

10

Correlation

10.1—The sample correlation coefficient *r*. The *correlation coefficient* is a measure of the closeness of relationship between two variables—more exactly, of the closeness of linear relationship. Table 10.1.1 and figure 10.1.1 show the heights of 11 brothers and sisters, drawn from a large family study by Pearson and Lee (1). Since there is no reason to think of one height as the dependent variable and the other as the independent variable, the heights are designated X_1 and X_2 instead of Y and X. To find the sample correlation coefficient denoted by r, compute Σx_1^2, Σx_2^2, and $\Sigma x_1 x_2$. Then,

$$r = \Sigma x_1 x_2 / \sqrt{(\Sigma x_1^2)(\Sigma x_2^2)} = 0.558$$

as shown in table 10.1.1. Roughly speaking, r is a quantitative expression of the commonly observed similarity among children of the same parents—the tendency of the taller sisters to have the taller brothers. In figure 10.1.1 the dots lie in a band extending from lower left to upper right instead of being scattered randomly over the whole field. The band is often shaped like an ellipse, with the major axis sloping upward toward the right when r is positive.

Figure 10.1.1 also shows the sample linear regressions AB of sister's height X_2 on brother's height X_1 and the sample linear regression CD of brother's height X_1 on sister's height X_2. The two lines are different. In mathematics, on the other hand, the equation by which we calculate X_2 when given X_1 is the same as the equation by which we calculate X_1 when given X_2. There are two different regressions because in correlation problems we deal with relationships that are not followed exactly. For any X_1 there is a whole population of values of X_2. The

TABLE 10.1.1
STATURE (IN.) OF BROTHER AND SISTER
(illustration taken from Pearson and Lee's sample of 1401 families)

Family Number	1	2	3	4	5	6	7	8	9	10	11
Brother, X_1	71	68	66	67	70	71	70	73	72	65	66
Sister, X_2	69	64	65	63	65	62	65	64	66	59	62

$n = 11$ $\quad \bar{X}_1 = 69 \quad \bar{X}_2 = 64 \quad \Sigma x_1^2 = 74 \quad \Sigma x_2^2 = 66 \quad \Sigma x_1 x_2 = 39$

$r = \Sigma x_1 x_2 / \sqrt{(\Sigma x_1^2)(\Sigma x_2^2)} = 39/\sqrt{(74)(66)} = 0.558$ Pearson and Lee's $r = 0.553$

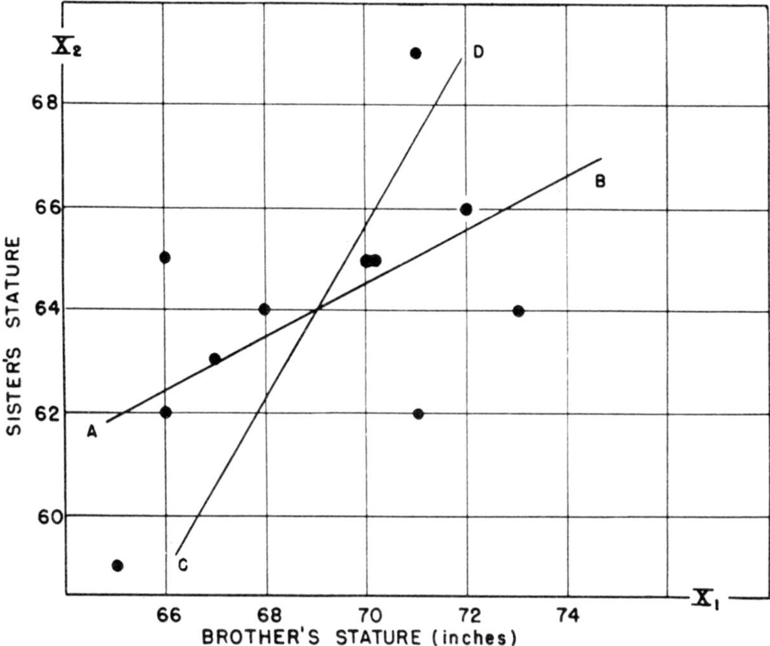

Fig. 10.1.1—Scatter (or dot) diagram of stature of 11 brother-sister pairs; $r = 0.558$.

population regression of X_2 on X_1 is the line that relates the means $\mu_{2.1}$ of these populations to X_1. Similarly, for each X_2 there is a population of values of X_1, and the population regression of X_1 on X_2 is the line relating their means $\mu_{1.2}$ to X_2. These two lines answer two different questions. They coincide only if we have an exact mathematical relationship.

EXAMPLE 10.1.1—Verify that $r = 0.91$ in the following pairs:

X_1	2	5	6	10	8	14	12	15	18	20
X_2	1	2	2	2	3	3	4	4	4	5

10.2—Properties of *r*. Two properties of r should be noted:

(*i*) The sample correlation coefficient r is a pure number without units or dimensions because the units of its numerator and denominator are both the products of the units in which X_1 and X_2 are measured. One useful consequence is that r can be computed from coded values of X_1 and X_2. No decoding is required.

(*ii*) Also, r always lies between -1 and $+1$ (see section 10.4). Positive values of r indicate a tendency of X_1 and X_2 to increase together. When r is negative, large values of X_1 are associated with small values of X_2.

To help you acquire some experience of the nature of r, a number of simple tables with the corresponding graphs are displayed in figure 10.2.1. In each of these tables $n = 9$, $\bar{X}_1 = 12$, $\bar{X}_2 = 6$, $\Sigma x_1^2 = 576$, $\Sigma x_2^2 = 144$. Only $\Sigma x_1 x_2$ changes

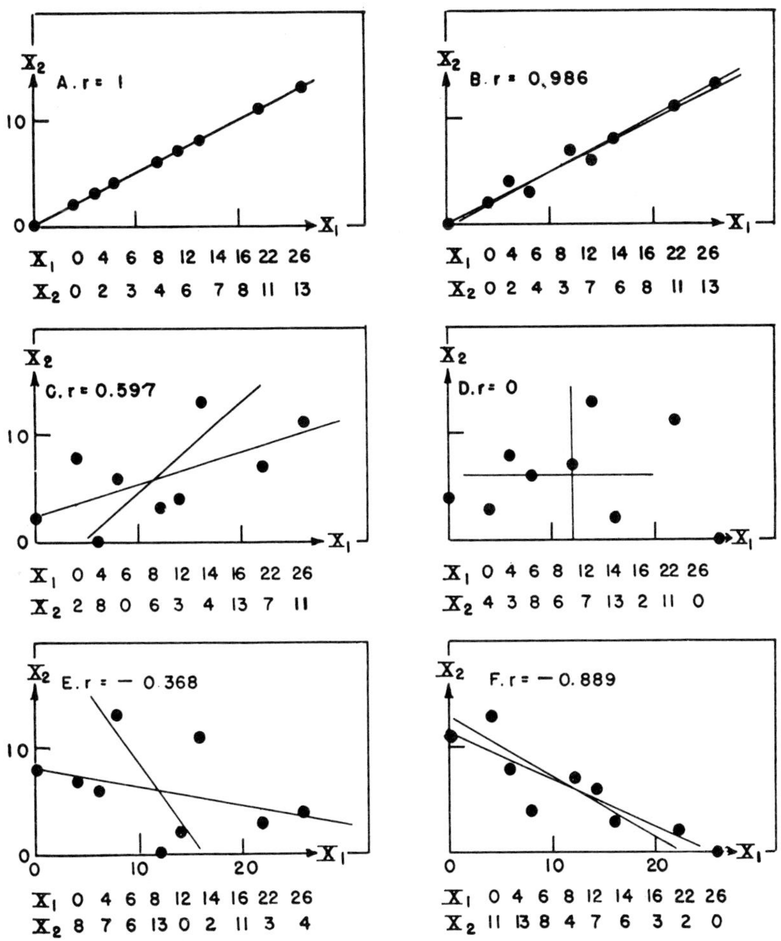

Fig. 10.2.1—Scatter diagrams with correlations ranging from 1 to −0.889.

and with it the value of r. Since $\sqrt{(\Sigma x_1^2)(\Sigma x_2^2)} = \sqrt{(576)(144)} = 288$, the correlation is easily evaluated in the several tables by calculating $\Sigma x_1 x_2$ and dividing by 288.

In A, the nine points lie on a straight line, the condition for $r = 1$. The two variables keep in perfect step, any change in one being accompanied by a proportionate change in the other. Graph B depicts some deviation from an exact relationship, the ellipse being long and thin with r slightly less than 1. In C, the ellipse widens; it approaches a circle in D where $r = 0$, denoting no relation between the two variables. Graphs E and F show negative correlations tending toward −1. To summarize, the thinness of the ellipse of points reflects the absolute size of r, while the inclination of the axis upward or downward shows its sign.

The larger correlations, either positive or negative, are fairly obvious from the graphs. It is not so easy to make a visual evaluation if the absolute value of r is less than 0.5; even the direction of inclination of the ellipse is elusive if r is between -0.3 and $+0.3$. Furthermore, in these small samples a single dot can make a great deal of difference. In D, for example, if the point (26, 0) were changed to (26, 9), r would be increased from 0 to 0.505. This emphasizes the fact that sample correlations from a bivariate population are quite variable if n is small. In assessing the value of r in a table, select some extreme values of one variable and note whether they are associated with extreme values of the other. If no such tendency can be detected, r is likely to be small.

EXAMPLE 10.2.1—In the following data, verify that $r = 0$.

X_1	0	1	2	3	4	5	6
X_2	0	5	8	9	8	5	0

Nevertheless, the data were constructed from an exact relationship, $X_2 = 6X_1 - X_1^2$. The example illustrates that r measures closeness of *linear* relationship only.

10.3—Bivariate normal distribution. The population correlation coefficient ρ and its sample estimate r are intimately connected with a bivariate population known as the bivariate normal distribution. This distribution is illustrated by table 10.3.1, which shows the joint frequency distributions of height X_1 and length of forearm X_2 for 348 men. The data are from the article by Galton (2) in 1888 in which the term "co-relation" was first proposed. As its name implies, the distribution is the natural extension of the normal distribution from one to two variates.

The theoretical distribution has the following properties. (Since Galton's data are a sample and since the relationship between height and length of forearm is not exactly bivariate normal, table 10.3.1 exhibits these properties only roughly.)

1. For each value of X_1 the corresponding values of X_2 follow a normal distribution. The means $\mu_{2.1}$ of these normal distributions lie on a straight line—the population regression of X_2 on X_1. Also, each distribution has the same standard deviation $\sigma_{2.1}$. In table 10.3.1 note that the sample means $\bar{X}_{2.1}$ lie fairly close to a straight line. The sample standard deviations $s_{2.1}$ for each X_1 vary around a mean of about 0.55.

2. Similarly, for each value of X_2 the corresponding X_1s are normally distributed. The means $\mu_{1.2}$ of these distributions also lie on a straight line and the standard deviations $\sigma_{1.2}$ are constant. The sample estimates $s_{1.2}$ suggest a value of about 1.7 for $\sigma_{1.2}$.

3. The marginal distributions of X_1 and X_2 are also normal. The variate X_1 follows $\mathcal{N}(\mu_1, \sigma_1^2)$, while X_2 follows $\mathcal{N}(\mu_2, \sigma_2^2)$.

4. The relative frequency of the bivariate normal distribution at the point (X_1, X_2) has the coefficient $1/(2\pi\sigma_1\sigma_2\sqrt{1 - \rho^2})$, followed by the exponential e with the exponent

$$-\frac{1}{2(1-\rho^2)}\left[\frac{(X_1-\mu_1)^2}{\sigma_1^2} - \frac{2\rho(X_1-\mu_1)(X_2-\mu_2)}{\sigma_1\sigma_2} + \frac{(X_2-\mu_2)^2}{\sigma_2^2}\right] \qquad (10.3.1)$$

TABLE 10.3.1

FREQUENCY OF PAIRS OF MEASUREMENTS OF HEIGHT AND LENGTH OF FOREARM. GALTON'S DATA

Length of Forearm X_2	Height, X_1 in.																		
	59–60	60–61	61–62	62–63	63–64	64–65	65–66	66–67	67–68	68–69	69–70	70–71	71–72	72–73	73–74	74–75	f_2	$\overline{X}_{1.2}$	$s_{1.2}$
21.0–21.5																1	1	74.4	. . .
20.5–21.0													1	1			2	72.0	0.71
20.0–20.5														1			1	72.4	. . .
19.5–20.0										2			1		2		5	71.0	2.51
19.0–19.5									2	4	6	11	8	4	2	1	38	70.6	1.62
18.5–19.0					1		2	6	8	7	15	13	2	1			55	68.8	1.77
18.0–18.5						3	8	15	28	14	25	5	2	2			102	68.0	1.67
17.5–18.0				2	1	2	12	18	15	7	2	1	1				61	66.7	1.62
17.0–17.5			1	3	6	11	10	7	7	3	1						49	65.4	1.80
16.5–17.0			1	5	6	5	4	1	1	1	1						25	64.4	1.97
16.0–16.5	1	1	1	3	2												8	62.0	1.41
15.5–16.0		1															1	60.4	. . .
Frequency, f_1	1	2	3	13	16	21	36	47	61	38	50	30	15	9	4	2	348		
Mean, $\overline{X}_{2.1}$	16.2	16.0	16.8	16.9	17.1	17.3	17.7	17.9	18.1	18.3	18.4	18.8	19.1	19.2	19.5	20.2	18.1		
SD, $s_{2.1}$	. . .	0.36	0.50	0.52	0.60	0.48	0.54	0.42	0.51	0.68	0.48	0.41	0.70	0.83	0.29	1.41	0.905		

This distribution has five parameters: μ_1, μ_2, σ_1, σ_2, and the population correlation coefficient ρ, of which r is an estimator.

5. In property 1, the population regression of X_2 on X_1 is the straight line

$$\mu_{2.1} = \mu_2 + (\rho\sigma_2/\sigma_1)(X_1 - \mu_1) \tag{10.3.2}$$

The residual standard deviation $\sigma_{2.1} = \sigma_2\sqrt{1 - \rho^2}$. Similarly, the population regression of X_1 for fixed X_2 in property 2 is the line

$$\mu_{1.2} = \mu_1 + (\rho\sigma_1/\sigma_2)(X_2 - \mu_2) \tag{10.3.3}$$

and the standard deviation $\sigma_{1.2} = \sigma_1\sqrt{1 - \rho^2}$.

A common expression for ρ can be derived from the mathematical form of the bivariate normal. Define the *population covariance* of X_1 and X_2 as

$$\text{Cov}(X_1, X_2) = \text{average value of } (X_1 - \mu_1)(X_2 - \mu_2) \tag{10.3.4}$$

Then

$$\rho = \text{Cov}(X_1, X_2)/(\sigma_1\sigma_2) \tag{10.3.5}$$

This form might have been suggested by considering the formula for r in large samples:

$$r = \Sigma(x_1x_2)/\sqrt{(\Sigma x_1^2)(\Sigma x_2^2)}$$

Divide numerator and denominator on the right by $n - 1$:

$$r = \Sigma(X_1 - \overline{X}_1)(X_2 - \overline{X}_2)/[(n - 1)s_1s_2] \tag{10.3.6}$$

As n becomes large, $\overline{X}_1$ and $\overline{X}_2$ tend to coincide with μ_1 and μ_2, respectively; s_1 and s_2 tend to σ_1 and σ_2; and division by $n - 1$ becomes equivalent to division by n. Hence, (10.3.6) becomes (10.3.5).

EXAMPLE 10.3.1—(*i*) In table 10.3.1, plot the $\overline{X}_{2.1}$ against the class midpoints of the heights for the classes from 62–63 in. height to 72–73 in. height. The classes with very high or low heights are omitted because they have such small sample sizes that their class means are unstable. As an illustration of the bivariate normal, note that the $\overline{X}_{2.1}$ lie very close to a straight line in this plot. (*ii*) Similarly, plot the $s_{2.1}$ values against X_1 and note that there is no consistent trend, reflecting the property that $\sigma_{2.1}$ is constant for the bivariate normal as X_1 varies.

EXAMPLE 10.3.2—The body and comb weights of ten 15-day-old chicks are given in example 9.7.1. In the notation of the present chapter the data are as follows:

Body weight, X_1 (g)	83	72	69	90	90	95	95	91	75	70
Comb weight, X_2 (mg)	56	42	18	84	56	107	90	68	31	48

For these data, $\overline{X}_1 = 83$ g, $\overline{X}_2 = 60$ mg, $\Sigma x_1^2 = 1000$, $\Sigma x_2^2 = 6854$, $\Sigma x_1x_2 = 2302$. Plot the points and the sample regressions of X_1 on X_2 and X_2 on X_1 in a form similar to figure 10.1.1. Calculate r. Ans. $r = 0.879$.

EXAMPLE 10.3.3—If $X_3 = a + bX_1$ and $X_4 = c + dX_2$, where a, b, c, and d are constants, prove that $r_{34} = r_{12}$. This result shows, as mentioned in section 10.2, that coding does not change the value of r.

10.4—Some uses of the correlation coefficient. The regression techniques described in chapter 9 and their extensions to multivariate problems are much more widely used than correlation techniques. The correlation coefficient merely estimates the degree of closeness of linear relationship between two variables Y and X, and the meaning of this concept is not easy to grasp. Often the interesting questions are, How much does Y change for a given change in X? What is the shape of the curve connecting Y and X? How accurately can Y be predicted from knowledge of X? These questions are handled by regression techniques. However, the correlation coefficient enters naturally into some important applications.

A useful property of r is obtained from the method of computing $s_{y.x}^2$ in a regression problem without finding any deviations. Recall from (9.3.7) and (9.3.9) that

$$\Sigma d_{y.x}^2 = (n - 2)s_{y.x}^2 = \Sigma y^2 - (\Sigma xy)^2/\Sigma x^2$$

Substituting $(\Sigma xy)^2 = r^2 \Sigma x^2 \Sigma y^2$, we have

$$\Sigma d_{y.x}^2 = (n - 2)s_{y.x}^2 = (1 - r^2)\Sigma y^2 \qquad (10.4.1)$$

Since $\Sigma d_{y.x}^2$ cannot be negative, this equation shows that r must lie between -1 and $+1$. Moreover, if r is ± 1, $\Sigma d_{y.x}^2 = 0$ and the sample points lie exactly on a line.

Result (10.4.1) provides another way of appraising the closeness of the relation between two variables. The original sample variance of Y, when no regression is fitted, is $s_y^2 = \Sigma y^2/(n - 1)$, while the variance of the deviations of Y from the linear regression is $(1 - r^2)\Sigma y^2/(n - 2)$ as shown above. Hence, the proportion of the variance of Y that is *not* associated with its linear regression on X is estimated by

$$s_{y.x}^2/s_y^2 = (n - 1)(1 - r^2)/(n - 2) \doteq (1 - r^2)$$

if n is at all large. Thus r^2 may be described approximately as *the estimated proportion of the variance of Y that can be attributed to its linear regression on X, while* $(1 - r^2)$ *is the proportion free from X*. The quantities r^2 and $(1 - r^2)$ are shown in table 10.4.1 for a range of values of r.

When r is 0.5 or less, only a minor portion of the variation in Y can be attributed to its linear regression on X. At $r = 0.7$, about half the variance of Y is associated with X, and at $r = 0.9$, about 80%. In a sample of size 200, an r of 0.2 would be significant at the 1% level but would indicate that 96% of the variation of Y was not explainable through its relation with X. A verdict of statistical significance merely shows that there is a linear relation with nonzero slope. Remember also that convincing evidence of an association, even though close, does not prove that X is the cause of the variation in Y. Evidence of causality must come from other sources.

TABLE 10.4.1
ESTIMATED PROPORTIONS OF THE VARIANCE OF Y ASSOCIATED AND NOT ASSOCIATED WITH X IN A LINEAR REGRESSION

	Proportion			Proportion	
r	Associated r^2	Not $1 - r^2$	r	Associated r^2	Not $1 - r^2$
±0.1	0.01	0.99	±0.6	0.36	0.64
± .2	.04	.96	± .7	.49	.51
± .3	.09	.91	± .8	.64	.36
± .4	.16	.84	± .9	.81	.19
±0.5	0.25	0.75	±0.95	0.90	0.10

An example of this type of application occurs in the classical agricultural experiments at the Rothamsted Experimental Station. Wheat was grown year after year on a given plot with the same plant foods and farming methods. What caused the yields Y on the plot to vary from year to year? First, the annual yields showed slow, long-term changes. On some plots the soil fertility was gradually increased and on others decreased by the plant foods given. Beyond this, the obvious sources to examine are different aspects of the weather during the growing seasons. An important part of the research consisted of estimating the proportions of the year-to-year variance in yields on the plot that were associated with long-term trends and with variations in rainfall, temperature, and sunshine during the growing seasons.

As another example of this type, a large sample study of the performance of school children of different ethnic groups in standard exams led to the Coleman Report (3), which was widely discussed. In this report the variance in exam scores among children in a given grade was partitioned into the proportions associated with the school attended, with objective aspects of the home (e.g., parents' education, reading material in the home), with the parents' interest in the child's progress, and with the child's own outlook. As we have seen, these proportions can all be estimated by means of r^2 and its extension to problems with more than one X variable.

A second major application of the correlation coefficient is to the investigation of the precision of measurements. In many types of research, particularly with human subjects, the problem of measuring the variable that we wish to study is formidable. Examples are when X stands for intelligence, amount of pain suffered, and aspects of personality. Often, as noted in section 9.14, we cannot measure X. What we can measure is a related variable $X' = X + e$, where e is an error of measurement, or sometimes an *error of specification,* when we realize that X' measures a concept somewhat different from X. The variable X' is called the fallible measurement and X the true measurement.

In the simplest situation in which X and e are uncorrelated and the errors e have population mean zero (i.e., the measuring process is unbiased),

$$\sigma_{X'}^2 = \sigma_X^2 + \sigma_e^2 \qquad (10.4.2)$$

Often the important quantity in appraising the quality of our measurement process is the ratio $\sigma_X^2/\sigma_{X'}^2$, that is, the fraction of variance of our fallible measurement that represents true variation in X as distinct from error or noise, as it is sometimes called. This ratio is called the *reliability* (13) of the measuring process.

One method of measuring reliability is to draw a random sample from the population under study and make two independent measurements X'_{i1}, X'_{i2} of each X_i. The correlation coefficient comes into the picture because if the *e*s and X are uncorrelated,

$$\text{Cov}(X'_{i1}, X'_{i2}) = \text{Cov}(X_i + e_{i1})(X_i + e_{i2}) = \sigma_X^2 \tag{10.4.3}$$

Also, X'_{i1} and X'_{i2} both have variance $\sigma_{X'}^2$. Hence ρ, the population correlation between X'_{i1} and X'_{i2}, is

$$\rho = \text{Cov}(X'_{i1}, X'_{i2})/\sigma_{X'}^2 = \sigma_X^2/\sigma_{X'}^2 \tag{10.4.4}$$

Thus, ρ is the reliability and the sample r estimates the reliability of the method of measurement.

This approach and variations of it are widely used in education and psychology. The concept of reliability has had to be modified to meet different situations. Suppose that a number of questions are given to a subject, either in an exam that measures skill or knowledge or in measuring some aspect of the subject's personality, perhaps aggressiveness or docility. One technique is to calculate the correlation r, called *split half* reliability, between the scores obtained by a subject on two random halves of a set of questions. If (10.4.3) holds, r estimates the reliability of half the exam. From this, the reliability of the whole exam can easily be estimated. In another type of reliability, *test-retest* reliability, two sets of questions (e.g., two exams) intended to measure the same skill are given to the subject on different occasions, usually not far apart in time. In test-retest reliability the term e may include more than pure error of measurement, since it reflects any time-variations in the subject's ability due to minor illness or need to concentrate on other problems, for example. For some uses the investigator may consider this kind of reliability more relevant to the purpose than split-half reliability. Other variants of the concept of reliability are also used.

EXAMPLE 10.4.1—The following sample with $n = 16$ was drawn from a bivariate normal distribution with $\rho = 0.7$, $\beta = 1$. The (X, Y) pairs are (0.1, 0.7), (1.0, 2.2), (1.1, 1.5), (1.7, 3.0), (1.8, 1.8), (2.0, 2.8), (2.1, 2.8), (2.3, 2.3), (2.6, 2.7), (2.7, 4.4), (2.9, 2.6), (3.4, 3.2), (4.4, 2.7), (4.4, 3.8), (4.9, 4.0), (6.1, 6.6). The sample correlation r was 0.701 and the sample regression of Y on X was $\hat{Y} = 0.258 + 0.988X$. Create a sample of size 8 by selecting the pairs with the four smallest X values and the four largest X values. For this sample, calculate r and the regression of Y on X. Ans. $r = 0.883$, substantially increased owing to the selection of extremes in X. Regression: $\hat{Y} = 0.313 + 0.928X$, changed only by random sampling errors. Selection on X gives a distorted estimate r of ρ.

EXAMPLE 10.4.2—Suppose that u_1, u_2, u_3 are independent standard normal variates and $X_1 = 3u_1 + u_2$, $X_2 = 3u_1 + u_3$. What is the correlation ρ between X_1 and X_2? Ans. 0.9. More generally, if $X_1 = fu_1 + u_2$, $X_2 = fu_1 + u_3$, then $\rho = f^2/(f^2 + 1)$. It can also be shown that X_1 and X_2 follow the bivariate normal so that this method will provide samples from a bivariate normal with any desired ρ. (For negative ρs, take $X_2 = -fu_1 + u_2$.)

EXAMPLE 10.4.3—Some measuring instruments give the correct measurement but are very expensive. One method of appraising a cheaper substitute measurement is to take n specimens, obtaining for each the correct measurement X_{1i} and the cheaper fallible measurement $X_{2i} = X_{1i} + e_i$. Show that if e_i and X_{1i} are uncorrelated, the square of the population correlation between X_1 and X_2 is the reliability of X_2.

EXAMPLE 10.4.4—If ρ is the reliability of the average score on a random half of the exam questions in the split-half method, show that the reliability of the average score on all the questions is $2\rho/(1 + \rho)$. (Assume, as in the simplest model, that the errors of measurement in the two halves are uncorrelated with each other and with the true score.)

10.5—Testing the null hypothesis, $\rho = 0$. Figure 10.5.1 shows the frequency distributions of r for samples of 8 drawn from bivariate normal populations with correlations 0 and 0.8. In both distributions, r lies between -1 and $+1$. The former is not far from normal. The reason for the pronounced skewness of the latter is not hard to see. Since the parameter is 0.8, sample values can exceed this by no more than 0.2 but may be as much as 1.8 less than the parameter value. Whenever there is a limit to the variation of a statistic at one end of the scale with practically none at the other, the distribution curve is likely to be asymmetrical. Of course, with increasing sample size this skewness tends to disappear. Samples of 400 pairs, drawn from a population with a correlation even as great as 0.8, have little tendency to range more than 0.05 on either side of the parameter. Consequently, the upper limit, unity, would not constitute a restriction and the distribution would be almost normal.

From the distribution of r when $\rho = 0$, table A 11 gives the 10%, 5%, 2%,

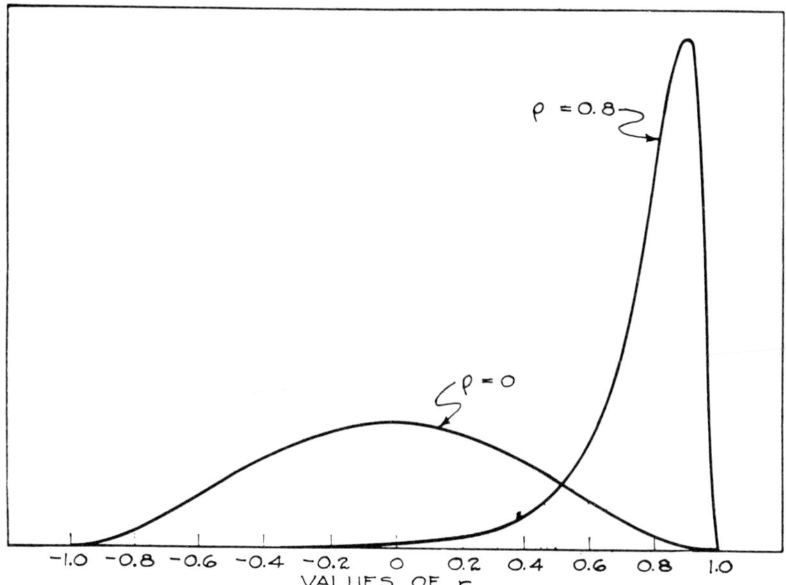

Fig. 10.5.1—Distribution of sample correlation coefficients in samples of 8 pairs drawn from two normally distributed bivariate populations having the indicated values of ρ.

and 1% significance levels of r. Note that the table is entered by the *degrees of freedom*, $n - 2$, in the residual mean square $s_{y.x}^2$ or $s_{x.y}^2$. As an illustration, consider the value $r = 0.597$ that was obtained from a sample of size 9 in diagram C of figure 10.2.1. For 7 df, the 5% value of r in table A 11 is 0.666. The observed r is not statistically significant and the null hypothesis is not rejected. This example throws light on the difficulty of graphic evaluation of correlations, especially when the number of degrees of freedom is small—they may be no more than accidents of sampling. Since the distribution of r is symmetrical when $\rho = 0$, the sign of r is ignored when making a two-tailed test. For one-tailed tests at the 5% and 1% levels, use the 10% and 2% levels in table A 11.

Among the following correlations, observe how conclusions are affected by both sample size and the size of r:

Pairs	df	r	Hypothesis, $\rho = 0$
20	18	0.60	reject at 1% level
100	98	.21	reject at 5% level
10	8	.60	not rejected
15	13	− .50	not rejected
500	498	−0.15	reject at 1% level

You now know two methods of testing for a linear relation between the variables Y and X. The first is to test the regression coefficient $b_{y.x}$ by calculating $t = b_{y.x}/s_b$ and reading the t table with $(n - 2)$ df. The second is the test of r. Fisher (14) showed that the two tests are identical. In fact, the table for r can be computed from the t table by means of the relation

$$t = b_{y.x}/s_b = r\sqrt{n - 2}/\sqrt{1 - r^2} \qquad (\text{df} = n - 2) \qquad (10.5.1)$$

(See example 10.5.1.) To illustrate, we found that the 5% level of r for 7 df is 0.666. Let us compute

$$t = 0.666\sqrt{7}/\sqrt{1 - 0.666^2} = 2.365$$

Reference to the t table shows that this is the 5% level of t for 7 df. In practice, use whichever test you prefer.

This relation raises a subtle point. The t test of b requires only that Y be normally distributed; the values of X need not be normal and may be selected by the investigator. On the other hand, we have stressed that r and ρ are intimately connected with random samples from the bivariate normal distribution. Fisher proved that in the particular case $\rho = 0$, the distribution of r is the same whether X is normal or not, provided that Y is normal.

EXAMPLE 10.5.1—To prove relation (10.5.1), which connects the t test of b with the test of r, you need three relations: (*i*) $b_{y.x} = rs_y/s_x$, (*ii*) $s_b = s_{y.x}/\sqrt{\Sigma x^2}$, (*iii*) $s_{y.x}^2 = (1 - r^2)\Sigma y^2/(n - 2)$, as shown in (10.4.1). Start with $t = b_{y.x}/s_b$ and make these substitutions to establish the result.

10.6—Confidence limits and tests of hypotheses about ρ. The methods given in this section, which apply when $\rho \neq 0$, require the assumption that the

(X, Y) or (X_1, X_2) pairs are a random sample from a bivariate normal distribution.

Table A 11 or the t table can be used only for testing the null hypothesis $\rho = 0$. They are unsuited for testing other null hypotheses, such as $\rho = 0.5$, or $\rho_1 = \rho_2$, or for making confidence statements about ρ. When $\rho \neq 0$ the distribution of r changes, becoming skew, as seen in figure 10.5.1.

A solution of these problems was provided by Fisher (4), who devised a transformation from r to a quantity z that was distributed almost normally with standard error approximately $\sigma_z = 1/\sqrt{n-3}$. The relation of z to r is given by

$$z = (1/2)[\ln(1 + r) - \ln(1 - r)]$$

Table A 12 (r to z) and A 13 (z to r) enable us to change from one to the other with sufficient accuracy. Following are some examples of the use of z:

1. *Set confidence limits to the value of ρ in the population from which a sample r has been drawn.* As an example, consider $r = -0.889$, based on 9 pairs of observations, figure 10.2.1F. From table A 12, $z = 1.417$ corresponds to $r = 0.889$. Since $n = 9$, $\sigma_z = 1/\sqrt{6} = 0.408$. Since z is distributed almost normally, independent of sample size, $z_{0.01} = 2.576$. For $P = 0.99$, we have as confidence limits for z,

$$1.417 - (2.576)(0.408) \leq z \leq 1.417 + (2.576)(0.408)$$
$$0.366 \leq z \leq 2.468$$

Using table A 13 to find the corresponding r and restoring the sign, the 0.99 confidence limits for ρ are given by

$$-0.986 \leq \rho \leq -0.350$$

Note two facts: (*i*) in small samples the estimate r is not accurate; and (*ii*) the limits are not equally spaced on either side of r, a consequence of its skewed distribution.

2. *Test the hypothesis that ρ has some particular value other than zero in the sampled population.* In a class exercise in sampling, $r = 0.28$ was observed in a sample of 50 pairs from $\rho = 0.5$. What is the probability of a larger deviation? For $r = 0.28$, $z = 0.288$, and for $\rho = 0.5$, $z = 0.549$. The difference, $0.549 - 0.288 = 0.261$, has a standard error $1/\sqrt{n-3} = 1/\sqrt{47} = 0.1459$. Hence, the normal deviate is $0.261/0.1459 = 1.80$, which does not reach the 5% level in a two-tailed test; the sample is not as unusual as a 1-in-20 chance.

3. *Test the hypothesis that two independent values r_1 and r_2 are estimates of the same ρ.* Convert each to z and test the significance of the difference between the two zs. For one method, two measurements of each object with independent errors of measurement give an estimated reliability $r_1 = 0.862$, based on $n_1 = 60$. Another method gives $r_2 = 0.720$, based on an independent sample with $n_2 = 49$. Does the first method have higher reliability? The details of the test are given in table 10.6.1.

The variance of $D = z_1 - z_2$ is the sum 0.0392 of their two individual

TABLE 10.6.1
TEST OF SIGNIFICANCE OF THE DIFFERENCE BETWEEN TWO CORRELATION COEFFICIENTS

Method	Sample Size	r	z	$1/(n-3)$
1	60	0.862	1.301	0.0175
2	49	0.720	0.908	0.0217
			$D = 0.393$	Sum = 0.0392

variances. Hence $D/\sigma_D = 0.393/0.198 = 1.98$, just about the 5% level in a two-tailed test. We conclude that the first method is more accurate.

4. *Test the hypothesis that several rs are estimates of the same* ρ. Lush's data (5) in table 10.6.2 on the correlation between initial weight and gain in weight of six lots of steers will serve as an example, though with these small sample sizes the power of the test is weak. First convert the r_i to z_i, recording also the terms $(n_i - 3)$ for each z_i Under H_0 the z_i are all estimates of the same mean μ but have different variances $\sigma_i^2 = 1/(n_i - 3)$. The test of significance is based on the result that if k normal deviates have the same mean μ but have different variances σ_i^2, the quantity

$$\sum_{i=1}^{k} w_i(z_i - \bar{z}_w)^2 = \sum_{i=1}^{k} w_i z_i^2 - \left(\sum_{i=1}^{k} w_i z_i\right)^2 \Big/ \sum_{i=1}^{k} w_i \tag{10.6.1}$$

is distributed as χ^2 with $(k - 1)$ df, where $w_i = 1/\sigma_i^2$ and $\bar{z}_w = \Sigma w_i z_i/\Sigma w_i$. In this application, $w_i = n_i - 3$ and

$$\chi^2 = \Sigma(n_i - 3)z_i^2 - [\Sigma(n_i - 3)z_i]^2/\Sigma(n_i - 3) \tag{10.6.2}$$

In table 10.6.2 the individual values of $(n_i - 3)z_i^2$ and $(n_i - 3)z_i$ are shown. They give

$$\chi^2 = 9.753 - 15.478^2/39 = 3.610$$

with 5 df. From the χ^2 table, $P = 0.61$, so H_0 is not rejected.

TABLE 10.6.2
TEST OF HYPOTHESIS OF COMMON ρ AND ESTIMATION OF ρ. CORRELATION BETWEEN INITIAL WEIGHT AND GAIN OF STEERS

Samples	n_i	$n_i - 3$	r_i	z_i	Weighted z $(n_i - 3)\,z_i$	Weighted Square $(n_i - 3)\,z^2$	Corrected z_i
1927 Herefords	4	1	0.929	1.651	1.651	2.726	1.589
1927 Brahmans	13	10	.570	0.648	6.480	4.199	0.633
1927 Backcrosses	9	6	.455	0.491	2.946	1.446	0.468
1928 Herefords	6	3	−.092	−0.092	−0.276	0.025	−0.055
1928 Brahmans	11	8	.123	0.124	0.992	0.123	0.106
1928 Backcrosses	14	11	0.323	0.335	3.685	1.234	0.321
Total	57	39			15.478	9.753	

5. *Combine independent estimates of ρ.* If it is decided to accept H_0, the best combined z value is $\bar{z}_w = 15.478/39 = 0.397$ from table 10.6.2. The combined estimate of ρ corresponding to $\bar{z}_w$ is $\hat{\rho} = 0.377$. The standard deviation of $\bar{z}_w$ is $1/\sqrt{\Sigma w_i} = 1/\sqrt{39}$. As in example 1 in this section, confidence limits are calculated in the z scale and then converted to the ρ scale.

Fisher pointed out that there is a small bias in z_i, each being too large by $\rho/[2(n - 1)]$. The bias may usually be neglected. It might be serious if large numbers of correlations were averaged, because the bias accumulates. If there is need to adjust for bias in table 10.6.2, $\hat{\rho} = 0.377$ may be substituted for ρ; then the approximate bias for each z may be deducted and the calculation of the average z repeated. Since this will decrease the estimated ρ, it is well to guess ρ slightly less than $\bar{r}$. For instance, it may be guessed that $\rho = 0.37$; then the correction in the first z is $0.37/[2(4 - 1)] = 0.062$, and the corrected z is $1.651 - 0.062 = 1.589$. The other corrected zs are in the last column of the table. The sum of the products,

$$\Sigma(n - 3)(\text{corrected } z_i) = 14.941$$

is divided by 39 to get the corrected mean value of the z_i, 0.383. The corresponding correlation is 0.365, only a minor change.

For tables of the distribution of r when $\rho \neq 0$, see reference (6).

EXAMPLE 10.6.1—(*i*) Set 95% confidence limits to the correlation 0.986, $n = 533$, between live and dressed weights of swine. (*ii*) What would have been the confidence limits if the number of swine had been 25? Ans. (*i*) (0.983, 0.988), (*ii*) (0.968, 0.994).

EXAMPLE 10.6.2—In four studies of the correlation between wing and tongue length in bees, Grout (7) found values of $r = 0.731, 0.354, 0.690$, and 0.740, each based on a sample of 44. Test the hypothesis that these samples are from a common ρ. Ans. $\chi^2 = 9.164$, df $= 3$, $P = 0.03$. In only about 3 trials per 100 would you expect such disagreement among four correlations drawn from a common population. One would like to know more about the discordant correlation 0.354 before drawing conclusions.

EXAMPLE 10.6.3—Estimate ρ in the population from which the three bee correlations, 0.731, 0.690, and 0.740, were drawn. Ans. 0.721.

EXAMPLE 10.6.4—Set 99% confidence limits on the foregoing bee correlation. Note: The combined z has standard error $1/\sqrt{123}$. The confidence limits for ρ are (0.590, 0.815).

10.7—Variance of a linear function. We have already encountered equations applicable to the variance of the sum $S = X_1 + X_2$ of two variates or the difference $D = X_1 - X_2$. If X_1 and X_2 are uncorrelated,

$$V(S) = V(D) = \sigma_1^2 + \sigma_2^2 \tag{10.7.1}$$

Then, in comparing paired with independent samples, we had the result in (6.13.1) that when X_1 and X_2 may be correlated,

$$\sigma_D^2 = \sigma_1^2 + \sigma_2^2 - 2\,\text{Cov}(X_1, X_2) \tag{10.7.2}$$

$$= \sigma_1^2 + \sigma_2^2 - 2\rho\sigma_1\sigma_2 \tag{10.7.3}$$

In the equation for the variance of a sum, the minus sign in the covariance term becomes plus. The corresponding sample equations are

$$s^2_{X_1 \pm X_2} = s_1^2 + s_2^2 \pm 2rs_1s_2 \qquad (10.7.4)$$

The most general equation of this type is for the variance of a *linear function* L of k variates, $L = \Sigma_{i=1}^{k} \lambda_i X_i$, where the coefficients λ_i are constants. In summation notation the equation is

$$\sigma_L^2 = \sum_{i=1}^{k} \lambda_i^2 \sigma_i^2 + 2 \sum_{i=1}^{k} \sum_{j>i}^{k} \lambda_i \lambda_j \text{Cov}(X_i, X_j) \qquad (10.7.5)$$

The covariance terms, $k(k - 1)/2$ in number, extend over every pair of variates. For example, for three variates,

$$\sigma_L^2 = \lambda_1^2\sigma_1^2 + \lambda_2^2\sigma_2^2 + \lambda_3^2\sigma_3^2 + 2\lambda_1\lambda_2\text{Cov}(X_1, X_2) + 2\lambda_1\lambda_3\text{Cov}(X_1, X_3) + 2\lambda_2\lambda_3\text{Cov}(X_2, X_3)$$

Equation (10.7.5) and particular cases arising from it are used repeatedly in statistical work. If the X_i are uncorrelated,

$$\sigma_L^2 = \sum_{i=1}^{k} \lambda_i^2 \sigma_i^2 \qquad (10.7.6)$$

The general equation (10.7.5) is proved by noting that the sampling error $L - \mu_L$ in L may be written $\Sigma\lambda_i(X_i - \mu_i)$. Expand the square of this quantity as a quadratic and substitute the average or expected value of each term. The result is the right-hand side of (10.7.5). But since the variance of L is $E(L - \mu_L)^2$ by definition, this proves (10.7.5).

An equation for the covariance of two linear functions of the same variables, $L = \Sigma\lambda_i X_i$ and $M = \Sigma\gamma_i X_i$, is sometimes useful:

$$\text{Cov}(L, M) = \sum_{i=1}^{k} \lambda_i \gamma_i \sigma_i^2 + \sum_{i=1}^{k} \sum_{j>i}^{k} (\lambda_i\gamma_j + \lambda_j\gamma_i)\text{Cov}(X_i, X_j) \qquad (10.7.7)$$

EXAMPLE 10.7.1—In table 10.1.1, subtract each sister's height from her brother's; then compute the sum of squares of deviations of the differences. Use the results given in table 10.1.1 to verify (10.7.4) as applied to differences.

EXAMPLE 10.7.2—You have a random sample of n pairs (X_{1i}, X_{2i}) in which the correlation between X_{1i} and X_{2i} is ρ. What is the correlation between the sample means $\bar{X}_1$ and $\bar{X}_2$? Ans. ρ, the same as the correlation between members of an individual pair.

EXAMPLE 10.7.3—The members of a sample, $X_1, X_2, \ldots, X_n$, all have variance σ^2 and every pair has correlation ρ. By (10.7.5), what is the variance of the sample mean $\bar{X}$? Ans. $\sigma_{\bar{X}}^2 = \sigma^2[1 + (n - 1)\rho]/n$. The standard formula σ^2/n for $\sigma_{\bar{X}}^2$ requires the members of the sample to be uncorrelated.

EXAMPLE 10.7.4—Given a random sample of n independent values X_i from a population, show from (10.7.7) that the linear functions $L = \Sigma l_i X_i$ and $M = \Sigma m_i X_i$ are uncorrelated if and only if $\Sigma l_i m_i = 0$.

EXAMPLE 10.7.5—In a sample of 300 ears of corn (8), the weight G of the grain had a standard deviation of 24.6 g and the weight C of the cob had a standard deviation of 4.2 g, while the correlation between these weights was $r = 0.691$. Find the standard deviation of the total ear weight $W = G + C$ and the correlation between G and W. Ans. $s_W = 27.7$ g, $r_{GW} = 0.993$.

EXAMPLE 10.7.6—Variables X_1 and X_2 have variances σ_1^2, σ_2^2 and correlation ρ. Show that the correlation between their sum $X_1 + X_2$ and their difference $X_1 - X_2$ is $(\sigma_1^2 - \sigma_2^2)/[(\sigma_1^2 + \sigma_2^2)^2 - 4\rho^2\sigma_1^2\sigma_2^2]^{1/2}$. The result—this correlation is zero for any value of ρ when $\sigma_1 = \sigma_2$—is used in the next section.

EXAMPLE 10.7.7—This example gives some practice in applying the rather complicated (10.7.5) and (10.7.7) for the variances and covariances of linear functions. For the variates X_1, X_2, X_3, we have $\sigma_1 = 1, \sigma_2 = 2, \sigma_3 = 4, \rho_{12} = 0.5, \rho_{13} = 0.2, \rho_{23} = -0.3$. Two linear functions are $L_1 = X_1 + X_2 + X_3$, $L_2 = 3X_1 - 2X_2 - X_3$. Find the standard deviations of L_1 and L_2 and the correlation between them. Ans. $\sigma_{L_1} = 4.45$, $\sigma_{L_2} = 3.82$, $\rho = -0.659$.

10.8—Comparison of two correlated variances in paired samples. In section 6.12 (p. 98) we test the null hypothesis that two *independent* estimates of variance, s_1^2 and s_2^2, are each estimates of the same unknown population variance σ^2. For a two-tailed test, calculate $F = s_1^2/s_2^2$, where s_1^2 is the larger of the two, and refer to table A 14.

This problem arises also when the two estimates s_1^2 and s_2^2 are correlated. For instance, in the sample of pairs of brothers and sisters (section 10.1.), we might wish to test whether brothers' heights X_1 are more or less variable than sisters' heights X_2. We can calculate s_1^2 and s_2^2, the variances of the two heights between families. But in our sample of 11 families the correlation between X_1 and X_2 was found to be $r = 0.558$. Although this did not reach the 5% level of r (0.602 for 9 df), the presence of a correlation was confirmed by Pearson and Lee's value of $r = 0.553$ for the sample of 1401 families from which our data were drawn. In another application, a specimen may be sent to two laboratories that make estimates X_1, X_2 of the concentration of a rare element contained in it. If a number of specimens are sent, we might wish to examine whether one laboratory gives more variability in results than the other.

The test to be described is valid for a sample of pairs of values X_1, X_2 that follows a bivariate normal. It holds for any value ρ of the population correlation between X_1 and X_2. If you are confident that ρ is zero the ordinary F test should be used, since it is slightly more powerful. When ρ is not zero, the F test is invalid.

The test is derived by an ingenious approach due to Pitman (9). Suppose that X_1 and X_2 have variances σ_1^2 and σ_2^2 and correlation ρ. Since X_1 and X_2 follow a bivariate normal, it is known that $D = X_1 - X_2$ and $S = X_1 + X_2$ also follow a bivariate normal. Now

$$\text{Cov}(DS) = \text{Cov}(X_1 - X_2)(X_1 + X_2) = \sigma_1^2 - \sigma_2^2 \tag{10.8.1}$$

since the two terms in $\text{Cov}(X_1X_2)$ cancel. Hence $\rho_{DS} = 0$ if $\sigma_1^2 = \sigma_2^2$.

Thus, the null hypothesis $\sigma_1^2 = \sigma_2^2$ can be tested by finding D and S for each pair, computing the sample correlation coefficient r_{DS}, and referring to table A 11. A significantly positive value of r_{DS} indicates $\sigma_1^2 > \sigma_2^2$, while a significantly negative one indicates $\sigma_1^2 < \sigma_2^2$.

We sometimes decide to make this test after computing s_1^2, s_2^2, and r but before finding D and S. By the following identity, r_{DS} can be calculated from $F = s_1^2/s_2^2$ and r.

$$r_{DS} = (F - 1)/\sqrt{(F + 1)^2 - 4r^2F} \qquad (10.8.2)$$

For example, in a sample of 173 boys aged 13–14, height had a standard deviation $s_1 = 5.299$, while leg length had $s_2 = 4.766$, both figures expressed as percentages of the sample means (10). The correlation between height and length was $r = 0.878$, a high value, as would be expected. To test whether height is relatively more variable than leg length, we have

$$F = (5.299/4.766)^2 = 1.236$$

and from (10.8.2),

$$r_{DS} = (0.236)/\sqrt{2.236^2 - 4(0.878^2)(1.236)} = 0.236/1.090 = 0.217$$

with df = 173 − 2 = 171. This value if r_{DS} is significant at the 1% level, since table A 11 gives the 1% level as 0.208 for 150 df.

The above test is two-tailed; for a one-tailed test, use the 10% and 2% levels in table A 11.

EXAMPLE 10.8.1—It is well known that a person's systolic and diastolic blood pressures are highly correlated and that the systolic pressure is more variable from person to person than the diastolic. The following data for a sample of 12 middle-aged males serves as an example of the test in this section (X_1 denotes systolic and X_2 diastolic pressure in millimeters).

X_1	100	105	110	110	120	120	125	130	130	150	170	195
X_2	65	65	75	70	78	80	75	82	80	90	95	90

Test the null hypothesis $\sigma_1 = \sigma_2$ by using (10.8.1) and (10.8.2) to get r_{DS} and verify that they agree. Ans. $r_{DS} = 0.939$, highly significant with df = 10, $r_{12} = 0.880$, $F = 8.65$.

10.9—Nonparametric methods: rank correlation. Often a bivariate population is far from normal. In some cases a transformation of the variables X_1 and X_2 brings their joint distribution close to the bivariate normal, making it possible to estimate ρ in the new scale. Failing this, methods of expressing the amount of correlation in nonnormal data by means of a parameter like ρ have not proceeded very far.

Nevertheless, we may still want to examine whether two variables are independent or whether they vary in the same or in opposite directions. For a test of the null hypothesis that there is no correlation, r may be used provided that *one* of the variables is normal. When neither variable seems normal, the best-known procedure is that in which X_1 and X_2 are both converted to rankings. If two judges each rank 12 abstract paintings in order of attractiveness, we may wish to know whether there is any degree of agreement among the rankings. Table 10.9.1 shows similar rankings of the condition of 7 rats after a period of deficient feeding. With data that are not initially ranked, the first step is to rank X_1 and X_2 separately.

TABLE 10.9.1
RANKING OF SEVEN RATS BY TWO OBSERVERS OF THEIR CONDITION AFTER THREE WEEKS ON A DEFICIENT DIET

Rat Number	Ranking by Observer 1	Ranking by Observer 2	Difference d	d^2
1	4	4	0	0
2	1	2	−1	1
3	6	5	1	1
4	5	6	−1	1
5	3	1	2	4
6	2	3	−1	1
7	7	7	0	0
			$\Sigma d = 0$	$\Sigma d^2 = 8$

$r_s = 1 - 6\Sigma d^2/[n(n^2 - 1)] = 1 - (6)(8)/[7(49 - 1)] = 0.857$

The *rank correlation coefficient,* due to Spearman (11) and usually denoted by r_s, is the ordinary correlation coefficient r between the *ranked* values X_1 and X_2. It can be calculated in the usual way as $\Sigma(x_1x_2)/\sqrt{(\Sigma x_1^2)(\Sigma x_2^2)}$. An easier method of computing r is given by

$$r_s = 1 - 6\Sigma D^2/[n(n^2 - 1)]$$

Its calculation is illustrated in table 10.9.1. Like r, the rank correlation can range in samples from −1 (complete discordance) to +1 (complete concordance).

For samples of 10 or fewer pairs, the significance levels of r_s are given in table A 11(*ii*). This table was constructed by arranging X_1 in rank order 1, 2, . . . , n. On the hypothesis of no correlation, the rank order of X_2 is assumed drawn at random from the $n!$ permutations of the ranks. Thus the test is nonparametric. In the rankings of the rats, $r_s = 0.857$ with 7 pairs. The correlation is significant at the 5% level but not at the 1%. For samples of more than 10 pairs, the null distribution of r_s is similar to that of r, and table A 11 is used for testing r_s. Remember that the degrees of freedom in table A 11 are two less than the number of pairs (size of sample). Thus, for $n = 11$ with r_s, look up df = 9 in table A 11.

Another measure of degree of concordance, closely related to r_s, is Kendall's τ (12). To compute this, rearrange the two rankings so that one of them is in the order 1, 2, 3, . . . , n. For table 10.9.1, putting observer 1 in this order, we have:

Rat Number	2	6	5	1	4	3	7
Observer 1	1	2	3	4	5	6	7
Observer 2	2	3	1	4	6	5	7

Taking each rank given by observer 2 in turn, count the smaller ranks to the *right* of it and add these counts. For the rank 2 given to rat 2 the count is 1, since only rat 5 has a smaller rank. The six counts are 1, 1, 0, 0, 1, 0, there being no need to count the extreme right rank. The total is $Q = 3$. Kendall's τ is

$$\tau = 1 - 4Q/[n(n-1)] = 1 - 12/42 = 5/7 = 0.714$$

Like r_s, τ lies between +1 (complete concordance) and −1 (complete disagreement). It takes a little longer to compute, but its frequency distribution on the null hypothesis is simpler and it can be extended to study partial correlation. For details, see (12).

The quantities r_s and τ can be used as a measure of ability to appraise or detect something by ranking. For instance, a group of subjects might each be given bottles containing four different strengths of a delicate perfume and asked to place the bottles in order of the concentration of perfume. If X_1 represents the correct ranking of the strengths and X_2 a subject's ranking, the value of r_s or τ for this subject measures, although rather crudely, success at this task. From the results for random or paired samples of men and women, we could investigate whether women are better at this task than men. The difference between $\bar{\tau}$ or $\bar{r}_s$ for women and men could be compared, approximately, by an ordinary t test. Transformation of each person's r_s or τ to z by table A 12 before applying the t test may improve the approximation by decreasing the skewness of the rs or τs.

TECHNICAL TERMS

bivariate normal distribution
correlation coefficient
error of specification
linear function
population covariance
rank correlation
reliability
scatter diagram
split-half reliability
test-retest reliability

REFERENCES

1. Pearson, K., and Lee, A. *Biometrika* 2 (1902–3):357.
2. Galton, F. *Proc. R. Soc. Lond.* 45 (1888):45.
3. Coleman, J. S., et al. 1966. *Equality of Educational Opportunity*. U.S. Government Printing Office.
4. Fisher, R. A. *Metron* 1 (1921):3.
5. Lush, J. L. *J. Agric. Res.* 42 (1931):853.
6. David, F. N. 1938. *Tables of the Correlation Coefficient*. Cambridge Univ. Press.
7. Grout, R. A. 1937. Iowa Agric. Exp. Stn. Res. Bull. 218.
8. Haber, E. S. Data, Iowa Agric. Exp. Stn.
9. Pitman, E. J. G. *Biometrika* 31 (1939):9.
10. Mumford, A. A., and Young, M. *Biometrika* 15 (1923):108.
11. Spearman, C. *Am. J. Psych.* 15 (1904):88.
12. Kendall, M. G. 1970. *Rank Correlation Methods*, 3rd ed. Charles Griffin, London.
13. Stanley, J. C. 1971. *Reliability in Educational Measurement*, 2nd ed. Edited by R. L. Thorndike, p. 356. Am. Counc. of Educ., Washington, D.C.
14. Fisher, R. A. *Biometrika* 10 (1915):507.

11

Analysis of frequencies in one-way and two-way classifications

11.1—Introduction. In chapter 7 on binomial data, the discussion is confined to cases in which the population contains only two classes of individuals and in which only one or two populations are sampled. We now extend the discussion to populations classified into more than two classes and to samples drawn from more than two populations. Section 11.2 considers the simplest situation in which the expected numbers in the classes are completely specified by the null hypothesis.

11.2—Single classifications with more than two classes. In crosses between two types of maize, Lindstrom (1) found four distinct types of plants in the second generation. In a sample of 1301 plants, there were

$$f_1 = 773 \text{ green}$$
$$f_2 = 231 \text{ golden}$$
$$f_3 = 238 \text{ green-striped}$$
$$f_4 = 59 \text{ golden-green-striped}$$

According to a simple type of Mendelian inheritance, the probabilities of obtaining these four types of plants are 9/16, 3/16, 3/16, and 1/16, respectively. We select this as the null hypothesis.

The χ^2 goodness of fit test in chapter 5 is applicable. Accordingly, we calculate the numbers of plants that would be expected in the four classes if the null hypothesis were true. These numbers, the deviations $f - F$, and the contributions to χ^2 are shown in table 11.2.1.

Since H_0 is completely specified, the number of degrees of freedom in χ^2 = (number of classes) $- 1 = 4 - 1 = 3$. Is χ^2 as large as 9.25 with 3 df a common event in sampling from the population specified by the null hypothesis 9 : 3 : 3 : 1, or is it rare? In table A 5, 9.25 is beyond the 5% point, near the 2.5% point. On this evidence the null hypothesis would be rejected.

With more than two classes, this χ^2 test is usually only a first step in the examination of the data. From the test we have learned that the deviations between observed and expected numbers are too large to be reasonably attributed to sampling fluctuations. But the χ^2 test does not tell us in what way

TABLE 11.2.1
A χ^2 TEST OF LINDSTROM'S DATA

f_i	F_i	$f_i - F_i$	$(f_i - F_i)^2/F_i$
773	(9/16)(1301) = 731.9	+41.1	2.31
231	(3/16)(1301) = 243.9	−12.9	0.68
238	(3/16)(1301) = 243.9	− 5.9	0.14
59	(1/16)(1301) = 81.3	−22.3	6.12
1301	1301.0	0.0	χ^2 = 9.25

the observed and expected numbers differ. For this, we look first at the individual deviations and their contributions to χ^2. The first class (green) gives a large positive deviation +41.1, but 2.31, its contribution to χ^2, is not particularly large. Among the other classes, the last class (golden-green-striped) gives the largest deviation, −22.3, and the largest contribution to χ^2, 6.12 out of a total of 9.25. Lindstrom commented that the deviations could be largely explained by a physiological cause, namely the weakened condition of the last three classes due to their chlorophyll abnormality. He pointed out in particular that the last class (golden-green-striped) was not very vigorous.

To illustrate this type of subsequent analysis, let us examine whether the significant value of 9.25 for χ^2 can be attributed to poor survivorship of the golden-green-striped class. To do this we perform two additional χ^2 tests: (*i*) a test of the null hypothesis that the numbers in the first three classes are in the Mendelian ratios 9:3:3 and (*ii*) a test of whether the last class has a frequency consistent with the Mendelian probability 1/16.

The 9:3:3 hypothesis is tested by a χ^2 test applied to the first three classes. The calculations appear in table 11.2.2.

In the first class, $F_1 = (0.6)(1242) = 745.2$, and so on. The value of χ^2 is now 2.70 with $3 - 1 = 2$ df. Table A 5 shows that the probability is about 0.25 of obtaining a χ^2 as large as this when there are 2 df.

We also test whether the last class (golden-green-striped) has a frequency of occurrence significantly less than would be expected from its Mendelian probability 1/16. For this we observe that 1242 plants fell into the first three classes, with total probability 15/16, as against 59 plants in the fourth class, with probability 1/16. The corresponding expected numbers are 1219.7 and 81.3 In this case the χ^2 test reduces to that given in section 7.7 for testing a theoretical binomial proportion. We have

TABLE 11.2.2
TEST OF THE MENDELIAN HYPOTHESIS IN THE FIRST THREE CLASSES

Class	f	Hypothetical Probability	F	$f - F$	$(f - F)^2/F$
Green	773	9/15 = 0.6	745.2	+27.8	1.04
Golden	231	3/15 = 0.2	248.4	−17.4	1.22
Green-striped	238	3/15 = 0.2	248.4	−10.4	0.44
Total	1242	15/15 = 1	1242.0	0.0	2.70

$$\chi^2 = \frac{(1242 - 1219.7)^2}{1219.7} + \frac{(59 - 81.3)^2}{81.3} = \frac{(+22.3)^2}{1219.7} + \frac{(-22.3)^3}{81.3} = 6.53$$

with 1 df. The significance probability is close to the 1% level.

To summarize, the high value of χ^2 obtained initially, 9.25 with 3 df, can be ascribed to a deficiency in the number of golden-green-striped plants, with the other three classes not deviating abnormally from the Mendelian probabilities. (As Lindstrom suggests, some deficiencies may also exist in the second and third classes relative to the first class, which would show up more definitely in a larger sample.)

This device of making comparisons among subgroups of the classes is useful in two situations. Sometimes, especially in exploratory work, the investigator has no clear ideas about the way in which the numbers in the classes will deviate from the initial null hypothesis; indeed, it may be considered likely that the first χ^2 test will support the null hypothesis. The finding of a significant χ^2 should be followed, as in the above example, by inspection of the deviations to see what can be learned from them. This process may lead to the construction of new hypotheses that are tested by further χ^2 tests among subgroups of the classes. Conclusions drawn from this analysis must be regarded as tentative, because the new hypotheses were constructed after seeing the data and should strictly be tested by gathering new data.

In the second situation the investigator has some specific ideas about the types of departure that the data are likely to show from the initial null hypothesis, that is, about the nature of the alternative hypothesis. The best procedure is then to construct tests aimed specifically at these types of departure. Often, the initial χ^2 test is omitted in this situation. This approach is illustrated in later sections.

When calculating χ^2 with more than 1 df, it is not worthwhile to make a correction for continuity. The exact distribution of χ^2 is still discrete but the number of different possible values of χ^2 is usually large, so the correction, when properly made, produces only a small change in the significance probability.

EXAMPLE 11.2.1—In 193 pairs of Swedish twins (2), 56 were of type *MM* (both male), 72 of type *MF* (one male, one female), and 65 of type *FF*. On the hypothesis that a twin is equally likely to be a boy or a girl and that the sexes of the two members of a twin pair are determined independently, the probabilities of *MM*, *MF*, and *FF* pairs are 1/4, 1/2, and 1/4, respectively. Compute the value of χ^2 and the significance probability. Ans. $\chi^2 = 13.27$ with 2 df, $P < 0.005$.

EXAMPLE 11.2.2—In the preceding example we would expect the null hypothesis to be false for two reasons. The probability that a twin is male is not exactly 1/2. This discrepancy produces only minor effects in a sample of size 193. Second, identical twins are always of the same sex. The presence of identical twins decreases the probability of *MF* pairs and increases the probabilities of *MM* and *FF* pairs. Construct χ^2 tests to answer the questions: (*i*) Are the relative numbers of *MM* and *FF* pairs (ignoring the *MF* pairs) in agreement with the null hypothesis? (*ii*) Are the relative numbers of twins of like sex (*MM* and *FF* combined) and unlike sex (*MF*) in agreement with the null hypothesis? Ans. (*i*) χ^2 (uncorrected) = 0.67 with 1 df, $P > 0.25$; (*ii*) $\chi^2 = 12.44$ with 1 df, P very small. The failure of the null hypothesis is due, as anticipated, to the identical twins of like sex.

EXAMPLE 11.2.3—In a study of the χ^2 distribution with 1 df, 230 samples from binomial distributions with known p were drawn and χ^2 was computed from each sample. The cumulative theoretical distribution was divided into seven classes. The observed and expected numbers of χ^2 values in the seven classes are as follows:

Observed	57	59	62	32	14	3	3	Total
Expected	57.5	57.5	57.5	34.5	11.5	9.2	2.3	230.0

Test whether the deviations of observed from expected numbers are of a size that occurs frequently by chance. Ans. $\chi^2 = 5.50$, df = 6, P about 0.5.

EXAMPLE 11.2.4—In the Lindstrom example in the text, we had χ_3^2 (3 df) = 9.25. It was followed by χ_2^2 (2 df) = 2.70, which compared the first three classes, and $\chi_1^2 = 6.53$, which compared the combined first three classes with the fourth class. Note that $\chi_2^2 + \chi_1^2 = 9.23$, while $\chi_3^2 = 9.25$. In examples 11.2.1 and 11.2.2, $\chi_2^2 = 13.27$, while the sum of the two 1-df χ^2s is $0.67 + 12.44 = 13.11$. When a χ^2 within each subgroup of a classification and a χ^2 that compares the total frequencies in the subgroups are computed, the degrees of freedom add up to the degrees of freedom in the initial χ^2, but the values of χ^2 do not add up to exactly the initial χ^2. They usually add up to a value that is fairly close to the initial χ^2 and worth noting as a clue to mistakes in calculation.

11.3—Single classifications with equal expectations. Often the null hypothesis specifies that all the classes have equal probabilities. In this case, χ^2 has a particularly simple form. As before, let f_i denote the observed frequency in the ith class and let $n = \Sigma f_i$ be the total size of sample. If there are k classes, the null hypothesis probability that a member of the population falls into any class is $p = 1/k$. Consequently, the expected frequency F_i in any class is $np = n/k = \bar{f}$, the mean of the f_i. Thus,

$$\chi^2 = \sum_{i=1}^{k} \frac{(f_i - F_i)^2}{F_i} = \sum_{i=1}^{k} \frac{(f_i - \bar{f})^2}{\bar{f}} \qquad (k - 1 \text{ df}) \qquad (11.3.1)$$

This test is applied to any new table of random numbers. The basic property of such a table is that each digit has a probability 1/10 of being chosen at each draw. To illustrate the test, the frequencies of the first 250 digits in the random number table A 1 are as follows:

Digit	0	1	2	3	4	5	6	7	8	9	Total
f_i	22	24	28	23	18	33	29	17	31	25	250

Only 17 sevens and 18 fours have appeared, as against 31 eights and 33 fives. The mean frequency $\bar{f} = 25$. Thus, by the usual shortcut method of computing the sum of squares of deviations, $\Sigma(f_i - \bar{f})^2$,

$$\chi^2 = (1/25)(22^2 + 24^2 + \ldots + 25^2 - 250^2/10) = 10.08 \qquad (9 \text{ df})$$

Table A 5 shows that the probability of a χ^2 as large as this lies between 0.5 and 0.3; χ^2 is not unusually large.

11.4—Test that Poisson samples have the same mean. This test can be related to the Poisson distribution. Suppose that the f_i are the numbers of occurrences of some rare event in a series of k independent samples. The null hypothesis is that the f_i follow Poisson distributions with the same mean μ. Then, as shown by Fisher, the quantity $\Sigma(f_i - \bar{f})^2/\bar{f}$ is distributed approximately as χ^2 with $(k - 1)$ df. To go a step further, the test can be interpreted as a comparison of the observed variance of the f_i with the variance expected from the Poisson distribution, in which the variance equals the mean μ, of which the sample estimate is $\bar{f}$. The observed variance among the f_i is $s^2 = \Sigma(f_i - \bar{f})^2/(k - 1)$. Hence

$$\chi^2 = (k - 1)(\text{observed variance})/(\text{Poisson variance})$$

This χ^2 test is sensitive in detecting the alternative hypothesis that the f_i follow independent Poisson distributions with *different* means μ_i. Under this alternative, the expected value of χ^2 may be shown to be approximately

$$E(\chi^2) \doteq (k - 1) + \sum_{i=1}^{k} (\mu_i - \bar{\mu})^2/\bar{\mu} \tag{11.4.1}$$

where $\bar{\mu}$ is the mean of the μ_i. If the null hypothesis holds, $\mu_i = \bar{\mu}$ and χ^2 has its usual average value $k - 1$. But any differences among the μ_i increase the expected value of χ^2 and tend to make it large. The test is sometimes called a variance test of the homogeneity of the Poisson distribution.

You now have two methods of testing whether a sample follows the Poisson distribution: the goodness of fit test of section 7.14 and the variance test of this section. If the members of the population actually follow Poisson distributions with *different* means, the variance test is more sensitive than the goodness of fit test. The goodness of fit test is a general purpose test, since *any* type of difference between the observed and expected numbers, if present in sufficient force, makes χ^2 large. But if something is known about the nature of the alternative hypothesis, we can often construct a different test that is more powerful for this type of alternative. The same remarks apply to the binomial distribution. A variance test for the binomial is given in section 11.7.

EXAMPLE 11.4.1—In 1951, the number of babies born with harelips in Birmingham, England, are quoted by Edwards (3) as follows:

Month	Jan.	Feb.	Mar.	Apr.	May	June	July	Aug.	Sept.	Oct.	Nov.	Dec.
Number	8	19	11	12	16	8	7	5	8	3	8	8

Test the null hypothesis that the probability of a baby with a harelip is the same in each month. Ans. $\chi^2 = 23.5$ with 11 df, P between 0.025 and 0.01. Strictly, the variable that should be examined in studies of this type is the ratio (number of babies with harelip)/(total number of babies born); even if this ratio is constant from month to month, the actual number of babies with harelip will vary if the total number born varies. Edwards points out that in these data the total number varies little and shows no relation to the variation in number with harelip. He proceeds to fit the above data by a periodic (cosine) curve, which indicates a maximum in March.

EXAMPLE 11.4.2—Leggatt (4) counted the number of seeds of the weed *potentilla* found in 98 quarter-ounce batches of the grass *Phleum praetense*. The 98 numbers varied from 0 to 7 and were grouped into the following frequency distribution. Test the null hypothesis that the weed seeds are distributed at random through the batches.

Number of seeds, y_j	0	1	2	3	4	5	6	7	Total
Number of batches, f_j	37	32	16	9	2	0	1	1	98

Calculate $\chi^2 = \Sigma f_j(y_j - \bar{y})^2/\bar{y}$. Ans. $\chi^2 = 145.4$ with 97 df. From table A 5, with 100 df, P is clearly less than 0.005. The high value of χ^2 is due primarily to the batches with 6 and 7 seeds.

EXAMPLE 11.4.3—Compute the significance probability in the preceding example by finding the normal deviate Z at the foot of table A 5. Ans. $Z = 3.16$, $P = 0.0008$. The correct probability, found from a larger table of χ^2, is $P = 0.0010$.

11.5—Additional tests. As in section 11.2, the χ^2 test for the Poisson distribution can be supplemented or replaced by other tests directed more specifically against the type of alternative hypothesis that the investigator has in mind. If it is desired to examine whether a rare meteorological event occurs more frequently in the summer months, we might compare the total frequency in June, July, and August with the total frequency in the rest of the year, the null hypothesis probabilities being very close to 1/4 and 3/4. If a likely alternative hypothesis is that an event shows a slow but steady increase or decrease in frequency over a period of nine years, construct a variate $X_i = 1, 2, 3, \ldots, 9$ or alternatively $-4, -3, -2, \ldots, +3, +4$ (making $\bar{X} = 0$), to represent the years. The average change in the f_i per year is estimated by the regression coefficient $\Sigma f_i x_i/\Sigma x_i^2$, where as usual $x_i = X_i - \bar{X}$. The value of χ^2 for testing this coefficient against the null hypothesis that there is no change is

$$\chi^2 = (\Sigma f_i x_i)^2/(\bar{f}\ \Sigma x_i^2) \qquad (1 \text{ df}) \qquad (11.5.1)$$

Another example is found in an experiment designed to investigate various treatments for the control of cabbage loopers (insect larvae)(5). Each treatment was tested on four plots. Table 11.5.1 shows the numbers of loopers counted on each plot for five of the treatments. The objective of the analysis is to examine whether the treatments produced differences in the average number of loopers per plot.

Since the sum of a number of independent Poisson variables also follows a Poisson distribution (section 7.14), we can compare the treatment totals by the Poisson variance test provided we can assume that the counts on plots treated alike follow the same Poisson distribution. To test this assumption, the χ^2 values

TABLE 11.5.1

NUMBER OF LOOPERS ON 50 CABBAGE PLANTS IN A PLOT

(four plots treated alike; five treatments)

Treatment	Loopers per Plot	Plot Total	Plot Mean	χ^2	df
1	11, 4, 4, 5	24	6.00	5.67	3
2	6, 4, 3, 6	19	4.75	1.42	3
3	8, 6, 4, 11	29	7.25	3.69	3
4	14, 27, 8, 18	67	16.75	11.39	3
5	7, 4, 9, 14	34	8.50	6.24	3
Total		173		28.41	15

for each treatment are computed in table 11.5.1. Although only one of the five χ^2 values is significant at the 5% level, their total (28.41, df = 15) gives P of about 0.02. This finding invalidates the use of the Poisson variance test for the comparison of treatment totals. Some additional source of variation within plots is present, which must be taken into account when investigating whether plot means differ from treatment to treatment. Problems of this type, which are common, are handled by the technique known as the analysis of variance. The analysis of these data is completed in example 12.13.3.

11.6—Two-way classifications: the 2 × C contingency table. We come now to data classified by two different criteria. The simplest case (the 2 × 2 table), in which each classification has only two classes, is discussed in section 7.10. The next simplest case occurs when one classification has only two classes, the other having $C > 2$ classes. In the example in table 11.6.1, leprosy patients were classified at the start of an experiment as to whether they exhibited little or much infiltration (a measure of a certain type of skin damage). They were also classified into five classes according to the change in their general health during a subsequent 48-week period of treatment (7). The patients did not all receive the same drugs, but since no differences in the effects of these drugs could be detected, the data were combined for this analysis. The table is called a *2 × 5 contingency table.*

The question at issue is whether the change in health is related to the initial degree of infiltration. The χ^2 test extends naturally to 2 × C tables. The overall proportion of patients with little infiltration is 144/196. On the null hypothesis of no relationship between degree of infiltration and change in health, we expect to find (18)(144)/196 = 13.22 patients with little infiltration and marked improvement, as against 11 observed. As in the 2 × 2 table, the rule for finding an expected number is (row total)(column total)/(grand total). The expected numbers F and the deviations $f - F$ are shown in table 11.6.2. Note that only four expected numbers need be calculated; the rest can be found by subtraction.

The value of χ^2 taken over the ten cells in the table is

$$\chi^2 = \Sigma(f - F)^2/F$$
$$= (-2.22)^2/13.22 + 2.22^2/4.78 + \ldots + (-2.18)^2/3.18 = 6.87$$

The number of degrees of freedom is $(R - 1)(C - 1)$, where R, C are the

TABLE 11.6.1
196 Patients Classified according to Change in Health and Degree of Infiltration

Degree of Infiltration	Change in Health: Improvement: Marked	Moderate	Slight	Stationary	Worse	Total
Little	11	27	42	53	11	144
Much	7	15	16	13	1	52
Total	18	42	58	66	12	196

TABLE 11.6.2
EXPECTED NUMBERS AND DEVIATIONS CALCULATED FROM TABLE 11.6.1

Degree of Infiltration	Improvement: Marked	Moderate	Slight	Stationary	Worse	Total
			Expected numbers, F			
Little	13.22	30.86	42.61	48.49	8.82	144.00
Much	4.78	11.14	15.39	17.51	3.18	52.00
Total	18.00	42.00	58.00	66.00	12.00	196.00
			Deviations, $f - F$			
Little	−2.22	−3.86	−0.61	+4.51	+2.18	0.00
Much	+2.22	+3.86	+0.61	−4.51	−2.18	0.00

numbers of rows and columns, respectively. In this example $R = 2$, $C = 5$ and we have 4 df. This rule for degrees of freedom is in line with the fact that when four of the deviations in a row are known, all the rest can be found. With $\chi^2 = 6.87$, df = 4, the probability lies between 0.25 and 0.10.

Although this test has not rejected the null hypothesis, the deviations show a systematic pattern. In the class of much infiltration, the observed numbers are higher than expected for patients showing any degree of improvement and lower than expected for patients classified as stationary or worse. The reverse is, of course, true for the class of little infiltration. Contrary to the null hypothesis, these deviations suggest that patients with much infiltration progressed on the whole better than those with little infiltration. This suggestion is studied further in section 11.8.

11.7—Test for homogeneity of binomial samples. In the preceding example we obtained a $2 \times C$ contingency table because the data were classified into 2 classes by one criterion and into C classes by a second criterion. Alternatively, we may have recorded some binomial variate $p_i = a_i/n_i$ in each of C independent samples, where i varies from 1 to C and n_i is the size of the ith sample. The objective now is to examine whether the true p_i vary from sample to sample.

A quicker method of computing χ^2 that is particularly appropriate in this situation was devised by Snedecor and Irwin (8). It is illustrated by the preceding example. Think of the columns in table 11.7.1 as representing $C = 5$ binomial samples.

First calculate the proportion $p_i = a_i/n_i$ of patients of much infiltration in each column, and the corresponding overall proportion $\bar{p} = A/N = 52/196 = 0.26531$. Then it may be shown that

$$\begin{aligned}\chi^2 &= (\Sigma p_i a_i - \bar{p}A)/(\bar{p}\bar{q}) \qquad (11.7.1)\\ &= [(0.3889)(7) + \ldots + (0.0833)(1)\\ &\quad - (0.26531)(52)]/[(0.26531)(0.73469)] = 6.88\end{aligned}$$

as before, with 4 df.

TABLE 11.7.1
ALTERNATIVE CALCULATION OF χ^2 FOR THE DATA IN TABLE 11.6.1

Degree of Infiltration	Improvement: Marked	Moderate	Slight	Stationary	Worse	Total
Little	11	27	42	53	11	144
Much, a_i	7	15	16	13	1	52, A
Total, n_i	18	42	58	66	12	196, N
$p_i = a_i/n_i$	0.3889	0.3571	0.2759	0.1970	0.0833	0.26531 ($\bar{p}$)

If p_i is the variable of interest, you will want to calculate these values in order to examine the results. Extra decimals should be carried to ensure accuracy in computing χ^2, particularly when the a_i are large.

The formula for χ^2 can be written, alternatively,

$$\chi^2 = \Sigma n_i(p_i - \bar{p})^2/(\bar{p}\bar{q}) \tag{11.7.2}$$

If the binomial estimates p_i are all based on the same sample size n, χ^2 becomes

$$\chi^2 = \sum_{i=1}^{C} (p_i - \bar{p})^2/(\bar{p}\bar{q}/n) = (C - 1)s_p^2/(\bar{p}\bar{q}/n) \tag{11.7.3}$$

In (11.7.2) and (11.7.3), χ^2 is essentially a comparison of the observed variance s_p^2 among the p_i with the variance $\bar{p}\bar{q}/n$ that the p_i would have if they were independent samples from the same binomial distribution. A high value of χ^2 indicates that the true proportions differ from sample to sample.

This test, sometimes called the variance test for homogeneity of the binomial distribution, has many applications. Different investigators may have estimated the same proportion in different samples and we wish to test whether the estimates agree, apart from sampling errors. In a study of an attribute in human families, where each sample is a family, a high value of χ^2 indicates that members of the same family tend to be alike with regard to this attribute, with variation from family to family.

When the p_i vary from column to column, as indicated by a high value of χ^2, the binomial formula $\sqrt{pq/N}$ underestimates the standard error of the overall proportion $\bar{p}$ for the combined sample. An approximation for the standard error of $\bar{p}$ in this situation is

$$se(\bar{p}) = (1/\bar{n})\sqrt{(\Sigma a_i^2 - 2p\,\Sigma a_i n_i + p^2\,\Sigma n_i^2)/[C(C-1)]} \tag{11.7.4}$$

where C is the number of samples and

$$p_i = a_i/n_i \qquad \bar{p} = \Sigma a_i/\Sigma n_i \qquad N = \Sigma n_i \qquad \bar{n} = N/C$$

EXAMPLE 11.7.1—The numbers of tomato plants attacked by spotted wilt disease were counted in each of 160 areas of 9 plants (6). In all, 261 plants were diseased out of $9 \times 160 = 1440$ plants. A binomial distribution with $n = 9$, $p = 261/1440$ was fitted to the distribution of numbers of diseased plants out of 9. The observed and expected numbers are as follows:

Diseased plants	0	1	2	3	4	5	6	7	Total
Observed frequency	36	48	38	23	10	3	1	1	160
Expected frequency	26.45	52.70	46.67	24.11	8.00	1.77	0.25	0.03	159.98

Perform the χ^2 goodness of fit test. Ans. $\chi^2 = 10.28$ with 4 df after combining classes 5–7, $P < 0.05$. The significant χ^2 is not surprising; diseased plants are seldom scattered at random in a field.

EXAMPLE 11.7.2—In a series of trials a set of r successes, preceded and followed by a failure, is called a *run* of length r. Thus the series *FSFSSSF* contains one run of successes of length 1 and one of length 3. If the probability of a success is p at each trial, the probability of a run of length r may be shown to be $p^{r-1}q$. In 207 runs of diseased plants in a field, the frequency distribution of lengths of run was as follows:

Length of run, r	1	2	3	4	5	Total
Observed frequency, f_r	164	33	9	1	0	207

The estimate of p from these data is $\hat{p} = (T - N)/T$, where $N = \Sigma f_r = 207$ is the total number of runs and $T = \Sigma r f_r$ is the total number of successes in these runs. Estimate p; fit the distribution, called the *geometric distribution;* and test the fit by χ^2. Ans. $\chi^2 = 0.96$ with 2 df, $P > 0.50$. Note: The expression $(T - N)/T$ used for estimating p is derived from a general method of estimation known as the method of maximum likelihood. The same estimate $\hat{p}$ is obtained by equating the average length of run T/N in this sample to its theoretical expected value $1/q$. The expected frequency of runs of length r is $N\hat{p}^{r-1}\hat{q}$.

EXAMPLE 11.7.3—In example 11.7.1 the observed frequencies exceed the expected frequencies at the extremes—with 0 and with 4 or more diseased plants—suggesting that the sample variance of $\hat{p}$ from area to area exceeds the binomial variance $\hat{p}\hat{q}/n$. (*i*) Perform the variance test for homogeneity of the binomial distribution. (*ii*) How much larger is the observed variance than the binomial variance? Hint: Since all areas have $n = 9$ plants, χ^2 is most easily computed as $[\Sigma f_i a_i^2 - (\Sigma f_i a_i)^2/N]/(n\bar{p}\bar{q})$, where f_i $(i = 0, 1, \ldots, 7)$ is the frequency of areas having a_i diseased plants and $N = 160$. This formula is derived from (11.7.3) as applied to grouped data. Ans. (*i*) $\chi^2 = 225.5$, df = 159, $Z = 3.43$, P very small. (*ii*) Observed variance is 0.0234, about 40% higher than the binomial variance, 0.0165.

EXAMPLE 11.7.4—Ten samples of 5 mice from the same laboratory were injected with the same dose of *typhimurium* (9). The numbers of mice dying (out of 5) were as follows: 3, 1, 5, 5, 3, 2, 4, 2, 3, 5. Test whether the proportion dying can be regarded as constant from sample to sample. Ans. $\chi^2 = 16.1$, df = 9, $P = 0.06$. Since the death rate is found so often to vary within the same laboratory, a standard agent is usually tested along with each new agent because comparisons made over time cannot be trusted.

EXAMPLE 11.7.5—Uniform doses of a toxin were injected into rats, the sizes of the samples being dictated by the numbers of animals available at the dates of injection. The sizes, the numbers of surviving rats, and the proportion surviving, are as follows:

Number in sample	40	12	22	11	37	20
Number surviving	9	2	3	1	2	3
Proportion surviving	0.2250	0.1667	0.1364	0.0909	0.0541	0.1500

Test the null hypothesis that the probability of survival is the same in all samples. Ans. $\chi^2 = 4.97$, df = 5, $P = 0.43$.

EXAMPLE 11.7.6—In another test with four samples of inoculated rats, χ^2 was 6.69, $P = 0.086$. Combine the values of χ^2 for the two tests. Ans. $\chi^2 = 11.66$, df = 8, $P = 0.17$.

EXAMPLE 11.7.7—Burnett (10) tried the effect of five storage locations on the viability of seed corn. In the kitchen garret, 111 kernels germinated among 120 tested; in a closed toolshed, 55 out of 60; in an open toolshed, 55 out of 60; outdoors, 41 out of 48; and in a dry garret, 50 out of 60. Calculate $\chi^2 = 5.09$, df = 4, $P = 28\%$.

EXAMPLE 11.7.8—In 13 families in Baltimore, the numbers of persons (n_i) and the numbers who had consulted a doctor during the previous 12 months (a_i) were as follows: 7, 0; 6, 0; 5, 2; 5, 5; 4, 1; 4, 2; 4, 2; 4, 2; 4, 0; 4, 0; 4, 4; 4, 0; 4, 0. Compute the overall percentage who had consulted a doctor and the standard error of the percentage. Note: One would expect the proportion who had seen a doctor to vary from family to family. Verify this by finding $\chi^2 = 35.6$, df = 12, $P < 0.005$. Consequently, (11.7.4) is used to estimate the standard error of $\bar{p}$. Ans. Percentage = $100\bar{p} = 30.5\%$, standard error = 10.5%. (These data were selected from a large sample for illustration.)

11.8—Ordered classifications. In the leprosy example of section 11.6, the classes (marked improvement, moderate improvement, slight improvement, stationary, worse) are an example of an *ordered classification.* Such classifications are common in the study of human behavior and preferences and whenever different degrees of some phenomenon can be recognized.

With a single classification of Poisson variables, the ordering might lead us to expect that if the null hypothesis $\mu_i = \mu$ does not hold, an alternative $\mu_1 \leq \mu_2 \leq \mu_3 \leq$, where the subscripts represent the order, should hold. For instance, if working conditions in a factory have been classified as excellent, good, fair, we might expect that if the number of defective articles per worker varies with working conditions, the order should be $\mu_1 \leq \mu_2 \leq \mu_3$. Similarly, with ordered columns in a $2 \times C$ contingency table, the alternative $p_1 \leq p_2 \leq p_3 \leq$ might be expected.

An approach used by numerous workers (7, 11, 12) is to attach a score to each class so that an ordered scale is created. To illustrate from the leprosy example, we assigned scores of 3, 2, 1, respectively, to the marked, moderate, and slight improvement classes, 0 to the stationary class, and -1 to the worse class. These scores are based on the judgment that the five classes constructed by the expert represent equal gradations on a continuous scale. We considered giving a score of $+4$ to the marked improvement class and -2 to the worse class, since the expert seemed to examine patients at greater length before assigning them to one of these extreme classes, but rejected this since our impression may have been erroneous.

Having assigned the scores, we may think of the leprosy data as consisting of two independent samples of 144 and 52 patients, respectively. (See table 11.8.1.) For each patient there is a discrete measure X of change in health, where X takes only the values 3, 2, 1, 0, -1. We can estimate the average change in health for each sample, with its standard error, and can test the null hypothesis that this average change is the same in the two populations. For this test the ordinary two-sample t test as applied to grouped data is used as an approximation. The calculations appear in table 11.8.1. On the X scale the average change in health is $+1.269$ for patients with much infiltration and $+0.819$ for those with little infiltration. The difference $\bar{D}$ is 0.450, with standard error ± 0.172 (194 df) computed in the usual way. The value of t is $0.450/0.172 = 2.616$, with $P < 0.01$. Contrary to the initial χ^2 test, this test reveals a significantly greater amount of progress for the patients with much infiltration.

The assignment of scores is appropriate when (*i*) the phenomenon in question is one that could be measured on a continuous scale if the instruments of measurement were good enough, and (*ii*) the ordered classification can be regarded as a kind of grouping of this continuous scale or as an attempt to

TABLE 11.8.1
ANALYSIS OF THE LEPROSY DATA BY ASSIGNED SCORES

Change in Health, X	Infiltration Cases, f	
	Little	Much
3	11	7
2	27	15
1	42	16
0	53	13
−1	11	1
Total Σf	144	52
ΣfX	118	66
$\overline{X} = \Sigma fX/\Sigma f$	0.819	1.269
ΣfX^2	260	140
$(\Sigma fX)^2/\Sigma f$	96.7	83.8
Σfx^2	163.3	56.2
df	143	51
s^2	1.142	1.102
Pooled s^2	1.131	

$$s_{\overline{D}}^2 = (1.131)(1/144 + 1/52) = 0.0296 \qquad s_{\overline{D}} = 0.172$$

$$t = \frac{\overline{D}}{s_{\overline{D}}} = \frac{1.269 - 0.819}{0.172} = 2.616 \qquad \text{df} = 194 \qquad P < 0.01$$

approximate the continuous scale by a cruder scale that is the best we can do in the present state of knowledge. A similar process occurs in many surveys. The householder is shown five specific income classes and asked to indicate the class within which the family income falls, without naming actual income.

The advantage in assigning scores is that the more flexible and powerful methods of analysis developed for continuous variables become available. One can begin to think of the sizes of the average differences between different groups in a study, consider whether the differences are important or minor, and compare the difference between groups A and B with that between groups E and F. Regressions of the group means $\overline{X}$ on a further variable Z can be worked out. The relative variability of different groups can be examined by computing s for each group.

This approach assumes that the standard methods of analysis of continuous variables, like the t test, can be used with an X variable that is discrete and takes only a few values. As noted in section 8.10 on scales with limited values, the standard methods appear to work well enough for practical use. However, heterogeneity of variance and correlation between s^2 and $\overline{X}$ are more frequently encountered because of the discrete scale. If most of the patients in a group show marked improvement, most of their Xs will be 3 and s^2 will be small. Pooling of variances should not be undertaken without examining the individual s^2. In the leprosy example the two s^2 values are 1.142 and 1.102 (table 11.8.1), and this difficulty is not present.

The chief objection to the assignment of scores is that the method is more or less arbitrary. Two investigators may assign different scores to the same set of data. In our experience, however, moderate differences between two scoring systems seldom produce marked differences in the conclusions drawn from the analysis. In the leprosy example, the alternative scores 4, 2, 1, 0, -2 give t = 2.549 as against t = 2.616 in the analysis in table 11.8.1. Some ordered classifications, however, present particular difficulty in scoring. If the degrees of injury to persons in accidents are recorded as slight, moderate, severe, disabling, and fatal, there seems no entirely satisfactory way of placing the last two classes on the same scale as the first three.

Several alternative principles have been used to construct scores. In studies of different populations of school children, K. Pearson (13) assumed that the underlying continuous variate was normally distributed in a standard population of school children. If the classes are regarded as a subdivision of this normal distribution, the class boundaries for the normal variate are easily found from the relative frequencies in the classes. The score assigned to a class is the mean of the normal variate within the class. A related approach due to Bross (14) also uses a standard population but does not assume normality. The score *(ridit)* given to a class is the relative frequency up to the midpoint of that class in the standard population. Fisher (15) has assigned scores so as to maximize the F ratio of treatments to experimental error as defined in section 12.5.

EXAMPLE 11.8.1—In the leprosy data, verify the value of t = 2.549 quoted for the scoring 4, 2, 1, 0, -2.

11.9—Test for a linear trend in proportions. When interest is centered on the proportions $p_i = a_i/n_i$ in a $2 \times C$ contingency table, there is another way of viewing the data. Table 11.9.1 shows the leprosy data with the assigned scores X_i, but in this case the variable that we analyze is p_i, the proportion of patients with much infiltration. The contention now is that if these patients have fared better than patients with little infiltration, the values of p_i should increase steadily as we move from the worse class ($X = -1$) towards the marked improvement class ($X = 3$). This is the case with the p_i in table 11.9.1.

If so, the regression coefficient of p_i on X_i should be a good test criterion. On the null hypothesis (no relation between p_i and X_i) each p_i is distributed about the same mean, estimated by $\bar{p}$, with variance $\bar{p}\bar{q}/n_i$. The regression coefficient b is calculated as usual except that each p_i must be weighted by the sample size n_i

TABLE 11.9.1

Testing a Linear Regression of p_i on the Score for Leprosy Data

Degree of Infiltration	Improvement					
	Marked	Moderate	Slight	Stationary	Worse	Total
Little	11	27	42	53	11	144
Much, a_i	7	15	16	13	1	52
Total, n_i	18	42	58	66	12	196, N
$p_i = a_i/n_i$	0.3889	0.3571	0.2759	0.1970	0.0833	0.2653, $\bar{p}$
Score, X_i	3	2	1	0	-1	

on which it is based. The numerator and denominator of b are computed as follows:

$$\begin{aligned}\text{num.} &= \Sigma n_i(p_i - \bar{p})(X_i - \bar{X}) = \Sigma n_i p_i X_i - (\Sigma n_i p_i)(\Sigma n_i X_i)/\Sigma n_i \\ &= \Sigma a_i X_i - (\Sigma a_i)(\Sigma n_i X_i)/N \\ &= 66 - (52)(184)/196 = 66 - 48.82 = 17.18 \\ \text{den.} &= \Sigma n_i X_i^2 - (\Sigma n_i X_i)^2/N \\ &= 400 - 184^2/196 = 400 - 172.8 = 227.2\end{aligned}$$

Then $b = 17.18/227.2 = 0.0756$. On H_0 its standard error is

$$s_b = \sqrt{(\bar{p}\bar{q}/\text{den.})} = \sqrt{(0.2653)(0.7347)/227.2} = 0.0293$$

The normal deviate for testing the null hypothesis $\beta = 0$ is therefore

$$Z = b/s_b = 0.0756/0.0293 = 2.580 \qquad P = 0.0098$$

Although it is not obvious at first sight, Yates (11) showed that this regression test is essentially the same as the t test in section 11.8 of the difference between the mean scores in the little and much infiltration classes. In this example the regression test gave $Z = 2.580$, while the t test gave $t = 2.616$ (194 df). The slight difference in numerical results arises because the two approaches use slightly different large sample approximations to the exact distributions of Z and t with these discrete data.

EXAMPLE 11.9.1—Armitage (12) quotes the following data by Holmes and Williams for the relation in children between size of tonsils and the proportion of children who are carriers of *Streptococcus pyogenes* in the nose.

Size of tonsils, $\bar{X}$	0	1	2	Total
Carriers a_i	19	29	24	72, A
Noncarriers	497	560	269	1326
Total, n_i	516	589	293	1398, N
Carrier-rate, p_i	0.0368	0.0492	0.0819	0.051502, $\bar{p}$

Calculate: (*i*) the normal deviate Z for testing the linear regression of the proportion of carriers on size of tonsils, (*ii*) the value of t for comparing the difference between the mean size of tonsils in carriers and noncarriers. Ans. (*i*) $Z = 2.681$, (*ii*) $t = 2.686$ with 1396 df.

EXAMPLE 11.9.2—As noted in example 7.11.1, Murphy and Abbey (20) compared the cancer rates in relatives of a sample of 200 women with breast cancer and in relatives of a control sample of 198 women without breast cancer. In such observational studies it is important to check that the cancer and control samples are similar with regard to extraneous characteristics (e.g., age, educational level) that might affect the amount of cancer reported in relatives. The following are the data on the educational levels attained by the cancer and control women.

	No School	Grade	High	College	Total
Cancer	1	18	65	116	200
Control	4	14	79	101	198

(*i*) Test by χ^2 whether the two groups can be considered to have the same distribution of scholastic levels. (*ii*) By assigning scores 0, 1, 2, 3 to the schooling levels, test whether the mean levels differ for the two samples. Ans. (*i*) $\chi^2 = 4.69$ (3 df), P about 0.2. (*ii*) Both t as in section 11.8 and b/s_b as in this section give $t = 0.305$, indicating no evidence of a real difference in average levels.

EXAMPLE 11.9.3—Fisher (16) applied χ^2 tests to the experiments conducted by Mendel in 1863 to test different aspects of this theory, as follows:

Trifactorial	$\chi^2 = 8.94$	17 df
Bifactorial	$\chi^2 = 2.81$	8 df
Gametic ratios	$\chi^2 = 3.67$	15 df
Repeated 2 : 1 test	$\chi^2 = 0.13$	1 df

Show that in random sampling the probability of obtaining a total χ^2 *lower* than that observed is less than 0.005 (use the χ^2 table). More accurately, the probability is less than 1 in 2000. Thus, the agreement of the results with Mendel's laws looks too good to be true. Fisher gives an interesting discussion of possible reasons.

11.10—The $R \times C$ contingency table. If each member of a sample is classified by one characteristic into R classes and by a second characteristic into C classes, the data may be presented in a table with R rows and C columns. The entry in any of the RC cells is the number of members of the sample falling into that cell. Strand and Jessen (17) classified a random sample of farms in Audubon County, Iowa, into three classes (owned, rented, mixed) according to the tenure status and into three classes (I, II, III) according to the level of the soil fertility (table 11.10.1).

Before drawing conclusions about the border totals for tenure status, this question is asked, Are the relative numbers of owned, rented, and mixed farms in this county the same at the three levels of soil fertility? This question might be phrased, Is the distribution of the soil fertility levels the same for owned, rented, and mixed farms? (If a little reflection does not make it clear that these two

TABLE 11.10.1

NUMBERS OF FARMS ON THREE SOIL FERTILITY GROUPS IN AUDUBON COUNTY, IOWA, CLASSIFIED ACCORDING TO TENURE

Soil		Owned	Rented	Mixed	Total
I	f	36	67	49	152
	F	36.75	62.92	52.33	
	$f - F$	−0.75	4.08	−3.33	
II	f	31	60	49	140
	F	33.85	57.95	48.20	
	$f - F$	−2.85	2.05	0.80	
III	f	58	87	80	225
	F	54.40	93.13	77.47	
	$f - F$	3.60	−6.13	2.53	
Total		125	214	178	517

$$\chi^2 = \sum \frac{(f - F)^2}{F} = \frac{(-0.75)^2}{36.75} + \ldots + \frac{2.53^2}{77.47} = 1.54 \qquad \text{df} = (R - 1)(C - 1) = 4$$

questions are equivalent, see example 11.10.1.) Sometimes the question is put more succinctly as, Is tenure status *independent* of fertility level?

The χ^2 test extends naturally to this situation.

$$\chi^2 = \Sigma(f - F)^2/F$$

where f is the observed frequency in any cell and F the frequency expected if the null hypothesis of independence holds. The rule for calculating the F values is obtained as follows.

In the population, let p_R be the probability that a farm falls in row R and p_C the probability that it falls in column C. Then on the null hypothesis of independence, the expected number of farms F in row R and column C will be np_Rp_C where n is total size of sample. Take the ratio (row total)/n as the estimate of p_R and the ratio (column total)/n as the estimate of p_C. Then

$$F = (\text{row total})(\text{column total})/n$$

for finding the expected cell values if H_0 is true.

As a check on arithmetic, note that the sum of the deviations $(f - F)$ in each row and in each column is zero. These facts dictate the number of degrees of freedom in χ^2. If we know $R - 1$ deviations in a column, the remaining deviation is known, since their sum is zero over the column. Similarly, if the deviations are known in $C - 1$ columns, they are known in the last column. Thus, knowledge of $(R - 1)(C - 1)$ deviations enables us to find all the deviations. Therefore, df $= (R - 1)(C - 1)$.

The calculation of χ^2 is given in the table. Since $P > 0.8$, the null hypothesis is not rejected.

When χ^2 is significant, the next step is to study the nature of the departure from independence in more detail. Examining the cells in which the contribution to χ^2 is greatest and taking note of the signs of the deviations $f - F$ furnish clues, but these are hard to interpret because the deviations in different cells are correlated. Computation of the percentage distribution of the row classification within each column, followed by a scrutiny of the changes from column to column, may be more informative. Further χ^2 tests may help. For instance, if the percentage distribution of the row classification appears the same in two columns, a χ^2 test for these two columns may confirm it. The two columns can then be combined for comparison with other columns. Examples 11.10.2–11.10.5 illustrate this approach.

EXAMPLE 11.10.1—Show that if the expected distribution of the column classification is the same in every row, the expected distribution of the row classification is the same in every column. For the ith row, let $F_{i1}, F_{i2}, \ldots, F_{iC}$ be the expected numbers in the respective columns. Let $F_{i2} = a_2F_{i1}$, $F_{i3} = a_3F_{i1}, \ldots, F_{iC} = a_cF_{i1}$. Then the numbers $a_2, a_3, \ldots, a_C$ must be the same in every row, since the expected distribution of the column classification is the same in every row. Now the expected row distribution in the first column is $F_{11}, F_{21}, \ldots, F_{R1}$. In the second column it is $F_{12} = a_2F_{11}$, $F_{22} = a_2F_{21}, \ldots, F_{R2} = a_2F_{R1}$. Since a_2 is a constant multiplier, this distribution is the same as in the first column and similarly for any other column.

EXAMPLE 11.10.2—In a study of the relation between blood type and disease, large samples of patients with peptic ulcer, patients with gastric cancer, and control persons free from these diseases were classified as to blood type (O, A, B, AB). In this example, the relatively small numbers of AB patients were omitted for simplicity. The observed numbers are as follows:

Blood Type	Peptic Ulcer	Gastric Cancer	Controls	Total
O	983	383	2892	4258
A	679	416	2625	3720
B	134	84	570	788
Total	1796	883	6087	8766

Compute χ^2 to test the null hypothesis that the distribution of blood types is the same for the three samples. Ans. $\chi^2 = 40.54$, 4 df, P very small.

EXAMPLE 11.10.3—To examine this question further, compute the percentage distribution of blood types for each sample, as shown below.

Blood Type	Peptic Ulcer	Gastric Cancer	Controls
O	54.7	43.4	47.5
A	37.8	47.1	43.1
B	7.5	9.5	9.4
Total	100.0	100.0	100.0

This suggests (*i*) there is little difference between the blood type distributions for gastric cancer patients and controls; (*ii*) peptic ulcer patients differ principally in having an excess of patients of type O. Going back to the frequencies in example 11.10.2, test the hypothesis that the blood type distribution is the same for gastric cancer patients and controls. Ans. $\chi^2 = 5.64$ (2 df.), P about 0.06.

EXAMPLE 11.10.4—Combine the gastric cancer and control samples. (*i*) Test whether the distribution of A and B types is the same in this combined sample as in the peptic ulcer sample (omit the O types). Ans. $\chi^2 = 0.68$ (1 df), $P > 0.7$. (*ii*) Test whether the proportion of O types versus A + B types is the same for the combined sample as for the peptic ulcer samples. Ans. $\chi^2 = 34.29$ (1 df), P very small. To sum up, the high value of the original 4 = df χ^2 is due primarily to an excess of O types among the peptic ulcer patients.

EXAMPLE 11.10.5—The preceding χ^2 tests may be summarized as follows:

O, A, B types in gastric cancer (g) and controls (c)	2 df	$\chi^2 = 5.64$
A, B types in peptic ulcer and combined (g, c)	1 df	$\chi^2 = 0.68$
O, A, and B types in peptic ulcer and combined (g, c)	1 df	$\chi^2 = 34.29$
Total	4 df	$\chi^2 = 40.61$

The total χ^2, 40.61, is close to the original χ^2, 40.54, because we have broken the original 4 df into a series of independent operations that account for all 4 df. The difference between 40.61 and 40.54, however, is not just a rounding error; the two quantities differ a little algebraically.

EXAMPLE 11.10.6—In the preceding examples we might have chosen the hypotheses to be tested as follows: *(i)* Test the null hypothesis that the ratio of A to B blood types is the same in the three disease types; *(ii)* test that the ratio of O to A and B combined is the same for gastric cancer patients and controls; (*iii*) test that the ratio of O types to A + B types is the same for the peptic ulcer patients as for the others (already done in example 11.10.4). Ans. (*i*) $\chi^2 = 1.01$ (2 df), $P > 0.50$. (*ii*) $\chi^2 = 5.30$ (1 df), P about 0.02. The three χ^2 values add to 40.60.

11.11—Sets of 2 × 2 tables. Sometimes the task is to combine the evidence from a number of 2 × 2 tables. The same two treatments or types of subject may have been compared in different studies and it is desired to summarize the combined data. Alternatively, the results of a single investigation are often subclassified by the levels of a factor or variable that is thought to influence the results. The data in table 11.11.1, made available by Martha Rogers, are of this type.

The data form part of a study of the possible relationship between complications of pregnancy of mothers and behavior problems in children. The comparison is between mothers of children in Baltimore schools who had been referred by their teachers as behavior problems and mothers of control children not so referred. Each mother's history of infant losses (for example, stillbirths) prior to the birth of the child was recorded. Since these loss rates increase with the birth order of the child, as table 11.11.1 shows, and since the two samples might not be comparable in the distributions of birth orders, the data were examined separately for three birth-order classes. This precaution is common.

Each of the three 2 × 2 tables is first inspected. None of the χ^2 values in a single table, shown at the right, approaches the 5% significance level. However, note that in all three tables the percentage of mothers with previous losses is higher in the problem children than in the controls. We seek a test sensitive in detecting a population difference that is consistently in one direction, although it may not show up clearly in the individual tables.

In the ith 2 × 2 table, let p_{i1}, p_{i2} be the population proportions of losses for problems and controls and let $\delta_i = p_{i1} - p_{i2}$. In applications of this type, two mathematical models have been used to describe how δ_i may be expected to change as p_{i2} varies over the range 0–100%. In one model, the difference between the two populations is assumed constant on a *logit* scale. The logit of a proportion p is ln (p/q), the symbol ln denoting the logarithm to base e. A constant difference on the logit scale therefore means that $\ln(p_{i1}/q_{i1}) - \ln(p_{i2}/q_{i2})$ is constant as p_{i2} varies. This model is widely used, partly because in numerous bodies of data the quantity (logit p_{i1} − logit p_{i2}) has been found to be more nearly constant as p_{i2} varies than $(p_{i1} - p_{i2})$ itself. For example, suppose that we are comparing the proportions of cases in which body injury is suffered

TABLE 11.11.1
A SET OF THREE 2 × 2 TABLES: NUMBERS OF MOTHERS WITH PREVIOUS INFANT LOSSES

Birth Order	Children	Mothers with Losses	Mothers with No losses	Total	% Loss	χ^2 (1 df)
2	Problems	20	82	$102 = n_{11}$	$19.6 = \hat{p}_{11}$	
	Controls	10	54	$64 = n_{12}$	$15.6 = \hat{p}_{12}$	
	Total	30	136	$166 = n_1$	$18.1 = \hat{p}_1$	0.42
3–4	Problems	26	41	$67 = n_{21}$	$38.8 = \hat{p}_{21}$	
	Controls	16	30	$46 = n_{22}$	$34.8 = \hat{p}_{22}$	
	Total	42	71	$113 = n_2$	$37.2 = \hat{p}_2$	0.19
5+	Problems	27	22	$49 \pm n_{31}$	$55.1 = \hat{p}_{31}$	
	Controls	14	23	$37 = n_{32}$	$37.8 = \hat{p}_{32}$	
	Total	41	45	$86 = n_3$	$47.7 = \hat{p}_3$	2.52

TABLE 11.11.2
SIZE OF DIFFERENCE $\delta = p_1 - p_2$ FOR A RANGE OF VALUES OF p_2

$p_2\%$	1	5	10	30	50	70	90	95	99
Constant logit	1.3	6.0	10.6	20.0	20.0	14.5	5.5	2.8	0.6
Constant Z	2.6	8.1	12.4	20.0	20.0	15.3	6.4	3.5	0.8

in auto accidents by seat-belt wearers and nonwearers. The accidents have been classified by severity of impact into mild, moderate, severe, and extreme, giving four 2 × 2 tables. Under the mild impacts, both p_{11} and p_{12} may be small and δ_1 also small, since injury rarely occurs with mild impact. Under extreme impact, p_{41} and p_{42} may both be close to 100%, making δ_4 also small. The large δs may occur in the two middle tables where the ps are nearer 50%. The differences may be found more nearly constant on a logit scale.

A second model postulates that the difference is constant on a *normal deviate* (Z) scale. The value of Z corresponding to any proportion p is such that the area of a standard normal curve to the left of Z is p. For instance, $Z = 0$ for $p = 0.5$, $Z = 1.282$ for $p = 0.9$, $Z = -1.282$ for $p = 0.1$.

To illustrate the meaning of a constant difference on these transformed scales, table 11.11.2 shows, as p_2 varies, the size of difference on the original percentage scale that corresponds to a constant difference on (a) the logit scale and (b) the normal deviate scale. The size of the difference was chosen to equal 20% at $p_2 = 50\%$. Note that (*i*) the differences diminish toward both ends of the p scale as in the seat-belt example and (*ii*) the two transformations do not differ greatly.

A test that is sensitive in detecting an overall difference when differences are constant either on a logit or on a Z scale was developed by Cochran (7). If $\hat{p}_i$ is the combined percentage in the ith table and

$$w_i = n_{i1}n_{i2}/(n_{i1} + n_{i2}) \qquad d_i = \hat{p}_{i1} - \hat{p}_{i2}$$

we compute

$$\Sigma w_i d_i / \sqrt{\Sigma w_i \hat{p}_i \hat{q}_i}$$

and refer to the normal table. This test gives proper weight to 2 × 2 tables with large and small samples and with different $\hat{p}$ values. For the data in table 11.11.1

TABLE 11.11.3
TEST FOR A CONSTANT DIFFERENCE ON A LOGIT OR A Z SCALE

Birth Order	w_i	d_i	$w_i d_i$	$\hat{p}_i$	$\hat{p}_i\hat{q}_i$	$w_i\hat{p}_i\hat{q}_i$
2	39.3	+0.040	+1.57	0.181	0.1482	5.824
3–4	27.3	+0.040	+1.09	.372	.2336	6.377
5+	21.1	+0.173	+3.65	0.477	0.2494	5.262
Total			+6.31			17.463

TABLE 11.11.4

MANTEL-HAENSZEL TEST FOR THE INFANT LOSS DATA IN TABLE 11.11.1

Birth Order	O	E_i	$n_{i1}n_{i2}c_{i1}c_{i2}/[n_i^2(n_i - 1)]$
2	20	18.43	5.858
3–4	26	24.90	6.426
5+	27	23.36	5.321
Total	73	66.69	17.605

$Z = (73 - 66.69 - 0.5)/\sqrt{17.605} = 1.38$

the computations are shown in table 11.11.3 (with the d_i in proportion to keep the numbers smaller). The test criterion is $6.31/\sqrt{17.463} = 1.51$.

There is another way of computing this test. In the ith table, let O_i be the observed number of problems losses and E_i the expected number under H_0. For birth order 2 (table 11.11.1), $O_1 = 20$ and $E_1 = (30)(102)/166 = 18.43$; Then $O_1 - E_1 = +1.57$, *which is the same as* w_1d_1. This result may be shown to hold in any 2×2 table. The test criterion can therefore be written

$$\Sigma(O_i - E_i)/\sqrt{\Sigma w_i\hat{p}_i\hat{q}_i}$$

This form of the test has been presented by Mantel and Haenszel (18, 19) with two refinements that are worthwhile when the ns are small. First, a correction for continuity is applied by subtracting 0.5 from the absolute value of $\Sigma(O_i - E_i)$. This version of the test is shown in table 11.11.4. The correction for continuity makes a noticeable difference even with samples of this size.

Second, the variance of w_id_i or $O_i - E_i$ on H_0 is not $w_i\hat{p}_i\hat{q}_i$ but the slightly larger quantity $n_{i1}n_{i2}\hat{p}_i\hat{q}_i/(n_{i1} + n_{i2} - 1)$. If the margins of the 2×2 table are n_{i1}, n_{i2}, c_{i1}, and c_{i2}, this variance can be computed as

$$n_{i1}n_{i2}\,c_{i1}c_{i2}/[n_i^2(n_i - 1)] \qquad (n_i = n_{i1} + n_{i2}) \qquad (11.11.1)$$

Table 11.11.4 shows the computation. The increase in variance has little effect in this problem; the standard error of Σw_id_i increases from $\sqrt{17.463} = 4.18$ (table 11.11.3) to $\sqrt{17.605} = 4.20$ (table 11.11.4).

TECHNICAL TERMS

logit
ordered classifications
$R \times C$ contingency table
ridits
scores
test for homogeneity

REFERENCES

1. Lindstrom, E. W. 1918. Cornell Agric. Exp. Stn. Mem. 13.
2. Edwards, A. W. F. *Ann. Hum. Gen.* 24 (1960):309.
3. Edwards, J. H. *Ann. Hum. Gen.* 25 (1961):89.
4. Leggatt, C. W. *Comptes rendus de l'association internationale d'essais de semences* 5 (1935):27.
5. Caffrey, D. J., and Smith, C. E. 1934. Bureau of Entomology and Plant Quarantine, USDA, Baton Rouge.

6. Cochran, W. G. *J. R. Stat. Soc. Suppl.* 3 (1936):49.
7. Cochran, W. G. *Biometrics* 10 (1954):417.
8. Snedecor, G. W., and Irwin, M. R. *Iowa State Coll. J. Sci.* 8 (1933):75.
9. Irwin, J. O., and Cheeseman, E. A. *J. R. Stat. Soc. Suppl.* 6 (1939):174.
10. Burnett, L. C. 1906. Master's thesis. Iowa State College.
11. Yates, F. *Biometrika* 35 (1948):176.
12. Armitage, P. *Biometrics* 11 (1955):375.
13. Pearson, K. *Biometrika* 5 (1905–6):105.
14. Bross, I. D. J. *Biometrics* 14 (1958):18.
15. Fisher, R. A. *Statistical Methods for Research Workers*. 1941. Oliver & Boyd, Edinburgh.
16. Fisher, R. A. *Ann. Sci.* 1 (1936):117.
17. Strand, N. V., and Jessen, R. J. 1943. Iowa Agric. Exp. Stn. Res. Bull. 315.
18. Mantel, N., and Haenszel, W. *J. Natl. Cancer Inst.* 22 (1959):719.
19. Mantel, N. *J. Am. Stat. Assoc.* 58 (1963): 690.
20. Murphy, D. P., and Abbey, H. 1959. *Cancer in Families*. Harvard Univ. Press, Cambridge.

12

One-way classifications; analysis of variance

12.1—Extension from two samples to many. Statistical methods for two independent samples are presented in chapter 6 but the needs of the investigator are seldom confined to the comparison of two samples only. For binomial data, the extension to more than two samples is made in the preceding chapter. We are now ready to do the same for measurement data.

First, recall the analysis used in the comparison of two samples. In the numerical example (section 6.8, p. 91), the comb weights of two samples of 11 chicks were compared, one sample having received sex hormone A, the other sex hormone C. Briefly, the principal steps in the analysis were as follows: (*i*) the mean comb weights $\overline{X}_1$, $\overline{X}_2$ were computed; (*ii*) the within-sample sum of squares of deviations Σx^2 with 10 df was found for each sample; (*iii*) a pooled estimate s^2 of the within-sample variance was obtained by adding the two values of Σx^2 and dividing by the sum of the degrees of freedom, 20; (*iv*) the standard error of the mean difference $\overline{X}_1 - \overline{X}_2$ was calculated as $\sqrt{2s^2/n}$, where $n = 11$ is the size of each sample; and finally, (*v*) a test of the null hypothesis $\mu_1 = \mu_2$ and confidence limits for $\mu_1 - \mu_2$ were given by the result that the quantity

$$[\overline{X}_1 - \overline{X}_2 - (\mu_1 - \mu_2)]/\sqrt{2s^2/n}$$

follows the t distribution with 20 df.

In the next section we apply this method to an experiment with four treatments, i.e., four independent samples.

12.2—An experiment with four samples. During cooking, doughnuts absorb fat in various amounts. Lowe (1) wished to learn if the amount absorbed depends on the type of fat used. For each of four fats, six batches of doughnuts were prepared, a batch consisting of 24 doughnuts. The data in table 12.2.1 are the grams of fat absorbed per batch, coded by deducting 100 g to give simpler figures. Data of this kind are called a *single* or *one-way* classification, each fat representing one class.

Before beginning the analysis, note that the totals for the four fats differ substantially—from 372 for fat 4 to 510 for fat 2. Indeed, there is a clear separation between the individual results for fats 4 and 2; the highest value given

TABLE 12.2.1
GRAMS OF FAT ABSORBED PER BATCH MINUS 100 G

Fat	1	2	3	4	Total
	64	78	75	55	
	72	91	93	66	
	68	97	78	49	
	77	82	71	64	
	56	85	63	70	
	95	77	76	68	
ΣX	432	510	456	372	1,770
$\overline{X}$	72	85	76	62	295
ΣX^2	31,994	43,652	35,144	23,402	134,192
$(\Sigma X)^2/n$	31,104	43,350	34,656	23,064	132,174
Σx^2	890	302	488	338	2,018
df	5	5	5	5	20

Pooled $s^2 = 2018/20 = 100.9$ $\quad s_{\overline{D}} = \sqrt{(2s^2/n)} = \sqrt{(2)(100.9)/6} = \pm 5.80$

by fat 4 is 70, while the lowest for fat 2 is 77. Every other pair of samples, however, shows some overlap.

Proceeding as in the case of two samples, we calculate for each sample the mean $\overline{X}$ and the sum of squares of deviations Σx^2, as shown in table 12.2.1. We then form a pooled estimate s^2 of the within-sample variance. Since each sample provides 5 df for Σx^2, the pooled $s^2 = 100.9$ has 20 df. Of course, this pooling involves the assumption that the variance between batches is the same for each fat. The standard error of the mean of any fat is $\sqrt{s^2/6} = 4.10$ g.

Thus far, the only new problem is that we have four means to compare instead of two. The comparisons of interest are not necessarily confined to the differences $\overline{X}_i - \overline{X}_j$ between pairs of means; their exact nature depends on the questions the experiment is intended to answer. For instance, if fats 1 and 2 were animal fats and fats 3 and 4 vegetable fats, we might be particularly interested in the difference $(\overline{X}_1 + \overline{X}_2)/2 - (\overline{X}_3 + \overline{X}_4)/2$. A rule for making planned comparisons of this nature is outlined in section 12.8, with further discussion in section 12.9.

Before considering the comparison of means, we present an alternative method of doing the preliminary calculations. This method, of great utility and flexibility, is known as the *analysis of variance* and was developed by Fisher in the 1920s. The analysis of variance performs two functions:

1. It is an elegant and slightly quicker way of computing the pooled s^2. In a single classification this advantage in speed is minor, but in the more complex classifications studied later, the analysis of variance is the only simple and reliable method of determining the appropriate pooled error variance s^2.

2. It provides an F test of the null hypothesis that the population means μ_1, μ_2, μ_3, μ_4 for the four fats are identical. This test is often useful in a preliminary inspection of the results and has many subsequent applications. You have already come across the F distribution in section 6.12, p. 98, in testing the null

hypothesis that two independent sample variances s_1^2, s_2^2 are estimates of the same σ^2 and also as applied to regression in section 9.1.

EXAMPLE 12.2.1—Here are some data selected for easy computation. Calculate the pooled s^2 and state how many degrees of freedom it has.

Sample number	1	2	3	4
	11	13	21	10
	4	9	18	4
	6	14	15	19

Ans. $s^2 = 21.5$ with 8 df.

12.3—Analysis of variance; model I (fixed effects). In the doughnut example, suppose for a moment that there are no differences between the average amounts absorbed for the four fats. In this situation, all 24 observations are distributed about a common mean μ with variance σ^2.

The analysis of variance develops from the fact that on this supposition we can make three different estimates of σ^2 from the data in table 12.2.1. Since we are assuming that all 24 observations come from the same population, we can compute the total sum of squares of deviations for the 24 observations as

$$64^2 + 72^2 + 68^2 \ldots + 70^2 + 68^2 - 1770^2/24 = 134{,}192 - 130{,}538 = 3654 \quad (12.3.1)$$

This sum of squares has 23 df. The mean square, $3654/23 = 158.9$, is the first estimate of σ^2.

The second estimate is the pooled s^2 already obtained. Within each fat we computed the sum of squares between batches (890, 302, etc.), each with 5 df. These sums of squares were added to give

$$890 + 302 + 488 + 338 = 2018 \quad (12.3.2)$$

This quantity is called the sum of squares *between batches within fats,* or more concisely the sum of squares *within fats*. The sum of squares is divided by its degrees of freedom, 20, to give the second estimate, $s^2 = 2018/20 = 100.9$.

For the third estimate, consider the means for the four fats, 72, 85, 76, and 62. These are also estimates of μ but have variances $\sigma^2/6$, since they are means of samples of 6. The sum of squares of deviations from their mean is

$$72^2 + 85^2 + 76^2 + 62^2 - 295^2/4 = 272.75 \qquad (3\text{ df})$$

The mean square, 272.75/3, is an estimate of $\sigma^2/6$. Consequently, if we multiply by 6 we have the third estimate of σ^2, namely 545.5. We can accomplish this by multiplying the sum of squares by 6, giving

$$6(72^2 + 85^2 + 76^2 + 62^2 - 295^2/4) = 1636.5 \qquad (12.3.3)$$

the mean square being $1636.5/3 = 545.5$.

Since the total for any fat is six times the fat means, this sum of squares can be computed from the fat totals as

$$\frac{432^2 + 510^2 + 456^2 + 372^2}{6} - \frac{1770^2}{24} = 132{,}174 - 130{,}538 = 1636 \qquad (12.3.4)$$

This sum of squares is called the sum of squares *between fats*.

Now list the degrees of freedom and the sums of squares in (12.3.3), (12.3.2), and (12.3.1) and the corresponding mean squares in table 12.3.1. Notice a new and important result: The degrees of freedom and the sums of squares for the two components (between fats and within fats) add to the corresponding total figures. These results hold in any single classification. The result for the degrees of freedom is not hard to verify. With a classes and n observations per class, the df = $(a - 1)$ for between fats, $a(n - 1)$ for within fats, and $(an - 1)$ for the total. But

$$(a - 1) + a(n - 1) = a - 1 + an - a = an - 1$$

The result for the sums of squares follows from an algebraic identity proved in the next section. Because of this relation, a common practice is to compute only the total sum of squares and the sum of squares between fats. The sum of squares within fats, leading to the pooled s^2, is obtained by subtraction.

It is time to state the assumptions underlying the analysis of variance for single classifications. A notation common in statistical papers is to use the subscript i to denote the class, where i takes on the values $1, 2, \ldots, a$. The subscript j designates the members of a class, j moving from 1 to n.

Within class i, the observations X_{ij} are assumed normally distributed about a mean μ_i with variance σ^2. The mean μ_i may vary from class to class, but σ^2 is assumed the same in all classes. We denote the mean of the a values of μ_i by μ and write $\mu_i = \mu + \alpha_i$. It follows, of course, that $\Sigma\alpha_i = 0$. Mathematically, the model may be written

$$X_{ij} = \mu + \alpha_i + \epsilon_{ij} \qquad [i = 1, \ldots, a; \quad j = 1, \ldots, n; \quad \epsilon_{ij} = \mathcal{N}(0, \sigma^2)]$$

TABLE 12.3.1
ANALYSIS OF VARIANCE FOR DOUGHNUT DATA

Source of Variation	Degrees of Freedom	Sum of Squares	Mean Squares
Between fats	$a - 1 = 3$	1636	545.3
Between batches within fats	$a(n - 1) = 20$	2018	100.9
Total	$an - 1 = 23$	3654	158.9

In words: *Any observed value is the sum of three parts: (i) an overall mean, (ii) a treatment or class deviation, and (iii) a random element from a normally distributed population with mean 0 and standard deviation σ.*

In this model (often called model I, the *fixed effects model*) the effects of the treatments or classes, measured by the parameters α_i, are regarded as fixed but unknown quantities to be estimated.

The random element ϵ_{ij} in the model represents the combined contribution of other influences on the unit's responsiveness. These may be variations from unit to unit in the effectiveness of the treatment, errors of measurement, or individual characteristics of the unit.

We have noted that on the null hypothesis $\mu_i = \mu$, the mean squares s_b^2 between classes and s_w^2 within classes are both estimates of σ^2. Consequences of the normality assumption in the model are that s_b^2 and s_w^2 are independent and that both follow the distribution described in section 5.7 for an estimate of variance. Hence, if the null hypothesis holds, s_b^2/s_w^2 follows the F distribution with $(a - 1)$ and $a(n - 1)$ df. Section 12.4 gives an illustration of the behavior of F when the null hypothesis does not hold.

Since the analysis of variance table is unfamiliar at first, the beginner should work a number of examples.

EXAMPLE 12.3.1—Compute the analysis of variance for the simple data in example 12.2.1. Verify that you obtain 21.5 for the pooled s^2, as found by the method of example 12.2.1.

Source of Variation	Degrees of Freedom	Sum of Squares	Mean Square
Between samples	3	186	62.0
Within samples	8	172	21.5
Total	11	358	32.5

EXAMPLE 12.3.2—As part of a larger experiment (2), three levels of vitamin B_{12} were compared, with each level being fed to three different pigs. The average daily gains in weight of the pigs (up to 75 lb liveweight) were as follows:

5 mg/lb B_{12}	10 mg/lb B_{12}	20 mg/lb B_{12}
1.52	1.63	1.44
1.56	1.57	1.52
1.54	1.54	1.63

Analyze the variance as follows:

Source of Variation	Degrees of Freedom	Sum of Squares	Mean Square
Between levels	2	0.0042	0.0021
Within levels	6	.0232	.0039
Total	8	0.0274	0.0034

Hint: If you subtract 1.00 from each gain (or 1.44 if you prefer) you will save time. Subtraction of a common figure from every observation does not alter results in the analysis of variance table.

EXAMPLE 12.3.3—The percentage of clean wool in seven bags was estimated by taking three batches at random from each bag. The percentages of clean wool in the batches were as follows:

1	2	3	4	5	6	7
41.8	33.0	38.5	43.7	34.2	32.6	36.2
38.9	37.5	35.9	38.9	38.6	38.4	33.4
36.1	33.1	33.9	36.3	40.2	34.8	37.9

Calculate the mean squares for bags (11.11) and batches within bags (8.22).

12.4—Effect of differences between population class means. What happens to F when the population means for the classes are different? To illustrate from a simple example in which you can easily verify the calculations, we drew (using a table of random normal deviates) six observations normally distributed with population mean $\mu = 5$ and $\sigma = 1$. They were arranged in three sets of two observations to simulate an experiment with $a = 3$ treatments and $n = 2$ observations per treatment.

The data and the analysis of variance appear as Case I at the top of table 12.4.1. In the analysis of variance table, the between-classes sum of squares is labeled treatments and the within-classes sum of squares is labeled error. This terminology is common in planned experiments. The mean squares, 0.83 for treatments and 1.12 for error, are both estimates of $\sigma^2 = 1$.

In Case II we subtracted 1 from each observation for treatment 1 and added 2 to each observation for treatment 3. This simulates an experiment with real differences in the effects of the treatments, the population means being $\mu_1 = 4$, $\mu_2 = 5$, $\mu_3 = 7$. In the analysis of variance, the error sum of squares and mean square are unchanged. This is not surprising, because the error sum of squares is the pooled Σx^2 *within* treatments and subtracting any constant from all the observations in a treatment has no effect on Σx^2. The treatments mean square has, however, increased from 0.83 in Case I to 7.26 in Case II.

Case III represents an experiment with larger differences between treatments. Each original observation for treatment 1 was reduced by 2, and each

TABLE 12.4.1

A SIMULATED EXPERIMENT WITH 3 TREATMENTS AND 2 OBSERVATIONS PER TREATMENT

Treatment				Analysis of Variance		
1	2	3		df	SS	MS
Case I. $\mu_1 = \mu_2 = \mu_3 = 5$						
4.6	3.3	6.3	Treatments	2	1.66	0.83
5.2	4.7	4.2	Error	3	3.37	1.12
9.8	8.0	10.5	Total	5	5.03	
Case II. $\mu_1 = 4, \mu_2 = 5, \mu_3 = 7$						
3.6	3.3	8.3	Treatments	2	14.53	7.26
4.2	4.7	6.2	Error	3	3.37	1.12
7.8	8.0	14.5	Total	5	17.90	
Case III. $\mu_1 = 3, \mu_2 = 5, \mu_3 = 9$						
2.6	3.3	10.3	Treatments	2	46.06	23.03
3.2	4.7	8.2	Error	3	3.37	1.12
5.8	8.0	18.5	Total	5	49.43	

observation for treatment 3 was increased by 4. The means are now $\mu_1 = 3$, $\mu_2 = 5$, $\mu_3 = 9$. As before, the error mean square is unchanged. The treatments mean square has increased to 23.03. Note that the samples for the three treatments have now moved apart so that there is no overlap.

When the means μ_i differ, it can be proved that the treatments mean square is an unbiased estimate of

$$\sigma^2 + n \sum_{i=1}^{a} (\mu_i - \bar{\mu})^2/(a - 1) \tag{12.4.1}$$

In Case II with $\mu_i = 4, 5, 7$, $\Sigma(\mu_i - \bar{\mu})^2$ is 4.67, while n and $a - 1$ are both 2 and $\sigma^2 = 1$, so (12.4.1) becomes $1 + 4.67 = 5.67$. Thus the treatments mean square, 7.26, is an unbiased estimate of 5.67. If we drew a large number of samples and calculated the treatments mean square for Case II for each sample, their average should be close to 5.67.

In Case III, $\Sigma(\mu_i - \bar{\mu})^2$ is 18.67, so the treatments mean square, 23.03, is an estimate of the population value, 19.67.

12.5—The *F* test. These results suggest that the quantity,

$$F = \frac{\text{treatments mean square}}{\text{error mean square}} = \frac{\text{mean square between classes}}{\text{mean square within classes}}$$

should be a good criterion for testing the null hypothesis that the population means are the same in all classes. The value of F should be around 1 when the null hypothesis holds and should become large when the μ_i differ substantially. The distribution was first tabulated by Fisher in the form $z = \ln \sqrt{F}$. In honor of Fisher, the criterion was named F by Snedecor (3). Fisher and Yates (4) designate F as the *variance ratio.*

Table A 14, part I, gives the 5% and 1% levels of F and part II the 25%, 10%, 2.5% and 0.5% levels. The degrees of freedom are ν_1 in the numerator, ν_2 in the denominator. For a one-way classification, $\nu_1 = a - 1$, $\nu_2 = a(n - 1)$.

In Case I in table 12.4.1, F is $0.83/1.12 = 0.74$. In Case II, F increases to $7.26/1.12 = 6.48$, and in Case III, to $23.03/1.12 = 20.56$. From table A 14, parts I and II, you will find that in Case II, F (which has 2 and 3 df) is significant at the 10% level but not at the 5% level. In Case III, F is significant at the 5% level.

Any set of computer programs for statistical use is likely to have a program for the analysis of a one-way classification. The output includes the analysis of variance, the value of F, and an indication of its significance level. For the doughnuts example, this output might appear as follows.

ANALYSIS OF VARIANCE

SOURCE	DF	SUM OF SQUARES	MEAN SQUARE	F RATIO	P
BETWEEN GROUPS	3	1636	545.3	5.40	0.0070
WITHIN GROUPS	20	2018	100.9		
TOTAL	23	3654			

The statistical significance of F can be indicated by giving, in another column to the right, the tail probability—the probability of a value of F larger than that observed if H_0 holds. For $\nu_1 = 3$, $\nu_2 = 20$, the 1% level of F in table A 14 is 4.94 and the 0.5% level is 5.82. The value of F is clearly significant at the 1% level, a more accurate tail probability being 0.0070. A preliminary conclusion is that the fats have different capabilities for being absorbed by doughnuts—preliminary, because we need to examine more closely the nature of the differences among batch means.

EXAMPLE 12.5.1—Four tropical feedstuffs were fed to a different lot of five baby chicks (9). The gains in weight were:

Lot 1	55	49	42	21	52
2	61	112	30	89	63
3	42	97	81	95	92
4	169	137	169	85	154

Analyze the variance and test the equality of the μs. Ans. Mean squares: (*i*) lots, 8745; (*ii*) chicks within lots, 722; $F = 12.1$. Since the sample F is far beyond the tabular 1% point, there is little doubt that the feedstuff populations have different μs.

EXAMPLE 12.5.2—In the wool data of example 12.3.3, test the hypothesis that the bags are all from populations with a common mean. Ans. $F = 1.35$, $F_{0.05} = 2.85$. There is not strong evidence against the hypothesis—the bags may all have the same percentage of clean wool.

EXAMPLE 12.5.3—In the vitamin B_{12} experiment of example 12.3.2, the mean gains for the three levels differ less than is to be expected from the mean square within levels. The value of F is 0.54 with 2 and 6 df. To discover how unusual is F as low as 0.54, find from table A 14 the probability of F as high as $1/0.54 = 1.85$ with 6 and 2 df; P is clearly higher than 0.25. With a very rare low F, inquiry into the experiment might reveal something needing correction.

12.6—Analysis of variance with only two classes. With only two classes, the F test is equivalent to the t test used in section 6.5 to compare the means of two independent samples. With two classes, the relation $F = t^2$ holds, as follows. If $\bar{X}_{1.}$, $\bar{X}_{2.}$ denote the means of the two classes,

$$t = \sqrt{n}(\bar{X}_{1.} - \bar{X}_{2.})/\sqrt{2s^2} \tag{12.6.1}$$

where s^2 is the pooled within-class mean square.

But if $\bar{X}_{..}$ is the mean of the whole sample,

$$\bar{X}_{1.} - \bar{X}_{..} = \bar{X}_{1.} - (\bar{X}_{1.} + \bar{X}_{2.})/2 = (\bar{X}_{1.} - \bar{X}_{2.})/2$$

while $(\bar{X}_{2.} - \bar{X}_{..})$ is the same quantity with a minus sign. Hence, the *mean square between classes* is

$$n\sum_{i=1}^{2} (\bar{X}_{i.} - \bar{X}_{..})^2 = n(\bar{X}_{1.} - \bar{X}_{2.})^2/2 \qquad \text{(1 df)} \tag{12.6.2}$$

From this result, $F = n(\bar{X}_{1.} - \bar{X}_{2.})^2/(2s^2) = t^2$, both having $2(n - 1)$ error df. In table A 14 you will find that any significance level of F with 1 and ν df is the square of the corresponding level of t with ν df. This result extends to the case where the two samples are of unequal sizes.

While it is a matter of choice whether F or t is used, the fact that we are nearly always interested in the size and direction of $\bar{X}_{1.} - \bar{X}_{2.}$ favors the t test.

EXAMPLE 12.6.1—Work out the analysis of variance of the experiment analyzed by t in table 6.8.1, and verify that $\sqrt{F} = t$. Ans.

Source of Variation	Degrees of Freedom	Sum of Squares	Mean Squares
Between samples	1	9,245	9245
Within samples	20	16,220	811
	$F = 9245/811 = 11.40$	$\sqrt{F} = 3.38 = t$	

EXAMPLE 12.6.2—Hansberry and Richardson (5) gave the percentages of wormy apples on two groups of 12 trees each. Group A, sprayed with lead arsenate, had 19, 26, 22, 13, 26, 25, 38, 40, 36, 12, 16, and 8% of apples wormy. Those of group B, sprayed with calcium arsenate and buffer materials, had 36, 42, 20, 43, 47, 49, 59, 37, 28, 49, 31, and 39% wormy. Compute the mean square within samples, 111.41 with 22 df, and that between samples, 1650.04 with 1 df. Then $F = 1650.04/111.41 = 14.8$. Next, test the significance of the difference between the sample means as in section 6.9. The value of t is $3.85 = \sqrt{14.8}$.

EXAMPLE 12.6.3—For $\nu_1 = 1$, $\nu_2 = 20$, verify that the 5% and 1% significance levels of F are the squares of those of t with 20 df.

12.7—Proof of the algebraic identity in the analysis of variance. Some readers may like to see a proof of the basic result that the sums of squares between and within classes add to the total sum of squares. In controlled experiments, as in the doughnuts example, the classes (fats) often have an equal number n of observations. But in many observational studies and in some experiments the classes have unequal numbers. The analysis of variance identity holds for this situation also. Let n_i be the number of observations in the ith class and let $N = \Sigma n_i$ be the total number of observations, while a is the number of classes.

If X_{ij} denotes the observation for the jth member of the ith class, the subscript i runs from 1 to a, while in the ith class, j runs from 1 to n_i. Denote the sum of class i by $X_{i.}$ and the mean by $\bar{X}_{i.}$. The grand total is $X_{..}$ and the overall mean, $\bar{X}_{..}$. The subscript dot denotes summation over the values of a subscript.

In this notation the *total sum of squares* of deviations of the observations from the sample mean is

$$\sum_{i=1}^{a} \sum_{j=1}^{n_i} X_{ij}^2 - X_{..}^2/N \qquad (12.7.1)$$

The sum of squares of deviations within the ith class is

$$\sum_{j=1}^{n_i} X_{ij}^2 - X_{i.}^2/n_i$$

On adding this quantity over all classes we get for the pooled *sum of squares within classes,*

$$\sum_{i=1}^{a}\sum_{j=1}^{n_i} X_{ij}^2 - \sum_{i=1}^{a} X_{i.}^2/n_i \tag{12.7.2}$$

The proof of the algebraic identity is only a matter of expressing the sum of squares between class means in the same form as (12.7.1) and (12.7.2). A result that we need is

$$N\overline{X}_{..} = X_{..} = \sum_{i=1}^{a} X_{i.} = \sum n_i \overline{X}_{i.} \tag{12.7.3}$$

Hence $\overline{X}_{..} = \Sigma n_i \overline{X}_{i.}/N$ is a *weighted* mean of the class means when sample sizes within classes are unequal. In calculating the sum of squares of deviations of the class means from the overall mean with unequal n_i, each $(\text{deviation})^2$ is weighted by the class size n_i. Hence, the sum of squares between classes is

$$\sum_{i=1}^{a} n_i(\overline{X}_{i.} - \overline{X}_{..})^2 = \sum n_i \overline{X}_{i.}^2 - 2\overline{X}_{..}\sum n_i \overline{X}_{i.} + N\overline{X}_{..}^2$$

$$= \sum n_i \overline{X}_{i.}^2 - N\overline{X}_{..}^2 \tag{12.7.4}$$

since $\Sigma n_i \overline{X}_{i.} = N\overline{X}_{..}$, by (12.7.3). Substituting $\overline{X}_{i.} = X_{i.}/n_i$ and $\overline{X}_{..} = X_{..}/N$ in (12.7.4), we get the *sum of squares between classes:*

$$\sum_{i=1}^{a} X_{i.}^2/n_i - X_{..}^2/N \tag{12.7.5}$$

It is clear that the between-classes sum of squares in (12.7.5) plus the within-classes sum of squares in (12.7.2) add to the total sum of squares in (12.7.1). Thus the identity is proved.

12.8—Planned comparisons among class means. The analysis of variance is only the first step in studying the results. The next step is to examine the class means and the sizes of differences among them.

Often, particularly in controlled experiments, the investigator plans the experiment in order to estimate a limited number of specific quantities. For instance, in part of an experiment on sugar beets, the three treatments (classes) were: (*i*) mineral fertilizers (PK) applied in April one week before sowing, (*ii*) PK applied in December before winter plowing, (*iii*) no minerals. The mean yields of sugar in cwt per acre were PK in April, $\overline{X}_1 = 68.8$; PK in December, $\overline{X}_2 = 66.8$; no PK, $\overline{X}_3 = 62.4$. The objective is to estimate two quantities:

average effect of PK: $(1/2)(\overline{X}_1 + \overline{X}_2) - \overline{X}_3 = 67.8 - 62.4 = 5.4$ cwt
April *minus* December application: $\overline{X}_1 - \overline{X}_2 = 2.0$ cwt

A rule for finding standard errors and confidence limits of estimates of this type is now given. Both estimates are linear combinations of the means, each mean being multiplied by a number. In the first estimate, the numbers are 1/2, 1/2, −1. In the second, they are 1, −1, 0, where we put 0 because $\overline{X}_3$ does not appear. Further, in each estimate, the sum of the numbers is zero. Thus,

$$1/2 + 1/2 + (-1) = 0 \qquad 1 + (-1) + 0 = 0 \tag{12.8.1}$$

Definition. Any linear combination

$$L = \lambda_1\overline{X}_1 + \lambda_2\overline{X}_2 + \ldots + \lambda_k\overline{X}_k \tag{12.8.2}$$

where the λs are fixed numbers is called a *comparison* of the treatment means or a *contrast* among the means if $\Sigma\lambda_i = 0$. The comparison may include all a treatment means ($k = a$) or only some of the means ($k < a$). The standard error of L is a particular case of the more general (10.7.5) in section 10.7.

Rule 12.8.1. The standard error of L is $\sqrt{\Sigma\lambda^2}(\sigma/\sqrt{n})$, and the estimated standard error is $\sqrt{\Sigma\lambda^2}(s/\sqrt{n})$ with degrees of freedom equal to those in s, where n is the number of observations in each mean $\overline{X}_i$. The contribution of L to the treatments sum of squares is therefore $nL^2/\Sigma\lambda^2$.

In the example the value of $s/\sqrt{n}$ was 1.37 with 24 df. Hence, for the average effect of PK, with $\lambda_1 = 1/2$, $\lambda_2 = 1/2$, $\lambda_3 = -1$, the estimated standard error is

$$\sqrt{(1/2)^2 + (1/2)^2 + (-1)^2}(1.37) = \sqrt{1.5}(1.37) = 1.68 \qquad (24\text{ df})$$

The value of t for testing the average effect of PK is $t = 5.4/1.68 = 3.2$, significant at the 1% level. Confidence limits (95%) are 5.4 ± (2.06)(1.68), or 1.9 and 8.9 cwt/acre.

For the difference between the April and December applications, with $\lambda_1 = 1$, $\lambda_2 = -1$, the estimated standard error is $\sqrt{2}$ (1.37) = 1.94. The difference is not significant at the 5% level, the confidence limits being 2.0 ± (2.06)(1.94) or −2.0 and +6.0.

Further examples of planned comparisons appear in the next two chapters. Common cases are the comparison of a no minerals treatment with minerals applied in four different ways; the comparison of different levels of the same ingredient, usually at equal intervals, where the purpose is to fit a curve that describes the relation between yield and the amount of the ingredient (section 14.10); and factorial experimentation, which is the subject of chapter 16.

Incidentally, when several different comparisons are being made, one or two of the comparisons may show significant effects even if the initial F test shows nonsignificance.

When reporting the results of a series of comparisons, it is not enough to make statements such as "the difference was not significant" or "the difference was significant at the 1% level." The reader should be given the sizes of the differences with their estimated standard errors. Most computer program printouts also give all the treatment means and their standard errors. These data enable the reader to estimate percentage differences (which are often more easily remembered), to judge whether the overall mean level was typical or unusually high or low, to make comparisons that did not interest the investigator, and to combine the results of this study with those of other studies—often important when trying to summarize the current state of knowledge on the effects of the treatments. The summary tables should contain an accurate description of the treatments, which is not always easy (for example, when "swimming lessons" or "psychiatric guidance when needed" is the treatment).

For the numerical example in this section, a report on the two comparisons might read: "Application of mineral fertilizers produced a significant average increase in sugar of 5.4 cwt/acre ($\pm$1.68). The yield of the April application exceeded that of the December application by 2.0 cwt ($\pm$1.94), but this difference was not significant."

12.9—Orthogonal comparisons. The two comparisons in the sugar beet example in section 12.8 have one other property. The sum of the products of corresponding coefficients in the two linear functions is zero. Thus, from (12.8.1), $(1/2)(1) + (1/2)(-1) + (-1)(0) = 0$.

Two comparisons with this property are called *orthogonal*. Orthogonal comparisons are uncorrelated and are independently distributed if the data are normal. Because of this independence they have the advantage in the summary of results that each deals with a distinctly different question and supplies different information. If we can find $a - 1$ planned comparisons with a treatments, every pair of which is orthogonal, discussion of each comparison is one way of presenting essentially all the information that the experiment possesses in a way that is logically tidy. An orthogonal set of $a - 1$ comparisons is not, however, unique; many sets can be constructed. But usually the set that the investigator had in mind when planning the experiment will be found the most informative, though some other way of looking at the data may suggest new ideas or other interpretations of the results.

Orthogonal comparisons have one further interesting consequence. For any comparison L, the quantity $Q = nL^2/\Sigma\lambda^2$ is a component with 1 df of the between-classes sum of squares, SSB, as noted in section 12.8. With two orthogonal comparisons $L_1 = \Sigma\lambda_{1i}\overline{X}_{i.}$ and $L_2 = \Sigma\lambda_{2i}\overline{X}_{i.}$, the quantities Q_1 and Q_2 are both components of SSB, while the remainder,

$$SSB - Q_1 - Q_2$$

is a component of SSB with $(a - 3)$ df. With a set of $a - 1$ mutually orthogonal comparisons, we have the identity

$$Q_1 + Q_2 + \ldots + Q_{a-1} \equiv SSB$$

The numerical example (effect of PK and time of application on sugar beets) gives an illustration. With $a = 3$ treatments we have a complete set of 2 orthogonal comparisons. Ignoring the multiplier n, you may verify that

$$SSB = 21.44 \qquad Q_1(\text{PK}) = 5.4^2/1.5 = 19.44 \qquad Q_2(\text{time}) = 2^2/2 = 2.00$$

In the analysis of variance the between-classes sum of squares can therefore be partitioned into a set of squares with 1 df, one for each comparison. Each comparison can be tested by an F test or the corresponding t test.

EXAMPLE 12.9.1—In an experiment in which mangolds were grown on acid soil (6), part of the treatments were: (*i*) chalk and (*ii*) lime, both applied at the rate of 21 cwt calcium oxide (CaO)/acre; and (*iii*) no liming. For good reasons, there were twice as many no lime plots as plots with chalk or with lime. Consequently, the comparisons of interest may be expressed algebraically as

Effect of CaO: $(1/2)(\overline{X}_1 + \overline{X}_2) - (1/2)(\overline{X}_3 + \overline{X}_4)$

where $\overline{X}_3$, $\overline{X}_4$ represent the two no lime classes; $\overline{X}_1$ represents chalk; $\overline{X}_2$ represents lime; and chalk minus lime, $\overline{X}_1 - \overline{X}_2$.

The mean yields were (tons per acre): chalk, 14.82; lime, 13.42; no lime, 9.74. The standard error of any $\overline{X}_i$ was ± 2.06 T with 25 df. Calculate the two comparisons and their standard errors and write a report on the results. Ans. Effect of CaO, 4.38 ± 2.06 T; chalk minus lime, 1.40 ± 2.91 T.

EXAMPLE 12.9.2—An experiment on sugar beets (7) compared times and methods of applying mixed artificial fertilizers (NPK). The mean yields of sugar (cwt per acre) were as follows: no artificials, $\overline{X}_1 = 38.7$; artificials applied in January (plowed), $\overline{X}_2 = 48.7$; artificials applied in January (broadcast), $\overline{X}_3 = 48.8$; artificials applied in April (broadcast), $\overline{X}_4 = 45.0$. Their standard error was ± 1.22 with 14 df. Calculate 95% confidence limits for the following comparisons:

(*i*) average effect of artificials, $(1/3)(\overline{X}_2 + \overline{X}_3 + \overline{X}_4) - \overline{X}_1$
(*ii*) January *minus* April application, $(1/2)(\overline{X}_2 + \overline{X}_3) - \overline{X}_4$
(*iii*) Broadcast *minus* January plowed, $\overline{X}_3 - \overline{X}_2$

Ans. (*i*) (5.8, 11.8), (*ii*) (0.6, 7.0), (*iii*) (−3.6, +3.8) cwt/acre.

EXAMPLE 12.9.3—One can encounter linear combinations of the means that are not comparisons as we have defined them. For instance, in early experiments on vitamin B_{12}, rats were fed on a B_{12}-deficient diet until they ceased to gain in weight. If we compared a single and a double supplement of B_{12} in an experiment with two treatments and measured the subsequent gains $\overline{X}_i$ in weight produced, it might be reasonable to calculate $\overline{X}_2 - 2\overline{X}_1$, which should be zero if the gain in weight is proportional to the amount of B_{12}. Here $\lambda_1 + \lambda_2 \neq 0$. The formula for the standard error still holds; the standard error is $\sqrt{5}\sigma/\sqrt{n}$ in this example.

EXAMPLE 12.9.4—In the experiment on the palatability of three types of proteins described in sections 1.9 and 1.10, the average palatability scores given by females (25 per treatment) to the three treatments were C, 0.24; liquid H, 1.12; solid H, 1.04. In the analysis of variance, the error mean square per person (72 df) was 1.391. Verify that the average effect of H *minus* C and the average effect of liquid H *minus* solid H are orthogonal comparisons. Perform t tests of each comparison. Ans. $t = 2.91$ ($p < 0.01$) for average effect of H, $t = 0.24$ (not significant) for liquid versus solid H.

EXAMPLE 12.9.5—(*i*) Verify that the three comparisons in example 12.9.2 are mutually orthogonal. (*ii*) If the experiment has three replications ($n = 3$ per treatment), show that the treatments and error sums of squares and mean squares in the analysis of variance can be presented as follows:

Source of Variation	df	Sum of Squares	Mean Squares	F
Treatments	3	202.38	67.46	15.11
Effect of artificials	1	174.24	174.24	39.02
January *minus* April	1	28.12	28.12	6.30
Broadcast *minus* plowed	1	0.02	0.02	0.00
Error	14	62.51	4.465	

(*iii*) Do the F values seem consistent with the confidence interval limits in the answers to example 12.9.2? Give your reasons.

12.10—Samples of unequal sizes. In planned experiments the samples from the classes are usually of equal sizes, but in nonexperimental studies the investigator may have little control over sample sizes. As before, X_{ij} denotes the jth observation from the ith class, $X_{i.}$ denotes the class total of the X_{ij}, while $X_{..} = \Sigma X_{i.}$ is the grand total. The size of the sample in the ith class is n_i and $N = \Sigma n_i$ is the total size of all samples.

Table 12.10.1 presents algebraic instructions for the analysis of variance. They follow from the algebraic identity in section 12.7. The F ratio s_b^2/s^2 has $(a - 1)$ and $(N - a)$ df. The standard error (SE) of the difference between the ith and kth class means with $(N - a)$ df is

$$SE\,(\overline{X}_{i.} - \overline{X}_{k.}) = \sqrt{s^2(1/n_i + 1/n_k)} \qquad (12.10.1)$$

The standard error of the comparison $\Sigma\lambda_i\overline{X}_{i.}$ is

$$SE\,(\Sigma\lambda_i\overline{X}_{i.}) = \sqrt{s^2\Sigma\lambda_i^2/n_i} \qquad (12.10.2)$$

With unequal n_i, F and t tests are more affected by nonnormality and heterogeneity of variances than with equal n (14). When in doubt as to whether within-class variances are equal, estimate the standard error of the comparison $\Sigma\lambda_i\overline{X}_{i.}$ as $\sqrt{\Sigma\lambda_i^2 s_i^2/n_i}$. An approximate number of degrees of freedom may be assigned to this standard error for t tests by an extension of Satterthwaite's rule (section 6.11, p. 97). For this extension, let $v_i = \lambda_i^2 s_i^2/n_i$ and let s_i^2 have ν_i df. (In a one-way classification, of course, $\nu_i = n_i - 1$.) The approximate degrees of freedom in the estimated variance $\Sigma v_i = \Sigma\lambda_i^2 s_i^2/n_i$ are

$$\text{df} = (\Sigma v_i)^2/(\Sigma v_i^2/\nu_i) \qquad (12.10.3)$$

As an example with unequal n_i, the public school expenditures per pupil per

TABLE 12.10.1
ANALYSIS OF VARIANCE WITH SAMPLES OF UNEQUAL SIZES

Source of Variation	df	Sum of Squares	Mean Square	F
Between classes	$a - 1$	$\Sigma X_{i.}^2/n_i - X_{..}^2/N$	s_b^2	s_b^2/s^2
Within classes	$N - a$	$\Sigma\Sigma X_{ij}^2 - \Sigma X_{i.}^2/n_i$	s^2	
Total	$N - 1$	$\Sigma\Sigma X_{ij}^2 - X_{..}^2/N$		

TABLE 12.10.2
PUBLIC SCHOOL EXPENDITURES PER PUPIL PER STATE (IN $1000)

	Northeast	Southeast	South Central	North Central	Mountain Pacific
	1.33	1.66	1.16	1.74	1.76
	1.26	1.37	1.07	1.78	1.75
	2.33	1.21	1.25	1.39	1.60
	2.10	1.21	1.11	1.28	1.69
	1.44	1.19	1.15	1.88	1.42
	1.55	1.48	1,15	1.27	1.60
	1.89	1.19	1.16	1.67	1.56
	1.88		1.26	1.40	1.24
	1.86		1.30	1.51	1.45
	1.99			1.74	1.35
				1.53	1.16
Total	17.63	9.31	10.61	17.19	16.58
Mean	1.763	1.330	1.179	1.563	1.507
n_i	10	7	9	11	11
s_i^2	0.1240	0.0335	0.0057	0.0448	0.0404

state in five regions of the United States in 1977 (22) are shown in table 12.10.2. Do the average expenditures differ from region to region? The analysis of variance (table 12.10.3) shows F highly significant.

The variance of a regional mean is σ^2/n_i. These variances differ from region to region, but since the n_i do not differ greatly, we use the *average* variance of a regional mean for t tests. This average is σ^2/n_h, where n_h is the harmonic mean of the n_i, in this case 9.33. Hence the estimated standard error of a regional mean is taken as $\sqrt{0.0515/9.33} = 0.0743$. With 43 df, the difference between two means required for significance at the 5% level is $\sqrt{2}(0.0743)(2.02) = 0.212$. Using this yardstick, average expenditure per pupil in the Northeast exceeds that in any other region except the North Central, while the North Central exceeds the southern regions.

Since the within-region values of s_i^2 look quite dissimilar, a safer procedure when comparing two class means is to estimate the standard error of the difference in means using only the data from these two classes. With this method the mean expenditures for the Northeast and Mountain Pacific regions no longer differ significantly.

EXAMPLE 12.10.1—How many degrees of freedom does Satterthwaite's method assign to the pooled mean square within regions in the F test in table 12.10.3? Ans. Since the pooled within-regions sum of squares is $\Sigma\nu_i s_i^2$, we take $v_i = \nu_i s_i^2$, $\Sigma v_i = 2.215$ in (12.10.3). We get df $= 2.215^2/0.1818 = 27$, instead of 43 df in the analysis of variance, which assumes that the σ_i^2 are all equal.

TABLE 12.10.3
ANALYSIS OF VARIANCE

Source	df	Sum of Squares	Mean Squares	F
Between regions	4	1.856	0.464	9.01
Within regions	43	2.215	0.0515	

12.11—Weighted linear regressions. The one-way classification with unequal n_i provides an introduction to weighted linear regression. The data in table 12.11.1, made available through the courtesy of B. J. Vos and W. T. Dawson, show the lethal doses of ouabain injected into cats at four different rates of injection. We will denote the lethal doses by Y_{ij}, the rates of injection by X_i.

The analysis of variance of Y, following the methods in the preceding section, is given in table 12.11.2.

The differences between rate groups are highly significant, the lethal doses increasing consistently as the rate of injection increases. A plot of $\overline{Y}_{i.}$ against X_i shows a linear relation up to $X_i = 4$, with perhaps some falling off at $X_i = 8$. We first fit a linear regression of $\overline{Y}_{i.}$ on X_i. Note that the n_i differ from group to group.

When the means $\overline{Y}_{i.}$ are based on different sample sizes n_i, the estimates a and b in the sample regression line $\hat{\overline{Y}}_{i.} = a + bX_i$ are

$$a = \overline{Y}_{..} - b\overline{X} \qquad (12.11.1)$$

where $\overline{Y}_{..}$ and $\overline{X}$ are *weighted* means: $\overline{Y}_{..} = \Sigma n_i \overline{Y}_{i.}/\Sigma n_i = \Sigma Y_{i.}/N = 1594/41 = 38.88$; $\overline{X} = \Sigma n_i X_i/N = 142/41 = 3.463$. If x_i, $\overline{y}_i$ are deviations from the weighted means,

$$b = \Sigma n_i x_i \overline{y}_{i.}/\Sigma n_i x_i^2 \qquad (12.11.2)$$

TABLE 12.11.1

LETHAL DOSE MINUS 50 UNITS OF U.S. STANDARD OUABAIN, BY SLOW INTRAVENOUS INJECTION IN CAT UNTIL THE HEART STOPS

	x_i = Rate of Injection, (mg/kg/min)/1045.75				
	1	2	4	8	Total
	5	3	34	51	
	9	6	34	56	
	11	22	38	62	
	13	27	40	63	
	14	27	46	70	
	16	28	58	73	
	17	28	60	76	
	20	37	60	89	
	22	40	65	92	
	28	42			
	31	50			
	31				
$\Sigma Y_{ij} = Y_{i.}$	217	310	435	632	1,594
n_i	12	11	9	9	41
$\overline{Y}_{i.}$	18.1	28.2	48.3	70.2	
ΣY_{ij}^2	4727	10,788	22,261	45,940	83,716

TABLE 12.11.2
ANALYSIS OF VARIANCE OF LETHAL DOSE OF OUABAIN IN CATS

Source of Variation	df	Sum of Squares	Mean Square	F
Between rates	3	16,093	5364	35.1**
Within rates	37	5,651	153	
Total	40	21,744		

where

$$\Sigma n_i x_i \bar{y}_{i.} = \Sigma X_i Y_{i.} - (\Sigma n_i X_i)(Y_{..})/N = 7633 - (142)(1594)/41 = 2112.3$$

$$\Sigma n_i x_i^2 = \Sigma n_i X_i^2 - (\Sigma n_i X_i)^2/N = 776 - 142^2/41 = 284.2$$

The weighted sum of squares of the $\bar{Y}_{i.}$ between rates of injection may now be subdivided into one component (with 1 df) that represents the regression of the $\bar{Y}_{i.}$ on the X_i and one component (with 2 df) that represents deviations from the regression. As in chapter 9, the sum of squares due to regression is $(\Sigma n_i x_i \bar{y}_i)^2/(\Sigma n_i x_i^2) = 2112.3^2/284.2 = 15{,}700$. This subdivision is shown in table 12.11.3.

In table 12.11.3 the deviations sum of squares, 393, is of course found by subtracting 15,700 from the sum of squares between rate groups, 16,093 in table 12.11.2. If we test the regression sum of squares against the deviations sum of squares, $F = 80.1$. The linear regression is significant even though the deviations sum of squares has only 2 df.

This table illustrates a further use of the F test that has many applications. If the linear model holds for the relation between $\bar{Y}_{i.}$ and X_i, it can be shown that the ratio of (deviations mean square)/(within-groups mean square) is distributed as F with 2 and 37 df. If the relation is not linear or if other sources of heterogeneity are present, the deviations sum of squares and F will tend to become large. Thus we have a test of the linearity of the regression of lethal dose on rate of injection. In this example $F = 1.28$, with $P > 0.25$ from table A 14, part II. The hypothesis of a linear relation is accepted. (A further test on these data is made in section 19.4.)

As mentioned in section 9.15, a weighted regression is sometimes used with a single sample. In the population model $Y_i = \alpha + \beta X_i + \epsilon_i$, the investigator may have reason to believe that $V(\epsilon_i)$ is not constant but is of the form $\lambda_i \sigma^2$, where λ_i is known, being perhaps a simple function of X_i. In this situation the general principle of least squares chooses a and b by minimizing

TABLE 12.11.3
SUBDIVISION OF RATES SUM OF SQUARES INTO REGRESSION AND DEVIATIONS

Source of Variation	df	Sum of Squares	Mean Squares	F
Linear regression	1	15,700	15,700	80.1
Deviations	2	393	196	1.28
Within rates	37	5,651	153	

$$\sum_{i=1}^{n} w_i(Y_i - a - bX_i)^2 \tag{12.11.3}$$

where $w_i = 1/\lambda_i$. The principle is to weight *inversely as the residual variances.* This principle leads to the estimates $b = \Sigma w_i x_i y_i / \Sigma w_i x_i^2$ and $a = \bar{y} - b\bar{x}$, where $\bar{y}$, $\bar{x}$ are weighted means and the deviations x_i, y_i are taken from the weighted means. Furthermore, if the model holds, the residual mean square

$$\Sigma w_i(Y_i - \hat{Y}_i)^2/(n - 2) \tag{12.11.4}$$

is an unbiased estimate of σ^2 with $(n - 2)$ df, while $V(b) = \sigma^2/\Sigma w_i x_i^2$. Thus the standard tests of significance and methods for finding confidence limits can be used if the λ_i are known.

EXAMPLE 12.11.1—The fact that in table 12.11.1 each rate is twice the preceding rate might suggest that a linear regression of the $\bar{Y}_{i.}$ on the logs of the rates or on the numbers 1, 2, 3, 4 was anticipated. Fit a weighted regression of the $\bar{Y}_{i.}$ on $X_i = 1, 2, 3, 4$. Does the reduction in sum of squares due to regression exceed 15,700? Ans. No. The reduction is practically the same—15,684. On this X-scale, however, the slope is steeper between $X = 3$ and $X = 4$.

12.12—Testing effects suggested by the data. The rule that a comparison L is declared significant at the 5% level if L/s_L exceeds $t_{0.05}$ is recommended for any comparisons that the experiment was designed to make. Sometimes in examining the treatment means we notice a combination that we did not intend to test but which seems unexpectedly large. If we construct the corresponding L, use of the t test for testing L/s_L is invalid, since we selected L for testing solely because it looked large.

Scheffé (11) provides a conservative test for this situation that works both for equal and unequal n_i. With a means, the rule is:

Declare $|L|/s_L$ significant only if it exceeds $\sqrt{(a - 1)F_{0.05}}$

where $F_{0.05}$ is the 5% level of F for degrees of freedom $\nu_1 = a - 1$, $\nu_2 = N - a$, which equals $a(n - 1)$ in classes with equal n. In more complex experiments, ν_2 is the number of error degrees of freedom provided by the experiment. Scheffé's test agrees with the t test when $a = 2$. It requires a substantially higher value of L/s_L for statistical significance when $a > 2$—for example, $\sqrt{(a - 1)F} = 3.08$ when $a = 5$ and 4.11 when $a = 10$ as against 1.96 for t in experiments with many error degrees of freedom.

Scheffé's test allows us to test any number of comparisons that are picked out by inspection. The probability of finding an erroneous significant result (Type I error) in any of these tests is at most 0.05. A test with this property is said to control the *experimentwise error rate* at 5%.

This test supplies the protection against Type I error that the investigator presumably wants when a comparison catches the eye simply because it looks unexpectedly large. The penalty in having a stiffer rule before declaring a result significant is, of course, loss of power relative to the t test, a penalty that the investigator may be willing to incur when faced with a result puzzling and unexpected. It has been suggested that control of the experimentwise error rate is appropriate in exploratory studies in which the investigator has no comparisons planned in advance but merely wishes to summarize what the data seem to suggest. Even in such studies, however, the analysis often concentrates on seeking some simple pattern or structure in the data that seems plausible rather than on picking out comparisons that look large. Thus, experimentwise control may give too much protection against Type I errors at the expense of the powers of any tests made.

Ability to control the probability of making *any* Type I errors when summarizing the results of an experiment is nevertheless attractive. With this approach, we know where we stand as regards liability to Type I errors, even when numerous tests are made. Those worried about the loss of power could adopt, say, 10% or 25% for the experimentwise error rate instead of 5%. However, the relative importance of Type I and Type II errors varies from study to study, and no general rule for balancing Type I against Type II errors has been widely accepted. Duncan (17) has developed a method for applications in which the investigator can judge the relative importance of losses due to Type I and Type II errors. We illustrate this approach for a one-tailed test in comparing two means, with $H_0: \mu_1 \le \mu_2$, $H_A: \mu_1 > \mu_2$. Let $\delta = \mu_1 - \mu_2$.

When H_0 holds ($\mu_1 \le \mu_2$), there is no loss of power if H_0 is accepted, but Duncan assumes a loss of amount $k_1|\delta|$ if it is rejected. If H_A holds ($\mu_1 > \mu_2$), there is no loss if H_0 is rejected but a loss of amount $k_2\delta$ if H_0 is accepted. Duncan's policy chooses the significance level that minimizes the average loss in repeated decisions made from tests of significance. In using this method the investigator must know the ratio $k = k_1/k_2$—the relative seriousness of Type I and Type II errors. In comparing every pair among a set of a class means, the optimum significance level of L/s_L for this method is found to depend on the number of means a, on the loss ratio k_1/k_2, and on the F ratio in the analysis of variance of the data into between and within classes. The value of F enters into the rule because F provides information as to whether real differences among class means are likely to be large or small. The a priori judgment of the investigator on this point may also be built into the rule. Tables of significance levels are given in (18). The decision levels of L/s_L range from values smaller than $t_{0.05}$ (when k is small and F is large) to values similar to those required by Scheffé's test (when k is large and Type I errors are relatively serious).

The problem of performing a number of tests of significance or of constructing a number of confidence intervals when analyzing a body of data is often called the *multiple comparisons* problem and has attracted a good deal of attention. The next section presents the particular case that has been most thoroughly discussed.

12.13—Inspection of all differences between pairs of means. Sometimes the classes are not ordered in any way, for example, varieties of a crop or different makes of gloves. In such cases the objective of the statistical analysis is often either (i) to rank the class means, noting whether the means ranked at the top are really different from one another and from the rest of the means, or (*ii*) to examine whether the class means all have a common μ or whether they fall into two or more groups having different μs. For either purpose, initial tests of the differences between some or all pairs of means may be helpful.

In the doughnut data in table 12.2.1, the means for the four fats (in increasing order) are as follows:

Fat	4	1	3	2	LSD	$Q_{0.05}s_{\bar{X}}$
Mean grams absorbed	62	72	76	85	12.1	16.2

The standard error of the difference between two means, $\sqrt{2s^2/n}$, is ± 5.80 with 20 df (table 12.2.1). The 5% value of t with 20 df is 2.086. Hence the difference between a specific pair of means is significant at the 5% level if it exceeds $(2.086)(5.8) = 12.1$.

The highest mean, 85 for fat 2, is significantly greater than the means 72 for fat 1 and 62 for fat 4. The mean 76 for fat 3 is significantly greater than the mean 62 for fat 4. None of the other three differences between pairs reaches 12.1. The two solid lines in the table above connect sets of means judged not significantly different. The quantity 12.1, which serves as a criterion, is called the *least significant difference* (LSD). Similarly, 95% confidence limits for the population difference between any pair of means are given by adding ± 12.1 to the observed difference.

Objections to use of the LSD in multiple comparisons have been raised for many years. Suppose that all the population means μ_i are equal so that there are no real differences. Five classes, for instance, have ten possible comparisons between pairs of means. The probability that at least one of the ten exceeds the LSD is bound to be greater than 0.05; it can be shown to be about 0.29. With ten means (45 comparisons among pairs), the probability of finding at least one significant difference is about 0.63 and with 15 means it is around 0.83.

When the μ_i are all equal, the LSD method still has the basic property of a test of significance, namely, about 5% of the tested differences will erroneously be declared significant. The trouble is that when many differences are tested in an experiment, some that appear significant are almost certain to be found. If these are reported and attract attention, the test procedure loses its valuable property of protecting the investigator against making erroneous claims.

Commenting on this issue, Fisher (8) wrote, "When the z test (i.e., the F test) does not demonstrate significance, much caution should be used before claiming significance for special comparisons." In line with this remark, investigators are sometimes advised to use the LSD method only if F is significant. This method has been called the *protected* LSD, or PSD method. If F is not significant, we declare no significant differences between means.

Among other proposed methods, the best known is one that replaces the LSD by a criterion based on the tables of the Studentized range, $Q = (\bar{X}_{max} - \bar{X}_{min})/s_{\bar{X}}$. Table A 15 gives the upper 5% levels of Q, i.e., the value exceeded in 5% of experiments. This value depends on the number of means, a, and the number ν of degrees of freedom in $s_{\bar{X}}$. Having read $Q_{0.05}$ from table A 15, we compute the difference D between two means that is required for 5% significance as $Q_{0.05}s_{\bar{X}}$.

For the doughnuts, $a = 4$, $\nu = 20$; we find $Q_{0.05} = 3.96$. Hence $D = Q_{0.05}s_{\bar{X}} = (3.96)(4.1) = 16.2$. Looking back at the table, only the difference between fats 2 and 4 is significant with this criterion. The dotted line in the table summarizes the results of the Q tests. With only two means, the Q method becomes identical with the LSD method. Otherwise, Q requires a larger difference for significance than LSD.

The Q method has the property that if we test some or all the differences between pairs of means, the probability that no erroneous claim of significance will be made is greater than or equal to 0.95. Similarly, the probability that all the confidence intervals $(\bar{X}_i - \bar{X}_j) \pm D$ will correctly include the difference $\mu_i - \mu_j$ is 0.95. The price paid for this increased protection is, of course, that fewer real differences $\mu_i - \mu_j$ are detected and confidence intervals are wider.

Scheffé's method of the previous section could be used instead of Q, since it applies to all comparisons. But for comparisons confined to pairs of means it requires larger differences for significance than Q and thus has less power with no better protection against Type I errors.

Strictly, the Q method requires samples of equal size n in the classes. With unequal n_i, several authors have recommended that $s_{\bar{X}}$ be calculated as $s/\sqrt{n_h}$, where n_h is the harmonic mean of the n_i. This approximation appears to keep experimentwise Type I errors close to 5%.

A variant of the Q method, the NK method, due to Newman (10) and Keuls (12), gives less protection in some cases (see example 12.13.5) but is more powerful (13); that is, the variant will detect real differences more frequently than the original Q method.

Arrange the means in ascending order. For the doughnut fats, these means are as follows:

Fat	4	1	3	2	*SD*
	62	72	76	85	±4.10 (20 df)

As before, first test the extreme difference of means, fat 2 − fat 4 = 23, against $D = 16.2$. Since the difference exceeds D, proceed to test fat 2 − fat 1 = 13 and fat 3 − fat 4 = 14 against the D value for $a = 3$, because these comparisons are differences between the highest and lowest of a group of *three* means. For $a = 3$, $\nu = 20$, Q is 3.58, giving $D = (3.58)(4.10) = 14.7$. Both the differences, 13 and 14, fall short of D. Consequently we stop; the difference between means of fats 2 and 4 is the only significant difference in the experiment. If fat 3 − fat 4 had been, say, 17, we would have declared this difference significant and next tested fat 3 − fat 1 and fat 1 − fat 4 against the D value for $a = 2$.

Whenever the highest and lowest of a group of means are found not significantly different in this method, we declare that none of the members of this group is distinguishable. This rule avoids logical contradictions in the conclusions. The Q method uses the Q value for a class for all tests.

If μ_i is the population mean of the ith class, several writers, (e.g., Balaam 20 and Carmer and Swanson 21), have studied the performances of the principal multiple comparison methods under different sets of μ_i, using Monte Carlo methods. In one set the μ_i are all equal, in others they are equal in subgroups, and in others they are all unequal in varying amounts. The proportion of incorrect conclusions drawn in individual comparisons (i.e., of Type I and Type II errors made) is usually the principal criterion used to compare the methods. As is not surprising in a range of sets with equal and unequal μ_i in which there are more opportunities for Type II than Type I error, the methods that increase power by less tight control of Type I errors score best in such studies. Duncan's, the LSD, and the protected LSD rank at the top, followed by the NK, Q, and S (Scheffé) methods. The exception is, of course, the set with all μ_i equal, in which the ranking is reversed. On balance, Carmer and Swanson like the protected LSD, which has good control of Type I errors, more power than the NK, Q, or S methods, and is easy to use. The Duncan method is a close competitor in their opinion—a refinement of the protected LSD that requires only a little more effort.

To sum up for this section, no method is uniformly best for all sets of data and cost considerations. In critical situations, try to judge the relative costs of the two kinds of mistakes and be guided by these costs. References (15) and (16) give good reviews of recent work on this problem.

EXAMPLE 12.13.1—In Case III of the constructed example in table 12.4.1, with $\mu_1 = 3$, $\mu_2 = 5$, $\mu_3 = 9$, the observed means are $\bar{X}_1 = 2.9$, $\bar{X}_2 = 4.0$, $\bar{X}_3 = 9.25$, with $SE = \sqrt{s^2/n} = 0.75$ (3 df). Test the three differences by (*i*) the LSD test, (*ii*) the Q test. Construct a confidence interval for each difference by each method. (*iii*) Do all the confidence intervals include $\mu_i - \mu_j$? Ans. (*i*) LSD = 3.37, $\bar{X}_3$ significantly greater than $\bar{X}_2$ and $\bar{X}_1$. (*ii*) Required difference = 4.43, same significant differences. (*iii*) Yes.

EXAMPLE 12.13.2—In example 12.5.1, the mean gains in weight of baby chicks under four feeding treatments were $\bar{X}_1 = 43.8$, $\bar{X}_2 = 71.0$, $\bar{X}_3 = 81.4$, $\bar{X}_4 = 142.8$, while $\sqrt{s^2/n} = 12.0$ with 16 df. Compare the means by the LSD and the Q methods. Ans. Both methods show that $\bar{X}_4$ differs significantly from any other mean. The LSD method also gives $\bar{X}_3$ significantly greater than $\bar{X}_1$.

EXAMPLE 12.13.3—Table 11.5.1 records the number of loopers (insect larvae) on 50 cabbage plants per plot after the application of five treatments to each of four plots. The numbers were:

Treatment	1	2	3	4	5
	11	6	8	14	7
	4	4	6	27	4
	4	3	4	8	9
	5	6	11	18	14
Total	24	19	29	67	34
Mean	6.00	4.75	7.25	16.75	8.50

The data at first were thought to be Poisson but the within-treatment variation was too large. (*i*). Verify that the analysis of variance is as follows:

Source of Variation	df	Sum of Squares	Mean Square	F
Between treatment	4	359.30	89.82	4.33
Within treatments	15	311.25	20.75	
Total	19	670.55		

(*ii*). By the protected LSD test, show that the only significant difference is that treatment 4 gave a higher mean number of loopers than any other treatment. Note: This analysis of variance can be criticised because the data are counts, not normal variates. In particular, the within-treatment ranges 7, 3, 7, 19, and 10 appear correlated with the treatment totals 24, 19, 29, 67, and 34, suggesting inequality of variances. However, if the data are transformed to square roots before analysis, a method often suggested for counts (section 15.11), the analysis gives the same conclusion.

EXAMPLE 12.13.4—In an experiment at Waseca, Minnesota, in 1932 (19), the mean yields of five varieties of barley were:

Variety	M	P	V	S	T	$s_{\bar{X}}$ (18 df)
Mean yield (bu/acre)	100.4	108.1	112.2	115.5	147.5	±2.899

Compare every pair of means by (*i*) the Studentized range or the Q method, (*ii*) the Newman-Keuls (NK) method. For each method, draw lines connecting means judged not significantly different. Ans. (*i*) the Q method joins P, V, S and M, P, V. (*ii*) the NK method joins P, V, S and M, P.

EXAMPLE 12.3.5—This example illustrates the type of data for which the NK method does not control Type I error as well as the Q method. With six means, suppose that (unknown to us) $\mu_1 = \mu_2$, $\mu_3 = \mu_4$, $\mu_5 = \mu_6$, and the three groups of means are widely separated so that any difference between sample means in different groups is almost certain to be significant. Verify that in NK tests we will end by testing *each* of $\bar{X}_2 - \bar{X}_1$, $\bar{X}_4 - \bar{X}_3$, and $\bar{X}_6 - \bar{X}_5$ by ordinary t tests. The probability that at least one of these gives an erroneous significant result is close to $3 \times 0.05 = 0.15$, which is the experimentwise error. Check that the Q test does not lead to this situation.

TECHNICAL TERMS

comparison
contrast
experimentwise error rate
fixed effects model
least significant difference
model I
multiple comparison
Newman-Keuls method
one-way classifications
orthogonal
protected LSD method
Studentized range
test of linearity

REFERENCES

1. Lowe, B. 1935. Data, Iowa Agric. Exp. Stn.
2. Richardson, R., et al. *J. Nutr.* 44 (1951):371.
3. Snedecor, G. W. 1934. *Analysis of Variance and Covariance.* Ames.
4. Fisher, R. A., and Yates, F. 1938. *Statistical Tables.* Oliver & Boyd, Edinburgh.
5. Hansberry, T. R., and Richardson, C. H. *Iowa State Coll. J. Sci.* 10 (1935):27.
6. Rothamsted Experimental Station Report. 1936, p. 289.
7. Rothamsted Experimental Station Report. 1937, p. 212.
8. Fisher, R. A. 1935. *The Design of Experiments.* Oliver & Boyd, Edinburgh.
9. Query. *Biometrics* 5 (1949):250.
10. Newman, D. *Biometrika* 31 (1939):20.
11. Scheffé, H. 1959. *The Analysis of Variance.* Wiley, New York.
12. Keuls, M. *Euphytica* 1 (1952):112.
13. Hartley, H. O. *Commun. Pure Appl. Math.* 8 (1955):47.
14. Box, G. E. P. *Ann. Math. Stat.* 25 (1954):290.
15. O'Neill, R. T., and Wetherill, G. B. *J. R. Stat. Soc.* B33 (1971):218.
16. Miller, R. G., Jr. *J. Am. Stat. Assoc.* 72 (1977):779.
17. Duncan, D. B. *Biometrics* 31 (1975):339.
18. Waller, R. A., and Duncan, D. B. *J. Am. Stat. Assoc.* 64 (1969):1484, and *Corrigenda* 67 (1972):253.
19. Immer, F. R.; Haynes, H. K.; and Powers, L. R. *J. Am. Soc. Agron.* 26 (1934):403.
20. Balaam, L. N. *Austral. J. Stat.* 5 (1963):62.
21. Carmer, S. G., and Swanson, M. R. *J. Am. Stat. Assoc.* 68 (1973):66.
22. Statistical Abstract of the United States. U.S. Bur. of Census 152 (1977).

13

The random effects model

13.1—Model II: random effects. With some types of single classification data, the model used and the objectives of the analysis differ from those under model I. Suppose that we wish to determine the average content of some chemical in a large population of leaves. We select a random sample of a leaves from the population. For each selected leaf, n independent determinations of the chemical content are made giving $N = an$ observations in all. The leaves are the classes, and the individual determinations are the members of a class.

In model II, the chemical content found for the jth determination from the ith leaf is written as

$$X_{ij} = \mu + A_i + \epsilon_{ij} \qquad (i = 1, \ldots, a;\ j = 1, \ldots, n) \qquad (13.1.1)$$

where $A_i = \mathcal{N}(0, \sigma_A^2)$, $\epsilon_{ij} = \mathcal{N}(0, \sigma^2)$.

The symbol μ is the mean chemical content of the population of leaves—the quantity to be estimated. The symbol A_i represents the difference between the chemical content of the ith leaf and the average content over the population. By including this term, we take into account the fact that the content varies from leaf to leaf. Every leaf in the population has its value of A_i, so we may think of A_i as a random variable with a distribution over the population. This distribution has mean 0, since the A_i are defined as deviations from the population mean. In the simplest version of model II, it is assumed in addition that the A_i are normally distributed with variance σ_A^2. Hence, we have written $A_i = \mathcal{N}(0, \sigma_A^2)$.

What about the term ϵ_{ij}? This term is needed because (*i*) the determination is subject to an error of measurement; and (*ii*) if the determination is made on a small piece of the leaf, its content may differ from that of the leaf as a whole. The ϵ_{ij} and the A_i are assumed independent. The further assumption $\epsilon_{ij} = \mathcal{N}(0, \sigma^2)$ is often made.

With model II the variance of an observation X_{ij} is $\sigma_A^2 + \sigma^2$. The two parts are called *components of variance.*

13.2—Relation between model II and model I. There are some similarities and some differences between model II and model I. In model I,

$$X_{ij} = \mu + \alpha_i + \epsilon_{ij} \qquad [\alpha_i \text{ fixed}, \epsilon_{ij} = \mathcal{N}(0, \sigma^2)] \qquad (13.2.1)$$

where $\Sigma\alpha_i = 0$.

Note the following points:

(*i*) The α_i are fixed quantities to be estimated; the A_i are random variables. As will be seen, their variance σ_A^2 is often of interest.

(*ii*) The null hypothesis $\alpha_i = 0$ is identical with the null hypothesis $\sigma_A = 0$, since in this event all the A_i must be zero. Thus, the F test holds also in model II, being now a test of the null hypothesis $\sigma_A = 0$.

(*iii*) We saw in (12.4.1) that when the null hypothesis is false, the mean square between classes under model I is an unbiased estimate of

$$E(MS \text{ between}) = \sigma^2 + n\Sigma\alpha_i^2/(a - 1) \tag{13.2.2}$$

An analogous result for model II is

$$E(MS \text{ between}) = \sigma^2 + n\sigma_A^2 \tag{13.2.3}$$

Neither result requires the assumption of normality.

(*iv*) In drawing repeated samples under model I, we always draw from the same set of classes with the same α_i. Under model II, we draw a *new* random sample of a leaves. A consequence is that the general distributions of F when the H_0 is false differ. With model I, this distribution, called the *noncentral F distribution,* is complicated; tables by Tang (1) and charts by Pearson and Hartley (2) are available. With model II, the probability that the observed variance ratio exceeds any value F_0 is simply the probability that the ordinary F exceeds $F_0/(1 + n\sigma_A^2/\sigma^2)$.

13.3—Use of model II in problems of measurement. To turn to an example of model II, the data for calcium in table 13.3.1 come from a large experiment (3) on the precision of estimation of the chemical content of turnip greens. To keep the example small, we have used only the data for $n = 4$ determinations on each of $a = 4$ leaves from a single plant. In the analysis of variance (table 13.3.1), the mean square between leaves s_L^2 is an unbiased estimate of σ^2 +

TABLE 13.3.1
CALCIUM CONCENTRATION IN TURNIP GREENS
(% dry weight)

Leaf	% Calcium				Sum	Mean
1	3.28	3.09	3.03	3.03	12.43	3.11
2	3.52	3.48	3.38	3.38	13.76	3.44
3	2.88	2.80	2.81	2.76	11.25	2.81
4	3.34	3.38	3.23	3.26	13.21	3.30

Source of Variation	df	Mean Square	Parameters Estimated
Between leaves	3	0.2961	$\sigma^2 + 4\sigma_A^2$
Determinations	12	0.0066	σ^2
$s^2 = 0.0066$ estimates σ^2		$s_A^2 = (0.2961 - 0.0066)/4 = 0.0724$ estimates σ_A^2	

$n\sigma_A^2 = \sigma^2 + 4\sigma_A^2$. Consequently, an unbiased estimate of σ_A^2 is

$$s_A^2 = (s_L^2 - s^2)/4 = (0.2961 - 0.0066)/4 = 0.0724$$

The quantity σ_A^2 is called the *component of variance* for leaves. The value $F = 0.2961/0.0066 = 44.9$ (highly significant with 3 and 12 df) is an estimate of $(\sigma^2 + 4\sigma_A^2)/\sigma^2$.

We now consider the questions: (*i*) How precisely has the mean calcium content been estimated? (*ii*) Can we estimate it more economically? With n determinations from each of a leaves, the sample mean $\bar{X}_{..}$ is, from (13.1.1) for model II,

$$\bar{X}_{..} = \mu + \bar{A}_{.} + \bar{\epsilon}_{..}$$

where $\bar{A}_{.}$ is the mean of a independent values of A_i (one for each leaf), and $\bar{\epsilon}_{..}$ is the mean of an independent ϵ_{ij}. Hence the variance of $\bar{X}_{..}$ as an estimate of μ is

$$V(\bar{X}_{..}) = \frac{\sigma_A^2}{a} + \frac{\sigma^2}{an} = \frac{\sigma^2 + n\sigma_A^2}{an} = \frac{\sigma^2 + 4\sigma_A^2}{16} \qquad (13.3.1)$$

In the analysis of variance, the mean square between leaves, 0.2961, is an unbiased estimate of $\sigma^2 + 4\sigma_A^2$. Hence $\hat{V}(\bar{X}_{..}) = 0.2961/16 = 0.0185$. This result is important. The estimated variance of the sample mean is the *between-classes (leaves) mean square divided by the total number of observations.* This mean square takes into account both the variation from leaf to leaf, with variance σ_A^2, and the error of measurement of a single leaf, with variance σ^2.

Suppose the experiment is to be redesigned, changing n and a to n' and a'. From (13.3.1), the variance of $\bar{X}_{..}$ becomes

$$V'(\bar{X}_{..}) = \frac{\sigma_A^2}{a'} + \frac{\sigma^2}{a'n'} \sim \frac{0.0724}{a'} + \frac{0.0066}{a'n'}$$

where $\sim$ means "is estimated by." Since the larger numerator is 0.0724, it seems clear that a' should be increased and n' decreased if this is possible without increasing the total cost of the experiment. If a determination costs 10 times as much as a leaf, our data cost $(10)(16) + 4 = 164$ units. The choice of $n' = 1$ and $a' = 15$ will cost $(10)(15) + 5 = 155$ units, about the same as our original data. For this new design our estimate of the variance of $\bar{X}_{..}$ is

$$\hat{V}'(\bar{X}_{..}) = 0.0724/15 + 0.0066/15 = 0.0053$$

The change reduces the variance of the mean from 0.0185 to 0.0053, i.e., to less than one-third, because the costly determinations with small variability have been utilized to sample more leaves whose variation is large.

In some applications, as in this example, the cost of selecting a' units at random from the population and of making n' independent measurements on each unit is of the form $C = c_a a' + c a' n'$. Here, c is the cost of a single measurement on the unit or on a random subsample from it, while c_a is the cost of selecting a unit. If the variance of the sample mean $\overline{X}_{..}$ is as given by (13.3.1), it may be shown that this variance is minimized for given total cost when

$$n' = \sqrt{c_a \sigma_A^2/(c\sigma^2)} = \sqrt{(1/10)(0.0724/0.0066)} = 1.05$$

Since n' must be an integer, this equation shows that $n' = 1$, i.e., one measurement on each sample leaf, is the best choice in our example. When an extensive sample is being planned in order to estimate the population mean, a pilot sample giving data like that in table 13.3.1 is sometimes selected first as a guide to an efficient plan for the main sample (though a' should be greater than 4 in the pilot sample to give a better estimate of σ_A^2/σ^2).

To mention other applications, n replications of each of a inbred lines may be grown in an experiment in plant breeding. The component σ_A^2 represents differences in yield due to differences in the genotypes (genetic characteristics) of the inbreds, while σ^2 measures the effect of nongenetic influences on yield. The ratio $\sigma_A^2/(\sigma_A^2 + \sigma^2)$ of genetic to total variance gives a guide to the possibility of improving yield by selection of particular inbreds. The same concepts are important in human family studies, both in genetics and the social sciences, where the ratio $\sigma_A^2/(\sigma_A^2 + \sigma^2)$ measures the proportion of the total variance that is associated with the family. The interpretation is more complex, however, since human families differ not only in genetic traits but also in environmental factors that affect the variables under study.

EXAMPLE 13.3.1—The following data were abstracted from records of performance of Poland China swine in a single inbred line at the Iowa Agricultural Experiment Station. Two boars were taken from each of four litters with common sire and fed a standard ration from weaning to about 225 lb. Here are the average daily gains:

Litter	1	2	3	4
Gains, boar 1	1.18	1.36	1.37	1.07
boar 2	1.11	1.65	1.40	0.90

Assuming that the litter variable is normally distributed, show that σ_A^2 differs significantly from zero ($F = 7.41$) and that the sample estimate of σ_A^2 is 0.0474.

EXAMPLE 13.3.2—In a wheat sampling survey in Kansas (4), two subsamples were taken at random from each of 660 fields. The analysis of variance for yield (bushels per acre) was as follows:

Source of Variation	Degrees of Freedom	Mean Square
Fields	659	434.52
Samples within fields	660	67.54

Estimate the components of variance. Ans. $s_A^2 = 183.49$, $s^2 = 67.54$.

EXAMPLE 13.3.3—Estimate the variance of the sample mean from the data in example 13.3.2 for (*i*) 1 subsample from each of 800 fields, (*ii*) 2 subsamples from each of 400 fields, (*iii*) 8 subsamples from each of 100 fields. Ans. (*i*) 0.313, (*ii*) 0.543, (*iii*) 1.919.

EXAMPLE 13.3.4—To prove the result (13.2.3), note that

$$\overline{X}_{i.} = A_i + \bar{\epsilon}_{i.}$$

The two terms are independent, with variances σ_A^2 and σ^2/n. Hence, $\bar{X}_{i.}$ has variance $(\sigma_A^2 + \sigma^2/n)$. The sample variance of the $\bar{X}_{i.}$, namely $\Sigma(\bar{X}_{i.} - \bar{X}_{..})^2/(n - 1)$, is therefore an unbiased estimate of this quantity. Multiply by n to get the mean square between classes, and (13.2.3) follows.

13.4—Structure of model II illustrated by sampling. It is easy to construct a model II experiment by sampling from known populations. One population can represent the within-class terms ϵ_{ij} with variance σ^2 and another the between-class effects A_i with variance σ_A^2. Then samples are drawn from each and the sums $\epsilon_{ij} + A_i$ are formed to represent the data. Table 13.4.1 shows two units from each of ten classes. The ϵ_{ij} are in columns 2 and 3, the A_i are in columns 4 and 5, and their sums are in columns 6 and 7.

The analysis of variance is given in table 13.4.1; then the components of variance are estimated. From the 20 observations we obtained estimates $s^2 = 96.5$ of $\sigma^2 = 100$ and $s_A^2 = 24.0$ of $\sigma_A^2 = 25$, the two components that were put into the data.

This example was chosen because of its unusually accurate estimates. An idea of ordinary variation is obtained from examination of the records of 25 samples drawn similarly in table 13.4.2. One is struck immediately by the great variability in the estimates of σ_A^2, some of them being negative! These latter occur whenever the mean square for classes is less than that for individuals. Clearly, one cannot hope for accurate estimates of σ^2 and σ_A^2 from such small samples.

EXAMPLE 13.4.1—In table 13.4.2, how many negative estimates of σ_A^2 would be expected? Ans. A negative estimate occurs whenever the observed $F < 1$. From section 13.2, the probability

TABLE 13.4.1

ILLUSTRATION OF RANDOM EFFECTS MODEL, WITH $a = 10$, $n = 2$. WITHIN-CLASS COMPONENTS HAVE MEAN 30, $\sigma^2 = 100$; BETWEEN-CLASS COMPONENTS HAVE MEAN 0, $\sigma_A^2 = 25$

Class Number, i	Components: Within class, ϵ_{ij}		Components: Between class, A_i		Sums		Class Totals
1	37	39	−1	−1	36	38	74
2	26	7	2	2	28	9	37
3	30	49	−1	−1	29	48	77
4	32	32	0	0	32	32	64
5	33	42	−4	−4	29	38	67
6	39	33	−10	−10	29	23	52
7	35	26	10	10	45	36	81
8	11	20	2	2	13	22	35
9	26	48	4	4	30	52	82
10	45	24	−2	−2	43	22	65

Source of Variation	df	Mean Square	Parameters Estimated
Between classes	9	144.6	$\sigma^2 + 2\sigma_A^2$
Within classes	10	96.5	σ^2
$s^2 = 96.5$ estimates $\sigma^2 = 100$		$s_A^2 = (144.6 - 96.5)/2 = 24.0$ estimates $\sigma_A^2 = 25$	

TABLE 13.4.2
ESTIMATES OF $\sigma_A^2 = 25$ AND $\sigma^2 = 100$ MADE FROM 25 SAMPLES DRAWN AS IN TABLE 13.4.1

Sample Number	Estimate of $\sigma_A^2 = 25$	Estimate of $\sigma^2 = 100$	Sample Number	Estimate of $\sigma_A^2 = 25$	Estimate of $\sigma^2 = 100$
1	60	127	14	56	112
2	56	104	15	−33	159
3	28	97	16	67	54
4	6	91	17	−18	90
5	18	60	18	33	65
6	− 5	91	19	−21	127
7	7	53	20	−48	126
8	− 1	87	21	4	43
9	0	66	22	3	145
10	−78	210	23	49	142
11	14	148	24	75	23
12	7	162	25	77	106
13	68	76			
			Mean	17.0	102.6

that the observed $F < 1$ is the probability that the ordinary $F < 1/(1 + 2\sigma_A^2/\sigma^2)$, or in this example, $F < 1/1.5 = 2/3$, where F has 9 and 10 df. A property of the F distribution is that this probability equals the probability that F with 10 and 9 df exceeds 3/2 (or 1.5). From table A 14, with $\nu_1 = 10$, $\nu_2 = 9$, we see that F exceeds 1.59 with $P = 0.25$. Thus about $(0.25)(25) = 6.2$ negative estimates are expected, as against 7 found in table 13.4.2.

13.5—Intraclass correlation. When the component $\sigma_A^2 > 0$, we have seen that members of the same class tend to act alike to some extent. To describe this property, an alternative form of model II supposes that the observations X_{ij} are all distributed about the same mean with the same variance σ_1^2 but that any two members of the same class have a correlation coefficient ρ_I, called the *intraclass correlation coefficient*. Two members in different classes are uncorrelated. Historically, this model antedates the analysis of variance.

The expected values of the mean squares in the analysis of variance under the two forms of model II are shown in table 13.5.1. The ratio of the expected values can be expressed either as $[1 + (n - 1)\rho_I]/(1 - \rho_I)$ or as $1 + n\sigma_A^2/\sigma^2$, in the previous notation. If you equate these expressions you will find that

$$\rho_I = \sigma_A^2/(\sigma_A^2 + \sigma^2) \tag{13.5.1}$$

This result might have been obtained by noting that with $X_{ij} = A_i + \epsilon_{ij}$, the

TABLE 13.5.1
EXPECTED VALUES OF MEAN SQUARES

Source	Mean Square	Expected Value
Between classes	s_b^2	$\sigma_1^2\,[1 + (n - 1)\,\rho_I] = \sigma^2 + n\sigma_A^2$
Within classes	s_w^2	$\sigma_1^2\,(1 - \rho_I) = \sigma^2$

covariance of X_{ij}, X_{ik} (two members of the same class) is σ_A^2, while each has variance $\sigma_A^2 + \sigma^2$. Their correlation is therefore $\sigma_A^2/(\sigma_A^2 + \sigma^2) = \rho_I$.

Thus the intraclass correlation is the proportion of the total variance of an observation that is associated with the class to which it belongs. This form of model II is appropriate when our primary interest is in the size of ρ_I as a measure of class homogeneity or in questions such as, Is intrafamily correlation higher for some variables than others or higher in some countries than in others?

Furthermore, this form of model II is more general than the components of variance model. If ρ_I is negative, note that s_b^2 has a *smaller* expected value than s_w^2. With $X_{ij} = A_i + \epsilon_{ij}$, where the two terms are independent, this cannot happen. But if, for instance, four young animals in a pen compete for an insufficient supply of food, the stronger animals may drive away the weaker and may regularly get most of the food. For this reason the variance in weight within pens may be larger than that between pens, this being a real phenomenon and not an accident of sampling. We say that there is a negative correlation ρ_I between the weights within a pen. One restriction on negative values of ρ_I is that ρ_I cannot be less than $-1/(n-1)$ because the expected value of s_b^2 must be greater than or equal to zero.

From the analysis of variance it is clear that $(s_b^2 - s_w^2)$ estimates $n\rho_I\sigma^2$, while $[s_b^2 + (n-1)s_w^2]$ estimates $n\sigma^2$. This suggests that as an estimate of ρ_I we take

$$r = (s_b^2 - s_w^2)/[s_b^2 + (n-1)s_w^2] \tag{13.5.2}$$

A method of calculating confidence limits for ρ_I is given in section 13.6.

The data on identical twins in table 13.5.2 illustrate a high positive intraclass correlation. The numbers of finger ridges are nearly the same for the two members of each pair but differ markedly among pairs. From the analysis of variance, the estimate of ρ_I $(n = 2)$ is

$$r_I = (817.31 - 14.29)/(817.31 + 14.29) = 0.966$$

TABLE 13.5.2
NUMBER OF FINGER RIDGES ON BOTH HANDS OF INDIVIDUALS IN 12 PAIRS OF FEMALE IDENTICAL TWINS
(data from Newman, Freeman, and Holzinger 5)

Pair	Finger Ridges of Individuals		Pair	Finger Ridges of Individuals		Pair	Finger Ridges of Individuals	
1	71	71	5	76	70	9	114	113
2	79	82	6	83	82	10	94	91
3	105	99	7	114	113	11	75	83
4	115	114	8	57	44	12	76	72

Source of Variation	Degrees of Freedom	Mean Square
Twin pairs	11	817.31
Individuals	12	14.29
$s^2 = 14.29$	$s_A^2 = 401.51$	$r_I = 0.966$

13.6—Confidence limits related to variance components. The limits most frequently wanted appear to be those for σ_A^2/σ^2 or $\rho_I = \sigma_A^2/(\sigma_A^2 + \sigma^2)$. If the observations are normally distributed, limits can be obtained from the F distribution. Let $F_0 = s_b^2/s_w^2$ be the sample ratio of the mean square between classes to that within classes. Since F_0 is distributed as $F(1 + n\sigma_A^2/\sigma^2)$, as noted in section 13.2, then apart from a 1-in-20 chance,

$$F_{0.975}(1 + n\sigma_A^2/\sigma^2) < F_0 < F_{0.025}(1 + n\sigma_A^2/\sigma^2) \qquad (13.6.1)$$

where $F_{0.975}$ and $F_{0.025}$ are the lower and upper 2.5% one-tailed levels of F with df $\nu_1 = (a - 1)$, $\nu_2 = a(n - 1)$. Furthermore,

$$F_{0.975}(\nu_1, \nu_2) = 1/[F_{0.025}(\nu_2, \nu_1)]$$

and can therefore be obtained from table A 14, part II.

Express the inequalities (13.6.1) in terms of inequalities about σ_A^2/σ^2. We get, apart from a 1-in-20 chance,

$$(1/n)(F_0/F_{0.025} - 1) < \sigma_A^2/\sigma^2 < (1/n)(F_0/F_{0.975} - 1) \qquad (13.6.2)$$

Expression (13.6.2) gives 95% confidence limits for σ_A^2/σ^2.

EXAMPLE 13.6.1—In estimating the amount of plankton in an area of sea, seven runs (called hauls) were made, with six nets on each run (6). The analysis of variance was as follows:

Source of Variation	df	Mean Square	Estimate of	F
Between hauls	6	0.1011	$\sigma^2 + 6\sigma_A^2$	4.86
Within hauls	35	0.0208	σ^2	

To estimate 90% confidence limits for σ_A^2/σ^2, use $F_{0.05}$ and $F_{0.95}$ in (13.6.2) instead of $F_{0.025}$ and $F_{0.975}$. We have $F_{0.05} = 2.37$, $F_{0.95} = 1/3.79$. Hence the 90% limits are

$$(1/6)(4.86/2.37 - 1) < \sigma_A^2/\sigma^2 < (1/6)(4.86 \times 3.79 - 1)$$

$$0.18 < \sigma_A^2/\sigma^2 < 2.90$$

If ϕ_L, ϕ_U are the confidence limits for σ_A^2/σ^2, the limits for $\rho_I = \sigma_A^2/(\sigma_A^2 + \sigma^2)$ at the same confidence probability are $\phi_L/(1 + \phi_L)$ and $\phi_U/(1 + \phi_U)$, in this example 0.15 and 0.74.

Assuming normality, approximate confidence limits for the variance component σ_A^2 have been given by several workers. We illustrate one method (7) from the turnip greens example (table 13.3.1) for which $n = 4$, $\nu_1 = 3$, $\nu_2 = 12$, $s_A^2 = 0.0724$, $s^2 = 0.0066$. Four entries from the F table are needed. The table of 5% significance levels determines a two-tailed 90% confidence interval with 5% on each tail. The 5% values of F needed are as follows:

$$F_{\nu_1,\nu_2} = F_{3,12} = 3.49 \qquad F_{\nu_1,\infty} = F_{3,\infty} = 2.60$$

$$F_{\nu_2,\nu_1} = F_{12,3} = 8.74 \qquad F_{\infty,\nu_1} = F_{\infty,3} = 8.53$$

$$F_O = \text{observed } F = 44.86$$

The confidence limits are given as multipliers of the quantity $s^2/n = 0.0066/4 = 0.00165$. The lower limit for σ_A^2 is

$$\hat{\sigma}_{AL}^2 = [(F_O - F_{\nu_1,\nu_2})/F_{\nu_1,\infty}](s^2/n) = [(44.86 - 3.49)/2.60](0.00165) = 0.026$$

As would be expected, the lower limit becomes zero if $F_O = F_{\nu_1,\nu_2}$, that is, if F is just significant at the 5% level. For the upper limit,

$$\hat{\sigma}_{AU}^2 = (F_O - 1/F_{\nu_2,\nu_1})F_{\infty,\nu_1}(s^2/n) = (44.86 - 1/8.74)(8.53)(0.00165) = 0.63$$

To summarize, the estimate $s_A^2 = 0.0724$, with 90% limits 0.026 and 0.63. The method assumes normality; with long-tailed distributions, wider limits are needed.

13.7—Random effects with samples of unequal sizes. This case occurs frequently in family studies in human and animal genetics and in the social sciences. Model II is unchanged:

$$X_{ij} = \mu + A_i + \epsilon_{ij} \qquad (i = 1, \ldots, a; j = 1, \ldots, n_i) \tag{13.7.1}$$

The new feature is that n_i, the size of sample in the ith class, varies from class to class. The total sample size is $N = \Sigma n_i$. The A_i and ϵ_{ij} are assumed independent, with means 0 and variances σ_A^2 and σ^2.

The method used to obtain confidence limits for σ_A^2/σ^2 in section 13.6 no longer applies, since the ratio s_b^2/s^2 is not distributed as a multiple of F when the n_i are unequal. An ingenious method of finding these confidence limits with unequal n_i has been given by Wald (8), though it is a little troublesome to apply (see example 13.7.3).

The computations for the analysis of variance and the F test of the null hypothesis $\sigma_A^2 = 0$ are the same as with fixed effects, as given in section 12.10. With equal n_i $(=n)$, the mean square between classes was found to be an unbiased estimate of $\sigma^2 + n\sigma_A^2$ (section 13.2). With unequal n_i, the corresponding expression is $\sigma^2 + n_0\sigma_A^2$, where

$$n_0 = [1/(a - 1)](N - \Sigma n_i^2/N) \tag{13.7.2}$$

The multiplier n_0 is always less than the arithmetic mean $\bar{n}$ of the n_i, although usually only slightly less.

Consequently, if s_b^2 and s^2 are the mean squares between and within classes, unbiased estimates of σ^2 and σ_A^2 are given by

$$\hat{\sigma}^2 = s^2 \qquad \hat{\sigma}_A^2 = (s_b^2 - s^2)/n_0 \tag{13.7.3}$$

However, there are some mathematical complexities with unequal n_i. The estimate $\hat{\sigma}_A^2$ in (13.7.3) is not fully efficient. Some Monte Carlo studies suggest that if $\sigma_A^2/\sigma^2 > 1$, a better estimate of σ_A^2 may be obtained from the *unweighted* sum of squares of deviations between classes, $\Sigma(\bar{X}_{i.} - \bar{X}_u)^2$, where $\bar{X}_u = \Sigma\bar{X}_i/a$.

If $s_b'^2 = \Sigma(\overline{X}_{i.} - \overline{X}_u)^2/(a - 1)$, its mean value is

$$E(s_b'^2) = \sigma_A^2 + (\sigma^2/a)(\Sigma 1/n_i) = \sigma_A^2 + \sigma^2/n_h \qquad (13.7.4)$$

Hence, the quantity $s_b'^2 - s^2/n_h$ is an alternative unbiased estimate of σ_A^2, where the harmonic mean $n_h = a/(\Sigma 1/n_i)$. These estimates of σ_A^2 have the awkward feature that they can take negative values; biased estimators that are always positive may be superior.

EXAMPLE 13.7.1—In research on artificial insemination of cows, a series of semen samples from a bull are tested for their ability to produce conceptions. The following data from a larger set supplied by G. W. Salisbury show the percentages of conceptions obtained from samples for six bulls. (Since the data are percentages based on slightly differing numbers of tests, the assumption that σ^2 is constant in these data is not quite correct.)

Bull, i	Percentages of Conceptions to Services for Successive Samples	n_i	$X_{i.}$	$\overline{X}_{i.}$
1	46, 31, 37, 62, 30	5	206	41.20
2	70, 59	2	129	64.50
3	52, 44, 57, 40, 67, 64, 70	7	394	56.29
4	47, 21, 70, 46, 14	5	198	39.60
5	42, 64, 50, 69, 77, 81, 87	7	470	67.14
6	35, 68, 59, 38, 57, 76, 57, 29, 60	9	479	53.22
Total		35	1876	321.95

Source	df	SS	MS	$E(MS)$
Between bulls	5	3772	754	$\sigma^2 + 5.67\sigma_A^2$
Within bulls	29	6750	233	σ^2

$s^2 = 233$ estimates σ^2 $\quad$ $(754 - 233)/5.67 = 92$ estimates σ_A^2

(*i*) Verify the analysis of variance, the value of n_0 from (13.7.2), and the estimates s^2 of σ^2 and $s_A^2 = 92$ of σ_A^2. (*ii*) Estimate σ_A^2 from the unweighted sum of squares by (13.7.4). Ans. $\hat{\sigma}_A^2 = 82$. Since method (*i*) estimates σ_A^2/σ^2 as only about 0.4, method (*i*) would probably be preferred over method (*ii*) as an estimator of σ_A^2.

EXAMPLE 13.7.2—In the preceding example we might consider either fixed or random effects of bulls, depending on the objectives. If these six bulls were available for an artificial insemination program, we would be interested in comparing the percentages of success of these specific bulls in a fixed effects analysis.

EXAMPLE 13.7.3—Wald's method of finding confidence limits for σ_A^2/σ^2 with unequal n_i runs as follows. From (13.7.1) we get, say,

$$V(\overline{X}_{i.}) = \sigma_A^2 + \sigma^2/n_i = \sigma^2(\phi + 1/n_i) = \sigma^2/w_i$$

where ϕ is the unknown σ_A^2/σ^2. Wald proved that if we knew ϕ and hence knew the w_i, the quantity

$$s_1^2 = \Sigma w_i(\overline{X}_{i.} - \overline{X}_w)^2/(a - 1)$$

would be an unbiased estimate of σ^2 and the ratio s_1^2/s^2 would be distributed as F with $(a - 1)$ and $(N - a)$ df. Note that s_1^2 depends on the value of ϕ. Thus, to get 90% confidence limits for ϕ by Wald's method, we have to find by trial and error two values ϕ_L and ϕ_U such that the corresponding s_1^2/s^2 are equal to $F_{0.05}$ and $F_{0.95}$, respectively. Incidentally, if s_b^2/s^2 in the ordinary analysis of variance happens to equal $F_{0.05}$, then $\phi_L = 0$, as we would expect.

In example 13.7.1 calculate the w_i for the trial value $\phi_U = 2$, and show that this gives $s_1^2/s^2 = 0.23$, which is very close to $F_{0.95}$ in this example. Thus the upper 90% limit for σ_A^2/σ^2 is close to 2.

EXAMPLE 13.7.4—Verify that if the n_i are equal, Wald's method reduces to that given in section 13.6 for σ_A^2/σ^2.

13.8—Extension to three stages of sampling. Each sample may be composed of *subsamples* and these in turn may be subsampled, etc. The repeated sampling and subsampling gives rise to *nested* or *hierarchal classifications,* as they are sometimes called.

In table 13.8.1 is an example from the turnip greens experiment cited earlier (3). The four plants were taken at random; then three leaves were randomly selected from each plant. From each leaf were taken two samples of 100 mg in which calcium was determined by microchemical methods. The immediate objective is to estimate the components of variance due to the sources of variation—plants, leaves of the same plant, and determinations on the leaves.

The calculations are given in table 13.8.1. The *total* sums of squares for determinations, leaves, and plants are first obtained by the usual formulas. The

TABLE 13.8.1

CALCIUM CONCENTRATION (%, DRY BASIS) IN $b = 3$ LEAVES FROM EACH OF $a = 4$ TURNIP PLANTS, $n = 2$ DETERMINATIONS/LEAF. ANALYSIS OF VARIANCE

Plant, i $i = 1, \ldots, a$	Leaf, ij $j = 1, \ldots, b$	Determinations, X_{ijk}		$X_{ij.}$	$X_{i..}$	$X_{...}$
1	1	3.28	3.09	6.37		
	2	3.52	3.48	7.00		
	3	2.88	2.80	5.68	19.05	
2	1	2.46	2.44	4.90		
	2	1.87	1.92	3.79		
	3	2.19	2.19	4.38	13.07	
3	1	2.77	2.66	5.43		
	2	3.74	3.44	7.18		
	3	2.55	2.55	5.10	17.71	
4	1	3.78	3.87	7.65		
	2	4.07	4.12	8.19		
	3	3.31	3.31	6.62	22.46	72.29

total size $= abn = (4)(3)(2) = 24$ determinations

$C = (X_{...})^2/(abn) = 72.29^2/24 = 217.7435$

Determinations: $\Sigma X_{ijk}^2 - C = 3.28^2 + \cdots + 3.31^2 - C = 10.2704$

Leaves: $\Sigma X_{ij.}^2/n - C = (6.37^2 + \cdots + 6.62^2)/2 - C = 10.1905$

Plants: $\Sigma X_{i..}^2/bn - C = (19.05^2 + \cdots + 22.46^2)/6 - C = 7.5603$

Leaves of same plant = leaves − plants = 10.1905 − 7.5603 = 2.6302

Determinations on same leaf = determinations − leaves = 10.2704 − 10.1905 = 0.0799

Source of Variation	Degrees of Freedom	Sum of Squares	Mean Square
Plants	3	7.5603	2.5201
Leaves in plants	8	2.6302	0.3288
Determinations in leaves	12	0.0799	0.0067
Total	23	10.2704	

sum of squares *between leaves of the same plant* is found by subtracting the sum of squares between plants from that between leaves, as shown. Similarly, the sum of squares *between determinations on the same leaf* is obtained by deducting the total sum of squares between leaves from that between determinations. This process can be extended to successive subsampling.

We will now prove that this subtraction method of computing the analysis of variance sums of squares is valid. The pooled sum of squares of deviations between determinations made on the same leaf is

$$\sum_i^a \sum_j^b \sum_k^n (X_{ijk} - \overline{X}_{ij.})^2 \equiv \sum_i^a \sum_j^b \sum_k^n X_{ijk}^2 - \sum_i \sum_j X_{ij.}^2/n \tag{13.8.1}$$

But if $C = X_{...}^2/(abn)$, the *total* sum of squares between determinations is

$$\sum_i^a \sum_j^b \sum_k^n X_{ijk}^2 - C \tag{13.8.2}$$

and the *total* sum of squares between leaves (on a single determination basis) is

$$\sum_i^a \sum_j^b X_{ij.}^2/n - C \tag{13.8.3}$$

Subtraction of (13.8.3) from (13.8.2) gives the desired (13.8.1). This argument holds at any stage of sampling. Of course the terms C may be omitted, since they cancel.

The extension of model II to three stages is

$$X_{ijk} = \mu + A_i + B_{ij} + \epsilon_{ijk} \tag{13.8.4}$$

with $i = 1, \ldots, a$; $j = 1, \ldots, b$; $k = 1, \ldots, n$; and $A_i = \mathcal{N}(0, \sigma_A^2)$, $B_{ij} = \mathcal{N}(0, \sigma_B^2)$, $\epsilon_{ijk} = \mathcal{N}(0, \sigma^2)$. The variables A_i, B_{ij}, and ϵ_{ijk} are all assumed independent. In this example A refers to plants and B to leaves.

From the model it follows that the expected values of the mean squares in the analysis of variance are as shown in table 13.8.2. Unbiased estimates of the three components of variance are obtained easily as shown under the analysis of variance.

TABLE 13.8.2
ANALYSIS OF VARIANCE OF TURNIP GREENS DATA

Source of Variation	df	Mean Square	Parameters Estimated
Plants	3	2.5201	$\sigma^2 + n\sigma_B^2 + bn\sigma_A^2$
Leaves in plants	8	0.3288	$\sigma^2 + n\sigma_B^2$
Determinations in leaves	12	0.0067	σ^2

$n = 2, b = 3, a = 4 \qquad s^2 = 0.0067$ estimates σ^2
$s_B^2 = (0.3288 - 0.0067)/2 = 0.1610$ estimates σ_B^2
$s_A^2 = (2.5201 - 0.3288)/6 = 0.3652$ estimates σ_A^2

Null hypotheses that may be tested are:

$$\sigma_A^2 = 0 \qquad F = \frac{2.5201}{0.3288} = 7.66 \text{ estimates } \frac{\sigma^2 + n\sigma_B^2 + bn\sigma_A^2}{\sigma^2 + n\sigma_B^2} \qquad \nu = 3, 8$$

$$\sigma_B^2 = 0 \qquad F = \frac{0.3288}{0.0067} = 49 \text{ estimates } \frac{\sigma^2 + n\sigma_B^2}{\sigma^2} \qquad \nu = 8, 12$$

For the first, with df $\nu_1 = 3$ and $\nu_2 = 8$, F is almost on its 1% point, 7.59; for the second, with df 8 and 12, F is far beyond its 1% point, 4.50. In the sampled population the percent calcium evidently varies both from leaf to leaf and from plant to plant.

As with a single subclassification (plants and leaves in section 13.3), the estimated variance of the sample mean per determination is given by the mean square *between plants* divided by the number of determinations. This estimated variance can be expressed in terms of the estimated components of variance from table 13.8.2, as follows:

$$s_{\bar{X}}^2 = \frac{2.5201}{24} = 0.105 = \frac{0.0067 + n(0.1610) + bn(0.3652)}{nab}$$

$$= \frac{0.0067}{nab} + \frac{0.1610}{ab} + \frac{0.3652}{a}$$

This expression suggests that more information per dollar may be obtained by decreasing n, the number of expensive determinations per leaf that have a small component, and increasing b or a, the numbers of leaves or plants. Plants presumably cost more than leaves, but the component is also larger.

Confidence limits for σ_A^2 and σ_B^2 are calculated by the method described in section 13.6.

EXAMPLE 13.8.1—From (13.8.4) for the model, show that (*i*) the variance of the sample mean is $(\sigma^2 + n\sigma_B^2 + bn\sigma_A^2)/(abn)$; and (*ii*) an unbiased estimate of it is given by the mean square between plants divided by anb, i.e., by $2.5201/24 = 0.105$, as stated in section 13.8.

EXAMPLE 13.8.2—If one determination were made on each of two leaves from each of ten plants, what is your estimate of the variance of the sample mean? Ans. 0.045.

EXAMPLE 13.8.3—With one determination on one leaf from each plant, how many plants must be taken to reduce $s_{\bar{X}}$ to 0.2? Ans. About 14. (This estimate is very rough, since the mean square between plants has only 3 df.)

13.9—Three stages with mixed model. In some applications of subsampling, the major classes have fixed effects to be estimated. An instance is an evaluation of the breeding value of a set of five sires in raising pigs. Each sire is mated to a random group of dams, each mating producing a litter of pigs whose characteristics are the criterion. The model is

TABLE 13.9.1
AVERAGE DAILY GAIN OF TWO PIGS OF EACH LITTER

Sire	Dam	Pig Gains		Sums		
1	1	2.77	2.38	5.15		
	2	2.58	2.94	5.52	10.67	
2	1	2.28	2.22	4.50		
	2	3.01	2.61	5.62	10.12	
3	1	2.36	2.71	5.07		
	2	2.72	2.74	5.46	10.53	
4	1	2.87	2.46	5.33		
	2	2.31	2.24	4.55	9.88	
5	1	2.74	2.56	5.30		
	2	2.50	2.48	4.98	10.28	51.48

Source of Variation	df	Mean Square	Parameters Estimated
Sires	4	0.0249	$\sigma^2 + n\sigma_B^2 + nb\kappa^2$
Dams—same sire	5	.1127	$\sigma^2 + n\sigma_B^2$
Pairs—same dam	10	0.0387	σ^2

$n = 2, b = 2 \quad s^2 = 0.0387$ estimates σ^2

$s_B^2 = (0.1127 - 0.0387)/2 = 0.0370$ estimates $\sigma_B^2 \quad$ 0 estimates κ^2

To test $\sigma_B^2 = 0$, $F = 0.1127/0.0387 = 2.91$, $F_{0.05} = 3.33$.

$$X_{ijk} = \mu + \alpha_i + B_{ij} + \epsilon_{ijk} \tag{13.9.1}$$

The α_i are constants ($\Sigma\alpha_i = 0$) associated with the sires but the B_{ij} and the ϵ_{ijk} are random variables corresponding to dams and offspring. Hence the model is called *mixed.*

Table 13.9.1 is an example with $b = 2$ dams for each sire and $n = 2$ pigs chosen from each litter for easy analysis (from records of the Iowa Agricultural Experiment Station). The calculations proceed exactly as in the preceding section. The only change is that in the mean square for sires, the term $nb\kappa^2$, where $\kappa^2 = \Sigma\alpha^2/(a - 1)$, replaces $nb\sigma_A^2$.

In a mixed model of this type, two points must be noted. From (13.8.4), the observed class mean may be written

$$\overline{X}_{i..} = \mu + \alpha_i + \overline{B}_{i.} + \bar{\epsilon}_{i..}$$

where $\overline{B}_{i.}$ is the average of b values of the B_{ij} and $\bar{\epsilon}_{i..}$ is the average of nb values of the ϵ_{ijk}. Thus the variance of $\overline{X}_{i..}$, considered as an estimate of $\mu + \alpha_i$, is

$$V(\overline{X}_{i..}) = \sigma_B^2/b + \sigma^2/(nb) = (\sigma^2 + n\sigma_B^2)/(nb)$$

The analysis of variance shows that the mean square *between dams of the same sire* is the relevant mean square, being an unbiased estimate of $\sigma^2 + n\sigma_B^2$. The standard error of a sire mean is $\sqrt{0.1127/4} = 0.168$ with 5 df. Second, the F ratio for testing the null hypothesis that all $\alpha_i = 0$ is the ratio 0.0249/0.1127. Since this ratio is substantially less than 1, there is no indication of differences between sires in these data.

13.10—Tests of homogeneity of variance. From time to time we have questioned whether two or more mean squares differ significantly. With two mean squares, a two-tailed F test is made as follows. Place the larger of s_1^2, s_2^2 in the numerator of F and read table A 14, part II, at the 2.5% level to obtain a 5% test. With a one-tailed test having $H_A:\sigma_1^2 > \sigma_2^2$, place s_1^2 in the numerator and read F at the usual 5% level; of course if $s_1^2 \leq s_2^2$ in this case, we accept H_0.

With $a > 2$ independent estimates of variance s_i^2, Bartlett (9) has provided a test. If the s_i^2 all have the same number of degrees of freedom ν, the test criterion, using logarithms to base e, is

$$M = \nu(a \ln \bar{s}^2 - \Sigma \ln s_i^2) \qquad (\bar{s}^2 = \Sigma s_i^2/a) \tag{13.10.1}$$

On the null hypothesis that each s_i^2 is an estimate of the same σ^2, the quantity M/C is distributed approximately as χ^2 with $(a - 1)$ df, where $C = 1 + (a + 1)/(3a\nu)$.

In table 13.10.1 this test is applied to the variances of grams of fat absorbed with four types of fat in the doughnut example of table 12.2.1. Here $a = 4$, $\nu = 5$. The value of M is 1.88, clearly not significant with 3 df. To illustrate the method, however, $\chi^2 = M/C = 1.74$ has also been computed.

When the degrees of freedom ν_i differ, as with samples of unequal sizes,

$$M = (\Sigma\nu_i) \ln \bar{s}^2 - \Sigma\nu_i \ln s_i^2 \qquad (\bar{s}^2 = \Sigma\nu_i s_i^2/\Sigma\nu_i)$$
$$C = 1 + \{1/[3(a - 1)]\}(\Sigma 1/\nu_i - 1/\Sigma\nu_i)$$
$$\chi^2 = M/C \qquad (a - 1)\text{df}$$

In table 13.10.2 this test is applied to the variances of the birth weights of five litters of pigs. Since $\bar{s}^2$ is the pooled variance (weighting by degrees of freedom), we have included a column of sums of squares as well as a column of reciprocals for finding C. The computations give $\chi^2 = 16.99$ with 4 df, showing that intralitter variances differ in these data.

The χ^2 approximation is less satisfactory if most of the df ν_i are less than 5. Special tables for this are given in the *Biometrika Tables* (10). This reference also gives the significance levels of s_{max}^2/s_{min}^2, the ratio of the largest to the smallest of the a variances. This ratio provides a quick test of homogeneity of variances that will often settle the issue, though usually less sensitive than Bartlett's test.

Unfortunately, Bartlett's test and the preceding test give too many signifi-

TABLE 13.10.1
BARTLETT'S TEST WHEN ALL ESTIMATES HAVE $\nu = 5$ DF

Fat	s_i^2	$\ln s_i^2$	
1	178	5.182	$M = (5)[4(4.614) - 18.081]$
2	60	4.094	$= 1.88 \qquad (\text{df} = 3)$
3	98	4.585	$C = 1 + (a + 1)/(3a\nu)$
4	68	4.220	$= 1 + 5/[(3)(4)(5)] = 1.083$
Total	404	18.081	
	$\bar{s}^2 = 100.9$	$\ln \bar{s}^2 = 4.614$	$\chi^2 = 1.88/1.083 = 1.74 \qquad P > 0.5$

TABLE 13.10.2

BARTLETT'S TEST FOR HOMOGENEITY OF VARIANCE WHEN s_i^2 HAVE DIFFERING DF ν_i

Litter	Sum of Squares $\nu_i s_i^2$	df ν_i	Mean Squares s_i^2	ln s_i^2	Reciprocals $1/\nu_i$
1	8.18	9	0.909	−0.095	0.1111
2	3.48	7	.497	−0.699	.1429
3	0.68	9	.076	−2.577	.1111
4	0.72	7	.103	−2.273	.1429
5	0.73	5	0.146	−1.924	0.2000
$a = 5$	13.79	37			0.7080

$\bar{s}^2 = \Sigma\nu_i s_i^2/\Sigma\nu_i = 13.79/37 = 0.3727$
$(\Sigma\nu_i) \ln \bar{s}^2 = (37)(-0.9870) = -36.519$
$M = (\Sigma\nu_i) \ln \bar{s}^2 - \Sigma\nu_i \ln s_i^2 = -36.519 - (-54.472) = 17.96$
$C = 1 + [1/(3)(4)]\,(0.7080 - 0.0270) = 1.057$
$\chi^2 = M/C = 17.96/1.057 = 16.99 \qquad (\text{df} = 4) \qquad P < 0.01$

cant results with observations that come from long-tailed distributions—distributions with positive kurtosis. An approximate test that is much less sensitive to nonnormality in the data has been given by Levene (11).

13.11—Levene's test of homogeneity of variance. As a measure of the variation within a class, Levene's test uses the average of the absolute deviations $S|X_{ij} - \bar{X}_{i.}|/n$ instead of the mean square of the deviations $s_i^2 = \Sigma(X_{ij} - \bar{X}_{i.})^2/(n-1)$. This avoidance of squaring makes the test criterion much less sensitive to long-tailed distributions. As an example, four independent samples with $n = 7$ were drawn from the t distribution with 3 df—a symmetrical long-tailed distribution—with the number 7 added to all observations to avoid negative observations. In this example, of course, we know that $H_0{:}\sigma_i^2 = \sigma^2$ is correct.

Table 13.11.1 shows the original data on the left and the absolute deviations $|X_{ij} - \bar{X}_{i.}|$ on the right. An observation in the data that catches the eye is the

TABLE 13.11.1

EXAMPLE OF LEVENE'S TEST OF HOMOGENEITY OF VARIANCE

	Data for Class				Absolute Deviations from Class Mean			
	1	2	3	4	1	2	3	4
	7.40	8.84	8.09	7.55	0.54	2.08	1.89	0.71
	6.18	6.69	7.96	5.65	0.68	0.07	1.76	1.19
	6.86	7.12	5.31	6.92	0.00	0.36	0.89	0.08
	7.76	7.42	7.39	6.50	0.90	0.66	1.19	0.34
	6.39	6.83	0.51	5.46	0.47	0.07	5.69	1.38
	5.95	5.06	7.84	7.40	0.91	1.70	1.64	0.56
	7.48	5.35	6.28	8.37	0.62	1.40	0.08	1.53
Total	48.02	47.31	43.38	47.85	4.12	6.34	13.14	5.79
Mean	6.86	6.76	6.20	6.84	0.589	0.906	1.877	0.827
s_i^2	0.500	1.630	7.325	1.100	0.095	0.668	3.214	0.302

TABLE 13.11.2
ANALYSIS OF VARIANCE OF MEAN DEVIATIONS

Source	df	Sum of Squares	Mean Squares	F
Between classes	3	6.773	2.258	2.11
Within classes	24	25.674	1.070	

value 0.51, which looks like a gross error, in class 3. Actually, 0.51 is not particularly unusual as the most extreme value for this long-tailed distribution.

Bartlett's test gives $\chi^2 = 11.22$, $P \doteq 0.01$, erroneously rejecting the null hypothesis (example 13.11.1). For Levene's test we perform an analysis of variance of the mean deviations in the right half of table 13.11.1. The class means, 0.589, 0.906, etc., are our estimates of the variability within the classes. Table 13.11.2 gives the analysis.

The F value, 2.11, indicates $P > 0.10$ with 3 and 24 df—not significant. The test is approximate because the absolute deviations are not normal, and the within-class s_i^2 suggest a much higher variance within class 3 than within other classes. In fact, Satterthwaite's rule, section 12.10, suggests that the within-class mean square should have 10 df, not 24.

EXAMPLE 13.11.1—Apply Bartlett's test to the within-class s_i^2 from the original data on the left half of table 13.11.1. Ans. $M = 11.995$, $C = 1.069$, $\chi^2 = 11.22$ (3 df), P about 0.01.

EXAMPLE 13.11.2—In the data on state expenditures per pupil in 1977, the within-class mean squares s_i^2 and the degrees of freedom in five regions of the United States were as follows: Northeast, $s_1^2 = 0.1240$, $\nu_1 = 9$; Southeast, 0.0335, 6; South Central, 0.0057, 8; North Central, 0.0448, 10; Mountain Pacific, 0.0404, 10. Apply Bartlett's test. Ans. $\chi^2 = 15.35$, df $= 4$, $P = 0.01$.

TECHNICAL TERMS

components of variance
hierarchal classification
homogeneity of variance
intraclass correlation
mixed effects
model II
nested classification
noncentral F distribution
random effects
three-stage sampling

REFERENCES

1. Tang, P. C. *Stat. Res. Mem.* 2 (1938):126.
2. Pearson, E. S., and Hartley, H. O. *Biometrika* 38 (1951):112.
3. South. Coop. Ser. Bull. 10. 1951.
4. King, A. J., and McCarty, D. E. *J. Mark.* 6 (1941):462.
5. Newman, H. H.; Freeman, F. N.; and Holzinger, K. J. 1937. *Twins.* Univ. of Chicago Press.
6. Winsor, C. P., and Clarke, G. L. *J. Mar. Res.* 3 (1940):1.
7. Boardman, T. J. *Biometrics* 30 (1974):251.
8. Wald, A. *Ann. Math. Stat.* 11 (1940):96.
9. Bartlett, M. S. *J. R. Stat. Soc.* 4 (1937):137.
10. Pearson, E. S., and Hartley, H. O. 1954. *Biometrika Tables for Statisticians,* vol. I. Cambridge Univ. Press, Tables 31 and 32.
11. Levene, H. 1960. In *Contributions to Probability and Statistics.* Stanford Univ. Press, Stanford, Calif., p. 278.

14

Two-way classifications

14.1—Introduction. When planning a controlled experiment, the experimenter often acquires the ability to predict roughly the behavior of the experimental material. In identical environments young male rats are known to gain weight faster than young female rats. In a machine that subjects five different pieces of cloth to simulated wearing, experience shows that the cloths placed in positions 4 and 5 will receive less abrasion than those in the other positions. Such knowledge can be used to increase the accuracy of an experiment. If a treatments are to be compared, experimental units are first arranged in groups of a. Units assigned to the same group should be as similar in responsiveness as possible. Each treatment is then allocated by randomization to one unit in each group. This produces a two-way classification, since any observation is classified by the treatment it receives and the group to which it belongs.

The name given to the group varies with the type of application. In agricultural field experiments, long experience has shown that plots near one another tend to give similar yields. The group will therefore often be a compact piece of land, called a *block*. The experimental plan is described as *randomized blocks*. Another name used for the group is *replication*—meaning a single trial or repetition of the comparison between the treatments. Many experiments on human subjects show considerable variation from one subject to another. Sometimes it is possible to give each treatment to every subject on different occasions—as when comparing different analgesics for the relief of chronic headaches or different rewards for performance in a repetitive task. The objective is to make the comparisons among treatments more accurate, since they are made within subjects. The groups would then probably be called "subjects." In the abrasion tests just mentioned, the groups would be "positions," all the pieces of cloth (treatments) being tested in each position. The name used often describes the classification employed in the grouping.

Two-way classifications are frequent in surveys and observational studies, also. We encounter an example in section 11.10 in which farms are classified by soil type and owner-tenant status. In a survey of family expenditures on food, classification of the results by size of family and income level is obviously relevant.

We first present an example to familiarize you with the standard computations needed to perform the analysis of variance and make any desired comparisons. Later, the mathematical assumptions are discussed.

14.2—Experiment in randomized blocks. Table 14.2.1 comes from an experiment (1) in which four seed treatments were compared with no treatment (check) on soybean seeds. The data are the number of plants that failed to emerge out of 100 planted in each plot. The two criteria of classification are treatments and replications (blocks).

The first steps are to find the treatment totals, the replication totals, the grand total, and the usual correction for the mean. The total sum of squares and the sum of squares for treatments are computed just as in a one-way classification. The new feature is that the sum of squares for replications is also calculated. The rule for finding this sum of squares is the same as for treatments. The sum of squares of the replication totals is divided by the number of observations in each replication (5) and the correction factor is subtracted. Finally, in the analysis of variance, we compute the line

residuals = total − replications − treatments

As shown later, the residuals mean square, 5.41 with 16 df, is an unbiased estimate of the error variance per observation.

The F ratio for treatments is $20.96/5.41 = 3.87$ with 4 and 16 df, significant at the 5% level. Actually, since this experiment has certain designed comparisons (discussed in the next section), the overall F test is not of great importance. Note that the replications mean square is more than twice the

TABLE 14.2.1
ANALYSIS OF VARIANCE OF A TWO-WAY CLASSIFICATION
(number of failures out of 100 planted soybean seeds)

	Replication						
Treatment	1	2	3	4	5	Total	Mean
Check	8	10	12	13	11	54	10.8
Arasan	2	6	7	11	5	31	6.2
Spergon	4	10	9	8	10	41	8.2
Semesan, Jr.	3	5	9	10	6	33	6.6
Fermate	9	7	5	5	3	29	5.8
Total	26	38	42	47	35	188	

Correction: $C = 188^2/25 = 1413.76$
Total SS: $8^2 + 2^2 + \ldots + 6^2 + 3^2 - C = 220.24$
Treatments SS: $(54^2 + 31^2 + \ldots + 29^2)/5 - C = 83.84$
Replications SS: $(26^2 + 38^2 + \ldots + 35^2)/5 - C = 49.84$

Source of Variation	df	Sum of Squares	Mean Square	F
Replications	4	49.84	12.46	
Treatments	4	83.84	20.96	3.87
Residuals (error)	16	86.56	5.41	
Total	24	220.24		

residuals mean square, indicating real differences between replication means and suggesting that the classification into blocks was successful in improving accuracy. A method of estimating the amount of gain in accuracy is presented in section 14.7.

EXAMPLE 14.2.1—In three species of citrus trees the ratio of leaf area to dry weight was determined for three conditions of shading (2).

Shading	Shamouti Orange	Marsh Grapefruit	Clementine Mandarin
Sun	112	90	123
Half shade	86	73	89
Shade	80	62	81

Compute the analysis of variance. Ans. Mean squares for shading and error, 942.1 and 21.8; F = 43.2 with 2 and 4 df. The shading was effective in decreasing the relative leaf area. Example 14.3.1 gives further discussion.

EXAMPLE 14.2.2—With only two treatments, the data reduce to two paired samples, previously analyzed by the t test in chapter 6. This t test is equivalent to the F test of treatments given in this section. Verify this result by performing the analysis of variance of the mosaic virus example in section 6.3 as follows:

	Degrees of Freedom	Sum of Squares	Mean Square
Replications (pairs)	7	575	82.2
Treatments	1	64	64.0
Error	7	65	9.29

$F = 6.89$, df $= 1, 7$; $\sqrt{F} = 2.63 = t$ as given in section 6.3.

14.3—Comparisons among means. The previous discussion of planned comparisons, orthogonal comparisons, and partitioning the treatments sum of squares (sections 12.8, 12.9) applies also to two-way classifications. To illustrate a planned comparison, we compare the mean number of failures for the check with the corresponding average for the four chemicals. From table 14.2.1 the means are: check, 10.8; Arasan, 6.2; Spergon, 8.2; Semesan, Jr., 6.6; and Fermate, 5.8. The comparison is, therefore,

$$10.8 - (6.2 + 8.2 + 6.6 + 5.8)/4 = 10.8 - 6.7 = 4.1 \qquad (14.3.1)$$

To find the standard error of this comparison, note that the experiment has five replications (blocks), with $s = \sqrt{5.41} = 2.326$ (16 df). Hence, by Rule 12.8.1 the required standard error is

$$(s/\sqrt{5})\sqrt{1^2 + 1/4^2 + 1/4^2 + 1/4^2 + 1/4^2} = (2.326/\sqrt{5})\sqrt{5/4} = 1.163$$

with 16 df. The value of t is $4.1/1.163 = 3.53$, highly significant. The 95% confidence limits for the average reduction in failure rate due to the chemicals are

$$4.1 \pm (2.120)(1.163) = 4.1 \pm 2.5 = 1.6, 6.6$$

The next step is to compare the means for the four chemicals. For this the discussion in section 12.12 (inspection of all differences between pairs of means) is relevant. From table A 15, the Studentized range Q test requires a difference of $(4.05)(2.326)/\sqrt{5} = 4.21$ for significance of any difference between means at the 5% level. None of the differences between pairs of chemicals approaches this value.

An alternative method is to start with an F test of the differences among the means of the four chemicals. From the totals in table 14.2.1 the sum of squares between chemicals is

$$(31^2 + 41^2 + 33^2 + 29^2)/5 - 134^2/20 = 16.60 \qquad (3 \text{ df})$$

The value of F is $16.60/[3(5.41)] = 1.02$, again indicating no sign of differences.

Since all the chemicals appear with a coefficient $-1/4$ in the comparison (14.3.1) between chemicals and check, it is not hard to show that any comparison among the means of the four chemicals is orthogonal to the comparison (14.3.1). It follows that the sum of squares (1 df) for the comparison (14.3.1) plus the sum of squares (16.60) among chemicals (3 df) together equal the treatments sum of squares. The sum of squares with 1 df is $nL^2/\Sigma\lambda^2 = (5)(4.1^2)/1.25 = 67.24$. As expected, $67.24 + 16.60 = 83.84$ = treatments sum of squares. In analysis of variance form, this partitioning of the treatments sum of squares might be reported as follows:

Source	df	Sum of Squares	Mean Square	F
Chemicals vs. check	1	67.24	67.24	12.42
Among chemicals	3	16.60	5.53	1.02
Residuals (error)	16	86.56	5.41	

EXAMPLE 14.3.1—In example 14.2.1 on the effect of shade on the leaf area-weight ratio, the treatment totals (3 replications) were calculated as follows:

	Sun 325	Half Shade 248	Shade 223
Effect of shade	+1	0	−1
Half shade vs. rest	−1/2	+1	−1/2

We might measure the effect of shade by the extreme comparison between sun and shade. We might also be interested in whether the effect of half shade is the simple average of the effects for sun and shade. (i) Show that the treatments sum of squares in the analysis of variance partitions as follows:

Source	df	Sum of Squares	Mean Square	F
Treatments	2	1884		
effect of shade	1	1734	1734	79.5
half shade vs. rest	1	150	150	6.9
Residuals (error)	4	87	21.8	

(ii) Verify that the effect of shade is significant at the 1% level, and with $F = 6.9$, the hypothesis that the effect of half shade is halfway between that of sun and shade is not quite rejected at the 5% level (P about 0.06). However, the treatment totals suggest that the results for half shade are quite close to those for shade. More data are needed to pinpoint the effect of half shade.

14.4—Notation and mathematical model. Table 14.4.1 gives an algebraic notation that has become standard for a two-way classification with one observation per cell of the table. The measurement obtained for the unit in the ith row and jth column is represented by X_{ij}. Row totals and means are denoted by $X_{i.}$ and $\bar{X}_{i.}$, while $X_{.j}$ and $\bar{X}_{.j}$ denote column totals and means. The overall mean is $\bar{X}_{..}$. Instructions for computing the analysis of variance appear in the table. Note that the number of degrees of freedom for residuals (used as error) is $(I - 1)(J - 1)$, the product of the numbers of degrees of freedom for rows and columns.

The two-way classification is often referred to as a *crossed classification*—the rows and columns crossing one another as seen in the table.

The mathematical model for a two-way classification is

$$X_{ij} = \mu + \alpha_i + \beta_j + \epsilon_{ij} \qquad (i = 1, \ldots, I; j = 1, \ldots, J) \qquad (14.4.1)$$

where μ represents the overall mean, the α_i stand for row effects, and the β_j for column effects. Note that since the parameter μ tells us the overall mean, the parameter α_i is needed only to inform us how much better or worse the mean of the ith row is than the overall mean. Hence, without restricting the scope of the model we can adopt the convention $\Sigma\alpha_i = 0$ and similarly $\Sigma\beta_j = 0$.

The model involves two basic assumptions:

1. The mathematical form $\mu + \alpha_i + \beta_j$ implies that row and column effects are *additive*. Apart from the experimental errors ϵ_{ij}, the difference in effect

TABLE 14.4.1

NOTATION FOR A TWO-WAY TABLE WITH I ROWS AND J COLUMNS

Rows $i = 1, \ldots, I$	Columns, $j = 1, \ldots, J$: 1	j	J	Total	Mean
1	X_{11}	X_{1j}	X_{1J}	$X_{1.}$	$\bar{X}_{1.}$
.					
i	X_{i1}	X_{ij}	X_{iJ}	$X_{i.}$	$\bar{X}_{i.}$
.					
I	X_{I1}	X_{Ij}	X_{IJ}	$X_{I.}$	$\bar{X}_{I.}$
Total	$X_{.1}$	$X_{.j}$	$X_{.J}$	$X_{..}$	
Mean	$\bar{X}_{.1}$	$\bar{X}_{.j}$	$\bar{X}_{.J}$		$\bar{X}_{..}$

Sum of Squares (SS)

Correction: $C = X_{..}^2/(IJ)$ Total: $TSS = \Sigma\Sigma X_{ij}^2 - C$

Rows: $RSS = (X_{1.}^2 + \ldots + X_{I.}^2)/J - C$

Columns: $CSS = (X_{.1}^2 + \ldots + X_{.J}^2)/I - C$

Residuals: $DSS = TSS - RSS - CSS$

Analysis of Variance

Source of Variation	df	Sum of Squares	Mean Square
Rows	$I - 1$	RSS	$RSS/(I - 1)$
Columns	$J - 1$	CSS	$CSS/(J - 1)$
Residuals (deviations)	$(I - 1)(J - 1)$	DSS	$DSS/(I - 1)(J - 1)$
Total	$IJ - 1$	TSS	

between row 2 and row 1 in column j is

$$(\mu + \alpha_2 + \beta_j) - (\mu + \alpha_1 + \beta_j) = \alpha_2 - \alpha_1 \tag{14.4.2}$$

That is, the row difference is the same in all columns. This assumption may or may not hold when we analyze data.

2. The residuals ϵ_{ij} in (14.4.1) are assumed to be independent random variables, normally distributed with mean 0 and variance σ^2. The ϵ_{ij} represent the extent to which the data depart from a strictly additive model. These departures may have several components. (*i*) They may represent natural variations among the units to which treatments are applied. Thus, in a randomized blocks experiment, treatment 2 may do particularly well in block or replication 3 because it is applied to an unusually good unit. (*ii*) They may represent errors of measurement of the X_{ij}. (*iii*) Replications have different levels of soil fertility, and some treatments may perform relatively better in high fertility replications than in low fertility replications. The residuals ϵ_{ij} are often called treatment × replication interactions, or more generally, *row × column interactions,* and are denoted by rows × columns in the analysis of variance.

In a controlled experiment arranged in randomized groups, the treatments are usually regarded as having fixed effects α_i in the sense of model I. Groups, blocks, or replications are generally thought of as random factors β_j. Ideally, they are intended to sample the kind of population to which we hope that the comparisons among the treatments will apply. Thus, in tests of analgesics for the relief of headaches we would like our subjects to be a random sample of the people with chronic headaches whom we are trying to help with analgesics. For many practical reasons, this ideal is seldom attained in experimentation. Nevertheless, the role of blocks, subjects, citrus trees, etc., used as replications in experiments is to sample a wider population so that it is appropriate to regard the β_j as random.

14.5—Method of estimation: least squares. Thus far we have proceeded as if the row and column means are the obvious statistics from which to estimate row and column effects. With more complex sets of data or more complex mathematical models, however, it may not be clear at first how to estimate the parameters in the model. We need some rule or method for making estimates that will perform well in such situations. The method used in standard analyses is the method of least squares. To explain the method, let $\hat{\mu}$, $\hat{\alpha}_i$, and $\hat{\beta}_j$ denote estimates of μ, α, and β made from the data. The predicted value of X_{ij} from the fitted model is then

$$\hat{X}_{ij} = \hat{\mu} + \hat{\alpha}_i + \hat{\beta}_j \tag{14.5.1}$$

The method of least squares chooses as estimates the values of $\hat{\alpha}$, $\hat{\beta}$, and $\hat{\mu}$ that minimize the sum of squares of the residuals, namely,

$$\sum_i \sum_j (X_{ij} - \hat{X}_{ij})^2 = \sum_i \sum_j (X_{ij} - \hat{\mu} - \hat{\alpha}_i - \hat{\beta}_j)^2 \tag{14.5.2}$$

Since we want the model to fit the data well, it seems wise to choose estimates that keep the values of $X_{ij} - \hat{X}_{ij}$ small. It is not obvious, however, that we should minimize the sum of *squares* of the residuals. This choice was made around 1800, partly because of mathematical convenience and partly because the resulting estimates perform well when the residuals are normal and independent.

The equations for the minimizing values are easily found by differential calculus. For the two-way classification, the equation for any parameter is written as follows. On the right side of the equals sign, put the total of all observations in whose model the parameter occurs. On the left side, put the value that the model predicts for this total. For instance, the mean μ appears in the model for *every* observation. The equation for finding $\hat{\mu}$ is therefore

$$\sum_i \sum_j (\hat{\mu} + \hat{\alpha}_i + \hat{\beta}_j) = X_{..}$$

Since by convention $\Sigma_i \hat{\alpha}_i$ and $\Sigma_j \hat{\beta}_j$ are zero, we get

$$IJ\hat{\mu} = X_{..} \qquad \hat{\mu} = \overline{X}_{..} \tag{14.5.3}$$

The parameter $\hat{\alpha}_i$ appears in the model only for observations that are in the ith row. Hence, on adding (14.5.1) over the ith row, we get

$$J\hat{\mu} + J\hat{\alpha}_i = X_{i.} \qquad \hat{\alpha}_i = \overline{X}_{i.} - \overline{X}_{..} \tag{14.5.4}$$

where the $\hat{\beta}_j$ disappear because $\hat{\beta}_1 + \ldots + \hat{\beta}_J = 0$. Similarly,

$$I\hat{\mu} + I\hat{\beta}_j = X_{.j} \qquad \hat{\beta}_j = \overline{X}_{.j} - \overline{X}_{..} \tag{14.5.5}$$

We have now verified that in the two-way classification the method of least squares leads to the overall mean and the row and column means as estimates of the corresponding effects. When all the assumptions in model (14.4.1) hold, the method of least squares gives estimates that are unbiased, have the smallest variance of all unbiased estimates that are linear in the observations, and are hard to improve. In section 9.5 we encountered the use of the method in fitting a straight line to data in regression analysis. The method has been criticized for data in which the ϵ_{ij} are not normally distributed or for data liable to gross errors, on the grounds that minimizing a sum of *squares* gives too much weight to extreme observations.

14.6—Deviations from the model. The least squares estimates are

$$\hat{\mu} = \overline{X}_{..} \qquad \hat{\alpha}_i = \overline{X}_{i.} - \overline{X}_{..} \qquad \hat{\beta}_j = \overline{X}_{.j} - \overline{X}_{..}$$

If we estimate any individual observation X_{ij} from the fitted model, the estimate is

$$\begin{aligned}\hat{X}_{ij} = \hat{\mu} + \hat{\alpha}_i + \hat{\beta}_j &= \overline{X}_{..} + (\overline{X}_{i.} - \overline{X}_{..}) + (\overline{X}_{.j} - \overline{X}_{..}) \\ &= \overline{X}_{i.} + \overline{X}_{.j} - \overline{X}_{..} \end{aligned} \tag{14.6.1}$$

Table 14.6.1 shows a set of observations X_{ij}, the estimates $\hat{X}_{ij}$, and the deviations $d_{ij} = X_{ij} - \hat{X}_{ij}$ of the observations from the estimates. For treatment 1 in replication 2, for instance, we have from table 14.6.1,

$$X_{12} = 29 \qquad \hat{X}_{12} = 35 + 26 - 29 = 32 \qquad d_{12} = -3$$

From (14.6.1),

$$d_{ij} = X_{ij} - \hat{X}_{ij} = X_{ij} - \bar{X}_{i.} - \bar{X}_{.j} + \bar{X}_{..} \tag{14.6.2}$$

The deviations d_{ij} have some important properties:

(*i*) Their sum is zero in any row or column.
(*ii*) Their sum of squares,

$$(-4)^2 + (+2)^2 + \ldots + (-3)^2 + (+1)^2 = 132$$

is equal to the residuals sum of squares in the analysis of variance in table 14.6.1. Thus the residuals sum of squares measures the extent to which the linear additive model fails to fit the data. This result is a consequence of a general algebraic identity:

$$\begin{aligned} \text{residuals } SS &= \sum_i \sum_j (X_{ij} - \bar{X}_{i.} - \bar{X}_{.j} + \bar{X}_{..})^2 \\ &= \sum_i \sum_j (X_{ij} - \bar{X}_{..})^2 - b\sum_i (\bar{X}_{i.} - \bar{X}_{..})^2 - a\sum_j (\bar{X}_{.j} - \bar{X}_{..})^2 \\ &= \text{total } SSD - \text{treatments } SSD - \text{replications } SSD \end{aligned} \tag{14.6.3}$$

This equation shows that the analysis of variance is a quick method of finding the sum of squares of the deviations of the observations from the fitted model.

To prove this result, note that since the d_{ij} sum to zero in every row and column, the sum of products of the d_{ij} with the row means, the column means, and the general mean is zero. Hence,

$$\Sigma\Sigma\, d_{ij}(\bar{X}_{i.} - \bar{X}_{..}) = 0 \qquad \Sigma\Sigma\, d_{ij}(\bar{X}_{.j} - \bar{X}_{..}) = 0 \qquad \Sigma\Sigma\, \bar{X}_{i.}(\bar{X}_{.j} - \bar{X}_{..}) = 0 \tag{14.6.4}$$

Now write

$$X_{ij} - \bar{X}_{..} = (\bar{X}_{i.} - \bar{X}_{..}) + (\bar{X}_{.j} - \bar{X}_{..}) + d_{ij}$$

Square each side and add over all observations. By (14.6.4), all three cross-product terms vanish. The squared terms give the analysis of variance identity (14.6.3).

Computer programs for the analysis of variance usually print the d_{ij} on request. This serves two purposes—it enables the investigator to look over the d_{ij}

TABLE 14.6.1
DEVIATIONS FROM THE MODEL IN A TWO-WAY CLASSIFICATION

Treatment	Replication 1	2	3	Mean $\bar{X}_{i.}$
1: X_{ij}	30	29	46	35
$\hat{X}_{ij}$	34	32	39	
d_{ij}	− 4	− 3	+ 7	
2: X_{ij}	35	34	33	34
$\hat{X}_{ij}$	33	31	38	
d_{ij}	+ 2	+ 3	− 5	
3: X_{ij}	31	30	32	31
$\hat{X}_{ij}$	30	28	35	
d_{ij}	+ 1	+ 2	− 3	
4: X_{ij}	16	11	21	16
$\hat{X}_{ij}$	15	13	20	
d_{ij}	+ 1	− 2	+ 1	
Mean $\bar{X}_{.j}$	28	26	33	29

Source of Variation	Degrees of Freedom	Sum of Squares	Mean Square
Replications	2	104	52
Treatments	3	702	234
Residuals	6	132	22

for gross errors or systematic departures from the linear model (as discussed in chapter 15) and it provides a check on the residuals sum of squares.

(*iii*) A third property of the deviations, which may be proved by substituting the values given by the model for X_{ij}, $\bar{X}_{i.}$, $\bar{X}_{.j}$, and $\bar{X}_{..}$ in (14.6.2), is

$$d_{ij} = \epsilon_{ij} - \bar{\epsilon}_{i.} - \bar{\epsilon}_{.j} + \bar{\epsilon}_{..} \tag{14.6.5}$$

Thus, if the additive model holds, each d_{ij} is a linear combination of the random errors. It may be shown with a rows and b columns that any d_{ij}^2 is an unbiased estimate of $(a - 1)(b - 1)\sigma^2/(ab)$. It follows that the residuals sum of squares is an unbiased estimate of $(a - 1)(b - 1)\sigma^2$, giving the basic result that the residuals mean square, with $(a - 1)(b - 1)$ df, is an unbiased estimate of σ^2.

EXAMPLE 14.6.1—Suppose that with $a = b = 2$, treatment and replication effects are *multiplicative*. Treatment 2 gives results 20% higher than treatment 1 and replication 2 gives results 10% higher than replication 1. With no random errors, the observations would be as shown on the left below.

Treatment	X_{ij}, Replication 1	2	$\hat{X}_{ij}$, Replication 1	2
1	1.00	1.10	0.995	1.105
2	1.20	1.32	1.205	1.315

Verify that the $\hat{X}_{ij}$ given by fitting the linear model are as shown above. Any d_{ij} is only ±0.005. The linear model gives a good fit to a multiplicative model when treatment and replication effects are small or moderate. If, however, treatment 2 gives a 100% increase and replication 2 a 50% increase, you will find $d_{ij} = \pm 0.125$, not so good a fit.

EXAMPLE 14.6.2—In table 14.6.1, verify that $\hat{X}_{33} = 35$, $d_{33} = -3$.

EXAMPLE 14.6.3—Perform an analysis of variance of the $\hat{X}_{ij}$ in table 14.6.1. Verify that the treatments and replications sums of squares are the same as for the X_{ij} but the residuals sum of squares is zero. Can you explain these results?

EXAMPLE 14.6.4—The result,

$$d_{ij} = \epsilon_{ij} - \bar{\epsilon}_{i.} - \bar{\epsilon}_{.j} + \bar{\epsilon}_{..}$$

shows that d_{ij} is a linear combination of the form $\Sigma\Sigma\,\lambda_{ij}\epsilon_{ij}$. By Rule 12.8.1, its variance is $\sigma^2\Sigma\Sigma\,\lambda_{ij}^2$. For d_{11}, for example, the λ_{ij} work out as follows:

Observations	Number of Terms	λ_{ij}
d_{11}	1	$(a-1)(b-1)/(ab)$
Rest of d_{1j}	$(b-1)$	$-(a-1)/(ab)$
Rest of d_{i1}	$(a-1)$	$-(b-1)/(ab)$
Rest of d_{ij}	$(a-1)(b-1)$	$+1/(ab)$

It follows that $\Sigma\Sigma\,\lambda_{ij}^2 = (a-1)(b-1)/(ab)$. Thus d_{11}^2 and similarly any d_{ij}^2 estimates $(a-1)(b-1)\sigma^2/(ab)$, as stated in the text.

14.7—Efficiency of blocking. When a randomized blocks experiment has been completed, the experimenter sometimes wants to estimate how effective the blocking was in increasing the precision of the experiment. The criterion chosen for blocking may not have been appropriate or the blocks chosen may have been troublesome to use. From the analysis of variance of a randomized blocks (RB) experiment, we can estimate the error variance that would have been obtained if the J replicates of any treatment had been assigned to the units at random in the plan known as the *completely randomized* (CR) design, for which the analysis of variance is a one-way classification into treatments and error.

Call the two error variances per unit σ_{RB}^2 and σ_{CR}^2. With randomized blocks the variance of a treatment mean is σ_{RB}^2/J. To get the same variance of a treatment mean in a completely randomized design, the number of replications n must satisfy the relation

$$\sigma_{CR}^2/n = \sigma_{RB}^2/J \quad \text{or} \quad n/J = \sigma_{CR}^2/\sigma_{RB}^2$$

For this reason the ratio $\sigma_{CR}^2/\sigma_{RB}^2$ is used to measure the *relative efficiency* of the blocking.

Table 14.7.1 shows how unbiased estimates of σ_{RB}^2 and σ_{CR}^2 are obtained from the analysis of variance of the results of the RB experiment. For an RB experiment, the residuals mean square s^2 is, of course, an unbiased estimate of σ_{RB}^2. For σ_{CR}^2 we would expect the corresponding estimate to be a combination of s_b^2 and s^2, since in that design comparisons between treatments are no longer restricted to be entirely within blocks. The combination giving an unbiased estimate (3, 4) is shown on the right in table 14.7.1.

Using the soybean seeds experiment (table 14.2.1) as an example, $s_b^2 = 12.46$, $s^2 = 5.41$, $I = J = 5$, and the relative efficiency of randomized blocks is

TABLE 14.7.1

ESTIMATES OF σ^2_{RB} AND σ^2_{CR} FROM THE RB ANALYSIS

Source	df	Mean Square	
Blocks	$J - 1$	s_b^2	$\hat{\sigma}^2_{RB} = s^2$
Treatments	$I - 1$		
Residuals	$(I - 1)(J - 1)$	s^2	$\hat{\sigma}^2_{CR} = \dfrac{(J-1)s_b^2 + J(I-1)s^2}{(IJ-1)}$

$$\hat{\sigma}^2_{CR}/\hat{\sigma}^2_{RB} = [4(12.46) + 20(5.41)]/[24(5.41)] = 1.22$$

If a CR plan had been used, about six replications instead of five would have been needed to obtain the same standard error of a treatment mean as with RB.

This comparison is not quite fair to the CR design, which would provide $\nu_{CR} = 20$ df for error as against $\nu_{RB} = 16$ with RB and would therefore require smaller values of t in calculating confidence intervals. To take account of the difference in error degrees of freedom, Fisher (5) suggests replacing the ratio $\hat{\sigma}^2_{CR}/\hat{\sigma}^2_{RB}$ by the ratio

$$(\nu_{RB} + 1)(\nu_{CR} + 3)(\hat{\sigma}^2_{CR}/\hat{\sigma}^2_{RB})/[(\nu_{RB} + 3)(\nu_{CR} + 1)]$$
$$= (16 + 1)(20 + 3)(1.22)/[(16 + 3)(20 + 1)] = 1.20$$

Fisher calls this ratio the *relative amount of information.* The adjustment for degrees of freedom has little effect except in small experiments.

EXAMPLE 14.7.1—In a randomized blocks experiment that compared four strains of Gallipoli wheat (6), the mean yields (pounds per plot) and the analysis of variance were as follows:

Strain	*A*	*B*	*C*	*D*
Mean yield	34.4	34.8	33.7	28.4

Source of Variation	Degrees of Freedom	Sum of Squares	Mean Square
Blocks	4	21.46	5.36
Strains	3	134.45	44.82
Error	12	26.26	2.19

(*i*) How many replications were there? (*ii*) Estimate s^2_{CR}/s^2_{RB}. (*iii*) Estimate the relative amount of information by Fisher's formula. Ans. (*ii*) 1.30, (*iii*) 1.26.

EXAMPLE 14.7.2—In example 14.7.1, verify that the LSD and the Q methods both show D inferior to the other strains but reveal no differences among the others.

14.8—Two-way classifications with *n* observations per cell. Two-way classifications with more than one observation in a cell are encountered quite frequently. An experiment may be completely randomized with n replications, but the IJ treatments may be a two-way classification, consisting of all combinations of I levels of one factor and J levels of another factor in what is known as factorial experimentation (chapter 16). For instance, three types of nitrogen fertilizer might be tested on each of four varieties of a crop. Examples also occur commonly in observational studies in which the observations have been classified by two variables that are of interest, though here n often varies from cell to cell.

As a second example, in some single-factor experiments in randomized blocks the data for a cell have to be obtained by sampling because it is too costly or not feasible to measure X_{ij} itself. An example is an experiment that compares soil fumigants intended to reduce the number of wireworms that attack roots in the top layers of the soil. Cores, perhaps 9 × 9 × 5 in., are driven in at selected sample points on a plot, the soil is removed, and the wireworms are counted in each sample. Each plot may have n sample observations.

Let the subscript k denote individual members of a cell. The model is

$$X_{ijk} = \mu + \alpha_i + \beta_j + \eta_{ij} + \epsilon_{ijk} \tag{14.8.1}$$

The interpretations of the terms η_{ij} and ϵ_{ijk} depend on the type of data. Consider first a completely randomized factorial experiment in which the treatments are combinations of fixed levels or variants of two factors A and B. The terms η_{ij} represent the fixed AB interactions and the terms ϵ_{ijk} the experimental errors. However, in a single-factor experiment in randomized blocks with subsampling, the terms η_{ij} represent random treatments × blocks interactions. The terms ϵ_{ijk} measure the sampling errors.

The analysis of variance contains an additional line—the within-cells sum of squares. It can be calculated directly as

$$\text{within-cells } SS = \sum_i \sum_j \sum_k (X_{ijk} - \bar{X}_{ij.})^2$$

and has $IJ(n - 1)$ df. Alternatively, we may calculate the total sum of squares between individual observations

$$\sum_i \sum_j \sum_k X_{ijk}^2 - X_{...}^2/(nIJ)$$

and subtract the sum of squares between all cell totals divided by n, namely,

$$\left[\sum_i \sum_j X_{ij.}^2 - X_{...}^2/(IJ)\right] \Big/ n$$

The sums of squares for rows, columns, and rows × columns are found as usual except that they are also divided by n if computed from cell totals so as to make them comparable with the within-cells sum of squares. This analysis is described as being on a *single-observation* basis. A report of the analysis should state whether sums of squares and mean squares refer to single observations or cell means or totals. Table 14.8.1 shows degrees of freedom and descriptions of the sources of variation when we have two factors in a completely randomized (CR)

TABLE 14.8.1
TWO-WAY ANALYSES OF VARIANCE WITH n OBSERVATIONS PER CELL

Two Factors (CR)	Single Factor (RB) with Subsampling	
Source of Variation	Source of Variation	df
A	Treatments	$I - 1$
B	Blocks	$J - 1$
$A \times B$ interactions	Experimental error	$(I - 1)(J - 1)$
Experimental error	Sampling error	$IJ(n - 1)$
Total	Total	$IJn - 1$

experiment and when we have one factor in randomized blocks (RB) with subsampling.

Discussion of the analysis of factorial experiments is deferred until chapter 16. In examples in which the within-cells mean square represents sampling errors, the analysis of variance enables us to appraise the effects of different amounts of sampling on the experimental error, as illustrated in example 14.8.1.

EXAMPLE 14.8.1—The data following are the numbers of wireworms in soil samples 6 × 6 × 5 in. from part of a randomized blocks experiment (7) on the effects of two fumigants C and S as against no fumigant 0 in reducing the numbers of wireworms. There were 4 samples per plot and 5 blocks (the experiment itself had 10 blocks). The blocks are numbered I, II, . . . , V.

Block	C		S		0		Total
I	5, 4, 5, 2	16	5, 5, 1, 2	13	12, 20, 8, 8	48	77
II	0, 9, 3, 3	15	6, 4, 5, 4	19	7, 4, 4, 5	20	54
III	4, 4, 3, 9	20	2, 9, 3, 7	21	9, 6, 7, 11	33	74
IV	7, 3, 5, 12	27	6, 4, 8, 4	22	12, 22, 17, 13	64	113
V	4, 9, 8, 6	27	2, 9, 7, 3	21	7, 8, 5, 9	29	77
	Total	105		96		194	395

Compute the analysis of variance on a single-sample basis. Ans.:

	df	Sum of Squares	Mean Square	F
Blocks	4	151.2		
Fumigants	2	293.4	146.7	5.98
Error	8	196.2	24.5	
Within plots	45	409.8	9.11	
Total	59	1050.6		

Note that comparisons of the effects of the fumigants can be made without using the between-samples sum of squares. This term is computed in experiments of this type as a guide to sampling strategy. With 4 samples per plot, sampling errors contribute an estimated variance 9.11 to the error variance of 24.5. If the sampling were doubled, the sampling error contribution would be halved, reducing the error variance to an estimated 19.9. This reduction is probably not worthwhile, since sorting the soil and counting the wireworms is laborious.

14.9—Balancing the order in which treatments are given. Experiments with living subjects, in which each subject receives all treatments in turn, are sometimes called *repeated measures* (or *repeated measurements*) experiments, the subject's response being measured each time a treatment is given. In some experiments of this type, there is reason to suspect systematic differences between successive measurements. For instance, the subjects may be nervous on the first occasion and respond relatively poorly, or if the measurements are onerous (e.g., performing some unattractive task) the subjects may not give their best efforts with treatments given later. In such circumstances a precaution is to plan the experiment so that each treatment is given first, second, . . . to an equal number of subjects. With this extra control, any systematic effects of the order in which measurements are made will not create bias in the comparison of the treatment means.

For such plans the number of subjects n must be a multiple of the number of treatments. With two treatments and n subjects, half the subjects, chosen at random, receive the order *AB* and the other half the order *BA*. With three treatments, each of the orders *ABC*, *BCA*, *CAB* can be given to $n/3$ subjects chosen at random from the n subjects, and so on.

We now face a three-way classification by treatments, subjects, and order. They are denoted by subscripts i, j, and k, respectively. If X_{ijk} represents a response measurement for the jth subject, the simplest model likely to be realistic is

$$X_{ijk} = \mu + \alpha_i + \beta_j + \gamma_k + \delta_{ij} + \epsilon_{ijk} \tag{14.9.1}$$

Here α_i, β_j, and γ_k represent the average effects of treatments, subjects, and orders. The term δ_{ij} represents the fact that the effect of a treatment may vary from subject to subject (treatment × subject interactions). The independent term ϵ_{ijk} includes measurement errors and nonsystematic variation of a subject from occasion to occasion. Note that this model does not allow for treatment × order or subject × order interactions; they would require terms η_{ik} and λ_{jk}, a larger experiment, and a different analysis of variance.

With model (14.9.1), the error variance of a treatment mean is $(\sigma_\delta^2 + \sigma_\epsilon^2)/n$. In the previous analysis of variance for a randomized blocks experiment, we subtracted from the total sums of squares of deviations (*SSD*) the *SSD*s for subjects (blocks) and treatments in order to compute the residual sum of squares from which the error mean square is obtained. With the additional control of order, we also subtract the order sum of squares, which is calculated in the same way as the treatments sum of squares. With I treatments and I orders, it also has $(I - 1)$ df. If there are J subjects (replications) the error mean square has $(I - 1)(J - 2)$ df.

Table 14.9.1, taken from an experiment conducted by Smith et al. (10) illustrates this analysis of variance. The experiment compared the effectiveness of three injections—placebo, 7.5 mg morphine, and 15 mg morphine/70 kg body weight—in enabling volunteer subjects to withstand pain produced experimentally on the arm by a tourniquet. There were three treatments and 36 subjects in a repeated measures experiment. One of the response measurements was the time of exposure at which the subject reported the pain as "moderately distressing." The treatment mean times and the analysis of variance appear in table 14.9.1. All six orders of the three treatments were used in this plan, each order being given to 6 subjects.

TABLE 14.9.1
TREATMENT MEANS AND ANALYSIS OF VARIANCE IN A REPEATED MEASURES EXPERIMENT

	Placebo	7.5 mg Morphine	15 mg Morphine	
Mean Time (min)	6.54	7.64	9.20	
Source	**df**	**Sum of Squares**	**Mean Squares**	***F***
Subjects	35	3357.6	95.9	15.0
Order	2	83.8	41.9	6.6
Treatments	2	129.0		
linear	1	127.7	127.7	19.9
quadratic	1	1.2	1.2	0.2
Residuals (error)	68	434.8	6.39	
Total	107	4005.2		

From the F ratios, note that use of subjects as replications greatly improved the precision and that control of order was also worthwhile, though this was less so for pain levels (not shown here) that were more severe than "moderately distressing." Since the response to the amount of morphine appeared to be linear, the treatments sum of squares was separated into its linear and quadratic components.

14.10—Latin squares. In the plans described in the preceding section, each treatment appeared once with each subject and an equal number of times in each order of presentation. With 5 treatments A, B, C, D, E and only 5 subjects, the plan might look as shown in table 14.10.1 after randomization. This plan is called a *5 × 5 Latin square.* Each treatment appears *once* in each row (subject) and *once* in each column (order). Consistent differences between subjects or between orders cancel when we take the differences between treatment means. Latin squares were first used in agricultural field experiments. There is often a gradient in fertility running parallel to one side of the field and sometimes gradients running parallel to both sides. A Latin square with rows and columns parallel to the two sides gives protection against such gradients. More generally, the Latin square controls two different sources of variation that may increase the experimental errors.

To construct a Latin square, write down any arrangement (systematically)

TABLE 14.10.1
A 5 × 5 LATIN SQUARE

	Order				
Subject	1	2	3	4	5
1	*E*	*D*	*C*	*A*	*B*
2	*C*	*B*	*A*	*E*	*D*
3	*D*	*A*	*B*	*C*	*E*
4	*B*	*C*	*E*	*D*	*A*
5	*A*	*E*	*D*	*B*	*C*

that has the Latin square property. An adequate randomization is to rearrange rows at random and assign treatments to letters at random.

The model for a Latin square experiment (model I) is

$$X_{ijk} = \mu + \alpha_i + \beta_j + \gamma_k + \epsilon_{ijk} \qquad [i, j, k = 1, \ldots, a; \quad \epsilon_{ijk} = \mathcal{N}(0, \sigma)]$$

where α, β, and γ indicate treatment, row, and column effects with the usual convention that their sums are zero. The assumption of additivity is carried a step further than with a two-way classification, since we assume the effects of all three factors to be additive. This assumption is a penalty the Latin square pays for its compactness. Interactions between row and column effects have been shown to bias the treatment comparisons. The $a \times a$ Latin square is in fact an incomplete three-way classification. Only a^2 out of the a^3 possible combinations of levels of rows, columns, and treatments are measured in a Latin square.

It follows from the model that a treatment mean $\bar{X}_{i..}$ is an unbiased estimate of $\mu + \alpha_i$, the effects of rows and columns canceling because of the symmetry of the design. The standard error of $\bar{X}_{i..}$ is $\sigma/\sqrt{a}$. The estimate $\hat{X}_{ijk}$ of the observation X_{ijk} made from the fitted linear model is

$$\hat{X}_{ijk} = \bar{X}_{...} + (\bar{X}_{i..} - \bar{X}_{...}) + (\bar{X}_{.j.} - \bar{X}_{...}) + (\bar{X}_{..k} - \bar{X}_{...})$$

Hence, the deviation from the fitted model is

$$d_{ijk} = X_{ijk} - \hat{X}_{ijk} = X_{ijk} - \bar{X}_{i..} - \bar{X}_{.j.} - \bar{X}_{..k} + 2\bar{X}_{...} \qquad (14.10.1)$$

As in the two-way classification, the error sum of squares in the analysis of variance is the sum of the d_{ijk}^2 and the error mean square is an unbiased estimate of σ^2, if the additive model holds.

Table 14.10.2 shows the field layout and yields of a 5×5 Latin square experiment on the effects of spacing on yields of millet plants (8). In the computations, the sums for rows and columns are supplemented by sums and means for treatments (spacings). Sums of squares for rows, columns, and spacings are calculated by the usual rules. These are subtracted from the total sum of squares to give the error sum of squares $(a - 1)(a - 2) = 12$ df.

This experiment is typical of many in which the treatments consist of a series of levels of a variable, in this case width of spacing. The objective is to determine the relation between the treatment mean yields, which we will now denote by $\bar{Y}_{i.}$, and width of spacing X_i. Inspection of the mean yields suggests that the relation may be linear, the yield decreasing steadily as spacing increases. The X_i, x_i, and $\bar{Y}_i$ are shown in table 14.10.3.

The regression coefficient of yield on width of spacing is

$$b = \frac{\Sigma(X_i - \bar{X})(\bar{Y}_{i.} - \bar{Y}_{..})}{\Sigma(X_i - \bar{X})^2} = \frac{\Sigma\, x_i \bar{Y}_{i.}}{\Sigma\, x_i^2} = -\frac{178.0}{40} = -4.45$$

the units being grams per inch increase in spacing. Note that b is a *comparison* among the treatment means, with $\lambda_i = x_i/\Sigma\, x_i^2$. From Rule 12.8.1, the standard

TABLE 14.10.2

YIELDS (G) OF PLOTS OF MILLET ARRANGED IN A LATIN SQUARE
(spacings: *A*, 2-in.; *B*, 4; *C*, 6; *D*, 8; *E*, 10)

	Column					
Row	1	2	3	4	5	Total
1	*B*: 257	*E*: 230	*A*: 279	*C*: 287	*D*: 202	1255
2	*D*: 245	*A*: 283	*E*: 245	*B*: 280	*C*: 260	1313
3	*E*: 182	*B*: 252	*C*: 280	*D*: 246	*A*: 250	1210
4	*A*: 203	*C*: 204	*D*: 227	*E*: 193	*B*: 259	1086
5	*C*: 231	*D*: 271	*B*: 266	*A*: 334	*E*: 338	1440
Total	1118	1240	1297	1340	1309	6304

Summary by Spacing

	A: 2 in.	*B*: 4 in.	*C*: 6 in.	*D*: 8 in.	*E*: 10 in.	
Total	1349	1314	1262	1191	1188	6304
Mean	269.8	262.8	252.4	238.2	237.6	252.2

Sums of Squares

Correction: $6{,}304^2/25 = 1{,}589{,}617$

Total *SS*: $257^2 + \ldots + 338^2 - 1{,}589{,}617 = 36{,}571$

Row *SS*: $(1255^2 + \ldots + 1.440^2)/5 - 1{,}589{,}617 = 13{,}601$

Column *SS*: $(1118^2 + \ldots + 1309^2)/5 - 1{,}589{,}617 = 6146$

Spacing *SS*: $(1349^2 + \ldots + 1188^2)/5 - 1{,}589{,}617 = 4156$

Error *SS*: 12,668

Source of Variation	Degrees of Freedom	Sum of Squares	Mean Square
Total	24	36,571	
Rows	4	13,601	3400
Columns	4	6,146	1536
Widths of spacing	4	4,156	1039
Error	12	12,668	1056

error of b is

$$s_b = \sqrt{s^2 \Sigma \lambda^2 / a} = \sqrt{s^2/(a \Sigma x^2)} = \sqrt{(1056)/[(5)(40)]} = 2.298$$

With 12 df, 95% confidence limits for the population regression are $+0.6$ and -9.5 g/in. increase. The linear decrease in yield is not quite significant, since the limits include 0.

TABLE 14.10.3

DATA FOR CALCULATING THE REGRESSION OF YIELD ON SPACING

Spacing, X_i (in.)	2	4	6	8	10
$x_i = X_i - \overline{X}$	-4	-2	0	2	4
$\overline{Y}_{i.}$ (g)	269.8	262.8	252.4	238.2	237.6

TABLE 14.10.4
ANALYSIS OF REGRESSION OF MEAN YIELD ON WIDTH OF SPACING
(millet experiment)

Source of Variation	df	Sum of Squares	Mean Square	F
Spacings (table 14.10.2)	4	4,156		
regression	1	3,960	3960	3.75
deviations	3	196	65	0.06
Error (table 14.10.2)	12	12,668	1056	

In the analysis of variance, the treatments sum of squares can be partitioned into a part representing the linear regression on width of spacing and a part representing the deviations of the treatment means from the linear regression. This partition provides new information. If the true regression of the means on width of spacing is linear, the deviations mean square should be an estimate of σ^2. If the true regression is curved, the deviations mean square is inflated by the failure of the fitted straight line to represent the curved relationship. Consequently, F = (deviations mean square)/(error mean square) tests whether the straight line is an adequate fit.

As given in section 12.8, the sum of squares for a comparison L is $nL^2/\Sigma\,\lambda^2$. In this application, $n = a = 5$, $L = b = -4.45$, $\lambda_i = x_i/\Sigma\, x_i^2$, $\Sigma\, \lambda_i^2 = 1/\Sigma x_i^2 = 1/40$, so that $nL^2/\Sigma\lambda_i^2 = (5)(-4.45)^2(40) = 3960$ with 1 df. The sum of squares for deviations from the regression is found by subtracting 3960 from the total sum of squares for spacings, 4156 (table 14.10.4). The F ratio for deviations is very small, 0.06, giving no indication that the regression is curved. The F for regression, 3.75, is not quite significant, this test being the same as the t test for b.

The results of this experiment are probably disappointing. In trying to discover the best width of spacing, an investigator hopes to obtain a *curved* regression, with reduced yields at the narrowest and widest spacings, so that the range of spacings straddles the optimum. As it is, assuming the linear regression real, all we have learned is that the best spacing may lie below 2 in.

Since the number of replications in the Latin square is equal to the number of treatments, the experimenter is ordinarily limited to eight or ten treatments in using this design. For four or less treatments, the degrees of freedom for error are fewer than desirable, $(a - 1)(a - 2) = (3)(2) = 6$ for the 4×4. This difficulty can be remedied by replicating the squares.

EXAMPLE 14.10.1—In experiments affecting the milk yield of dairy cows the great variation among cows requires large numbers of animals for evaluating moderate differences. Efforts to apply several treatments successively to the same cow are complicated by the decreasing milk flow as the lactation proceeds and by carry-over effects of a treatment into the following period. The effort was made to control these difficulties by the use of several pairs of orthogonal Latin squares (9), the columns representing cows and the rows successive periods during lactation. The treatments were A = roughage, B = limited grain, C = full grain.

For this example, a single square is presented. The entries are total nutrient consumption (lb) for a 6-wk period. Compute the analysis of variance.

Period	Cow 1	Cow 2	Cow 3
I	*A*: 608	*B*: 885	*C*: 940
II	*B*: 715	*C*: 1087	*A*: 766
III	*C*: 844	*A*: 711	*B*. 832

Source of Variation	df	Sum of Squares	Mean Square	*F*
Periods	2	5,900	2,950	1.2
Cows	2	47,214	23,607	9.7
Treatments	2	103,436	51,718	21.4
Error	2	4,843	2,422	
Total	8	161,393		

Treatment means (lb): *A*, 695.0; *B*, 810.7; *C*, 957.0

EXAMPLE 14.10.2—Compute the expected values $\hat{X}_{ijk}$ and the residuals d_{ijk} from the Latin square model for the data in example 14.10.1. Verify that $\Sigma\Sigma\Sigma\, d_{ijk}^2$ is close to 4843. (Carry one decimal place in the residuals.)

EXAMPLE 14.10.3—The following 5 × 5 Latin square (7) tests four soil fumigants against no fumigation, O. The data are the total wireworms per plot from samples (9 × 9 × 5 in.).

P6	O3	N29	K8	M17
M8	K13	O18	N12	P16
O15	M13	K7	P10	N28
N14	P11	M13	O22	K7
K7	N26	P24	M14	O20

Perform the analysis of variance and summarize the results. Ans. Mean squares: rows, 24.6; columns, 78.9; fumigants, 119.0 (all having 4 df); error 28.0 (12 df). The *F* ratio for fumigants is significant ($P < 0.025$). By the protected LSD test (5% level), only *K*, which reduces the wireworms by nearly 50%, does significantly better than no fumigation, O.

TECHNICAL TERMS

additive effects
block
completely randomized design
crossed classification
interactions
Latin square
multiplicative effects
randomized blocks
relative efficiency
repeated measures experiment
replications
two-way classification

REFERENCES

1. Porter, R. H. 1936. Cooperative Soybean Seed Treatment Trials. Iowa State Coll. Seed. Lab.
2. Monselise, S. P. *Palest. J. Bot.* 8 (1951):1.
3. Cochran, W. G., and Cox, G. M. 1957. *Experimental Designs,* 2nd ed. Wiley, New York.
4. Kempthorne, O. 1952. *Design and Analysis of Experiments.* Wiley, New York.
5. Fisher, R. A. 1935–51. *The Design of Experiments.* Oliver & Boyd, Edinburgh.
6. Forster, H. C., and Vasey, A. J. *Austral. J. Dept. Agric.* 30 (1932):35.
7. Ladell, W. R. S. *Ann. Appl. Biol.* 25 (1938):341.
8. Li, H. W.; Meng, C. J.; and Liu, T. N. *J. Am. Soc. Agron.* 28 (1936):1.
9. Cochran, W. G.; Autrey, K. M.; and Cannon, C. Y. *J. Dairy Sci.* 24 (1941):937.
10. Smith, G. M., et al. *J. Pharm. Exp. Ther.* 163 (1968):468.

15

Failures in the assumptions

15.1—Introduction. In standard analyses of variance the model specifies that the effects of the different factors (treatments, rows, columns, etc.) are additive and the residuals from the model are normally and independently distributed with the same variance. It is unlikely that these ideal conditions are ever exactly realized in practice. While minor failures in the assumptions do not greatly disturb the conclusions drawn from standard analyses, the data analyst is responsible for checking that the principal assumptions seem reasonably well satisfied in the data being analyzed. In this chapter some advice is given on the detection and handling of more serious failures. In this activity, electronic computers and computer programs have been of great help; analyses and plots that aid in checking a model or in deciding how to cope with nonconformity are now available.

Types of failure are classified into missing data, presence of gross errors, lack of independence of errors, unequal error variances due to the nature of the treatments, nonadditivity, heterogeneity of variances, and nonnormality.

15.2—The problem of missing data. Accidents often result in loss of data. Crops are destroyed on some plots, some subjects do not take all the treatments as planned, and mistakes are made in the application of some treatments or in recording the results. In the standard methods for handling missing data, it is assumed that missing items are due to mishaps or mistakes and not to failure of a treatment. To put it another way, any missing observation X_{ij} is assumed to follow the same mathematical model as the observations that are present.

Thus the mathematical model for the plan remains the same, but certain of the X_{ij} are missing. Since the assumptions of the model apply to the X_{ij} that are present, the correct analysis is to apply to them the method of least squares. Missing items, however, destroy the symmetry and simplicity of the analysis, which becomes a good deal more complex if several X_{ij} are missing. Most computer programs with the following titles handle the analysis of incomplete data as well as of complete data:

General analysis of variance
Analysis of variance and covariance
Multiple linear regression

TABLE 15.2.1
YIELDS OF FOUR STRAINS OF WHEAT IN FIVE RANDOMIZED BLOCKS (LB/PLOT) WITH ONE MISSING VALUE

	Block					
Strain	1	2	3	4	5	Total
A	32.3	34.0	34.3	35.0	36.5	172.1
B	33.3	33.0	36.3	36.8	34.5	173.9
C	30.8	34.3	35.3	32.3	35.8	168.5
D	. . .	26.0	29.8	28.0	28.8	112.6
Total	96.4	127.3	135.7	132.1	135.6	627.1

Analysis of Variance (with 25.4 inserted)

Source of Variation	Degrees of Freedom	Sum of Squares	Mean Squares
Blocks	4	35.39	
Strains	3	171.36	57.12 (45.79)
Error	11	17.33	1.58
Total	18	224.08	

In their output these programs usually provide the analysis of variance, with the correct F test for treatments and the correct residuals mean square and degrees of freedom. The least squares estimates of the treatment effects with their standard errors are usually available also. If comparisons on or among treatment means are to be made, check that the program will provide standard errors for these comparisons or instructions for computing them—estimated treatment means often become correlated when data are missing.

In a one-way classification, the effect of missing values is merely to reduce sample sizes in the affected classes. The analysis is handled correctly by the methods in section 12.10 for one-way classifications with unequal sample sizes in the classes.

With a single missing value in a two-way classification, we obtain most of what we want by inserting the least squares estimate of the missing value in the vacant cell and analyzing the complete data. This method gives

least squares estimate of every treatment mean
correct residual (error) sum of squares

If the missing observation is in row i and column j, insert the value (16)

$$\hat{X}_{ij} = (IX_{i.} + JX_{.j} - X_{..})/[(I - 1)(J - 1)] \tag{15.2.1}$$

where the totals are taken over the values that are *present*.

As an example, table 15.2.1 shows the yields in an experiment on four strains of Gallipoli wheat in which we have supposed that the yield X_{41} for strain D in block 1 is missing. From the totals in table 15.2.1, the value to be inserted is

$$[4(112.6) + 5(96.4) - 627.1]/[(3)(4)] = 25.4$$

This value is entered in the table as the yield of the missing plot. All sums of squares in the analysis of variance are then computed as usual. However, the degrees of freedom in the total and error sums of squares are both reduced by 1, since there are actually only 18 df for the total and 11 for the error sums of squares.

For comparisons among treatment means, the estimated standard error of the mean of the treatment with the missing value is

$$\frac{s}{\sqrt{J}}\sqrt{1+\frac{I}{(I-1)(J-1)}}=\frac{s}{\sqrt{(J-1)}}\sqrt{1+\frac{1}{J(I-1)}} \qquad (15.2.2)$$

As might be expected, this is slightly larger than $s/\sqrt{J-1}$. The standard error of a mean with no missing value is $s/\sqrt{J}$ as usual. For example, the standard error of the difference between the means for strains C and D may be verified to be ± 0.859.

The treatments (strains) mean square in the analysis of variance is somewhat inflated. To correct for this bias, subtract from the *mean square* the quantity

$$\frac{[X_{.j}-(I-1)\hat{X}_{ij}]^2}{I(I-1)^2}=\frac{[96.4-(3)(25.4)]^2}{(4)(3)(3)}=11.33$$

Then $57.12 - 11.33 = 45.79$ is the correct treatments mean square.

For the $I \times I$ Latin square with the value X_{ijk} missing, insert

$$\hat{X}_{ijk}=[I(X_{i..}+X_{.j.}+X_{..k})-2X_{...}]/[(I-1)(I-2)] \qquad (15.2.3)$$

Deduction from treatments *mean square* for bias is

$$[X_{...}-X_{.j.}-X_{..k}-(I-1)X_{i..}]^2/[(I-1)^3(I-2)^2] \qquad (15.2.4)$$

The estimated standard error of the mean of the treatment with a missing value is

$$\frac{s}{\sqrt{I-1}}\sqrt{1+\frac{2}{I(I-2)}} \qquad (15.2.5)$$

15.3—More than one missing value. Methods of estimating more than one missing value may be illustrated by the small randomized blocks experiment in table 15.3.1, in which X_{31} and X_{22} are missing. One method is to guess X_{22}, insert this guessed value, and obtain an estimate $\hat{X}_{31(1)}$ of X_{31} by (15.2.1). Then insert $\hat{X}_{31(1)}$ and try to obtain a better estimate $\hat{X}_{22(1)}$ of X_{22} by (15.2.1). Then insert $\hat{X}_{22(1)}$ and reestimate X_{31} by (15.2.1). Continue this iterative process until it does not change either value. Convergence is usually quick.

TABLE 15.3.1
RANDOMIZED BLOCKS EXPERIMENT WITH TWO MISSING VALUES

Treatments	Blocks 1	2	3	Total
A	6	5	4	15
B	15	X_{22}	8	23
C	X_{31}	15	12	27
Total	21	20	24	65

With a program that analyzes complete data rapidly, Healy and Westmacott (1) give a more general iterative method. First, insert guesses $X_{ij(1)}$ for *all* missing values. Analyze the complete data. As second estimates $X_{ij(2)}$, use the predictions $\hat{X}_{ij}$ for the missing values from this analysis. Since most programs print residuals $d_{ij} = X_{ij(1)} - \hat{X}_{ij}$, we can obtain $\hat{X}_{ij(2)}$ as

$$\hat{X}_{ij(1)} - d_{ij}$$

Continue this process until the residual sum of squares shows no appreciable change. Convergence is slow unless the original guesses are very good.

Rubin (2) gives a noniterative method. With m missing values a program is required that will invert an $m \times m$ matrix. Matrix operations are extensively used in fitting multiple linear regressions. The steps in Rubin's method are:

1. Complete the data by inserting for every missing value the mean yield of the observations that are present ($65/7 = 9.286$ in table 15.3.1). Analyze the completed data and obtain the residual d_{ij} at each missing value. For table 15.3.1 these residuals will be found to be $d_{31} = -3.619$, $d_{22} = -1.952$. This pair of values is called a *row vector* in matrix terminology.

2. For each missing value in turn, construct data consisting of 1 at the missing value and 0 everywhere else. The two sets of data for table 15.3.1 are as follows:

For X_{31}				Means
	0	0	0	0
	0	0	0	0
	1	0	0	1/3
Means	1/3	0	0	1/9

$d_{31} = 4/9$, $d_{22} = +1/9$

For X_{22}				Means
	0	0	0	0
	0	1	0	1/3
	0	0	0	0
Means	0	1/3	0	1/9

$d_{31} = +1/9$, $d_{22} = 4/9$

Analyze each set and find d_{31}, d_{22} for each set.

3. Arrange these d_{ij} in a 2×2 matrix and invert it. The matrix is

$$\begin{bmatrix} 4/9 & 1/9 \\ 1/9 & 4/9 \end{bmatrix}$$

The inverse of a 2×2 matrix is easily calculated as follows:

$$\begin{array}{cc} \text{Matrix} & \text{Inverse} \\ \begin{bmatrix} C_{11} & C_{12} \\ C_{21} & C_{22} \end{bmatrix} & \begin{bmatrix} C_{22}/D & -C_{21}/D \\ -C_{12}/D & C_{11}/D \end{bmatrix} \end{array}$$

where $D = C_{11}C_{22} - C_{12}C_{21}$. With our data, $D = 15/81$ and the inverse is shown below.

$$\begin{array}{cc} \text{Matrix} & \text{Inverse} \\ \begin{bmatrix} 4/9 & +1/9 \\ +1/9 & 4/9 \end{bmatrix} & \begin{bmatrix} 12/5 = \ \ 2.4 & -3/5 = -0.6 \\ 3/5 = -0.6 & 12/5 = -2.4 \end{bmatrix} \end{array}$$

(With more than 2 missing values, this is the place at which a program that inverts a matrix is needed.)

4. Premultiply this inverse matrix by the row vector $(-3.619, -1.952)$ obtained in step 1. For this operation the rule is that the row terms are multiplied by corresponding column terms in the matrix and added. The result is the vector

$$(-3.619)(\ \ 2.4) + (-1.952)(-0.6) = -7.514$$
$$(-3.619)(-0.6) + (-1.952)(\ \ 2.4) = -2.513$$

5. The correct least squares estimates of the missing values are obtained by subtracting these adjustments from our first estimates, which were the overall mean:

$$\hat{X}_{31} = 9.286 - (-7.514) = 16.80$$
$$\hat{X}_{22} = 9.286 - (-2.513) = 11.80$$

These are the values to which the iterative methods should converge and that we insert to complete table 15.3.1.

6. Table 15.3.2 shows the complete data from table 15.3.1 and the analysis of variance. This analysis gives the least squares estimate of any treatment mean (shown in table 15.3.2) and the correct residual sum of squares. As regards degrees of freedom, we subtract m (in this case 2) from the total and residuals (error) degrees of freedom.

TABLE 15.3.2
COMPLETED DATA AND ANALYSIS OF VARIANCE

Treatments	Blocks 1	2	3	Total	Mean
A	6	5	4	15	5.0
B	15	11.8	8	34.8	11.6
C	16.8	15	12	43.8	14.6
Total	37.8	31.8	24	93.6	

Source of Variation	Degrees of Freedom	Sum of Squares	Mean Square
Blocks	2	31.92	
Treatments	2	144.72	72.36
Residuals	2	6.40	3.20
Total	6	183.04	

In this analysis of variance, the treatments sum of squares and mean squares are biased upwards. To obtain the correct treatments sum of squares, we calculate how much the residuals sum of squares would increase from its present value of 6.40 if there were no treatment effects. For this, we reanalyze the seven observations in table 15.3.1 as a one-way classification with unequal numbers, the blocks being the classes. The new residuals sum of squares will be found to be 122.50 with 4 df. Subtract from this 6.40 with 2 df. The difference, 122.50 − 6.40 = 116.10 with 4 − 2 = 2 df, is the correct treatments sum of squares. The F ratio is 58.05/3.20 = 18.1 with 2 and 2 df.

15.4—Extreme observations. A measurement may be read, recorded, or transcribed wrongly; the instrument on which it is made may have become faulty; or the treatment for this measurement may have been wrongly applied. A major error in an experiment greatly distorts the mean of the treatment involved. By inflating the error variance, it affects conclusions about other treatments as well. The principal safeguards are vigilance in carrying out the operations, the measurements, and the recording, plus eye inspection of the data before analysis.

If a figure in the data looks suspicious, an inquiry about this observation should be made. Sometimes the inquiry shows a mistake and also reveals the correct value for the observation, which is then used in the analysis. Sometimes it is certain that a gross error was made, but there is no way of finding the correct value. In this event, omit the erroneous observation and use the analysis for data with one missing value. (In such cases, check that the source of the gross error did not affect other observations also.)

Numerous tables have been constructed to guide the investigator as to how rare an extreme observation would be if the data were normal and followed the assumed model. The extreme observation is often called an *outlier*.

SINGLE SAMPLE. For a sample from a normal distribution, table *A* 16(*i*) shows the 5% and 1% levels of the criterion

$$\text{MNR} = \frac{\max|x|}{\sqrt{\Sigma x^2}} = \frac{\max|X - \overline{X}|}{\sqrt{\Sigma(X - \overline{X})^2}}$$

Stefansky (3) has called a statistic of this type a *maximum normed residual* (MNR), the residuals for a single sample being the deviations $X_i - \overline{X}$. Grubbs (4) has tabulated significance levels of the equivalent criterion $T = \max|x|/s = \sqrt{n-1}(\text{MNR})$, sometimes called the *extreme Studentized deviate* (ESD).

EXAMPLE 1—In table 12.2.1, the grams of fat absorbed by doughnuts in the replicates of fat 1 are, in increasing order: 56, 64, 68, 72, 77, 95. The value 95 looks a bit out of line. Is a deviation of this size very unusual? We find $\overline{X} = 72$, $\Sigma(X - \overline{X})^2 = 890$. Hence, $\text{MNR} = |95 - 72|/\sqrt{890} = 0.771$, $n = 6$. Evidently a deviation of this size or larger occurs more than 5% of the time under normal conditions, the 5% level [table A 16(*i*)] being 0.844.

For the same purpose (testing an outlier in a normal sample), W. J. Dixon gives a test based on order statistics that is very quick and convenient. Table A 17 gives 5% and 1% levels. Arrange the sample in increasing order, X_1 being the smallest and X_n the highest value. The best test criterion varies with the sample size and is given in Table A 17. For our example, with $n = 6$, the criterion is

$$(X_6 - X_5)/(X_6 - X_1) = (95 - 77)/(95 - 56) = 18/39 = 0.462$$

This falls short of the 5% level, 0.560, in agreement with the MNR. Both tests make $P > 0.10$. If the suspected outlier is the *lowest* observation, the criterion is $10\ (X_2 - X_1)(X_6 - X_1)$.

Two suspicious outliers may be tested by repeated use of either of these tests. First, test the most extreme value; then test the second most extreme value in the sample of size $(n - 1)$ formed by omitting the most extreme value. *Make both tests* no matter what the verdict of the first test is, since a second outlier may mask a first outlier if both lie on the same side. Consider the data $(n = 10)$:

$$-1.88,\ -0.85,\ -0.50,\ -0.42,\ -0.26,\ 0.02,\ 0.24,\ 0.44,\ 4.74,\ 5.86$$

For the most extreme value, Dixon's criterion is $(X_{10} - X_9)/(X_{10} - X_2) = (5.86 - 4.74)/[5.86 - (-0.85)] = (5.86 - 4.74)/(5.86 + 0.85) = 0.17$—nowhere near the 5% level. But the second extreme gives $(4.74 - 0.44)/(4.74 + 0.85) = 0.769$. For $n = 9$, the 1% level of Dixon's criterion is 0.635. Thus, *both* are regarded as outliers. The MNR method gives the same results.

Since we are now making more than one test, the levels in tables A 16(*i*) and A 17 must be changed to keep the probability of finding no spurious outlier at the 5% or 1% levels. Rosner (5) has given the 5%, 1%, and 0.5% levels for the ESD (MNR) method with two outliers.

15.5—Suspected outliers in one-way or two-way classifications. If X_{ij} is the *j*th observation in the *i*th class in a one-way classification, its residual from the standard model is $d_{ij} = X_{ij} - \overline{X}_{i.}$, since $\hat{X}_{ij} = \overline{X}_{i\cdot}$. If we expect that within-class variances will be heterogeneous, we may decide to test a large residual by comparing it only with other residuals in the same class. Then table

A 16(i) or A 17 can be used, since we are dealing with a single sample. If within-class variances look homogeneous, however, a more powerful test is obtained by defining the MNR as

$$\text{MNR} = \max|d_{ij}| \Big/ \sqrt{\sum_i \sum_j d_{ij}^2} = \max|d_{ij}| / \sqrt{\text{residual } SS} \qquad (15.5.1)$$

Regarding the distribution of this MNR, any d_{ij} is correlated with the other d_{ik} in its class but uncorrelated with the d_{lm} in other classes. Its significance levels in table A 16 lie between those in column (i) MNR and those in column (ii) MNR, which would apply if *all* d_{ij} were mutually independent. In this use of table A 16, note that n is the total number of observations over *all* classes. The significance levels are usually much closer to column (ii) than to column (i). With 5 classes and 4 in a class, we read table A 16 at $n = 20$.

In a two-way classification with one observation per cell, the residuals that we examine when checking on the presence of extreme values are

$$d_{ij} = X_{ij} - \overline{X}_{i.} - \overline{X}_{.j} + \overline{X}_{..}$$

Table A 18 gives the 5% and 1% levels of the MNR $= \max|d|\sqrt{\Sigma\Sigma d_{ij}^2}$ for a two-way classification, computed by Stefansky (3). The table is indexed by the number of rows and columns in the two-way classification. As an example of its use, we take an experiment (6) in which the gross error is an impossible value that should have been spotted at once and rejected but was not. The data are ratios of dry weight to wet weight of wheat grain, which of course cannot exceed 1. In table 15.5.1 the MNR test detects this outlier, 1.04 ($P < 0.01$).

If an observation is clearly aberrant but no explanation can be found, it is not obvious what is best to do. Occasionally, the aberrant observation can be repeated; a subject who gave an aberrant reading under treatment 3 might take the treatment again on a later occasion with only a minor departure from the experimental plan. If it is decided to treat the outlier as a missing observation, this decision and the value of the outlier should be reported. The investigator

TABLE 15.5.1

RATIO OF DRY TO WET GRAIN IN AN EXPERIMENT WITH FOUR TREATMENTS, FOUR BLOCKS

	Data Nitrogen Applied				Residuals Nitrogen Applied			
Block	None	Early	Middle	Late	None	Early	Middle	Late
1	0.72	0.73	0.73	0.79	0.021	−0.079	0.011	0.046
2	.72	0.78	.72	.72	.029	− .021	.009	− .016
3	.70	1.04	.76	.76	− .071	.159	− .031	− .056
4	0.73	0.76	0.74	0.78	0.021	−0.059	0.011	0.026

Residual $SS = \Sigma\Sigma d_{ij}^2 = 0.0497$ max $|d| = 0.159$

MNR $= 0.159/\sqrt{0.0497} = 0.713$ (1% level $= 0.665$)

should also analyze the data with the outlier present to learn which conclusions are affected by presence or absence of the outlier.

Sometimes an investigator knows from past experience that occasional wild observations occur, though the process is otherwise stable. Except in such cases, statisticians warn against automatic rejection rules based on tests of significance, particularly if there appear to be several outliers. The apparent outliers may reflect distributions of the observations that are skew or have long tails and are better handled by methods being developed for nonnormal distributions. Ferguson (7) has noted that the tests for skewness and kurtosis (sections 5.13, 5.14) are good tests for detecting the presence of numerous outliers.

Grubbs (4) gives an excellent review of techniques for detecting outliers.

15.6—Correlations between the errors. If care is not taken, an experiment may be conducted in a way that induces positive correlations between the errors for different replicates of the same treatment. In an industrial experiment, all the replications of a given treatment might be processed at the same time by the same technicians to cut down the chance of mistakes or to save money. Any differences that exist between the batches of raw materials used with different treatments or in the working methods of the technicians create positive correlations within treatments.

In the simplest case these situations are represented mathematically by supposing an intraclass correlation ρ_I between any pair of errors within the same treatment. In the absence of real treatment effects, the mean square between treatments is an unbiased estimate of $\sigma^2 [1 + (n - 1)\rho_I]$, where n is the number of replications, while the error mean square is an unbiased estimate of $\sigma^2(1 - \rho_I)$, as pointed out in section 13.5. The F ratio is an estimate of $[1 + (n - 1)\rho_I]/(1 - \rho_I)$. With ρ_I positive, this ratio can be much larger than 1; for instance, with $\rho_I = 0.2$ and $n = 6$, the ratio is 2.5. Thus, positive correlations among the errors within a treatment vitiate the F test, giving too many significant results. The disturbance affects t tests also and may be major.

15.7—The role of transformations. In the standard analysis of variance the model assumes that the effects of the different factors (rows, columns, or treatments) are additive and that the residual errors are normally and independently distributed with the same variance. In data with small effects and limited range (e.g., with the highest observation less than 1.5 times the lowest), departures from these assumptions do not greatly disturb the conclusions drawn from a standard analysis. In data of wider range in which one or more of the basic assumptions fails, a transformation of the observations X_{ij} into another scale often allows the standard analysis to be used as an adequate approximation.

If the trouble is that the error variance of X_{ij} depends on the expected value of X_{ij}, Bartlett (8) showed how to find a transformation that decreases such heterogeneity of variance. If the trouble is nonadditivity of row and column effects, Tukey (9) showed how to seek a transformation such that effects in the transformed scale are approximately additive. Transformations that bring data closer to normality have also been studied. Not surprisingly, the individual transformations that handle heterogeneity of variance, nonadditivity, and nonnormality are usually different. We first present the transformations for nonadditivity and for heterogeneity of variance.

15.8—A test for nonadditivity. Tukey's procedure is useful in a number of ways: (*i*) to help decide, by means of a test for nonadditivity, if a transformation is necessary; (*ii*) to suggest a suitable transformation; and (*iii*) to learn if the transformation has produced additivity.

The method applies to transformations of the form $Y = X^p$, in which X is the original scale and we seek a power p of X such that effects are additive in the scale of $Y = X^p$. Thus, $p = 1/2$ represents a square root transformation and $p = -1$ a reciprocal or inverse transformation, analyzing $1/X$ instead of X. For values of p near zero, X^p behaves like ln X, suggesting a log transformation. In fact, $(X^p - 1)/p$ becomes equal to ln X when p is very small. For instance, for $X = 5$ and $p = 0.01$, $(X^p - 1)/p = 1.62$, while ln $X = 1.61$.

In the additive model, the parameters α_i and β_j represent deviations of the row and column means from the general mean. If nonadditivity is not too complex, a model that approximates quite a range of types of nonadditivity is one in which, apart from residual errors, X_{ij} is a *quadratic* function of α_i and β_j. A quadratic function can be written

$$X_{ij} = \mu + \alpha_i + \beta_j + a\alpha_i^2 + b\beta_j^2 + c\alpha_i\beta_j \tag{15.8.1}$$

Note that $a\alpha_i^2$ and $b\beta_j^2$, like α_i and β_j, represent *additive* row and column effects. The only nonadditive term in (15.8.1) is $c\alpha_i\beta_j$.

To sum up: Recall that the residuals d_{ij} in the analysis of variance of the X_{ij} are the deviations of the X_{ij} from an additive model. If we actually have a quadratic nonadditive model, the d_{ij} would be expected to have a linear regression on the product $d_i d_j = (\overline{X}_{i.} - \overline{X}_{..})(\overline{X}_{.j} - \overline{X}_{..})$, since the latter is the sample estimate of $\alpha_i\beta_j$.

Tukey gave two further valuable results: (*i*) The regression coefficient of d_{ij} on $d_i d_j$ is approximately $B = (1 - p)/\overline{X}_{..}$. Hence, the power p to which X_{ij} must be raised to produce additivity is estimated by $\hat{p} = (1 - B\overline{X}_{..})$. Anscombe and Tukey (17) warn, however, that a body of data rarely determines p closely. Try several values of p, including any suggested by subject matter knowledge.

(*ii*) Tukey gives a test of the null hypothesis that the population value of B is zero. For this, we perform a test of the regression coefficient B. From the residual sum of squares in the analysis of variance of X we subtract a term with 1 df that represents the regression of d_{ij} on $d_i d_j$. Table 15.8.1 shows this partition of the residual sum of squares and the resulting F test of B. The table uses the results: $\Sigma\Sigma d_{ij} d_i d_j = (\Sigma d_i^2)(\Sigma d_j^2)$ and $\Sigma\Sigma(d_i d_j)^2 = (\Sigma d_i^2)(\Sigma d_j^2)$. The F value has 1 and $(IJ - I - J)$ df.

Table 15.8.2 illustrates this test for artificial data in which we know the

TABLE 15.8.1
TUKEY'S TEST FOR NONADDITIVITY

Source	Degrees of Freedom	Sum of Squares	Mean Square
Nonadditivity, NA	1	$(\Sigma\Sigma d_i d_j X_{ij})^2/[(\Sigma d_i^2)(\Sigma d_j^2)]$	s_{NA}^2
Remainder, REM	$(I-1)(J-1)-1$	by subtraction	s_{REM}^2
Residuals	$(I-1)(J-1)$	$\Sigma\Sigma d_{ij}^2$	$(F = s_{NA}^2/s_{REM}^2)$

transformation that restores additivity. The data in table 15.8.2(*i*) have exactly additive row and column effects. Table 15.8.2(*ii*) gives the *squares* of the values in (*i*); these squares are the X_{ij} to be analyzed. In table 15.8.2(*iii*) we find the residuals d_{ij}, the d_i, and the d_j, while (*iv*) presents the plot of the d_{ij} against the products $d_i d_j$. Evidently the points lie close to a straight line, with a slope near 0.2.

The residuals d_{ij} and the products $d_i d_j$ are as follows:

d_{ij}	0.16	0.03	−0.17	0.02	0.01	−0.01	−0.17	−0.02	0.20
$d_i d_j$	0.734	0.153	−0.877	0.137	0.028	−0.163	−0.864	−0.180	1.032

The regression coefficient B of d_{ij} on $d_i d_j$ is 0.196. Since $\bar{X}_{..} = 2.58$ (table 15.8.2), we find $\hat{p} = 1 - (0.196)(2.58) = 0.49$, suggesting that $X^{0.49} \doteq \sqrt{X}$ is the additive scale, as we know. This example happens to be one in which the

TABLE 15.8.2
ILLUSTRATION OF TRANSFORMATION FOR NONADDITIVITY

(*i*) Additive data

1	1.2	1.5
1.3	1.5	1.8
1.7	1.9	2.2

(*ii*) Squares (X_{ij})

				Means
	1	1.44	2.25	1.56
	1.69	2.25	3.24	2.39
	2.89	3.61	4.84	3.78
Means	1.86	2.43	3.44	2.58

(*iii*) d_{ij}, d_i, and d_j

				d_i
	+.16	+.03	−.17	−1.02
	+.02	+.01	−.01	− .19
	−.17	−.02	+.20	+1.20
d_j	−.72	−.15	+.86	

(*iv*) Plot of d_{ij} against $d_i\,d_j$

d_{ij} .2 .1 −.1 −.2 ; −1 −.5 .5 1 $d_i d_j$

technique works exceptionally well, but the advice of Anscombe and Tukey mentioned earlier in this section should be heeded.

EXAMPLE 15.8.1—For the additive data in table 15.8.2(i), the values of e^X are as follows:

	X_{ij}			Means
	2.72	3.32	4.48	3.51
	3.67	4.48	6.05	4.73
	5.47	6.69	9.03	7.06
Means	3.95	4.83	6.52	5.10

For these data a log or ln transformation ($p = 0$) will, of course, restore additivity. You will find that $B = \Sigma d_i d_j/[(\Sigma d_i^2)(\Sigma d_j^2)] = 1.00$, so $\hat{p} = 0$.

15.9—Application to an experiment. The data in table 15.9.1 were taken from a larger experiment (10). Each entry is the geometric mean of insect (macrolepidoptera) catches in a trap on three successive nights. The periods are different three-night slices cut from the distribution of the population over time. The objective was to find a scale on which period and trap effects are additive. The steps in a convenient method of making the calculations for an additivity test are as follows.

TABLE 15.9.1
MACROLEPIDOPTERA CATCHES BY THREE TRAPS IN FIVE PERIODS
(calculations for test of additivity)

	Trap			Sum	Mean		
Period	1	2	3	$X_{i.}$	$\overline{X}_{i.}$	d_i	$w_i = \Sigma X_{ij} d_j$
1	19.1	50.1	123.0	192.2	64.1	−74.9	14,025
2	23.4	166.1	407.4	596.9	199.0	+60.0	51,096
3	29.5	223.9	398.1	651.5	217.2	+78.2	47,543
4	23.4	58.9	229.1	311.4	103.8	−35.2	28,444
5	16.6	64.6	251.2	332.4	110.8	−28.1	32,243
Sum $X_{.j}$	112.0	563.6	1408.8	2084.4			$\Sigma w_i =$ 173,351
Mean $\overline{X}_{.j}$	22.4	112.7	281.8		139.0		
d_j	−116.6	−26.2	+142.8			0.0	

(i) Find $d_i = X_{i.} - \overline{X}_{..}$ and $d_j = \overline{X}_{.j} - \overline{X}_{..}$, both adding exactly to zero.
(ii) $w_1 = (19.1)(-116.6) + (50.1)(-26.2) + (123.0)(+142.8) = 14{,}025$
$w_5 = (16.6)(-116.6) + (64.6)(-26.2) + (251.2)(+142.8) = 32{,}243$
Check: $173{,}351 = (112.0)(-116.6) + (563.6)(-26.2) + (1408.8)(+142.8)$
$N = \Sigma w_i d_i = (14{,}025)(-74.9) + \ldots + (32{,}243)(-28.1) = 3.8259 \times 10^6$
(iii) $\Sigma d_i^2 = (-74.9)^2 + \ldots + (-28.1)^2 = 17{,}354$
$\Sigma d_j^2 = (-116.6)^2 + \ldots + (+142.8)^2 = 34{,}674$
$D = (\Sigma d_i^2)(\Sigma d_j^2) = (17{,}354)(34{,}674) = 601.7 \times 10^6$
(iv) SS for nonadditivity $= N^2/D = (3.8259)^2(10^{12})/[(601.7)(10^6)] = 24{,}327$

(*i*) Calculate $d_i = \overline{X}_{i.} - \overline{X}_{..}$ and $d_j = \overline{X}_{.j} - \overline{X}_{..}$, rounding if necessary so that both sets add *exactly* to zero.

(*ii*) Compute $w_i = \Sigma_j X_{ij} d_j$ and record them in the extreme right column. Then find N, the numerator of B:

$$N = \sum_i w_i d_i = \Sigma\Sigma X_{ij} d_i d_j$$

(*iii*) The denominator D of B is $(\Sigma d_i^2)(\Sigma d_j^2)$. Thus, $B = N/D$.

(*iv*) The contribution of nonadditivity to the error sum of squares of X is N^2/D with 1 df. It is tested by an F test against the remainder of the residual sum of squares, which has $[(r - 1)(c - 1) - 1]$ df. The test is made in table 15.9.2.

The hypothesis of additivity is untenable ($P < 0.01$). What type of transformation is suggested by this test?

$$B = N/D = 3.8259/601.7 = 0.006358$$

$$\hat{p} = 1 - B\overline{X}_{..} = 1 - (0.006358)(139.0) = 1 - 0.88 = 0.12$$

The test suggests $X^{0.1}$, which behaves very much like log X.

Two further checks on the data may be made. The first is the test for nonadditivity in the log scale, which checks whether a log transformation of the data seems to have restored additivity. You will find that in logs, F for nonadditivity has been reduced to 3.23, which is nonsignificant (example 15.9.1). Also, note that in table 15.9.1 the large effects are those due to traps. Within the traps the ranges of the data in the original scale are trap 1, 12.9; trap 2, 173.8; and trap 3, 284.4, indicating nonhomogeneity of residual variances. The standard deviation increases steadily as the trap mean increases. In the log scale this relation disappears, the ranges becoming (in natural logs) 0.57, 1.50, and 1.20. Thus the log transformation has also helped homogeneity of variance.

TABLE 15.9.2
ANALYSIS OF VARIANCE AND TEST OF ADDITIVITY

	df	Sum of Squares	Mean Square	F
Periods	4	52,066	13,016	3.4
Traps	2	173,333	86,660	22.7
Residuals	8	30,607	3,826	
nonadditivity	1	24,327	24,327	27.1
remainder	7	6,280	897	

EXAMPLE 15.9.1—Transform the data in table 15.9.1 to logs, perform the analysis, and test for nonadditivity. Ans. In natural logs, we found the following analysis of variance:

	df	Sum of Squares	Mean Square	F
Periods	4	2.249	0.562	6.12
Traps	2	15.357	7.678	83.64
Residuals	8	0.735	0.0918	
nonadditivity	1	0.232	0.232	3.23
remainder	7	0.503	0.0719	
Total		18.341		

Note also that the F values for periods and traps have substantially increased in the log scale.

EXAMPLE 15.9.2—The data in example 14.2.1 on the effect of shade on leaf area/leaf weight in citrus trees provide a simple example of a nonadditivity test for a 3 × 3 table. Ans. For nonadditivity, $F = 5.66$ with 1 and 3 df, giving P about 0.10.

EXAMPLE 15.9.3—In a randomized blocks experiment with 2 treatments and n replications forming a $2 \times n$ table, (*i*) prove that the nonadditivity test is a test for a linear regression of the differences $X_{1j} - X_{2j}$ between the two treatments on the mean levels $\bar{X}_{.j}$ of replication j. (*ii*) If b is this regression coefficient, prove that $B = 2b/(\bar{X}_{1.} - \bar{X}_{2.})$. The following data from table 6.3.1, p. 87, provide an illustration, the X_{ij} being the numbers of lesions on half leaves inoculated with two strengths of tobacco mosaic virus.

	Replications							
Treatments	1	2	3	4	5	6	7	8
1	31	20	18	17	9	8	10	7
2	18	17	14	11	10	7	5	6
Mean	24.5	18.5	16.0	14.0	9.5	7.5	7.5	6.5
$X_{1j} - X_{2j}$	13	3	4	6	−1	1	5	1

By the method of table 15.9.1, the nonadditivity test gives the following partition of the residual (error) sum of squares for the regression of d_{ij} on $d_i d_j$.

	df	Sum of Squares	Mean Square	F
Nonadditivity	1	38.1	38.1	8.5
Remainder	6	26.9	4.48	
Residuals	7	65.0		

The significant F value seems to be due mainly to the first replication. (*iii*) Verify that you can reproduce this analysis of variance by computing the regression of the treatment differences $X_{1j} - X_{2j}$ on the means $\bar{X}_{.j}$. Remember that the residuals sum of squares is $\Sigma(x_{1j} - x_{2j})^2/2$.

15.10—Variance-stabilizing transformations. If σ_X^2 is a known function of the mean μ of X, say $\sigma_X^2 = \phi(\mu)$, a transformation of the data that makes the variance almost independent of the mean is obtained by an argument based on calculus. Let the transformation be $Y = f(X)$ and let $f'(X)$ denote the derivative of $f(X)$ with respect to X. By a one-term Taylor expansion,

$$Y \doteq f(\mu) + f'(\mu)(X - \mu)$$

To this order of approximation, the mean value $E(Y)$ of Y is $f(\mu)$, since $E(X - \mu) = 0$. With the same approximation, the variance of Y is

$$E[Y - f(\mu)]^2 \doteq [f'(\mu)]^2 E(X - \mu)^2 = [f'(\mu)]^2 \sigma_X^2 = [f'(\mu)]^2 \phi(\mu)$$

Hence, to make the variance of Y independent of μ, we choose $f(\mu)$ so that the term on the extreme right above is a constant. This makes $f(\mu)$ the indefinite integral of $d\mu/\sqrt{\phi(\mu)}$. For the Poisson distribution, this gives $f(\mu) = \sqrt{\mu}$, i.e., $Y = \sqrt{X}$. For the binomial, the method gives $Y = \text{arc sin } \sqrt{p}$, that is, Y is the angle whose sine is $\sqrt{p}$. If $\sigma_X^2 \propto \mu^2$, $Y = \ln X$ or $\log X$. When $f(X)$ has been chosen in this way, the value of the constant variance on the transformed scale is obtained approximately by finding $[f'(\mu)]^2\phi(\mu)$. For the Poisson, with $\phi(\mu) = \mu$, $f(\mu) = \sqrt{\mu}$, we have $f'(\mu) = (1/2)\sqrt{\mu}$, so $[f'(\mu)]^2\phi(\mu) = 1/4$. The variance on the transformed scale is 1/4.

15.11—Square root transformation for counts. Counts of rare events, such as numbers of defects or of accidents, tend to be distributed approximately in Poisson fashion. A transformation to $\sqrt{X}$ is often effective; the variance on the square root scale will be close to 0.25. If some counts are small, $\sqrt{X + 1}$ or $\sqrt{X} + \sqrt{X + 1}$ stabilizes the variance more effectively (11).

The square root transformation can also be used with counts in which it appears that the variance of X is *proportional* to the mean of X, that is $\sigma_X^2 = k\bar{X}$. For a Poisson distribution of errors $k = 1$, but we often find k larger than 1, indicating that the distribution of errors has a variance greater than that of the Poisson.

An example is the record of poppy plants in oats (8) shown in table 15.11.1, where the numbers are large. The ranges increase steadily as the treatment means increase. Table 15.11.1 shows that the ranges do not increase as fast as the means. However, the ratio of range to $\sqrt{\text{mean}}$ shows no trend, suggesting a square root transformation.

TABLE 15.11.1
NUMBER OF POPPY PLANTS IN OATS
(plants per 3 3/4 ft^2)

	Treatment				
Block	*A*	*B*	*C*	*D*	*E*
1	438	538	77	17	18
2	442	422	61	31	26
3	319	377	157	87	77
4	380	315	52	16	20
Mean, M	395	413	87	38	35
Range, R	123	223	105	71	59
$R/\sqrt{M}$	6.2	11.0	11.2	11.5	10.0
R/M	0.31	0.54	1.21	1.87	1.68

TABLE 15.11.2
SQUARE ROOTS OF THE POPPY NUMBERS IN TABLE 11.15.1

Block	*A*	*B*	*C*	*D*	*E*
1	20.9	23.2	8.8	4.1	4.2
2	21.0	20.5	7.8	5.6	5.1
3	17.9	19.4	12.5	9.3	8.8
4	19.5	17.7	7.2	4.0	4.5
Mean	19.8	20.2	9.1	5.8	5.6
Range	3.1	5.5	5.3	5.3	4.6

Source of Variation	Degrees of Freedom	Sum of Squares	Mean Square
Blocks	3	22.65	
Treatments	4	865.44	216.36
Error	12	48.69	4.06

In table 15.11.2 the square roots of the numbers are recorded and analyzed. The ranges in the several treatments are now similar. That there are differences among treatments is obvious; it is unnecessary to compute F. The 5% LSD value is 3.09, suggesting that D and E are superior to C, while the C, D, E group is much superior to A and B in reducing the numbers of undesired poppies.

Investigators sometimes like to report results in the original scale, feeling that they will be more easily understood on this scale. The means in the square root scale are reconverted to the original scale by squaring. This gives $19.8^2 = 392$ plants for A, $20.2^2 = 408$ plants for B, and so on. These values are slightly lower than the original means, 395 for A, 413 for B, etc., because the mean of a set of square roots is less than the square root of the original mean. As a rough correction (22) for this discrepancy, add to each reconverted mean (19.8^2, etc.) the quantity $(n - 1)s^2/n$, where n is the number of replications, 4, and s^2 is the error mean square in square roots, 4.06. In this example we add $(3)(4.06)/4 = 3$.

Since transformations change the meaning of additivity, the data in table 15.11.2 were tested for nonadditivity. The F value for nonadditivity did not reach the 5% level.

EXAMPLE 15.11.1—The numbers of wireworms counted in the plots of a Latin square (19) following soil fumigations in the previous year were:

	1		2		3		4		5	
1	*P*	3	*O*	2	*N*	5	*K*	1	*M*	4
2	*M*	6	*K*	0	*O*	6	*N*	4	*P*	4
3	*O*	4	*M*	9	*K*	1	*P*	6	*N*	5
4	*N*	17	*P*	8	*M*	8	*O*	9	*K*	0
5	*K*	4	*N*	4	*P*	2	*M*	4	*O*	8

Since these numbers are so small, transform to $\sqrt{X + 1}$. The first number, 3, becomes $\sqrt{3 + 1} = 2$, etc. Compute an analysis of variance on the transformed scale. Ans. Mean square for treatments, 1.4457; for error, 0.3259; $F = 4.44$ ($P < 0.025$).

EXAMPLE 15.11.2—Calculate the Studentized range $D_{0.05} = 1.15$ and show that K gave significantly fewer wireworms than M, N, and O.

EXAMPLE 15.11.3—Estimate the average numbers of wireworms per plot for the several treatments. Ans. With no bias correction: K, 0.99; M, 6.08; N, 6.40; O, 5.55; P, 4.38. To make the bias correction, add 0.26, giving $K = 1.25$; $M = 6.34$, etc.

EXAMPLE 15.11.4—If the error variance of X in the original scale is k times the mean of X and if effects are additive in the square root scale, it can be shown that the true error variance in the square root scale is approximately $k/4$. Thus, the value of k can be estimated from the analysis in the square root scale. If k is close to 1, the distribution of errors in the original scale may be close to the Poisson distribution. In example 15.11.1, k is about $4(0.3259) = 1.3$, suggesting that most of the variance in the original scale is of the Poisson type. With the poppy plants (table 15.11.2), k is about 16, indicating a variance much greater than the Poisson.

EXAMPLE 15.11.5—The square roots might be tried in the analysis of the cabbage looper data (table 11.5.1, example 12.13.3). The conclusions are unchanged.

15.12—Arc sine transformation for proportions. The arc sine transformation, also called the *angular* transformation, was developed for binomial proportions. If a_{ij} successes out of n are obtained in the jth replicate of the ith treatment, the proportion $\hat{p}_{ij} = a_{ij}/n$ has variance $p_{ij}(1 - p_{ij})/n$. By means of table A 21, due to C. I. Bliss, we replace $\hat{p}_{ij}$ by the angle whose sine is $\sqrt{\hat{p}_{ij}}$. In the angular scale, proportions near zero or one are spread out so as to increase their variance. If all the error variance is binomial, the error variance in the angular scale is about $821/n$. The transformation does not remove inequalities in variance arising from differing values of n. If the ns vary widely, a weighted analysis in the angular scale is advisable.

With $n < 50$, a zero proportion should be counted as $1/(4n)$ and a 100% proportion as $(n - 1/4)/n$ before transforming to angles. This empirical device, suggested by Bartlett (20), improves the equality of variance in the angles. A more accurate transformation for small n has been tabulated by Mosteller and Youtz (11).

Angles may also be used with proportions that are subject to other sources of variation in addition to the binomial, if it is thought that the variance of $\hat{p}_{ij}$ is some multiple of $p_{ij}(1 - p_{ij})$. However, since this product varies little for p_{ij} lying between 30% and 70%, the angular transformation is scarcely needed if nearly all the observed $\hat{p}_{ij}$ lie in this range. In fact, this transformation is unlikely to produce a noticeable change in the conclusions unless the $\hat{p}_{ij}$ range from near zero to over 30% or from below 70% to 100%.

Table 15.12.1, taken from a larger randomized blocks experiment (12), shows the percentages of unsalable ears of corn, the treatments being a control A and three mechanical methods of protecting against damage by corn earworm larvae. The value of n, about 36, was not constant but its variations were fairly small and are ignored. Note that the percents range from 2.1% to 55.5%.

In the analysis of variance of the angles (table 15.12.1), the error mean square was 36.4. Since $821/n = 821/36 = 22.8$, some variation in excess of the binomial may be present. The F value for treatments is large. The 5% LSD for comparing two treatments is 7.4. Treatments B, C, and D were all superior to the

TABLE 15.12.1
PERCENTAGE OF UNSALABLE EARS OF CORN

	Block							
Treatments	1	2	3	4	5	6		
A	42.4	34.3	24.1	39.5	55.5	49.1		
B	33.3	33.3	5.0	26.3	30.2	28.6		
C	8.5	21.9	6.2	16.0	13.5	15.4		
D	16.6	19.3	16.6	2.1	11.1	11.1		
	Angle = Arc Sin $\sqrt{\text{Proportion}}$						Mean	%
A	40.6	35.8	29.4	38.9	48.2	44.5	39.6	40.6
B	35.2	35.2	12.9	30.9	33.3	32.3	29.9	24.9
C	17.0	27.9	14.4	23.6	21.6	23.1	21.3	13.2
D	24.0	26.1	24.0	8.3	19.5	19.5	20.2	11.9

Analysis of Variance in Angles

	Degrees of Freedom	Sum of Squares	Mean Square
Blocks	5	359.8	
Treatments	3	1458.5	486.2
Error	15	546.1	36.4
Total	23	2364.4	

control *A*, while *C* and *D* were superior to *B*. The angle means are retranslated to percents at the right side of the table.

15.13—Logarithmic transformation. Logarithms are used to stabilize the variance if the standard deviation in the original scale varies directly as the mean, in other words, if the coefficient of variation is constant. There are mathematical reasons why this type of relation between standard deviation and mean is likely to be found when the effects are *proportional* rather than *additive*—for example, when treatment 2 gives results consistently 23% higher than treatment 1 rather than results higher by say 18 units. In this situation the log transformation may bring about both additivity of effects and equality of variance. If some zero values of X occur, $\log(X + 1)$ is often used.

The plankton catches (18) of table 15.13.1 yielded nicely to the log transformation. Logs to base 10 were used. The original ranges and means for the four kinds of plankton were nearly proportional, the ratios of range to mean being 0.99, 0.87, 0.79, and 1.00. After transformation the ranges were almost equal and uncorrelated with the means.

Transforming back, the estimated mean numbers for the four kinds of plankton caught are antilog 2.802 = 634; 1663; 30,200; and 9162. These are *geometric means*.

The means of the logs differ significantly for the four kinds of plankton. The standard deviation of the logarithms is $\sqrt{0.0070} = 0.084$, and the antilogarithm of this number is 1.21. Quoting Winsor and Clark (18, p. 5), "Now a

TABLE 15.13.1
ESTIMATED NUMBERS OF FOUR KINDS OF PLANKTON CAUGHT IN SIX HAULS WITH EACH OF TWO NETS

	Estimated Numbers				Logarithms			
Haul	I	II	III	IV	I	II	III	IV
1	895	1520	43,300	11,000	2.95	3.18	4.64	4.04
2	540	1610	32,800	8,600	2.73	3.21	4.52	3.93
3	1020	1900	28,800	8,260	3.01	3.28	4.46	3.92
4	470	1350	34,600	9,830	2.67	3.13	4.54	3.99
5	428	980	27,800	7,600	2.63	2.99	4.44	3.88
6	620	1710	32,800	9,650	2.79	3.23	4.52	3.98
7	760	1930	28,100	8,900	2.88	3.29	4.45	3.95
8	537	1960	18,900	6,060	2.73	3.29	4.28	3.78
9	845	1840	31,400	10,200	2.93	3.26	4.50	4.01
10	1050	2410	39,500	15,500	3.02	3.38	4.60	4.19
11	387	1520	29,000	9,250	2.59	3.18	4.46	3.97
12	497	1685	22,300	7,900	2.70	3.23	4.35	3.90
Mean	671	1701	30,775	9,396	2.802	3.221	4.480	3.962
Range	663	1480	24,400	9,440	0.43	0.39	0.36	0.41

Analysis of Variance of Logarithms

Source of Variation	Degrees of Freedom	Sum of Squares	Mean Square
Kind of plankton	3	20.2070	6.7357
Haul	11	0.3387	0.0308
Residuals	33	0.2300	0.0070

deviation of 0.084 in the logarithms of the catch means that the catch has been multiplied (or divided) by 1.21. Hence we may say that one standard deviation in the logarithm corresponds to a percentage standard deviation, or coefficient of variation, of 21% in the catch."

15.14—Nonadditivity in a Latin square. In a Latin square, suppose that X_{ijk} can be expressed as a quadratic function of α_i, β_j, γ_k when row, column, and treatment effects are not additive. Then by analogy with the argument given in section 15.8, we would expect the residuals d_{ijk} from the additive model to have a linear regression on the quantity

$$d_i d_j + d_i d_k + d_j d_k \tag{15.14.1}$$

Moreover, Tukey (13) has shown that the regression coefficient is approximately $(1 - p)/\bar{X}_{...}$, where p is the power of the X_{ijk} that restores additivity. We could, therefore, compute and test the regression of d_{ijk} on expression (15.14.1).

We shall, instead, illustrate an alternative method of doing the computations, due to Tukey (13), which works well with a computer program that finds estimated values and residuals quickly. Table 15.14.1 comes from an experiment on monkeys (21), the raw data being the number of responses to auditory or visual stimuli administered under five conditions ($A, \ldots, E$). Each pair of

TABLE 15.14.1

LOGS OF NUMBERS OF RESPONSES BY PAIRS OF MONKEYS UNDER FIVE STIMULI
(test of additivity in a Latin square)

Pair		Week 1		2		3		4		5	$\overline{X}_{i..}$
1	*B*	1.99	*D*	2.25	*C*	2.18	*A*	2.18	*E*	2.51	2.222
$\hat{X}_{ijk}$		2.022		2.268		2.220		2.084		2.518	
d_{ijk}		−0.032		−0.018		−0.040		0.098†		−0.008	
U_{ijk}		37		3		0		17		92	
2	*D*	2.00	*B*	1.85	*A*	1.79	*E*	2.14	*C*	2.31	2.018
		1.950		1.932		1.852		2.152		2.206	
		0.052†		−0.082		−0.062		−0.012		0.104	
		70		80		132		4		0	
3	*C*	2.17	*A*	2.10	*E*	2.34	*B*	2.20	*D*	2.40	2.242
		2.132		2.082		2.348		2.178		2.472	
		0.038		0.018		−0.006†		0.022		−0.072	
		7		18		18		1		66	
4	*E*	2.41	*C*	2.47	*B*	2.44	*D*	2.53	*A*	2.44	2.458
		2.456		2.462		2.366		2.526		2.482	
		−0.046		0.010†		0.074		0.004		−0.042	
		58		61		23		97		71	
5	*A*	1.85	*E*	2.32	*D*	2.21	*C*	2.05	*B*	2.25	2.136
		1.862		2.248		2.176		2.162		2.234	
		−0.012		0.072		0.034		−0.112		0.018†	
		125		1		2		3		0	
$\overline{X}_{.j.}$		2.084		2.198		2.192		2.220		2.382	2.215
		A		*B*		*C*		*D*		*E*	
$\overline{X}_{..k}$		2.072		2.146		2.236		2.278		2.344	

†Adjusted to make the deviations add to zero over every row, column, and treatment.

monkeys received one type of stimulus per week, the order from week to week being determined by the randomized columns of the Latin square.

The standard deviation of the number of responses was discovered to be almost directly proportional to the mean, so the counts were transformed to logs. Each entry in the table is the mean of the log counts for the two members of a pair. Are the effects of stimuli, weeks, and monkey pairs reasonably close to additive? The steps follow.

1. Find the row, column, and treatment means, as shown, and the fitted values $\hat{X}_{ijk}$ by the additive model

$$\hat{X}_{ijk} = \overline{X}_{i..} + \overline{X}_{.j.} + \overline{X}_{..k} - 2\overline{X}_{...}$$

For E in row 2, column 4,

$$\hat{X}_{245} = 2.018 + 2.220 + 2.344 - 2(2.215) = 2.152$$

2. Find the residuals $d_{ijk} = X_{ijk} - \hat{X}_{ijk}$ as shown, adjusting if necessary so that the sums are zero over every row, column, and treatment. Adjusted values are denoted by † in table 15.14.1.

3. Construct the 25 values of a variate $U_{ijk} = c_1(\hat{X}_{ijk} - c_2)^2$, where c_1 and c_2 are any two convenient constants. We took $c_2 = \overline{X}_{...} = 2.215$, which is often suitable, and $c_1 = 1000$, so the Us are mostly between 0 and 100. For B in row 1, column 1,

$$U_{112} = 1000(2.022 - 2.215)^2 = 37$$

4. Calculate the regression coefficient of the d_{ijk} on the residuals of the U_{ijk}. Since $\Sigma d_{ijk}\hat{U}_{ijk}$ is easily shown to be zero, the numerator of the regression coefficient is

$$N = \Sigma d_{ijk}U_{ijk} = (-0.032)(37) + \ldots + (0.018)(0) = -20.356$$

The denominator D is the *residual* or *error* sum of squares of the U_{ijk}. It is found by performing the ordinary Latin square analysis of the U_{ijk}. The value of D is 22,330.

5. To perform the test for additivity, find the sum of squares, 0.0731, of the d_{ijk}, which equals the error sum of squares of the X_{ijk}. The contribution due to nonadditivity is $N^2/D = (-20.356)^2/22{,}330 = 0.0186$. Finally, compare the mean square for nonadditivity with the remainder mean square.

	df	Sum of Squares	Mean Square	F
Error SS	12	0.0731		
Nonadditivity	1	.0186	0.0186	3.76 ($P = 0.08$)
Remainder	11	0.0545	0.00495	

The value of P is 0.08—a little low, though short of the 5% level. Since the interpretations are not critical (examples 15.14.1, 15.14.2), the presence of slight nonadditivity should not affect them.

The above procedure applies also in more complex classifications. Note that if we expand the quadratic $c_1(\hat{X}_{ijk} - \overline{X}_{...})^2$ in U_{ijk}, the coefficient of terms like $(\overline{X}_{i..} - \overline{X}_{...})(\overline{X}_{.j.} - \overline{X}_{...})$ is $2c_1$. Hence the regression coefficient B of sections 15.8 and 15.9 is $B = 2c_1N/D$. If a power transformation is needed, the suggested power is as before $p = 1 - B\overline{X}_{...}$.

EXAMPLE 15.14.1—Analyze the variance of the logarithms of the monkey responses. The analysis follows:

	df	Sum of Squares	Mean Square	F
Monkey pairs	4	0.5244	0.1311	
Weeks	4	.2294	.0574	9.5
Stimuli	4	.2313	.0578	9.6
Error	12	0.0725	0.00604	

EXAMPLE 15.14.2—Test all differences among the stimuli means in table 15.14.1, using the LSD method. Ans. $E > A, B, C$; $D > A, B$; $C > A$.

EXAMPLE 15.14.3—Calculate the sum of squares in table 15.14.1 due to the regression of log response on weeks. It is convenient to code the weeks as $X = -2, -1, 0, 1, 2$. Then, taking the weekly means as Y, $\Sigma xy = 0.618$ and $(\Sigma xy)^2/\Sigma x^2 = 0.03819$. In the analysis of variance table, the sum of squares due to regression is $5(0.03819) = 0.1910$. The line for weeks in example 15.14.1 may now be separated into two parts:

Linear regression	1	0.1910	0.1910
Deviations from regression	3	0.0384	0.0128

Comparing the mean squares with error, it is seen that deviations are not significant—most of the sum of squares for weeks is due to regression. Control of the order in which the stimuli were given was important in this experiment, the weeks mean square giving an F of 9.5.

15.15—Simultaneous study of different effects of a transformation. For the kinds of failures in the standard assumptions that are handled by transformation of the scale in which the data are analyzed, the transformation that restores additivity will usually differ from the transformation that improves homogeneity of variance or induces near-normality, as we noted earlier. Given this conflict, there is unlikely to be an ideal transformation for a given body of data. In view of the simplicity and efficiency of an additive model, much can be said for giving primary attention to removal of nonadditivity while keeping an eye on heterogeneity of variance, as illustrated in previous sections. Or if the original scale reveals a relation between variance and mean but no serious nonadditivity, one might try a variance-stabilizing transformation, checking that nonadditivity has not been worsened thereby.

A more balanced strategy is available to the investigator who has access to computer programs that have graphic display facilities and can perform many analyses of variance quickly and cheaply. The technique is to analyze the data in each of a whole series of powers p of the observations. From the results of these analyses, plot each criterion that is considered important against the power p. For instance, if the data provide these tests, we might choose as criteria the F value in a test of nonadditivity, the value of Bartlett's χ^2 or Levene's F for testing heterogeneity of variance, a criterion for testing skewness, and the F ratio for treatments. The values of these criteria are plotted against the power p in the same figure. From this, a choice of the best compromise value of p is made.

For an illustration we are indebted to N. R. Draper and W. G. Hunter (14). Their data consisted of a 3×4 classification with four observations per cell. Earlier the data had been analyzed using maximum likelihood methods by G. E. P. Box and D. R. Cox (15), who proposed and illustrated this approach to the study of transformations. The data are the survival times (in 10-hr units) of animals. Forty-eight animals were divided at random into 12 groups of 4 animals each. A group received one of the 12 combinations of 3 poisons and 4 treatments in a completely randomized design. The analysis of variance of the data uses the following notation.

Source	df	Sum of Squares	Mean Square	F ratio
Poisons, P	2	SSP	MSP	$p = MSP/s^2$
Treatments, T	3	SST	MST	$t = MST/s^2$
Residuals, $P \times T$	6	SSI	MSI	$i = MSI/s^2$
Within groups	36	SSW	s^2	

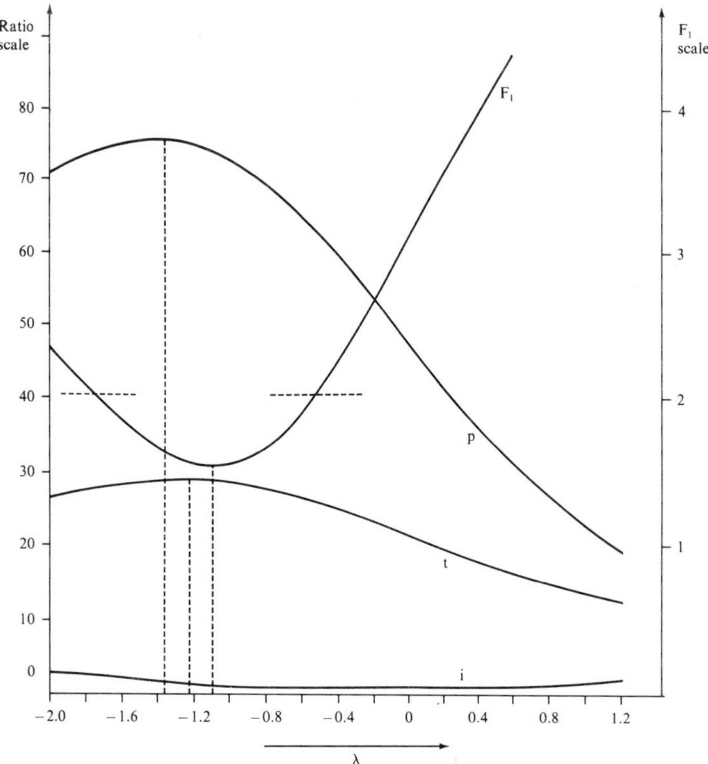

Fig. 15.15.1—A reproduction, by permission, of Figure 1 in Transformations: Some Examples Revisited, *Technometrics* 11 (1969):25, by N. R. Draper and W. G. Hunter.

Draper and Hunter consider four criteria: (*i*) the F ratio i (interactions), which they regard as a measure of the nonadditivity of the effects of poisons and treatments and which they would like to keep small; (*ii*) and (*iii*) the F ratios p and t (for P and T), which measure the power of the experiment in detecting the additive effects of poisons and treatments and which they would like to maximize; and (*iv*) Levene's ratio F_1 for testing the heterogeneity of within-group variances, which they would like to have not much above 1.

Draper and Hunter use λ instead of our p as the power in the transformation of the X_{ijk}. Figure 15.15.1 plots i, p, t, and F_1 against λ for values of λ from -2 to $+1.2$. (The 0 on the left scale for i, p, and t should be on the abscissa, not beside the i curve.) They note that $\lambda = -0.6$ minimizes nonadditivity but does not seem a good choice for p, t, or F_1. Moreover, nonadditivity is not a serious problem here, the F ratio MSI/s^2 being only 1.87 in the original scale. They select $\lambda = -1$ (reciprocals) as a good compromise value. Box and Cox also chose this value from their analysis, noting that the reciprocal scale has a simple interpretation as the *rate of dying*. As the analyses by Box and Cox (15) showed, the effects of this transformation were to increase markedly the F ratios for the additive effects (from 23.3 to 72.9 for P and from 13.8 to 28.4 for T), to reduce but probably not eliminate heterogeneity of within-cell variance, and to reduce the F for $P \times T$ interactions from 1.88 to 1.09.

Clearly, this technique has a number of advantages. The investigator can appraise the original scale with regard to each criterion, examine the effects of a range of powers, and judge whether a satisfactory compromise power can be selected or whether any chosen power still leaves some failure in the assumptions that require a more complex analysis.

TECHNICAL TERMS

arc sine transformation
extreme Studentized deviate (ESD)
maximum normed residual (MNR)
missing data
outlier
row vector
variance-stabilizing transformation

REFERENCES

1. Healy, M., and Westmacott, M. *Appl. Stat.* 5 (1956):203.
2. Rubin, D. B. *Appl. Stat.* 21 (1972):136.
3. Stefansky, W. *Technometrics* 14 (1972):469.
4. Grubbs, F. E. *Technometrics* 11 (1969):1.
5. Rosner, B. *Technometrics* 17 (1975):221.
6. Cochran, W. G. *Biometrics* 3 (1947):22.
7. Ferguson, T. S. *Rev. Inst. Int. Stat.* 3 (1961):29.
8. Bartlett, M. S. *J. R. Stat. Soc.* 3 (1936):68.
9. Tukey, J. W. *Biometrics* 5 (1949):232.
10. Williams, C. B. *Bull. Entomol. Res.* 42 (1951):513.
11. Mosteller, F., and Youtz, C. *Biometrika* 48 (1961):433.
12. Cochran, W. G. *Ann. Math. Stat.* 11 (1940):344.
13. Tukey, J. W. *Biometrics* 11 (1955):111.
14. Draper, N. R., and Hunter, W. G. *Technometrics* 11 (1969):23.
15. Box, G. E. P., and Cox, D. R. *J. R. Stat. Soc.* 26 (B) (1964):211.
16. Yates, F. *Emp. J. Exp. Agric.* 1 (1933):129.
17. Anscombe, F. J., and Tukey, J. W. *Technometrics* 5 (1963):141.
18. Winsor, C. P., and Clarke, G. L. *J. Mar. Res.* 3 (1940):1.
19. Cochran, W. G. *Emp. J. Exp. Agric.* 6 (1938):157.
20. Bartlett, M. S. *Biometrics* 3 (1947):39.
21. Butler, R. A. *J. Exp. Psychol.* 48 (1954):19.
22. Quenouille, M. H. 1953. *The Design and Analysis of Experiments.* Charles Griffin, London.

16

Factorial experiments

16.1—Introduction. A common problem in research is investigating the effects of each of a number of variables, or *factors,* on some response Y. Suppose a company in the food industry proposes to market a cake mix from which a cake can be made by adding water and then baking. The company must determine the best kind of flour and the correct amounts of fat, sugar, liquid (milk or water), eggs, baking powder, and flavoring, as well as the best oven temperature and the proper baking time. Nine factors are present, any one of which may affect the palatability and the keeping quality of the cake to a noticeable degree. Similarly, a research program designed to learn how to increase the yields of the principal cereal crop in a country is likely to try to measure the effects on yield of different amounts of nitrogen, phosphorus, and potassium when added as fertilizers to the soil. Problems of this type occur frequently in industry; with complex chemical processes there can be as many as 10 to 20 factors that may affect the final product.

In earlier times the advice was sometimes given to study one factor at a time, a separate experiment being devoted to each factor. Later, Fisher (1) pointed out that important advantages are gained by combining the study of several factors in the same *factorial experiment.* Factorial experimentation is highly efficient, because every observation supplies information about all the factors included in the experiment. Also, factorial experimentation is a systematic method of investigating the relationships between the effects of different factors.

16.2—The single-factor approach. To illustrate the difference between the one-factor-at-a-time approach and the factorial approach, consider an investigator who has two factors, A and B, to study. For simplicity, suppose that only two *levels* of each factor, say a_1, a_2 and b_1, b_2 are to be compared. In a cake mix, a_1, a_2 might be two types of flour and b_1, b_2 two amounts of flavoring. Four replications are considered sufficient by the investigator.

In the single-factor approach, the first experiment is a comparison of a_1 with a_2. The level of B is kept constant in the first experiment; the investigator must decide what this constant level is to be. We shall suppose that B is kept at b_1; the choice does not affect our argument. The two treatments in the first experiment may be denoted by the symbols a_1b_1 and a_2b_1, replicated four times.

The effect of A, that is, the mean difference $\overline{a_2b_1} - \overline{a_1b_1}$, is estimated with variance $2\sigma^2/4 = \sigma^2/2$.

The second experiment compares b_2 with b_1. If a_2 performed better that a_1 in the first experiment, the investigator is likely to use a_2 as the constant level of A in the second experiment (again, this choice is not vital to the argument). Thus, the second experiment compares a_2b_1 with a_2b_2 in four replications and estimates the effect of B with variance $\sigma^2/2$.

In the two single-factor experiments, 16 observations have been made, and the effects of A and B have each been estimated with variance $\sigma^2/2$.

But suppose that someone else, interested in these factors, hears that experiments on them have been done and asks the investigator, "In my work, I have to keep A at its lower level, a_1. What effect does B have when A is at a_1?" Obviously, the investigator cannot answer this question, since the effect of B was measured only when A was held at its higher level. Another person might ask, "Is the effect of A the same at the two levels of B?" Once again the investigator has no answer, since A was tested at only one level of B.

16.3—The factorial approach. In the factorial approach, the investigator compares all treatments that can be formed by combining the levels of the different factors. The four such treatment combinations are a_1b_1, a_2b_1, a_1b_2, a_2b_2. Note that each replication of this experiment supplies *two* estimates of the effect of A. The comparison $a_2b_2 - a_1b_2$ estimates the effect of A when B is held constant at its higher level, while the comparison $a_2b_1 - a_1b_1$ estimates the effect of A when B is held constant at its lower level. These effects have been called *simple effects*. The average of these two estimates is called the *main effect* of A, the adjective *main* being a reminder that this average is taken over the levels of the other factor. In terms of our definition of a comparison or contrast (section 12.8) the main effect of A may be expressed as

$$L_A = (1/2)(a_2b_2) + (1/2)(a_2b_1) - (1/2)(a_1b_2) - (1/2)(a_1b_1) \qquad (16.3.1)$$

where a_2b_2 denotes the yield given by the treatment combination a_2b_2 (or the average yield if the experiment has r replications) and so on. By Rule 12.8.1 the variance of L_A is

$$(\sigma^2/r)[(1/2)^2 + (1/2)^2 + (1/2)^2 + (1/2)^2] = \sigma^2/r$$

If the investigator uses two replications (8 observations), the main effect of A is estimated with variance $\sigma^2/2$.

Now consider B. Each replication furnishes two estimates, $a_2b_2 - a_2b_1$ and $a_1b_2 - a_1b_1$, of the effect of B. The main effect of B is the comparison

$$L_B = (1/2)(a_2b_2) + (1/2)(a_1b_2) - (1/2)(a_2b_1) - (1/2)(a_1b_1) \qquad (16.3.2)$$

With two replications of the factorial experiments (8 observations), L_B *like* L_A *has variance* $\sigma^2/2$.

Thus, the factorial experiment requires only 8 observations, as against 16 by the single-factor approach, to estimate the effects of A and B with the same variance $\sigma^2/2$. With three factors, the factorial experiment requires only 1/3 as many observations, with four factors only 1/4, and so on. These striking gains in efficiency occur because every observation, like a_2b_1, $a_1b_2c_2$, or $a_2b_1c_1d_2$, is used in the estimate of the effect of every factor. In the single-factor approach, as noted, an observation supplies information only about the effect of one factor.

What about the relationship between the effects of the factors? The factorial experiment provides a separate estimate of the effects of A at each level of B, though these estimates are less precise than the main effect of A, with variance σ^2. The question, Is the effect of A the same at the two levels of B? can be examined by means of the comparison:

$$(a_2b_2 - a_1b_2) - (a_2b_1 - a_1b_1) \qquad (16.3.3)$$

This expression measures the difference between the effect of A when B is at its higher level and the effect of A when B is at its lower level. If the question is, Does the level of A influence the effect of B? the relevant comparison is

$$(a_2b_2 - a_2b_1) - (a_1b_2 - a_1b_1) \qquad (16.3.4)$$

Notice that (16.3.3) and (16.3.4) are identical. The expression is called the AB *two-factor interaction.* In this, the combinations a_2b_2 and a_1b_1 receive a plus sign, the combinations (a_2b_1) and (a_1b_2) a minus sign.

Because of its efficiency and comprehensiveness, factorial experimentation is extensively used in research programs, particularly in industry. One limitation is that a factorial experiment is usually larger and more complex than a single-factor experiment. The potentialities of factorial experimentation in clinical medicine, for example, have not been fully exploited because it is usually difficult to find enough suitable patients to compare more than two or three treatment combinations.

16.4—Analysis of the 2^2 factorial experiment. In analyzing the results of a 2^2 factorial, the commonest procedure is to look first at the two main effects and the two-factor interaction. If the interaction seems absent, we need only report the main effects, with some assurance that each effect holds at either level of the other variate.

A compact notation for describing treatment combinations is also common. The presence of a letter *a* or *b* denotes one level of the factor in question, while the absence of the letter denotes the other level. Thus, a_2b_2 becomes *ab*, and a_1b_2 becomes *b*. The combination a_1b_1 is denoted by the symbol (1). In this notation, table 16.4.1 shows how to compute the main effects and the interaction from the treatment *totals* over *r* replications.

The quantities $[A]$, $[B]$, and $[AB]$ given by the sums of products in table 16.4.1 are called the *factorial effect totals.* The *factorial effect mean*—the main effect—of A is

$$[A]/(2r) = [(ab) - (b) + (a) - (1)]/(2r)$$

TABLE 16.4.1
CALCULATION OF MAIN EFFECTS AND INTERACTION IN A 2^2 FACTORIAL

Factorial Effect	Multiplier for Treatment Total (1)	a	b	ab	Divisor for Mean	Contribution to Treatments SS
A	−1	1	−1	1	$2r$	$[A]^2/(4r)$
B	−1	−1	1	1	$2r$	$[B]^2/(4r)$
AB	1	−1	−1	1	$2r$	$[AB]^2/(4r)$

Use of the same divisor ($2r$) for the AB interaction mean, as in table 16.4.1, is a common convention. For interpretation it seems more natural, however, to divide by r, since the AB mean then estimates the difference between the effect of a when b is present and its effect when b is absent.

A preliminary examination of the data can proceed as follows. By the analysis of variance appropriate to the design of the experiment, calculate the error mean squares, s^2. Then calculate the factorial effect means as in table 16.4.1. Their standard errors are $s/\sqrt{r}$. From these results you can judge the sizes and the statistical significance of the main effects and the interaction. If there seems no sign of interaction, the report of results can concentrate on the main effects.

In table 16.4.1, note that the three comparisons $[A]$, $[B]$, and $[AB]$ are orthogonal. The treatments sum of squares in the analysis of variance can therefore be broken down into single components $[A]^2/(4r)$, $[B]^2/(4r)$, and $[AB]^2/(4r)$, each with 1 df.

The case where no interaction appears is illustrated by an experiment (3) on the fluorometric determination of the riboflavin content of dried collard leaves (table 16.4.2). The two factors were A, the size of sample (0.25 g, 1.00 g) from which the determination was made, and B, the effect of the inclusion of a permanganate-peroxide clarification step in the determination. This was a randomized blocks design replicated on three successive days.

The usual analysis of variance into replications, treatments, and error is computed. Then the factorial effect totals for A, B, and AB are calculated from the treatment totals, using the multipliers given in table 16.4.2. Their squares are divided by $4r$, or 12, to give the contributions to the treatments sum of squares. The P value corresponding to the F ratio 13.02/8.18 for interaction is about 0.25; we shall assume interaction absent. Consequently, attention can be concentrated on the main effects. The permanganate step produced a large reduction in the estimated riboflavin concentration. The effect of sample size was not quite significant.

Instead of subdividing the treatments sum of squares and making F tests, one can proceed directly to compute the factorial effect means. These are obtained by dividing the effect totals by $2r$, or 6, and are shown in table 16.4.2 beside the effect totals. The standard error of an effect mean is $\sqrt{s^2/r} = \sqrt{2.73} = 1.65$. The t tests of the effect means are of course the same as the F tests in the analysis of variance. Use of the effect means has the advantage of showing the magnitude and direction of the effects.

TABLE 16.4.2
APPARENT RIBOFLAVIN CONCENTRATION (MCG/G) IN COLLARD LEAVES

Replication	Without *B* 0.25 g	Without *B* 1.00 g	With *B* 0.25 g	With *B* 1.00 g	Total
1	39.5	38.6	27.2	24.6	129.9
2	43.1	39.5	23.2	24.2	130.0
3	45.2	33.0	24.8	22.2	125.2
Total	127.8	111.1	75.2	71.0	

	(1)	*a*	*b*	*ab*	Factorial Effect Total	Factorial Effect Mean	Factorial Effect *SE*
Sample Size, *A*	−1	1	−1	1	−20.9	− 3.5	± 1.65
Permanganate, *B*	−1	−1	1	1	−92.7	−15.4	
Interaction, *AB*	1	−1	−1	1	12.5	2.1	

Source of Variation	df	Sum of Squares	Mean Square	*P*
Replications	2	3.76		
Treatments	(3)	(765.53)		
sample size	1	$(-20.9)^2/12 =$ 36.40	36.40	0.08
permanganate	1	$(-92.7)^2/12 =$ 716.11	716.11	< 0.01
interaction	1	$12.5^2/12 =$ 13.02	13.02	0.25
Error	6	49.08	8.18	

The principal conclusion from this experiment was: "In the fluorometric determination of riboflavin of the standard dried collard sample, the permanganate–hydrogen peroxide clarification step is essential. Without this step, the mean value is 39.8 mcg/g, while with it the more reasonable mean of 24.4 is obtained." These data are discussed further in example 16.5.3.

16.5—The 2^2 factorial when interaction is present. When interaction is present, the results of a 2^2 experiment require more detailed study. If both main effects are large, an interaction that is significant but much smaller than the main effects may imply merely that there is a minor variation in the effect of *A* according as *B* is at its higher or lower level, and vice versa. In this event, reporting of the main effects may still be an adequate summary. But in most cases we must revert to a report based on the 2 × 2 table.

Table 16.5.1 contains the results (slightly modified) of a 2^2 experiment in a completely randomized design. The factors were vitamin B_{12} (0, 5 mg) and antibiotics (0, 40 mg) fed to swine; the response was the average daily gain in weight. A glance at the totals for the four treatment combinations suggests that with no antibiotics, B_{12} had little or no effect (3.66 versus 3.57), apparently because intestinal flora utilized the B_{12}. With antibiotics present to control the flora, the effect of the vitamin was marked (4.63 versus 3.10). Looking at the table the other way, the antibiotics alone decreased gain (3.10 versus 3.57), perhaps by suppressing intestinal flora that synthesize B_{12}; but with B_{12} added,

TABLE 16.5.1
AVERAGE DAILY GAIN IN WEIGHT OF SWINE (LB)

Antibiotics	0		40 mg	
B_{12}	0	5 mg	0	5 mg
	1.30	1.26	1.05	1.52
	1.19	1.21	1.00	1.56
	1.08	1.19	1.05	1.55
Total (±0.105)	3.57	3.66	3.10	4.63

Source	Degrees of Freedom	Sum of Squares	Mean Squares
Treatments	3	0.4124	0.1375
Error	8	0.0293	0.00366

the antibiotics produced a gain by decreasing the activities of unfavorable flora (4.63 versus 3.66).

A summary of the results of this experiment is therefore presented in a 2 × 2 table of the means of the four treatment combinations. It displays the simple effects of the two factors:

Antibiotics	0	40 mg	Difference (±0.049)
B_{12}, 0	1.19	1.03	−0.16
B_{12}, 5 mg	1.22	1.54	+0.32
Difference (±0.049)	+0.03	+0.51	

The se (simple effect) of each difference is $\sqrt{2s^2/3} = \pm 0.049$. The decrease due to antibiotics when B_{12} is absent and the increases to each additive when the other is present are all clearly significant.

If we begin, instead, by calculating factorial effects, as shown in table 16.5.2, we learn from the effect means that there is a significant interaction at the 1% level (0.240 ± 0.035). This immediately directs attention back to the four individual treatment totals or means in order to study the nature of the interaction and seek an explanation. The main effects both happen to be significant but are of no interest.

EXAMPLE 16.5.1—Yates (2) pointed out that the concept of factorial experimentation can be applied to gain accuracy when weighing objects on a balance with two pans. Suppose that two objects are to be weighed and that in any weighing the balance has an error distributed about 0 with

TABLE 16.5.2
CALCULATION OF FACTORIAL EFFECT TOTALS AND MEANS

	(1)	*a*	*b*	*ab*	Factorial Effect		
Total	3.57	3.66	3.10	4.63	Total	Mean	*SE*
B_{12}	−1	1	−1	1	1.62	0.270**	
Antibiotics	−1	−1	1	1	0.50	0.083*	±0.035
Interaction	1	−1	−1	1	1.44	0.240**	

variance σ^2. If the two objects are weighed separately, the balance estimates each weight with variance σ^2. Instead, both objects are placed in one pan, giving an estimate y_1 of the sum of the weights. Then the objects are placed in different pans, giving an estimate y_2 of the difference between the weights. Show that the quantities $(y_1 + y_2)/2$ and $(y_1 - y_2)/2$ give estimates of the individual weights with variance $\sigma^2/2$.

EXAMPLE 16.5.2—Show how to conduct the weighings of four objects so that the weight of each object is estimated with variance $\sigma^2/4$. Hint: First weigh the sum of the objects; then refer to table16.5.2, in which (1), *a*, *b*, *ab* represent the four objects and $-$ and $+$ represent the two pans.

EXAMPLE 16.5.3—Our use of the riboflavin data in section 16.4 as an example with no interaction might be critized on two grounds: (1) a *P* value of 0.25 in the test for interaction in a small experiment suggests the possibility of an interaction that a larger experiment might reveal and (2) perhaps the effects are multiplicative in these data. If you analyze the logs of the data in table 16.4.1, you will find that the *F* value for interaction is now only 0.7. Thus the assumption of zero interaction seems better grounded on a log scale than on the original scale.

EXAMPLE 16.5.4—In a randomized blocks experiment on sugar beets in Iowa with four replicates, the treatment totals for the numbers of surviving plants were as follows: no treatment, 904; superphosphate (P), 1228; potash (K), 943; and (P + K), 1246. The error mean square was 1494 (9 df). (*i*) Write a brief report on the results. (*ii*) Calculate the treatments sum of squares and the sum of squares due to P, K, and the PK interaction, verifying that they add to the treatments sum of squares. Ans. (*i*) The only effect was a highly significant increase of about 34% in plant number due to P. This result is a surprise, since P seldom shows this effect. (*ii*) Sums of squares: P, 24,571; K, 203; PK, 28.

EXAMPLE 16.5.5—If (A), (B), and (AB) denote the factorial effect means and M the general mean, show that we can recapture the means (1), (a), (b), and (ab) of the individual treatment combinations by the following equations:

$$(ab) = M + (1/2)[(A) + (B) + (AB)]$$

$$(a) = M + (1/2)[(A) - (B) - (AB)]$$

$$(b) = M + (1/2)[-(A) + (B) - (AB)]$$

$$(1) = M + (1/2)[-(A) - (B) + (AB)]$$

These equations also provide estimates of the treatment-combination means when we make assumptions about the factorial effects. Thus, if (AB) is considered negligible, we estimate (a) as $M + (1/2)[(A) - (B)]$.

16.6—The general two-factor experiment. Leaving the special case of two levels per factor, we now consider the general arrangement with *a* levels of the first factor and *b* levels of the second. As before, the layout of the experiment may be completely randomized, randomized blocks, or any other standard plan.

With *a* levels, the main effects of *A* in the analysis of variance now have $(a - 1)$ df, while those of *B* have $(b - 1)$ df. Since there are *ab* treatment combinations, the treatments sum of squares has $(ab - 1)$ df. Consequently, there remain

$$(ab - 1) - (a - 1) - (b - 1) = ab - a - b + 1 = (a - 1)(b - 1) \text{ df}$$

which may be shown to represent the *AB* interactions. In the 2×2 factorial, in which the *AB* interaction had only 1 df, the comparison corresponding to this degree of freedom was called *the AB* interaction. In the general case, the AB interaction represents a set of $(a - 1)(b - 1)$ independent comparisons. They can be subdivided into single comparisons in many ways.

In subdividing the AB sum of squares, the investigator is guided by the questions at hand when planning the experiment. Any comparison among the levels of A is estimated independently at each of the b levels of B. For a comparison that is of particular interest, the investigator may wish to examine whether the level of B affects these estimates. The sum of squares of deviations of the estimates, with the appropriate divisor, is a component of the AB interaction, with $(b - 1)$ df, which may be isolated and tested against the error mean square. Incidentally, since the main effect of A represents $a - 1$ independent comparisons, these components of the AB interaction jointly account for $(a - 1)(b - 1)$ df and will be found to sum to the sum of squares for AB.

As an illustration, the data in table 16.6.1 show the gains in weight of male rats under six feeding treatments in a completely randomized experiment. The factors were: A (3 levels): source of protein—beef, cereal, pork; and B (2 levels): level of protein—high, low.

Often the investigator has decided in advance how to subdivide the comparisons among the treatment means. In more exploratory situations, we often start with a breakdown of the treatments sum of squares into the sum of squares for A, B, and AB; see table 16.6.1. Looking at the main effects of A, the three sources of protein show no differences in average rates of gain ($F = 0.6$); but there is a clear effect of level of protein ($F = 14.8$), the gain being about 18% larger with the high level.

TABLE 16.6.1
GAINS IN WEIGHT (G) OF RATS UNDER SIX DIETS

	High Protein			Low Protein		
	Beef	Cereal	Pork	Beef	Cereal	Pork
	73	98	94	90	107	49
	102	74	79	76	95	82
	118	56	96	90	97	73
	104	111	98	64	80	86
	81	95	102	86	98	81
	107	88	102	51	74	97
	100	82	108	72	74	106
	87	77	91	90	67	70
	117	86	120	95	89	61
	111	92	105	78	58	82
Total	1000	859	995	792	839	787

Source of Variation	df	Sum of Squares	Mean Square	F
Treatments	5	4,613.0		
A (source of protein)	2	266.5	133.2	0.6
B (level of protein)	1	3,168.3	3168.3	14.8**
AB	2	1,178.2	589.1	2.7
Error	54	11,585.7	214.6	

For AB, the value of F is 2.7, between the 10% and the 5% level. In the general two-factor experiment and in more complex factorials it often happens that a few of the comparisons comprising the main effects have substantial interactions, while the majority of the comparisons have negligible interactions. Consequently, the F test of the AB interaction sum of squares as a whole is not a good guide as to whether interactions can be ignored. It is wise to look over the two-way table of treatment totals or means before concluding that there are no interactions, particularly if F is larger than 1.

Another working rule tested by experience in a number of areas is that large main effects are more likely to have interactions than small ones. Consequently, we look particularly at the effects of B, level of protein. From the treatment totals in table 16.6.1 we see that high protein gives large gains over low protein for beef and pork but only a small gain for cereal. This suggests a breakdown into: (1) cereal versus the average of beef and pork and (2) beef versus pork. This subdivision is natural, since beef and pork are animal sources of protein and cereal is a vegetable source, and would probably be planned from the beginning in this type of experiment.

Table 16.6.2 shows how this breakdown is made by means of five single comparisons. Study the coëfficients for each comparison carefully and verify that the comparisons are mutually orthogonal. In the lower part of the table, the divisors required to convert the squares of the factorial effect totals into sums of squares in the analysis of variance are given. Each divisor is n times the sum of squares of the coefficients in the comparison ($n = 10$). As anticipated, the interaction of the animal versus vegetable comparison with level of protein is significant at the 5% level. There is no sign of a difference between beef and pork at either level.

TABLE 16.6.2

SUBDIVISION OF THE *SS* FOR MAIN EFFECTS AND INTERACTIONS

	High Protein			Low Protein			
Comparisons (treatment totals)	Beef 1000	Cereal 859	Pork 995	Beef 792	Cereal 839	Pork 787	Factorial Effect Total
Level of protein	+1	+1	+1	−1	−1	−1	436
Animal vs. vegetable	+1	−2	+1	+1	−2	+1	178
Interaction with level	+1	−2	+1	−1	+2	−1	376
Beef vs. pork	+1	0	−1	+1	0	−1	10
Interaction with level	+1	0	−1	−1	0	+1	0

Comparison	Divisor for *SS*	Degrees of Freedom	Sum of Squares	Mean Square
Level of protein	60	1	3168.3**	
Animal vs. vegetable	120	1	264.0	
Interaction with level	120	1	1178.1*	
Beef vs. pork	40	1	2.5	
Interaction with level	40	1	0.0	
Error		54		214.6

The principal results can be summarized in the following 2 × 2 table of mean rat gains in weight per week (g).

Level of Protein	Source of Protein: Animal	Vegetable	Difference	*SE*
High	99.8	85.9	+13.9*	±5.67
Low	79.0	83.9	− 4.9	±5.67
Difference	+20.8**	+ 2.0		
SE	± 4.6	± 6.5		

As a consequence of the interaction, the animal proteins gave substantially greater gains in weight than cereal protein at the high level but showed no superiority to cereal protein at the low level.

16.7—Polynomial response curves. The levels of a factor frequently represent increasing amounts X of some substance. It may then be of interest to examine whether the response Y to the factor has a linear relation to the amount X. An example has already been given in section 14.10, in which the linear regression of yield of millet on width of spacing of the rows was worked out for a Latin square experiment. If the relation between Y and X is curved, a more complex mathematical expression is required to described it. Sometimes the form of this expression is suggested by subject matter knowledge. Failing this, a polynomial in X is often used as a descriptive equation.

With equally spaced levels of X, auxiliary tables (table A19) are available that facilitate the fitting of these polynomials. An introduction is given here to enable them to be used in the analysis of factorial experiments. The tables are based essentially on an ingenious coding of the values of X, X^2, and so on, which makes the successive power orthogonal.

With three levels, the values of X are coded as −1, 0, + 1 so that they sum to zero. If Y_1, Y_2, Y_3 are the corresponding response *totals* over n replicates, the linear regression coefficient b_1 is $\Sigma XY/(n\Sigma X^2)$ or $(Y_3 - Y_1)/(2n)$. The values of X^2 are 1, 0, 1. Subtracting their mean, 2/3, so that they add to zero gives 1/3, −2/3, 1/3. Multiplying by 3 to obtain whole numbers, we have coefficients 1, −2, 1. In its coded form, this variable is $X_2 = 3X^2 - 2$. The regression coefficient of Y on X_2 is $b_2 = \Sigma X_2 Y/(n\Sigma X_2^2)$ or $(Y_3 - 2Y_2 + Y_1)/(6n)$. The equation for the parabola fitted to the level *means* of Y is

$$\hat{Y} = \bar{Y} + b_1 X + b_2 X_2 \tag{16.7.1}$$

With four levels, the X are coded −3, −1, +1, +3 so that they are whole numbers adding to zero. The values of X^2 are 9, 1, 1, 9 with mean 5. Subtracting the mean gives +4, −4, −4, +4, which we divide by 4 to give the coefficients +1, −1, −1, +1 for the parabolic component. These components represent the variable $X_2 = (X^2 - 5)/4$. The fitted parabola has the same form as (16.7.1), where

$$b_1 = (3Y_4 + Y_3 - Y_2 - 3Y_1)/(20n) \qquad b_2 = (Y_4 - Y_3 - Y_2 + Y_1)/(4n)$$

TABLE 16.8.2
LINEAR AND QUADRATIC COMPONENTS FOR EACH VARIETY IN COWPEA EXPERIMENT

	4″	8″	12″	Totals for Components	
Linear	−1	0	+1		
Quadratic	+1	−2	+1	Linear	Quadratic
Variety I	190	203	223	33	7
Variety II	249	234	209	−40	−10
Variety III	224	257	292	68	2
Total	663	694	724	SE ± 11.9	±20.6

VS (variety × spacing) interactions. The table of treatment combination totals immediately above shows an upward trend in yield with wider spacing for varieties I and III but an opposite trend with variety II. This presumably accounts for the large *VS* mean square and warns that no useful overall statements can be made from the main effects.

To examine the trends of yield Y on spacing X, the linear and quadratic components are calculated for each variety. The factorial effect totals for these components are shown with their standard errors in table 16.8.2. Note the following results, from table 16.8.2, given by t tests:

(*i*) As anticipated, the linear slopes are positive for varieties I and III and negative for variety II.

(*ii*) The linear trend for each variety is significant at the 1% level, while no variety shows any sign of curvature.

(*iii*) If the upward trends for varieties I and III are compared, the trend for variety III will be significantly greater.

To summarize, the varieties have different linear trends on spacing. Apparently I and III have heavy vegetative growth that requires more than 12 in. spacing for maximum yield. In further experiments the spacings tested for varieties I and III should differ from those for II.

As an illustration, table 16.8.3 shows how the sums of squares for the linear and quadratic terms add to the combined sum of squares for spacings and interactions in table 16.8.1.

EXAMPLE 16.8.1—In the variety × spacing experiment, verify the statement that the linear regression of yield on width of spacing is significantly greater for variety III than for variety I.

TABLE 16.8.3
CONTRIBUTIONS TO SUMS OF SQUARES

	Linear	Quadratic
Variety I	$33^2/[(4)(2)] = 136.12^{**}$	$7^2/[(4)(6)] = 2.04$
II	$(-40)^2/[(4)(2)] = 200.00^{**}$	$(-10)^2/[(4)(6)] = 4.17$
III	$68^2/[(4)(2)] = 578.00^{**}$	$2^2/[(4)(6)] = 0.17$
Total	914.12	6.38

Verification: $914.12 + 6.38 = 155.06 + 765.44$ ($= S + VS$), df $= 6$

EXAMPLE 16.8.2—If the primary interest in this experiment were in comparing the varieties when each has its highest yielding spacing, we might compare the totals 223 (I), 249 (II), and 292 (III). Show that the optimum for III exceeds the others at the 1% level.

16.9—An example with both factors quantitative. We turn now to a 3 × 4 experiment in which there is regression in each factor. The data are from the Foods and Nutrition Section of the Iowa Agricultural Experiment Station (6). The object was to learn about losses of ascorbic acid in snap beans stored at 3 temperatures for 4 periods, each 2 weeks longer than the preceding. The beans were all harvested under uniform conditions before eight o'clock one morning. They were prepared and quick-frozen before noon of the same day. Three packages were assigned at random to each of the 12 treatments and all packages were stored at random positions in the locker, a completely randomized design with three replications.

The totals of three ascorbic acid determinations are recorded in table 16.9.1. The error mean square was 0.706 with 24 df. It is clear that the concentration of ascorbic acid decreases with higher storage temperatures and with storage time except at 0°. For practical purposes one might conclude simply that to preserve the ascorbic acid (vitamin C), it is best to store at 0°F; at this temperature there is no loss of vitamin C for storage up to 8 weeks.

As an exercise in fitting a response surface, we shall examine in more detail the combined effects of storage length and temperature. It looks from table 16.9.1 as if the rate of decrease of ascorbic acid with increasing temperature is not linear and is not the same for the several storage periods. We first calculate (table 16.9.2) the linear and quadratic temperature components T_L and T_Q at each storage period, the multipliers being $(-1, 0, +1)$ and $(+1, -2, +1)$.

For the *main* effects of temperature, the sums of squares for the analysis of variance are as follows:

TABLE 16.9.1
SUM OF THREE ASCORBIC ACID DETERMINATIONS (MG/100 G) FOR EACH OF 12 TREATMENTS IN A 3 × 4 FACTORIAL EXPERIMENT ON SNAP BEANS

	Weeks of Storage				
Temperature, °F	2	4	6	8	Sum
0	45	47	46	46	184
10	45	43	41	37	166
20	34	28	21	16	99
Total	124	118	108	99	449

	df	Sum of Squares	Mean Square	F
Temperature, T	2	334.39	167.2	236.8
Two-week period, P	3	40.53	13.5	19.1
Interaction, TP	6	34.05	5.7	8.0
Error	24		0.706	

TABLE 16.9.2
LINEAR AND QUADRATIC TEMPERATURE COMPONENTS

Weeks of Storage	2	4	6	8	Total
Linear, T_L	−11	−19	−25	−30	−85
Quadratic, T_Q	−11	−11	−15	−12	−49

$$T_L = \frac{(-85)^2}{(12)(2)} = 301.04^{**} \qquad T_Q = \frac{(-49)^2}{(12)(6)} = 33.35^{**}$$

(Note that each temperature total is taken over 12 observations.) Both sums of squares (1 df) are highly significant with respect to the error sum of squares, 0.706. Evidently quality decreases with accelerated rapidity as the temperature increases. The two sums of squares add up, of course, to the sum of squares for temperature, 334.39.

Are the linear and quadratic regressions the same for all storage periods? To answer this question, calculate the interactions of these components with periods. The sums of squares are

$$T_LP = \frac{(-11)^2 + \ldots + (-30)^2}{(3)(2)} - \frac{(-85)^2}{(4)(3)(2)} = 33.46^{**} \qquad (3 \text{ df})$$

$$T_QP = \frac{(-11)^2 + \ldots + (-12)^2}{(3)(6)} - \frac{(-49)^2}{(4)(3)(2)} = 0.59 \qquad (3 \text{ df})$$

Note the divisors; for a single period, T_L has a divisor (3)(2) and T_Q a divisor (3)(6). Note also that the two subtraction terms are the overall T_L and T_Q, 301.04 and 33.35, respectively. We conclude that the linear terms T_L change with storage period but that the quadratic terms T_Q, for which $F = 0.59/[(3)(0.706)] = 0.28$, do not.

Table 16.9.2 suggests that this change in T_L may be a linear decrease with storage period. To examine this, we calculate the linear, quadratic, and cubic components of the regression of T_L on storage period (table 16.9.3). The sums of squares in table 16.9.3 confirm this suggestion.

The sum of squares for the T_QP interactions, 0.59, is so small that there seems no point in finding the corresponding components. You may wish to verify that $T_QP_L = 0.14$, $T_QP_Q = 0.12$, and $T_QP_C = 0.34$.

TABLE 16.9.3
COMPONENTS OF REGRESSION OF T_L ON STORAGE PERIOD

Weeks T_L	2 −11	4 −19	6 −25	8 −30	Comparison	Divisor for SS	Sum of Squares	F
Linear, T_LP_L	−3	−1	+1	+3	−63	(3)(2)(20)	33.08**	46.9
Quadratic, T_LP_Q	+1	−1	−1	+1	3	(3)(2)(4)	0.38	0.5
Cubic, T_LP_C	−1	+3	−3	+1	−1	(3)(2)(20)	0.01	0.01
					Sum = sum of squares for T_LP =		33.47	

TABLE 16.9.4
ANALYSIS OF VARIANCE OF ASCORBIC ACID IN SNAP BEANS

Source of Variation	Degrees of Freedom	Sum of Squares	Mean Square
Temperature	(2)	(334.39)	
T_L	1		301.04**
T_Q	1		33.35**
Period	(3)	(40.53)	
P_L	1		40.14**
P_Q	1		0.25
P_C	1		0.14
Interaction	(6)	(34.05)	
T_LP_L	1		33.08**
T_LP_Q	1		0.38
T_LP_C	1		0.01
T_QP_L	1		0.14
T_QP_Q	1		0.12
T_QP_C	1		0.34
Error	24		0.706

Turning to the main effects of storage periods, the storage totals in table 16.9.1 indicate a linear decrease in ascorbic acid concentration with each additional 2 weeks of storage. The sums of squares, P_L, 40.14**; P_Q, 0.25; and P_C, 0.14, agree with this conclusion.

Table 16.9.4 shows the complete breakdown of the treatments sum of squares in the analysis of variance.

In summary, T_L and T_Q show that the relation of ascorbic acid to temperature is parabolic, the rate of decline increasing as storage time lengthens (T_LP_L). The regression on storage period is linear, sloping downward more rapidly as temperature increases. In fact, we noted in table 16.9.1 that at the coldest temperature, 0°F, there is no decline in amount of ascorbic acid in up to 8 weeks of storage.

16.10—Fitting the response surface. These results can be expressed as a mathematical relation between ascorbic acid Y, storage temperature T, and weeks of storage W. As we have seen, we require terms in T_L, T_Q, P_L, and T_LP_L to describe the relation adequately. It is helpful to write down these polynomial coefficients for each of the 12 treatment combinations, as shown in table 16.10.1.

For the moment, think of the mathematical relation as having the form

$$\hat{Y} = \bar{Y} + b_1X_1 + b_2X_2 + b_3X_3 + b_4X_4$$

where $\hat{Y}$ is the predicted ascorbic acid *total* over three replications, while $X_1 = T_L$, $X_2 = T_Q$, $X_3 = P_L$, and $X_4 = T_LP_L$. The regression coefficient $b_i = \Sigma X_iY/\Sigma X_i^2$. The quantities ΣX_iY, which were all obtained in the earlier analysis, are given at the end of table 16.10.1, as well as the divisors ΣX_i^2. Hence, the

TABLE 16.10.1
CALCULATION OF THE RESPONSE SURFACE

°F	Weeks	Y Totals	$T_L = 0.1(T-10)$	$T_Q = 3T_L^2 - 2$	$P_L = W - 5$	$T_LP_L = 0.1(T-10)(W-5)$	$\hat{Y}$
0	2	45	−1	+1	−3	+3	45.53
	4	47	−1	+1	−1	+1	45.84
	6	46	−1	+1	+1	−1	46.16
	8	46	−1	+1	+3	−3	46.47
10	2	45	0	−2	−3	0	45.75
	4	43	0	−2	−1	0	42.92
	6	41	0	−2	+1	0	40.08
	8	37	0	−2	+3	0	37.25
20	2	34	+1	+1	−3	−3	33.73
	4	28	+1	+1	−1	−1	27.74
	6	21	+1	+1	+1	+1	21.76
	8	16	+1	+1	+3	+3	15.77
$\Sigma X_i Y$		449	−85	−49	−85	−63	
Divisor for b		12	8	24	60	40	

relation is as follows:

$$\hat{Y} = 37.417 - 10.625X_1 - 2.042X_2 - 1.417X_3 - 1.575X_4 \qquad (16.10.1)$$

Since the values of the X_i are given in table 16.10.1, the predicted values $\hat{Y}$ are easily computed for each treatment combination. For example, for 0°F and 2 weeks storage,

$$\hat{Y} = 37.417 - (10.625)(-1) - 2.042(+1) - 1.417(-3) - (1.575)(+3)$$
$$= 45.53$$

as shown in the right-hand column of table 16.10.1.

By decoding, we can express the prediction equation (16.10.1) in terms of T (°F) and W (weeks). You may verify that the relations between $X_1(T_L)$, $X_2(T_Q)$, $X_3(P_L)$, $X_4(T_LP_L)$, and T and W are as given at the top of table 16.10.1. After making these substitutions and dividing by 3 so that the prediction refers to the ascorbic acid *mean* per treatment combination, we obtain

$$\hat{\bar{Y}} = 15.070 + 0.316T - 0.02042T^2 + 0.0528W - 0.05250TW \qquad (16.10.2)$$

Geometrically, a relation of this type is called a *response surface*, since we have now a relation in three dimensions Y, T, and W. With quantitative factors, the summarization of the results by a response surface has proved highly useful, particularly in industrial research. If the objective of the research is to maximize Y, the equation shows the combinations of levels of the factors that give responses close to the maximum. Further accounts of this technique, with experimental plans specifically constructed for fitting response surfaces, are given in (7) and (8). The analysis in this example is based on (6).

A word of warning. In the example we fitted a multiple regression of Y on four variables X_1, X_2, X_3, X_4. The methods by which the regression coefficients b_i were computed apply only if the X_i are mutually orthogonal, as is the case in table 16.10.1. General methods are presented in chapter 17.

EXAMPLE 16.10.1—Check that (16.10.2) gives the predicted $\hat{Y}$ values in table 16.10.1 for (*i*) $T = 0$, $W = 6$; (*ii*) $T = 10$, $W = 4$; (*iii*) $T = 20$, $W = 8$, noting that the $\hat{Y}$ are for totals over three replications.

EXAMPLE 16.10.2—Find the residuals $(\hat{Y} - Y)$ and their sum of squares divided by 3 in table 16.10.1. State which terms in the analysis of variance in table 16.9.4 are components of this sum of squares, and verify that the total sum of squares for these terms $= \Sigma(\hat{Y} - Y)^2/3$. Ans. The terms are P_Q, P_C, and all interactions except $T_L P_L$; $SS = 1.38$.

16.11—Three-factor experiments: the 2^3. The experimenter often wants to estimate the effects of three or more factors in a common environment. The simplest arrangement is that of three factors each at two levels, the $2 \times 2 \times 2$ or 2^3 factorial experiment. The eight treatment combinations may be tried in any of the common experimental designs.

The data in table 16.11.1 are taken from a 2^3 factorial experiment (7, p. 160), in which two of the factors appeared to have negligible effects. The crop was mangolds; the three factors of interest were

sulphate of ammonia (A): none and 0.6 cwt nitrogen/acre
agricultural salt (S): none and 5 cwt/acre
dung (D): none and 10 T/acre

The plot yields in table 16.11.1 have been converted to tons per acre. There were four replications in randomized blocks. For the treatment combinations we have adopted the compact notation in section 16.4.

For a first look at the results we calculate the factorial effect totals. There

TABLE 16.11.1
YIELDS OF MANGOLDS ROOTS (T/ACRE) IN A 2^3 FACTORIAL

	Treatment Combinations								
Block	(1)	*a*	*s*	*as*	*d*	*ad*	*sd*	*asd*	Block Total
1	19.2	20.6	18.9	25.3	20.8	26.8	22.2	27.7	181.5
2	15.5	16.9	20.2	27.6	18.5	17.8	18.6	28.6	163.7
3	17.0	19.5	16.7	29.1	20.1	18.6	22.3	28.7	172.0
4	11.7	21.9	20.7	25.4	19.2	19.0	21.1	28.5	167.5
Total	63.4	78.9	76.5	107.4	78.6	82.2	84.2	113.5	684.7

	Degrees of Freedom	Sum of Squares	Mean Square
Blocks	3	22.1	7.37
Treatments	7	484.2	69.17
Error	21	106.4	5.07

TABLE 16.11.2
FACTORIAL EFFECT TOTALS EXPRESSED AS COMPARISONS (2^3 FACTORIAL)

Factorial Effect	Treatment Totals (1) 63.4	*a* 78.9	*s* 76.5	*as* 107.4	*d* 78.6	*ad* 82.2	*sd* 84.2	*asd* 113.5	Effect Total	Sum of Squares
A	−1	+1	−1	+1	−1	+1	−1	+1	79.3**	196.5
S	−1	−1	+1	+1	−1	−1	+1	+1	78.5**	192.6
AS	+1	−1	−1	+1	+1	−1	−1	+1	41.1**	52.8
D	−1	−1	−1	−1	+1	+1	+1	+1	32.3*	32.6
AD	+1	−1	+1	−1	−1	+1	−1	+1	−13.5	5.7
SD	+1	+1	−1	−1	−1	−1	+1	+1	−4.7	0.7
ASD	−1	+1	+1	−1	+1	−1	−1	+1	10.3	3.3
									$SE = \pm 12.7$	484.2

are three main effects, *A*, *S*, and *D*; three two-factor interactions, *AS*, *AD*, and *SD*; and a *three-factor interaction, ASD*. The comparisons that give the factorial effect totals from the treatment totals appear in table 16.11.2. The coefficients for the main effects and the two-factor interactions are the same as in a 2^2 factorial. For the main effect of *A*, the total for combinations *a*, *as*, *ad*, and *asd* receive a plus sign, while (1), *s*, *d*, and *sd* receive a minus sign. For the *AS* interaction, the combinations (1), *as*, *d*, and *asd* receive a plus and *a*, *s*, *ad*, and *sd* receive a minus.

The new term is the three-factor interaction *ASD*. From table 16.11.2 the *AS* interaction (apart from its divisor) is estimated at the higher level of *D* as $d + asd - ad - sd$. An independent estimate at the lower level of *D* is $(1) + as - a - s$. The sum of these quantities is the factorial effect total for *AS*. Their difference,

$$a + s + d + asd - (1) - as - ad - sd \qquad (16.11.1)$$

might be described as measuring the effect of *D* on the *AS* interaction. By the symmetry of (16.11.1), it equally measures the effect of *S* on the *AD* interaction and of *A* on the *SD* interaction. It is called the factorial effect total for *ASD*. A significant three-factor interaction is a sign that the corresponding three-way table of treatment means must be examined for the interpretation of the results. Fortunately, three-factor interactions are often negligible except in experiments that have very large main effects.

From the error mean square 5.07 in table 16.11.1, the standard error of any factorial effect total is $\sqrt{(32)(5.07)} = \pm 12.7$. The main effects of *A* and *S* and the positive *AS* interaction in table 16.11.2 are significant at the 1% level. The response to dung is significant at the 5% level, but no interaction involving dung appears to be present.

For a report on the results we therefore state that dung gave an increase of $32.3/16 = 2.02$ T/acre in mangolds roots and present the two-way *AS* table of mean yields as follows.

Agricultural Salt	Sulphate of Ammonia		Increase
	None	0.6 cwt/acre	
None	17.75	20.14	2.39
5 cwt/acre	20.09	27.61	7.52
Increase	2.34	7.47	

The amounts of sulphate of ammonia and agricultural salt given appear to have been about equally effective and to have interacted positively. Each fertilizer increases yields by about 2.3 T/acre in the absence of the other and by about 7.5 T/acre in the presence of the other.

The right-hand column of table 16.11.2 shows the contribution of each factorial effect to the treatments sum of squares in the analysis of variance. With r replications, the square of any factorial effect total is divided by $r2^n$, in this case 32, to obtain the sum of squares.

The 2^n factorial experiment has proved a powerful research weapon in many fields. For more complete discussion with examples, see (7, 8, 10).

16.12—Yates' algorithm. Yates (10) has given simple rules for finding the factorial effect totals in a 2^n factorial from the totals for the individual treatment combinations. Advantages of these rules are that (*i*) one does not have to remember or work out the combination of plus and minus signs needed to give each factorial effect total as we did in table 16.11.2; (*ii*) the rules hold for any number of factors at two levels; and (*iii*) the rules are very easily programmed in a computer, the term *algorithm* applying to this use.

We shall illustrate the algorithm on the mangolds data in table 16.11.2. We start with the eight totals for the individual treatment combinations, (table 16.12.1). To get the first four numbers in column (1), add the treatment totals in

TABLE 16.12.1
ILLUSTRATION OF YATES' ALGORITHM

Treatment Combination	Total	Column			Effect Total
		(1)	(2)	(3)	
(1)	63.4	142.3	326.2	684.7	*G*†
a	78.9	183.9	358.5	79.3	*A*
s	76.5	160.8	46.4	78.5	*S*
as	107.4	197.7	32.9	41.1	*AS*
d	78.6	15.5	41.6	32.3	*D*
ad	82.2	30.9	36.9	−13.5	*AD*
sd	84.2	3.6	15.4	−4.7	*SD*
asd	113.5	29.3	25.7	10.3	*ASD*

†Grand total for the experiment.

successive pairs; for example,

$$63.4 + 78.9 = 142.3, \ldots, 84.2 + 113.5 = 197.7$$

To get the last four numbers in column (1), subtract the *first* member of each pair from the *second;* thus

$$78.9 - 63.4 = 15.5, \ldots, 113.5 - 84.2 = 29.3$$

Column (2) is formed from column (1) exactly as column (1) was formed from the treatment totals, first adding in pairs and then subtacting within pairs. Column (3) is obtained similarly from column (2). With three factors, we stop at column (3). The figures in this column are the grand total and the factorial effect totals for A, S, AS, D, AD, SD, and ASD, which agree with those shown in table 16.11.2. With n factors we stop at column n.

16.13—Three-factor experiments: a 2 × 3 × 4. This section illustrates briefly the general method of analysis for a three-factor experiment. The factors were lysine (4 levels), methionine (3 levels), and soybean meal (2 levels of protein) as food supplements to corn in feeding male pigs (9). The experiment was a 2 × 3 × 4 factorial in two randomized blocks. Table 16.13.1 gives the original analysis of variance into blocks, treatments, and error. The individual observations are not shown.

With factors at more than 2 levels, it is usually best to start with a breakdown of the treatments sum of squares into main effects and interactions as a guide to the summary tables that need to be examined in writing the report. The treatment totals and detailed manual computations are given in table 16.13.2, followed by the complete analysis of variance. Manual computing instructions are as follows.

1. For each pair of factors, form a two-way table of sums. From the $L \times M$ table (table A), obtain the total sum of squares (11 df) and the sum of squares for L and M. The sum of squares for the LM interactions is found by subtraction. The $M \times P$ table (table B) supplies the sum of squares for M (already obtained), for P, and for the MP interactions by subtraction. The $L \times P$ table (table C) supplies the sum of squares for the LP interactions.

2. From the sum of squares for treatments in table 16.13.1 subtract the

TABLE 16.13.1
ORIGINAL ANALYSIS OF VARIANCE

	Degrees of Freedom	Sum of Squares	Mean Squares
Blocks	1	0.1334	
Treatments	23	1.2755	0.5546
Error	23	0.6319	0.0275

TABLE 16.13.2

TREATMENT TOTALS (TWO REPLICATIONS) FOR AVERAGE DAILY GAINS OF PIGS FED SUPPLEMENTARY LYSINE, METHIONINE, PROTEIN

	Methionine, *M* (%)	0		0.25		0.50	
Lysine, *L* (%)	Protein, *P* (%)	12	14	12	14	12	14
0		2.08	2.97	2.08	2.49	2.06	2.91
0.05		2.30	3.08	2.24	2.58	2.08	3.03
0.10		2.35	2.46	2.75	2.61	2.53	2.85
0.15		2.22	2.09	2.52	2.81	2.49	2.89

Summary Table *A*

Methionine	Lysine 0	0.05	0.10	0.15	Total
0	5.05	5.38	4.81	4.31	19.55
0.025	4.57	4.82	5.36	5.33	20.08
0.050	4.97	5.11	5.38	5.38	20.84
Total	14.59	15.31	15.55	15.02	60.47

Computations ($C = 60.47^2/48 = 76.1796$):

Entries are sums of 2 levels of protein: $5.05 = 2.08 + 2.97$, etc.
Total in *A*: $(5.05^2 + \ldots + 5.38^2)/4 - C = 0.3496$.
Lysine, *L*: $(14.59^2 + \ldots + 15.02^2)/12 - C = 0.0427$.
Methionine, *M*: $(19.55^2 + 20.08^2 + 20.84^2)/16 - C = 0.0526$.
LM: $0.3496 - (0.0427 + 0.0526) = 0.2543$.

Summary Table *B*

Methionine	Protein 12	14	Total
0	8.95	10.60	19.55
0.025	9.59	10.49	20.08
0.050	9.16	11.68	20.84
Total	27.70	32.77	60.47

Entries are sums of 4 levels of lysine: $8.95 = 2.08 + 2.30 + 2.35 + 2.22$, etc.
Total in *B*: $(8.95^2 + \ldots + 11.68^2)/8 - C = 0.6702$.
Protein, *P*: $(27.70^2 + 32.77^2)/24 - C = 0.5355$.
MP: $0.6702 - (0.5355 + 0.0526) = 0.0821$.

Summary Table *C*

Protein	Lysine 0	0.05	0.10	0.15	Total
12	6.22	6.62	7.63	7.23	27.70
14	8.37	8.69	7.92	7.79	32.77
Total	14.59	15.31	15.55	15.02	60.47

Entries are sums of 3 levels of methionine: $6.22 = 2.08 + 2.08 + 2.06$, etc.
Total in *C*: $(6.22^2 + \ldots + 7.79^2)/6 - C = 0.8181$.
LP: $0.8181 - (0.5355 + 0.0427) = 0.2399$.
LMP: $1.2756 - (0.0427 + 0.0526 + 0.5355 + 0.2543 + 0.0821 + 0.2399) = 0.0685$.

TABLE 16.13.2.—*Continued*

Analysis of Variance				
Source of Variation	df	Sum of Squares	Mean Square	F
Replications	1	0.1334		
Lysine, L ($l = 4$)	3	.0427	0.0142	0.5
Methionine, M ($m = 3$)	2	.0526	.0263	1.0
Protein, P ($p = 2$)	1	.5355	.5355**	19.5
LM	6	.2543	.0424	1.5
LP	3	.2399	.0800	2.9
MP	2	.0821	.0410	1.5
LMP	6	.0685	.0114	0.4
Error ($r = 2$)	23	0.6319	0.0275	

sum of squares for L, M, P, LM, MP, and LP to obtain that for the LMP three-factor interaction.

In the analysis of variance the only significant mean square is that for soybean mean (P). Furthermore, the sums of squares for L, M, MP, and LMP are all so small that no single comparison isolated from them could reach 5% significance. But LP, which is almost significant at the 5% level, deserves further study.

In the $L \times P$ summary table C, the differences between 14% and 12% protein are 2.15, 2.07, 0.29, 0.56 at the successive levels of L. These figures suggest that the beneficial effect of the higher level of soybean meal decreases as more lysine is added. (This effect is not surprising, since soybean meal contains lysine and in the complete experiment of eight replications, lysine had a significant effect at the lower level of soybean meal.) By applying the multipliers -3, -1, $+1$, $+3$ to these differences, we obtain the factorial effect total for L_LP_L as -6.55. The divisor for its square is $(6)(2)(20) = 240$, so the F value to test L_LP_L is $6.55^2/[(240)(0.0275)] = 6.50$ with 1 and 23 df, giving P less than 0.025. Therefore this component of LP is significant.

The sum of squares for LM (6 df) is also large enough to contain at least one significant comparison, even though there is no sign of main effects of L or M. In summary table A the differences between the highest and the lowest levels of M are -0.08, -0.27, 0.57, 1.07 at the four levels of L, with standard error $\pm\sqrt{(8)(0.027)} = \pm 0.453$; the value 1.07 is significant.

To sum up, the higher level of soybean meal increased the gain by $2.15/6 = 0.358$ or 28% when no lysine was added, but this increase declined sharply at the higher levels of lysine. Methionine had little effect except at the highest level of lysine, at which level 0.050% methionine increased gain by 0.268 or 21%. The direct effects of lysine were small.

In conclusion, with factors at more than two levels there may be several ways of reporting main effects and interactions, so the summary of results requires careful thought and, as we have seen, a good deal of computation.

16.14—Expected values of mean squares. In the analysis of variance of a factorial experiment, the expected values of the mean squares for main effects and interactions can be expressed in terms of components of variance that are part of the mathematical model underlying the analysis. These formulas have two principal uses. They show how to obtain unbiased estimates of error for the comparisons that are of interest. In studies of variability they provide estimates of the contributions made by different sources to the variance of a measurement.

Consider a two-factor $A \times B$ experiment in a completely randomized design, with a levels of A, b levels of B, and n replications. The observed value for the kth replication of the ith level of A and the jth level of B is

$$X_{ijk} = \mu + \alpha_i + \beta_j + (\alpha\beta)_{ij} + \epsilon_{ijk} \tag{16.14.1}$$

where $i = 1, \ldots, a$; $j = 1, \ldots, b$; $k = 1, \ldots, n$. (If the plan is in randomized blocks or a Latin square, further parameters are needed to specify block, row, or column effects.)

The parameters α_i and β_j represent effects of the levels of the respective factors. If the effects are described as *fixed,* we mean that the levels of the factors in the experiment are the only population levels of interest to us. If the effects are called *random,* the corresponding α_i or β_j are assumed drawn at random from an infinite population with mean 0, variance σ_A^2 or σ_B^2. The $(\alpha\beta)_{ij}$ are the two-factor interaction effects. They are random if either A or B is random, with mean 0 and variance σ_{AB}^2. The residuals ϵ_{ijk} have mean 0, variance σ^2.

By the main effect of the ith level of A we mean the average value of X_{ijk} taken over the population of levels of B and of the residual errors. The main effects of A are the comparisons of the main effects of the levels of A.

The values of the main effects of A depend on whether the other factor B is fixed or random. To illustrate, let A represent two fertilizers (a_1 and a_2) and B two fields. Experimental errors ϵ are assumed negligible, and results are as follows:

	a_1	a_2	$a_2 - a_1$
Field 1	10	17	+7
Field 2	18	13	−5
Mean	14	15	+1

Our estimate of the main effect of A is $15 - 14 = 1$. When B is *fixed* this estimate is *exact,* since it is averaged over the only two fields (levels of B) in our population and experimental errors are assumed negligible. If B is *random* our estimate of the main effect of A is still $15 - 14 = 1$, but the population value of this main effect is the average difference between levels a_2 and a_1 over a population of which these two fields are a random sample. The estimate is no longer exact but has a standard error (with 1 df) that may be computed as $[7 - (-5)]/2 = 6$. Note that this standard error is derived from the AB interaction, whose effect total is $[7 - (-5)] = 12$.

To sum up, the estimates of the main effects of A from the experiment are the same whether B is fixed or random; but the population quantities being estimated are not the same, and consequently different standard errors are required.

We turn now to the general two-factor experiment in (16.14.1). The sample mean of the ith level of A is

$$\overline{X}_{i..} = \mu + \alpha_i + \overline{\beta} + (\overline{\alpha\beta})_{i.} + \overline{\epsilon}_{i..} \qquad (16.14.2)$$

where $\overline{\beta} = (\beta_1 + \ldots + \beta_b)/b$, $(\overline{\alpha\beta})_{i.} = [(\alpha\beta)_{i1} + \ldots + (\alpha\beta)_{ib}]/b$ and $\overline{\epsilon}_{i..}$ is the average of nb independent values of ϵ.

Consider first B fixed. When we average $\overline{X}_{i..}$ over the levels of B and the $\overline{\epsilon}_{i..}$, the only change in (16.4.2) is that the term in $\overline{\epsilon}_{i..}$ vanishes.

The true main effects of A are therefore the differences of the quantities $[\alpha_i + (\overline{\alpha\beta})_{i.}]$ from level to level of A. In this case it is customary, for simplicity of notation, to redefine the parameter α_i as $\alpha_i' = \alpha_i + (\overline{\alpha\beta})_i$. Thus with B fixed, it follows from (16.14.2) that

$$\overline{X}_{i..} - \overline{X}_{...} = \alpha_i' - \overline{\alpha}' + \overline{\epsilon}_{i..} - \overline{\epsilon}_{...} \qquad (16.14.3)$$

From this relation the expected value of the mean square for A is

$$E(A) = E\left[\frac{nb\Sigma(\overline{X}_{i..} - \overline{X}_{...})^2}{a-1}\right] = \frac{nb\Sigma(\alpha_i' - \overline{\alpha}')^2}{a-1} + \sigma^2 \qquad (16.14.4)$$

The quantity $\Sigma(\alpha_i' - \overline{\alpha}')^2/(a-1)$ is the quantity previously denoted by κ_A^2.

If A is random and B is fixed, repeated sampling involves drawing a fresh set of a levels of the factor A in each experiment, retaining the same set of b levels of B. In $E(A)$, κ_A^2 in (16.14.4) is an unbiased estimate of σ_A^2, the population variance of α_i'. Hence, with A random and B fixed,

$$E(A) = nb\sigma_A^2 + \sigma^2$$

Now consider B random. Consider (16.14.2):

$$\overline{X}_{i..} = \mu + \alpha_i + \overline{\beta} + (\overline{\alpha\beta})_{i.} + \overline{\epsilon}_{i..}$$

In each new sample we draw fresh values of β_j and of $(\alpha\beta)_{ij}$ so that $\overline{\beta}$ and $(\overline{\alpha\beta})_{i.}$ change from sample to sample. However, since the population means of $\overline{\beta}$, $(\overline{\alpha\beta})_{i.}$, and $\overline{\epsilon}_{i..}$ are all zero, the population mean of $\overline{X}_{i..}$ is $\mu + \alpha_i$. Consequently, the population variance of the main effects of A is defined as $\kappa_A^2 = \Sigma(\alpha_i - \overline{\alpha})^2/(a-1)$ if A is fixed or as the variance σ_A^2 of the αs if A is random. But since

$$\overline{X}_{i..} - \overline{X}_{...} = \alpha_i - \overline{\alpha} + (\overline{\alpha\beta})_{i.} - (\overline{\alpha\beta})_{..} + \overline{\epsilon}_{i..} - \overline{\epsilon}_{...}$$

the expected value of the mean square of A now involves σ_{AB}^2 as well as σ^2. It follows that when B is random,

$$E(A) = nb\kappa_A^2 + n\sigma_{AB}^2 + \sigma^2 \qquad (A \text{ fixed})$$
$$E(A) = nb\sigma_A^2 + n\sigma_{AB}^2 + \sigma^2 \qquad (A \text{ random})$$

The preceding results are particular cases of a more general formula (22). Suppose that the population of levels of B contains B' levels of which b are chosen at random for the experiment. By definition, the main effects of A are the comparisons among the quantities $[\alpha_i + \underline{(\overline{\alpha\beta})_{i.}}]$, where the interaction term is averaged over B' levels. In the estimate $[\alpha_i + (\overline{\alpha\beta})_{i.}]$, the interaction is averaged over only b levels. Hence, with A random,

$$E(A) = nb\sigma_A^2 + n\left(\frac{B' - b}{B'}\right)\sigma_{AB}^2 + \sigma^2 \qquad (16.14.5)$$

This case can occur in practice. Suppose that a company owns B' factories or stores and carries out experiments to guide some policy decision in a random sample of b factories or stores. The population of interest is that of the B' units.

If $b = B'$, the term in σ_{AB}^2 vanishes and we regard factor B as fixed. As B' tends to infinity, the coefficient of σ_{AB}^2 tends to n, factor B being random. If A is fixed, σ_A^2 becomes κ_A^2 in (16.14.5).

The AB mean square is derived from the sum of squares of the terms $(X_{ij.} - \overline{X}_{i..} - \overline{X}_{.j.} + \overline{X}_{...})$. From the model, this term is

$$(\alpha\beta)_{ij} - (\overline{\alpha\beta})_{i.} - (\overline{\alpha\beta})_{.j} + (\overline{\alpha\beta})_{..} + \bar{\epsilon}_{ij.} - \bar{\epsilon}_{i..} - \bar{\epsilon}_{.j.} + \bar{\epsilon}_{...}$$

Unless both A and B are fixed, the interaction term in the above is a random variable from sample to sample, giving

$$E(AB) = n\sigma_{AB}^2 + \sigma^2$$

With both factors fixed, σ_{AB}^2 is replaced by κ_{AB}^2. Table 16.14.1 summarizes this series of results.

This analysis and table 16.14.1 lead to the conclusion that if B is a fixed factor, we should test A against the experimental error mean square even if there is a substantial AB interaction. On the other hand, some writers (11, 12) have argued that the test of A against the experimental error in the presence of an AB interaction seldom answers a question that is of much interest or importance. The preceding 2 × 2 table with fertilizers A and fields B was intended as an example with a large AB interaction and a small main effect of A. When would it be of interest to regard B as fixed and test the main effect of A against experimental error? One such case might be if the two fertilizers cost the same

TABLE 16.14.1

Expected Values of Mean Squares in a Two-Factor Experiment

Mean Squares	Fixed Effects	Random Effects	Mixed Model: A fixed, B Random
A	$\sigma^2 + nb\kappa_A^2$	$\sigma^2 + n\sigma_{AB}^2 + nb\sigma_A^2$	$\sigma^2 + n\sigma_{AB}^2 + nb\kappa_A^2$
B	$\sigma^2 + na\kappa_B^2$	$\sigma^2 + n\sigma_{AB}^2 + na\sigma_B^2$	$\sigma^2 + na\sigma_B^2$
AB	$\sigma^2 + n\kappa_{AB}^2$	$\sigma^2 + n\sigma_{AB}^2$	$\sigma^2 + n\sigma_{AB}^2$
Error	σ^2	σ^2	σ^2

TABLE 16.14.2
EXPECTED VALUES OF MEAN SQUARES IN A THREE-WAY FACTORIAL

Mean Squares	Expected Values: A Fixed, B and C Random
A	$\sigma^2 + n\sigma^2_{ABC} + nc\sigma^2_{AB} + nb\sigma^2_{AC} + nbc\kappa^2_A$
B	$\sigma^2 + na\sigma^2_{BC} + nac\sigma^2_B$
C	$\sigma^2 + na\sigma^2_{BC} + nab\sigma^2_C$
AB	$\sigma^2 + n\sigma^2_{ABC} + nc\sigma^2_{AB}$
AC	$\sigma^2 + n\sigma^2_{ABC} + nb\sigma^2_{AC}$
BC	$\sigma^2 + na\sigma^2_{BC}$
ABC	$\sigma^2 + n\sigma^2_{ABC}$
Error	σ^2

amount, if the two fields were the only fields on which this crop is grown, and if the farmer is trying to decide whether it is more profitable to put a_1 on both fields or a_2 on both fields. An obvious question is, Why consider this decision? Surely he would make still more money by using a_2 on field 1 and a_1 on field 2. This comment is in line with the approach used earlier in this chapter, which is to turn attention to simple effects rather than to main effects when there are substantial interactions. Of course, situations occur in which, for cogent reasons, only a single level of A is to be chosen and the proposed test of significance applies to the decision that is faced. The discussion of this issue by Elston and Bush (21) is helpful.

From the algebraic analysis, general rules are available for factors A, B, C, D, . . . at levels a, b, c, d, . . . with n replications of each treatment combination. Any factors may be fixed or random. In presenting these rules, the symbol U denotes the factorial effect in whose mean square we are interested (for instance, U may be the main effect of A, the BC interaction, or the ACD interaction).

Rule 16.14.1. The expected value of the mean square for U contains a term in σ^2 and a term in σ^2_U. It also contains a variance term for any interaction in which (*i*) *all* the letters in U appear and (*ii*) *all* the other letters in the interaction represent random effects.

Rule 16.14.2. The coefficient of the term in σ^2 is 1. The coefficient of any other variance is n times the product of all letters a, b, c, . . . that do *not* appear in the set of capital letters A, B, C, . . . specifying the variance.

Table 16.14.2 illustrates the rules for three factors with A fixed and B and C random—a case of frequent interest. Note from table 16.14.2 that no single mean square in the analysis of variance is an appropriate denominator for an F test of the main effects of A. Approximate F tests can be constructed. On H_0: $\kappa^2_A = 0$, table 16.14.2 shows that

$$E(A) = E(AC) + E(BC) - E(ABC)$$

A suggested test criterion (8) is

$$F' = [(A) + (ABC)]/[(AC) + (BC)]$$

where (A) denotes the mean square for A, and so on. The approximate degrees of freedom for the numerator, by Satterthwaite's rule (section 6.11) are

$$\nu_1 = [(A) + (ABC)]^2/[(A)^2/\nu_A + (ABC)^2/\nu_{ABC}]$$

and similarly for ν_2.

An alternative test criterion is, of course (18),

$$F'' = (A)/[(AC) + (BC) - (ABC)] \qquad (16.14.6)$$

This might have more power but has the criticism that Satterthwaite's approximation to the numbers of degrees of freedom does not work so well when the linear function contains a negative term, as occurs in (16.14.6). This criticism may be less important if (ABC) is relatively small and has numerous degrees of freedom.

EXAMPLE 16.14.1—Show that in a trifactorial experiment with factors A and C fixed and factor B random, consequences of rules 16.14.1 and 16.14.2 are that appropriate denominators for F tests of A, C, and AC are their interaction mean squares with the random factor B.

16.15—Split-plot or nested design. It is sometimes necessary or convenient to test one factor on one size of unit and test a second factor on a smaller size of unit. For example, three sources of vitamin might be compared by trying them on three males of the same litter, replicating the experiment on 20 litters. It would be a randomized blocks design with high precision, providing 20 replications. Superimposed on this could be some experiment with the litters as units. Four types of housing could be tried, one litter to each type, thus allowing 5 replications with 12 df for error. The main treatments (housings) would not be compared as accurately as the subtreatments (sources of vitamin) for two reasons—less replication is provided and litter differences contribute to the error for evaluating the housing effects. Nevertheless, some information about housing may be obtained at little extra expense and any interaction between housing and vitamin will be accurately evaluated.

In experiments on varieties or fertilizers on small plots, cultural practices with large machines may be tried on whole groups of the smaller plots, each group containing all the varieties. (Irrigation is one practice that demands large areas per treatment.) The series of cultural practices is usually replicated only a small number of times but the varieties are repeated on every cultural plot. Experiments of this type are called *split-plot,* the cultural *main plot* being split into smaller varietal *subplots*.

This design is also common in industrial research. Comparisons among relatively large machines, or comparisons of different conditions of temperature and humidity under which machines work, are *main plot* treatments, while adjustments internal to the machines are *subplot* treatments. Since the word *plot* is inappropriate in such applications, the designs are often called *nested.*

The essential feature of a nested design is that there are two sizes of experimental unit. Each large unit or main unit contains, say, t subunits. One factor M is applied to the main units in a standard randomized design. Within any main unit, the t levels of a second factor T are assigned at random to one of the t subunits. A consequence is that the error mean square per subunit for the factor M is usually larger than those for T and TM.

Figure 16.15.1 shows two replicates of a split-plot experiment on alfalfa in which the main treatments were varieties, the subtreatments being four dates of final cutting (13). The first two cuttings were common to all plots, with the second on July 27, 1943. The third harvests were: none, September 1 (S1), September 20 (S20), October 7 (O7). Yields (six replications) in 1944 are recorded in table 16.15.1. Such an experiment is usually evaluated by several seasons' yields.

In the analysis of variance the data form a trifactorial, with varieties (V) and dates (D) as fixed factors and blocks (B) as random factors. It follows from Rule 16.14.1 (see example 16.14.1) that appropriate error mean squares for V, D, and VD are their interactions with blocks. The VB interactions are derived from comparisons between main plot totals; the DB and VDB interactions are

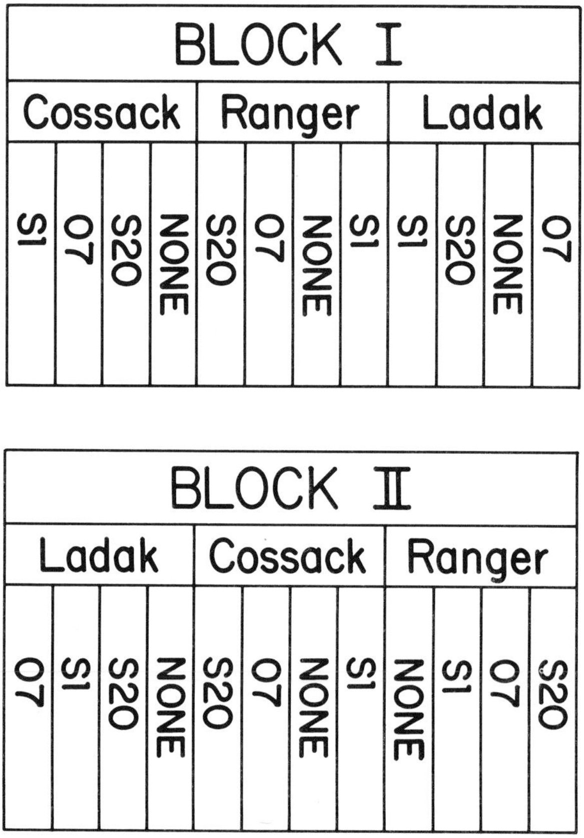

BLOCK I											
Cossack				Ranger				Ladak			
S1	O7	S20	NONE	S20	O7	NONE	S1	S1	S20	NONE	O7

BLOCK II											
Ladak				Cossack				Ranger			
O7	S1	S20	NONE	S20	O7	NONE	S1	NONE	S1	O7	S20

Fig. 16.15.1—First two blocks of a split-plot experiment on alfalfa, illustrating random arrangement of main and subplot treatments.

TABLE 16.15.1

YIELDS OF THREE VARIETIES OF ALFALFA (T/ACRE) IN 1944 FOLLOWING THIRD DATE OF CUTTING IN 1943

Variety	Date	Block 1	2	3	4	5	6
Ladak	None	2.17	1.88	1.62	2.34	1.58	1.66
	S1	1.58	1.26	1.22	1.59	1.25	0.94
	S20	2.29	1.60	1.67	1.91	1.39	1.12
	O7	2.23	2.01	1.82	2.10	1.66	1.10
		8.27	6.75	6.33	7.94	5.88	4.82
Cossack	None	2.33	2.01	1.70	1.78	1.42	1.35
	S1	1.38	1.30	1.85	1.09	1.13	1.06
	S20	1.86	1.70	1.81	1.54	1.67	0.88
	O7	2.27	1.81	2.01	1.40	1.31	1.06
		7.84	6.82	7.37	5.81	5.53	4.35
Ranger	None	1.75	1.95	2.13	1.78	1.31	1.30
	S1	1.52	1.47	1.80	1.37	1.01	1.31
	S20	1.55	1.61	1.82	1.56	1.23	1.13
	O7	1.56	1.72	1.99	1.55	1.51	1.33
		6.38	6.75	7.74	6.26	5.06	5.07

Variety	Date of Cutting: None	S1	S20	O7	Total
Ladak	11.25	7.84	9.98	10.92	39.99
Cossack	10.59	7.81	9.46	9.86	37.72
Ranger	10.22	8.48	8.90	9.66	37.26
Total	32.06	24.13	28.34	30.44	114.97

Date	Block 1	2	3	4	5	6	Total
None	6.25	5.84	5.45	5.90	4.31	4.31	32.06
S1	4.48	4.03	4.87	4.05	3.39	3.31	24.13
S20	5.70	4.91	5.30	5.01	4.29	3.13	28.34
O7	6.06	5.54	5.82	5.05	4.48	3.49	30.44
Total	22.49	20.32	21.44	20.01	16.47	14.24	114.97

from comparisons between subplots in the same main plot. Often, the *DB* and *VDB* sums of squares are pooled to give the subplot error. But since date of cutting effects are large in this experiment, we have reason to expect that the *DB* mean square will be larger than the *VDB* mean square, as was the case.

The analysis of variance is given in table 16.15.2. All sums of squares except that for *DVB* are obtained from the data in table 16.15.1; the sum of squares for *DVB* is found by subtraction. The sums of squares are on a subplot basis. Thus the squares of variety totals are divided by 24, since each total is taken over 24 subplots, and so on. Note that the main plot error mean square was 0.1362 and the subplot error mean square was 0.0367 for *D* and 0.0236 for *DV*.

The only significant differences were those among dates, S1 and S20 giving

TABLE 16.15.2
ANALYSIS OF VARIANCE (ON A SUBPLOT BASIS) OF A SPLIT-PLOT EXPERIMENT

Source	df	Sum of Squares	Mean Square	F
Main plot comparisons				
varieties, V	2	0.1781	0.0890	0.65
blocks, B	5	4.1499	.8300	
main plot error, VB	10	1.3622	0.1362	
Subplot comparisons				
dates, D	3	1.9625	0.6542**	17.82
date × variety, DV	6	0.2105	.0351	1.49
date × block, DB	15	0.5500	.0367	
date × variety × block, DVB	30	0.7086	0.0236	
Total	71	9.1218		

smaller 1944 yields. This was not unexpected. The last 1943 cutting should be either early enough to allow restoration of depleted root reserves or late enough so that little depletion will occur. The absence of DV interactions is a little surprising—Ladak is slow to renew growth after cutting and might have reacted differently from the other varieties.

In writing up the results from the table of treatment means (table 16.15.3), care is required to give the correct standard errors for different comparisons. The standard error of a date mean is of course $\sqrt{0.0367/18} = \pm 0.0452$, while that of a variety mean is $\sqrt{0.1362/24} = \pm 0.0753$. For a single mean in the 3 × 4 table, the standard error $\sqrt{0.0236/6} = \pm 0.0627$ applies to comparisons that are part of the variety-date interactions and to comparisons among dates for a single variety (e.g., none versus S1 for Ladak). Some comparisons, for example those among varieties for a given date, require a standard error that is more complex, as described in (8).

Experimenters sometimes split the subplots, accommodating three factors on different sizes of unit. The statistical methods are a natural extension of those given here. Example 16.15.1 gives an illustration, which naturally involves a good deal of computation.

EXAMPLE 16.15.1—An $I \times S \times F$ split split-plot experiment on corn tested 3 rates of planting (stands, S), with 3 levels of fertilizer (F), in irrigated (I) and nonirrigated plots (14). The main plots, irrigated or not, were in 4 randomized blocks of 2 plots. Each plot had 3 subplots with 10,000, 13,000, and 16,000 plants/acre. Finally, each subplot had 3 parts, fertilized with 60, 120, and 180 lb of nitrogen (N). Yields given in the table are in bushels per acre. Calculate the analysis of

TABLE 16.15.3
TREATMENT MEANS (T/ACRE) AND STANDARD ERRORS

	Date ($\pm\sqrt{0.0236/6} = \pm 0.0627$)				
Variety	None	S1	S20	O7	Means
Ladak	1.88	1.31	1.66	1.82	1.67 $\pm \sqrt{(0.1362/24)} = 0.0753$
Cossack	1.76	1.30	1.58	1.64	1.57
Ranger	1.70	1.41	1.48	1.61	1.55
Means	1.78	1.34	1.57	1.69	1.60
		$\pm\sqrt{0.0367/18} = \pm 0.0452$			

variance. In this experiment the different components of the subplot and sub-subplot errors have been pooled, rather than testing each mean square against its interaction with blocks (B). In the table, (SF, 11) denotes 10,000 plants/acre with 60 lb N, and so on.

Treatments	Nonirrigated Blocks					Irrigated Blocks				
SF	1	2	3	4	Sum	1	2	3	4	Sum
11	90	83	85	86	344	80	102	60	73	315
12	95	80	88	78	341	87	109	104	114	414
13	107	95	88	89	379	100	105	114	114	433
Sum	292	258	261	253	1064	267	316	278	301	1162
21	92	98	112	79	381	121	99	90	109	419
22	89	98	104	86	377	110	94	118	131	453
23	92	106	91	87	376	119	123	113	126	481
Sum	272	302	307	252	1134	350	316	321	366	1353
31	81	74	82	85	322	78	136	119	116	449
32	92	81	78	89	340	98	133	122	136	489
33	93	74	94	83	344	122	132	136	133	523
Sum	266	229	254	257	1006	298	401	377	385	1461
Total	831	789	822	862	3204	915	1033	976	1052	2976

Ans. Mean squares in analysis of variance:

	df	Mean Square		df	Mean Square
Main plots			Sub-subplots		
blocks, B	3	64.8	fertilizer, F	2	988.7**
irrigation, I	1	8277.6*	IF	2	476.7**
error, IB	3	470.6	SF	4	76.2
Subplots			ISF	4	58.7
stand, S	2	879.2	error, FB, IFB,		
IS	2	1373.5*	SFB, $ISFB$	36	86.4
error, SB, ISB	12	232.3			

EXAMPLE 16.15.2—Irrigation gave an increase of about 24% in corn yield. The substantial IS and IF interactions direct attention to the $I \times S$ and $I \times F$ two-way tables of totals:

	$S1$	$S2$	$S3$	$F1$	$F2$	$F3$
Not irrigated	1064	1134	1006	1047	1058	1099
Irrigated	1162	1353	1461	1183	1356	1437
Standard error	$\sqrt{(12)(232.3)} = \pm 52.8$			$\sqrt{(12)(86.4)} = \pm 32.2$		

Clearly, neither stand nor fertilizer affected yield materially on the nonirrigated plots, but they appear to have had roughly linear effects on the irrigated plots. Since irrigated plots with their higher yields may have had higher experimental errors than found in the analysis of variance given in example 16.15.1, analyze the irrigated plots separately as a split-plot experiment. Verify the following results:

	df	Mean Square		df	Mean Square
Blocks, B	3	424	Fertilizer, F		
Stand, S			linear	1	2688**
linear	1	3725**	quadratic	1	118
quadratic	1	96	SF	4	92
Error, SB	6	316	Error, FB, SFB	18	137

With irrigation, the highest stand tested, 16,000 plants/acre, gave about 25% higher yield than 10,000 plants/acre. Higher stands and amounts of fertilizer might be worth testing.

16.16—Experiments with repeated measurements. As noted in section 14.9, the term *repeated measurements experiment* sometimes denotes a randomized blocks experiment in which the subject forms the block, receiving each treatment in turn so that the subject is measured repeatedly. A second use occurs in the study of the long-term effects of treatments. Each subject receives only one treatment but is subsequently measured repeatedly to study how its effects change as time elapses.

Suppose that the experiment has 5 treatments in a randomized design, with 1 treatment to a subject and 4 subsequent measurements per subject. In the analysis we may want to compare the average effects and the time trends in response for different treatments. However, while totals on different subjects may be regarded as independent, successive measurements on the same subject are usually positively correlated. This problem is often handled by noting its similarity to a split-plot experiment, in which main plot totals are independent but results for subplot treatments may be thought of as correlated because they lie in the same main plot. Hence, in a repeated measurements experiment we might apply a main plot analysis to treatment totals and a subplot analysis to time differences and treatment × time interactions. This method is often satisfactory.

Box (15) pointed out, however, that this analysis requires the assumption that every pair of subplot times has the same correlation ρ. In a split-plot experiment, the randomization makes this assumption reasonable for the subplot treatments. But with repeated measurements we may expect the correlation between times 1 and 2 to be higher than that between times 1 and 3, which are farther apart, and still higher than that between times 1 and 4. For repeated measurements experiments Box gives a test, based on multivariate theory, as to whether the assumptions made in a split-plot analysis appear to hold. If they do not, several authors (16, 17) have given multivariate methods of analysis that do not require these assumptions.

When the assumptions do not hold or when we are uncertain that they do, univariate analysis of treatment × time interactions can still be conducted. Calculate for every subject any time comparison that interests us (e.g., a linear time trend) and analyze these data by the same method as for subject totals. Thus, for any single degree of freedom component of treatment × times, we use its interaction with blocks as error, avoiding pooling of errors that is not valid if ρ is not constant.

For illustration, the data in table 16.16.1 are taken from an experiment by Haber and Snedecor (19) to compare the effects of various cutting treatments on asparagus. Planting was in 1927 and cutting began in 1929. One plot in each block was cut until June 1 in each year, others to June 15, July 1, and July 15. The yields are for the four succeeding years, 1930, 1931, 1932, and 1933. The yields are the weights cut to June 1 in every plot, irrespective of later cuttings in some of them. This weight is a measure of vigor, and the objective is to compare the relative effectiveness of the different harvesting plans.

A glance at the four-year totals (5706, 5166, 4653, 3075) leaves little doubt that prolonged cutting decreased the vigor. The cutting totals were separated into linear, quadratic, and cubic components of the regression on duration of cutting. The significant quadratic component may be due mainly to the sharp drop from July 1 to July 15.

TABLE 16.16.1

Weight (oz) of Asparagus Cut before June 1 from Plots with Various Cutting Treatments

Blocks	Year	Cutting Ceased: June 1	June 15	July 1	July 15	Total
1	1930	230	212	183	148	773
	1931	324	415	320	246	1,305
	1932	512	584	456	304	1,856
	1933	399	386	255	144	1,184
		1465	1597	1214	842	5,118
2	1930	216	190	186	126	718
	1931	317	296	295	201	1,109
	1932	448	471	387	289	1,595
	1933	361	280	187	83	911
		1342	1237	1055	699	4,333
3	1930	219	151	177	107	654
	1931	357	278	298	192	1,125
	1932	496	399	427	271	1,593
	1933	344	254	239	90	927
		1416	1082	1141	660	4,299
4	1930	200	150	209	168	727
	1931	362	336	328	226	1,252
	1932	540	485	462	312	1,799
	1933	381	279	244	168	1,072
		1483	1250	1243	874	4,850
Total		5706	5166	4653	3075	18,600

	Degrees of Freedom	Sum of Squares	Mean Square
Blocks	3	30,170	
Cuttings	(3)	(241,377)	
linear	1		220,815**
quadratic	1		16,835*
cubic	1		3,727
Error	9		2,429

Coming to time trends, we show the annual treatment totals in table 16.16.2. The large annual variations in yield make a simple summary difficult. However, it looks as if differences between treatments are increasing as time goes on. Consequently, we examined for different treatments the linear component of the regression of yield on years. On each plot we multiplied the yields in the four years by -3, -1, $+1$, and $+3$ and added. These regressions (with an

TABLE 16.16.2

Annual Treatment Totals (oz)

Year	June 1	June 15	July 1	July 15	Total
1930	865	703	755	549	2872
1931	1360	1325	1241	865	4791
1932	1996	1939	1732	1176	6843
1933	1485	1199	925	485	4094

TABLE 16.16.3
ANALYSIS OF THE LINEAR COMPONENT OF THE REGRESSION OF YIELD ON YEARS

	Cutting Ceased				
Blocks	June 1	June 15	July 1	July 15	Total
1	695†	691	352	46	1784
2	566	445	95	−41	1065
3	514	430	315	28	1287
4	721	536	239	86	1582
Total	2496	2102	1001	119	5718

	Degrees of Freedom	Sum of Squares	Mean Square
Blocks	3	3,776	
Cuttings	(3)	43,633	14,544**
linear	1	42,354**	
quadratic	1	744	
cubic	1	536	
Error	9	2,236	248

†695 = 3(399) + 512 − 324 − 3(230), from table 16.16.1.

appropriate divisor) measure the average improvement in yield per year. From the treatment totals in table 16.16.3, it is evident that the annual improvement is greatest for the June 1 cutting and declines approximately linearly with later cutting times, as shown by the analysis of variance in table 16.16.3. Evidently, each additional two weeks of cutting decreased the annual improvement in yield up to June 1 by about the same amount. Other features of this experiment are discussed more fully by Snedecor and Haber (19, 20).

TECHNICAL TERMS

factorial effect mean
factorial effect total
factorial experimentation
main effect
nested design
repeated measurements experiments
response curve
response surface
simple effect
split-plot design
Yates' algorithm

REFERENCES

1. Fisher, R. A. 1935–51. *The Design of Experiments.* Oliver & Boyd, Edinburgh.
2. Yates, F. *J. R. Stat. Soc. Suppl.* 2 (1935):210.
3. Southern Cooperative Series Bulletin 10. 1951, p. 114.
4. Rothamsted Experimental Station Report. 1937, p. 218.
5. Saunders, A. R. 1939. Union of South Africa Dept. Agric. For. Sci. Bull. 200.
6. Snedecor, G. W. *Proc. Int. Stat. Conf.* 3 (1947):440.
7. Daniel, C. 1976. *Applications of Statistics to Industrial Experimentation.* Wiley, New York.
8. Cochran, W. G., and Cox, G. M. 1957. *Experimental Designs.* Wiley, New York.
9. Iowa Agricultural Experiment Station. 1952. Anim. Husb. Swine Nutr. Exp. 577.
10. Yates, F. 1937. "The Design and Analysis of Factorial Experiments." *Commonw. Bur. Soil Sci. Tech. Commun.* 35.
11. Nelder, J. A. *J. R. Stat. Soc.* 140 (A) (1977):48.

12. Cox, D. R. *Biometrika* 45 (1958):69.
13. Wilsie, C. P. 1944. Iowa State Coll. Agric. Exp. Stn.
14. Porter, R. H. 1943. Cooperative Soybean Seed Treatment Trials, Iowa State Univ. Seed Lab.
15. Box, G. E. P. *Biometrics* 6 (1950):362.
16. Danford, M. B.; Hughes, H. M.; and McNee, R. C. *Biometrics* 16 (1960):547.
17. Cole, J. W. L., and Grizzle, J. E. *Biometrics* 22 (1966):810.
18. Hudson, J. D., and Krutchkoff, R. G. *Biometrika* 55 (1968):431
19. Haber, E. S., and Snedecor, G. W. *Am. Soc. Hort. Sci.* 48 (1946):481.
20. Snedecor, G. W., and Haber, E. S. Biometrics Bull. 2 (1946):61.
21. Elston, R. C., and Bush, N. *Biometrics* 20 (1964):681.
22. Cornfield, J., and Tukey, J. W. *Ann. Math. Stat.* 27 (1956):907

17

Multiple linear regression

17.1—Introduction. Chapter 9 treats the regression of Y on a single X-variable. Often, however, two or more Xs are available to give additional information by means of a multiple regression on the Xs. In this chapter we consider *multiple linear regression* (MLR), in which the regression is linear in the Xs. Among the principal uses of multiple regression are:

1. Constructing an equation in the Xs that gives the best predictions of the values of Y. With many Xs it may involve selecting the subset that seems best. In predicting future weather conditions at an airport, for example, over 50 X-variables may be available to measure different aspects of the present weather pattern at neighboring weather stations. Many of these variables may contribute little or nothing to improved prediction in the presence of other variables.
2. Discovering which variables are related to Y, rating them if possible in the order of their importance, and considering the interpretation and the implications of the sizes and signs of the regression coefficients.

Multiple linear regression is a complex subject. The steps in the calculations multiply rapidly as the number of Xs increases. The calculations may not cause much difficulty, since they are usually done by standard electronic programs. However, understanding what a regression coefficient in a multiple regression means, and what it does not mean, is not easy. Fortunately, much can be learned about the calculations and the pitfalls in interpretation by studying a regression on two X-variables.

In a two-variable example we first present the calculations for fitting the sample regression line and for testing the sample regression coefficients b_1 and b_2 plus other results usually given in the output of a standard computer program.

17.2—Two independent variables. With only one X-variable, the sample values of Y and X can be plotted in figures that show both the regression line and the distributions of the individual values of Y about the line. But if Y depends partly on X_1 and partly on X_2 for its value, solid geometry instead of plane is required. Any observation now involves three numbers—the values of Y, X_1, and X_2. The pair (X_1, X_2) can be represented by a point on graph paper. The values of Y corresponding to this point are on a vertical axis perpendicular to the graph paper. In the population these values of Y form a frequency distribution, so we must try to envisage a frequency distribution of Y on each vertical axis. Each frequency distribution has a mean—the mean value of Y for specified X_1, X_2. The surface determined by these means is the *regression surface.* In this chapter the surface is a *plane,* since only linear regressions on X_1 and X_2 are being studied.

The mathematical model is

$$Y = \alpha + \beta_1 X_1 + \beta_2 X_2 + \epsilon \qquad \epsilon = \mathcal{N}(0, \sigma^2) \tag{17.2.1}$$

For given X_1 and X_2, the values of Y vary about the regression plane in a normal distribution with mean 0 and variance σ^2, sometimes denoted by $\sigma^2_{Y.12}$ as a reminder that it is the variance of the deviations from the regression on X_1, X_2. The mean value of Y for given X_1, X_2 is the population regression plane

$$\mu_{Y.12} = \alpha + \beta_1 X_1 + \beta_2 X_2 \tag{17.2.2}$$

Given a sample of n values (Y, X_1, X_2), the sample regression (prediction equation) is

$$\hat{Y} = a + b_1 X_1 + b_2 X_2 \tag{17.2.3}$$

By the method of least squares, the values of a, b_1, and b_2 are chosen so as to minimize $\Sigma(Y - \hat{Y})^2$, the sum of the squares of the n differences between the actual and predicted Y values. As in the case of a single X-variable, the estimates a, b_1, b_2, and $\hat{Y}$ are unbiased and have the smallest standard errors of any unbiased estimates that are linear expressions in the Ys. When we minimize $\Sigma(Y - a - b_1 X_1 - b_2 X_2)^2$, the minimizing value of a is

$$a = \overline{Y} - b_1 \overline{X}_1 - b_2 \overline{X}_2 \tag{17.2.4}$$

If we write $Z_i = Y_i - b_1 X_{1i} - b_2 X_{2i}$, the least squares value of a is an application of the result that $\Sigma(Z_i - a)^2$ is minimized by choosing $a = \overline{Z}$. The quantity a is of course the intercept of the sample line on the Y-axis. If we substitute for a in (17.2.3), it becomes

$$\Sigma(Y - \hat{Y})^2 = \Sigma(y - b_1 x_1 - b_2 x_2)^2 \tag{17.2.5}$$

For b_1 and b_2 we have to solve two equations, often called the *normal equations* (though the word *normal* here has no connection with the normal distribution):

$$b_1(\Sigma x_1^2) + b_2(\Sigma x_1 x_2) = (\Sigma x_1 y) \tag{17.2.6}$$

$$b_1(\Sigma x_1 x_2) + b_2(\Sigma x_2^2) = (\Sigma x_2 y) \tag{17.2.7}$$

For the solutions, first calculate the difference D and then b_1 and b_2:

$$D = (\Sigma x_1^2)(\Sigma x_2^2) - (\Sigma x_1 x_2)^2 \tag{17.2.8}$$

$$b_1 = [(\Sigma x_1 y)(\Sigma x_2^2) - (\Sigma x_2 y)(\Sigma x_1 x_2)]/D \quad (17.2.9)$$

$$b_2 = [(\Sigma x_2 y)(\Sigma x_1^2) - (\Sigma x_1 y)(\Sigma x_1 x_2)]/D \quad (17.2.10)$$

The numerical example (table 17.2.1) comes from an investigation (1) of the sources from which corn plants in various Iowa soils obtain their phosphorus. The concentrations in parts per million (ppm) of inorganic (X_1) and organic (X_2) phosphorus in the soils and the phosphorous content Y of corn grown in the soils were measured for 17 soils. The sample means, sums of squares, and cross products of deviations are also given.

Substitution in (17.2.8)–(17.2.10) gives

$$D = (1519.3)(2888.5) - 835.7^2 = 3{,}690{,}104$$

$$b_1 = [(1867.4)(2888.5) - (757.5)(835.7)]/3{,}690{,}104 = 1.2902 \quad (17.2.11)$$

$$b_2 = [(757.5)(1519.3) - (1867.4)(835.7)]/3{,}690{,}104 = -0.1110 \quad (17.2.12)$$

From (17.2.3), $a = 76.18 - (1.2902)(11.07) + (0.1110)(41.18) = 66.47$. The sample regression equation is therefore

TABLE 17.2.1

INORGANIC PHOSPHORUS X_1, ORGANIC PHOSPHORUS X_2, AND ESTIMATED PLANT-AVAILABLE PHOSPHORUS Y IN 17 IOWA SOILS AT 20°C (PPM)

Soil Sample	X_1	X_2	Y	$\hat{Y}$	$Y - \hat{Y}$
1	0.4	53	64	61.1	2.9
2	0.4	23	60	64.4	−4.4
3	3.1	19	71	68.4	2.6
4	0.6	34	61	63.5	−2.5
5	4.7	24	54	69.9	−15.9
6	1.7	65	77	61.4	15.6
7	9.4	44	81	73.7	7.3
8	10.1	31	93	76.1	16.9
9	11.6	29	93	78.2	14.8
10	12.6	58	51	76.3	−25.3
11	10.9	37	76	76.4	−0.4
12	23.1	46	96	91.2	4.8
13	23.1	50	77	90.7	−13.7
14	21.6	44	93	89.5	3.5
15	23.1	56	95	90.1	4.9
16	1.9	36	54	64.9	−10.9
17	29.9	51	99	99.4	−0.4
Total	188.2	700	1295	1295.2	−0.2
Mean	11.07	41.18	76.18		

$\Sigma x_1^2 = 1519.3$ $\quad \Sigma x_1 x_2 = 835.7$ $\quad \Sigma x_2^2 = 2888.5$

$\Sigma x_1 y = 1867.4$ $\quad \Sigma x_2 y = 757.5$ $\quad \Sigma y^2 = 4426.5$

$$\hat{Y} = 66.47 + 1.2902X_1 - 0.110X_2 \qquad (17.2.13)$$

From this equation the predicted value $\hat{Y}$ is estimated for each soil sample in table 17.2.1. For instance, for soil 1,

$$\hat{Y} = 66.47 + (1.2902)(0.4) - (0.1110)(53) = 61.1 \text{ ppm}$$

The observed value $Y = 64$ ppm deviates by $+2.9$ ppm from the regression estimate. The residuals $Y - \hat{Y}$ in table 17.2.1 measure the sizes of the errors in the regression predictions of the sample points.

The predictions and the deviations are considered further in later sections. Inspection of the $Y - \hat{Y}$ values in table 17.2.1 shows, however, that the soil phosphorous readings do not predict Y very accurately; seven samples have prediction errors exceeding ± 10 ppm.

17.3—The analysis of variance and the F test. In the multiple regression model (17.2.1), the deviations of the Ys from the population regression plane have mean 0, variance σ^2. An unbiased estimate of σ^2 can be shown to be $\Sigma(Y - \hat{Y})^2/(n - k - 1)$, where n is the sample size and k the number of β parameters estimated in fitting the regression. In the example, $n = 17$ with two β parameters β_1, β_2, giving $n - k - 1 = 14$. (The fitted parameter α accounts for the -1 in the divisor.)

The deviations or residuals sum of squares $\Sigma(Y - \hat{Y})^2$ can be computed in two ways. If the individual deviations have been tabulated, as in the last column of table 17.2.1, their sum of squares is added directly, giving $\Sigma(Y - \hat{Y})^2 = 2101.1$.

If the deviations $Y - \hat{Y}$ have not been found, a quicker method follows. Substituting for a in (17.2.2) we get

$$\hat{Y} = \overline{Y} + b_1x_1 + b_2x_2 \qquad (17.3.1)$$

Since the sample means of x_1 and x_2 are both zero, the sample mean of the fitted values $\hat{Y}$ is $\overline{Y}$. Write $\hat{y} = \hat{Y} - \overline{Y}$. Putting deviations $d = Y - \hat{Y}$, we have

$$y = Y - \overline{Y} = \hat{y} + d \qquad (17.3.2)$$

Two important results (proved at the end of this section) follow. First,

$$\Sigma y^2 = \Sigma \hat{y}^2 + \Sigma d^2 \qquad (17.3.3)$$

This result states that the sum of squares of the deviations of the Ys from their mean splits into two parts: (*i*) the sum of squares of deviations of the fitted values from their mean and (*ii*) the sum of squares of deviations d from the fitted values. The sum of squares $\Sigma\hat{y}^2$ is often called *the sum of squares due to regression.*

TABLE 17.3.1
ANALYSIS OF VARIANCE OF PHOSPHOROUS DATA

Source of Variation	df	Sum of Squares	Mean Square	F
Regression	2	2325.2	1162.6	7.75**
Deviations	14	2101.3	150.1	
Total	16	4426.5	276.7	

The second result is

$$\Sigma\hat{y}^2 = b_1\Sigma x_1 y + b_2\Sigma x_2 y \tag{17.3.4}$$

From (17.3.3) and (17.3.4) the value of $\Sigma d^2 = \Sigma(Y - \hat{Y})^2$ is

$$\Sigma d^2 = \Sigma y^2 - b_1\Sigma x_1 y - b_2\Sigma x_2 y \tag{17.3.5}$$
$$= 4426.5 - (1.2902)(1867.4) - (-0.1110)(757.5) = 2101.3$$

By this method, Σd^2 can be found without calculating any individual deviations. Agreement of the two methods of finding Σd^2 (apart from rounding errors) is an excellent check on the regression computations.

The mean square of the deviations is $s^2 = 2101.3/14 = 150.1$, the sample estimate of the residual σ^2. These calculations are summarized in the analysis of variance table 17.3.1.

Sometimes an investigator is not at all confident initially that any of the Xs are related to Y. In this event, a test of the null hypothesis $\beta_1 = \beta_2 = 0$ is helpful. This test is made in the analysis of variance table from the result that if $\beta_1 = \beta_2 = 0$, the ratio (regression mean square)/(deviations mean square) $= 1162.6/150.1 = 7.75$ is distributed as F with 2 and 14 df. This F value, 7.75, is significant at the 1% level, rejecting the null hypothesis.

The remainder of this section is devoted to proofs of the basic results (17.3.3) and (17.3.4). Recall that

$$\hat{y} = \hat{Y} - \bar{Y} = b_1x_1 + b_2x_2 \qquad y = \hat{y} + d$$
$$d = y - b_1x_1 - b_2x_2$$

Start with the normal equations,

$$b_1\Sigma x_1^2 + b_2\Sigma x_1x_2 = \Sigma x_1 y$$
$$b_1\Sigma x_1x_2 + b_2\Sigma x_2^2 = \Sigma x_2 y$$

These may be rewritten in the form

$$\Sigma x_1(y - b_1x_1 - b_2x_2) = \Sigma x_1 d = 0 \tag{17.3.6}$$

$$\Sigma x_2(y - b_1x_1 - b_2x_2) = \Sigma x_2 d = 0 \tag{17.3.7}$$

These results show that the deviations d have zero sample correlations with any X-variable. This is not surprising, since d represents the part of Y that is not linearly related either to X_1 or to X_2.

Multiply (17.3.6) by b_1 and (17.3.7) by b_2 and add. Then

$$\Sigma(b_1x_1 + b_2x_2)d = \Sigma\hat{y}d = 0 \qquad (17.3.8)$$

Hence, the deviations are also uncorrelated with the predicted values. Now

$$\Sigma y^2 = \Sigma(\hat{y} + d)^2 = \Sigma\hat{y}^2 + 2\Sigma\hat{y}d + \Sigma d^2 = \Sigma\hat{y}^2 + \Sigma d^2$$

using (17.3.8). This proves the first result (17.3.3). To obtain the second result, we have

$$\Sigma\hat{y}^2 = \Sigma(b_1x_1 + b_2x_2)^2 = b_1^2\Sigma x_1^2 + 2b_1b_2\Sigma x_1x_2 + b_2^2\Sigma x_2^2$$

Reverting to the normal equations, multiply the first one by b_1, the second by b_2, and add:

$$b_1^2\Sigma x_1^2 + 2b_1b_2\Sigma x_1x_2 + b_2^2\Sigma x_2^2 = b_1\Sigma x_1y + b_2\Sigma x_2y$$

This establishes (17.3.4), the shortcut method of computing the reduction $\Sigma\hat{y}^2$ in the sum of squares due to regression.

The correlation coefficient between Y and $\hat{Y}$ is called the *multiple correlation coefficient* between Y and the Xs. From (17.3.5) and (17.3.8) it follows that $\Sigma y\hat{y} = \Sigma\hat{y}^2$, so the sample value R is always ≥ 0. Furthermore,

$$R^2 = (\Sigma y\hat{y})^2/[(\Sigma y^2)(\Sigma\hat{y}^2)] = (\Sigma\hat{y}^2)^2/[(\Sigma y^2)(\Sigma\hat{y}^2)] = \Sigma\hat{y}^2/\Sigma y^2$$

In the analysis of variance table 17.3.1, $R^2 = 2325.2/4426.5 = 0.53$.

Since $\Sigma\hat{y}^2 = R^2\Sigma y^2$, we have $\Sigma d^2 = (1 - R^2)\Sigma y^2$. Hence in the analysis of variance table, $F = (n - k - 1)R^2/[k(1 - R^2)]$. Thus the F table supplies a test of significance of the multiple correlation coefficient.

17.4—Extension of the analysis of variance. Suppose that we have fitted a regression on k X-variables but think it possible that only $p < k$ of the variables contribute to the regression. We might express this idea by the null hypothesis that $\beta_{p+1}, \beta_{p+2}, \ldots, \beta_k$ are all zero, making no assumptions about $\beta_1, \beta_2, \ldots, \beta_p$. The analysis of variance in section 17.3 is easily extended to provide an F test of this null hypothesis.

The method is as follows. Compute the sample regression of Y on all k X-variables. Then compute the sample regression of Y on $X_1, X_2, \ldots, X_p$ alone, that is, on the variables not being tested. Construct the analysis of variance in table 17.4.1.

In table 17.4.1, R_p and R_k denote the reductions in Σy^2 due to the regressions on p and k Xs. If R_k is substantially greater than R_p, this suggests

TABLE 17.4.1
ANALYSIS OF VARIANCE FOR TESTING A SUBSET OF $(k - p)$ βs

Source of Variation	df	Sum of Squares	Mean Square
Reduction to $X_1, \ldots, X_p$	p	R_p	
Additional reduction to $X_{p+1}, \ldots, X_k$	$k - p$	$R_k - R_p$	$(R_k - R_p)/(k - p)$
Residuals	$n - K - 1$	$\Sigma y^2 - R_k$	s^2
Total	$n - 1$	Σy^2	

that some of the last $(k - p)$ of the βs are nonzero. We must remember, however, that the k-variable regression has $(k - p)$ extra parameters at our disposal for increasing R_k. It has been shown that if $\beta_{p+1}, \ldots, \beta_k$ are all zero, then (*i*) $(R_k - R_p)/(k - p)$ is an unbiased estimate of the residual σ^2, just as s^2 is; and (*ii*) $(R_k - R_p)/[(k - p)s^2]$ is distributed as F with $(k - p)$ and $(n - k - 1)$ df.

To illustrate, the two-variable numerical example in section 17.3 was taken from a three-variable regression, with X_3 a different type of organic phosphorus from X_2. The reduction in Σy^2 was $R_3 = 2339.3$. The sample regression coefficients were

$$b_1 = 1.3014 \qquad b_2 = 0.1303 \qquad b_3 = 0.0228$$

We will test $H_0{:}\beta_2 = \beta_3 = 0$, namely that the organic phosphorous measures are unrelated to the plant-available phosphorus. For this, we compute $R_1 = 2295.3$, $R_3 = 2339.3$, and set up the following analysis of variance:

Source	df	Sum of Squares	Mean Square	F
Reduction to X_1	1	2295.3		
Additional reduction to X_2, X_3	2	44.0	22.0	0.14
Residuals	13	2087.2	160.6	
Total	16	4426.5		

The reduction to X_1 when it is alone in the regression is, of course, $(\Sigma x_1 y)^2/\Sigma x_1^2 = 1867.4^2/1519.3 = 2295.3$, using the results in table 17.2.1. The F value for testing $\beta_2 = \beta_3 = 0$ is 0.14 (P about 0.84), so these two forms of organic phosphorus show no relation to plant-available phosphorus.

Coming back to our two-variable example in sections 17.2 and 17.3, it is clear that we can use this method to make F tests of the two null hypotheses, (*i*) $\beta_1 = 0$ and (*ii*) $\beta_2 = 0$ (see example 17.4.2). In the next section we see how to make t tests of these hypotheses and to construct confidence limits for the β_i.

EXAMPLE 17.4.1—The following data come from a study of the relation between the number of moths caught on successive nights in a light trap and two weather variates. The variates are: Y = log(number of moths + 1), X_1 = maximum temperature (previous day) in °F, X_2 = average wind speed. The sums of squares and products of deviations are

$$\Sigma x_1^2 = 14.03 \qquad \Sigma x_1 x_2 = 1.99 \qquad \Sigma x_2^2 = 2.07$$

$$\Sigma x_1 y = 2.07 \qquad \Sigma x_2 y = -0.64 \qquad \Sigma y^2 = 3.552$$

where Σy^2 has 72 df. (*i*) Calculate b_1 and b_2. Do their signs look reasonable? (*ii*) Calculate the analysis of variance and test $H_0:\beta_1 = \beta_2 = 0$. (*iii*) By the extended analysis of variance, perform separate F tests of b_1 and b_2. Ans. (*i*) $b_1 = 0.222$, $b_2 = -0.522$. (*ii*) Analysis of variance:

Source	df	Sum of Squares	Mean Square	F
Regression on X_1, X_2	2	0.793	0.396**	10.05
Residuals	70	2.759	0.0394	
Total	72	3.552		

(*iii*) Extended analysis of variance:

Source	df	Sum of Squares	Mean Square	F
X_1	1	0.305		
X_2 after X_1	1	0.488	0.488**	12.4
Residuals	70	2.759	0.0394	
X_2	1	0.198		
X_1 after X_2	1	0.595	0.595**	15.1
Residuals	70	2.759	0.0394	

Both b_1 and b_2 are clearly significant in the presence of the other X-variable.

Several points in this example are worth noting. The F value for testing X_1 when it is included first, namely $0.305/0.0394 = 7.7$, differs from $F = 15.1$ when X_1 is tested after including X_2. This property is characteristic of multiple regression. Similarly, b_1 changes from 0.148 to 0.222 and b_2 from -0.309 to -0.522 when the other X is included in the regression. The bs both increased in absolute size when the other variable was included because the correlations of X_1 and X_2 with Y have opposite signs, yet X_1 and X_2 are positively correlated. The results in this example are used in example 17.6.1.

EXAMPLE 17.4.2—In the phosphorous example in tables 17.2.1 and 17.3.1, perform F tests of b_1 and b_2. Ans. For b_1, $F = 14.2$**; for b_2, $F = 0.2$. This type of organic phosphorus seems unrelated to plant-available phosphorus.

EXAMPLE 17.4.3—Sometimes the relation between Y and the Xs is not linear but can be made approximately linear by a transformation of the variable before analysis. Let W be the weight of a hen's egg, L its maximum length, and B its maximum breadth. Schrek (3) noted that if the shape of an egg is an ellipse rotated about its major axis (the length), the relation should be $W = cLB^2$, where c is a constant. Thus in the log scale the relation should be $\log W = a + \beta_1 \log L + \beta_2 \log B$, where $\beta_1 \doteq 1$, $\beta_2 \doteq 2$. Following are the measurements from 10 eggs. To provide easier data for calculation, we have taken $Y = 10(\log W - 1.6)$, $X_1 = 10(\log L - 1.7)$, $X_2 = 10(\log B - 1.6)$.

X_1	X_2	Y	X_1	X_2	Y
0.51	0.42	1.10	0.10	0.62	0.76
.79	.38	1.03	1.32	0.72	2.23
.52	.47	0.66	0.99	0.48	1.37
.10	.37	0.33	1.00	0.96	2.42
0.12	0.87	1.37	0.73	1.05	2.42
			6.18	6.34	13.69

$\Sigma x_1^2 = 1.625$ $\quad \Sigma x_1 x_2 = 0.196$ $\quad \Sigma x_2^2 = 0.573$

$\Sigma x_1 y = 2.017$ $\quad \Sigma x_2 y = 1.397$ $\quad \Sigma y^2 = 5.091$

(*i*) Calculate the sample regression and the analysis of variance. (*ii*) Compare the actual and predicted weights for the first and last eggs. Ans. (*i*) $\hat{Y} = 0.988X_1 + 2.100X_2 - 0.573$. Clearly, $\hat{Y} =$

$X_1 + 2X_2 - 0.517$ would do about as well. In the analysis of variance, $R_2 = 4.926$, $s^2 = 0.0236$ (7 df), $F = 104.4$. (*ii*) First egg, $W = 51.3$, $\hat{W} = 48.0$ (this prediction was the poorest); last egg, $W = 69.5$, $\hat{W} = 68.4$.

17.5—Variances and covariances of the regression coefficients. In chapter 9, with only one X-variable, we had $b = \Sigma xy/\Sigma x^2$ and $V(b) = \sigma^2/\Sigma x^2$. With two X-variables we have, from (17.2.9),

$$b_1 = [(\Sigma x_1 y)(\Sigma x_2^2) - (\Sigma x_2 y)(\Sigma x_1 x_2)]/D \tag{17.5.1}$$

where

$$D = (\Sigma x_1^2)(\Sigma x_2^2) - (\Sigma x_1 x_2)^2 = 3{,}690{,}104 \tag{17.5.2}$$

The variances and the covariance of b_1 and b_2 are

$$V(b_1) = \sigma^2(\Sigma x_2^2)/D \qquad \text{Cov}(b_1, b_2) = -\sigma^2(\Sigma x_1 x_2)/D$$
$$V(b_2) = \sigma^2(\Sigma x_1^2)/D \tag{17.5.3}$$

With k Xs, these quantities are usually denoted by $V(b_i) = \sigma^2 c_{ii}$, $\text{Cov}(b_i b_j) = \sigma^2 c_{ij}$. Note that b_1 and b_2 are correlated if x_1 and x_2 are correlated in the sample, and their correlation has the opposite sign from that of x_1 and x_2. For the phosphorous data from table 17.2.1 and (17.5.2),

$$V(b_1) = 0.00078277\sigma^2 \qquad \text{Cov}(b_1, b_2) = -0.00022647\sigma^2$$
$$V(b_2) = 0.00041172\sigma^2 \tag{17.5.4}$$

For the estimated variances and covariance we replace σ^2 by the residual mean square s^2 from the fitted regression with $(n - k - 1)$ df. Since $s^2 = 150.1$, from (17.5.4)

$$s^2(b_1) = 0.1175 \qquad s^2(b_2) = 0.0618 \qquad \text{Cov}(b_1, b_2) = -0.0340$$

This set of numbers is called the *covariance matrix* of the b_i. In computer printouts it usually appears in one of the two forms below:

	1	2
1	0.1175	−0.0340
2	−0.0340	0.0618

	1	2
1	0.1175	
2	−0.0340	0.0618

The numbers 1, 2 denote the X-variates. The covariance matrix is needed when we are interested in comparisons among the bs. The variance for a linear function of the bs, $\Sigma L_i b_i$, is

$$V\left(\sum_i L_i b_i\right) = \sum_i L_i^2 s^2(b_i) + 2\sum_i \sum_{j>i} L_i L_j \,\text{Cov}(b_i b_j) \tag{17.5.5}$$

To test the significance of b_1 and b_2, we have $b_1 = 1.2902$, $b_2 = -0.1110$ (from section 17.2). Hence, for b_1, $t = 1.2902/\sqrt{0.1175} = 3.76$** with 4 df. For b_2, $t = -0.1110/\sqrt{0.0618} = -0.45$ (not significant). These tests are identical with the F tests given by the extended analysis of variance (see example 17.4.2, where you should find $F = t^2$).

The relation between a sample regression coefficient in univariate and multivariate regressions is made clearer by the following results. The sample regression of X_1 on X_2 is

$$\hat{X}_1 = \overline{X}_1 + \frac{\Sigma x_1 x_2}{\Sigma x_2^2} x_2$$

Consider the variate $X_{1.2}$, the deviation of X_1 from its regression on X_2:

$$X_{1.2} = X_1 - \hat{X}_1 = x_1 - \frac{\Sigma x_1 x_2}{\Sigma x_2^2} x_2 \tag{17.5.6}$$

Let us find the regression coefficient of Y on $X_{1.2}$, omitting X_2 so that it is a univariate regression. Now, from (17.5.6),

$$\Sigma y x_{1.2} = \Sigma x_1 y - \frac{(\Sigma x_2 y)(\Sigma x_1 x_2)}{\Sigma x_2^2} = \frac{(\Sigma x_1 y)(\Sigma x_2^2) - (\Sigma x_2 y)(\Sigma x_1 x_2)}{\Sigma x_2^2} \tag{17.5.7}$$

Also,

$$\Sigma x_{1.2}^2 = \Sigma x_1^2 - \frac{(\Sigma x_1 x_2)^2}{\Sigma x_2^2} = \frac{D}{\Sigma x_2^2} \tag{17.5.8}$$

Hence the univariate regression coefficient of Y on $X_{1.2}$ is, by (17.5.1),

$$\frac{\Sigma y x_{1.2}}{\Sigma x_{1.2}^2} = \frac{(\Sigma x_1 y)(\Sigma x_2^2) - (\Sigma x_2 y)(\Sigma x_1 x_2)}{D} = b_1$$

To summarize, in a bivariate regression b_1 is the *univariate* regression coefficient of Y, not on X_1 but on the deviations $x_{1.2}$ of X_1 from its regression on X_2. This important result extends to any number of X-variables. With k Xs, b_1 is the *univariate* regression coefficient of Y on $X_{1.23\ldots k}$, the deviation of X_1 from its regression on all the *other* X-variates. The meaning of a regression coefficient therefore depends on the other Xs that are present in the regression. The coefficient b_i measures the linear regression of Y on the part of X_i that is uncorrelated with any of the other X-variates.

Since in our example b_1 is the regression of Y on $X_{1.2}$, it follows by the univariate formulas that $V(b_1) = \sigma^2/\Sigma x_{1.2}^2 = \sigma^2(\Sigma x_2^2)/D$, from (17.5.8). This agrees with the result for $V(b_1)$ in (17.5.3). With k X-variables, $V(b_1) = \sigma^2/\Sigma x_{1.23\ldots k}^2$.

17.6—Regression calculations by the sweep operator. In fitting multiple regression equations, many readers will use one of the standard computer programs and need not be concerned with calculation techniques. However, some knowledge of the calculations helps the understanding of multiple regression. We give here a simple method (4) of calculating the *b*s, the residual sum of squares, and the covariance matrix of the *b*s. This method is suitable for a pocket calculator or extends naturally to any number of *X*-variables, and is also adapted for computer programming.

Notation. Since the sweep operator treats Y just as it does the Xs, it helps to think of Y as X_3 in our phosphorous example, or, more generally, as X_{k+1}. The sums of squares and products Σx_i^2, $\Sigma x_i x_j$, $\Sigma x_i y$ are denoted by the slightly more compact S_{ii}, S_{ij}, $S_{i,k+1}$. In (17.5.3) the variances and the covariance of the *b*s are found by applying certain multipliers to the residual σ^2. The multiplier for finding $\text{Cov}(b_i b_j)$ is denoted by c_{ij} and that for $V(b_i)$ is c_{ii}.

Precoding. In a single-variable regression, $c_{11} = 1/\Sigma x_1^2 = 1/\ S_{11}$. In multiple regression, c_{ii} is of the order of magnitude of $1/S_{ii}$, with c_{ij} intermediate between $1/S_{ii}$ and $1/S_{jj}$. Thus, in a regression problem with $S_{11} = 20{,}874$ and $S_{22} = 0.00982$, c_{11} is of the order of 0.00004 and c_{22} is of the order of 100. These variations make it harder to retain the same number of significant figures throughout a calculation and to keep the correct number of zeros after the decimal place. If this problem seems to be present when we see the S_{ij}, precoding Y and the Xs is worth considering.

When necessary, code each variable so that S_{ii} lies between 0.1 and 10. This can always be done by dividing X_i by a power of 10, say 10^p, where p may be positive or negative. If X_i is divided by 10^p and X_j by 10^q, then S_{ii} is divided by 10^{2p}, S_{ij} by 10^{p+q}, S_{jj} by 10^{2q}. In the phosphorous example we had $S_{11} = 1519.3$. If we divide X_1 by 10^2, S_{11} becomes 0.15193. The advantage is that the coded *c*s are usually not far from 1, so the same number of decimal places can be carried for each variable.

Decoding. Suppose that X_i has been divided by 10^p, X_j by 10^q, Y by 10^r. To recover results in the original units when the calculations are finished:

1. Divide the c_{ii} by 10^{2p}; divide the c_{ij} by 10^{p+q}.
2. Divide the b_i by 10^{p-r}.
3. Multiply the residual sum of squares of Y by 10^{2r}.

In the phosphorous example we divide X_1, X_2, and Y each by 10^2. Hence, in decoding, the c_{ij} are divided by 10^4, the b_i are unchanged, and $\Sigma d_{y.12}^2$ is multiplied by $10^4 = 10{,}000$.

Calculations. For X_1, X_2, and Y (after coding), set out the sums of squares and products S_{ij} in an arrangement called a 3×3 matrix:

$$\begin{bmatrix} 0.15193 & 0.08357 & 0.18674 \\ 0.08357 & 0.28885 & 0.07575 \\ 0.18674 & 0.07575 & 0.44265 \end{bmatrix} \qquad (17.6.1)$$

Down the diagonal we have $S_{11} = \Sigma x_1^2$, $S_{22} = \Sigma x_2^2$, $S_{33} = \Sigma y^2$. Note that the sum of product terms $S_{12} = S_{21}$, etc., appear twice, symmetrically placed about the diagonal.

We now perform an operation known as sweeping on the first row and column, which leads to the following 3 × 3 matrix:

SWP[1]

$$\begin{bmatrix} -6.58198 & 0.55006 & 1.22912 \\ 0.55006 & 0.24288 & -0.02697 \\ 1.22912 & -0.02697 & 0.21312 \end{bmatrix} \tag{17.6.2}$$

The rules for sweeping with respect to row and column i are as follows:

1. *The diagonal term* S_{ii} becomes $-1/S_{ii}$. Thus 0.15193 is replaced by $-1/0.15193 = -6.58198$.

2. *Other terms in the same row or column* become $S_{ij}/S_{ii} = S_{ji}/S_{ii}$. The term 0.08357 becomes $0.08357/0.15193 = 0.55006$.

3. *Other diagonal terms* become $S_{jj} - S_{ij}^2/S_{ii}$. From 0.28885 we get $0.28885 - 0.08357^2/0.15193 = 0.24288$.

4. *Terms* S_{jk} *not in the same row, column, or diagonal* become $S_{jk} - S_{ji}S_{ik}/S_{ii}$. Here there is only one such term, $S_{23} = S_{32} = 0.7575$, which becomes $0.07575 - (0.18674)(0.08357)/0.15193 = -0.02697$.

Verify the effect of the sweeping on S_{33} and S_{13}. The procedure on a pocket calculator is quickly learned with the aid of one's fingers and the reminder: "diagonal term S_{ii}, same row, other diagonal terms, remaining terms."

5. *The next step (the last with two X-variates)* is to sweep the SWP[1] matrix on row and column 2. This gives

SWP[1]SWP[2] = SWP[1, 2]

$$\begin{bmatrix} -7.82772 & 2.26474 & 1.29020 \\ 2.26474 & -4.11726 & -0.11104 \\ 1.29020 & -0.11104 & 0.21013 \end{bmatrix} \tag{17.6.3}$$

For instance, $-4.11726 = -1/0.24288$; $-0.11104 = -0.02697/0.24288$; $-7.82773 = -6.58198 - 0.55006^2/0.24288$; $1.29020 = 1.22912 - (0.55006)(-0.02697)/0.24288$.

The results of the SWP[1]SWP[2] operations in (17.6.3) may be shown to be

$$\begin{bmatrix} -c_{11} & -c_{12} & b_1 \\ -c_{12} & -c_{22} & b_2 \\ b_1 & b_2 & \Sigma d_{y.12}^2 \end{bmatrix} \tag{17.6.4}$$

On decoding (17.6.3), we get

$$\begin{bmatrix} c_{11} & c_{12} & b_1 \\ c_{12} & c_{22} & b_2 \\ b_1 & b_2 & \Sigma d^2_{y.12} \end{bmatrix} = \begin{bmatrix} 0.00078277 & -0.00022647 & 1.2902 \\ -0.00022647 & 0.00041173 & -0.1104 \\ 1.2902 & -0.1104 & 2101.3 \end{bmatrix} \quad (17.6.5)$$

Checking back, you will find that these results agree with the b_i in (17.2.11, 17.2.12), the c_{ij} in (17.5.4), and the deviations sum of squares in table 17.3.1. These results give $s^2 = \Sigma d^2_{y.12}/(n - k - 1) = 2101.3/14 = 150.2$. From this we get the sample covariance matrix of the b_i on multiplying by the c_{ij}. Note also that before coding, the diagonal term 4426.5 in matrix (17.6.1) is Σy^2 with 16 df. With $\Sigma d^2_{y.12}$ from (17.6.5) we have the analysis of variance, the F test of $H_0{:}\beta_1 = \beta_2 = 0$, and the value of multiple R^2.

6. *With k X-variates,* the rules of sweeping with respect to a row and column remain the same. To obtain matrix (17.6.4) with k Xs, sweep in succession with respect to rows 1, 2, . . . , k. The number of individual steps increases very rapidly with k. Rows 1, 2, . . . , k may be swept in any order, giving the same result after k steps.

When we have swept on rows 1, 2, . . . , $p < k$ in a k-variate regression, the numbers $b_1, b_2, \ldots, b_p$ in the right-hand column of the matrix SWP[1, 2, . . . , p] are the regression coefficients in a regression of Y on X_1, $X_2, \ldots, X_p$, omitting the remaining X-variables. Also, at this stage the diagonal term $c_{k+1,k+1} = \Sigma d^2_{y.12\ldots p}$, so that by the extended analysis of variance an F test of the $H_0{:}\beta_{p+1}, \ldots, \beta_k = 0$ is provided from SWP[1, 2, . . . , p] and SWP[1, 2, . . . , k].

EXAMPLE 17.6.1—(*i*) In example 17.4.1, perform the SWP[1]SWP[2] operations. Note that no coding is needed. Verify that the resulting matrix SWP[1, 2] provides the b_i and the analysis of variance. (*ii*) Find the covariance matrix of the b_i. Perform t tests of b_1 and b_2 and verify that they agree with the F tests of the bs in example 17.4.1. Ans. (*i*) With five decimal places carried, rounded to four at the end,

SWP[1]SWP[2] = SWP[1, 2]

$$\begin{bmatrix} -0.0825 & 0.0793 & 0.2216 \\ 0.0793 & -0.5594 & -0.5222 \\ 0.2216 & -0.5222 & 2.7588 \end{bmatrix}$$

(*ii*) $t_1 = 3.887$, $t_2 = -3.517$.

EXAMPLE 17.6.2—In matrix (17.6.1), perform SWP[2], then SWP[1]. Verify that this gives the same result as SWP[1]SWP[2].

EXAMPLE 17.6.3—The following data (supplied by Dr. Gene M. Smith) come from a class of 66 students of nursing: X_1 = rank in high school (high values meaning high ranks), X_2 = score on a verbal aptitude test, X_3 = measure of strength of character, $Y = X_4$ = score in an exam on theory. The S_{ij} (65 df) are given below. (*i*) Verify (apart from rounding differences) the results of applying SWP[1, 2, 3].

S_{ij}

$$\begin{bmatrix} 24,633 & 2212 & 5,865 & 925.3 \\ 2,212 & 7760 & 2,695 & 745.9 \\ 5,865 & 2695 & 28,432 & 1537.8 \\ 925.3 & 745.9 & 1,537.8 & 670.3 \end{bmatrix}$$

SWP[1, 2, 3]

$$\begin{bmatrix} -0.4337 \times 10^{-4} & 0.0957 \times 10^{-4} & 0.0803 \times 10^{-4} & 0.02064 \\ 0.0957 \times 10^{-4} & -1.3536 \times 10^{-4} & 0.1086 \times 10^{-4} & 0.07542 \\ 0.0803 \times 10^{-4} & 0.1086 \times 10^{-4} & -0.3786 \times 10^{-4} & 0.04268 \\ 0.02064 & 0.07533 & 0.04264 & 529.3 \end{bmatrix}$$

(*ii*) Show that the regression coefficients and their standard errors are

$$b_1 = 0.0206 \pm 0.0192 \qquad b_2 = 0.0754 \pm 0.0340 \qquad b_3 = 0.0427 \pm 0.0180$$

Which X-variables are related to performance in theory? (*iii*) Show that $F = 5.50$ for the three-variable regression. What is the P value? (*iv*) Verify that $R^2 = 0.210$. (*v*) Test the correlation between rank in school and verbal aptitude. Ans. (*ii*) X_2, X_3; (*iii*) $P < 0.005$; (*v*) $r = 0.160$ (not significant).

EXAMPLE 17.6.4—When you are familiar with the sweep method, you will find it necessary to use only the numbers in the leading diagonal and above it. In this example, apply the method algebraically to the S_{ij} so as to verify that SWP[1, 2] produces the results claimed in (17.6.4). Remember that in the S_{ij} notation, $D = S_{11}S_{22} - S_{12}^2$, $b_1 = (S_{1y}S_{22} - S_{2y}S_{12})/D$, $b_2 = (S_{2y}S_{11} - S_{1y}S_{12})/D$, $c_{11} = S_{22}/D$, $c_{12} = -S_{12}/D$, $c_{22} = S_{11}/D$. (Instead of X_3 we use the original Y notation.) Show that SWP[1] gives in the leading diagonal and above it:

$$\begin{matrix} -1/S_{11} & S_{12}/S_{11} & S_{1y}/S_{11} \\ & D/S_{11} & (S_{2y}S_{11} - S_{1y}S_{12})/S_{11} = b_2 D/S_{11} \\ & & S_{yy} - S_{1y}^2/S_{11} \end{matrix}$$

while SWP[1, 2] produces

$$\begin{matrix} -(1 + S_{12}^2/D)/S_{11} & S_{12}/D = -c_{12} & S_{1y}/S_{11} - [S_{12}(S_{2y}S_{11} - S_{1y}S_{12})/(S_{11}D)] \\ & -S_{11}/D = -c_{22} & b_2 \\ & & S_{yy} - S_{1y}^2/S_{11} - b_2^2 D/S_{11} \end{matrix}$$

After simplification of the terms in row 1, column 1, and column 3, we get

SWP[1, 2]

$$\begin{matrix} -S_{22}/D = c_{11} & -c_{12} & (S_{1y}/S_{22} - S_{2y}S_{12})/D = b_1 \\ & -c_{22} & b_2 \\ & & S_{yy} - S_{1y}^2/S_{11} - b_2^2 D/S_{11} \end{matrix}$$

In the (3, 3) cell, the term $-S_{1y}^2/S_{11}$ is the reduction in Σy^2 due to the regression on X_1 alone. Furthermore, since $V(b_2) = c_{22}\sigma^2 = \sigma^2 S_{11}/D$, the term $-b_2^2 D/S_{11}$ is the additional reduction in the sum of squares due to X_2 after fitting X_1. Thus the (3, 3) cell gives $\Sigma d_{y.12}^2$.

17.7—Statistics in computer printouts. In computer programs for multiple regression, the results available and the way in which they are presented and labeled vary with the program. However, in most programs the items in figure 17.7.1 are available, either automatically or if requested. As an illustration we have used the two X-variable phosphorous regression; a computer printout

REGRESSION TITLE....... PLANT PHOSPHORUS

NUMBER OF CASES 17

VARIABLE	MEAN	ST. DEV.	C.V.
1 INORGANIC	11.07	9.745	0.880
2 ORGANIC	41.18	13.436	0.326
3 PLANT AVAILABLE	76.18	16.633	0.218

DEPENDENT VARIABLE 3

CORRELATION MATRIX

	1	2	3
1	1.000		
2	0.399	1.000	
3	0.720	0.212	1.000

MULTIPLE R-SQUARE 0.525

ANALYSIS OF VARIANCE

SOURCE	SUM OF SQUARES	DF	MEAN SQUARE	F RATIO	P (TAIL)
REGRESSION	2325.2	2	1162.6	7.75	0.005
RESIDUAL	2101.3	14	150.1		

VARIABLE	COEFF	S.E.	STD. REG COEFF	T	P (2 TAIL)
INTERCEPT	66.47				
INORGANIC 1	1.2902	0.343	0.756	3.76	0.002
ORGANIC 2	−0.1110	0.249	−0.090	−0.45	0.660

COVARIANCE MATRIX OF REG COEFF

	1	2
1	0.1175	
2	−0.0340	0.0618

Fig. 17.7.1—Statistics given in most computer program printouts.

usually carries several more decimal places than given here. All the statistics in figure 17.7.1 have been computed in previous sections except where mentioned.

First are the sample size (number of cases); the means; standard deviations; and the coefficients of variation of X_1, X_2, and $Y = X_3$. Then comes the correlation matrix of these three variables (not calculated previously). In this, the off-diagonal figures are the sample correlation coefficients r_{12}, r_{13}, r_{23}. These figures sometimes reveal a discrepancy in calculations and can alert the user to a computing problem. Algebraically, the multiple R^2 must be at least as great as the maximum r_{iy}^2. Hence, if one or two r_{iy} exceed 0.9, we know that R^2 must be high. The phosphorous example has $r_{1y}^2 = 0.518$ and $R^2 = 0.525$; we learn that X_2 scarcely makes a difference in the presence of X_1. Furthermore, if r_{12}, r_{13} are very high in a multiple regression, $\Sigma x_{1.23...}^2$ is likely to be small, so $V(b_1) = \sigma^2/\Sigma x_{1.23...}^2$ will be large and b_1 will be poorly determined. We may find in this case that $V(b_2)$ is very large also if X_1 is in the regression. Disentangling the effects of X_1 and X_2 on Y may not be feasible in this situation. Finally, when several rs are very high, there is a risk of severe loss of decimal accuracy in the calculations and extra decimal places must be carried.

The remaining printout material presents R^2; the analysis of variance and the F test; the estimates a, b_1, b_2, with t tests of b_1 and b_2; and the covariance matrix of the bs. The only new figures are the STD. REG COEFF—the standard regression coefficients. If b_i' denotes b_i in standard units, $b_i' = b_i s_{x_i}/s_y$. For example, for X_1, $b_1' = (1.2902)(9.745)/16.633 = 0.756$. These coefficients were developed in earlier attempts to compare the strengths of the relations between different Xs and Y. The quantity b_i estimates the average change in Y per unit increase in X_i (with other Xs unchanged). But if X_1 varies much more in the population than X_2, the population variation in Y could be due much more to variations in X_1 than to X_2 (which does not vary much), even if $b_1 < b_2$. The rescaling removes the effect of this difference in variability but does not remove the difficulty that the meaning and the value of a regression coefficient depend on the other Xs in the regression.

17.8—Examination of residuals. On request, most computer programs will print the residuals $Y - \hat{Y}$ and the predicted values $\hat{Y}$. Some will print supplementary plots that help the user to judge whether the assumptions made in a regression analysis are satisfied in the data. Among the plots are:

1. $Y - \hat{Y}$ against $\hat{Y}$
2. $Y - \hat{Y}$ against each X_i
3. $Y - \hat{Y}$ on normal probability paper

1. The plot against $\hat{Y}$ gives a rough check on constancy of residual variances as Y changes. Probably the most common departure is finding that the $|Y - \hat{Y}|$ values increase with $\hat{Y}$, suggesting that σ^2 increases with Y. Possible remedies are a weighted regression, with weights some reasonable function of $\hat{Y}$, or a variance-stabilizing transformation for Y if this does not distort linearity.

2. In plots against each X_i we look primarily for signs of nonlinearity in the

regression. A simple type of curvature that is caused by a term in X_i^2 is suggested when most residuals have one sign when X_i is low or high and the opposite sign with intermediate X_i. Addition of a term in X_i^2 or use of a nonlinear function of X_i in place of X_i should improve the fit.

3. With signs of marked skewness or kurtosis (sections 5.13, 5.14) in the probability plot, a transformation of Y nearer to normality may help. Another possibility is to give up the least squares method of fitting, which has low efficiency when the residuals are long-tailed. Mosteller and Tukey (5) discuss and illustrate alternative methods of fitting that are designed to be more robust. One method makes repeated use of weighted least squares, the weights depending on the absolute sizes of the residuals in the preceding fit. Up to a certain size, smaller residuals receive almost equal weight, but unusually large residuals are weighted down severely.

The hints received from the preceding plots are frequently not very clear or easy to interpret, particularly with multiple regression in small samples, but they should indicate extreme failures in the assumptions that need attention.

A suspiciously large residual may be tested approximately by the method in section 9.11. Suppose the suspect Y value is Y_m. When Y_m is one of the observations from which the regression was computed, the standard error of $Y_m - \hat{Y}_m$ is $\sigma/\sqrt{g}$, where

$$g = 1 - 1/n - \sum_i^k \sum_j^k c_{ij}x_i x_j = 1 - 1/n - c_{11}x_1^2 - 2c_{12}x_1x_2 - c_{22}x_2^2 \quad (17.8.1)$$

for two X-variables, where x_i, x_j refer to the mth observation. Furthermore, the quantity

$$s'^2 = \left[\sum^n (Y - \hat{Y})^2 - (Y_m - \hat{Y}_m)^2/g\right] \Big/ (n - k - 2) \quad (17.8.2)$$

can be shown to be the mean square of the deviations obtained if the suspect Y_m is omitted when fitting the regression. If Y_m were chosen randomly, the quantity

$$t = (Y_m - \hat{Y}_m)/(s'\sqrt{g})$$

would follow the t distribution with $(n - k - 2)$ df, in this case 13 df. The largest deviation in table 17.2.1 is -25.3 for observation 10. The value of g for this deviation is found to be 0.8345, while $\Sigma(Y - \hat{Y})^2$ is 2101.3 (table 17.3.1) with 14 df. Hence, from (17.8.2),

$$s'^2 = (1/13)(2101.3 - 25.3^2/0.8345) = 102.6$$

$$t = 25.3/\sqrt{(102.6)(0.8345)} = 2.73$$

The P value from the t table (A 4) is 0.017. Multiplying P by $n = 17$ to make a rough allowance for the fact that we selected the largest of 17 deviations, we get $nP = 0.29$, so this deviation is not unusual.

17.9—Standard errors of predicted values. When the objective is to predict the population regression plane or the value of Y for a new specimen, the formulas for the standard errors of the predicted values are natural extensions of those in sections 9.8–9.10 for univariate regression. The predicted values themselves for either application are

$$\hat{Y} = \overline{Y} + \sum_{i}^{k} b_i(X_i - \overline{X}_i) \tag{17.9.1}$$

where the X_i have the values relevant to the problem. We might, for instance, predict the average weight of 11-year-old boys of specified height X_1, chest girth X_2, and leg length X_3. The estimated variance of $\hat{Y}$ when predicting the population regression value μ for this set of Xs is

$$s_{\hat{\mu}}^2 = s^2\left(1/n + \sum_{i}^{k} c_{ii}x_i^2 + 2\sum_{i}^{k}\sum_{j>i}^{k} c_{ij}x_ix_j\right) \tag{17.9.2}$$

To illustrate the calculation, suppose that we were predicting the average plant-available phosphorus for soils with $X_1 = 4$, $X_2 = 24$. For this regression, previous calculations gave $s = 12.25$, $n = 17$, $c_{11} = 0.0007828$, $c_{22} = 0.0004117$, $c_{12} = -0.0002265$, $x_1 = -7.1$, $x_2 = -17.2$. Hence

$$s_{\hat{\mu}} = (12.25)[1/17 + (0.0007828)(7.1^2) + (0.0004117)(17.2^2)$$
$$+ 2(-0.0002265)(-7.1)(-17.2)]^{1/2} = \pm 4.97 \text{ ppm}$$

If $\hat{Y}$ is used to predict the value of Y for an individual *new* member of the population (not one in the regression calculations), we have, instead of (17.9.2),

$$s_{\hat{Y}}^2 = s^2\left(1 + 1/n + \sum_{i}^{k} c_{ii}x_i^2 + 2\sum_{i}^{k}\sum_{j>i}^{k} c_{ij}x_ix_j\right) \tag{17.9.3}$$

The confidence limits $\hat{Y} \pm t_\alpha s_{\hat{Y}}$ apply to a *single* prediction. As noted in section 9.10, if predictions are being made from the same line for n' new specimens, the wider confidence limits $\hat{Y} \pm t_{\alpha/n'} s_{\hat{Y}}$ must be used to ensure a confidence probability $1 - \alpha$ for all n' predictions.

EXAMPLE 17.9.1—If Y_i is one of the observations from which the regression was computed, the variance of $Y_i - \hat{Y}$ is, from (17.8.1),

$$\sigma^2(1 - 1/n - c_{11}x_2^2 - c_{22}x_2^2 - 2c_{12}x_1x_2)$$

If this expression is added over all the n sample values, we get

$$\sigma^2(n - 1 - c_{11}\Sigma x_1^2 - c_{22}\Sigma x_2^2 - 2c_{12}\Sigma x_1x_2)$$

From the equations (17.5.3) for the cs, show that the above equals $\sigma^2(n + 3)$. This is one way of seeing that $\Sigma(Y - \hat{Y})^2$ has $(n - 3)$ df when there are 2 xs, or more generally, $(n - k - 1)$ df.

EXAMPLE 17.9.2—For a soil with the relatively high values of $X_1 = 30$, $X_2 = 50$, calculate the standard errors of (*i*) the prediction of the population Y, (*ii*) the prediction of Y for a new individual soil. The values of the cs and of s are given following (17.9.2). Ans. (*i*) ± 6.65 ppm, (*ii*) ± 13.94 ppm.

EXAMPLE 17.9.3—What is the standard error of the predicted value of Y for a new individual soil with $X_1 = 30$ if X_1 alone is included in the regression prediction? Ans. ± 13.62 ppm, slightly smaller than when X_2 is included.

EXAMPLE 17.9.4—In the eggs example, 17.4.3, find the standard error of log W for the predicted value for a new egg that has the sample average values of X_1 and X_2. Note that $Y = 10(\log W - 1.6)$ and that s^2 is given in the answer to example 17.4.3. Ans. ± 0.0161.

17.10—Interpretation of regression coefficients. In observational studies, multiple regression analyses are used extensively in attempts to disentangle and measure the effects of different X-variables on some response Y. However, there are important limitations on what can be learned by this technique. In a multiple regression, as we have seen, b_1 is the univariate regression of Y on the deviations of X_1 from its linear regression on $X_2, X_3, \ldots, X_k$. When the X-variables are highly intercorrelated, both the observed size of b_1 and its meaning may be puzzling. To illustrate, table 17.10.1 shows three variables: Y = gross national product, X_1 = total number employed, X_2 = total population (excluding institutions) aged 14 years or over. The data are for the 16 years from 1947 to 1962. These data were taken from Longley's (6) study of the performance of computer programs in multiple regression when there are some high intercorrelations. Our data have been rounded. High correlations are $r_{1y} = 0.983$, $r_{2y} = 0.986$, $r_{12} = 0.953$.

We might expect total employment (X_1) and the GNP (Y) to be highly correlated, as they are ($r_{1y} = 0.983$), with $b_1' = 27.6$ billion dollars per million workers when X_1 alone is fitted. However, the correlation between the GNP and the total population over 14 is slightly higher still ($r_{2y} = 0.986$), which is puzzling. Also, when both variates are fitted, b_1' drops to $b_1 = 13.50$ (less than half). Thus the size of a regression coefficient is a slippery quantity when

TABLE 17.10.1

REGRESSION OF Y = GNP (BILLIONS OF DOLLARS) ON X_1 = TOTAL EMPLOYMENT (MILLIONS) AND X_2 = POPULATION AGED 14 AND OVER (MILLIONS) (1947–1962).

No.	X_1	X_2	Y	$\hat{Y}$	No.	X_1	X_2	Y	$\hat{Y}$
1	60.3	108	234	246.4	9	66.0	117	396	393.4
2	61.1	109	259	265.0	10	67.9	119	419	434.6
3	60.2	110	258	260.6	11	68.2	120	443	446.4
4	61.2	112	285	289.7	12	66.5	122	445	439.0
5	63.2	112	329	316.7	13	68.7	123	483	476.5
6	63.6	113	347	329.9	14	69.6	125	503	504.2
7	65.0	115	365	364.3	15	69.3	128	518	523.5
8	63.8	116	363	355.9	16	70.6	130	555	556.6

$\Sigma X_1^2 = 186.33$ $\Sigma x_1 x_2 = 341.625$ $\Sigma x_2^2 = 689.9375$

$\Sigma x_1 y = 5{,}172{,}625$ $\Sigma x_2 y = 9978.19$ $\Sigma y^2 = 148{,}535$

intercorrelations are high; its value and even its sign can change according to the other Xs that are entered.

The prediction equation is

$$\hat{Y} = 13.5026X_1 + 7.7766X_2 - 1407.642$$

with $s_{Y.12} = \pm 9.172$, about 2.3%. Both b_1 and b_2 are highly significant. Thus, predictions of the GNP are quite good, but it is not easy to interpret the bs, particularly b_2, in any causal sense.

Incidentally, the residuals $Y - \hat{Y}$ calculated from table 17.10.1 show the typical pattern that suggests some year-to-year correlation. Four negative values are followed by five positive values, with three negatives at the end.

In the phosphorous study, on the other hand, the regression coefficient b_1 on inorganic phosphorus changed very little from the univariate to the trivariate regression, giving us more confidence in its numerical value. In seeking data, it helps if the investigator can find a set in which the intercorrelations among the xs are small.

17.11—Omitted *X*-variables. Another difficulty is that the fitted regression may not include some X-variables that are present in the population regression. These variables may be thought to be unimportant, not feasible to measure, or unknown to the investigator. To keep the algebra simple, we replace the omitted terms $\Sigma\beta_j X_j$ by a single term $\beta_o X_o$ that stands for their joint effect (where the subscript o denotes "omitted").

To take the easiest case, suppose that a regression of Y on X_1 alone is fitted to a random sample, when the correct model is

$$Y = \alpha + \beta_1 X_1 + \beta_o X_o + \epsilon'$$

where the residual ϵ' has mean zero and is uncorrelated with X_1 and X_o.

In repeated samples in which the Xs are fixed,

$$E(b_1) = E\frac{\Sigma x_1 y}{\Sigma x_1^2} = \frac{\beta_1 \Sigma x_1^2 + \beta_o \Sigma x_1 x_o}{\Sigma x_1^2} = \beta_1 + \beta_o b_{o1} \qquad (17.11.1)$$

where b_{o1} is the sample linear regression of X_o on X_1. (If X_o has a linear regression on X_1 in the population and we average further over random samples, we get $E(b_1) = \beta_1 + \beta_o \beta_{o1}$.)

In a bivariate regression in which a third variable X_o is omitted,

$$E(b_1) = \beta_1 + \beta_o b_{o1.2} \qquad (17.11.2)$$

in samples with fixed Xs. (See example 17.11.2.) Here $b_{o1.2}$ is the sample regression coefficient of X_o on X_1 after allowing for the effects of X_2. It is usually hard to form a judgment about the amount of bias from this source.

For example, an agronomist might try to estimate the effects of nitrogen and phosphorous fertilizers on the yield of a common farm crop by taking a sample of farms. On each field the crop yield Y at the most recent harvest and the amounts X_1, X_2 of N and P per acre applied in that field are recorded. If, however, substantial amounts of fertilizer are used mainly by the more competent farmers, the fields on which X_1 and X_2 have high values will, in general, have better soil, more potash fertilizer, superior drainage and tillage, more protection against insect and crop damage, and so on. If $\beta_o X_o$ denotes the combined effect of these variables on yield, X_o will be positively correlated with X_1 and X_2, so $b_{o1.2}$ will be positive. Further, β_o will be positive if these practices increase yields. Thus the regression coefficients b_1 and b_2 will overestimate the increase in yield caused by additional amounts of N and P. This type of overestimation is likely to occur whenever the beneficial effects of an innovation in some process are being estimated by regression analysis, if the more capable operators are the innovators.

In planning studies in which the sizes of the regression coefficients are of primary interest, a useful precaution is to include in the regression any X-variable that seems likely to have a material effect on Y, even though this variable is not of direct interest. Note from (17.11.2) that no contribution to the bias in b_1 comes from β_2, since X_2 was included in the regression. Another strategy is to find, if possible, a source population in which X-variables not of direct interest have only narrow ranges of variation. The effect is to decrease $b_{o1.2}$ (see example 17.11.1) and hence lessen the bias in b_1. It also helps if the study is repeated in diverse populations that are subject to different X_o variables. Finding stable values for b_1 and b_2 gives reassurance that the biases are not major.

Incidentally, when the purpose is to find a regression formula that predicts Y accurately, the bias in b_1 may actually be advantageous. If the unknown variables in X_o are good predictors of Y and are stably related to X_1, the regression value of b_1 is trying to improve the prediction by capitalizing on these relationships. This can be seen from an artificial example. Suppose that the correct model is $Y = 1 + 3X_o$. This implies that in the correct model (*i*) X_1 is useless as a predictor, since $\beta_1 = 0$; (*ii*) if X_o could be measured it would give perfect predictions, since the model has no residual term ϵ'. In the data (table 17.11.1), we have constructed an X_1 that is highly correlated with X_o. You may check that the prediction equation based on the regression of Y on X_1,

TABLE 17.11.1
PREDICTION FROM A REGRESSION MODEL WITH AN OMITTED VARIABLE

Observation	X_o	$Y = 1 + 3X_o$	X_1	$\hat{Y}_1$	$Y - \hat{Y}_1$
1	1	4	0	2.5	+1.5
2	2	7	2	9.5	−2.5
3	4	13	3	13.0	0.0
4	6	19	5	20.0	−1.0
5	7	22	5	20.0	+2.0
Total	20	65	15	65.0	0.0
Mean	4	13	3	13.0	0.0
$\Sigma x_1^2 = 18$	$\Sigma x_1 y = 63$	$\Sigma x_o x_1 = 21$	$b_1 = 63/18 = 3.5$	$b_{o1} = 21/18 = 7/6$	

$$\hat{Y}_1 = 2.5 + 3.5X_1$$

gives good, although not perfect, predictions. Since $\beta_o = 3$, $b_{o1} = 7/6$, and $b_1 = 7/2$, the relation $b_1 = \beta_o b_{o1}$ is also verified.

EXAMPLE 17.11.1—When there are omitted variables (denoted by X_o), the bias they create in b_1 depends both on the size β_o of their effect on Y and on the extent to which X_o varies. Let $Y = X_1 + X_o$ so that $\beta_1 = \beta_o = 1$. In sample 1, X_1 and X_o have the same distribution. Verify that $b_1 = 2$. In sample 2, X_1 and X_o still have a perfect correlation but the variance of X_o is greatly reduced. Verify that b_1 is now 1.33, giving a much smaller bias. Of course, steps that reduce the correlation between X_1 and X_o are also helpful.

Sample 1			Sample 2		
X_1	X_o	Y	X_1	X_o	Y
−6	−6	−12	−6	−2	−8
−3	−3	−6	−3	−1	−4
0	0	0	0	0	0
0	0	0	0	0	0
9	9	18	9	3	12
0	0	0	0	0	0
$\Sigma x_1^2 = 126$	$\Sigma x_1 y = 252$		$\Sigma x_1^2 = 126$	$\Sigma x_1 y = 168$	

EXAMPLE 17.11.2—Suppose that we fit a bivariate regression of Y on X_1, X_2 but that the correct model is

$$Y = \alpha + \beta_1 X_1 + \beta_2 X_2 + \beta_o X_o + \epsilon \qquad \epsilon \sim (0, \sigma^2) \tag{1}$$

Assume the Xs fixed in repeated samples. Let the sample bivariate regression of X_o on X_1, X_2 be

$$\hat{X}_o = a_{o.12} + b_{o1.2}X_1 + b_{o2.1}X_2$$

Substitute $d_{o.12} + \hat{X}_o$ for X_o in (1). (*i*) Show that the population regression in equation (1) above may then be written, in terms of X_1, X_2, and $d_{o.12}$,

$$Y = \alpha + \beta_o a_{o.12} + (\beta_1 + \beta_o b_{o1.2})X_1 + (\beta_2 + \beta_o b_{o2.1})X_2 + \beta_o d_{o.12} \tag{2}$$

It follows that if a trivariate sample regression of Y on X_1, X_2, and $d_{o.12}$ is fitted, the sample regression coefficients b_1', b_2', and b_o', say, are unbiased estimates of the population regression coefficients in equation (2). (*ii*) Write down the normal equations for b_1' and b_2' in this trivariate regression and verify that they are exactly the same as the normal equations for b_1 and b_2 in a bivariate regression of Y on X_1 and X_2 with X_o omitted. This shows that in repeated samples with fixed Xs, b_1 is an unbiased estimate of $\beta_1 + \beta_2 b_{o1.2}$, and similarly for b_2.

17.12—Effects possibly causal. In many problems the variables X_1 and X_2 are thought to have causal effects on Y. We would like to learn how much Y will be increased or decreased by a given change ΔX_1 in X_1. The estimate of this amount suggested by a bivariate regression equation is $b_1 \Delta X_1$. As we have seen, this quantity is actually an estimate of $(\beta_1 + \beta_o b_{o1.2})\Delta X_1$. Further, while we may be able to impose a change of amount ΔX_1 in X_1, we may be unable to control other consequences of this change. These consequences may include changes ΔX_2 in X_2 and ΔX_o in X_o. Thus the real effect of a change ΔX_1 may be

$$\beta_1 \Delta X_1 + \beta_2 \Delta X_2 + \beta_o \Delta X_o \tag{17.12.1}$$

whereas our estimate of this amount, which assumes that ΔX_1 can be changed without producing a change in X_2 and ignores the unknown variables, approximates $(\beta_1 + \beta_o b_{o1.2})\Delta X_1$. If enough is known about the situation, a more realistic mathematical model can be constructed, perhaps involving a system of equations or path analysis (7, 8). In this way a better estimate of (17.12.1) might be made, but estimates of this type are always subject to hazard. As Box (9) has remarked in a discussion of this problem in industrial work, "To find out what happens to a system when you interfere with it you have to interfere with it (not just passively observe it)."

To sum up, when it is important to find some way of increasing or decreasing Y, multiple regression analyses provide indications as to which X-variables might be changed to accomplish this end. Our advance estimates of the effects of such changes on Y, however, may be wrong by substantial amounts. If these changes are to be imposed, we should plan whenever feasible a direct study of the effects of the changes on Y so that false starts can be corrected quickly.

In controlled experiments these difficulties can be largely overcome. The investigator is able to impose the changes (treatments) whose effects are desired and to obtain direct measurements of them. The extraneous and unknown variables represented by X_o are present just as in observational studies. But the device of randomization plus replication (10, 11) makes X_o in effect independent of X_1 and X_2 in the probability sense. Thus X_o acts like the residual term ϵ in the standard regression model and the assumptions of this model are more nearly satisfied. Of course, if the effects of X_o are large, the deviations mean square, which is used as the estimate of error, will be large, and the experiment may be too imprecise to be useful. A large error variance should lead the investigator to study the uncontrolled sources of variation to find a way of doing more accurate experimentation.

In this connection, it would be interesting to compare data in which the same effect (presumably causal) has been estimated by regression methods in observational surveys and in controlled, randomized experiments. Can we learn something from cases in which the estimates agree and cases in which they disagree? Data that are strictly comparable are naturally hard to find. Yates and Boyd (12) report a comparison of estimates of the effect of farmyard manure (FYM) on the yields of potatoes. In randomized experiments the estimated effect of FYM on yield was an average increase of 2.8 T/acre when no inorganic fertilizers were applied (234 experiments) and 1.4 T/acre when inorganics were also applied (132 experiments). In a sample survey the estimated effect of FYM was an average *decrease* in yield of 0.2 T/acre after adjustment for other variables thought relevant. The authors' guess about the reason for this wide discrepancy in estimates is that specialist potato growers who get high yields often have few livestock and little or no FYM to apply. Livestock farmers, on the other hand, have plenty of FYM to apply to potatoes if they grow any. But they are not particularly skillfull at growing potatoes—it is not their livelihood. Thus, in this explanation the survey farms with large amounts of FYM do not get very high potato yields because high yields are not of major importance to such farmers.

17.13—Relative importance of different *X*-variables. In a multiple regression analysis the question may be asked, Which X variables are most important in determining Y? Usually no unique or fully satisfactory answer can be given, though several approaches have been tried. Consider first the situation in which the objective is to predict Y or to "explain" the variation in Y. The problem would be fairly straightforward if the X-variables were independent. From the model

$$Y = \alpha + \beta_1 X_1 + \beta_2 X_2 + \ldots + \beta_k X_k + \epsilon$$

we have, in the population,

$$\sigma_Y^2 = \beta_1^2\sigma_1^2 + \beta_2^2\sigma_2^2 + \ldots + \beta_k^2\sigma_k^2 + \sigma^2$$

where σ_i^2 denotes the variance of X_i. The quantity $\beta_i^2\sigma_i^2/\sigma_Y^2$ measures the fraction of the variance of Y attributable to its linear regression on X_i. This fraction can reasonably be regarded as a measure of the relative importance of X_i. With a random sample from this population, the quantities $b_i^2\Sigma x_i^2/\Sigma y^2$ are sample estimates of these fractions. (In small samples a correction for bias might be advisable, since $b_i^2\Sigma x_i^2/\Sigma y^2$ is not an unbiased estimate of $\beta_i^2\sigma_i^2/\sigma_Y^2$.)

As mentioned in section 17.7, the square roots of these quantities, $b_i\sqrt{\Sigma x_i^2/\Sigma y^2}$, are called the *standard regression coefficients* and have sometimes been used as measures of relative importance, with the Xs ranked in order of the sizes of these coefficients (ignoring sign).

When the Xs are correlated, however, the contribution of an X to Σy^2 depends on the order in which this X is introduced into the model. The analysis of variance in table 17.13.1 comes from the regression of Y = GNP on X_1 = total number employed and X_2 = total population over 14 in table 17.10.1. The contribution of X_1 to Σy^2 drops from 143,535 to 3132 when it is introduced second.

Occasionally, the investigator's question may be a different one: Is X_1 alone a better predictor of Y than X_2 alone? (This question may arise if the bivariate

TABLE 17.13.1

EFFECT OF THE ORDER IN WHICH AN *X*-VARIABLE IS BROUGHT INTO THE MODEL

Source	df	Sum of Squares	Source	df	Sum of Squares
Regression on			Regression on		
X_1	1	143,535	X_2	1	144,309
X_2 after X_1	1	3,846	X_1 after X_2	1	3,132
Deviations	13	1,094	Deviations	13	1,094
Total	15	148,535	Total	15	148,535

prediction from X_1 and X_2 is little better than that from a single one of the Xs and measurement is expensive, so that it seems wise to use only one X. A test for this problem was given by Hotelling (13) for two X-variables and extended by Williams (14) to more than two.

In some applications there may be a rational way of deciding the order in which the Xs should be brought into the regression so that their contributions to Σy^2 sum to the correct combined contribution. In his early studies of variation in the yields of wheat grown continuously on the same plots for many years at Rothamsted, Fisher (2) postulated the sources of variation in the following order: (*i*) a steady increase or decrease in level of yield, measured by a linear regression on time; (*ii*) other slow changes in yields through time, represented by a polynomial in time with terms in T^2, T^3, T^4, T^5; (*iii*) the effect of total annual rainfall on the deviations of yields from the temporal trend; (*iv*) the effect of the distribution of rainfall throughout the growing season on the deviations from the preceding regression.

Finally, if the purpose is to learn how to change Y in some population by changing some X-variable, the investigator might estimate the sizes ΔX_1, ΔX_2, etc., of the changes that can be imposed on X_1 and X_2 in this population by a given expenditure of resources. Then the variables can be rated in the order of the sizes of $b_i \Delta X_i$ in absolute terms, these being the estimated amounts of change that will be produced in Y. As we have seen in the preceding section, this approach has numerous pitfalls.

17.14—Selection of variates for prediction. A problem related to that in the preceding section arises when a regression is constructed for prediction. Often some of the X-variables contribute little or nothing to the accuracy of the predictions. For instance, we may start with 11 X-variables, but a suitable choice of 3 of them may give the best predictions. The problem is to decide how many variables to retain and which ones.

The most thorough approach is to work out the regression of Y on every subset of the k X-variables, that is, on each X-variable singly, on every pair of X-variables, on every triplet, and so on. With 11 X-variables, the number of regressions to be computed is $2^{11} - 1$, or 2047. Fortunately, algorithms have been developed for computing regressions very quickly and for selecting the best of a group of regressions according to the criteria without having to work out all members of the group. For example, program P9R in the BMDP *P*-series of programs (1977) (15) can handle 27 X-variables, selecting the best 5 or 10, say, at the option of the statistician.

As a criterion by which to rate different subsets, the value of multiple R^2 = regression sum of squares$/\Sigma y^2$ has sometimes been used. This has the disadvantage, however, that R^2 always increases as more Xs are added to the regression, so that R^2 is not suitable for comparing a subset having 3 X-variables with one having, say, 7 X-variables. A criterion that is free from this disadvantage is the residual mean squares $s_p^2 = \Sigma(Y - \hat{Y})^2/(n - p - 1)$ when the subset contains p Xs. An equivalent criterion is one called the *adjusted* $R^2 = 1 - s_p^2/s_y^2$, where, as usual, $s_y^2 = \Sigma y^2/(n - 1)$.

However, in comparing subgroups with different ps, the residual mean squares s_p^2 and adjusted R^2 have a similar defect for which Mallows (16) proposed a slightly different criterion. When $\hat{Y}$ is actually used to predict for new specimens, the variance of a prediction contains terms

$$\sigma^2\left(1/n + \sum_i^p \sum_j^p c_{ij}x_i x_j\right) \tag{17.14.1}$$

due to sampling errors in a and the b_i', as shown in (17.9.3). The sum of such terms over the n sample values is identically $(p + 1)\sigma^2$. (You were asked to prove an equivalent result for $p = 2$ as part of example 17.9.1.) This sampling error contribution becomes larger as p increases. Consequently, a subset with $p = 3$ may make better predictions than one with $p = 7$ even if the residual $s_3^2 > s_7^2$ and the s_p^2 criterion chooses $p = 7$.

For comparing subsets with different values of $p < k$, Mallows' criterion C_p is essentially the total mean square error of predictions made for new Ys from the chosen regression at all the sample sets of values of the Xs. If the residual variance σ^2 from the population regression were known, we could use as the criterion

$$\Sigma d_{y.p}^2 + (p + 1)\sigma^2$$

where $\Sigma d_{y.p}^2$ is the residual sum of squares from the regression on the chosen subset of p variables and $(p + 1)\sigma^2$ is the expected increase in the total mean square error due to sampling errors in a and the bs that affect the predictions. Mallows uses the quantity

$$[\Sigma d_{y.p}^2 - (n - p - 1)\sigma^2] + (p + 1)\sigma^2 = \Sigma d_{y.p}^2 - (n - 2p - 2)\sigma^2 \tag{17.14.2}$$

In this, the expected value of the first term is zero if none of the Xs omitted from the subset contributes to the prediction, since $E(\Sigma d_{y.p}^2) = (n - p - 1)\sigma^2$ in that case. But if, in keeping p small, we leave out some Xs that do contribute, $E(\Sigma d_{y.p}^2)$ exceeds zero, so this term is a contribution due to possible biases in predictions from the subset.

Usually s_k^2, the residual mean square from the regression on all k Xs, is the best available estimate of σ^2 in (17.14.2). In addition, Mallows divides by s_k^2, so that C_p is

$$C_p = \frac{(n - p - 1)s_p^2}{s_k^2} - (n - 2p - 2) = \frac{\Sigma d_{y.p}^2}{s_k^2} - (n - 2p - 2) \tag{17.14.3}$$

In most presentations of C_p, a is counted as one of the regression coefficients, so their p denotes our $p + 1$ and their subtraction term is $n - 2p$ in C_p. In this notation, $E(C_p) = p$ when $E(s_p^2) = \sigma^2$ so that there is no bias.

When C_p and s_p^2 differ in their subset rankings, C_p chooses a subset with smaller p. For instance, in the regression with four variables for which results for all subsets were worked out by Hald (17), $n = 13$, s_4^2 (number of Xs omitted) =

5.990. The best four subsets according to the s_p^2 and C_p criteria were as follows:

Subset	s_p^2	Subset	C_p
1, 2, 4	5.330	1, 2	2.67
1, 3, 4	5.347	1, 2, 4	3.01
1, 2	5.791	1, 3, 4	3.03
1, 4	7.476	1, 4	5.48

The criterion C_p chooses 1, 2, while s_p^2 chooses either 1, 2, 4 or 1, 3, 4, which perform about equally well. For further illustrations of the use of C_p, see (18).

EXAMPLE 17.14.1—In studies of the fertilization of red clover by honeybees (22), it was desired to learn the effects of various lengths of the insects' proboscises. The measurement is difficult, so a pilot experiment was performed to determine more convenient measurements that might be highly correlated with proboscis length. Three measurements were tried on 44 bees with the results indicated:

$n = 44$	Dry Weight X_1 (mg)	Length of Wing X_2 (mm)	Width of Wing X_3 (mm)	Length of Proboscis Y (mm)
Mean	13.10	9.61	3.28	6.59
	Sum of Squares and Products			
	X_1	X_2	X_3	Y
X_1	16.6840	1.9279	0.8240	1.5057
X_2		0.9924	0.3351	0.5989
X_3			0.2248	0.1848
Y				0.6831

Which subset of the Xs is best for predicting Y, using as criterion (*i*) s_p^2, (*ii*) C_p? (*iii*) What is the reliability of the best predictor considered as a measuring instrument? Ans. (*i*) For s_p^2 it is practically a tie between X_2, with $s_p^2 = 0.007659$, and (X_1, X_2), with $s_p^2 = 0.007656$. (*ii*) C_p chooses X_2 in preference to (X_1, X_2), the C_p values being 1.58 and 2.57. (*iii*) With either X_2 or (X_1, X_2) the residual s^2 is 0.0077, whereas s_y^2, the variance in the sample between the lengths of the bees' proboscises, is $0.6831/43 = 0.0159$. Errors in predicting Y give a reliability of $0.0159/0.0236 = 67\%$ in this sample. For the purpose at hand, this measurement would not be considered accurate. Furthermore, the estimated reliability should be reduced a little to allow for sampling errors in the bs.

17.15—Step-up and step-down methods. These methods are designed to find a good subset by computing only a small fraction of the regressions on all subsets. At its simplest, the step-up method is often called *forward selection.* It begins with the regressions of Y on $X_1, \ldots, X_k$ taken singly. The X-variable for which the regression has the highest F value and therefore gives the smallest residual mean square is selected. One proviso is that this F must *exceed* some boundary value chosen in advance. The best choice of this boundary is not clear. The 5% significance level has been used, but this is not a problem in testing significance. Values of $F = 2, 3,$ or 4 might be tried. Suppose that X_1 is selected.

Next, all $k - 1$ bivariate regressions in which X_1 appears are worked out. The variate giving the greatest additional reduction in sum of squares after fitting X_1 is selected, provided that its F exceeds the boundary. Call this variate X_2. All trivariate regressions including both X_1 and X_2 are then computed, and the variate that makes the greatest additional reduction in sum of squares is selected. At each stage, the computer program prints the F values for all Xs not yet included in the regression. It is then easy to spot the largest F and see whether it exceeds the boundary. The process stops when no X_i not yet selected gives an F exceeding the boundary.

When there are substantial intercorrelations among the Xs, this method has a weakness. For example, X_1 might be chosen first; when X_2, X_5, and X_6 have been subsequently chosen, the additional reduction in sum of squares due to X_1 after X_2, X_5, and X_6 might have an F less than the boundary. The following is a check against this effect. In the step at which X_6 is chosen, the computer prints F values for the additional reductions due to X_1 after X_2, X_5, and X_6; to X_2 after X_1, X_5, and X_6; and to X_5 after X_1, X_2, and X_6. If one of these Fs is below the boundary, the X in question is dropped. These Fs are sometimes labeled "F to remove." The boundary chosen for removal is sometimes smaller than that for inclusion.

Note that this method essentially uses s_p^2 as a criterion. It does not claim to give the best subset but has often worked well, particularly when intercorrelations are modest.

An alternative method, the step-down, works first the regression on all k Xs, then on the k regressions from which X_i has been omitted in turn. The X with the smallest F is dropped, provided that this F falls *below* a predetermined boundary. The process continues until no X not yet dropped gives an F below the boundary. This method is sometimes called *backward selection.* The step-up method usually involves less computing, particularly if the chosen p is small relative to k, and is probably preferable with the check that is mentioned.

17.16—Partial correlation. In a sample of 18-year-old college freshmen, the variables measured might be height, weight, blood pressure, basal metabolism, economic status, aptitude, etc. One purpose might be to examine whether aptitude (Y) is linearly related to the physiological measurements. If so, the regression methods of the preceding sections would apply. But the objective might be to study the correlations among such variables as height, weight, blood pressure, basal metabolism, etc., among which no variables can be specified as independent or dependent. In that case, *partial correlation* methods are appropriate.

You may recall that the ordinary correlation coefficient is closely related to the bivariate normal distribution. With more than two variables, an extension of this distribution called the *multivariate normal distribution* (19) forms the basic model in correlation studies. A property of the multivariate normal model is that any variable has a linear regression on the other variables (or on any subset of the other variables), with deviations that are normally distributed. Thus the assumptions made in multivariate regression studies hold for a multivariate normal population.

If there are three variables, there are three simple correlations among them,

ρ_{12}, ρ_{13}, ρ_{23}. The *partial correlation coefficient* $\rho_{12.3}$ is the correlation between variables 1 and 2 in a cross section of individuals *all having the same value of variable 3;* the third variable is held constant so that only 1 and 2 are involved in the correlation. In the multivariate normal model, $\rho_{12.3}$ is the same for every value of variable 3.

A sample estimate $r_{12.3}$ of $\rho_{12.3}$ can be obtained by calculating the deviations $d_{1.3}$ of variable 1 from its sample regression on variable 3. Similarly, find $d_{2.3}$. Then $r_{12.3}$ is the simple correlation coefficient between $d_{1.3}$ and $d_{2.3}$. The idea is to measure that part of the correlation between variables 1 and 2 that is not simply a reflection of their relations with variable 3. It may be shown that $r_{12.3}$ satisfies the following formula:

$$r_{12.3} = \frac{r_{12} - r_{13}r_{23}}{\sqrt{(1 - r_{13}^2)(1 - r_{23}^2)}} \qquad (17.16.1)$$

Table A 11 is used to test the significance of $r_{12.3}$. Enter it with $(n - 3)$ degrees of freedom instead of $(n - 2)$ as for a single correlation coefficient.

In Iowa and Nebraska, a random sample of 142 older women was drawn for a study of nutritional status (20). Three of the variables were age (A), blood pressure (B), and cholesterol concentration in the blood (C). The three simple correlations were

$$r_{AB} = 0.3332 \qquad r_{AC} = 0.5029 \qquad r_{BC} = 0.2495$$

Since high blood pressure might be associated with above average amounts of cholesterol in the walls of blood vessels, it is interesting to examine r_{BC}. But it is evident that both B and C increase with age. Are they correlated merely because of their common association with age or is there a real relation at every age? The effect of age is eliminated by calculation

$$r_{BC.A} = \frac{0.2495 - (0.3332)(0.5029)}{\sqrt{(1 - 0.3332^2)(1 - 0.5029^2)}} = 0.1005$$

With df $= 142 - 3 = 139$, this correlation is not significant.

Confidence limits may be found by the methods of section 10.6. With $r = 0.1005$, z in table A 12 is 0.1008. For its standard error we take $1/\sqrt{n - 4}$ instead of $1/\sqrt{n - 3}$, since age has been eliminated. Hence the 95% one-tailed upper confidence limit in z is $0.1008 + 1.645/\sqrt{138} = 0.241$, giving $\rho = 0.24$. The partial correlation, if present, is evidently small.

Formula (17.6.1) may be proved by the sweep operator. Suppose we scale each X so that $\Sigma x_1^2 = \Sigma x_2^2 = \Sigma x_3^2 = 1$. This scaling will not affect correlation coefficients, which are dimensionless. But it makes $\Sigma x_1 x_2 = r_{12}$, and so on, and turns the covariance matrix into the correlation matrix. On the left below we show the correlation matrix and on the right the effect of sweeping on row and column 3, so that the scaled x_1 and x_2 become $d_{1.3}$ and $d_{2.3}$, the deviations from their regression on X_3.

Correlation Matrix			Effect of SWP[3]		
1	r_{12}	r_{13}	$(1 - r_{13}^2)$	$(r_{12} - r_{13}r_{23})$	r_{13}
r_{12}	1	r_{23}	$(r_{12} - r_{13}r_{23})$	$(1 - r_{23}^2)$	r_{23}
r_{13}	r_{23}	1	r_{13}	r_{23}	-1

Hence, from SWP[3], $r_{12.3} = (r_{12} - r_{13}r_{23})/\sqrt{(1 - r_{13}^2)(1 - r_{23}^2)}$ as in (17.16.1).

With four variables the partial correlation coefficient between variables 1 and 2 can be computed after eliminating the effects of the other variables, 3 and 4. The formulas are, respectively,

$$r_{12.34} = \frac{r_{12.4} - r_{13.4}r_{23.4}}{\sqrt{(1 - r_{13.4}^2)(1 - r_{23.4}^2)}}$$

$$r_{12.34} = \frac{r_{12.3} - r_{14.3}r_{24.3}}{\sqrt{(1 - r_{14.3}^2)(1 - r_{24.3}^2)}} \qquad (17.16.2)$$

the two formulas being identical.

To test this quantity in table A 11, use $(n - 4)$ df.

EXAMPLE 17.16.1—Brunson and Willier (21) examined the correlations among ear circumference E, cob circumference C, and number of rows of kernels K calculated from measurements of 900 ears of corn:

$$r_{EC} = 0.799 \qquad r_{EK} = 0.570 \qquad r_{CK} = 0.507$$

Among the ears having the same kernel number, what is the correlation between E and C? Ans. $r_{EC.K} = 0.720$.

EXAMPLE 17.16.2—Among ears of corn having the same circumference, is there any correlation between C and K? Ans. $r_{CK.E} = 0.105$. The correlation is small but gives $P < 0.01$, since m is large.

EXAMPLE 17.16.3—In a random sample of 54 Iowa women (20), the intake of two nutrients was determined together with age and the concentration of cholesterol in the blood. If P symbolizes protein, F fat, A age, and C cholesterol, the correlations are as follows:

	A	P	F
P	−0.4865		
F	−0.5296	0.5784	
C	0.4737	−0.4249	−0.3135

What is the correlation between age and cholesterol independent of the intake of protein and fat? Ans.

$$r_{AC.PF} = \frac{r_{AC.F} - r_{AP.F}r_{CP.F}}{\sqrt{(1 - r_{AP.F}^2)(1 - r_{CP.F}^2)}} = \frac{0.3820 - (-0.2604)(-0.3145)}{\sqrt{(1 - 0.2604^2)(1 - 0.3145^2)}} = 0.3274$$

EXAMPLE 17.16.4—Set up the data in example 17.16.3 in a 4×4 correlation matrix and find $r_{AC.PF}$ by performing SWP[P, F].

TECHNICAL TERMS

adjusted R^2
backward selection
correlation matrix
covariance matrix
forward selection
Mallows' C_p
multiple correlation coefficient
multiple linear regression
normal equations
omitted variable
partial correlation coefficient
precoding
regression plane
standard regression coefficient
step-down method
step-up method

REFERENCES

1. Eid, M. T., et al. 1954. Iowa Agric. Exp. Stn. Res. Bull. 406.
2. Fisher, R. A. *Philos. Trans.* B (1934):89.
3. R. Schrek. *Hum. Biol.* 14 (1942):95.
4. Beaton, A. E. 1964. "The Use of Special Matrix Operators in Statistical Calculus." Ph.D. dissertation. Harvard University.
5. Mosteller, F., and Tukey, J. W. 1977. *Data Analysis and Regression.* Addison-Wesley, Reading, Mass.
6. Longley, J. W. *J. Am. Stat. Assoc.* 62 (1967):819.
7. Wright, S. *Biometrics* 16 (1960):189.
8. Duncan, O. D. *Am. J. Sociol.* 72 (1966):1.
9. Box, G. E. P. *Technometrics* 8 (1966):625.
10. Kempthorne, O. 1952. *Design and Analysis of Experiments.* Wiley, New York.
11. Cochran, W. G., and Cox, G. M. 1957. *Experimental Designs,* 2nd ed. Wiley, New York.
12. Yates, F., and Boyd, D. A. Br. Agric. Bull. 4 (1951):206.
13. Hotelling, H. *Ann. Math. Stat.* 11 (1940):271.
14. Williams, E. J. 1952. *Regression Analysis.* Wiley, New York.
15. BMDP. *Biomedical Computer Programs, P*-series. 1977. Univ. California Press, Berkeley.
16. Mallows, C. L. *Technometrics* 15 (1973):661.
17. Hald, A. 1952. *Statistical Theory with Engineering Applications.* Wiley, New York.
18. Daniel, C., and Wood, F. S. 1971. *Fitting Equations to Data.* Wiley-Interscience, New York.
19. Mood, A. M., and Graybill, F. A. 1963. *Introduction to the Theory of Statistics,* 2nd ed. McGraw-Hill, New York.
20. Swanson, P. P., et al. *J. Gerontol.* 10 (1955):41.
21. Brunson, A. M., and Willier, J. G. *J. Am. Soc. Agron.* 21 (1929):912.
22. Grout, R. A. 1937. Iowa Agric. Exp. Stn. Res. Bull. 218.

18

Analysis of covariance

18.1—Introduction. The analysis of covariance is a technique that combines the features of analysis of variance and regression. In a one-way classification, the typical analysis of variance model for the value Y_{ij} of the jth observation in the ith class is

$$Y_{ij} = \alpha_i + e_{ij}$$

where the α_i represent the population means of the classes and the e_{ij} are the residuals. But suppose that on each unit we have also measured another variable X_{ij} that is linearly related to Y_{ij}. It is natural to set up the model,

$$Y_{ij} = \alpha_i + \beta(X_{ij} - \overline{X}_{..}) + \epsilon_{ij}$$

where β is the regression coefficient of Y on X. This model is typical for the analysis of covariance. If X and Y are closely related, we may expect this model to fit the Y_{ij} values better than the original analysis of variance model. That is, the residuals ϵ_{ij} should be in general smaller than the e_{ij}.

The model extends easily to more complex situations. With a two-way classification, as in a randomized blocks experiment, the model is

$$Y_{ij} = \mu + \alpha_i + \rho_j + \beta(X_{ij} - \overline{X}_{..}) + \epsilon_{ij}$$

With a one-way classification and two auxiliary variables X_{1ij} and X_{2ij}, both linearly related to Y_{ij}, we have

$$Y_{ij} = \alpha_i + \beta_1(X_{1ij} - \overline{X}_{1..}) + \beta_2(X_{2ij} - \overline{X}_{2..}) + \epsilon_{ij}$$

The analysis of covariance has numerous uses:

1. *To increase precision in randomized experiments.* In such applications the covariate X is a measurement taken on each experimental unit before the treatments are applied that predicts to some degree the final response Y on the unit. In the earliest application suggested by Fisher (1), the Y_{ij} were the yields of tea bushes in an experiment. An important source of error is that, by the luck of the draw, some treatments will have been allotted to a more productive set of

bushes than others. The X_{ij} were the yields of the bushes in a period before treatments were applied. Since the relative yields of tea bushes show a good deal of stability from year to year, the X_{ij} serve as predictors of the inherent yielding abilities of the bushes. By adjusting the treatment mean yields so as to remove these differences in yielding ability, we obtain a lower experimental error and more precise comparisons among the treatments. This use of covariance is probably the commonest.

2. *To adjust for sources of bias in observational studies.* An investigator is studying the relation between obesity in workers and the physical activity required in their occupations. Measures of obesity Y_{ij} are obtained from samples of workers from each of a number of occupations. The age X_{ij} of each worker is also recorded and differences between the mean ages of the workers in different occupations are noted. If obesity is linearly related to age, differences found in obesity among different occupations may be due in part to these age differences. Consequently the investigator introduces the term $\beta(X_{ij} - \bar{X}_{..})$ into the model in the hope of removing a possible source of bias in the comparison among occupations. However, difficulties are connected with this use of covariance; see sections 18.6, 18.7.

3. *To throw light on the nature of treatment effects in randomized experiments.* In an experiment on the effects of soil fumigants on nematodes, which attack some farm crops, significant differences between fumigants were found both in the numbers of nematode cysts X_{ij} and in the yields Y_{ij} of the crop. The question is raised, Can the differences in yields be ascribed to the differences in numbers of nematodes? One way of examining this question is to see whether treatment differences in yields remain or whether they shrink to insignificance after adjusting for the regression of yields on nematode numbers.

4. *To study regressions in multiple classifications.* For example, an investigator is studying the relation between expenditure per student in schools (Y) and per capita income (X) in large cities. If data for a large number of cities for each of four years are available, the investigator may want to examine whether the relation is the same in different sections of the country or whether it remains the same from year to year. Sometimes the question is whether the relation is straight or curved.

18.2—Covariance in a completely randomized experiment. We begin with a simple example of the use of covariance in increasing precision in randomized experiments. With a completely randomized design, the data form a one-way classification, the treatments being the classes. In the model

$$Y_{ij} = \alpha_i + \beta(X_{ij} - \bar{X}_{..}) + \epsilon_{ij} \qquad (18.2.1)$$

the α_i represent the effects of the treatments. The observed mean for the ith treatment is

$$\bar{Y}_{i.} = \alpha_i + \beta(\bar{X}_{i.} - \bar{X}_{..}) + \bar{\epsilon}_i$$

Thus $\bar{Y}_{i.}$ is an unbiased estimate of $\alpha_i + \beta(\bar{X}_{i.} - \bar{X}_{..})$. It follows that as an estimate of α_i we use

$$\alpha_i = \bar{Y}_{i.} - \beta(\bar{X}_{i.} - \bar{X}_{..})$$

the second term on the right being the adjustment introduced by the covariance analysis. The adjustment accords with common sense. For instance, suppose we were told that in the previous year the tea bushes receiving treatment 1 yielded 20 lb more than the average over the experiment. If the regression coefficient of Y on X was 0.4, meaning that each pound of increase in X corresponds to 0.4 lb of increase in Y, we would decrease the observed Y mean by (0.4)(20) = 8 lb to make treatment 1 more comparable to the other treatments. In this illustration the 0.4 is β and the 20 is $(\bar{X}_{i.} - \bar{X}_{..})$.

The problem of estimating β from the results of the experiment remains. In a single sample you may recall that the regression coefficient is estimated by $b = \Sigma xy/\Sigma x^2$ and that the reduction in sum of squares of Y due to the regression is $(\Sigma xy)^2/\Sigma x^2$. These results continue to hold in multiple classifications (completely randomized, randomized blocks, and Latin square designs) except that β *is estimated from the error line in the analysis of variance.* We may write $b = E_{xy}/E_{xx}$. The error sum of squares of X in the analysis of variance, E_{xx}, is familiar, but the quantity E_{xy} is new. It is the error sum of products of X and Y. A numerical example will clarify it.

The data in table 18.2.1 were selected from a larger experiment on the use of drugs in the treatment of leprosy at the Eversley Childs Sanitarium in the Philippines. On each patient six sites on the body at which leprosy bacilli tend to congregate were selected. The variate X, based on laboratory tests, is a score representing the abundance of leprosy bacilli at these sites before the experiment began. The variate Y is a similar score after several months of treatment. Drugs A and B are antibiotics, while drug C is an inert drug included as a control. Ten patients were selected for each treatment for this example.

The first step is to compute the analysis of sums of squares and products, shown in the table. In the columns headed Σx^2 and Σy^2, we analyze X and Y in the usual way into "between drugs" and "within drugs." For the Σxy column, make the corresponding analysis of the products of X and Y, as follows:

$$\text{total: } (11)(6) + (8)(0) + \ldots + (12)(20) - (322)(237)/30 = 731.2$$

$$\text{between drugs: } \frac{(93)(53) + (100)(61) + (129)(123)}{10} - \frac{(322)(237)}{30} = 145.8$$

The within-drugs sum of products, 585.4, is found by subtraction. Note that any of these sums of products may be either positive or negative. The within-drugs (error) sum of products, 585.4, is the quantity we call E_{xy}, while the error sum of squares of X, 593.0, is E_{xx}.

The reduction in the error sum of squares Y due to the regression is E_{xy}^2/E_{xx} with 1 df. The deviations mean square, 16.05 with 26 df, provides the estimate of

TABLE 18.2.1
SCORES FOR LEPROSY BACILLI BEFORE (X) AND AFTER (Y) TREATMENT

	A		*B*		*C*		Overall	
	X	Y	X	Y	X	Y	X	Y
	11	6	6	0	16	13		
	8	0	6	2	13	10		
	5	2	7	3	11	18		
	14	8	8	1	9	5		
	19	11	18	18	21	23		
	6	4	8	4	16	12		
	10	13	19	14	12	5		
	6	1	8	9	12	16		
	11	8	5	1	7	1		
	3	0	15	9	12	20		
Total	93	53	100	61	129	123	322	237
Mean	9.3	5.3	10.0	6.1	12.9	12.3	10.73	7.90

Analysis of Sums of Squares and Products

Source	df	Σx^2	Σxy	Σy^2
Total	29	665.9	731.2	1288.7
Between drugs	2	72.9	145.8	293.6
Within drugs (error)	27	593.0	585.4	995.1
Reduction due to regression	1		$(585.4)^2 / 593.0 =$	577.9
Deviations from regression	26			417.2

deviations mean square = 417.2/26 = 16.05

error. The original error mean square of Y is $995.1/27 = 36.86$. The regression has produced a substantial reduction in the error mean square.

The next step is to compute b and the adjusted means. We have $b = E_{xy}/E_{xx} = 585.4/593.0 = 0.987$. The adjusted means, as follow,

$$A\colon\ \overline{Y}_{1.} - b(\overline{X}_{1.} - \overline{X}_{..}) = 5.3 - (0.987)(9.3 - 10.73) = 6.71$$

$$B\colon\ \overline{Y}_{2.} - b(\overline{X}_{2.} - \overline{X}_{..}) = 6.1 - (0.987)(10.0 - 10.73) = 6.82$$

$$C\colon\ \overline{Y}_{3.} - b(\overline{X}_{3.} - \overline{X}_{..}) = 12.3 - (0.987)(12.9 - 10.73) = 10.16$$

have improved the status of C, which happened to receive initially a set of patients with somewhat high scores.

For tests of significance or confidence limits relating to the adjusted means, the error variance is derived from the mean square $s^2_{y.x} = 16.05$ with 26 df. Algebraically, the difference between the adjusted means of the ith and the jth treatments is

$$D = \overline{Y}_{i.} - \overline{Y}_{j.} - b(\overline{X}_{i.} - \overline{X}_{j.}) \qquad (18.2.2)$$

The estimated variance of D is

$$s^2_D = s^2_{y.x}[2/n + (\overline{X}_{i.} - \overline{X}_{j.})^2/E_{xx}] \qquad (18.2.3)$$

where n is the sample size per treatment. The second term on the right is an allowance for the sampling error of b.

For a comparison of the adjusted means of treatments A and C, $n = 10$ and $\overline{X}_A - \overline{X}_C = 9.3 - 12.9 = -3.6$, so

$$s_D^2 = (16.05)(1/5 + 3.6^2/593.0) = 3.561 \tag{18.2.4}$$

Thus, $t = (6.71 - 10.16)/1.887 = -1.83$ with 26 df, which is not significant at the 5% level.

Note that since the same regression coefficient b is used to adjust all means, the adjusted means are slightly correlated. This explains why we calculated D and s_D^2 directly. Similarly, a comparison or contrast among the adjusted means may be expressed as $\Sigma L_i \overline{Y}_i - b(\Sigma L_i \overline{X}_i)$, where $\Sigma L_i = 0$. Its estimated variance is therefore

$$s_{y.x}^2[\Sigma L_i^2/n + (\Sigma L_i \overline{X}_i)^2/E_{xx}] \tag{18.2.5}$$

As an example, let us test the difference between the adjusted mean 6.765 of the two active drugs A and B and the adjusted mean 10.16 of the inert drug C. The L_i are 1/2, 1/2, −1, and $\Sigma L_i \overline{X}_i = -3.25$, with variance

$$s^2 = (16.05)[1.5/10 + (-3.25)^2/593.0] = 2.693$$

The t value is $-3.395/1.641 = -2.07$, just significant at the 5% level.

A standard computer program is likely to give either standard errors or t tests of the difference between every pair of adjusted means and, on request, the same statistics for any specified comparisons. If a program is not available, the following approximation saves some time in computing standard errors. It works well enough if (*i*) at least 20 df are in the error line of the analysis of variance and (*ii*) the treatments mean square for X is nonsignificant, as it should be since the Xs were measured before treatments were assigned. In such cases an average value of s_D^2 may be used. By an algebraic identity (2) the average value of s_D^2, taken over every pair of treatments, is

$$\overline{s_D^2} = (2/n)\, s_{y.x}^2(1 + t_{xx}/E_{xx}) \tag{18.2.6}$$

where t_{xx} is the treatments *mean square* for X. More generally, we may regard

$$s'^2 = s_{y.x}^2(1 + t_{xx}/E_{xx})$$

as the effective error mean square per observation when computing the error variance for any comparison among the adjusted treatment means.

With this s'^2 the adjusted means are regarded as independent when calculating standard errors for comparisons among adjusted means. For example, in this experiment $t_{xx} = 72.9/2 = 36.5$ (from table 18.2.1) and $E_{xx} = 593.0$, giving $t_{xx}/E_{xx} = 0.0616$. Hence,

$$s'^2 = (16.05)(1.0616) = 17.04 \qquad s' = 4.128$$

With 10 replicates this gives $\overline{s_D} = 4.128\sqrt{0.2} = 1.846$ for the difference between two treatment means. For $A - C$ we previously found $s_D = 1.887$ from (18.2.4).

After completing a covariance analysis, the experimenter may ask, Is it worthwhile? The efficiency of the adjusted means relative to the unadjusted means is estimated by the reciprocal of the ratio of the corresponding effective error mean squares:

$$\frac{s_y^2}{s'^2} = \frac{s_y^2}{s_{y.x}^2(1 + t_{xx}/E_{xx})} = \frac{36.86}{17.04} = 2.16$$

Covariance with 10 replicates per treatment gives nearly as precise estimates as the unadjusted means with 21 replicates.

In experiments like this, in which X measures the same quantity as Y (score for leprosy bacilli), an alternative to covariance is to use $Y - X$, the *change* in the score, as the measure of treatment effect. The error mean square for $Y - X$ is obtained from table 18.2.1 as

$$(E_{yy} - 2E_{xy} + E_{xx})/27 = [995.1 - 2(585.4) + 593.0]/27 = 15.46$$

compared with 17.04 for covariance. In this experiment, use of $Y - X$ is slightly more efficient than the use of covariance as well as faster computationally because $b = 0.987$ is so close to 1. The value of $Y - X$ was the recommended variable for analysis in the larger experiment from which these data were selected. In many experiments, $Y - X$ is inferior to covariance and may also be inferior to Y if the correlation between X and Y is low.

Why is b estimated from the error or residuals line in the analysis of variance? With a covariance adjustment, the measurement used to estimate and compare treatment effects is $Y - bX$. Thus the analysis of covariance is essentially an analysis of variance of the quantity $Y - bX$. Suppose that the value of b has not yet been chosen. The error sum of squares of $Y - bX$ is

$$E_{yy} - 2bE_{xy} + b^2E_{xx}$$

Complete the square on b. The error sum of squares becomes

$$E_{xx}(b - E_{xy}/E_{xx})^2 + E_{xy} - E_{xy}^2/E_{xx} \qquad (18.2.7)$$

By the method of least squares, b is selected so as to minimize the error sum of squares of $Y - bX$. From (18.2.7), this happens when $b = E_{xy}/E_{xx}$. The minimum error sum of squares is $E_{yy} - E_{xy}^2/E_{xx}$.

EXAMPLE 18.2.1—This example uses Eden's data (14) of the yields in two successive periods of 16 plots of tea bushes in Ceylon. Fisher (1) introduced the analysis of covariance by means of this example. Previous to the use of covariance, the yields X in the first period might be used to form blocks (as below) so that X varies as little as possible within a block. If X is a good predictor of the yields Y in the second period, these should be good blocks for Y. How much does blocking on X reduce the error variance of Y per plot? Does a covariance adjustment within blocks reduce $s^2_{y.x}$ still further? Since no real treatments were applied, you may treat the data as a one-way classification into blocks. The number 80 has been subtracted from each yield.

	1		2		3		4		Total	
	X	Y	X	Y	X	Y	X	Y	X	Y
	8	10	11	5	22	27	29	34		
	8	1	14	13	22	13	30	26		
	8	12	14	13	25	26	35	31		
	11	12	16	22	29	34	38	41		
Total	35	35	55	53	98	100	132	132	320	320

Ans. With no blocking, error mean square = $\Sigma y^2/15$ = 136.0; with blocking, the within-blocks s^2_y = 48.0; with a further covariance adjustment, $s^2_{y.x}$ = 27.3.

18.3—The *F* test of the adjusted means. An F test can be made of the null hypothesis that there are no differences among the adjusted treatment means—the α_i in model (18.2.1) are all equal. This test is a straightforward application of the F test in the extended analysis of variance described in section 17.4. Recall that in the multiple regression $Y = \alpha + \Sigma^k \beta_i X_i$, we wish to test the null hypothesis that $\beta_{p+1}, \beta_{p+2}, \ldots, \beta_k$ were all zero. If the null hypothesis holds, the simpler model $Y = \alpha + \Sigma^p \beta_i X_i$ fits as well as the more general alternative hypothesis model. Expressed in more general terms, the prescription for making the F test is as follows.

1. Fit the null hypothesis (NH) model and find the reduction R_{NH} in Σy^2 due to this model. The deviations sum of squares $D_{\mathrm{NH}} = \Sigma y^2 - R_{\mathrm{NH}}$ can be used instead.
2. Fit the more general alternative hypothesis (AH) model and find R_{AH} and D_{AH}. Let s^2_{AH} be the mean square of the deviations from this model.
3. If the null hypothesis holds,

$$F = (R_{\mathrm{AH}} - R_{\mathrm{NH}})/(\nu_1 s^2_{\mathrm{AH}}) = (D_{\mathrm{NH}} - D_{\mathrm{AH}})(\nu_1 s^2_{\mathrm{AH}}) \qquad (18.3.1)$$

is distributed as F with ν_1, ν_2 df. Here, ν_1 is the difference between the numbers of fitted parameters in the alternative and null hypotheses and ν_2 is the number of degrees of freedom in s^2_{AH}. Large values of F suggest that the null hypothesis does not hold.

In our example (covariance in a one-way classification), the null hypothesis model (no treatment effects) is

$$Y_{ij} = \mu + \beta(X_{ij} - \overline{X}_{..}) + \epsilon_{ij}$$

This equation states simply that Y has a linear regression on X. Hence,

$$D_{NH} = (E_{yy} + T_{yy}) - (E_{xy} + T_{xy})^2/(E_{xx} + T_{xx}) \quad (18.3.2)$$

where T_{yy}, etc., refer to the treatments in the analysis of variance. In our case, $E_{yy} + T_{yy} = \Sigma y^2$, and (18.3.2) could be put more simply as $D_{NH} = \Sigma y^2 - (\Sigma xy)^2/\Sigma x^2$. But (18.3.2) holds in randomized blocks and Latin squares also. Furthermore,

$$D_{AH} = E_{yy} - E_{xy}^2/E_{xx} \quad (18.3.3)$$

The number of fitted parameters is 2 in the null hypothesis and 4 in the alternative hypothesis, so $\nu_1 = 2$, or $a - 1$ with a treatments.

In table 18.3.1 the test is made for the leprosy example. The first step is to form a line for treatments + error. Following (18.3.2) we then subtract $731.2^2/665.9 = 802.9$ from 1288.7 to give 485.8, the sum of squares of deviations D_{NH}. From 485.8 we subtract $417.2 = D_{AH}$, giving 68.6, which we divide by the difference in degrees of freedom, 2. The F ratio is $34.3/16.05 = 2.14$ with 2 and 26 df; it lies between the 25% and 10% levels.

If calculations are done by a standard computer program, the F test in table 18.3.1 may be presented in more familiar analysis of variance form, somewhat as

SOURCE	D.F.	SUM OF SQ.	MEAN SQ.	F	TAIL P
EQUALITY OF ADJ. MEANS	2	68.6	34.3	2.14	0.14
RESIDUAL ERROR	26	417.2	16.05		

18.4—Multiple regression and multiple classification analysis. By introducing dummy X-variables, sometimes called *indicator variables,* we can express the mathematical models for analyses of variance and covariance as models for multiple regressions. For manual calculations of analyses of variance and covariance, this device does not save time. But it has the advantage that a standard computer program for multiple linear regressions can perform many analyses of variance and covariance, including applications more complex than any described in this book.

As an illustration we show how the analysis of table 18.3.1 is expressed as a multiple regression analysis. In the original notation of the problem, our

TABLE 18.3.1

ANALYSIS OF COVARIANCE F TEST (ONE-WAY CLASSIFICATION) LEPROSY DATA

						Deviations from Regressions		
	df	Σx^2	Σxy	Σy^2	R	df	SS	MS
Treatments, T	2	72.9	145.8	293.6				
Error, E	27	593.0	585.4	995.1	577.9	26	417.2	16.05
$T + E$	29	665.9	731.2	1288.7	802.9	28	485.8	
						2	68.6	34.3

variables are X_{ij} and Y_{ij}, where the suffix $i = 1, 2, 3$ denotes the class (in this case the treatment) and $j = 1, \ldots, 10$ denotes the members of the class. In multiple regression notation, we have a *single* sample of size 30, the members of the sample being denoted by the subscript $u = 1, 2, \ldots, 30$. Since we will be bringing in two dummy variables, X_{1u} and X_{2u}, we denote our original X_{ij} by X_{3u} and Y_{ij} by X_{4u}.

Let X_{1u} have the value 1 for every observation receiving treatment B and 0 for every other observation. Let X_{2u} have the value 1 for every observation receiving treatment C and 0 for every other observation. We now show that these dummy variables enable us to represent the analyses of variance and covariance models as multiple regressions and find the values of the regression coefficients β_1 and β_2 in terms of our original problem.

Consider first the analysis of variance of $X_{ij} = X_{3u}$. For a one-way classification the notations for the analysis of variance (ANOVA) and the multiple linear regression (MLR) are as follows:

$$\text{ANOVA: } X_{ij} = \alpha_i + \epsilon_{ij} \qquad (i = 1, 2, 3; j = 1, 2, \ldots, 10) \tag{18.4.1}$$

$$\text{MLR: } X_{3u} = \alpha + \beta_1 X_{1u} + \beta_2 X_{2u} + \epsilon_u \qquad (u = 1, 2, \ldots, 30) \tag{18.4.2}$$

Inserting the values given to X_{1u} and X_{2u}, we show the two notations for any member of class 1 (treatment A), class 2 (treatment B), and class 3 (treatment C):

	ANOVA	MLR	Values of Parameters
Class 1	$X_{1j} = \alpha_1 + \epsilon_{1j}$	$X_{3u} = \alpha + \epsilon_u$	$\alpha = \alpha_1$
Class 2	$X_{2j} = \alpha_2 + \epsilon_{2j}$	$X_{3u} = \alpha + \beta_1 + \epsilon_u$	$\beta_1 = \alpha_2 - \alpha_1$
Class 3	$X_{3j} = \alpha_3 + \epsilon_{3j}$	$X_{3u} = \alpha + \beta_2 + \epsilon_u$	$\beta_2 = \alpha_3 - \alpha_1$

Check carefully the MLR column. In class 1, both dummy Xs are 0, so the only parameter in the MLR model is the constant term α. It follows that in terms of our original ANOVA model, $\alpha = \alpha_1$; that is, the constant term measures the mean of class 1, or treatment A. In class 2, $X_{1u} = 1, X_{2u} = 0$; so the MLR model coincides with the ANOVA model if $\alpha + \beta_1 = \alpha_2$, from which we get $\beta_1 = \alpha_2 - \alpha_1$. Thus the regression coefficient β_1 measures the difference between the means of class 2 and class 1. Similarly, the MLR model and the ANOVA model are the same in class 3 if $\beta_3 = \alpha_2 - \alpha_1$, the mean of class 3 *minus* the mean of class 1. Thus, when the MLR analysis is performed on the sample data, the quantities a, $(a + b_1)$, and $(a + b_2)$ are estimates of the means for the three treatments. Naturally, some practice is needed in selecting dummy variables so that the MLR and ANOVA models are identical and in expressing the β_i in terms of parameters that we want to estimate.

For the analysis of covariance in table 18.2.1 we have $X_{4u} \equiv Y_{ij}$, and the model in the two notations becomes

$$\text{Covariance: } Y_{ij} = \alpha_i + \beta(X_{ij} - \overline{X}_{..}) + \epsilon_{ij} \tag{18.4.3}$$

$$\text{MLR: } X_{4u} = \alpha + \beta_1 X_{1u} + \beta_2 X_{2u} + \beta_3(X_{3u} - \overline{X}_3) + \epsilon_u \tag{18.4.4}$$

As before, the models are the same if $\alpha = \alpha_1$, $\beta_1 = \alpha_2 - \alpha_1$, $\beta_2 = \alpha_3 - \alpha_1$, and $\beta_3 = \beta$ except that now the α_i represent *adjusted* means in the covariance model.

Table 18.4.1 shows the data in table 18.2.1 as they would be written for multiple regression analysis. The data are now a single sample of size 30.

In the MLR notation, table 18.4.2 shows the covariance matrix: $\Sigma_u x_{iu}^2$ and $\Sigma_u x_{iu}x_{ju}$, calculated from table 18.4.1. For instance, $\Sigma x_{1u}^2 = 10 - 10^2/30 = 20/3$. Similarly, $\Sigma x_{1u}x_{2u} = 0 - 10^2/30 = -10/3$. The results of applying SWP[1], SWP[1, 2], and SWP[1, 2, 3] to this matrix are also shown.

Look first at SWP[1, 2], which represents the analysis of variance of the one-way classification. In column 3, the figures +0.7 and +3.6 are the differences $\overline{X}_B - \overline{X}_A$, $\overline{X}_C - \overline{X}_A$ in table 18.2.1; similar figures for Y are in column 4. With $X_{3u} = X_{ij}$, the mean $\overline{X}_A$ in regression notation is

$$\overline{X}_A = a = \overline{X}_3 - b_1\overline{X}_1 - b_2\overline{X}_2 = [322 - (0.7)(10) - (3.6)(10)]/30 = 9.3$$

The error sums of squares and products for X_{ij} and Y_{ij} appear in rows 3 and 4 of columns 3 and 4 of SWP[1, 2].

SWP[1, 2, 3] shows the results of adjustment by the linear regression on X_{ij}. From MLR model (18.4.4), the adjusted mean of $Y = X_4$ for treatment A is

$$a = \overline{X}_4 - b_1\overline{X}_1 - b_2\overline{X}_2 - b_3(0) = [237 - (0.109)(10) - (3.446)(10)]/30$$
$$= 6.715$$

For B and C the adjusted means are $6.715 + 0.109 = 6.824$ and $6.715 + 3.446 = 10.161$, in agreement with the values for A, B, and C above (18.2.2). The residual mean square $s^2 = 417.235/26 = 16.05$ as before. Rows and columns 1, 2 in SWP[1, 2, 3] give the multipliers of s^2 that provide the

TABLE 18.4.1
TABLE 18.2.1 IN MULTIPLE REGRESSION NOTATION

No.	X_{1u}	X_{2u}	X_{3u}	X_{4u}	No.	X_{1u}	X_{2u}	X_{3u}	X_{4u}
1	0	0	11	6	16	1	0	8	4
2	0	0	8	0	17	1	0	19	14
3	0	0	5	2	18	1	0	8	9
4	0	0	14	8	19	1	0	5	1
5	0	0	19	11	20	1	0	15	9
6	0	0	6	4	21	0	1	16	13
7	0	0	10	13	22	0	1	13	10
8	0	0	6	1	23	0	1	11	18
9	0	0	11	8	24	0	1	9	5
10	0	0	3	0	25	0	1	21	23
11	1	0	6	0	26	0	1	16	12
12	1	0	6	2	27	0	1	12	5
13	1	0	7	3	28	0	1	12	16
14	1	0	8	1	29	0	1	7	1
15	1	0	18	18	30	0	1	12	20
					Total	10	10	322	237

TABLE 18.4.2
MULTIPLE REGRESSION ANALYSIS BY THE SWEEP METHOD

	Covariance Matrix, $\Sigma x_i x_j$				SWP[1]			
	1†	2	3	4	1	2	3	4
1	20/3	−10/3	−22/3	−18	−0.15	−0.5	−1.1	−2.7
2	−10/3	20/3	65/3	44	−0.5	5.0	18.0	35.0
3	−22/3	65/3	665.9	731.2	−1.6	18.0	657.8	711.4
4	−18	44	731.2	1288.7	−2.7	35.0	711.4	1240.1
	SWP[1, 2]				SWP[1, 2, 3]			
1	−0.2	−0.1	+0.7	+0.8	−0.2008	−0.1042	0.0012	0.109
2	−0.1	−0.2	+3.6	+7.0	−0.1042	−0.2218	0.0061	3.446
3	+0.7	+3.6	593.0	585.4	0.0012	0.0061	−0.0017	0.987
4	+0.8	7.0	585.4	995.1	0.109	3.446	0.987	417.2

†Denotes the variables $X_1, \ldots, X_4 = Y$.

covariance matrix of b_1 and b_2. For example, $V(b_2) = (0.2218)(16.05) = 3.560$, as given in (18.2.4). Other comparisons among the adjusted treatment means can be expressed as linear functions of b_1 and b_2.

With access to a computer multiple regression program, one can construct dummy Xs so that the analysis of variance or covariance wanted becomes a multiple linear regression and interpret the estimates a, b_i in terms of the quantities to be estimated.

EXAMPLE 18.4.1—The sweep method does not directly give the F test of section 18.3. How is it obtained from table 18.4.2? Ans. From the covariance matrix, $D_{\text{NH}} = 1288.7 - 731.2^2/665.9 = 485.8$ with 28 df. Since table 18.4.2 gives $D_{\text{AH}} = 417.2$ from SWP[1, 2, 3], the F test in table 18.3.1 follows immediately.

EXAMPLE 18.4.2—From table 18.4.2, perform a t test of the difference between the adjusted mean of A and B and that of C. Ans. Since $b_1 = \hat{\alpha}_2 - \hat{\alpha}_1$ and $b_2 = \hat{\alpha}_3 - \hat{\alpha}_1$ in our original model, the comparison that we seek is $(\hat{\alpha}_1 + \hat{\alpha}_2)/2 - \hat{\alpha}_3 = -(b_2 - b_1/2)$. From columns 1 and 2 of SWP[1, 2, 3], the multiplier of $s^2_{y.x}$ to give the desired variance is $0.2218 - 0.1042 + 0.2008/4 = 0.1678$. Hence, $V(b_2 - b_1/2) = (0.1678)(16.05) = 2.693$ and $t = -3.392/1.641 = -2.07$ as in section 18.2.

EXAMPLE 18.4.3—Verify the numbers −22/3 and 44 in columns 3 and 4 of the covariance matrix in table 18.4.2.

EXAMPLE 18.4.4—Many sets of indicator variables can be used to express the classes in an analysis of variance or covariance in multiple linear regression notation. With t treatments, a convenient notation is to let the dummy x_{iu} (for $i = 1, 2, \ldots, t - 1$) have the value +1 in treatment i, the value −1 in treatment t, and the value 0 in all other treatments. With this notation, $\overline{X}_i = 0$ if all treatments have the same number of replicates.

The analysis of variance (ANOVA) and multiple linear regression (MLR) notations are as follows:

ANOVA: $E(Y_{ij}) = \mu + \alpha_i$ in treatment i
MLR: $E(Y_u) = \alpha + \Sigma\beta_i X_{iu}$

With the usual ANOVA convention $\Sigma\alpha_i = 0$, show that $\alpha = \mu$ and $\beta_i = \alpha_i$ $(i = 1, 2, \ldots, t - 1)$.

18.5—Covariance in a two-way classification. As before, the regression coefficient is estimated from the error (treatments × blocks) line in the analysis of sums of squares and products, and the F test of the adjusted treatment means is made by recomputing the regression from the treatments + error lines, following the procedure in section 18.3. To put it more generally, the regression coefficient is estimated from the rows × columns line, and either the adjusted row means or the adjusted column means may be tested. Two examples from experiments are presented to illustrate points that arise in applications.

The data in table 18.5.1 are from an experiment on the effects of two drugs on mental activity (11). The mental activity score was the sum of the scores on seven items in a questionnaire given to each of 24 volunteer subjects. The treatments were morphine, heroin, and a placebo (an inert substance) given in subcutaneous injections. On different occasions, each subject received each drug in turn. The mental activity was measured before taking the drug (X) and at

TABLE 18.5.1

MENTAL ACTIVITY SCORES BEFORE (X) AND TWO HR AFTER (Y) A DRUG

	Morphine		Heroin		Placebo		Total	
Subject	X	Y	X	Y	X	Y	X	Y
1	7	4	0	2	0	7	7	13
2	2	2	4	0	2	1	8	3
3	14	14	14	13	14	10	42	37
4	14	0	10	0	5	10	29	10
5	1	2	4	0	5	6	10	8
6	2	0	5	0	4	2	11	2
7	5	6	6	1	8	7	19	14
8	6	0	6	2	6	5	18	7
9	5	1	4	0	6	6	15	7
10	6	6	10	0	8	6	24	12
11	7	5	7	2	6	3	20	10
12	1	3	4	1	3	8	8	12
13	0	0	1	0	1	0	2	0
14	8	10	9	1	10	11	27	22
15	8	0	4	13	10	10	22	23
16	0	0	0	0	0	0	0	0
17	11	1	11	0	10	8	32	9
18	6	2	6	4	6	6	18	12
19	7	9	0	0	8	7	15	16
20	5	0	6	1	5	1	16	2
21	4	2	11	5	10	8	25	15
22	7	7	7	7	6	5	20	19
23	0	2	0	0	0	1	0	3
24	12	12	12	0	11	5	35	17
Total	138	88	141	52	144	133	423	273

	df	Σx^2	Σxy	Σy^2
Between subjects	23	910	519	558
Between drugs	2	1	5	137
Error	46	199	−16	422
Total	71	1110	508	1117

1/2, 2, 3, and 4 hr after. The response data (Y) in table 18.5.1 are those at 2 hr after. As a common precaution in these experiments, eight subjects took morphine first, eight took heroin first, and eight took the placebo first, and similarly on the second and third occasions. These data show no apparent effect of the order in which drugs were given, and the order is ignored in the analysis of variance presented here.

In planning this experiment two sources of variation were recognized. The first source is consistent differences in level of mental activity between subjects. This source was removed from the experimental error by the device of having each subject test all three drugs so that comparisons between drugs are made within subjects. Second, the subjects' levels change from time to time—they feel sluggish on some occasions and unusually alert on others. Insofar as these differences are measured by the pretest mental activity score on each occasion, the covariance analysis should remove this source of error.

As it turned out, the covariance was ineffective in this experiment. The error regression coefficient is actually slightly negative ($b = -16/199$) and showed no sign of statistical significance. Consequently, comparison of the drugs is best made from the 2-hr readings alone in this case. Incidentally, covariance would have been effective in removing differences in mental activity between subjects, since the between subjects $b = 519/910$ is positive and strongly significant.

Unlike the leprosy example in section 18.2, the use of the change in score, 2 hr − pretest, would have been unwise as a measure of the effects of the drugs. From table 18.5.1 the error sum of squares for $Y - X$ is $422 + 199 - 2(-16) = 653$, which is substantially larger than the sum of squares, 422, for Y alone.

The second example, table 18.5.2, illustrates another issue (3). The experiment compared the yields Y of six varieties of corn. There was some variation from plot to plot in number of plants (stand). If this variation is caused by differences in fertility in different plots and if higher plant numbers result in higher yields per plot, increased precision will be obtained by adjusting for the covariance of yield on plant number. The plant numbers in this event serve as an index of the fertility levels of the plots. But if some varieties characteristically have higher plant numbers than others through a greater ability to germinate or to survive when the plants are young, the adjustment for stand distorts the yields because it is trying to compare the varieties at some average plant number level that the varieties do not attain in practice.

With this in mind, look first at the F ratio for varieties in X (stand). From table 18.5.2 the mean squares are: varieties 9.17 and error 7.59, giving $F = 1.21$. The low value of F gives some assurance that the variations in stand are mostly random and that adjustment for stand will not introduce bias.

In the analysis, note the use of the variety + error line in computing the F test of the adjusted means. The value of F is $645.38/97.22 = 6.64$, highly significant with 5 and 14 df. The adjustment produced a striking decrease in the error mean square, from 583.5 to 97.2, and an increase in F from 3.25 to 6.64.

The adjusted means are found to be A, 191.8; B, 191.0; C, 193.1; D, 219.3; E, 189.6; and F, 213.6. From (18.2.7), the average standard error of the difference between two adjusted means is 7.25 with 14 df. By the protected LSD method (section 12.13), the two highest yielding varieties D and F are not

TABLE 18.5.2
STAND (X) AND YIELD (Y) (POUNDS FIELD WEIGHT OF EAR CORN) OF SIX VARIETIES OF CORN. COVARIANCE IN FOUR RANDOMIZED BLOCKS

	1		2		3		4		Total	
Varieties	X	Y	X	Y	X	Y	X	Y	X	Y
A	28	202	22	165	27	191	19	134	96	692
B	23	145	26	201	28	203	24	180	101	729
C	27	188	24	185	27	185	28	220	106	778
D	24	201	28	231	30	238	30	261	112	931
E	30	202	26	178	26	198	29	226	111	804
F	30	228	25	221	27	207	24	204	106	860
Total	162	1166	151	1181	165	1222	154	1225	632	4794

					Deviations from Regression		
Source of Variation	df	Σx^2	Σxy	Σy^2	df	Sum of Squares	Mean Square
Total	23	181.33	1485.00	18,678.50			
Blocks	3	21.67	8.50	436.17			
Varieties	5	45.83	559.25	9,490.00			
Error	15	113.83	917.25	8,752.33	14	1361.07	97.22
Variety + error	20	159.66	1476.50	18,242.33	19	4587.99	
For testing adjusted means					5	3226.92	645.38**

significantly different but they are significantly superior to all the others, which do not differ significantly among themselves.

In some cases, plant numbers might be influenced partly by fertility variations and partly by basic differences between varieties. The possibility of a partial adjustment has been considered by H. F. Smith (4).

EXAMPLE 18.5.1—Verify the adjusted means in the corn experiment and carry through the protected LSD test of the differences between pairs.

EXAMPLE 18.5.2—Estimate the efficiency of the covariance adjustments, as in section 18.2. Ans. 5.55.

EXAMPLE 18.5.3—As an alternative to covariance, could we analyze the yield per plant (Y/X) as a means of removing differences in plant numbers? Ans. This is satisfactory if the relation between Y and X is a straight line through the origin. But b is often substantially less than the mean yield per plant because when plant numbers are high, competition between plants reduces the yield per plant. If this happens, the use of Y/X overcorrects for stand. In the corn example, $b = 8.1$ and the overall yield per plant is $4794/632 = 7.6$, in good agreement; yield per plant would give results similar to covariance. Of course, yield per plant should be analyzed if there is direct interest in this quantity.

EXAMPLE 18.5.4—The following data are the yields (Y) in bushels per acre and the percents of stem canker infection (X) in a randomized blocks experiment comparing four lines A, B, C, D of soybeans (5).

Blocks	*A* X	*A* Y	*B* X	*B* Y	*C* X	*C* Y	*D* X	*D* Y	Total X	Total Y
1	19.3	21.3	10.1	28.3	4.3	26.7	14.0	25.1	47.7	101.4
2	29.2	19.7	34.7	20.7	48.2	14.7	30.2	20.1	142.3	75.2
3	1.0	28.7	14.0	26.0	6.3	29.0	7.2	24.9	28.5	108.6
4	6.4	27.3	5.6	34.1	6.7	29.0	8.9	29.8	27.6	120.2
Total	55.9	97.0	64.4	109.1	65.5	99.4	60.3	99.9	246.1	405.4

By looking at some plots with unusually high and unusually low values of X, note that there seems a definite negative relation between Y and X. Before removing this source of error by covariance, check that the lines do not differ in the amounts of infection. The analysis of sums of squares and products is as follows:

	df	Σx^2	Σxy	Σy^2
Blocks	3	2239.3	−748.0	272.9
Treatments, T	3	14.3	10.2	21.2
Error, E	9	427.0	−145.7	66.0
$T + E$	12	441.3	−135.5	87.2

(*i*) Perform the F test of the adjusted means. (*ii*) Find the adjusted means and test the differences among them. (*iii*) Estimate the efficiency of the adjustments. Ans. (*i*) $F = 4.79^*$; df = 3, 8. (*ii*) A, 23.77; B, 27.52; C, 25.19; D, 24.87. By the protected LSD test, B significantly exceeds C and D, but the others do not differ significantly. (*iii*) 3.56. Strictly, a slight correction to this figure should be made for the reduction in degrees of freedom from 9 to 8.

18.6—Use of regression adjustments in observational studies. As noted in chapter 1, the objective in many observational studies is to compare the effects of different treatments or programs applied to groups of subjects, but the investigator lacks the power to use randomization in determining which subjects receive a specific treatment. Often the groups select themselves. Men and women decide how many cigarettes they smoke daily; drivers of autos decide whether to wear seat belts. Sometimes the groups are selected by another agency, as with groups of persons judged eligible for welfare or of children entered into a program for the disadvantaged.

In addition to differing in the treatments to be compared, such groups may show systematic differences in other variables that are related to the response variable Y. Consequently, the investigator should guard against ascribing to the treatments differences in response that are due in whole or in part to these extraneous variables. If they are quantitative (X-variables), their values can often be measured and recorded for every subject. By analysis of variance and covariance the investigator can then examine whether there are systematic differences in X among the treatment groups and whether X is linearly related to Y. If so, it is natural to adjust the treatment means of Y by an analysis of covariance on X.

Note that the primary objective of the adjustments is to remove or reduce biases due to systematic differences in the treatment means of X. In randomized experiments, on the other hand, the purpose of covariance adjustments is to increase precision, as we have seen.

Under the right conditions, covariance adjustment can remove all the bias. Suppose that each study group is a random sample from some population. For the jth member of group i, let the regression model be

$$Y_{ij} = \alpha_i + \beta X_{ij} + \epsilon_{ij} \quad (18.6.1)$$

where ϵ_{ij} has mean 0, variance σ^2 for given X_{ij}. The parameter α_i represents the average effect of the ith treatment. Let η_i be the mean value of X_{ij} in the ith population. For simplicity, the discussion is confined to the most common situation with two groups—the extension to k groups is usually straightforward. From (18.6.1), averaging over the populations,

$$E(\overline{Y}_1 - \overline{Y}_2) = (\alpha_1 - \alpha_2) + \beta(\eta_1 - \eta_2) \quad (18.6.2)$$

Thus the initial bias due to the systematic difference in X is $\beta(\eta_1 - \eta_2)$.

If we adjust $\overline{Y}_1 - \overline{Y}_2$ by a covariance analysis, the adjusted estimate is

$$\overline{Y}_{1a} - \overline{Y}_{2a} = (\overline{Y}_1 - \overline{Y}_2) - b(\overline{X}_1 - \overline{X}_2) \quad (18.6.3)$$

Averaging first over repeated samples with fixed values of X, we get $E[b(\overline{X}_1 - \overline{X}_2)] = \beta(\overline{X}_1 - \overline{X}_2)$. Then, averaging over repeated random samples from the two populations,

$$\begin{aligned} E[(\overline{Y}_1 - \overline{Y}_2) - b(\overline{X}_1 - \overline{X}_2)] &= (\alpha_1 - \alpha_2) + \beta(\eta_1 - \eta_2) - \beta(\eta_1 - \eta_2) \\ &= (\alpha_1 - \alpha_2) \end{aligned} \quad (18.6.4)$$

Thus, covariance adjustment removes all the bias if we have random samples and the regression of Y on X is linear with the same slope in each population. With random samples the latter two assumptions can be checked in part from the sample data: Section 18.9 presents a test for differences among the bs in different groups. If the initial bias in X is large, however, one weakness of this approach is that the standard error of the adjusted estimate $\overline{Y}_{1a} - \overline{Y}_{2a}$ becomes large because of the term $(\overline{X}_1 - \overline{X}_2)^2/E_{xx}$ in its variance, due to errors in b [see (18.2.3)]. For example, in a related application, H. F. Smith (4) found that the unadjusted difference in Y between two groups was 145 ± 19.6, while the adjusted difference was −114 ± 105, "with almost all its standard error derived from uncertainty about the regression."

18.7—Problems with regression adjustments. Two additional assumptions not yet mentioned must also hold if the adjustment is to remove all the bias. First, X must be measured without error. In some studies the investigator can be confident that this is so, but in fields such as social science, education, and psychology the relevant Xs are often hard to either define or measure—for example, X may be intelligence, amount of parental care, or nervousness. Here, the measurement error may be what is called an error of specification—the tools of measurement are measuring the wrong quantity. Second, there must be no omitted variable Z that enters into the population regression and has different means in the two groups. Before considering the effects of measurement errors in X and of omitted variables, we take up one other topic.

Suppose we conclude that the regression lines have different slopes in the two populations. The model becomes

$$Y_{ij} = \alpha_i + \beta_i X_{ij} + \epsilon_{ij} \tag{18.7.1}$$

It follows that for any given value of X (the same in the two populations),

$$E(Y_{1j} - Y_{2j} | X_{1j} = X_{2j} = X) = (\alpha_1 - \alpha_2) + (\beta_1 - \beta_2)X \tag{18.7.2}$$

What do we make of (18.7.2)? One interpretation is that the difference in the effects of the two treatments depends on the level of X. Sometimes, however, this conclusion does not seem reasonable from what is known about the treatments. Another interpretation is that something is wrong with the model (18.7.1). If the first interpretation is accepted, our principal objective may be to estimate the *average* difference between the treatments taken over one of the populations. For example, Belsen (6) used covariance adjustments to estimate the effect of four BBC television programs called "Bon Voyage," intended to teach useful French words and phrases to viewers about to make a first trip to France. Group 1 consisted of viewers of the program and group 2 was a control group of nonviewers. Here, the quantity to be estimated is obviously the mean over population 1 of viewers or potential viewers.

From (18.7.2) with $X = \eta_1$, this quantity is

$$E(\bar{Y}_1 - \bar{Y}_2 | X = \eta_1) = (\alpha_1 - \alpha_2) + \beta_1\eta_1 - \beta_2\eta_1 \tag{18.7.3}$$

But the unadjusted mean difference gives, from (18.7.1),

$$E(\bar{Y}_1 - \bar{Y}_2) = \alpha_1 - \alpha_2 + \beta_1\eta_1 - \beta_2\eta_2 \tag{18.7.4}$$

Comparing (18.7.4) with (18.7.3), we see that the bias in the unadjusted difference is $-\beta_2(\eta_1 - \eta_2)$. Hence, an unbiased estimate of the mean over population 1 is

$$(\bar{Y}_1 - \bar{Y}_2) - b_2(\bar{X}_1 - \bar{X}_2) \tag{18.7.5}$$

The slope *in sample* 2 is used for the covariance adjustment, as Belsen recommended.

We now revert to the standard covariance model (18.6.1):

$$Y_{ij} = \alpha_i + \beta X_{ij} + \epsilon_{ij}$$

Regarding errors of measurement or specification, suppose that instead of X we measure $X' = X + e$, where $E(e|X) = 0$ and e, X are normal and independent. The covariance adjustment to $\bar{Y}_1 - \bar{Y}_2$ is now $-b'(\bar{X}'_1 - \bar{X}'_2)$, where b' is the sample regression coefficient of Y on X'. Note that in (9.14.1) the population regression of Y on X' has slope $\beta' = \beta\sigma_X^2/(\sigma_X^2 + \sigma_e^2)$. The ratio $R = \sigma_X^2/(\sigma_X^2 + \sigma_e^2)$ is called the *reliability coefficient* of the measurement. Over repeated samples with the same values of X, the average value of $-b'(\bar{X}_1' - \bar{X}_2') = -\beta'(\bar{X}_1 - \bar{X}_2)$. When we average further over all random samples, we see that adjustment reduces the bias by $\beta'(\eta_1 - \eta_2) = \beta R(\eta_1 - \eta_2)$. Since the initial bias in $\bar{Y}_1 - \bar{Y}_2$ was $\beta(\eta_1 - \eta_2)$, adjustment leaves a residual bias of amount $\beta(1 - R)(\eta_1 - \eta_2)$. If R is relatively small owing to errors of specification, most of the initial bias may remain after adjustment.

Campbell and Erlebacher (8) point out that this removal of only part of the initial bias by a fallible covariate results in underestimating the effects of programs designed to help disadvantaged or backward children. Suppose that such a program is given to a backward group 1. Group 2 is a normal nonbackward group used as a control. In what is called a before-after study, let X' (subject to error) be a measure of initial ability in a child before the program starts, while Y measures final ability. The parameter α_2 in the covariance model (18.6.1) measures the average effect of the natural maturation of the child, while α_1 measures maturation plus the effect of the program. Clearly, in this kind of study, $\eta_1 < \eta_2$, since group 1 is backward initially. The average residual bias in the estimate of $\alpha_1 - \alpha_2$ is $\beta(1 - R)(\eta_1 - \eta_2)$, which is negative if $\beta > 0$, as we would expect. Thus the effect of the program is underestimated, perhaps seriously if R is moderate and $\eta_2 - \eta_1$ is large.

With some types of measurement, the error e has correlation ρ with the correct value X. Under normality, the relation between β' and β then becomes (9)

$$\beta' = \beta(\sigma_X^2 + \rho\sigma_e\sigma_X)/(\sigma_X^2 + \sigma_e^2 + 2\rho\sigma_e\sigma_X) \tag{18.7.6}$$

For given σ_e and σ_X, positive values of ρ are found to decrease β'/β, so the covariance adjustment removes a smaller fraction of the initial bias in $\bar{Y}_1 - \bar{Y}_2$. Negative values of ρ increase the ratio β'/β. In fact, if $\rho = -\sigma_e/\sigma_X$, you may verify that $\beta' = \beta$ and the adjustment removes all the bias; this is the Berkson case mentioned at the end of section 9.14. If $\rho < -\sigma_e/\sigma_X$, then $\beta' > \beta$ and the adjustment overcorrects; the residual and initial biases have opposite signs.

If little is known about the size of the error of measurement e and its relation to X, it is uncertain how much confidence to place in the adjusted estimates of $\alpha_1 - \alpha_2$. When some information about the joint distribution of e and X is available, attempts have been made (12, 13) to use it to obtain better estimates of β and more trustworthy adjustments for bias.

Now consider variables omitted in the adjustment process. Suppose that the model is

$$Y_{ij} = \alpha_i + \beta X_{ij} + \gamma Z_{ij} + \epsilon_{ij} \qquad (i = 1, 2) \tag{18.7.7}$$

but we adjust only for the regression of Y on X. If Z_{ij} has means ν_1, ν_2 in the two populations, the initial bias in the estimate of $\alpha_1 - \alpha_2$ is

$$E(\bar{Y}_1 - \bar{Y}_2) - (\alpha_1 - \alpha_2) = \beta(\eta_1 - \eta_2) + \gamma(\nu_1 - \nu_2) \tag{18.7.8}$$

Assume that Y, X, Z have trivariate normal distributions in the two populations. Equation (17.11.1) in section 17.11 shows that under model (18.7.7) the regression coefficient b of Y on X is an unbiased estimate of

$$\beta'' = \beta + \gamma\beta_{ZX}$$

where β_{ZX} is the within-population regression of Z on X. Hence, the average value of the covariance adjustment $-b(\bar{X}_1 - \bar{X}_2)$ is $-\beta''(\eta_1 - \eta_2) = -(\beta + \gamma\beta_{ZX})(\eta_1 - \eta_2)$. From (18.7.8), the residual bias in $\hat{\alpha}_1 - \hat{\alpha}_2$ after adjustment is therefore

$$\gamma[(\nu_1 - \nu_2) - \beta_{ZX}(\eta_1 - \eta_2)] \qquad (18.7.9)$$

With this model the adjustment removes all the bias due to X and attempts to reduce also the bias due to Z by taking advantage of the regression of Z on X. The ratio $(\nu_1 - \nu_2)/(\eta_1 - \eta_2)$ can be described as the slope of the between-populations regression of Z on X. Hence, from (18.7.9), adjustment on X removes all the bias due to Z if the between-population and within-population regression coefficients of Z on X are equal. However, if the between-population regression is of opposite sign from the within-population regression, adjustment on X increases the bias due to Z. For comparison of the performance of linear regression on X and matching the samples on X, see (10).

To sum up, linear regression adjustments can be highly effective in removing or reducing an initial bias in the comparison of two groups or they can leave a residual bias after adjustment that is large enough to distort the conclusions badly. The result depends on the relation between the model assumed in making the regression adjustment and the model that actually holds. When planning a study and when preparing to report results, try to estimate, if possible, how large the initial bias is. This can be done in studies in which the response variable Y is measured *before* as well as after treatment differences between groups have been introduced. From the pretreatment measurements it is worth looking at both $(\bar{X}_1 - \bar{X}_2)/s_{\bar{X}}$ and $(\bar{X}_1 - \bar{X}_2)/\bar{X}$, dividing by some measure of average level. For instance, we might judge from $(\bar{X}_1 - \bar{X}_2)/s_X$ that the initial bias is large enough to make tests of significance of the final difference unreliable. But if $(\bar{X}_1 - \bar{X}_2)/\bar{X}$ is, say, 5% and we find a posttreatment difference of 35%, we may conclude that the treatment effect was large, despite signs of some initial bias.

Furthermore, is anything known about measurement and specification errors in X from which one might make guesses about β'/β and the fraction of initial bias removed by adjustment? In some studies there are many omitted variables that might influence Y but would be costly to include. Which, if any, seem to have large enough values of γ (regression of Y on Z) to justify special efforts to include them? From judgments like these, the investigator may form an opinion as to how firmly the final conclusions seem to be established.

EXAMPLE 18.7.1—With errors of measurement in X_1', suppose that $\sigma_X^2 = 8$, $\sigma_e^2 = 2$. If e and X are uncorrelated, $\beta'/\beta = 0.8$. From (18.7.7), what are the values of β'/β if $\rho = 0.4, 0.6, -0.4, -0.6$? Ans. 0.73, 0.70, 0.94, 1.08.

18.8—Covariance in interpreting differences between classes. In either experiments or observational studies we may have data in which significant differences between treatments or groups are found *both* in the response variate Y and in a covariate X thought to be causally related to Y. As mentioned in section 18.1, we may be interested in examining whether the treatment differences in Y can be ascribed to the differences in X. We compute an analysis of covariance and make an F test of the differences between the adjusted treatment means in Y. If F has become nonsignificant, we may be inclined to conclude that apart from random errors the differences in Y are due entirely to differences in X.

Remember, however, as noted at the end of section 18.6, that when there are large differences between the group means for X, the standard errors of the adjusted differences in Y become large. A consequence is that the power of the F test of the adjusted means, on which we are basing our conclusion, becomes weak. Calculation of confidence intervals for the adjusted differences helps. From these intervals we may learn that the adjusted differences have *not* been shown to be almost certainly small but merely to be poorly estimated. With wide differences between the X means, we also ask if we can trust the covariance adjustments because they involve an element of unverifiable extrapolation. This element is not taken into account in the confidence interval calculations, which assume that the model holds all the way.

H. F. Smith (4), who drew attention to these points, illustrated them from an experiment in which alfalfa was prepared as field-cured hay, barn-cured hay, and silage, and fed to cows in an experiment in six randomized blocks. The variable Y was the vitamin A potency per pound of butterfat produced by the cow. The three methods of curing produce large differences in the carotene content X per ration of hay and it is well known that animals make vitamin A from carotene. A question of interest was, Do the curing methods affect the vitamin A apart from their effects on carotene contents of the forage? The analysis was made in logs.

Table 18.8.1 shows the treatment means for X and Y and the analyses of

TABLE 18.8.1
CAROTENE CONTENT (X) AND VITAMIN A POTENCY PER POUND OF BUTTERFAT (Y) FOR TYPES OF ALFALFA FORAGE (ON A LOG SCALE)

	Field Hay		Barn Hay		Silage	
	X	Y	X	Y	X	Y
Means	146.3	109.7	336.2	133.5	796.5	254.3

					Deviations from Regression	
	df	Σx^2	Σxy	Σy^2	df	SS
Treatments, T	2	1,341,320	308,411	72,181		
Error, E	10	29,930	11,922	11,578	9	6829
$T + E$	12	1,371,250	320,333	83,759	11	8927

TABLE 18.8.2
F TESTS OF DIFFERENCES BETWEEN TREATMENT MEANS

	Unadjusted				Adjusted			
	df	*SS*	*MS*	*F*	df	*SS*	*MS*	*F*
Treatments	2	72,181	36,090	31.2	2	2098	1049	1.38
Error	10	11,578	1,158		9	6829	759	

Σx^2, Σxy, and Σy^2. You may verify that the F tests of the unadjusted and adjusted treatment means are as shown in table 18.8.2. Note that the covariance adjustment reduced F from 31.2 to 1.38, suggesting that the treatment effects on vitamin A can be fully explained by their effects on the carotene in the forage. However, let us find confidence limits for the difference between the adjusted means for silage and field hay, which showed the largest unadjusted Y difference, 144.6. With $b = 0.39833$, the difference between the adjusted means is $144.6 - (0.39833)(650.2) = -114.4$. From (18.2.3), the error variance of the adjusted difference is $(759)(0.333 + 650.9^2/29{,}930) = 10{,}997$. With 9 df, the resulting 95% confidence limits for silage *minus* field hay are $(-352, +123)$. Relative to field hay, silage could have had a strong depressing effect or a moderate positive effect $(+123)$ on vitamin A in addition to its influence on the carotene content. The adjusted Y means are in fact not reliable enough to determine whether there was *any* additional effect of the treatments. As H. F. Smith suggests, addition of carotene in a new experiment so that different hays are compared at the same level of carotene intake by the cows would give a more reliable comparison of any additional effects that are present.

18.9—Comparison of regression lines. Frequently the relation between Y and X is studied in samples obtained by different workers, or in different environments, or at different times. In summarizing these results, the question arises, Can the regression lines be regarded as the same? If not, in what respects do they differ? This issue is also relevant as a check when the analysis of covariance is used, as in previous sections, to compare adjusted means in experiments or observational studies. The standard model that we use for these adjustments assumes that the slopes and the deviations mean squares are homogeneous in different classes.

A numerical example provides an introduction to handling these questions. The example has only two samples, but the techniques extend naturally to more than two samples. In a survey to examine relationships between the nutrition and the health of women in the Middle West (7), the concentration of cholesterol in the blood serum was determined on 56 randomly selected subjects in Iowa and 130 in Nebraska. Table 18.9.1 contains subsamples from the survey data. Graphs of the data from each state suggest that the regression of cholesterol concentration on age is linear, which will be assumed in this discussion.

The purpose is to examine whether the linear regressions of cholesterol on age are the same in Iowa and Nebraska. They may differ in slope, in elevation,

TABLE 18.9.1
AGE (X) AND CONCENTRATION OF CHOLESTEROL (MG/100 ML) (Y) IN THE BLOOD SERUM OF IOWA AND NEBRASKA WOMEN

Iowa, $n = 11$				Nebraska, $n = 19$					
X	Y	X	Y	X	Y	X	Y	X	Y
33	201	54	259	18	137	42	214	56	197
39	182	58	189	19	189	43	223	58	262
41	112	65	249	21	191	44	190	58	257
46	181	71	224	30	140	44	173	63	337
49	121	76	339	31	159	47	196	67	356
52	228			33	177	51	225	70	261
								78	241
Total		584	2285					873	4125
		$\bar{X}_I = 53.1$	$\bar{Y}_I = 207.7$	$\bar{X}_N = 45.9$		$\bar{Y}_N = 217.1$			

or in the residual variances $\sigma_{y.x}^2$. The most convenient approach is to compare the residual variances first, then the slopes, and then the elevations. In terms of the model, we have

$$Y_{ij} = \alpha_i + \beta_i X_{ij} + \epsilon_{ij}$$

where $i = 1, 2$ denotes the two states. We first compare the residual variances σ_1^2 and σ_2^2; next β_1 and β_2; and finally the elevations of the lines, α_1 and α_2.

If no computer program is available, record separately the within-class sums of squares and products for each state, as shown in table 18.9.2 on lines 1 and 2. The next step is to find the residual sum of squares from regression for each state, as on the right in lines 1 and 2. The residual mean squares, 2392 and 1581, are compared by the two-tailed F test or, with more than two samples, by Bartlett's test (section 13.10). If heterogeneous variances were evident, this might be pertinent information in itself. In this example, $F = 1.51$ with 9 and 17

TABLE 18.9.2
COMPARISON OF REGRESSION LINES. CHOLESTEROL DATA

		df	Σx^2	Σxy	Σy^2	Reg. Coeff.	Deviations from Regression df	SS	MS
	Within								
1	Iowa	10	1829	5,922	40,698	3.24	9	21,524	2392
2	Nebraska	18	5565	14,026	62,226	2.52	17	26,875	1581
3							26	48,399	1862
4	Pooled, W	28	7394	19,948	102,924	2.70	27	49,107	1819
5				Difference between slopes			1	708	708
6	Between, B	1	355	−466	613				
7	$W + B$	29	7749	19,482	103,537		28	54,557	
8				Between adjusted means			1	5,450	5450

Comparison of slopes: $F = 708/1862 = 0.38$ (df = 1, 26) not significant.
Comparison of elevations: $F = 5450/1819 = 3.00$ (df = 1, 27) not significant.

df, giving P a value greater than 0.40 in a two-tailed test. The residual mean squares show no sign of a real difference.

Assuming homogeneity of residual variances, we now compare the slopes or regression coefficients, 3.24 for Iowa and 2.52 for Nebraska. The test is made by the extended analysis of variance by which we obtained the F test of differences between the adjusted class means in section 18.3. Add the degrees of freedom and the sum of squares of deviations from the individual regressions, recording these sums in line 3. The mean square 1862 is the quantity called s_{AH}^2 in section 18.3—the residual mean square obtained on the alternative hypothesis, when *separate* regressions are fitted in each state. In line 4 we add the sums of squares and products, obtaining the pooled slope 2.70 and the sum of squares 49,107, representing deviations D_{NH} from a model in which a *single* pooled slope is fitted. The difference, $D_{NH} - D_{AH} = 49{,}107 - 48{,}399 = 708$ (line 5) with 1 df, measures the contribution of the difference between the two regression coefficients to the sum of squares of deviations. If there were k coefficients, this difference would have $(k - 1)$ df. The corresponding mean square is compared with the within-states mean square, 1862, by the F test. In these data, $F = 708/1862 = 0.38$, df $= 1, 26$, supporting the assumption that the slopes do not differ.

Algebraically, the difference 708 in the sum of squares may be shown to be $\Sigma_1 \Sigma_2 (b_1 - b_2)^2/(\Sigma_1 + \Sigma_2)$, where Σ_1, Σ_2 are the values of Σx^2 for the two states. With more than two states the difference is $\Sigma w_i (b_i - \bar{b})^2$, where $w_i = 1/\Sigma_i$ and the pooled slope $\bar{b}$ is $\Sigma w_i b_i / \Sigma w_i$. The sum of squares of deviations of the bs is a weighted sum because the variances of the b_i, namely $\sigma_{y.x}^2/\Sigma_i$, depend on the values of Σx^2.

If the sample regressions were found to differ significantly, the investigation might be ended. Interpretation would involve the question, Why? The final question about the elevations of the population regression lines usually has little meaning unless the lines are parallel.

Assuming parallel lines and homogeneous variance, we write the model as

$$Y_{ij} = \alpha_i + \beta X_{ij} + \epsilon_{ij}$$

where $i = 1, 2$ denotes the state. It remains to test the null hypothesis $\alpha_1 = \alpha_2$. The least squares estimates of α_1 and α_2 are $\hat{\alpha}_1 = \bar{Y}_1 - b\bar{X}_1$ and $\hat{\alpha}_2 = \bar{Y}_2 - b\bar{X}_2$. Hence, this test is identical to the test of the hypothesis that the adjusted means of the Ys are the same in the two states. This is, of course, the F test of the difference between adjusted treatment means that was made in section 18.3. It is made in the usual way in lines 4 to 8 in table 18.9.2. Line 4 gives the pooled within-states sums of squares and products, while line 6 shows the between-states sums of squares and products. In line 7 these are combined, just as we combined error and treatments in section 18.3. A deviations sum of squares, 54,557, is obtained from line 7 and the deviations sum of squares in line 4 is subtracted to give 5450, the sum of squares between adjusted means. We find $F = 3.00$; df $= 1, 27$; P about 0.10. In the original survey the difference was smaller than in these subsamples. The investigators felt justified in combining the two states for further examination of the relation between age and cholesterol.

With a computer program, the printout might summarize the F tests of slopes and adjusted means as follows:

ANALYSIS OF VARIANCE	D.F.	SUM OF SQ.	MEAN SQ.	F	TAIL AREA P
EQUALITY OF SLOPES	1	708	708	0.38	0.54
RESIDUAL ERROR	26	48,399	1862		
EQUALITY OF ADJ. MEANS	1	5450	5450	3.00	0.095
RESIDUAL ERROR	27	49,107	1819		

EXAMPLE 18.9.1—As would be expected, states with higher per capita incomes (pci) X tend to spend more per pupil in public school expenditures Y. The average within-region regression coefficient (0.218) represents about an additional $220 per pupil for each additional $1000 pci. From the following data for 1976 and 1977, test whether the regression coefficient varies from region to region (pci values and expenditures are divided by 1000).

Region	df	Σx^2	Σxy	Σy^2	b	$\overline{X}/1000$
Northeast	9	4.898	1.901	1.116	0.338	6.54
Southeast	6	2.647	0.688	0.201	0.260	5.84
South Central	8	2.576	0.162	0.046	0.063	5.52
North Central	10	4.905	1.023	0.448	0.209	6.27
Mountain Pacific	10	5.128	0.629	0.404	0.123	6.24
Total	43	20.154	4.403	2.214		

Ans. $F = 2.43$ with 4, 38 df; P about 0.065. You will also find some heterogeneity of residual mean squares by Bartlett's test. As a consequence, P should be slightly higher than 0.065. The coefficient b is highest (0.338) where income is highest and lowest (0.063) where income is lowest, but the three other regions do not show this relation to pci.

18.10—Multiple covariance. With two or more independent variables there is no change in the theory beyond the addition of extra terms in X. For analyses of the results of experiments with two X-variates, we need:

1. The regression coefficients b_1 and b_2, calculated from the sums of squares and products in the error line appropriate to the type of experiment.
2. The adjusted treatment means.
3. The sum of squares and mean square s^2 of deviations of Y from the error regression on x_1 and x_2.
4. Standard errors and t tests of differences between the adjusted means and of comparisons among them. For this, the c_{ij} multipliers (section 17.5) that give the variances and covariances of the bs are very useful, since any comparison is a linear function of the bs with known coefficients.

We may also want:

5. An F test of the null hypothesis that there are no differences among the adjusted means.
6. The t tests of b_1 and b_2 and an F test of the reduction in error sum of squares due to the regression.

The method is illustrated for a one-way classification by the average daily gains of pigs in table 18.10.1. Presumably they are predicted at least partly by the ages

TABLE 18.10.1

INITIAL AGE (X_1), INITIAL WEIGHT (X_2), AND RATE OF GAIN (Y) OF 40 PIGS
(four treatments in lots of equal size)

	Treatment 1			Treatment 2		
	Initial Age, X_1 (days)	Weight, X_2 (lb)	Gain, $Y = X_3$ (lb/day)	Initial Age, X_1 (days)	Weight X_2 (lb)	Gain, $Y = X_3$ (lb/day)
	78	61	1.40	78	74	1.61
	90	59	1.79	99	75	1.31
	94	76	1.72	80	64	1.12
	71	50	1.47	75	48	1.35
	99	61	1.26	94	62	1.29
	80	54	1.28	91	42	1.24
	83	57	1.34	75	52	1.29
	75	45	1.55	63	43	1.43
	62	41	1.57	62	50	1.29
	67	40	1.26	67	40	1.26
Total	799	544	14.64	784	550	13.19
Means	79.9	54.4	1.46	78.4	55.0	1.32
	Treatment 3			Treatment 4		
	78	80	1.67	77	62	1.40
	83	61	1.41	71	55	1.47
	79	62	1.73	78	62	1.37
	70	47	1.23	70	43	1.15
	85	59	1.49	95	57	1.22
	83	42	1.22	96	51	1.48
	71	47	1.39	71	41	1.31
	66	42	1.39	63	40	1.27
	67	40	1.56	62	45	1.22
	67	40	1.36	67	39	1.36
Total	749	520	14.45	750	495	13.25
Means	74.9	52.0	1.44	75.0	49.5	1.32

Sums of Squares and Products

	df	Σx_1^2	$\Sigma x_1 x_2$	Σx_2^2
Treatments	3	187.70	160.15	189.08
Error	36	4548.20	2877.40	4876.90
Total	39	4735.90	3037.55	5065.98
	df	$\Sigma x_1 y$	$\Sigma x_2 y$	Σy^2
Treatments	3	1.3005	1.3218	0.1776
Error	36	5.6230	26.2190	0.8452
Total	39	6.9235	27.5408	1.0228

and weights at which the pigs were started in the experiment, which compared four feeds.

This experiment is an example of a technique in experimental design known as *balancing*. The assignment of pigs to the four treatments was not made by strict randomization. Instead, pigs were allotted so that the means of the four lots agreed closely in both X_1 and X_2. An indication of the extent of the balancing can be seen by calculating the F ratios for treatments/error from the analyses of variance of X_1 and X_2, given in table 18.10.1. These Fs are 0.50 for X_1 and 0.47 for X_2, both well below 1.

If X_1 and X_2 are linearly related to Y, this balancing produces a more accurate comparison among the Y means. One complication is that, since the variance within treatments is greater than that between treatments for X_1 and X_2, the same happens to some extent for Y. Consequently, in the analysis of variance of Y the error mean square is an overestimate and the F test of Y gives too few significant results. However, if the covariance model holds, the analysis of covariance will give an unbiased estimate of error and a correct F test for the adjusted means of Y. The situation is interesting in that, with balancing, the reason for using covariance is to obtain a proper estimate of error rather than to adjust the Y means. If perfect balancing were achieved, the adjusted Y means would be the same as the unadjusted means.

The first step is to calculate the error and total (treatments + error) sums of squares and products shown in table 18.10.1. Then the equations for the bs are solved for the error line. We find

$$b_1 = -0.0034542 \qquad b_2 = 0.0074142$$

The c_{ij} multipliers are

$$c_{11} = 0.0003508 \qquad c_{12} = -0.0002070 \qquad c_{22} = 0.0003272$$

Deviations sum of squares = 0.6702 (34 df); $s^2 = 0.0197$. The standard errors of b_1 and b_2 are

$$s_{b_1} = \sqrt{s^2 c_{11}} = 0.00263 \qquad s_{b_2} = \sqrt{s^2 c_{22}} = 0.00254$$

It follows that b_2 is definitely significant but b_1 is not. In practice, we might decide to drop X_1 (age) at this stage and use the regression of Y on X_2 alone. But for illustration, we shall retain both variables.

If an F test of the adjusted means is wanted, make a new calculation of b_1 and b_2 from the treatments + error lines, in this case the total line. The results are $b_1 = -0.0032903$, $b_2 = 0.0074093$. The sum of squares for deviations from the regression is 0.8415 (37 df). The F test is made in table 18.10.2. It gives $F = 2.90$, $P < 0.05$.

The adjusted Y means are computed in table 18.10.3. In our notation, $\overline{Y}_i$,

TABLE 18.10.2

DEVIATIONS FROM REGRESSION AND F TEST IN ANALYSIS OF COVARIANCE

Source of Variation	df	Sum of Squares	Mean Square	F
Residuals (NH)	37	$D_{NH} = 0.8415$		
Residuals (AH)	34	$D_{AH} = 0.6702$	0.0197	
Difference, $D_{NH} - D_{AH}$	3	0.1713	0.0571	2.90

TABLE 18.10.3
CALCULATION OF ADJUSTED MEANS

Treatment	1	2	3	4	Multiplier
$\overline{Y}_i$	1.46	1.32	1.44	1.32	1
$\overline{X}_{1i} - \overline{X}_1$	+2.9	+1.4	−2.1	−2.0	$0.00345 = -b_1$
$\overline{X}_{2i} - \overline{X}_2$	+1.7	+2.3	−0.7	−3.2	$-0.00741 = -b_2$
$\overline{Y}_{\text{adj}}$	1.46	1.31	1.44	1.34	

$\overline{X}_{1i}$, $\overline{X}_{2i}$ denote the means of Y, X_1, and X_2 for the ith treatment, while $\overline{X}_1$, $\overline{X}_2$ denote the overall means of X_1 and X_2. Thus, for treatment 4,

$$\overline{Y}_{\text{adj}} = \overline{Y}_4 - b_1(\overline{X}_{14} - \overline{X}_1) - b_2(\overline{X}_{24} - \overline{X}_2)$$

$$= 1.32 + 0.00345(-2.0) - 0.00741(-3.2) = 1.34$$

There is practically no change from unadjusted to adjusted means because of the balancing. However, comparison of $s_y^2 = 0.0235$ with $s^2 = s_{y.12}^2 = 0.0197$ shows that the error variance of the Y means would have been overestimated by about 20% without the covariance analysis. By the LSD test, the adjusted means of treatments 1 and 3 are significantly higher than that for treatment 2, while treatment 4 occupies an intermediate position.

Computer programs for the analysis of covariance are likely to give the t values for the differences between every pair of adjusted means. If other comparisons are wanted, check that the printout provides their standard errors or results that lead to them.

By an extension of (18.2.7) for one X-variate, an average error variance can be used as an approximation for any comparison among adjusted treatment means. The approximation works well enough if treatments do not affect the Xs and with at least 20 df for error. The approximate error mean square per observation is

$$s'^2 = s^2(1 + c_{11}t_{11} + 2c_{12}t_{12} + c_{22}t_{22}) \tag{18.10.1}$$

where the ts are the treatments *mean squares and product* for the Xs (found from the data in table 18.10.1). In these data, $s'^2 = (0.0197)(1.020) = 0.0201$. The approximate standard error of the difference between any two adjusted treatment means is $\sqrt{2s'^2/10} = \pm 0.0634$. Different treatment means are regarded as independent with this approximation.

TECHNICAL TERMS

adjusted means
analysis of covariance
balancing
between-populations regression
dummy variables
indicator variables
within-groups regression

REFERENCES

1. Fisher, R. A. *Statistical Methods for Research Workers,* 4th ed. Oliver & Boyd, Edinburgh, §49.1.
2. Finney, D. J. Biometrics Bull. 2 (1946):53.
3. Sprague, G. F. 1952. Iowa Agric. Exp. Stn. data.
4. Smith, H. Fairfield. *Biometrics* 13 (1957):282.
5. Crall, J. M. 1949. Iowa Agric. Exp. Stn. data.
6. Belsen, W. A. *Appl. Stat.* 5 (1956).
7. Swanson, P. P., et al. *J. Gerontol.* 10 (1955):41.
8. Campbell, D. T., and Erlebacher, A. 1970. *Compensatory Education: A National Debate, 3.* J. Hellmuth. Brunner-Mazel, New York, p. 185.
9. Cochran, W. G. *Technometrics* 10 (1968):637.
10. Cochran, W. G., and Rubin, D. B. *Sankhya* A35 (1973):417.
11. Smith, G. M., and Beecher, H. T. *J. Pharm. Exp. Ther.* 136 (1962):47.
12. DeGracie, J. S., and Fuller, W. A. *J. Am. Stat. Assoc.* 67 (1972):930.
13. Hidiroglou, M. A.; Fuller, W. A.; and Hickman, R. D. 1978. *SUPER CARP*—a program designed for the analysis of survey data, including data subject to measurement error. Tapes available for purchase from Stat. Lab., Iowa State University, Ames.
14. Eden, T. *J. Agric. Sci.* 21 (1931):547.

19

Nonlinear relations

19.1—Introduction. Although linear regression is adequate for many needs, some variables are not connected by so simple a relation. The discovery of a precise description of the relation between two or more quantities is one problem of *curve fitting,* known as *curvilinear regression.* From this general view the fitting of the straight line is a special case, the simplest and the most useful.

Figure 19.1.1 shows four common nonlinear relations. Part (a) is the compound interest law or *exponential growth curve,* $W = A(B^X)$, where we have written W in place of our usual Y. If $B = 1 + i$, where i is the annual rate of interest, W gives the amount to which a sum of money A will rise if left at compound interest for X years. As we shall see, this curve also represents the way in which some organisms grow at certain stages. The curve shown in (a) has $A = 1$.

If $B < 1$, this curve assumes the form shown in (b). It is often called an *exponential decay curve,* the value of W declining to zero from its initial value A as X increases. The decay of emissions from a radioactive element follows this curve.

The curve in (c) is $W = A - B\rho^X$, with $0 < \rho < 1$. This curve rises from the value $A - B$ when $X = 0$ and steadily approaches a maximum value A, called the asymptote, as X becomes large. The curve is known by various names. In agriculture it has been called *Mitscherlich's law,* from a German chemist (11) who used it to represent the relation between the yield W of a crop (grown in pots) and the amount of fertilizer X added to the soil in the pots. In chemistry it is sometimes called the *first order reaction* curve. The name *asymptotic regression* is also used.

Curve (d), the *logistic growth law,* has played a prominent part in the study of human populations. This curve gives a remarkably good fit to the growth of the U.S. population, as measured in the decennial censuses from 1790 to 1940.

In this chapter we illustrate the fitting of three types of curves: (1) certain nonlinear curves, like those in figure 19.1.1, (a) and (b), that can be reduced to straight lines by a transformation of the W or the X scale; (2) the polynomial in X, which often serves as a good approximation; (3) nonlinear curves, like figure 19.1.1, (c) and (d), that require more complex methods of fitting.

EXAMPLE 19.1.1—The fit of the logistic curve of the U.S. census populations (excluding Hawaii and Alaska) for the 150 year period from 1790 to 1940 is an interesting example, both of the striking accuracy of the fit and of its equally striking failure when extrapolated to give population forecasts for 1950 and 1960. The curve (fitted by Pearl, Reed, and Kish 1) is

$$\hat{W} = \frac{184.00}{1 + (66.69)(10^{-0.1398X})}$$

where $X = 1$ in 1790, and one unit in X represents 10 yr so that $X = 16$ in 1940. The table below shows the actual census population, the estimated population from the logistic, and the error of estimation.

Year	Population Actual	Population Estimated	$W - \hat{W}$	Year	Population Actual	Population Estimated	$W - \hat{W}$
1790	3.9	3.7	+0.2	1880	50.2	50.2	0.0
1800	5.3	5.1	+0.2	1890	62.9	62.8	+0.1
1810	7.2	7.0	+0.2	1900	76.0	76.7	−0.7
1820	9.6	9.5	+0.1	1910	92.0	91.4	+0.6
1830	12.9	12.8	+0.1	1920	105.7	106.1	−0.4
1840	17.1	17.3	−0.2	1930	122.8	120.1	+2.7
1850	23.2	23.0	+0.2	1940	131.4	132.8	−1.4
1860	31.4	30.3	+1.1	1950	150.7	143.7	+7.0
1870	38.6	39.3	−0.7	1960	178.5	152.9	+25.6

Note how poor the 1950 and 1960 forecasts are. The forecast from the curve is that the U.S. population will never exceed 184 million; the actual 1970 population was over 200 million. The postwar baby boom and improved health services are two responsible factors.

19.2—The exponential growth curve.

A characteristic of some of the simpler growth phenomena is that the rate of increase at any moment is proportional to the size already attained. During one phase in the growth of a culture of bacteria, the numbers of organisms follow such a law. The relation is nicely illustrated by the dry weights of chick embryos at ages 6 to 16 days (2) recorded in table 19.2.1. The graph of the weights in figure 19.2.1 ascends with greater rapidity as age increases, the regression equation being of the form

$$W = (A)(B^X) \tag{19.2.1}$$

where A and B are constants to be estimated. Applying logarithms to the equation,

$$\log W = \log A + (\log B)X \tag{19.2.2}$$

$$Y = \alpha + \beta X \tag{19.2.3}$$

where $Y = \log W$, $\alpha = \log A$, and $\beta = \log B$. If $\log W$ instead of W is plotted against X, the graph will be linear.

The values of $Y = \log W$ (to base 10) appear in the fourth column of the table and are plotted opposite X in the figure. The regression equation, computed as usual from columns X and Y in the table, is $\hat{Y} = 0.1959X - 2.689$. The

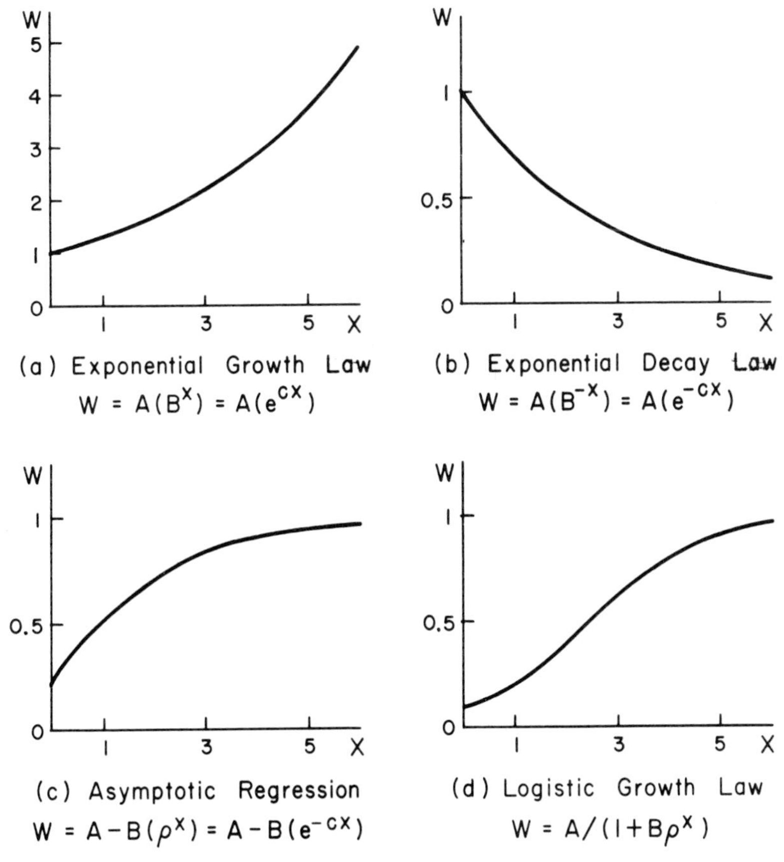

Fig. 19.1.1—Four common nonlinear curves.

TABLE 19.2.1

DRY WEIGHTS (GRAMS) OF CHICK EMBRYOS FROM AGES 6 TO 16 DAYS

Age X	Weight W	$\hat{W}$	Log W Y	$\hat{Y}$	$Y-\hat{Y}$
6	0.029	0.031	−1.538	−1.514	−0.024
7	0.052	0.048	−1.284	−1.318	+ .034
8	0.079	0.076	−1.102	−1.122	+ .020
9	0.125	0.119	−0.903	−0.926	+ .023
10	0.181	0.186	−0.742	−0.730	− .012
11	0.261	0.292	−0.583	−0.534	− .049
12	0.425	0.458	−0.372	−0.339	+ .033
13	0.738	0.721	−0.132	−0.142	+ .010
14	1.130	1.130	0.053	+0.053	.000
15	1.882	1.774	0.275	+0.249	+ .026
16	2.812	2.786	0.449	+0.445	−0.001

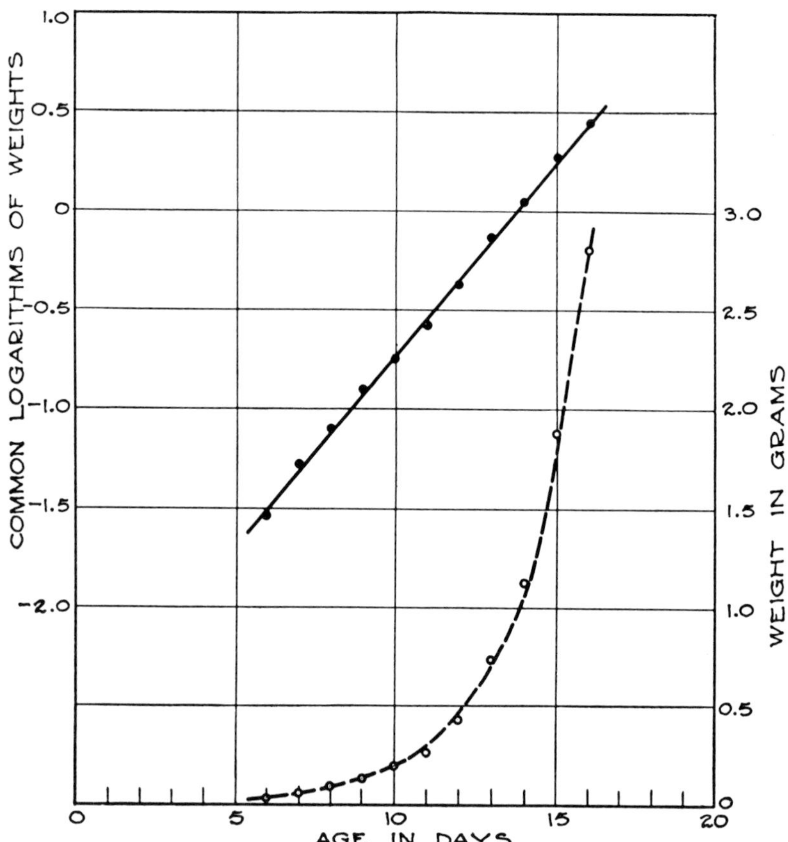

Fig. 19.2.1—Dry weights of chick embryos at ages 6–16 days with fitted curves. Weight scale: $W = 0.002046(1.57)^X$; logarithmic scale: $Y = 0.1959X - 2.689$.

regression line fits the data points with unusual fidelity, the correlation between Y and X being 0.9992. The conclusion is that the chick embryos, as measured by dry weight, are growing in accord with the exponential law; the logarithm of the dry weight increases at the estimated uniform rate of 0.1959 g per day.

Often the objective is to learn whether the data follow the exponential law. The graph of log W against X helps in making an initial judgment on this question and may be sufficient to settle the point. If so, the use of *semilogarithmic* graph paper avoids the necessity for finding the logarithms of W. The horizontal rulings on this graph paper are drawn to such a scale that the plotting of the original data results in a straight line if the data follow the exponential growth law. A more thorough method of testing whether the relation between log W and X is linear is given in section 19.3.

For those who know some calculus, the law that the rate of increase at any stage is proportional to the size already attained is described mathematically by the equation

$$\frac{dW}{dX} = cW$$

where c is the constant *relative* rate of increase. This equation leads to the relation

$$\ln W = \log_e W = \log_e A + cX \tag{19.2.4}$$

or $W = Ae^{cX}$, where $e = 2.718$ is the base of natural logarithms. Relation 19.2.4 is exactly the same as (19.2.3),

$$\log_{10} W = \alpha + \beta X$$

except that it is expressed in logs to base e instead of to base 10.

Since $\ln W = \log_e W = (\log_{10} W)(\log_e 10) = 2.3026 \log_{10} W$, it follows that $c = 2.3026\beta$. For the chick embryos, the relative rate of growth is $(2.3026)(0.1959) = 0.451$ g/day/g.

To convert the equation $\log \hat{W} = 0.1959X - 2.689$ into the original form, we have

$$\hat{W} = (0.00205)(1.57)^X$$

where $0.00205 = 10^{-2.689}$, while $1.570 = 10^{0.1959}$. Many pocket calculators have keys that produce 10^x and e^x for any x. In the exponential growth form,

$$\hat{W} = (0.00205)e^{0.451X}$$

Other relations that may be fitted by simple transformations of the W or the X variable are $W = 1/(\alpha + \beta X)$; $W = \alpha + \beta \ln X$; and $W = \alpha X^{\beta}$, where the power β must be estimated from the data. This relation leads to $\ln W = \ln \alpha + \beta \ln X$. For plotting this relation, paper with both vertical and horizontal logarithmic scales is available.

The transformation of variables so that a nonlinear relation becomes a straight line is a simple method of fitting but involves assumptions that should be noted. For the exponential growth curve, we are assuming that the population model is of the form

$$Y = \ln W = \alpha + \beta X + \epsilon \tag{19.2.5}$$

where the residuals ϵ are independent with zero means and constant variance. The assumption of constant variance in the Y scale seems reasonable with the check embryo data, as it often does with exponential growth curves. Application of the usual tests of significance to α and β involves a further assumption that the ϵs are normally distributed.

Sometimes it may seem more realistic, from our knowledge of the process, to assume that residuals are normal and have constant variance *in the original W scale*. Then we postulate a population model

$$W = AB^X + d \tag{19.2.6}$$

where A and B stand for population parameters and the residuals d are $\mathcal{N}(0, \sigma^2)$. If model (19.2.6) holds, the ϵs in model (19.2.5) will not be normal and their variances will change as X changes. Given model (19.2.6), the least squares method estimates A and B by minimizing $\Sigma(W - AB^X)^2$ taken over the sample values. This minimization produces nonlinear equations in A and B that must be solved by successive approximations. A method of fitting such equations is given in section 19.7. Such methods require insertion of initial estimates A_1, B_1 of A and B. The linear fit to log W usually provides good initial estimates that sometimes need very little change.

EXAMPLE 19.2.1—J. W. Gowen and W. C. Price counted the number of lesions of *Aucuba mosaic* virus developing after exposure to X rays for various times (data made available through courtesy of the investigators).

Minutes exposure	0	3	7.5	15	30	45	60
Count in hundreds	271	226	209	108	59	29	12

Plot count as the ordinate; then plot its logarithm. Derive the regression $Y = 2.432 - 0.02227X$, where Y is the logarithm of the count and X is minutes exposure.

EXAMPLE 19.2.2—Repeat the fitting of the last example using natural logarithms. Verify the fact that the rate of decrease of hundreds of lesions per minute per hundred is $(2.3026)(0.02227) = 0.05128$.

EXAMPLE 19.2.3—If the meaning of relative rate is not clear, try this approximate method of computing it. The increase in weight of the chick embryo during day 13 is $1.130 - 0.738 = 0.392$ g; that is, the average rate during this period is 0.392 g/day. But the average weight during the same period is $(1.130 + 0.738)/2 = 0.934$ g. The relative rate, or rate of increase of each gram, is therefore $0.392/0.934 = 0.42$ g/day/g. This differs from the average obtained in the whole period from 6 to 16 days, 0.451, partly because the weights and increases in weights are subject to sampling variation and partly because the correct relative rate is based on weight and increase in weight at any instant of time, not on day averages.

19.3—The second degree polynomial. Faced by nonlinear regression, one often has no knowledge of a theoretical equation to use. With a simple type of curvature, however, the second degree polynomial,

$$\hat{Y} = a + bX + cX^2 \tag{19.3.1}$$

is often found to fit the data satisfactorily. Relations like (19.3.1) are fitted by the methods of multiple regression. The calculations proceed exactly as in chapter 17, X and X^2 being the two independent variates. Terms in $\sqrt{X}$, log X, or $1/X$ might have been added instead of X^2 if the data had required it.

To illustrate the method and some of its applications, we present data on wheat yield and protein content (3) in table 19.3.1 and figure 19.3.1. The

TABLE 19.3.1
PERCENTAGE PROTEIN CONTENT (Y) AND YIELD (X) OF WHEAT

Yield† X	Square X^2	Y	$(X-20)^2$	Yield X	Square X^2	Y	$(X-20)^2$
5	25	16.2	225	22	484	12.6	4
8	64	14.2	144	24	576	11.6	16
10	100	14.6	100	26	676	11.0	36
11	121	18.3	81	30	900	9.8	100
14	196	13.2	36	32	1,024	10.4	144
16	256	13.0	16	34	1,156	10.9	196
17	289	13.0	9	36	1,296	12.2	256
17	289	13.4	9	38	1,444	9.8	324
18	324	10.6	4	43	1,849	10.7	529
20	400	12.8	0	421	11,469	238.3	2229

	1	2	3
1	2140.53	101,492	−339.23
2		5,043,754	−14,725.41
3			88.05

†Bushels per acre.

investigator wished to estimate the protein content for various yields. We also test the significance of the departure from linearity.

If we take $X_1 = X$, $X_2 = X^2$, and $X_3 = Y$, we find the matrix of sums of squares and products of deviations from the mean as shown at the foot of table 19.3.1. The sample regression

$$Y = 18.676 - 0.4367X + 0.005869X^2 \tag{19.3.2}$$

is plotted in the figure. At small values of X the plot is nearly linear, but the

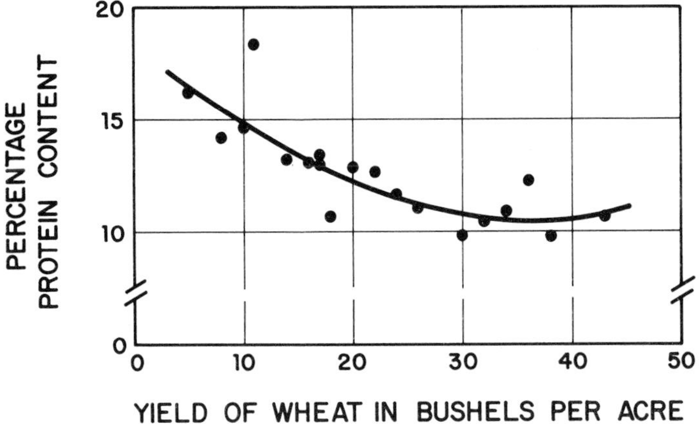

Figure 19.3.1—Regression of protein content on yield of wheat in 19 plots.

TABLE 19.3.2
Test of Significance of Departure from Linearity

Source of Variation	df	Sum of Squares	Mean Square	F
Linear regression	1	$R_1 = 53.76$		
Quadratic term	1	$R_2 - R_1 = 7.98$	7.98	4.85*
Deviations from quadratic	16	$\Sigma y^2 - R_2 = 26.31$	1.644	

downward slope decreases as X increases and at about $X = 37$ the slope begins to turn upward.

The test of departure from linearity may be made by the usual extension of the analysis of variance. The reduction R_2 in sums of squares due to the regression of Y on X and X^2 was found to be 61.74. From the data in table 19.3.1, the reduction R_1 due to a linear regression on X is $339.23^2/2140.53 = 53.76$. The deviations sum of squares from the quadratic was $(88.05 - 61.74) = 26.31$ with $19 - 3 = 16$ df. The analysis of variance and the test appear in table 19.3.2.

The parabolic term is just significant ($P = 0.043$). If several Y values have been measured at the same value of X, a test as to whether the data show consistent deviations from the quadratic relation can be made as described in section 19.4.

With certain ranges of values of X, the sample correlation between X and X^2 is uncomfortably high, requiring extra decimals to be carried in writing and solving the normal equations. In table 19.3.1, for example, the correlation is 0.977. The risk of a high correlation is avoided by choosing some rounded number a near $\overline{X}$ and taking $X_1 = X$, $X_2 = (X - a)^2$ instead of $X_2 = X^2$. Since $\overline{X} = 22.2$ in table 19.3.1, we choose $a = 20$ for illustration. The values of $X_2 = (X - 20)^2$ are given in table 19.3.1. The correlation between X_1 and X_2 has now dropped to 0.580. In example 19.3.1 the regression of Y on X and $(X - 20)^2$ is to be calculated.

Parabolic regressions can work well for estimation and interpolation within the range of the data even when we are sure the correct relation is not strictly parabolic. As always in regression, linear or curved, extrapolation beyond this range is risky. Looking at the chick embryo data in figure 19.2.1, for instance, one might be tempted by the excellent fit to assume the same growth rate before day 6 and after day 16. Actually, sharp breaks in the rate of growth occurred at both these days.

EXAMPLE 19.3.1—From table 19.3.1, work out the matrix of sums of squares and sums of products for the regression of Y on $X_1 = X$, $X_2 = (X - 20)^2$. (*i*) Show that the sample regression is $\hat{Y} = 16.327 - 0.2020X + 0.005869(X - 20)^2$. (*ii*) Show that this result agrees with $\hat{Y}$ in (19.3.2).

EXAMPLE 19.3.2—The test of significance of departure from linear regression in table 19.3.2 may also be used to examine whether a rectifying transformation of the type illustrated in section 19.2 has produced a straight-line relationship. Apply this test to the chick embryo data in table 19.2.1 by fitting a parabola in X to log weights Y. Verify that the parabola is $Y = -2.783162 + 0.214503X - 0.000846X^2$ and that the test works out as follows:

Source of Variation	df	Sum of Squares	Mean Square	F
Linear regression	1	$R_1 = 4.221057$		
Quadratic regression	1	$R_2 - R_1 = 0.000614$	0.000614	0.76
Deviations	8	0.006480	0.000810	

No indication of curvature appears in the log scale.

19.4—Data having several *Y*s for each *X*. If several values of Y have been measured for each X, the adequacy of a fitted polynomial can be tested more thoroughly. Suppose that for $X = X_i$, a group of n_i values of Y_{ij} are available. We then have a one-way classification. In section 12.11 (p. 230), a test was made of the adequacy of the linear model,

$$\overline{Y}_{i.} = \alpha + \beta X_i + \bar{\epsilon}_i \tag{19.4.1}$$

To recapitulate, with a classes the weighted sum of squares between class means, $\Sigma n_i(\overline{Y}_{i.} - \overline{Y}_{..})^2$, was divided into 1 df representing the linear regression on x_i and $(a - 2)$ df representing deviations of the class means from the linear regression. If the linear model holds, the ratio (deviations mean square)/(pooled within-classes mean square) is distributed as F with $(a - 2)$ and $(N - a)$ df, where $N = \Sigma n_i$. This test was illustrated in table 12.11.2 by the analysis of variance of data giving to cats lethal doses of ouabain injected at four different rates.

We can now take this test a stage further. The sum of squares for deviations from the linear regression is divided into 1 df for the quadratic term in a regression of $\overline{Y}_i$ on X_i and $(a - 3)$ df representing deviations from a quadratic regression. The point is that if linear regression does not hold and the relation between Y and X displays a simple type of curvature, the 1 df mean square for the quadratic term will reveal this curvature much more clearly than the deviations mean square with $(a - 2)$ df in the preceding paragraph. This method leads to the analysis of variance in table 19.4.1.

In examining table 19.4.1, first test $F = s_d^2/s^2$ with $(a - 3)$ and $(N - a)$ df. When will F become large? *Either* if the population regression is nonlinear but is not adequately represented by a quadratic (for instance, it might be a third degree curve or a curve with a periodic feature) *or* if other sources of variation

TABLE 19.4.1

Analysis of Variance for Tests of Linear Regression

Source of Variation	Degrees of Freedom	Mean Square
Linear regression of $\overline{Y}_{i.}$ on X_i	1	s_1^2
Quadratic term in regression	1	s_2^2
Deviations of $\overline{Y}_{i.}$ from quadratic	$a - 3$	s_d^2
Pooled within classes	$N - a$	s^2
Total	$N - 1$	

are constant within a group but vary from group to group. Consequently, if F is significant, look at the plot of $\overline{Y}_{i.}$ against X_i to see whether a higher degree polynomial or a different mathematical relationship is suggested. Examination of the deviations of the $\overline{Y}_{i.}$ from the fitted quadratic is also helpful. If no systematic trend is found, the most likely explanation is that some extra between-group variation has entered the data, e.g., perhaps the measurements in different groups were made at different times or in different hospitals.

If s_d^2/s^2 is clearly nonsignificant, form the pooled mean square s_p^2 of s_d^2 and s^2. Then test $F = s_2^2/s_p^2$ with 1 and $(N - 3)$ df as a test of curvature.

This procedure is illustrated by the ouabain data from table 12.11.1. There are four classes, with rates $X_i = 1, 2, 4, 8$ (mg/kg/min)/1045.95. The within-classes mean square was 153 with 37 df. For fitting the quadratic regression, we chose $X_{1i} = X_i$, $X_{2i} = (X_i - 4)^2$, $X_{3i} = \overline{Y}_i$. The class means and the matrix of weighted sums of squares and sums of products for fitting the weighted regression of the $\overline{Y}_i$ on X_{1i} and X_{2i} are given in table 19.4.2.

Note that the sums of squares and sums of products are weighted, since the n_i are unequal. For example,

$$\Sigma n_i x_{1i}\bar{y}_i = \Sigma X_{1i}Y_{i.} - (\Sigma n_i X_{1i})(Y_{..})/N = 7633 - (142)(1594)/41 = 2112$$

$$\Sigma n_i x_{1i}x_{2i} = \Sigma n_i X_{1i}X_{2i} - (\Sigma n_i X_{1i})(\Sigma n_i X_{2i})/N = 332.8$$

The final analysis of variance appears in table 19.4.3.

The mean square 11 for the deviations from the quadratic is much lower than the within-groups mean square, though not unusually so with only 1 df in the numerator. The pooled average of these two mean squares is 149, with 38 df. For the test of curvature, $F = 382/149 = 2.56$ with 1 and 38 df, lying between the 25% and the 10% level. We conclude that the results may be consistent with a linear relation in the population.

TABLE 19.4.2

MEAN LETHAL DOSE (MINUS 50 UNITS) OF OUABAIN BY INJECTION IN CATS AT FOUR DIFFERENT RATES

$X_{1i} = X_i$ (rate)	1	2	4	8	$\Sigma n_i X_{1i} = 142$
$X_{2i} = (X_i - 4)^2$	9	4	0	16	$\Sigma n_i X_{2i} = 296$
$\overline{Y}_{i.}$	18.1	28.2	48.3	70.2	
n_i	12	11	9	9	$N = \Sigma n_i = 41$
$Y_{i.} = n_i\overline{Y}_{i.}$	217	310	435	632	$Y_{..} = \Sigma n_i \overline{Y}_i = 1594$

Matrix of Sums of Squares and Products

	1	2	3
1	284.2	322.8	2,112
2		1315.0	1,797
3			21,744

TABLE 19.4.3
TESTS OF DEVIATIONS FROM LINEAR AND QUADRATIC REGRESSION

Source of Variation	df	Sum of Squares	Mean Square	F
Linear regression on X	1	15,700	15,700	
Quadratic regression on X	1	382	382	2.56
Deviations from quadratic	1	11	11	0.07
Pooled within classes	37	5,651	153	
Total	40	21,744		

EXAMPLE 19.4.1—The following data, selected from Swanson and Smith (4) to provide an example with equal n, show the total nitrogen content Y (g/100 cc plasma) of rat blood plasma at nine ages X (days).

Age of Rat	25	37	50	60	80	100	130	180	360
	0.83	0.98	1.07	1.09	0.97	1.14	1.22	1.20	1.16
	.77	.84	1.01	1.03	1.08	1.04	1.07	1.19	1.29
	.88	.99	1.06	1.06	1.16	1.00	1.09	1.33	1.25
	.94	.87	0.96	1.08	1.11	1.08	1.15	1.21	1.43
	.89	.90	0.88	0.94	1.03	0.89	1.14	1.20	1.20
	0.83	0.82	1.01	1.01	1.17	1.03	1.19	1.07	1.06
Total	5.14	5.40	5.99	6.21	6.52	6.18	6.86	7.20	7.39

A plot of the Y totals against X shows that (*i*) the Y values for $X = 100$ are abnormally low and require special investigation, (*ii*) the relation is clearly curved. Omit the data for $X = 100$ and test the deviations from a parabolic regression against the within-groups mean square. Ans. $F = 1.4$.

19.5—Test of departure from linear regression in covariance analysis. As in other correlation and regression work, it is necessary in linear covariance adjustment to check that the regression is linear. Recall that in the standard types of layout—one-way classifications, two-way classifications (randomized blocks), and Latin squares—the regression of Y on X is computed from the residual or error line in the analysis of variance. A graphic method of checking on linearity, which is often sufficient, is to plot the residuals of Y from the analysis of variance model against the corresponding residuals of X, looking for signs of curvature.

The numerical method of checking is to add a term in X^2 to the model. Writing $X_1 = X, X_2 = X^2$, work out the residual or error sums of squares of Y, X_1, and X_2 and the error sums of products of X_1X_2, YX_1, and YX_2, as illustrated in section 18.10 for a one-way classification. From these data, compute the test of significance of departure from linear regression as in table 19.3.2.

If the regression is found to be curved, the treatment means are adjusted for the parabolic regression. The calculations follow the method given in section 18.10.

19.6—Orthogonal polynomials. The method of fitting a quadratic to data can be extended to polynomials of higher degree, using some method or program for fitting multiple linear regressions. Also, the BMDP *P*-series of computer programs (10) contains a program for polynomial regression. It performs the calculations by creating polynomials X_i $(i = 1, 2, 3, \ldots, p)$ that are mutually orthogonal, thus avoiding the possibility that different powers of X are very highly correlated. In this program, polynomials of degree 1, 2, 3, . . . are fitted in succession. With each polynomial the output gives the regression coefficients, their standard errors, t values, and the residual mean square, from which the user can judge whether a polynomial of sufficiently high degree has been fitted. To start this program, the user specifies the highest degree p of polynomial that the program should try. For example, the user might have specified $p = 6$ but after seeing the output might judge that $p = 3$ fits about as well as $p = 6$.

If the values of X are equally spaced, the manual fitting of the polynomial

$$Y = b_0 + b_1X + b_2X^2 + b_3X^3 + \ldots$$

is simplified by the use of tables of orthogonal polynomials. The essential step is to replace X^i $(i = 1, 2, 3, \ldots)$ by a polynomial of degree i in X, which we will call X_i. The coefficients in these polynomials are chosen so that

$$\Sigma X_i = 0 \qquad \Sigma X_iX_j = 0 \qquad (i \neq j)$$

where the sums are over the n values of X in the sample. The different polynomials are *orthogonal* to one another. Explicit formulas for these polynomials are given later in this section.

Instead of calculating the polynomial regression of Y on X in the form above, we calculate it in the form

$$Y = B_0 + B_1X_1 + B_2X_2 + B_3X_3 + \ldots$$

which may be shown to give the same fitted polynomial. Because of the orthogonality of the X_i, we have

$$B_0 = \overline{Y} \qquad B_i = \Sigma X_iY/\Sigma X_i^2 \qquad (i = 1, 2, 3, \ldots)$$

The values of the X_i and of ΣX_i^2 are provided in the tables, making the computation of B_i simple. Further, the reductions in $\Sigma(Y - \overline{Y})^2$ due to the successive terms in the polynomial are given by

$$(\Sigma X_1Y)^2/\Sigma X_1^2 \qquad (\Sigma X_2Y)^2/\Sigma X_2^2 \qquad (\Sigma X_3Y)^2/\Sigma X_3^2$$

and so on. Thus it is easy to check whether the addition of a higher power in X to the polynomial produces a marked reduction in the residual sum of squares.

Tables of the first five polynomials are given in (5) up to $n = 75$ and of the first six in (6) up to $n = 52$. Table A 19 shows these polynomials up to $n = 12$.

For illustration, a polynomial will be fitted to the chick embryo data, though, as we saw in section 19.2, these data are probably more aptly fitted as an exponential growth curve. In this example, Y rather than W denotes the weights.

Table 19.6.1 shows the weights Y and the values of X_1, X_2, X_3, X_4, X_5 for $n = 11$ read from table A 19. To save space, most tables give the X_i values only for the upper half of the values of X. In our sample these are the values from $X =$ 11 to $X = 16$. The method of writing down the X_i for the lower half of the sample is seen in table 19.6.1. For the terms of *odd* degree, X_1, X_3, and X_5, the signs are changed in the lower half; for terms of *even* degree, X_2 and X_4, the signs remain the same.

We shall suppose that the objective is to find the polynomial of lowest degree that seems an adequate fit. Consequently, the reduction in sum of squares is tested as each successive term is added. At each stage, calculate $\Sigma X_i Y$ and $B_i = \Sigma X_i Y / \Sigma X_i^2$ (shown in table 19.6.1) and the reduction in sum of squares, $(\Sigma X_i Y)^2 / \Sigma X_i^2$ (table 19.6.2). For the linear term, the F value is $6.078511/0.232177 = 26.2$. The succeeding F values for the quadratic and cubic terms are even larger, 59.9 and 173.4. For the X_4 (quartic) term, F is 10.3, significant at the 5% but not at the 1% level. The fifth degree term, however, has an F less than 1. As a precautionary move we should check the sixth degree term also, but for this illustration we will stop and conclude that a fourth degree polynomial is a satisfactory fit.

For graphing the polynomial, the estimated values $\hat{Y}$ for each value of X are easily computed from table 19.6.1:

$$\hat{Y} = B_0 + B_1 X_1 + B_2 X_2 + B_3 X_3 + B_4 X_4$$

TABLE 19.6.1

FITTING A FOURTH DEGREE POLYNOMIAL TO CHICK EMBRYO WEIGHTS

Age X, (days)	Dry Weight, Y (g)	X_1	X_2	X_3	X_4	X_5	$\hat{Y}_4$
6	0.029	−5	15	−30	6	−3	0.026
7	0.052	−4	6	6	−6	6	0.056
8	0.079	−3	−1	22	−6	1	0.086
9	0.125	−2	−6	23	−1	−4	0.119
10	0.181	−1	−9	14	4	−4	0.171
11	0.261	0	−10	0	6	0	0.265
12	0.425	1	−9	−14	4	4	0.434
13	0.738	2	−6	−23	−1	4	0.718
14	1.130	3	−1	−22	−6	−1	1.169
15	1.882	4	6	−6	−6	−6	1.847
16	2.812	5	15	30	6	3	2.822
ΣX_i^2		110	858	4,290	286	156	
λ_i		1	1	5/6	1/12	1/40	
$\Sigma X_i Y$	7.714	25.858	39.768	31.873	1.315	−0.254	
B_i	0.701273	0.235073	0.046349	0.007430	0.004598		

TABLE 19.6.2
REDUCTIONS IN SUM OF SQUARES DUE TO SUCCESSIVE TERMS

Source	Degrees of Freedom	Sum of Squares	Mean Square	F
Total: $\Sigma(Y - \overline{Y})^2$	10	8.168108		
Reduction to linear	1	6.078511		
Deviations from linear	9	2.089597	0.232177	26.2
Reduction to quadratic	1	1.843233		
Deviations from quadratic	8	0.246364	0.030796	59.9
Reduction to cubic	1	0.236803		
Deviations from cubic	7	0.009561	0.001366	173.4
Reduction to quartic	1	0.006046		
Deviations from quartic	6	0.003515	0.000586	10.3
Reduction to quintic	1	0.000414		
Deviations from quintic	5	0.003101	0.000620	0.7

Note that $B_0 = \overline{Y} = 0.701273$. At $X = 6$,

$$\hat{Y} = 0.701273 - 5(0.235073) + 15(0.046349) - 30(0.007430) + 6(0.004598) = 0.026$$

To express the polynomial as an equation in the original X (= 1, 2, . . . , n) and its powers is more tedious. We need formulas giving the X_i in terms of X. In developing polynomial tables, Fisher started with a slightly different set of polynomials ξ_i. They were orthogonal but did not take integral values at $X = 1, 2, \ldots, n$. Then Fisher found by inspection the multiplier λ_i that would make $X_i = \lambda_i \xi_i$ the smallest possible set of integers, making manual calculations easier. The values of the λ_i are shown in table 19.6.1, table A 19, and references (5) and (6).

The first step in expressing the polynomial in the original X is to multiply each B_i in table 19.6.1 by the corresponding λ_i, giving

$$B'_1 = 0.235073 \qquad B'_2 = 0.046349 \qquad B'_3 = 0.006192 \qquad B'_4 = 0.0003832$$

These are the regression coefficients of Y on the ξ_i, so

$$\hat{Y} = \overline{Y} + B'_1\xi_1 + B'_2\xi_2 + B'_3\xi_3 + B'_4\xi_4 \tag{19.6.1}$$

Second, if $x = X - \overline{X}$, as usual, and $\xi_1 = x$, the equations for higher ξ_i in terms of x are

$$\xi_2 = x^2 - (n^2 - 1)/12 \qquad \xi_3 = x^3 - (3n^2 - 7)x/20$$

$$\xi_4 = x^4 - (3n^2 - 13)x^2/14 + 3(n^2 - 1)(n^2 - 9)/560$$

$$\xi_5 = x^5 - 5(n^2 - 7)x^3/18 + (15n^4 - 230n^2 + 407)x/1008$$

where n is the size of sample. The remaining task is to substitute for ξ_i in terms of x and then in terms of X in (19.6.1). For the quartic in table 19.6.1, the algebra works out as

$$\hat{Y} = 0.265373 + 0.124855x + 0.036769x^2 + 0.006192x^3 + 0.0003832x^4$$

$$= 0.7099 - 0.47652X + 0.110636X^2 - 0.010669X^3 + 0.0003832X^4$$

As a check, we again find $\hat{Y} = 0.026$ at $X = 6$.

EXAMPLE 19.6.1—Six points on the cubic, $Y = 9X - 6X^2 + X^3$, are (0, 0), (1, 4), (2, 2), (3, 0), (4, 4), (5, 20). Carry through the computations for fitting a linear, a quadratic, and a cubic regression. Verify that there is no residual sum of squares after fitting the cubic and that the polynomial values at that stage are exactly the Ys.

EXAMPLE 19.6.2—The method of constructing orthogonal polynomials can be illustrated by finding X_1 and X_2 when $n = 6$.

(1) X	(2) $\xi_1 = X - \overline{X}$	(3) $X_1 = 2\xi_1$	(4) ξ_2	(5) $X_2 = (3/2)\xi_2$
1	−5/2	−5	10/3	5
2	−3/2	−3	− 2/3	−1
3	−1/2	−1	− 8/3	−4
4	1/2	1	− 8/3	−4
5	3/2	3	− 2/3	−1
6	5/2	5	10/3	5

Start with $X = 1, 2, 3, 4, 5, 6$, with $\overline{X} = 7/2$. Verify that the values of $\xi_1 = x = X - \overline{X}$ are shown in column (2). Since the ξ_1 are not whole numbers, take $\lambda_1 = 2$, giving $X_1 = 2\xi_1$, column (3). To find ξ_2, write

$$\xi_2 = \xi_1^2 - b\xi_1 - c$$

which is a quadratic in X. We want $\Sigma\xi_2 = 0$; then

$$\Sigma\xi_1^2 - b\Sigma\xi_1 - nc = 0 \qquad 35/2 - 6c = 0 \qquad c = 35/12$$

Further, we want $\Sigma\xi_1\xi_2 = 0$, giving

$$\Sigma\xi_1^3 - b\Sigma\xi_1^2 - c\Sigma\xi_1 = 0 \qquad b\Sigma\xi_1^2 = 0 \qquad b = 0$$

Hence, $\xi_2 = \xi_1^2 - 35/12$. Verify the ξ_2 values in column (4). To convert these values to integers, multiply by $\lambda_2 = 3/2$.

19.7—A general method of fitting nonlinear regressions. Suppose the population relation between Y and X is of the form

$$Y_i = f(\alpha, \beta, \gamma, X_i) + \epsilon_i \qquad (i = 1, 2, \ldots, n)$$

where f is a regression function containing X_i and the parameters α, β, γ. (There may be more than one X-variable.) If the residuals ϵ_i have zero means and constant variance, the least squares method of fitting the regression is to estimate the values of the α, β, γ by minimizing

$$\sum_{i=1}^{n} [Y_i - f(\alpha, \beta, \gamma, X_i)]^2 \tag{19.7.1}$$

This section presents a general method of carrying out the calculations. The details require a knowledge of partial differentiation but the approach is simple.

The difficulty arises not because of nonlinearity in X_i but because of nonlinearity in one or more of the parameters α, β, γ. The parabola $\alpha + \beta X + \gamma X^2$ is fitted by the ordinary methods of multiple linear regression because it is linear in α, β, and γ. Consider the asymptotic regression, $\alpha - \beta(\gamma^X)$. If the value of γ were known in advance, we could write $X_1 = \gamma^X$. The least squares estimates of α and β would then be given by fitting an ordinary linear regression of Y on X_1. When γ must be estimated from the data, however, the methods of linear regression cannot be applied.

The first step in the general method is to obtain initial estimates a_1, b_1, c_1 of the final least square estimates $\hat{\alpha}$, $\hat{\beta}$, $\hat{\gamma}$. Good estimates are important if calculations must be done manually. For the common nonlinear functions, various techniques for finding first estimates have been developed, sometimes graphic, sometimes by special studies of the problem. Next, we use *Taylor's theorem.* This states that if $f(\alpha, \beta, \gamma, X)$ is continuous in α, β, and γ and if $(\alpha - a_1)$, $(\beta - b_1)$ and $(\gamma - c_1)$ are small,

$$f(\alpha, \beta, \gamma, X_i) \doteq f(a_1, b_1, c_1, X_i) + (\alpha - a_1)f_a + (\beta - b_1)f_b + (\gamma - c_1)f_c$$

The symbol $\doteq$ means "is approximately equal to." The symbols f_a, f_b, f_c denote the partial derivatives of f with respect to α, β, and γ, respectively, evaluated at the point a_1, b_1, c_1. For example, in the asymptotic regression,

$$f(\alpha, \beta, \gamma, X_i) = \alpha - \beta(\gamma^{X_i})$$

we have

$$f_a = 1 \qquad f_b = -c_1^{X_i}; \qquad f_c = -b_1 X_i (c_1^{X_i - 1})$$

Since a_1, b_1, and c_1 are known, the values of f, f_a, f_b, and f_c can be calculated for each member of the sample, where we have written f for $f(a_1, b_1, c_1, X_i)$. From Taylor's theorem, the original regression relation $Y_i = f(\alpha, \beta, \gamma, X_i) + \epsilon_i$ may therefore be written, approximately,

$$Y_i \doteq f + (\alpha - a_1)f_a + (\beta - b_1)f_b + (\gamma - c_1)f_c + \epsilon_i \tag{19.7.2}$$

Now write

$$Y_{\text{res}} = Y - f \qquad X_1 = f_a \qquad X_2 = f_b \qquad X_3 = f_c \tag{19.7.3}$$

From equation (19.7.3),

$$Y_{\text{res}} \doteq (\alpha - a_1)X_1 + (\beta - b_1)X_2 + (\gamma - c_1)X_3 + \epsilon_i \tag{19.7.4}$$

The variate Y_{res} is the residual of Y from the first approximation, f. The relation (19.7.4) represents an ordinary linear regression on Y_{res} on the variates X_1, X_2, X_3; the regression coefficients are $(\alpha - a_1)$, $(\beta - b_1)$, and $(\gamma - c_1)$. If the relation (19.7.4) held exactly instead of approximately, the computation of the sample regression of Y_{res} on X_1, X_2, X_3 would give the regression coefficients $(\hat{\alpha} - a_1)$, $(\hat{\beta} - b_1)$, and $(\hat{\gamma} - c_1)$, from which the correct least squares estimates $\hat{\alpha}$, $\hat{\beta}$, and $\hat{\gamma}$ would be obtained at once.

Since relation (19.7.4) is approximate, the fitting of this regression yields *second* approximations a_2, b_2, and c_2 to $\hat{\alpha}$, $\hat{\beta}$, and $\hat{\gamma}$, respectively. We then recalculate f, f_a, f_b, and f_c at point a_2, b_2, c_2, finding new Y_{res} values and new variates X_1, X_2, and X_3. The sample regression of this Y_{res} on X_1, X_2, and X_3 gives the regression coefficients $(a_3 - a_2)$, $(b_3 - b_2)$, and $(c_3 - c_2)$ from which third approximations a_3, b_3, c_3 to $\hat{\alpha}$, $\hat{\beta}$, $\hat{\gamma}$ are found, and so on.

If the process is effective, the sum of squares of the residuals, ΣY_{res}^2, should decrease steadily at each stage, called an *iteration,* the decreases becoming small as the least squares solution is approached. In practice, the calculations are stopped when the decrease in ΣY_{res}^2 and the changes in a, b, and c are considered small enough to be negligible. The mean square residual is

$$s^2 = \Sigma Y_{\text{res}}^2/(n - k)$$

where k is the number of parameters that have been estimated (in our example, $k = 3$). With nonlinear regression, s^2 is not an unbiased estimate of σ^2, though it tends to become unbiased as n becomes large.

An approximation to the covariance matrix of the final estimates is obtained from the c_{ij} multipliers in the regression calculation at the last iteration. Thus, $SE(\hat{\alpha}) \doteq s\sqrt{c_{11}}$; $SE(\hat{\beta}) \doteq s\sqrt{c_{22}}$; $SE(\hat{\gamma}) \doteq s\sqrt{c_{33}}$; $\text{Cov}(\hat{\alpha}, \hat{\beta}) \doteq s^2 c_{12}$. Approximate confidence limits for α are given by $\hat{\alpha} \pm ts\sqrt{c_{11}}$, where t has $(n - 3)$ df.

At some iterations, the residual sum of squares may be found to increase rather than decrease. One suggestion (13) for this problem is to halve the size of the change made in each parameter from its value in the previous iteration; that is, recalculate f and $\Sigma(Y - f)^2$ at the values $\hat{\alpha} = (a_i + a_{i-1})/2$, $\hat{\beta} = (b_i + b_{i-1})/2$, etc., instead of at a_i, b_i, etc. If this change now makes the residual sum of squares decrease, continue from the new $\hat{\alpha}$, $\hat{\beta}$, etc. If not, try halving again. Another method of successive approximation that appears to work well with many functions f is due to Marquardt (14, 15).

19.8—Fitting an asymptotic regression. As an example, we will calculate the second approximation when fitting an asymptotic regression (with ρ in place of γ):

$$f(\alpha, \beta, \gamma, X) = \alpha + \beta(\rho^X) \tag{19.8.1}$$

If $0 < \rho < 1$ and β is negative, this curve has the form shown in figure 19.1.1(c), rising from the value $\alpha + \beta$ at $X = 0$ to the asymptote α as X becomes large. If $0 < \rho < 1$ and β is positive, the curve declines from the value $\alpha + \beta$ at $X = 0$ to an asymptote α when X is large.

Since the function is nonlinear only in the parameter ρ, the method of successive approximation described in the preceding section simplifies a little. Let r_1 be a first approximation to ρ. By Taylor's theorem,

$$\alpha + \beta(\rho^X) \doteq \alpha + \beta(r_1^X) + \beta(\rho - r_1)(Xr_1^{X-1})$$

Write $X_0 = 1, X_1 = r_1^X, X_2 = Xr_1^{X-1}$. If we fit the sample regression

$$\hat{Y} = aX_0 + bX_1 + cX_2 \tag{19.8.2}$$

it follows that a, b are second approximations to the least squares estimates $\hat{\alpha}, \hat{\beta}$ of α and β in (19.8.1), while $c = b(r_2 - r_1)$ so

$$r_2 = r_1 + c/b \tag{19.8.3}$$

is the second approximation to $\hat{\rho}$.

The commonest case is that in which the values of X change by unity. Denote the corresponding Y values by $Y_0, Y_1, Y_2, \ldots, Y_{n-1}$. Note that the value of X corresponding to Y_0 need not be 0. For $n = 4$, 5, 6, and 7, good first approximations to ρ, due to Patterson (7), are as follows:

$n = 4$ $\quad r_1 = (4Y_3 + Y_2 - 5Y_1)/(4Y_2 + Y_1 - 5Y_0)$

$n = 5$ $\quad r_1 = (4Y_4 + 3Y_3 - Y_2 - 6Y_1)/(4Y_3 + 3Y_2 - Y_1 - 6Y_0)$

$n = 6$ $\quad r_1 = (4Y_5 + 4Y_4 + 2Y_3 - 3Y_2 - 7Y_1)/(4Y_4 + 4Y_3 + 2Y_2 - 3Y_1 - 7Y_0)$

$n = 7$ $\quad r_1 = (Y_6 + Y_5 + Y_4 - Y_2 - 2Y_1)/(Y_5 + Y_4 + Y_3 - Y_1 - 2Y_0)$

In a later paper (8), Patterson gives improved first approximations for sample sizes from $n = 4$ to $n = 12$. The value of r_1 in (8), obtained by solving a quadratic equation, is remarkably good in our experience.

In an illustration given by Stevens (9), table 19.8.1 shows six consecutive readings of a thermometer at half-minute intervals after lowering it into a refrigerated hold. From Patterson's formula (above) for $n = 6$, we find $r_1 = -104.2/(-188.6) = 0.552$. Taking $r_1 = 0.55$, compute the sample values of X_1 and X_2 and insert them in table 19.8.1. The matrix of sums of squares and products $\Sigma X_i X_j$ is shown in the table, where X_3 denotes Y. Note that the $\Sigma X_i X_j$ are not deviations from the sample means, since $X_0 \equiv 1$ has been inserted as a dummy X for a. (We could have used deviations from the sample means, obtaining a 3×3 matrix, but little time would have been saved.) Under the $\Sigma X_i X_j$ appear the results of the sweep operation, namely the c_{ij}, the regression

TABLE 19.8.1
DATA FOR FITTING AN ASYMPTOTIC REGRESSION

Time, X (1/2 min)	$X_3 =$ Temp., Y (°F)	$X_1 = 0.55^X$	$X_2 = X(0.55^{X-1})$	$\hat{Y}_2$	$Y_{res} = Y - \hat{Y}_2$
0	57.5	1.00000	0	57.544	−0.044
1	45.7	0.55000	1.00000	45.525	+0.175
2	38.7	0.30250	1.10000	38.892	−0.193
3	35.3	0.16638	0.90750	35.231	+0.069
4	33.1	0.09151	0.66550	33.211	−0.111
5	32.2	0.05033	0.45753	32.096	+0.104
Total	242.5	2.16072	4.13053		+0.001

	$\Sigma X_i X_j$ 0	1	2	3
0	6	2.16072	4.13053	242.5
1		1.43260	1.11767	104.86457
2			3.68578	157.06527
3				10,270.97000
0	$c_{00} = 1.62101$	$c_{01} = -1.34608$	$c_{02} = -1.40843$	$a = 30.723$
1		$c_{11} = 2.03212$	$c_{12} = 0.89229$	$b = 26.821$
2			$c_{22} = 1.57912$	$c = 0.05024$
3				$\Sigma d_y^2 = 0.09724$

coefficients, and the residual Σd_y^2:

$$a = 30.723 \qquad b = 26.821 \qquad c = b(r_2 - r_1) = 0.05024 \tag{19.8.4}$$

Hence $r_2 = r_1 + c/b = 0.55 + 0.05024/26.821 = 0.55187$. The second approximation to the curve is therefore

$$\hat{Y}_2 = 30.723 + 26.821(0.55187)^X \tag{19.8.5}$$

To see how much the residual sum of squares has been reduced, we find it for the first two approximations. The first approximation is

$$\hat{Y}_1 = a_1 + b_1(0.55^X) = a_1 + b_1 X_1 \tag{19.8.6}$$

where a_1 and b_1 are given by the linear regression of Y on X_1. In the preceding calculations, a_1 and b_1 were not computed, since they were not needed in finding the second Taylor approximation. From the regression of Y on X_1 we find $a_1 = 30.768$, $b_1 = 25.793$. By finding $\hat{Y}_1$ from (19.8.6) we get $\Sigma(Y - \hat{Y}_1)^2 = 0.0988$. For the second approximation we compute the powers of $r_2 = 0.55187$ and hence find $\hat{Y}_2$ by (19.8.5). The values of $\hat{Y}_2$ and $Y - \hat{Y}_2$ are shown in table 19.8.1. The sum of squares of residuals is 0.0977. Further approximations lead to a minimum of 0.0972. Incidentally, table 19.8.1 gives the value of Σd_y^2 from the Taylor approximation to $\hat{Y}_2$ as 0.09724, indicating that the Taylor approximation is good in this case.

The residual mean square for the second approximation is $s^2 = 0.0977/3 = 0.0326$ with $(n - 3) = 3$ df. Approximate standard errors for the estimated parameters are:

$$SE(a_2) = s\sqrt{c_{00}} = \pm 0.23 \qquad SE(b_2) = s\sqrt{c_{11}} = \pm 0.26$$

$$SE(r_2) = s\sqrt{c_{22}}/b_2 = 0.226/26.82 = \pm 0.0084$$

Strictly speaking, the values of the c_{ii} should be calculated for $r = 0.55187$ instead of $r = 0.55$, but the above results are close enough. Furthermore, since $r_2 - r_1 = c_2/b_2$, a better approximation to the standard error of r_2 may be given by the large sample formula for the standard error of a ratio:

$$SE(r_2) = s[c_{22} - 2(r_2 - r_1)c_{12} + (r_2 - r_1)^2 c_{11}]^{1/2}/b_2 \qquad (19.8.7)$$

However, since at the last iteration the term $r_i - r_{i-1}$ should be very small, the term c_{22} under the square root in (19.8.7) nearly always dominates, bringing us back to $s\sqrt{c_{22}}/b_2$. In this example, (19.8.7) gives $SE(r_2) = \pm 0.0085$.

In a computer program the user normally specifies the maximum number of iterations to be tried and may have to specify the initial estimates of the parameters except possibly with a common nonlinear function like an asymptotic regression, for which the program supplies initial values. The program may also carry a stopping rule; for example, it may stop when the residual SS $(\Sigma Y^2_{\text{res}})$ changes by less than a specified amount on each of a number of successive iterations. At the iteration with the smallest ΣY^2_{res}, the output is likely to include the c_{ij} or the corresponding correlation matrix, the residual mean square $\Sigma^2_{\text{res}}/(n - k)$, the parameter estimates with their approximate standard errors and t values, and the value of $\hat{Y}$ and of the residual $Y - \hat{Y}$ for each observation. For examples of the input and output with different programs, see (10, 16).

EXAMPLE 19.8.1—In a chemical reaction the amount of nitrogen pentoxide decomposed at various times after the start of the reaction was as follows (12):

Time, T	2	3	4	5	6	7
Amount decomposed, Y	18.6	22.6	25.1	27.2	29.1	30.1

Fit an asymptotic regression. We obtained $\hat{Y} = 33.802 - 26.698(0.753)^T$, with residual $SS = 0.105$.

EXAMPLE 19.8.2—Stevens (9) has remarked that when ρ is between 0.7 and 1, the asymptotic regression curve is closely approximated by a second degree polynomial. The asymptotic equation $Y = 1 - 0.9(0.8)^X$ takes the following values:

X	0	1	2	3	4	5	6
Y	0.100	0.280	0.424	0.539	0.631	0.705	0.764

Fit a parabola by orthogonal polynomials and observe how well the values of Y agree.

TECHNICAL TERMS

asymptotic regression
exponential growth curve
iteration
logistic growth curve
Mitscherlich's law
nonlinear regression
orthogonal polynomials
semilogarithmic paper
Taylor's theorem

REFERENCES

1. Pearl, R.; Reed, L. J.; and Kish, J. F. *Science* 92 (1940):486.
2. Penquite, R. 1936. Ph.D. dissertation, Iowa State College.
3. Metzger, W. H. *J. Am. Soc. Agron.* 27 (1935):653.
4. Swanson, P. P., and Smith, A. H. *J. Biol. Chem.* 97 (1932):745.
5. Fisher, R. A., and Yates, F. 1957. *Statistical Tables,* 5th ed. Oliver & Boyd, Edinburgh.
6. Pearson, E. S., and Hartley, H. O. *Biometrika Tables for Statisticians,* vol. 1. Cambridge Univ. Press.
7. Patterson, H. D. *Biometrics* 12 (1956):323.
8. Patterson, H. D. *Biometrika* 47 (1960):177.
9. Stevens, W. L. *Biometrics* 7 (1951):247.
10. BMDP. *Biomedical Computer Programs, P*-series. 1977. Univ. California Press, Berkeley.
11. Mitscherlich, E. A. *Landw. Jahrb.* 38 (1909):537.
12. Reed, L. J., and Theriault, E. J. *J. Phys. Chem.* 35 (1931):950.
13. Jennrich, R. I., and Sampson, P. F. *Technometrics* 10 (1968):63.
14. Marquardt, D. W. *J. Soc. Ind. Appl. Math.* 11 (1963):431.
15. Meeter, D. A. 1966. Non-linear least squares (GASTHAUS). Univ. Wisconsin Computing Center.
16. Daniel, C., and Wood, F. S. 1971. *Fitting Equations to Data.* Wiley, New York.

20

Two-way tables with unequal numbers and proportions

20.1—Introduction. For one reason or another the numbers of observations in the individual cells (subclasses) of a multiple classification may be unequal. This is the situation in many observational studies in which the investigator classifies the sample according to the factors or variables of interest, exercising no control over the way in which the numbers fall. With a one-way classification, the handling of the unequal numbers case is discussed in section 12.10. In this chapter we present methods for analyzing a two-way classification. The related problem of analyzing a proportion in a two-way table is taken up also.

The complications introduced by unequal subclass numbers can be illustrated by a simple example. Two diets were compared on samples of 10 rats. As it happened, 8 of the 10 rats on diet 1 were females, while only 2 of the 10 rats on diet 2 were females. Table 20.1.1 shows the subclass totals for gains in weight and the subclass numbers. The 8 females on diet 1 gained a total of 160 units, and so on.

From these data we obtain the row totals and means and the column totals and means. From the row means, it looks as if diet 2 had a very slight advantage over diet 1, 23 against 22. In the column means, males show greater gains than females, 26 against 19.

The subclass means per rat tell a different story.

	Female	Male
Diet 1	20	30
Diet 2	15	25

Diet 1 is superior by 5 units in both females and males. Further, males gain 10 units more than females under both diets, as against the estimate of 7 units obtained from the overall means.

Why do the row and column means give such distorted results? Clearly it is because of the inequality in the subclass numbers. The poorer feed, diet 2, had an excess of the faster growing males. Similarly, the comparison of male and female means is biased because most of the males were on the inferior diet.

To attempt to compute the analysis of variance by elementary methods also

TABLE 20.1.1

Total Gains in Weight and Subclass Numbers (Artifical Data)

		Females	Males	Sums	Means
Diet 1	totals	160	60	220	22
	numbers	8	2	10	
Diet 2	totals	30	200	230	23
	numbers	2	8	10	
Sums	totals	190	260	450	
	numbers	10	10	20	
Means		19	26		22.5

leads to difficulty. From table 20.1.1 the sum of squares between subclasses is correctly computed as

$$\frac{160^2}{8} + \frac{60^2}{2} + \frac{30^2}{2} + \frac{200^2}{8} - \frac{450^2}{20} = 325 \qquad (3\text{ df})$$

The sum of squares for diets, $(230 - 220)^2/20$, is 5 and that for sex, $(260 - 190)^2/20$, is 245, leaving an interaction sum of squares of 75. But from the cell means there is obviously no interaction; the difference between the diet means is the same for males as for females. In a correct analysis the interaction sum of squares should be zero.

20.2—Methods of attack. For the analysis of a two-way table with fixed row and column effects, the following approach is suggested:

1. First, test for interactions. This test is usually made by fitting an additive model to the cell means, namely,

$$\bar{X}_{ij.} = \mu + \alpha_i + \beta_j + \bar{\epsilon}_{ij} \qquad (20.2.1)$$

where $\bar{X}_{ij.}$ is the mean of the n_{ij} observations in the ith row and jth column. Since the residuals $\bar{X}_{ij} - \hat{\bar{X}}_{ij}$ of the cell means from the predicted additive values represent row $\times$ column interactions, an F test is made of the ratio $\Sigma\Sigma n_{ij}(\bar{X}_{ij} - \hat{\bar{X}}_{ij})^2/[(I-1)(J-1)s^2]$, where s^2 is the within-cells mean square. Unfortunately, with unequal n_{ij} the fitting of the additive model (20.2.1) requires the solution of a set of linear equations like those in fitting a multiple regression. When interactions are large, inspection of the cell means plus a few elementary tests may reveal the presence of interactions without fitting the additive model, as illustrated in section 20.3.

2. If interactions are substantial, study the row effects separately in each column and vice versa, with a view to understanding the nature of the interactions and writing a summary of the results. Overall row and column effects as given by model (20.2.1) are often of little interest, since the effect of each factor depends on the level of the other.

3. If interactions appear negligible, the summary of results is written from

TABLE 20.2.1
EXAMPLE OF PROPORTIONAL SUBCLASS NUMBERS

	Females			Males			Sums		
	Totals		Numbers	Totals		Numbers	Totals		Numbers
Diet 1	240		12	120		4	360		16
Means		20			30			22.5	
Diet 2	45		3	25		1	70		4
Means		15			25			17.5	
Sums	285		15	145		5	430		20
Means		19.0			29.0			21.5	

Analysis of Variance (correction term $C = 430^2/20 = 9245$)

	Degrees of Freedom	Sum of Squares	
Rows	1	$360^2/16 + 70^2/4 - C$	= 80
Columns	1	$145^2/5 + 285^2/15 - C$	= 375
Interactions	1	by subtraction	= 0
Between subclasses	3	$120^2/4 + \ldots + 45^2/3 - C$	= 455

the least squares estimates a_i, b_j of the row and column effects in the additive model (20.2.1). If a computer program is being used to fit the additive model, the program should give the values of the a_i and b_j and their covariance matrix, which will be needed for summarizing the results as well as for an analysis of variance and F tests of overall row and column effects and interactions. If the calculations must be done manually, remember that the additive model can be fitted and the test for interactions can be made by simple methods when the cell numbers n_{ij} are (*i*) equal, (*ii*) equal within any row or within any column, or (*iii*) proportional—that is, in the same proportion within any row. If actual cell numbers can be approximated reasonably well by one of these cases, an approximate analysis is obtained by using the actual cell means but replacing the cell numbers n_{ij} by the approximations.

The fact that elementary methods of analysis still apply when the cell numbers are proportional is illustrated in table 20.2.1. Here the cell means are exactly the same as in table 20.1.1 but males and females are now in the ratio 1:3 in each diet, with 4 males and 12 females on diet 1 and 1 male and 3 females on diet 2. Note that the overall row means show a superiority of 5 units for diet 1, just as the cell means do. Similarly, the overall column means show that the males gained 10 units more per animal than females. In the analysis of variance, the interactions sum of squares, found by subtraction, is now identically zero.

20.3—Inspection of cell means. Despite unequal numbers in the cells, the presence of interactions can often be detected and a report on the results written by elementary methods. In an experiment (1), three strains of mice were inoculated with three isolations (i.e., different types) of the mouse typhoid organism. The n_{ij} and the $\overline{X}_{ij.}$ (mean days-to-death) are shown for each cell in table 20.3.1. The pooled within-cells mean square is 5.015 with 774 df.

TABLE 20.3.1

CELL NUMBERS AND MEAN DAYS-TO-DEATH IN THREE STRAINS OF MICE INOCULATED WITH THREE TYPES OF TYPHOID BACILLUS

Type of Organism		Strain of Mice: *RI*	*Z*	*Ba*
9*D*	n_{ij}	34	31	33
	$\overline{X}_{ij.}$	4.0000	4.0323	3.7576
1*C*	n_{ij}	66	78	113
	$\overline{X}_{ij.}$	6.4545	6.7821	4.3097
*DSC*1	n_{ij}	107	133	188
	$\overline{X}_{ij.}$	6.6262	7.8045	4.1277

From inspection of the cell means, strain *Ba* looks about equally susceptible to all three types of organism, surviving 4.1 days on the average. Strains *RI* and *Z* also survived type 9*D* about 4 days but were more resistant to types 11*C* and *DSC*1, surviving over 6 days. In fact the survivals fall into three groups, combinations (*RI*, 11*C*), (*RI*, *DSC*1) and (*Z*, 11*C*) surviving around 6.6 days, while *Z* survived type *DSC*1 7.8 days. The differences among the three groups of survival times are significant. Thus the difference (6.4545 − 4.3097) = 2.1448 has estimated variance (5.015)/(1/66 + 1/113) = 0.120 or standard error ±0.346, while the difference (7.8045 − 6.7821) = 1.0225 has estimated variance (5.015)(1/133 + 1/78) = 0.1020, giving a standard error ±0.319, both differences being clearly significant. The preceding statements might suffice without further analysis for a report on the results in the table.

However, to look at part of the interactions sum of squares as an exercise, we show the comparison $(\overline{X}_{DSC1} + \overline{X}_{11C} - 2\overline{X}_{9D})$ for each strain of mice in table 20.3.2. This comparison should bring out the contrast between strains *RI*, *Z*, and *Ba*.

Beside each comparison is given the multiplier of σ^2 needed to obtain its variance. Now any comparison among the quantities 5.0807, 6.5220, and 0.9222 is a component of the interactions. The component that our earlier comments suggest examining is 5.0807 + 6.5220 − 2 × 0.9222 = 9.7583, with variance (5.015) (0.142 + 0.149 + 4 × 0.135) = (5.015) (0.831) = 4.167. The standard error is ±2.041, leading to a t value of 9.7583/2.041 = 4.78, highly significant. The contribution of this comparison to the interactions sum of squares is $9.7583^2/0.831 = 114.6$ with 1 df.

TABLE 20.3.2

COMPARISONS OF THREE STRAINS OF MICE INOCULATED WITH THREE TYPES OF TYPHOID BACILLUS

Strain	$\overline{X}_{DSC1} + \overline{X}_{11C} - 2\overline{X}_{9D}$	Variance/σ^2	w_i
RI	5.0807	1/107 + 1/66 + 4/34 = 0.142	7.042
Z	6.5220	1/133 + 1/78 + 4/31 = 0.149	6.711
Ba	0.9222	1/188 + 1/113 + 4/33 = 0.135	7.407

EXAMPLE 20.3.1—The figures w_i in the right column of table 20.3.2 are the reciprocals of the variance multipliers. If Y_i ($i = 1, 2, 3$) denote the comparisons in table 20.3.2, their contribution to the interactions sum of squares is $\Sigma w_i(\overline{Y}_i - \overline{Y}_w)^2$ with 2 df. Show that this sum of squares is 120.9, not much larger than the sum of squares 114.6 that measured the most striking contrast in the Y_i.

20.4—Unweighted analysis of cell means. In this method, the cell means $\overline{X}_{ij.}$ are treated as if they were all based on the same number of observations. Experience suggests that if the ratio of the largest to the smallest n_{ij} is less than 2, this approximation is adequate in making an F test of the interactions mean square and in testing comparisons among row and column means when interactions are negligible.

Under the additive model with fixed row and column effects,

$$\overline{X}_{ij.} = \mu + \alpha_i + \beta_j + \bar{\epsilon}_{ij.} \tag{20.4.1}$$

where $\bar{\epsilon}_{ij.}$ is the mean of n_{ij} random deviations. Hence if σ^2 is the variance per observation, the variance of $\overline{X}_{ij.}$ is σ^2/n_{ij}. With I rows and J columns, the average variance of the IJ cell means is

$$(\sigma^2/IJ)(1/n_{11} + 1/n_{12} + \ldots + 1/n_{IJ}) = \sigma^2/n_h \tag{20.4.2}$$

where n_h is called the *harmonic mean* of the n_{ij}.

In the unweighted analysis of variance of the cell means, multiply all sums of squares and mean squares by n_h to make them comparable with the within-cells mean square $s^2 = \Sigma\Sigma\Sigma(X_{ijk} - \overline{X}_{ij.})^2/\Sigma\Sigma(n_{ij} - 1)$. When examining main effects, we regard each row mean as having standard error $\pm s/\sqrt{bn_h}$ and each column mean as having standard error $\pm s/\sqrt{an_h}$.

Our example is somewhat artificial. The data were taken by Kutner (2) from an experiment (3) by Afifi and Azen, in which four drugs were tested on dogs with three different experimentally induced diseases. Each cell of the 4×3 table had 6 dogs. Kutner discarded 14 dogs at random to obtain an example with unequal n_{ij} for discussion of some points in methodology. The response variable is the increase in systolic blood pressure (mm Hg). Table 20.4.1 shows the n_{ij} and the cell means $\overline{X}_{ij.}$. The cell numbers n_{ij} range from 3 to 6, most values being intermediate. The average variance of the 12 cell means is

$$(\sigma^2/12)(1/6 + 1/5 + \ldots + 1/5) = 0.215\sigma^2 = \sigma^2/4.65$$

Every cell mean will be assumed to be based on 4.65 observations.

The within-cells sum of squares is 5081 with 46 df. Table 20.4.2 shows the analysis of variance. The sums of squares and mean squares for drugs, diseases, and interactions, from an unweighted analysis of the cell means, have been multiplied by 4.65 to make them comparable with the within-cells sums of squares and mean squares.

There are no signs of interactions. Differences between drugs are highly significant, while differences between diseases do not approach the 5% level of F. The standard error of any drug mean in table 20.4.2 is taken as $\sqrt{110.5/(3 \times 4.65)} = \pm 2.81$. From the means in table 20.4.1 it is clear that the

TABLE 20.4.1
INCREASE IN SYSTOLIC PRESSURE (MM HG) UNDER DRUGS GIVEN TO DOGS WITH THREE DIFFERENT DISEASES

Disease		Drugs 1	2	3	4	Total	Mean
1	$\bar{X}_{ij.}$	29.3	28.0	16.3	13.6	87.2	21.8
	n_{ij}	6	5	3	5		
2	$\bar{X}_{ij.}$	28.2	33.5	4.4	12.8	78.9	19.7
	n_{ij}	4	4	4	6		
3	$\bar{X}_{ij.}$	20.4	18.2	8.5	14.2	61.3	15.3
	n_{ij}	5	6	4	4		
Total		77.9	79.7	29.2	40.6	227.4	
Mean		26.0	26.6	9.7	13.5	18.95	
SE			±2.81				

drugs fall into two groups, drugs 1 and 2 giving much greater increases in systolic pressure than drugs 3 and 4. You may verify that the sum of squares for the comparison 1 + 2 − 3 − 4 is 2897 with 1 df. The remaining 2 df in the drugs sum of squares give an F slightly under 1.

Note that it is possible to calculate the correct standard error of any single comparison among row means, column means, and individual cells, since all such comparisons are linear functions of the individual cell means. For example, the estimated variance of the comparison between the means of drugs 3 and 4 in table 20.4.1 is (110.5) (1/3 + 1/5 + . . . + 1/5)/9 = 16.58, giving a standard error of ±4.07, as against 3.97 by the approximation. This approach loses the simplicity of the unweighted means method but might occasionally be useful.

EXAMPLE 20.4.1—Although the n_{ij} in table 20.3.1 are far from equal, calculate an unweighted analysis of variance of the unweighted cell means so as to give a rough F test of the interactions mean square. The harmonic mean of the n_{ij} is 59.6. If the sums of squares in the analysis of variance are multiplied by 59.6, show that the following analysis is obtained.

Source	df	Sum of Squares	Mean Square	F
Types	2	530		
Strains	2	447		
Interactions	4	191	47.8	9.5**
Within cells	774		5.015	

The least squares analysis gives $F = 8.8$ in the test for interactions.

TABLE 20.4.2
ANALYSIS OF VARIANCE OF UNWEIGHTED CELL MEANS

Source	df	Sum of Squares	Mean Squares	F
Drugs	3	3090	1030.0	9.3**
Diseases	2	407	203.5	1.8
Interactions	6	715	119.1	1.1
Within cells	46	5081	110.5	

20.5—Analysis by proportional numbers. Sometimes the n_{ij} appear to be approximately proportional to their row and column totals, that is, $n_{ij} \doteq n_{i.}n_{.j}/n_{..}$. In this event, a good approximation to the least squares analysis is obtained by using the actual cell means $\bar{X}_{ij.}$ but replacing the n_{ij} by proportional numbers $n'_{ij} = n_{i.}n_{.j}/n_{..}$. An example is provided in table 20.5.1 by a sample survey of farm tenancy in an Iowa county (4). It was found that farmers had about the same proportions of owned, rented, and mixed farms in three soil fertility classes, as tested by χ^2 (section 11.10). Each cell of table 20.5.1 shows first the n_{ij}, then the mean corn acres $\bar{X}_{ij.}$ per farm, and underneath this the proportional numbers n'_{ij}. For example, for owners in soil class I, $n'_{11} = (152)(125)/517 = 36.8$. The agreement between actual and porportional numbers is good.

In calculating the new row and column totals $\Sigma n'\bar{X}$ that will be used, no rounding has been done. The total sum of squares between cells is

$$(36.8)(32.7^2) + (62.9)(55.2^2) + \ldots + (77.5)(40.1^2) - (23{,}008.31^2)/517$$

The sum of squares between soil classes is calculated from the new soil class totals:

$$7321.82^2/152 + 6584.22^2/140 + 9102.27^2/225 - 23{,}008.31^2/517$$

In the analysis of variance there are large differences between the mean corn acres per farm for different tenure groups but there is no sign of interactions. The average corn acres are 51.1 for renters, 45.1 for those with

TABLE 20.5.1
CORN ACRES CLASSIFIED BY TENURE AND SOIL PRODUCTIVITY
(AUDUBON COUNTY, IOWA)

Soil Class		Owner		Renter		Mixed		Σn	Total Acres $\Sigma n'\bar{X}$	Mean Acres
I	$n, \bar{X}$	36	32.7	67	55.2	49	50.6	152	7,321.82	48.2
	n'	36.8		62.9		52.3		152		
II	$n, \bar{X}$	31	36.0	60	53.4	49	47.1	140	6,584.22	47.0
	n'	33.8		58.0		48.2		140		
III	$n, \bar{X}$	58	30.1	87	46.8	80	40.1	225	9,102.27	40.5
	n'	54.4		93.1		77.5		225		
	n	125		214		178		517		
Total acres, $\Sigma n'\bar{X}$		4057.60		10,926.36		8024.95			23,008.31	
Mean acres		32.5		51.1		45.1				44.5

Analysis of Variance Using Proportional Numbers

Source of Variation	df	Sum of Squares	Mean Squares	F
Soils	2	6,626	3,313	4.0*
Tenures	2	27,381	13,690	16.5**
Interactions	4	901	225	0.3
Within cells	508		830	

mixed tenure, and 32.5 for owners, all three means being significantly different. For example, the standard error of the difference 51.1 − 45.1 = 6.0 is $\sqrt{830(1/214 + 1/178)} = \pm 2.92$. Furthermore, soil classes I and II have significantly more corn than soil class III.

20.6—Least squares fitting of the additive model. If the individual observations in a two-way table are independent with residual variance σ^2, the additive model for the cell means is

$$\bar{X}_{ij.} = \mu + \alpha_i + \beta_j + \bar{\epsilon}_{ij.} \tag{20.6.1}$$

where $\bar{\epsilon}_{ij.}$ are independent with variance σ^2/n_{ij}. The least squares analysis finds estimates m, a_i, and b_j that minimize the weighted sum of squares of residuals

$$\sum_i \sum_j n_{ij}(\bar{X}_{ij.} - m - a_i - b_j)^2 \tag{20.6.2}$$

This analysis also provides an F test of the interactions mean square.

As noted previously, the parameter a_i is needed only to tell how much the mean of row i differs from the overall mean and similarly for b_j with respect to columns. Hence, with I rows and J columns, there are $I + J - 1$ parameters to be estimated. A method for manual fitting of this model by least squares was first given by Yates (5) in 1934. However, manual calculation becomes very tedious when I and J both exceed 3, so use of a computer program that supplies the output needed is desirable.

From the 3 × 3 soils × tenures data in the preceding section (table 20.5.1), we shall illustrate a method of least squares calculation that can be adapted to computer programming. By introducing dummy X variables, the fitting of the additive model is expressed as the fitting of a multiple linear regression as described in section 18.4. In the regression calculations, the sweep method of section 17.6 is used. With a 3 × 3 table 5 parameters must be estimated, so 5 dummy Xs are needed.

In table 20.6.1 the 3 × 3 soils × tenures cell means are set out as a single sample of size 9. There are several simple methods for constructing the dummy Xs so that the additive model is expressed as a multiple linear regression, $Y = \gamma_0 X_0 + \gamma_1 X_1 + \ldots + \gamma_4 X_4$. In our 3 × 3 table, let $X_0 = 1$ in every cell. The variate X_1 is given the value 1 for all members of the *second row* and the value 0 throughout the other rows. Similarly, $X_2 = 0$ except in the third row, in which $X_2 = 1$. As for the columns, $X_3 = 1$ for all members of the *second column* and 0 elsewhere, while $X_4 = 1$ in the third column and 0 elsewhere. In the notation of table 20.6.1 it follows that

$$E(\bar{X}_{11.}) = \gamma_0 \qquad E(\bar{X}_{12.}) = \gamma_0 + \gamma_3 \qquad E(\bar{X}_{31.}) = \gamma_0 + \gamma_2$$

and so on. Verify that the Xs in table 20.6.1 satisfy these rules.

This method extends naturally to an $I \times J$ table. The variate $X_0 = 1$ in all

TABLE 20.6.1
SOILS × TENURES TABLE 20.5.1 IN MULTIPLE LINEAR REGRESSION FORM

IJ	n_{ij}	$\overline{X}_{ij.}$	X_0	X_1	X_2	X_3	X_4	$n_{ij}\overline{X}_{ij.}$
11	36	32.7	1	0	0	0	0	1,177.2
12	67	55.2	1	0	0	1	0	3,698.4
13	49	50.6	1	0	0	0	1	2,479.4
21	31	36.0	1	1	0	0	0	1,116.0
22	60	53.4	1	1	0	1	0	3,204.0
23	49	47.1	1	1	0	0	1	2,307.9
31	58	30.1	1	0	1	0	0	1,745.8
32	87	46.8	1	0	1	1	0	4,071.6
33	80	40.1	1	0	1	0	1	3,208.0
	517							23,008.3

Uncorrected Sums of Squares and Products

	0	1	2	3	4	5
0	517	140	225	214	178	23,008.3
1		140	0	60	49	6,627.9
2			225	87	80	9,025.4
3				214	0	10,974.0
4					178	7,995.3
5						1,059,815

cells. The row variates are $X_1, X_2, \ldots, X_{I-1}$, where $X_{I-1} = 1$ in row I and 0 elsewhere. The column variates extend from X_I to X_{I+J-2}.

The cell total corn acreages $n_{ij}\overline{X}_{ij.}$ and the grand total are shown in the right column of table 20.6.1. At the bottom of the table are the weighted sums of squares and products of the dummy Xs and the independent variable $\overline{X}_{ij.} = Y$, which we call X_5 for the application of the sweep method. For example, two sums of products under table 20.6.1 are

$$\Sigma n X_0 X_2 = 58 + 87 + 80 = 225$$

$$\Sigma n X_3 X_5 = 3698.4 + 3204.0 + 4071.6 = 10{,}974.0$$

Before performing the sweep calculations, it is important to express the regression coefficients γ_i in terms of the parameters μ, α_i, β_j of model (20.6.1) and to check that the γ_i will provide the comparisons we want. Expressing $E(\overline{X}_{11.})$, etc., in the two sets of notations,

$$E(\overline{X}_{11.}) = \gamma_0 = \mu + \alpha_1 + \beta_1 \qquad E(\overline{X}_{21.}) = \gamma_0 + \gamma_1 = \mu + \alpha_2 + \beta_1$$

$$E(\overline{X}_{31.}) = \gamma_0 + \gamma_2 = \mu + \alpha_3 + \beta_1$$

Hence, $\gamma_1 = \alpha_2 - \alpha_1$, $\gamma_2 = \alpha_3 - \alpha_1$, and similarly, $\gamma_3 = \beta_2 - \beta_1$, $\gamma_4 = \beta_3 - \beta_1$. Satisfy yourself also that any comparison among the α_i can be expressed as a linear function of γ_1 and γ_2, while γ_3 and γ_4 perform the same function for the β_j.

Thus, in order to examine row and column effects, we need estimates

$g_1, \ldots, g_4$ of $\gamma_1, \ldots, \gamma_4$, with their covariance matrix. Note that the remaining parameter γ_0 represents a kind of average level and that performing SWP[0] in table 20.6.1 converts the sums of squares and products for the other dummy *X*s into sums of squares and products of deviations from the sample means, which are shown in table 20.6.2 for variables $X_1, X_2, \ldots, X_5 = Y$. Under this matrix are the results of performing SWP[1, 2, 3, 4], namely the c_{ij} that give the covariance matrix of the g_i; the g_i themselves; and the residual or interactions sum of squares, 890.

As for the analysis of variance, table 20.6.2 shows that the total sum of squares between cells (8 df) is 35,866 and the interactions sum of squares (4 df) is 890, leaving 34,976 (4 df) for the main effects of soils and tenures. Since soil and tenure effects are not orthogonal because of the unequal n_{ij} (as pointed out in section 20.1), the sums of squares for the main effects can be subdivided into either of the following two separate ways:

soils *SS* (unadjusted) + tenures *SS* (adjusted)

tenures *SS* (unadjusted) + soils *SS* (adjusted)

The sums of squares that must be used for *F* tests of the soils and tenures effects are the *adjusted* sums of squares, since the unadjusted soils sum of squares, for example, will contain hidden effects of tenure differences. This method of making *F* tests is another application of the extended analysis of variance (section 17.4).

Algebraically, the unadjusted soils sum of squares is

$$X_{1.}^2/n_{1.} + X_{2.}^2/n_{2.} + X_{3.}^2/n_{3.} - X_{..}^2/n_{..}$$

From table 20.6.1, this quantity is found to be 7759. It can also be found by subtracting the residual sum of squares after SWP[1, 2]—not shown here—from the sum of squares of deviations (8 df) of the cell means, 35,866 in table 20.6.2. The unadjusted sum of squares for tenures is found from table 20.6.1 to

TABLE 20.6.2
SUMS OF SQUARES AND PRODUCTS OF DEVIATIONS FROM THE MEANS

	1	2	3	4	5
1	102.0890	−60.9284	2.0503	0.7988	397.41
2		127.0793	−6.1335	2.5338	−987.88
3			125.4197	−73.6789	1,450.25
4				116.6789	73.68
5					35,866
			c_{ij}		$g_i = \hat{\gamma}_i$
	0.01373	0.00659	−0.0007	−0.00028	−1.1653
		0.01105	.00042	−.00002	−7.6862
			0.01270	.00801	18.5649
				0.01362	12.5254
					interactions *SS* = 890

be 28,438. These results lead to the *adjusted* sum of squares for tenures and soils and to the analysis of variance in table 20.6.3.

Both main effects are significant and there are no indications of any interactions. Incidentally, with a standard computer program, the output analysis of variance may show only the *adjusted* sums of squares and mean squares for rows and columns needed for the F tests. Note also that in the approximate methods of analysis by equal or proportional cell numbers, the sums of squares for rows and columns are approximations to the *adjusted* sums of squares in the least squares analysis. In the proportional numbers analysis in table 20.5.1, the tenures sum of squares and soils sum of squares were 27,381 and 6626, respectively, as against 27,217 and 6538 for the corresponding adjusted sums of squares in the least squares analysis.

As regards tenure effects, you may verify from the g_i and c_{ij} in table 20.6.2 that the mean differences in corn acreage are renter − owner, 18.6 ± 3.25; mixed − owner, 12.5 ± 3.36; renter − mixed, 6.0 ± 2.92. All differences are statistically significant.

Least squares estimates of the tenure means $m + b_1$, $m + b_2$, $m + b_3$ can be obtained in several ways. One is to impose the condition that the weighted mean of these quantities equals the observed mean 23,008.3/517. This condition may be written:

$$517m + 125b_1 + 214b_2 + 178b_3 = 23{,}008.3$$

Since $g_3 = b_2 - b_1$, $g_4 = b_3 - b_1$, then $b_2 = b_1 + g_3$, $b_3 = b_1 + g_4$. This leads to the equation that gives the value of $m + b_1$:

$$517(m + b_1) + 214g_3 + 178g_4 = 23{,}008.3 \qquad m + b_1 = 32.507$$

Hence, the estimated mean corn acres are

owner, 32.51   renter, 51.07   mixed, 45.03

The soil class means may be found similarly.

EXAMPLE 20.6.1—The artificial 2 × 2 table 20.1.1 provides an easy example of the method of least squares calculation given in this section. Introduce dummy X variables X_0, X_1, X_2 as in this

TABLE 20.6.3
ANALYSIS OF VARIANCE OF CORN ACREAGE DATA

Source of Variation	df	Sum of Squares	Mean Squares	F
Tenures (unadjusted)	2	28,438		
Soils (adjusted)	2	6,538	3,269	3.9*
Soils (unadjusted)	2	7,759		
Tenures (adjusted)	2	27,217	13,601	16.4**
Interactions	4	890	222	0.3
Within cells	508		830	

section. Then by applying the sweep method, show that in the weighted regression of $\overline{X}_{ij.} = X_3$ on X_1 and X_2, adjusted for X_0, the c_{ij}, g_i, and interactions sums of squares are as follows:

	1	2	3
	c_{ij}		g_i
1	0.3125	−0.1875	−5
2		0.3135	10
3			0

The rats gain 5 units more on diet 1 than on diet 2, males gain 10 units more than females, and the interactions sum of squares is 0.

EXAMPLE 20.6.2—In the approximate analysis of the corn acreage data in table 20.5.1 by proportional numbers, the difference between the mean acreages for soil class II and for soil class III was just significant at the 5% level. From tables 20.5.1 and 20.6.2, calculate the estimates of this mean difference and its standard error as given by (*i*) proportional numbers, (*ii*) least squares. Ans. Soil class II − soil class III = (*i*) 6.5 ± 3.10, (*ii*) 6.52 ± 3.10. The close agreement may be expected, since the cell numbers are so nearly proportional.

EXAMPLE 20.6.3—With only two rows or two columns, the F test of the interactions mean squares can be made quickly by a pocket calculator before deciding whether estimates of row and column effects should be based on the additive model. The following data show the mean six-week gains in weight (grams − 100) of male and female Wistar rats in four successive generations. The within-cells mean square is 409 with 141 df.

	Male		Female			
Generation	n_{1j}	$\overline{X}_{1j}$	n_{2j}	$\overline{X}_{2j}$	$W_j = n_{1j}n_{2j}/(n_{1j} + n_{2j})$	$D_j = \overline{X}_{1j} - \overline{X}_{2j}$
1	21	77.0	27	9.5	11.8	67.5
2	15	61.5	25	14.1	9.4	47.4
3	12	55.7	23	7.9	7.9	47.2
4	7	71.0	19	5.1	5.1	64.2
					34.2	$\Sigma W_i D_i = 1942.36$

The sex × generations interactions measure how much the sex differences D_j vary from generation to generation. The variance of D_j is $\sigma^2(1/n_{1j} + 1/n_{2j})$ so that its weight W_j is the reciprocal of this quantity. The interactions sum of squares (3 df) is therefore

$$\Sigma W_j D_j^2 - (\Sigma W_j D_j)^2/\Sigma W_j = 3189 \qquad MS = 1063$$

Hence, $F = 1063/409 = 2.60$, just short of the 5% level. The sex difference looks greater in generations 1 and 4 than in generations 2 and 3, but there is no prior reason to anticipate that the sex difference would change from generation to generation. It seems reasonable to go ahead with the additive model, while reporting the possibility of some interaction or extraneous variation.

20.7—Analysis of proportions in two-way tables. In chapter 11 we discuss methods of analysis for a binomial proportion. Sections 11.7–11.9 deal with a set of C proportions arranged in a one-way classification. Two-way tables in which the entry in every cell is a sample proportion are also common. Examples are sample survey results giving the percentage of voters who stated their intention to vote the Democratic ticket, classified by the age and income level of the voter, or a study of the proportion of patients with blood group O in a large hospital, classified by sex and type of illness.

With I rows and J columns, the data consist of IJ independent values of a binomial proportion $p_{ij} = a_{ij}/n_{ij}$. The data resemble those in preceding sections, but instead of a sample of n_{ij} continuous measurements X_{ijk} ($k = 1, 2, \ldots, n_{ij}$) in the (i, j) cell, we have a single binomial proportion p_{ij} in the cell. The questions of interest are usually the same in the binomial and continuous cases. We want to examine how the p_{ij} vary over the cells of the table. We may try an analysis in terms of main effects and interactions or a more ad hoc analysis.

From the viewpoint of theory, the analysis of proportions presents more difficulties than that of normally distributed variables. Few exact small sample results are available. Approximate methods that have been used depend on one of the following approaches.

1. Regard p_{ij} as a normally distributed variable with variance $p_{ij}q_{ij}/n_{ij}$, using the weighted methods of analysis in preceding sections, with weights $w_{ij} = n_{ij}/(p_{ij}q_{ij})$ and p_{ij} replacing $\overline{X}_{ij}$.
2. Transform the p_{ij} to equivalent angles y_{ij} (section 15.12), and treat the y_{ij} as normally distributed. Since the variance of y_{ij} for any p_{ij} is approximately $821/n_{ij}$, this method has the advantage that the analysis of variance of the y_{ij} is unweighted if the n_{ij} are constant. As we have seen, this transformation is frequently used in randomized blocks experiments in which the measurement is a proportion.
3. Transform p_{ij} to its logit, $Y_{ij} = \ln(p_{ij}/q_{ij})$. The estimated variance of Y_{ij} is approximately $1/(n_{ij}p_{ij}q_{ij})$, so in a logit analysis Y_{ij} is given a weight $n_{ij}p_{ij}q_{ij}$.

The assumptions involved in these approaches probably introduce little error in the conclusions if the observed numbers of successes and failures, $n_{ij}p_{ij}$ and $n_{ij}q_{ij}$, exceed 20 in every cell. Various small sample adjustments have been prepared to extend the validity of the methods.

When all p_{ij} lie between 25% and 75%, the results given by the three approaches seldom differ materially. If, however, the p_{ij} cover a wide range from almost 0 to 50% or beyond, there are reasons for expecting that row and column effects are more likely to be additive on a logit scale than on the original p scale. To repeat an example cited in section 11.11, suppose that the data are the proportions of cases in which the driver of the car suffered injury in automobile accidents, classified by severity of impact (rows) and by whether the driver wore a seat belt or not (columns). Under very mild impacts p is likely to be close to 0 for both wearers and nonwearers, with little if any difference between the two columns. At the other end, under extreme impacts, p will be near 100% whether a seat belt was worn or not, with again a small column effect. The beneficial effects of the belts, if any, will be revealed by the accidents that show intermediate proportions of injuries. The situation is familiar in biological assay, in which the toxic or protective effects of different agents are being compared. It is well known that two agents cannot be effectively compared at concentrations for which p is close to 0 or 100%; instead, the investigator aims at concentrations yielding p around 50%.

Thus, in the scale of p, row and column effects cannot be strictly additive over the whole range. The logit transformation pulls out the scale near 0 and 100% so that the scale extends from $-\infty$ to $+\infty$. In the logit analysis row and column effects may be additive, while in the p scale for the same data we find interactions that are entirely a consequence of the scale. The angular transformation occupies an intermediate position. As with logits, the scale is stretched at the ends but the total range remans finite, from 0° to 90°.

To summarize, with an analysis in the original scale it is easier to think about the meaning and practical importance of effects in this scale. The advantage of angles is the simplicity of the computations if the n_{ij} are equal or nearly so. Logits may permit an additive model to be used in tables showing large effects.

The preceding analyses use *observed weights;* the weight $w = n/(pq)$ attached to a proportion is computed from the observed value of p. Similarly, the observed weight $w_{ij} = n_{ij}p_{ij}q_{ij}$ for logit p_{ij} is the reciprocal of $[1/a_{ij} + 1/(n_{ij} - a_{ij})]$. This quantity is easily found with a pocket calculator that has reciprocals. When fitting the model we could instead use *expected weights* $n/(\hat{p}\hat{q})$ or $n\hat{p}\hat{q}$, where $\hat{p}$ is the estimated value given by the model. This approach requires a series of iterations. We guess first approximations to the weights and fit the model, obtaining second approximations to the $\hat{p}$. From these the weights are recomputed and the model fitted again, giving third approximations to the $\hat{p}$ and the w, and so on until no appreciable change occurs in the results.

With some adjustments due to Fisher, this set of calculations gives successive approximations to the *maximum likelihood* estimates of the $\hat{p}$ (6). Like the method of least squares, the method of maximum likelihood is a general method of finding good estimates of parameters in models. It reduces essentially to the method of least squares when the observations are normally distributed but does not depend on normality, requiring only that the probability distribution of the observations be known. It is therefore more general than least squares.

However, Berkson (7) found in a series of typical examples that the weighted least squares solutions using observed weights gave estimates and tests of significance that agree closely with the maximum likelihood values. With a model $\hat{Y}$ expressed in the logit scale, he calls his method the *minimum logit* χ^2 *method.* The estimates of the parameters in the model are chosen to minimize

$$\sum_i \sum_j n_{ij}p_{ij}q_{ij}(Y_{ij} - \hat{Y}_{ij})^2$$

where $Y_{ij} = \ln(p_{ij}/q_{ij}) = \text{logit } p_{ij}$ and $p_{ij} = a_{ij}/n_{ij}$.

20.8—Analysis in the p scale: a 3 × 2 table. In this example, inspection of the individual proportions indicates interactions that are due to the nature of the factors and would not be removed by a logit transformation. The data are the proportions of *children* with emotional problems in a study of family medical care (8), classified by the number of children in the family and by whether both,

TABLE 20.8.1
PROPORTION OF CHILDREN WITH EMOTIONAL PROBLEMS

Parents with Problems	Number of Children in Family	
	1–2	3–4
Both	$p = 33/57 = 0.579$	$p = 15/38 = 0.395$
One	$p = 18/54 = .333$	$p = 17/55 = .309$
Neither	$p = 10/37 = 0.270$	$p = 9/32 = 0.281$

one, or neither *parent* was recorded as having emotional problems, as shown in table 20.8.1.

In families having one or neither parent with emotional problems the four values of p are close to 0.3, any differences being easily accountable by sampling errors. Thus there is no sign of a relation to number of children in the family or the parents' status when neither or only one parent has emotional problems. When both parents have problems, there is a marked increase in p in the smaller families to 0.579 and a modest increase in the larger families to 0.395. Thus, inspection suggests that the proportion of children with emotional problems is increased when both parents have problems, and that this increase is less in the larger families.

It is not obvious what to write in the report of results. The difference (0.579 − 0.333) is clearly significant. But the corresponding value 0.395 with 3–4 children is not significantly higher than 0.309 nor is it significantly lower than 0.579. The tests are consistent with the suggestions made about the smaller families but do not give positive support to the statements about families with 3–4 children. Reference (8) presents additional data bearing on the scientific issue. In data of this type, nothing seems to be gained by transformation to logits.

20.9—Analysis of logits in a two-way table: calculations. When fitting an additive model in the logit scale, the minimum logit χ^2 method should be an adequate approximation. In this we minimize

$$\Sigma w_{ij}(Y_{ij} - \hat{Y}_{ij})^2 \qquad (20.9.1)$$

TABLE 20.9.1
MOTHERS CLASSIFIED BY SCHOOL STATUS AND BIRTH ORDER OF CHILD (3 × 2 × 2 TABLE)

	Number of Mothers with					
	Losses a	No Losses $n - a$	Losses a	No Losses $n - a$	Losses a	No Losses $n - a$
Birth order	2		3–4		5+	
Problems	20	82	26	41	27	22
Controls	10	54	16	30	14	23

where $Y_{ij} = \ln (p_{ij}/q_{ij}) = \ln [a_{ij}/(n_{ij} - a_{ij})]$ and the weight w_{ij} is the reciprocal of $[1/a_{ij} + 1/(n_{ij} - a_{ij})]$. The term "minimum logit χ^2" is used because, since the w_{ij} are approximations to the true weights, the quantity (20.9.1) is distributed approximately as χ^2 with $(I - 1)(J - 1)$ df if the additive model fits. Thus the χ^2 table is used when testing the interactions sums of squares. There is no within-cells sum of squares in the analysis of binomial proportions.

When the cell numbers a or $n - a$ are small, various adjustments to Y_{ij} and w_{ij} have been suggested to reduce biases that arise in the transformation to logits. From work by Gart and Zweifel (9) we suggest taking

$$Y = \ln [a + 1/2)(n - a + 1/2)]$$

$$V(Y) = 1/(a + 1/2) + 1/(n - a + 1/2) \qquad w = 1/[V(Y)]$$

when cell numbers are less than 10.

The calculations required to fit the additive model in the logit scale are the same as those in the weighted least squares fitting in section 20.6. As a simple example, the data in table 11.11.1 were chosen. The data were part of a study of a possible relation between losses of children during pregnancies and school behavior problems of subsequent live children in Baltimore. Mothers were classified according to the birth order of the live child and to whether the child was a problem child or a nonproblem control child in school. The variate to be analyzed was the proportion of mothers with previous infant losses. The comparison of primary interest was whether these proportions differed for mothers of problems and control children.

Section 11.11 presents a test of the average difference in proportions without transformation to logits. The test criterion, essentially the same as the Mantel-Haenszel criterion, was constructed so that it would be efficient if the effects of birth order and the problems minus controls differences were additive on a logit scale. By fitting an additive model in logits to these data, we get a χ^2 test of this additivity assumption, as well as estimates of the birth order and problems-controls main effects if additivity appears to hold. The cell numbers in the 2 × 2 × 3 table are given in table 20.9.1. For a look at the percent losses for problems and controls, refer to table 11.11.1, p. 211.

Table 20.9.2 shows the logits, $Y = \ln[a/(n - a)]$; the weights, $w = a(n - a)/n$; and the dummy Xs used to convert the additive model into a weighted multiple regression line, $\Sigma\gamma_i X_i$. As before, $X_0 = 1$ in all six cells. The remaining Xs were constructed by a different method from that in section 20.6. If there are i rows, X_i $(i = 1, 2, \ldots, I - 1) = 1$ in row i, $X_i = -1$ in the last (Ith) row, and $X_i = 0$ in all other rows. With only two rows, $X_1 = 1$ in row 1 and $X_1 = -1$ in row 2. The same method is followed in the columns for X_2 and X_3. Verify these Xs in table 20.9.2. With this system of Xs, the relation between the γs and the μ, α_i, and β_j in the analysis of variance notation is seen by expressing the model values for Y_{11}, Y_{12}, and Y_{13} in the two forms:

$E(Y_{11})$	$E(Y_{12})$	$E(Y_{13})$
$\mu + \alpha_1 + \beta_1$	$\mu + \alpha_1 + \beta_2$	$\mu + \alpha_1 + \beta_3$
$\gamma_0 + \gamma_1 + \gamma_2$	$\gamma_0 + \gamma_1 + \gamma_3$	$\gamma_0 + \gamma_1 - (\gamma_2 + \gamma_3)$

TABLE 20.9.2
ADDITIVE MODEL IN MULTIPLE REGRESSION NOTATION

(i, j)	$Y = \text{logit } p$	w	X_0	X_1	X_2	X_3	wY
11	−1.411	16.1	1	1	1	0	−22.7171
12	−0.455	15.9	1	1	0	1	− 7.2345
13	+0.205	12.1	1	1	−1	−1	+ 2.4805
21	−1.686	8.4	1	−1	1	0	−14.1624
22	−0.629	10.4	1	−1	0	1	− 6.5416
23	−0.496	8.7	1	−1	−1	−1	− 4.3152

Uncorrected Sums of Squares and Products

X_0	X_1	X_2	X_3	$X_4 = Y$
71.6	16.6	3.7	5.5	−52.4903
	71.6	4.3	2.1	− 2.4519
		45.3	20.8	−35.0448
			47.1	−11.9414
				65.98684

TABLE 20.9.3
RESULTS OF SWP[0, 1, 2, 3] APPLIED TO TABLE 20.9.2

	c_{ij}			
	1	2	3	Estimates
1	0.01482	−0.00126	0.00029	$\hat{\gamma}_1 = 0.18252$
2		0.02782	−0.01220	$\gamma_2 = -0.81510$
3			0.02678	$\gamma_3 = 0.18558$
				$\chi^2 = 0.8455$ (2 df)

It follows that $\beta_1 = \gamma_2$, $\beta_2 = \gamma_3$, $\beta_3 = -(\gamma_2 + \gamma_3)$, so this notation implies the common convention: $\beta_1 + \beta_2 + \beta_3 = 0$. Similarly, $\alpha_1 = \gamma_1$, $\alpha_2 = -\gamma_1$ (the same convention applied to two classes). The regression coefficient γ_0 (a measure of average level) is estimated from its normal equation:

$$71.6\hat{\gamma}_0 + 16.6\hat{\gamma}_1 + 3.7\hat{\gamma}_2 + 5.5\hat{\gamma}_3 = -52.4903 \tag{20.9.2}$$

These multipliers 71.6, 16.6, etc., are weighted sums of squares and products shown in table 20.9.2.

Table 20.9.3 shows the c_{ij}; the estimates $\hat{\gamma}_1$, $\hat{\gamma}_2$, and $\hat{g}_3$; and the weighted sum of squares of deviations from the additive model, which were found by calculating SWP[0, 1, 2, 3].

20.10—Results of the analysis. The value of the residual, $\chi^2 = 0.845$ with 2 df, gives no evidence for doubting the additive model. This gives support to the test criterion used in section 11.11, which implied additivity. In tests of significance or confidence limits for main effects, we treat the within-cell individual variance σ^2 as unity and use the normal and χ^2 distributions as an approximation instead of the t and F distributions. The task of principal interest is to test $\hat{\gamma}_1 = 0.1825$ (the estimated mean difference, problems minus controls) on the logit scale. From c_{11} in table 20.9.3, the standard error of $\hat{\gamma}_1$ is $\sqrt{c_{11}} = \sqrt{0.01482} = \pm0.1217$. The resulting normal deviate is $0.1825/0.1217 = 1.50$, with P about 0.13. Incidentally, the test in section 11 gave a normal deviate 1.51, practically the same. The increase in frequency of previous infant losses seen in mothers of behavior problems is not significant.

The estimates of the birth order effects are $b_1 = -0.81510$, $b_2 = +0.18558$, $b_3 = +0.62952$. Tests of differences among the bs are of little interest, since previous loss rates naturally increase with birth order. As an exercise, the reader may wish to verify that the standard error of $b_1 - b_2$ is ±0.2811, while that of $b_2 - b_3$ is $\sqrt{c_{22} + 4c_{23} + 4c_{33}} = \pm0.2935$.

If desired, a partition of χ^2 values into contributions due to rows, columns, and interactions can be obtained. As with a continuous response variable, the relevant χ^2 contributions are those due to each main effect after adjusting for the other main effects and they will not sum to the total χ^2 between cells. From table 20.9.3, the adjusted χ^2 for problems minus controls is $\gamma_1^2/c_{11} = 0.1825^2/0.01482 = 2.247$. The adjusted χ^2 value for birth order effects can be found by subtraction. The regression on X_0 and X_1 represents a fitting of effects of the mean and of rows. The residual χ^2 after fitting X_0 and X_1, found by SWP[0, 1], was 26.112 (4 df), as against 0.845 (2 df) after fitting birth order effects as well. Hence, the χ^2 value for birth order effects (adjusted) is 25.267. This gives for the χ^2 contributions:

	df	χ^2
Problems vs. controls (adj.)	1	2.247
Birth orders (adj.)	2	25.267
Interactions	2	0.845

EXAMPLE 20.10.1—(*i*) Find the value of $\hat{\gamma}_0$ from equation (20.9.2). (*ii*) By using table 20.9.2, find the value $\hat{Y}$ predicted in each cell by the additive model. For instance, $\hat{Y}_{23} = \hat{\gamma}_0 - \hat{\gamma}_1 - \hat{\gamma}_2 - \hat{\gamma}_3$. (*iii*) By calculation, $\Sigma w(Y - \hat{Y})^2$, verify the χ^2 value 0.845 for deviations from the model (apart from rounding errors). Ans. (*i*) $\hat{\gamma}_0 = -0.747570$. (*ii*) The predicted and residual logit values in the 6 cells are:

(i, j)	Y	$\hat{Y}$	$Y - \hat{Y}$	w
11	−1.411	−1.38009	−0.03091	16.1
12	−0.455	−0.37941	− .07559	15.9
13	+0.205	+0.06453	+ .14047	12.1
21	−1.686	−1.74525	+ .05925	8.4
22	−0.629	−0.74457	+ .11557	10.4
23	−0.496	−0.30063	−0.19537	8.7

(*iii*) $\chi^2 = 16.1(0.03091^2) + 15.9(0.07559^2) + \ldots = 0.845$, as before.

EXAMPLE 20.10.2—From the $\hat{Y}$ values in example 20.10.1 we can reverse the logit transformation and predict the cell numbers in table 20.9.1 from the additive model in logits. Since $\hat{Y} = \text{logit } \hat{p} = \ln(\hat{p}/\hat{q})$, it follows that $\hat{p} = e^{\hat{Y}}/(1 + e^{\hat{Y}})$. For example, for cell (1, 1) we find $\hat{Y}_{11} = -1.380$ and $\hat{p}_{11} = 0.2010$. Hence the predicted number of problem losses for birth order 2 is $\hat{p}_{11}n_{11} = 0.2010 \times 102 = 20.50$. Find the predicted number of control losses for birth order 5+. Ans. 15.74.

EXAMPLE 20.10.3—If p is the population proportion of successes, its logit is $\ln(p/q)$. If a binomial sample of size n gives a successes, the adjusted logit suggested in section 20.9, namely $\ln[(a + 1/2)/(n - a + 1/2)]$, is a relatively unbiased estimate of $\ln(p/q)$ for expectations np and nq as low as 3. Furthermore, its suggested variance,

$$V = 1/w = 1/[(a + 1/2) + 1/(n - a + 1/2)]$$

is also relatively unbiased. For $n = 10$, $p = 0.3$, $np = 3$, $nq = 7$, the values of the binomial probability P of getting a successes and of Y and V are shown below for each value of a as an illustration. The true logit is $\ln(3/7) = -0.8473$. Verify that (i) the mean value of Y is -0.8497 and (ii) the correct variance of Y is 0.4968, while the mean value of V is 0.5164, about 4% too large.

	P	Y	V	Y^2
0	0.0282	−3.046	2.095	9.278
1	.1211	−1.846	0.772	3.408
2	.2335	−1.224	0.518	1.498
3	.2668	−0.762	0.419	0.581
4	.2001	−0.367	0.376	0.135
5	.1029	0.000	0.364	0.000
6	.0368	0.367	0.376	0.135
7	.0090	0.762	0.419	0.581
8	.0014	1.224	0.518	1.498
9	.0001	1.846	0.772	3.408
10	0.0000	3.046	2.095	9.278

Berkson (7) gives a number of examples of the minimum logit χ^2 method, comparing the estimates and χ^2 values with those given by the maximum likelihood method. Grizzle, Starmer, and Koch (10) have extended the χ^2 method to a broader range of problems, with examples; their paper requires familiarity with matrix notation. For contingency tables in which *all* classifications have more than two classes, a general method that has attracted much attention (11, 12) is to fit the kind of model used in the analysis of factorial experiments to the natural logarithms of the cell numbers, sometimes called a *loglinear model.* This model is a natural extension of the model used here for analysis in the logit scale.

To see this, suppose we have a three-way classification. The first classification has two classes, numbered 1 and 2. Rows (i) and columns (j) denote the second and third classifications. Let a_{1ij}, a_{2ij} denote the cell numbers for classes 1 and 2 in row i, column j. In our notation, $p_{ij} = a_{1ij}/(a_{1ij} + a_{2ij})$, while logit $p_{ij} = \ln a_{1ij}/a_{2ij} = \ln a_{1ij} - \ln a_{2ij}$.

Now the full loglinear model can be written:

$$\ln a_{1ij} = \mu + \delta_1 + \alpha_i + \beta_j + (\delta\alpha)_{1i} + (\delta\beta)_{1j} + (\alpha\beta)_{ij} + (\delta\alpha\beta)_{1ij}$$

$$\ln a_{2ij} = \mu + \delta_2 + \alpha_i + \beta_j + (\delta\alpha)_{2i} + (\delta\beta)_{2j} + (\alpha\beta)_{ij} + (\delta\alpha\beta)_{2ij}$$

In the loglinear notation, δ_1, α_i, and β_j denote main effects; $(\delta\alpha)_{1i}$, $(\delta\beta)_{1j}$, and $(\alpha\beta)_{ij}$ denote two-factor interactions; and $(\delta\alpha\beta)_{1ij}$ denote three-factor interactions. If there are no three-factor interactions, it follows that

$$\text{logit } p_{ij} = \ln a_{1ij} - \ln a_{2ij} = (\delta_1 - \delta_2) + [(\delta\alpha)_{1i} - (\delta\alpha)_{2i}] + [(\delta\beta)_{1j} - (\delta\beta)_{2j}]$$

This relation can be written, in different notation,

$$\text{logit } p_{ij} = \mu' + \alpha'_i + \beta'_j$$

Thus, when one classification has two classes, an additive model in logits corresponds to a loglinear model with no three-factor interactions.

TECHNICAL TERMS

adjusted sums of squares
loglinear model
maximum likelihood
minimum logit χ^2 method
proportional numbers

REFERENCES

1. Gowen, J. W. *Am. J. Hum. Gen.* 4 (1952):285.
2. Kutner, M. H. *Am. Stat.* 28, 3 (1974):98.
3. Afifi, A. A., and Azen, S. P. 1972. *Statistical Analysis: A Computer Oriented Approach.* Academic Press, New York.
4. Strand, N., and Jessen, R. J. 1943. Iowa Agric. Exp. Stn. Res. Bull. 315.
5. Yates, F. *J. Am. Stat. Assoc.* 29 (1934):51.
6. Cochran, W. G. *Ann. Math. Stat.* 11 (1940):335.
7. Berkson, J. *Biometrics* 24 (1968):75.
8. Silver, G. A. 1963, *Family Medical Care.* Harvard Univ. Press, Cambridge, Mass., table 59.
9. Gart, J. J., and Zweifel, J. R. *Biometrika* 54 (1967):181.
10. Grizzle, J. E.; Starmer, C. F.; and Koch, G. G. *Biometrics* 25 (1969):489.
11. Bishop, Y. M. M.; Fienberg, S. E.; and Holland, P. W. 1975. *Discrete Multivariate Analysis: Theory and Practice.* MIT Press, Cambridge, Mass.
12. Haberman, S. J. 1974. *The Analysis of Frequency Data.* Univ. Chicago Press.

21

Sample surveys

21.1—Introduction. At the beginning of this book it is pointed out that one of the principal applications of statistical methods is to sample surveys, in which information about a specific population is obtained by selecting and measuring a sample of the members of the population. Three examples were described briefly: the nationwide sample from which the Census Bureau publishes monthly estimates of the number of employed and unemployed members of the labor force; a sample of waybills taken by the Chesapeake & Ohio Railroad in appraising whether sampling could be used to estimate the money due the railroad when the Chesapeake & Ohio is used for only a portion of the freight trip; and a sample of 100 farmers in Boone County, Iowa, taken to estimate the proportion of farmers in the county who had sprayed their cornfields to control the European corn borer.

Statistical bureaus in countries in Western Europe and in this country began to try sampling as a means of saving time and money toward the end of the nineteenth century. Acceptance of sampling took some time, but applications gradually spread as sampling techniques were developed and better understood. Nowadays, most published data except for decennial population counts is collected from samples.

Two simple methods for drawing a sample are introduced in chapter 1. One method is to leave the selection entirely to chance. The members of the population are first listed and numbered. If the members of the sample are to be selected one at a time, use of a table of random digits guarantees that at each draw any member of the population not already drawn has an equal chance of being selected for the sample. The method is called *random sampling without replacement,* or *simple random sampling.*

Simple random sampling is intuitively fair and free from distortion—every member of the population is equally likely to appear in the sample. Its weakness is that it does not use any relevant information or judgment that we have about the nature of the population—such as that people in one part of a city are wealthier than those in another or that farmers in the north of the county may be more likely to spray than those in the south. One method of using such knowledge is *stratified random sampling.* From this knowledge we try to divide the population into subpopulations or strata that are internally more homogeneous. Then we draw a sample separately from each stratum. The appeal of early applications of stratification was that it would make the sample more representative of the population. By selecting the same proportion of the members of each stratum, in the method known as *proportional stratification,* we guarantee that the sample has the correct population proportion of rich and poor members instead of leaving this matter to chance, as simple random sampling does. But Neyman (7) showed in 1934 that sometimes the deliberate selection of different proportions in different strata can give more accurate estimates than proportional stratification without introducing bias. As noted in section 1.6, this was the method used in the sample of waybills.

This chapter introduces some of the principal methods used for selecting a sample and estimating population characteristics from the sample data. We begin with examples of simple and stratified random sampling.

21.2—An example of simple random sampling. The population consists of $N = 6$ members, denoted by the letters *a* to *f*. The six values of the quantity being measured are as follows: *a*, 1; *b*, 2; *c*, 4; *d*, 6; *e*, 7; *f*, 16. The total for this population is 36. A sample of three members is to be drawn to estimate this total.

How good an estimate of the population total do we obtain by simple random sampling? We are not quite ready to answer this question. Although we know how the sample is to be drawn, we have not yet discussed how the population total is to be estimated from the results of the sample. Since the sample contains three members and the population contains six members, the simplest procedure is to multiply the sample total by 2, and this procedure will be adopted. Any sampling plan contains two parts—a rule for drawing the sample and a rule for making the estimates from the results of the sample.

We can now write down all possible samples of size 3, make the estimate from each sample, and see how close these estimates lie to the true value of 36. There are 20 different samples. Their results appear in table 21.2.1, where the successive columns show the composition of the sample, the sample total, the

TABLE 21.2.1
RESULTS FOR ALL POSSIBLE SAMPLE RANDOM SAMPLES OF SIZE THREE

Sample	Sample Total	Estimate of Population Total	Error of Estimate	Sample	Sample Total	Estimate of Population Total	Error of Estimate
abc	7	14	−22	*bcd*	12	24	−12
abd	9	18	−18	*bce*	13	26	−10
abe	10	20	−16	*bcf*	22	44	+ 8
abf	19	38	+ 2	*bde*	15	30	− 6
acd	11	22	−14	*bdf*	24	48	+12
ace	12	24	−12	*bef*	25	50	+14
acf	21	42	+ 6	*cde*	17	34	− 2
ade	14	28	− 8	*cdf*	26	52	+16
adf	23	46	+10	*cef*	27	54	+18
aef	24	48	+12	*def*	29	58	+22
				Average	18	36	0

estimated population total, and the error of estimate (estimate *minus* true value).

Some samples, e.g., *abf* and *cde*, do very well, while others like *abc* give poor estimates. Since we do not know in any individual instance whether we will be lucky or unlucky in the choice of a sample, we appraise any sampling plan by looking at its *average* performance.

The average of the errors of estimate (taking account of their signs) is called the *bias* of the estimate (or, more generally, of the sampling plan). A positive bias implies that the sampling plan gives estimates that are on the whole too high; a negative bias, too low. From table 21.2.1 it is evident that this plan gives unbiased estimates, since the average of the 20 estimates is exactly 36 and consequently the errors of estimate add to 0. With simple random sampling this result holds for any population and any size of sample. Unbiased estimates are a desirable feature of a sampling plan, but a plan that gives a small bias is not ruled out of consideration it has other attractive features.

As a measure of the accuracy of the sampling plan we use the *mean square error* (*MSE*) of the estimates taken about the true population value.

$$MSE = \Sigma(\text{error of estimate})^2/20 = 3504/20 = 175.2$$

The divisor 20 is used instead of 19, because the errors are measured from the true population value. To sum up, this plan gives an estimate of the population total that is unbiased and has a standard error $\sqrt{175.2} = 13.2$. This standard error amounts to 37% of the true population total; evidently the plan is not very accurate for this population.

21.3—An example of stratified random sampling. Suppose that before planning the sample we expect that f will give a much higher value than any other member in the population. How can we use this information? Clearly the

TABLE 21.3.1
RESULTS FOR ALL POSSIBLE STRATIFIED RANDOM SAMPLES WITH THE UNEQUAL SAMPLING FRACTIONS DESCRIBED IN TEXT

Sample	Sample Total in Stratum II, T_2	Estimate $16 + 2.5T_2$	Error of Estimate
abf	3	23.5	−12.5
acf	5	28.5	− 7.5
adf	7	33.5	− 2.5
aef	8	36.0	0.0
bcf	6	31.0	− 5.0
bdf	8	36.0	0.0
bef	9	38.5	+ 2.5
cdf	10	41.0	+ 5.0
cef	11	43.5	+ 7.5
def	13	48.5	+12.5
Average		36.0	0.0

estimate from the sample will depend to a considerable extent on whether f falls in the sample. This statement can be verified from table 21.2.1; every sample containing f gives an overestimate and every sample without f gives an underestimate.

The best plan is to be sure that f appears in every sample. We can do this by dividing the population into two parts or *strata*. Stratum I, which consists of f alone, is completely measured. In stratum II, containing a, b, c, d, and e, we take a simple random sample of size 2 to keep the total sample size equal to 3.

Some forethought is needed in deciding how to estimate the population total. To use twice the sample total, as was done previously, gives too much weight to f and always produces an overestimate of the true total. We can handle this problem by treating the two strata separately. For stratum I we know the correct total, which is 16, since we always measure f. For stratum II, where 2 members are measured out of 5, the natural procedure is to multiply the sample total in that stratum by 5/2 or 2.5. Hence the appropriate estimate of the population total is 16 + 2.5 × (sample total in stratum II).

These estimates are shown for the 10 possible samples in table 21.3.1. Again, we note that the estimate is unbiased. Its mean square error is

$$\Sigma(\text{error of estimate})^2/10 = 487.50/10 = 48.75$$

The standard error is 7.0 or 19% of the true total—a marked improvement over the standard error of 13.2 obtained with simple random sampling.

This sampling plan is called *stratified random sampling with unequal sampling fractions*. The last part of the title denotes the fact that stratum I is completely sampled and stratum II is sampled at a rate of 2 units out of 5, or 40%. Stratification allows us to divide the population into subpopulations or strata that are less variable than the original population and to sample different parts of the population at different rates when this seems advisable.

EXAMPLE 21.3.1—In the preceding example, suppose you expect that both e and f will give high values. You decide that the sample shall consist of e, f, and one member drawn at random from a, b, c, d. Show how to obtain an unbiased estimate of the population total and show that the standard error of this estimate is 7.7. (This sampling plan is not as accurate as the plan in which f alone was placed in a separate stratum, because the actual value for e is not very high.)

EXAMPLE 21.3.2—If previous information suggests that f will be high; d and e moderate; and a, b, and c small, we might try stratified sampling with three strata. The sample consists of f, either d or e, and one chosen from a, b, and c. Work out the unbiased estimate of the population total for each of the six possible samples and show that its standard error is 3.9, much better than that given by our two strata plan.

21.4—Probability sampling. The preceding examples are intended to introduce *probability sampling*. This general name is given to sampling plans in which

(i) every member of the population has a known probability of being included in the sample (*ii*) the sample is drawn by some method of random selection consistent with these probabilities

(*iii*) we take account of these probabilities of selection in making the estimates from the sample

Note that the probability of selection need not be equal for all members of the population; it is sufficient that these probabilities be known. In the example in section 21.2, each member of the population had an equal chance of being in the sample and each member of the sample received an equal weight in estimating the population total. But in the example in section 21.3, member f was given a probability 1 of appearing in the sample, as against 2/5 for the rest of the population. This inequality in the probabilities of selection was compensated for by assigning a weight 5/2 to the other members when making the estimate. The use of unequal probabilities produces a substantial gain in precision for some types of populations (see section 21.8).

Probability sampling has some important advantages. By probability theory it is possible to study the biases and the standard errors of the estimates from different sampling plans. In this way much has been learned about the scope, advantages, and limitations of each plan. This information helps greatly in selecting a suitable plan for a particular sampling job. As will be seen later, most probability sampling plans also enable the standard error of the estimate and confidence limits for the true population value to be computed from the results of the sample. Thus, when a probability sample has been taken, we have some idea as to how accurate the estimates are.

Probability sampling is by no means the only way of selecting a sample. One alternative method is to ask someone who has studied the population to point out average or typical members and then confine the sample to these members. When the population is highly variable and the sample is small, this method often gives more accurate estimates than probability sampling. Another method is to restrict the sampling to those members that are conveniently accessible. If bales of goods are stacked tightly in a warehouse, it is difficult to get at the inside bales of the pile and one is tempted to confine attention to the outside bales. In many biological problems it is hard to see how a workable probability sample can be devised, for instance, as in estimating the number of houseflies in a town, field mice in a wood, or plankton in the ocean.

One drawback of these alternative methods is that when the sample has been obtained, there is no way to determine how accurate the estimate is. Members of the population picked as typical by an expert may be more or less atypical. Outside bales may or may not be similar to interior bales. Probability sampling formulas for the standard error of the estimate or for confidence limits do not apply to these methods. Consequently, it is wise to use probability sampling unless it is clearly not feasible or prohibitively expensive.

In the following sections we give the formulas for the standard errors of the estimates from simple and stratified random sampling.

21.5—Standard errors for simple random sampling. If Y_i ($i = 1, 2, \ldots, N$) denotes the variable being studied, the standard deviation, S, of the population is defined as

$$S = \sqrt{\Sigma(Y_i - \overline{Y})^2/(N - 1)} \qquad (21.5.1)$$

where $\overline{Y}$ is the population mean of the Y_i and the sum Σ is taken over all sampling units in the population. The symbol S is used instead of σ because the population is finite and in (21.5.1) we have divided by $N - 1$ instead of N.

Since $\overline{Y}$ denotes the population mean, we shall use $\overline{y}$ to denote the sample mean. In a simple random sample of size n, the standard error of $\overline{y}$ is (2):

$$\sigma_{\overline{y}} = (S/\sqrt{n})\sqrt{1 - \phi} \qquad (21.5.2)$$

where $\phi = n/N$ is the *sampling fraction,* i.e., the fraction of the population that is included in the sample.

The term $\sigma/\sqrt{n}$ or in sample survey notation $S/\sqrt{n}$ is already familiar to you; it is the usual formula for the standard error of a sample mean. The factor $\sqrt{1 - \phi}$ is known as the *finite population correction.* It enters because we are sampling from a population of finite size N instead of from an infinite population as assumed in the usual theory. Note that this term makes the standard error zero when $n = N$, as it should do, since we have then measured every unit in the population. In practical applications the finite population correction is close to 1 and can be omitted when n/N is less than 10%, i.e., when the sample includes less than 10% of the population.

This result is very remarkable. In a large population with a fixed amount of variability (a given value of S), the standard error of the mean depends mainly on the size of sample and only to a minor extent on the fraction of the population sampled. For a given S, the mean of a sample of 100 is almost as precise when the population size is 200,000 as when the population size is 20,000 or 2000. Some people intuitively feel that one cannot possibly get accurate results from a sample of 100 from a population of 200,000, because only a tiny fraction of the population has been measured. Actually, whether the sampling plan is accurate or not depends primarily on the size of $S/\sqrt{n}$. This shows why sampling can bring about a great reduction in the amount of measurement needed.

For the *estimated* standard error of the sample mean we have

$$s_{\overline{y}} = (s/\sqrt{n})\sqrt{1 - \phi} \qquad (21.5.3)$$

where s is the standard deviation of the sample, calculated in the usual way.

If the sample is used to estimate the population *total* of the variable under study, the estimate is $N\overline{y}$ and its estimated standard error is

$$s_{N\overline{y}} = (Ns/\sqrt{n})\sqrt{1 - \phi} \qquad (21.5.4)$$

In simple random sampling for attributes, where every member of the sample is classified into one of two classes, we take

$$s_p = \sqrt{pq/n}\sqrt{1 - \phi} \qquad (21.5.5)$$

where p is the proportion of the sample that lies in one of the classes. Suppose that 50 families are picked at random from a list of 432 families who have telephones and that 10 of the families report they are listening to a certain radio program. Then $p = 0.2$, $q = 0.8$, and

$$s_p = \sqrt{(0.2)(0.8)/50}\,\sqrt{1 - 50/432} = 0.053$$

If we ignore the finite population correction, we find $s_p = 0.057$.

The formula for s_p holds only *if each sampling unit is classified as a whole into one of the two classes*. If your sampling unit is a group of elements and you are classifying individual elements within each group, a different formula for s_p must be used. For instance, in estimating the percentage of diseased plants in a field from a sample of 360 plants, the formula above holds if the plants were selected independently and at random. To save time in the field, however, we might have chosen 40 areas, each consisting of 3 plants in each of 3 neighboring rows. With this method the area (a group or cluster of plants) is the sampling unit. If the distribution of disease in the field were extremely patchy, it might happen that every area had either all plants diseased or no plants diseased. In this event the sample of 40 areas would be no more precise than a sample of 40 independently chosen plants, and we would be deceiving ourselves if we thought that we had a binomial sample of 360 plants.

The correct procedure for computing s_p in this case is simple. Calculate p separately for each sampling unit and apply formula (21.5.3) for continuous variates to these ps. That is, if p_i is the percentage diseased in the ith area, the sample standard deviation is

$$s = \sqrt{\Sigma(p_i - p)^2/(n - 1)}$$

where n is now the number of areas (cluster units). Then, by (21.5.3),

$$s_p = (s/\sqrt{n})\sqrt{1 - \phi}$$

For instance, suppose that the numbers of diseased plants in the 40 areas were as given in table 21.5.1. The standard deviation of the numbers of diseased plants in this sample is 2.331. Since the *proportions* of diseased plants in the 40 areas are found by dividing the numbers in table 21.5.1 by 9, the standard deviation of the proportions is $s = 2.331/9 = 0.259$. Hence (assuming N large),

$$s_p = s/\sqrt{n} = 0.259/\sqrt{40} = 0.041$$

For comparison, the result given by the binomial formula is worked out.

TABLE 21.5.1
NUMBERS OF DISEASED PLANTS (OUT OF 9) IN EACH OF 40 AREAS

2	5	1	1	1	7	0	0	3	2	3	0	0	0	7	0	4	1	2	6
0	0	1	4	5	0	1	4	2	6	0	2	4	1	7	3	5	0	3	6

grand total = 99

From the total in table 21.5.1, $p = 99/360 = 0.275$. The binomial formula is

$$s_p = \sqrt{pq/360} = \sqrt{(0.275)(0.725)/360} = 0.024$$

giving an overoptimistic notion of the precision of p.

EXAMPLE 21.5.1—If a sample of 4 from the 16 townships of a county has a standard deviation 45, show that the standard error of the mean is 19.5.

EXAMPLE 21.5.2—In the example in section 21.2, $N = 6$, $n = 3$, and the values for the 6 members of the population were 1, 2, 4, 6, 7, and 16. The formula for the true standard error of the estimated population total is

$$\sigma_{N\bar{y}} = (Ns/\sqrt{n})\sqrt{1 - n/N}$$

Verify that this formula agrees with the result, 13.2, which we found by writing down all possible samples.

EXAMPLE 21.5.3—A simple random sample of size 100 is taken to estimate some proportion (e.g., the proportion of males) whose value in the population is close to 1/2. Work out the standard error of the sample proportion p when the size of the population is (*i*) 200, (*ii*) 500, (*iii*) 1000, (*iv*) 10,000, (*v*) 100,000. Note how little the standard error changes for N greater than 1000.

EXAMPLE 21.5.4—Show that the coefficient of variation of the sample mean is the same as that of the estimated population total.

EXAMPLE 21.5.5—In simple random sampling for attributes, show that the standard error of p for given N and n is greatest when p is 50%, but that the coefficient of variation of p is largest when p is very small.

21.6—Size of sample. At an early stage in the design of a sample, the question, How large a sample do I need? must be considered. Although a precise answer may not be easy to find (for reasons that will appear), there is a rational method of attack on the problem. At present we assume simple random sampling and ignore n/N.

Clearly, we want to avoid making the sample so small that the estimate is too inaccurate to be useful. Equally, we want to avoid taking a sample that is too large so that the estimate is more accurate than we require. Consequently, the first step is to decide how large an error we can tolerate in the estimate. This demands careful thinking about the use to be made of the estimate and the consequences of a sizable error. The figure finally reached may be to some extent arbitrary, yet after some thought samplers often find themselves less hesitant about naming a figure than they expected.

The next step is to express the allowable error in terms of confidence limits. Suppose that L is the allowable error in the sample mean and we are willing to take a 5% chance that the error will exceed L. In other words, we want to be reasonably certain that the error will not exceed an amount $\pm L$. Remembering that the 95% confidence limits computed from a sample mean, assumed approximately normally distributed, are

$$\bar{y} \pm 2S/\sqrt{n}$$

we put $L = 2S/\sqrt{n}$. This gives, for the required sample size,

$$n = 4S^2/L^2 \qquad (21.6.1)$$

To use this relation, we must have an estimate of the population standard deviation S. Often a good guess can be made from the results of previous samplings of this population or of similar populations. For example, an experimental sample was taken in 1938 to estimate the yield per acre of wheat in certain districts of North Dakota (7). For a sample of 222 fields, the variance of the yield per acre from field to field was $s^2 = 90.3$ (in bu^2). How many fields are indicated if we wish to estimate the true mean yield within ± 1 bu, with a 5% risk that the error will exceed 1 bu? Then

$$n = 4s^2/L^2 = 4(90.3)/1^2 = 361 \text{ fields}$$

If this estimate were being used to plan a sample in a later year, it would be regarded as tentative, since the variance between fields might change from year to year.

In default of previous estimates, Deming (3) has pointed out that σ can be estimated from a knowledge of the highest and lowest values in the population and a rough idea of the shape of the distribution. If h = highest − lowest, then $\sigma = 0.29h$ for a uniform (rectangular) distribution, $\sigma = 0.24h$ for a symmetrical distribution shaped like an isosceles triangle, and $\sigma = 0.21h$ for a skew distribution shaped like a right triangle.

If the quantity to be estimated is a binomial proportion, the allowable error L for 95% confidence probability is

$$L = 2\sqrt{pq/n}$$

The sample size required to attain a given limit of error L is therefore

$$n = 4pq/L^2 \qquad (21.6.2)$$

In this formula, p, q, and L may be expressed either as proportions or as percentages, provided they are all expressed in the same units. The result necessitates an advance estimate of p. If p is likely to lie between 35% and 65%, the advance estimate can be quite rough, since the product pq varies little for p lying between these limits. If, however, p is near 0% or 100%, accurate determination of n requires a close guess about the value of p.

If the computed value of n is found to be more than 10% of the population size N, a revised value n' that takes proper account of the finite population fraction is obtained from the relation

$$n' = n/(1 + \phi) \qquad (21.6.3)$$

For example, casual inspection of a batch of 480 seedlings indicates that about 15% are diseased. Suppose we wish to know the size of sample needed to determine p, the percent diseased, to within $\pm 5\%$, apart from a 1-in-20 chance.

Formula 21.6.2 gives $n = 4(15)(85)/25 = 204$ seedlings. At this point we might decide that it would be as quick to classify every seedling as to plan a sample that is over 40% of the whole batch. If we decide on sampling, we make a revised estimate n':

$$n' = \frac{n}{1 + \phi} = \frac{204}{1 + 204/480} = 143$$

The preceding formulas assume simple random sampling, which has only limited use in practice. When a more complex plan such as stratified random sampling is employed, a useful quantity known as the design effect of the plan enables simple random sampling formulas to be used more extensively. Kish (5) defines the *design effect* (deff) of a complex plan as the ratio of the variance of the estimate given by the complex plan to the variance of the estimate given by a simple random sample of the same size. The design effects of many plans in common use can be estimated from their sample results. Suppose that a plan has given deff = 2 in recent applications. If we want confidence limits $\pm L$ when using this plan, we first calculate the sample size needed with a simple random sample. Then, if the finite population correction is negligible, we multiply the sample size by 2, or in general by deff, for use with the more complex plan.

EXAMPLE 21.6.1—A simple random sample of houses is to be taken to estimate the percentage of houses that are unoccupied. The estimate is desired to be correct to within $\pm 1\%$, with 95% confidence. One advance estimate is that the percentage of unoccupied houses will be about 6%; another is that it will be about 4%. What sizes of sample are required on these two forecasts? What size would you recommend?

EXAMPLE 21.6.2—The total number of rats in the residential part of a large city is to be estimated with an error of not more than 20%, apart from a 1-in-20 chance. In a previous survey, the mean number of rats per city block was 9 and the sample standard deviation was 19 (the distribution is extremely skew). Show that a simple random sample of around 450 blocks should suffice.

EXAMPLE 21.6.3—West (9) quotes the following data for 556 full-time farms in Seneca County, New York.

	Mean	Standard Deviation per Farm
Acres in corn	8.8	9.0
Acres in small grains	42.0	39.5
Acres in hay	27.9	26.9

If a coefficient of variation of up to 5% can be tolerated, show that a random sample of about 240 farms is required to estimate the total acreage of each crop in the 556 farms with this degree of precision. (Note that the finite population correction must be used.) This example illustrates a result that has been reached by several different investigators; with small farm populations such as counties, a substantial part of the whole population must be sampled to obtain accurate estimates.

21.7—Standard errors for stratified random sampling. The three steps in stratified random sampling are:

1. The population is divided into a number of parts, called *strata*.
2. A random sample is drawn independently in each part.
3. As an estimate of the population mean, we use

$$\bar{y}_{st} = \Sigma N_h \bar{y}_h / N \tag{21.7.1}$$

where N_h is the total number of sampling units in the hth stratum, $\bar{y}_h$ is the sample mean in the hth stratum, and $N = \Sigma N_h$ is the size of the population. Note that we must know the values of the N_h (i.e., the sizes of the strata) in order to compute this estimate.

Stratification is commonly employed in sampling plans for several reasons. Differences between the strata means in the population do not contribute to the sampling error of the estimate $\bar{y}_{st}$; it arises solely from variations among sampling units that are in the same stratum. If we can form strata so that a heterogeneous population is divided into parts, each of which is fairly homogeneous, we may expect a substantial gain in precision over simple random sampling. In stratified sampling, we can choose the size of sample to be taken from any stratum. This freedom of choice gives us scope to do an efficient job of allocating resources to the sampling within strata. Furthermore, when different parts of the population present different problems of listing and sampling, stratification enables these problems to be handled separately. For this reason hotels and large apartment houses are frequently placed in a separate stratum in a sample of the inhabitants of a city.

We now consider the estimate from stratified sampling and its standard error. For the population mean, estimate (21.7.1) may be written

$$\bar{y}_{st} = (1/N)\Sigma N_h \bar{y}_h = \Sigma W_h \bar{y}_h$$

where $W_h = N_h/N$ is the relative *weight* attached to the stratum. Note that the sample means $\bar{y}_h$ in the respective strata are weighted by the sizes N_h of the strata. The arithmetic mean of the sample observations is no longer the estimate of the population mean except with proportional stratification. If n_h/N_h = constant = n/N, as in proportional stratification, it follows that $W_h = N_h/N = n_h/n$ so that in (21.7.1) the estimate $\bar{y}_{st}$ becomes

$$\bar{y}_{st} = \Sigma W_h \bar{y}_h = \Sigma n_h \bar{y}_h / n = \bar{y} \tag{21.7.2}$$

since $\Sigma n_h \bar{y}_h$ is the total of all observations in the sample.

Since a simple random sample is drawn in each stratum, (21.5.2) gives

$$V(\bar{y}_h) = S_h^2(1 - \phi_h)/n_h \tag{21.7.3}$$

where $\phi_h = n_h/N_h$ is the sampling fraction in the hth stratum. Also, since sampling is independent in different strata and the W_h are known numbers,

$$V(\bar{y}_{st}) = \sum_h W_h^2 S_h^2(1 - \phi_h)/n_h \tag{21.7.4}$$

For the estimated standard error of $\bar{y}_{st}$, this gives

$$s(\bar{y}_{st}) = \sqrt{\sum_h W^2 s_h^2 (1 - \phi_h)/n_h} \qquad (21.7.5)$$

where s_h^2 is the sample variance in the hth stratum.

In the example to be presented, the L strata were of equal size, so $W_h = 1/L$; and proportional allocation was used, giving $n_h = n/L$, $\phi_h = n/N = \phi$. In this case, (21.7.5) reduces to $\sqrt{\Sigma s_h^2/nL}\sqrt{1 - \phi}$, or

$$s(\bar{y}_{st}) = (s_w/\sqrt{n})\sqrt{1 - \phi}$$

where s_w^2 is the average within-stratum mean square in the analysis of variance of the sample data as a one-way classification.

The data in table 21.7.1 come from an early investigation by Clapham (1) of the feasibility of sampling for estimating the yields of small cereal plots. A rectangular plot of wheat was divided transversely into three equal strata. Ten samples, each a meter length of a single row, were chosen by simple random sampling from each stratum. The problem is to compute the standard error of the estimated mean yield per meter of row.

In this example, $s_w = \sqrt{240.4} = 15.5$ and $n = 30$. Since the sample is only a negligible part of the whole plot, n/N is negligible and

$$s(\bar{y}_{st}) = s_w/\sqrt{n} = 15.5/\sqrt{30} = 2.83 \text{ g}$$

How effective was the stratification? In the analysis of variance, the mean square between strata is over four times as large as that within strata. This is an indication of real differences in level of yield from stratum to stratum. It is possible to go further and estimate what the standard error of the mean would have been if simple random sampling had been used without any stratification. With simple random sampling, the corresponding formula for the standard error of the mean is

$$s_{\bar{y}} = s/\sqrt{n}$$

where s is the ordinary sample standard deviation. In the sample under discussion, s is $\sqrt{295.3}$ (from the *total* mean square in table 21.7.1). Hence, as an estimate of the standard error of the mean under simple random sampling we might take $s_{\bar{y}} = \sqrt{295.3}/\sqrt{30} = 3.14$ g, as compared with 2.83 g for stratified

TABLE 21.7.1

ANALYSIS OF VARIANCE OF A STRATIFIED RANDOM SAMPLE

(wheat grain yields, g/m)

Source of Variation	Degrees of Freedom	Sum of Squares	Mean Square
Between strata	2	2073	1036.5
Within strata	27	6491	240.4
Total	29	8564	295.3

random sampling. Stratification has reduced the standard error by about 100%. The *design effect* of the stratified plan, as described in section 21.6, is deff = 240.4/295.3 = 0.81.

This comparison is not quite correct for the reason that the value of s was calculated from the results of a stratified sample and not, as it should have been, from the results of a simple random sample. The approximate method that we used is close enough, however, when stratification is proportional and at least ten sampling units are drawn from every stratum.

EXAMPLE 21.7.1—In the example of stratified sampling given in section 21.3, show that the estimate that we used for the population total was $N\bar{y}_{st}$. From (21.7.3), verify that the variance of the estimated population total is 48.75, as found directly in section 21.3. (Note that stratum I makes no contribution to this variance because $n_h = N_h$ in that stratum.)

21.8—Choice of sample sizes in the strata. It is sometimes thought that in stratified sampling we should sample the same fraction from every stratum; i.e., we should make n_h/N_h the same in all strata, using proportional allocation. A more thorough analysis of the problem shows, however, that the *optimum* allocation is to take n_h proportional to $N_h S_h/\sqrt{c_h}$, where S_h is the standard deviation of the sampling units in the hth stratum and c_h is the cost of sampling per unit in the hth stratum. This method of allocation gives the smallest standard error of the estimated mean $\bar{y}_{st}$ for a given total cost of taking the sample. The rule tells us to take a larger sample, as compared with proportional allocation, in a stratum that is unusually variable (S_h large) and a smaller sample in a stratum where sampling is unusually expensive (c_h large). Looked at in this way, the rule is consistent with common sense. The rule reduces to proportional allocation when the standard deviation and the cost per unit are the same in all strata.

To apply the rule, advance estimates of the relative standard deviations and of the relative costs in different strata are needed. These estimates need not be highly accurate; rough estimates often give results satisfactorily near the optimum allocation. When a population is sampled repeatedly, estimates can be obtained from the results of previous samplings. Even when a population is sampled for the first time, it is sometimes obvious that some strata are more accessible to sampling than others. In this event it pays to hazard a guess about the differences in costs. In other situations we are unable to predict with any confidence which strata will be more variable or more costly, or we think that any such differences will be small. Proportional allocation is then used.

Disproportionate sampling pays large dividends when the principal variable being measured has a highly skewed or asymmetrical distribution. Usually such populations contain a few sampling units that have large values for this variable and many units that have small values. Variables that are related to the sizes of economic institutions are often of this type, for instance, the total sales of grocery stores, the number of patients per hospital, the amounts of butter produced by creameries, family incomes, and prices of houses.

With populations of this type, stratification by size of institution is highly effective and the optimum allocation is likely to be much better than propor-

TABLE 21.8.1
DATA FOR TOTAL REGISTRATIONS PER SENIOR COLLEGE OR UNIVERSITY, ARRANGED IN FOUR STRATA

Stratum: Number of Students per Institution	Number of Institutions N_h	Total Registration for the Stratum	Mean per Institution $\bar{Y}_h$	Standard Deviation per Institution S_h
Less than 1000	661	292,671	443	236
1000–3000	205	345,302	1,684	625
3000–10,000	122	672,728	5,514	2,008
Over 10,000	31	573,693	18,506	10,023
Total	1019	1,884,394		3,860

tional allocation. As an illustration, table 21.8.1 shows data for the number of students per institution in a population consisting of the 1019 senior colleges and universities in the United States. The data, which apply mostly to the 1952–1953 academic year, might be used as background information for planning a sample designed to give a quick estimate of total registration in some future year. The institutions are arranged in four strata according to size.

Note that the 31 largest universities (about 3% in number) have 30% of the students, while the smallest group (which contains 65% of the institutions) contributes only 15% of the students. Note also that the within-stratum standard deviation S_h increases rapidly with increasing size of institution.

Table 21.8.2 shows the calculations needed for choosing the optimum sample sizes within strata. We are assuming equal costs per unit within all strata. The products $N_h S_h$ are formed and added over all strata. Then the relative sample sizes, $N_h S_h / \Sigma N_h S_h$, are computed. These ratios when multiplied by the intended sample size n give the sample sizes in the individual strata.

As a consequence of the large standard deviation in the stratum with the largest universities, the rule requires 37% of the sample to be taken from this stratum. Suppose we are aiming at a total sample size of 250. The rule then calls for (0.37)(250) or 92 universities from this stratum, although the stratum contains only 31 universities in all. With highly skewed populations, as here, the optimum allocation may demand 100% sampling, or even more, of the largest institutions. When this situation occurs, a good procedure is to take 100% of the

TABLE 21.8.2
CALCULATIONS FOR OBTAINING THE OPTIMUM SAMPLE SIZES IN INDIVIDUAL STRATA

Stratum: Number of Students	Number of Institutions N_h	$N_h S_h$	Relative Sample Sizes $N_h S_h / \Sigma N_h S_h$	Actual Sample Sizes	Sampling Rate (%)
Less than 1000	661	155,996	0.1857	65	10
1000–3000	205	128,125	.1526	53	26
3000–10,000	122	244,976	.2917	101	83
Over 10,000	31	310,713	0.3700	31	100
Total	1019	839,810	1.0000	250	

large stratum and employ the rule $n_h \propto N_h S_h$ to distribute the remainder of the sample over the other strata. Following this procedure, we include in the sample all 31 of the largest institutions, leaving 219 to be distributed among the first three strata. In the first stratum, the size of sample is

$$219[0.1857/(0.1857 + 0.1526 + 0.2917)] = 65$$

The allocations (second column from the right of table 21.8.2) call for over 80% sampling in the second largest group of institutions (101 out of 122) but only a 10% sample of the small colleges. In practice we might decide for administrative convenience to take a 100% sample in the second largest group as well as in the largest.

Is the optimum allocation much superior to proportional allocation? From tables 21.8.1 and 21.8.2 and the sampling error formulas, we can calculate the standard errors of the estimated population totals $N\bar{y}_{st}$ or $N\bar{y}$ by stratification with optimum allocation or with proportional allocation, and by simple random sampling. These standard errors are:

Sampling Plan	$s(\hat{Y})$
Simple random sampling	216,000
Stratification, proportional allocation	107,000
Stratification, optimum allocation	26,000

The reduction in the standard error due to stratification and the further reduction due to optimum allocation are both striking.

If every unit lies in one or the other of two classes, (e.g., sprayed, not sprayed), the estimate of the population proportion p_{st} in one of the classes is

$$p_{st} = \Sigma W_h p_h \tag{21.8.1}$$

where p_h is the sample proportion in stratum k and $W_h = N_h/N$ as before. To obtain the estimated standard error of p_{st}, substitute $p_h q_h$ for s_h^2 in (21.7.5).

For the optimum choice of sample sizes within strata, take n_h proportional to $N_h \sqrt{p_h q_h/c_h}$. If c_h is about the same in all strata, this rule implies that the fraction sampled, n_h/N_h, should be proportional to $\sqrt{p_h q_h}$. Now the quantity $\sqrt{pq}$ changes little as p ranges from 25% to 75%. Consequently, proportional allocation is nearly optimal if the strata proportions lie in this range.

EXAMPLE 21.8.1—From the data in table 21.8.1, verify the standard error of 107,000 reported for the estimated total registration as given by a stratified random sample with $n = 250$ and proportional allocation.

EXAMPLE 21.8.2—A sample of 692 families in Iowa was taken to determine among other things the percentage of families with vegetable gardens in 1943. The families were classified into three strata—urban, rural nonfarm, and farm—because these groups were expected to show differences in the frequency and size of vegetable gardens. The sampling fraction was roughly the same in all strata, a sample of 1 per 1000 being aimed at. The values of W_h, n_h, and the numbers and percentages with gardens are as follows:

Stratum	W_h	Sample Size n_h	Number with Gardens	Percent with Gardens
Urban	0.445	300	218	72.7
Rural nonfarm	0.230	155	147	94.8
Farm	0.325	237	229	96.6
Total	1.000	692	594	85.8

The finite population corrections can, of course, be ignored. (*i*) Calculate the estimated population percent p_{st} with gardens and give its standard error. (*ii*) If the costs c_h are constant, find the optimum sample sizes in the strata and the resulting $s(p_{st})$. Assume the sample p_h are the same as those in the population. Note that the optimum $n_h = nW_h\sqrt{p_h q_h}/\Sigma W_h\sqrt{p_h q_h}$, where $q_h = 100 - p_h$ when p_h is expressed in percent. (*iii*) Estimate approximately the value of $s(p)$ given by a simple random sample with $n = 692$. Ans. (*i*) $p_{st} = 85.6\%$, $SE = \pm 1.27\%$. (*ii*) Optimum $n_h = 445, 115, 132$; $SE(p_{st}) = \pm 1.17\%$. (*iii*) $SE \doteq \sqrt{(85.8)(14.2)/692} = \pm 1.33$. The gain in precision due to stratification and the further gain due to optimum allocation are both modest. The deff factors are 0.91 and 0.77 for stratification with proportional and optimal allocation.

21.9—Systematic sampling. To draw a 10% sample from a list of 730 cards, we might select a random number between 1 and 10, say 3, and pick every tenth card thereafter, i.e., the cards numbered 3, 13, 23, and so on, ending with card number 723. A sample of this kind is known as a *systematic sample,* since the choice of its first member, 3, determines the whole sample.

Systematic sampling has two advantages over simple random sampling. It is easier to draw, since only one random number is required, and it distributes the sample more evenly over the listed population. It has a built-in stratification. In our example, cards 1–10, 11–20, etc., in effect form strata, one sampling unit being drawn from each stratum. Systematic sampling differs, however, from stratified random sampling in that the unit from the stratum is not drawn at random; in our example, this unit is always in the third position. Systematic sampling often gives substantially more accurate estimates than simple random sampling and has become popular, for example, with samples taken regularly for inspection and control of quality in mass production.

Systematic sampling has one disadvantage and one potential disadvantage. It has no reliable method of estimating the standard error of the sample mean (the formula for stratified sampling cannot be used, since only one unit is drawn per stratum). Some formulas work well for particular types of populations but cannot be trusted for general application. However, systematic sampling is often a part of a more complex sampling plan such as two-stage sampling in which unbiased estimates of the sampling errors can be obtained.

A potential disadvantage is that if the population contains a periodic type of variation and if the interval between successive units in the sample happens to equal the wavelength or a multiple of it, the sample may be badly biased. For instance, a systematic sample of the houses in a city might contain too many (or too few) corner houses; a systematic sample of families listed by name might contain too many male heads of households or too many children. These situations can be guarded against by changing the random start number frequently.

EXAMPLE 21.9.1—In estimating mean per capita income per state, we might list the 48 U.S. states (excluding Hawaii and Alaska) in order from east to west, putting neighboring states near one another in the sequence, and draw a systematic sample of 1 in 4, with $n = 12$ states. For 1976 incomes (in $1000s), the following data give the four systematic samples.

Sample													Total
1	5.4	6.5	6.5	5.4	6.4	6.1	6.4	6.3	5.4	6.2	7.3	5.6	73.5
2	6.0	7.4	7.3	5.4	6.3	5.1	6.2	6.2	5.4	5.2	7.2	5.7	73.4
3	5.5	7.1	7.0	5.1	7.4	4.6	6.5	5.4	5.1	5.8	6.8	6.7	73.0
4	6.6	7.3	6.3	5.6	7.0	5.4	6.0	4.8	5.7	6.5	6.3	5.5	73.0

The population total is 292.9, the mean $\overline{Y}$ per state is 6102, and the variance S^2 is 0.5632 with 47 df. Compare the standard errors of the estimated mean per capita income per state as given by (*i*) the systematic sample, (*ii*) a simple random sample of 12 states, (*iii*) a stratified random sample with 12 strata and 1 unit per stratum. Note that for the systematic sample, $V(\overline{y}_{sy}) = \Sigma(\overline{y}_{sy} - \overline{Y})^2/4$. Ans. (*i*) $SE(\overline{y}_{sy}) = \pm 0.0190$, (*ii*) $SE(\overline{y}) = \pm 0.1876$, (*iii*) $SE(\overline{y}_{st}) = \pm 0.1212$. Why the systematic sample does so much better than stratified random sampling is puzzling.

21.10—Two-stage sampling. Consider the following miscellaneous group of sampling problems: (1) a study of the vitamin A content of butter produced by creameries, (2) a study of the protein content of wheat in the wheat fields in an area, (3) a study of red blood cell counts in a population of men aged 20–30, (4) a study of insect infestation of the leaves of the trees in an orchard, and (5) a study of the number of defective teeth in third-grade children in the schools of a large city. What do these investigations have in common? First, in each study an appropriate sampling unit suggests itself naturally—the creamery, the field of wheat, the individual man, the tree, and the school. Second, and this is the important point, in each study the chosen sampling units can be *subsampled* instead of being measured completely. Indeed, subsampling is essential in the first three studies. Obviously, we cannot examine *all* the blood in a man in order to make a complete count of his red cells. In the insect infestation study, it might be feasible, although tedious, to examine *all* leaves on any selected tree. If the insect distribution is spotty, however, we would probably decide to take only a small sample of leaves from any selected tree so as to include more trees.

This type of sampling is called *two-stage sampling.* The first stage is the selection of a sample of *primary sampling units*—the creameries, wheat fields, and so on. The second stage is the taking of a subsample of *second-stage units,* or *subunits,* from each selected primary unit.

As illustrated by these examples, the two-stage method is sometimes the only practicable way the sampling can be done. Even with a choice between subsampling the units and measuring them completely, two-stage sampling gives the sampler greater scope, since both the size of the sample of primary units and the size of the sample that is taken from a primary unit can be chosen by the sampler. In some applications an important advantage of two-stage sampling is that it facilitates the problem of listing the population. Often it is relatively easy to obtain a list of the primary units but difficult or expensive to list all the subunits.

Listing is an important problem that we have not discussed. To use probability sampling, we must have in effect a complete list of the sampling units in the population in order to select a sample according to our randomized plan.

In the national sample mentioned in section 1.3 for estimating unemployment, the primary unit in urban areas is what is called a standard metropolitan area; in rural areas it is a county or a group of small contiguous counties. These units have all been defined and listed. No list of all the households or families in the country exists. To list the trees in an orchard and draw a sample of them is usually simple, but the problem of making a random selection of the leaves on a tree may be very troublesome. With two-stage sampling this problem is faced only for those trees that are in the sample. No complete listing of all leaves in the orchard is required.

Assume for simplicity that the primary units are of equal size. The population contains N_1 primary units, each containing N_2 second-stage units or subunits. A random sample of n_1 primary units is drawn. From each selected primary unit, n_2 subunits are drawn at random. If the sampling fractions n_1/N_1 and n_2/N_2 are small, we can apply to our results the random effects model (section 13.3) for a one-way classification, the primary units being the classes. Considered as an estimate of the population mean, the observation on any subunit is the sum of two independent terms. One term, associated with the primary unit, has the same value for all subunits in the primary unit and varies from one primary unit to another with variance s_1^2. The second term measures differences between subunits in the same primary unit and has variance s_2^2.

The sample as a whole contains n_1 independent values of the first term and $n_1 n_2$ independent values of the second term. Hence the variance of the sample mean per subunit is

$$V(\bar{y}) = s_1^2/n_1 + s_2^2/(n_1 n_2) \tag{21.10.1}$$

Furthermore, as shown in section 13.3, the two components of variance s_1^2 and s_2^2 can be estimated from an analysis of variance of the sample results, as given in table 21.10.1. It follows from table 21.10.1 that an unbiased sample estimate of $V(\bar{y})$ in (21.10.1) is

$$\hat{V}(\bar{y}) = s_1^2/(n_1 n_2) = \Sigma(\bar{y}_i - \bar{y})^2/[n_1(n_1 - 1)] \tag{21.10.2}$$

When n_1/N_1 is negligible, it can be shown that this very simple result holds also (*i*) if the second-stage sampling fraction n_2/N_2 is not negligible; (*ii*) if the second-stage variance differs from one primary unit to another; and (*iii*) if the second-stage samples are drawn systematically, provided that the random start is chosen independently in each sample primary unit.

As pointed out in section 13.3, the analysis of variance (table 21.10.1) is

TABLE 21.10.1
ANALYSIS OF VARIANCE FOR A TWO-STAGE SAMPLE (SUBUNIT BASIS)

Source of Variation	df	Mean Square	Expected Value
Between primary units (p.u.)	$n_1 - 1$	s_1^2	$S_2^2 + n_2 S_1^2$
Between subunits within p.u.	$n_1(n_2 - 1)$	s_2^2	S_2^2
	$\hat{S}_1^2 = (s_1^2 - s_2^2)n_1$	$\hat{S}_2^2 = s_2^2$	

TABLE 21.10.2
ANALYSIS OF VARIANCE OF SUGAR PERCENT OF BEETS (ON A SINGLE BEET BASIS)

Source of Variation	df	Mean Square	Expected Value
Between plots (primary units)	80	2.9254	$S_2^2 + 10S_1^2$
Between beets (subunits) within plots	900	2.1374	S_2^2
$\hat{S}_1^2 = (2.9254 - 2.1374)/10 = 0.0788$		$\hat{S}_2^2 = 2.1374$	

useful as a guide in choosing values of n_1 and n_2 for future samples. Table 21.10.2 gives the analysis of variance in a study by Immer (6), whose object was to develop a sampling technique for the determination of the sugar percentage in sugar beets in field experiments. Ten beets were chosen from each of 100 plots in a uniformity trial; the plots were the primary units. The sugar percentage was obtained separately for each beet. To simulate conditions in field experiments, the between-plots mean square was computed as the mean square between plots within blocks of 5 plots. This mean square gives the experimental error variance that would apply in a randomized blocks experiment with five treatments.

Hence, if a new experiment is to consist of n_1 replications with n_2 beets sampled from each plot, the predicted variance of a treatment mean is, from the variance estimates in table 21.10.2

$$s_{\bar{y}}^2 = 0.0788/n_1 + 2.1374/(n_1 n_2)$$

We shall illustrate three of the questions that can be tackled from these data. How accurate are the treatment means in an experiment with 6 replications and 5 beets per plot? For this experiment we would expect

$$s_{\bar{y}} = \sqrt{0.0788/6 + 2.1374/30} = 0.29\%$$

The sugar percentage figure for a treatment mean would be correct to within $\pm(2)$ (0.29) or 0.58%, with 95% confidence, assuming $\bar{y}$ approximately normally distributed.

If the standard error of a treatment mean is not to exceed 0.2%, what combinations of n_1 and n_2 are allowable? We must have

$$0.0788/n_1 + 2.1374/(n_1 n_2) \leq 0.2^2 = 0.04$$

You can verify that with 4 replications ($n_1 = 4$), there must be 27 beets per plot; with 8 replications, 9 beets per plot are sufficient; and with 10 replications, 7 beets per plot. As one would expect, the intensity of subsampling decreases as the intensity of sampling is increased. The total size of sample also decreases from 108 beets when $n_1 = 4$ to 70 beets when $n_1 = 10$.

Some surveys entail a cost c_1 of selecting and getting access to a primary unit to sample it and a cost c_2 of selecting and measuring each sample subunit. Thus, apart from overhead cost, the cost of taking the sample is

$$C = c_1 n_1 + c_2 n_1 n_2 \qquad (21.10.3)$$

In section 13.3 it is noted that for a given total cost, the value of n_2 that minimizes $V(\bar{y})$ is

$$n_2 = \sqrt{c_1 S_2^2/(c_2 S_1^2)} \qquad (21.10.4)$$

With the sugar beets, $\sqrt{S_2^2/S_1^2} = \sqrt{2.1374/0.0788} = 5.2$, giving $n_2 = 5.2\sqrt{c_1/c_2}$.

In this study, cost data were not reported. If c_1 were to include the cost of the land and the field operations required to produce one plot, it would be much greater than c_2. Evidently a fairly large number of beets per plot would be advisable. In practice, factors other than the sugar percentage determinations must also be considered in deciding on costs and number of replications in sugar beet experiments.

21.11—Selection with probability proportional to size. In many important sampling problems the natural primary sampling units are of unequal sizes. Schools, hospitals, and factories all contain different numbers of individuals. A sample of the houses in a town may use blocks as first-stage units, the number of houses per block ranging from 0 to around 40. In national surveys the primary unit is often an administrative area—a county or a metropolitan district. A relatively large unit of this type cuts down travel costs and makes supervision and control of the field work more manageable.

When primary units vary in size, Hansen and Hurwitz (8) pointed out the advantages of selecting primary units with probabilities proportional to their sizes. To illustrate, consider a population with $N = 3$ schools having 600, 300, and 100 children. A 5% sample of 50 children is to be taken to estimate the population mean per child for some characteristic. The means per child in the three schools are $\bar{Y}_1 = 2$, $\bar{Y}_2 = 4$, $\bar{Y}_3 = 1$. Hence, the population mean per child is

$$\bar{\bar{Y}} = [(600)(2) + (300)(4) + (100)(1)]/1000 = 2.5 \qquad (21.11.1)$$

For simplicity, suppose that only one school is chosen, the 50 children are drawn at random from the selected school, and the variation in Y between children in the same school is negligible. Thus the mean $\bar{y}$ of any sample is equal to the mean of the school from which it is drawn.

In selecting the school with probability proportional to size (pps), the three schools receive probabilities 0.6, 0.3, and 0.1, respectively, of being drawn. We shall compare the mean square error of the estimate given by this method with that given by selecting the schools with equal probabilities. Table 21.11.1 contains the calculations.

If the first school is selected, the sample estimate is in error by $2.0 - 2.5 = -0.5$, and so on. These errors and their squares appear in the two right-hand columns of table 21.11.1. In repeated sampling with probability proportional to size, the first school is drawn 60% of the time, the second school 30%, and the third school 10%. The mean square error of the estimate is therefore

$$MSE_{\text{pps}} = (0.6)(0.25) + (0.3)(2.25) + (0.1)(2.25) = 1.05$$

TABLE 21.11.1
SELECTION OF A SCHOOL WITH PROBABILITY PROPORTIONAL TO SIZE

School	Children	Probability of Selection π_i	Mean per Child $\bar{Y}_i$	Error of Estimate $\bar{Y}_i - \bar{\bar{Y}}$	$(\bar{Y}_i - \bar{\bar{Y}})^2$
1	600	0.6	2	−0.5	0.25
2	300	0.3	4	+1.5	2.25
3	100	0.1	1	−1.5	2.25
Population	1000	1.0	2.5		

If, alternatively, the schools are drawn with equal probability, the mean square error is

$$MSE_{eq} = (1/3)(0.25 + 2.25 + 2.25) = 1.58$$

which is about 50% higher than that given by pps selection.

The reason it usually pays to select large units with higher probabilities is that the population mean depends more on the means of the large units than on those of the small units, as (21.11.1) shows. The large units are therefore likely to give better estimates when most of the variation is between primary units.

You may ask, Does the result in our example depend on the choice or the order of the means (2, 4, 1) assigned to schools 1, 2, and 3? The answer is yes. With means 4, 2, 1, you will find $MSE_{pps} = 1.29$ and $MSE_{eq} = 2.14$, the latter being 66% higher. Over the six possible orders of the numbers 1, 2, 4, the ratio MSE_{eq}/MSE_{pps} varies from 0.93 to 2.52. However, the ratio of the averages $\overline{MSE}_{eq}/\overline{MSE}_{pps}$ taken over all six possible orders does not depend on the numbers 1, 2, 4. With N_1 primary units in the population, the ratio is

$$\frac{\overline{MSE}_{eq}}{\overline{MSE}_{pps}} = \frac{(N_1 - 1) + N_1 \sum^{N} (\pi_i - \bar{\pi})^2}{(N_1 - 1) - N_1 \sum^{N} (\pi_i - \bar{\pi})^2} \qquad (21.11.2)$$

where π_i is the probability of selection (relative size) of the ith school. Clearly, the ratio exceeds one unless all π_i are equal, that is, all schools are the same size. In our example, this ratio is found to equal 1.47.

In two-stage sampling with primary units of unequal sizes, a simple method is to select n_1 primary units with probability proportional to size and take *an equal number of subunits* (e.g., children) in every selected primary unit, as in our illustration. This method gives every subunit in the population an equal chance of being in the sample. The method used in the sample (section 1.3) from which unemployment figures are estimated is an extension of this method to more than two stages of sampling. The sample mean $\bar{y}$ per subunit is an unbiased estimate of the population mean. If the n_1 primary units are drawn with replacement, an unbiased estimate of the variance of $\bar{y}$ is

$$s_{\bar{y}}^2 = \sum^{n_1} (\bar{y}_i - \bar{y})^2/[n_1(n_1 - 1)] \tag{21.11.3}$$

where $\bar{y}_i$ is the mean of the subsample from the ith primary unit.

We have illustrated only the simplest case. Some complications arise when we select units without replacement. Often the sizes of the units are not known exactly and have to be estimated in advance. Considerations of cost or of the structure of variability in the population may lead to the selection of units with unequal probabilities that are proportional to some quantity other than sizes. For details, see the references. In extensive surveys, multistage sampling with unequal probabilities of selection of primary units is the commonest method in current practice.

21.12—Ratio estimates. The *ratio estimate* is a different way of estimating population totals (or means) that is useful in many sampling problems. Suppose that you have taken a sample to estimate the population total Y of a variable y_i and that a complete count of the population was made on some previous occasion. Let x_i denote the value of the variable on the previous occasion. You might then compute the ratio

$$\hat{R} = \Sigma y_i/\Sigma x_i = \bar{y}/\bar{x}$$

where the sums are taken over the sample. This ratio is an estimate of the present level of the variate relative to that on the previous occasion. On multiplying the ratio by the known population total X on the previous occasion (8), you obtain the ratio estimate $\hat{Y}_R = RX = (\bar{y}/\bar{x})X$ of the population total of Y. Clearly, if the relative change is about the same on all sampling units, the estimate of the population total will be a good one.

The ratio estimate can also be used when x_i is some other kind of supplementary variable. The conditions for a successful application of this estimate are that the ratio y_i/x_i should be relatively constant over the population and the population total X should be known. Consider an estimate of the total amount of a crop, just after harvest, made from a sample of farms in a region. For each farm in the sample we record the total yield y_i and the total acreage x_i of that crop. In this case the ratio $\hat{R} = \Sigma y_i/\Sigma x_i$ is the sample estimate of the mean yield per acre. This is multiplied by the total acreage of the crop in the region, which would have to be known accurately from some other source. This estimate will be precise if the mean yield per acre varies little from farm to farm.

In large samples the estimated standard error of the ratio estimate $\hat{Y}_R$ of the population total from a simple random sample of size n is approximately

$$s(\hat{Y}_R) = N\sqrt{\frac{\Sigma(y_i - \hat{R}x_i)^2}{n(n-1)}}\sqrt{1 - \phi} \tag{21.12.1}$$

TABLE 21.12.1
1970 AND 1960 POPULATIONS (MILLIONS) OF SIX LARGE CITIES

	1	2	3	4	5	6	Total
1970 population, y_i	3.36	2.82	7.90	1.95	1.51	1.23	$18.77 = Y$
1960 population, x_i	3.55	2.48	7.78	2.00	1.67	0.94	$18.42 = X$

The ratio estimate is not always more precise than the simpler estimate $N\bar{y}$ (number of units in population × sample mean). It has been shown that in large samples the ratio estimate is more precise only if ρ, the correlation coefficient between Y and X, exceeds $cv(x)/[2cv(y)]$. Consequently, ratio estimates must not be used indiscriminately, although in appropriate circumstances they produce large gains in precision.

Sometimes the purpose of the sampling is to estimate a ratio, e.g., ratio of dry weight to total weight or ratio of clean wool to total wool. The estimated standard error of the estimate in large samples is then

$$s(R) = \frac{1}{\bar{x}}\sqrt{\frac{\Sigma(y_i - \hat{R}x_i)^2}{n(n-1)}}\sqrt{1-\phi}$$

This formula has already been given (in a different notation) at the end of section 19.8 in fitting an asymptotic regression.

As an illustration in which the ratio estimate works well, table 21.12.1 shows the 1970 and 1960 populations of the six U.S. cities with 1970 populations over 1 million.

Suppose that we have to estimate the 1970 total population of $N = 6$ cities from a simple random sample of $n = 2$ cities. The 1970 populations range from 1.23 million to 7.90 million; but while some cities have declined and some increased since 1960, the 1970/1960 ratios are relatively stable.

The estimate based on the sample mean is $\hat{Y} = N\bar{y} = 6\bar{y}$. From (21.5.2) for the standard error of $\bar{y}$, the variance of $\hat{Y}$ is

$$V(\hat{Y}) = N^2s^2(1-\phi)/n = (36)(6.1057)/3 = 73.268$$

The standard error of this estimate is ±8.56, giving a coefficient of variation of ±46%—very inaccurate.

The ratio estimate $\hat{Y}_R = (\bar{y}/\bar{x})X = 18.42\bar{y}/\bar{x}$ is slightly biased. Since no exact formula for its mean square error is known, table 21.12.2 presents the estimates from all 15 simple random samples of size 2. From these we calculate the mean square error as $\Sigma(\hat{Y}_R - Y)^2/15$. The 15 estimates $N\bar{y}$ are also shown for comparison.

Note that the bias in $\hat{Y}_R$, +0.05, is trivial and that the 15 ratio estimates $\hat{Y}_R$ range from 17.18 to 21.81, as against a range from 8.22 to 33.78 for the 15 unbiased estimates $N\bar{y}$. You may verify that $MSE(\hat{Y}_R) = 1.201$, giving a coefficient of variation of 5.8% and a deff value of only 0.016 relative to $\hat{Y}$.

TABLE 21.12.2
RATIO ESTIMATES $\hat{Y}_R = (\bar{y}/\bar{x})X$ AND ESTIMATES $\hat{Y} = N\bar{y}$

Sample Units	Sample Totals 1960	1970	$\hat{Y}_R$	$\hat{Y}$	Sample Units	Sample Totals 1960	1970	$\hat{Y}_R$	$\hat{Y}$
1, 2	6.03	6.18	18.88	18.54	2, 6	3.42	4.05	21.81	12.15
1, 3	11.33	11.26	18.31	33.78	3, 4	9.78	9.85	18.55	29.55
1, 4	5.55	5.31	17.62	15.93	3, 5	9.45	9.41	18.34	28.23
1, 5	5.22	4.87	17.18	14.61	3, 6	8.72	9.13	19.29	27.39
1, 6	4.49	4.59	18.83	13.77	4, 5	3.67	3.40	17.37	10.38
2, 3	10.26	10.72	19.25	32.16	4, 6	2.94	3.18	19.92	9.54
2, 4	4.48	4.77	18.38	14.31	5, 6	2.61	2.74	19.34	8.22
2, 5	4.15	4.33	19.22	12.99	Mean			18.82	18.77

21.13—Nonsampling errors. In many surveys, especially surveys dealing with human subjects and institutions, sources of error other than those due to sampling affect the estimates. The most common are probably missing data. In a survey taken by mail, only 30% of those to whom questionnaires are sent may reply. In an interview survey made by visiting a sample of names and addresses, perhaps 10% of the people may not be home and a further 4% refuse or are unable to answer the questions.

With missing data our sample is smaller than planned, but a bigger problem is that we often have reason to believe that the misses (the nonrespondents) differ systematically from the respondents. Consequently, our sample of respondents is biased, though evidence about the size of this bias may naturally be hard to obtain. To illustrate from an oversimplified model, suppose that our field method (mail, telephone, household interview) can reach a proportion w_1 of the population but fails to get replies from a proportion w_0; and that the two subpopulations have means $\bar{Y}_1$, $\bar{Y}_0$ for the variable being measured. Our sample of size n_1 is a random sample of the respondents. The mean square error of the sample mean $\bar{y}_1$ is then

$$MSE(\bar{y}_1) = E(\bar{y}_1 - \bar{Y})^2 = E(\bar{y}_1 - w_1\bar{Y}_1 - w_0\bar{Y}_0)^2$$

$$= E[(\bar{y}_1 - \bar{Y}_1) + w_0(\bar{Y}_1 - \bar{Y}_0)]^2 = s_1^2/n_1 + w_0^2(\bar{Y}_1 - \bar{Y}_0)^2$$

ignoring the finite population correction. With large samples, the bias term may dominate this mean square error and our sampling error formulas may seriously underestimate it and our real errors.

There are two strategies for attacking this problem. One is to use field methods that reduce w_0, for instance by insisting that at least three or four attempts be made to reach and obtain answers from any sample member. Alternatively, if supplementary information can be obtained about nonrespondents that indicates to some extent how they differ from respondents, another strategy is to use a different estimate that takes this knowledge into account.

For example, suppose that males differ markedly from females in their replies to one question. A planned sample of 1000 has 487 males and 513 females, this being the proportion in the population. Responses are obtained from 410 males (84%) and 492 females (96%). Instead of the sample mean, we use the estimate

$$\hat{\bar{Y}} = 0.487\bar{y}_m + 0.513\bar{y}_f$$

If the males who did not respond have relatively little bias as compared with those who did respond, this estimate should be almost free from bias. Another approach uses available knowledge about nonrespondents to substitute or *impute* estimates of the responses that they would have given. The assumptions in this approach are similar to those made in substituting for a missing value in randomized blocks from knowledge of the treatment and block corresponding to the missing observation.

Errors of measurement, including those introduced in classifying and coding the responses for analysis, are another source of inaccuracy. Sometimes the question is poorly worded and has different meanings for different subjects. Pretests and revisions of the questionnaire help here. In summary, the objective in planning and conducting a survey should be to minimize the total error, not just the sampling error. This involves the difficult job of allocating resources among reduction of sampling errors, missing data, and errors of measurement, and of deciding from what we know how best to use these resources for each purpose.

21.14—Further reading. The general books on sample surveys that have become standard (2, 3, 4, 5, 10) involve roughly the same level of mathematical difficulty and knowledge of statistics. Reference (3) is oriented toward applications in business and (10) toward those in agriculture. Another good book for agricultural applications, at a lower mathematical level, is (11).

Useful short books are (12), an informal, popular account of some of the interesting applications of survey methods; (13), which conducts the reader painlessly through the principal results in probability sampling at about the mathematical level of this chapter; and (14), which discusses the technique of constructing interview questions.

Books and papers have also begun to appear on some of the common specific types of application. For sampling a town under U.S. conditions, with the block as primary sampling unit, references (15) and (16) are recommended. Reference (17), intended primarily for surveys by health agencies to check on the immunization status of children, gives instructions for the sampling of attributes in local areas, while (18) deals with the sampling of hospitals and patients. Much helpful advice on the use of sampling in agricultural censuses is found in (19), and (20) presents methods for reducing errors of measurement.

TECHNICAL TERMS

design effect (deff)
finite population correlation
imputing
listing
nonsampling errors
probability proportional to size (pps)

probability sampling
proportional stratification
ratio estimates
stratified random sampling
systematic sampling
two-stage sampling

REFERENCES

1. Clapham, A. R. *J. Agric. Sci.* 19 (1929):214.
2. Cochran, W. G. 1967. *Sampling Techniques,* 3rd ed. Wiley, New York.
3. Deming, W. Edwards. 1960. *Sample Design in Business Research.* Wiley, New York.
4. Hansen, M. H.; Hurwitz, W. N.; and Madow, W. G. 1953. *Sample Survey Methods and Theory.* Wiley, New York.
5. Kish, L. 1965. *Survey Sampling.* Wiley, New York.
6. Immer, F. R. *J. Agric. Res.* 44 (1932):633.
7. Neyman, J. *J. R. Stat. Soc.* 97 (1934):558.
8. Hansen, M. H., and Hurwitz, W. N. *Ann. Math. Stat.* 14 (1943):333.
9. West, Q. M. 1951. Mimeographed Report. Cornell Univ. Agric. Exp. Stn.
10. Yates, F. 1960. *Sampling Methods for Censuses and Surveys,* 3rd ed. Griffin, London.
11. Sampford, M. R. 1962. *An Introduction to Sampling Theory.* Oliver & Boyd, Edinburgh.
12. Slonim, M. J. 1960. *Sampling in a Nutshell.* Simon & Schuster, New York.
13. Stuart, A. 1962. *Basic Ideas of Scientific Sampling.* Griffin, London.
14. Payne, S. L. 1951. *The Art of Asking Questions.* Princeton Univ. Press.
15. Woolsey, T. D. 1956. *Sampling Methods for a Small Household Survey.* Public Health Monographs 40.
16. Kish, L. *Am. Soc. Rev.* 17 (1952):761.
17. Serfling, R. E., and Sherman, I. L. 1965. *Attribute Sampling Methods.* U.S. Government Printing Office, Washington, D.C.
18. Hess, I.; Riedel, D. C.; and Fitzpatrick, T. B. 1975. *Probability Sampling of Hospitals and Patients,* 2nd ed. Univ. Michigan, Ann Arbor.
19. Zarcovich, S. S. 1965. *Sampling Methods and Censuses.* FAO, Rome.
20. Zarcovich, S. S. 1966. *Quality of Statistical Data.* FAO, Rome.

Appendix

LIST OF APPENDIX TABLES

NOTES

Interpolation. In analyses of data and in working the examples in this book, use of the nearest entry in any appendix table is accurate enough in most cases. The following examples illustrate linear interpolation, which will sometimes be needed.

1. Find the 5% significance level of χ^2 for 34 df. For $P = 0.050$, table A 5 gives

df	30		34		40
χ^2	43.77		?		55.76
distance		0.4		0.6	

Calculate $(34 - 30)/(40 - 30) = 0.4$. Thus, 34 is 0.4 of the way from 30 to 40. The required χ^2 value is therefore

$$(0.4)(55.76) + (0.6)(43.77) = 48.57$$

Note that 0.4 multiplies χ^2_{40}, not χ^2_{30}.

2. An analysis gave an F value of 2.04 for 3 and 18 df. Find the significance probability. For 3 and 18 df, table A 14, part II, gives the following entries:

P	0.25	?	0.10
F	1.49	2.04	2.42

Calculate $(2.04 - 1.49)/(2.42 - 1.49) = 0.55/0.93 = 0.59$. By the method in the preceding example,

$$P = (0.59)(0.10) + (0.41)(0.25) = 0.16$$

3. With a pocket calculator that has $\ln(x)$ and e^x keys, interpolation against $\ln P$ is usually more accurate than interpolation against P in finding P for a given value of F, t, χ^2, and so on, or the value of the variate for given P. The difference between interpolation against $\ln P$ and P is minor in the middle of the table, as in example 2, but can be important toward the extremes. As an

example, find the value of t with 11 df that has two-tailed probability $0.05/12 = 0.0042$. In table A 4 we find:

P	0.005	0.0042	0.001
t	3.497	?	4.437

We get the same answer in the ln scale if we multiply the P values by any number, so for simplicity we take $P' = 5$, 4.2, and 1. We then have

$\ln P'$	1.609	1.435	0

Since $1.435/1.609 = 0.89$, we get

distance	0.11	0.89

The required t is therefore

$$(0.89)(3.497) + (0.11)(4.437) = 3.600$$

Finding square roots. Table A 22 is a table of square roots. To save space the entries jump by 0.02 instead of 0.01, but interpolation will rarely be necessary. With very large or very small numbers, mistakes in finding square roots are common. The following examples should clarify the procedure.

Step	(1)	(2)	(3)	(4)
Number	Mark Off	Column Read	Reading	Square Root
6028.0	60,28.0	$\sqrt{10n}$	7.76	77.6
397.2	3,97.2	$\sqrt{n}$	1.99	19.9
46.38	46.38	$\sqrt{10n}$	6.81	6.81
0.194	0.19,4	$\sqrt{10n}$	4.40	0.440
0.000893	0.00,08,93	$\sqrt{n}$	2.99	0.0299

In step (1), mark off the digits *in twos* to the right or left of the decimal point. Step (2) tells which column of the square root table is to be read. With 3,97.2 and 0.00,08,93 read the $\sqrt{n}$ column, because there is a *single* digit (3 or 8) to the left of the first comma that has any nonzero digits to its left. If there are *two* digits to the left of the first comma, as in 60,28.0, read the $\sqrt{10n}$ column. Step (3) gives the reading, taken directly from the nearest entry in table A 22.

The final step (4) finds the actual square roots. Consider, first, numbers greater than 1. If column (1) has no comma to the left of the decimal, as with 46.38, the square root has one digit to the left of the decimal. If column (1) has one comma to the left of the decimal, as with 60,28.0 and 3,97.2, the square root has two digits to the left of the decimal, and so on. With numbers smaller than 1, replace any pair 00 to the right of the decimal by a single 0. Thus, the square root of 0.00,08,93 is 0.0299 as shown. The square root of 0.00,00,08,93 is 0.00299.

TABLE A 1
Ten Thousand Randomly Assorted Digits

	00–04	05–09	10–14	15–19	20–24	25–29	30–34	35–39	40–44	45–49
00	54463	22662	65905	70639	79365	67382	29085	69831	47058	08186
01	15389	85205	18850	39226	42249	90669	96325	23248	60933	26927
02	85941	40756	82414	02015	13858	78030	16269	65978	01385	15345
03	61149	69440	11286	88218	58925	03638	52862	62733	33451	77455
04	05219	81619	10651	67079	92511	59888	84502	72095	83463	75577
05	41417	98326	87719	92294	46614	50948	64886	20002	97365	30976
06	28357	94070	20652	35774	16249	75019	21145	05217	47286	76305
07	17783	00015	10806	83091	91530	36466	39981	62481	49177	75779
08	40950	84820	29881	85966	62800	70326	84740	62660	77379	90279
09	82995	64157	66164	41180	10089	41757	78258	96488	88629	37231
10	96754	17676	55659	44105	47361	34833	86679	23930	53249	27083
11	34357	88040	53364	71726	45690	66334	60332	22554	90600	71113
12	06318	37403	49927	57715	50423	67372	63116	48888	21505	80182
13	62111	52820	07243	79931	89292	84767	85693	73947	22278	11551
14	47534	09243	67879	00544	23410	12740	02540	54440	32949	13491
15	98614	75993	84460	62846	59844	14922	48730	73443	48167	34770
16	24856	03648	44898	09351	98795	18644	39765	71058	90368	44104
17	96887	12479	80621	66223	86085	78285	02432	53342	42846	94771
18	90801	21472	42815	77408	37390	76766	52615	32141	30268	18106
19	55165	77312	83666	36028	28420	70219	81369	41943	47366	41067
20	75884	12952	84318	95108	72305	64620	91318	89872	45375	85436
21	16777	37116	58550	42958	21460	43910	01175	87894	81378	10620
22	46230	43877	80207	88877	89380	32992	91380	03164	98656	59337
23	42902	66892	46134	01432	94710	23474	20423	60137	60609	13119
24	81007	00333	39693	28039	10154	95425	39220	19774	31782	49037
25	68089	01122	51111	72373	06902	74373	96199	97017	41273	21546
26	20411	67081	89950	16944	93054	87687	96693	87236	77054	33848
27	58212	13160	06468	15718	82627	76999	05999	58680	96739	63700
28	70577	42866	24969	61210	76046	67699	42054	12696	93758	03283
29	94522	74358	71659	62038	79643	79169	44741	05437	39038	13163
30	42626	86819	85651	88678	17401	03252	99547	32404	17918	62880
31	16051	33763	57194	16752	54450	19031	58580	47629	54132	60631
32	08244	27647	33851	44705	94211	46716	11738	55784	95374	72655
33	59497	04392	09419	89964	51211	04894	72882	17805	21896	83864
34	97155	13428	40293	09985	58434	01412	69124	82171	59058	82859
35	98409	66162	95763	47420	20792	61527	20441	39435	11859	41567
36	45476	84882	65109	96597	25930	66790	65706	61203	53634	22557
37	89300	69700	50741	30329	11658	23166	05400	66669	48708	03887
38	50051	95137	91631	66315	91428	12275	24816	68091	71710	33258
39	31753	85178	31310	89642	98364	02306	24617	09609	83942	22716
40	79152	53829	77250	20190	56535	18760	69942	77448	33278	48805
41	44560	38750	83635	56540	64900	42912	13953	79149	18710	68618
42	68328	83378	63369	71381	39564	05615	42451	64559	97501	65747
43	46939	38689	58625	08342	30459	85863	20781	09284	26333	91777
44	83544	86141	15707	96256	23068	13782	08467	89469	93842	55349
45	91621	00881	04900	54224	46177	55309	17852	27491	89415	23466
46	91896	67126	04151	03795	59077	11848	12630	98375	52068	60142
47	55751	62515	21108	80830	02263	29303	37204	96926	30506	09808
48	85156	87689	95493	88842	00664	55017	55539	17771	69448	87530
49	07521	56898	12236	60277	39102	62315	12239	07105	11844	01117

TABLE A 1—*Continued*

	50–54	55–59	60–64	65–69	70–74	75–79	80–84	85–89	90–94	95–99
00	59391	58030	52098	82718	87024	82848	04190	96574	90464	29065
01	99567	76364	77204	04615	27062	96621	43918	01896	83991	51141
02	10363	97518	51400	25670	98342	61891	27101	37855	06235	33316
03	86859	19558	64432	16706	99612	59798	32803	67708	15297	28612
04	11258	24591	36863	55368	31721	94335	34936	02566	80972	08188
05	95068	88628	35911	14530	33020	80428	39936	31855	34334	64865
06	54463	47237	73800	91017	36239	71824	83671	39892	60518	37092
07	16874	62677	57412	13215	31389	62233	80827	73917	82802	84420
08	92494	63157	76593	91316	03505	72389	96363	52887	01087	66091
09	15669	56689	35682	40844	53256	81872	35213	09840	34471	74441
10	99116	75486	84989	23476	52967	67104	39495	39100	17217	74073
11	15696	10703	65178	90637	63110	17622	53988	71087	84148	11670
12	97720	15369	51269	69620	03388	13699	33423	67453	43269	56720
13	11666	13841	71681	98000	35979	39719	81899	07449	47985	46967
14	71628	73130	78783	75691	41632	09847	61547	18707	85489	69944
15	40501	51089	99943	91843	41995	88931	73631	69361	05375	15417
16	22518	55576	98215	82068	10798	86211	36584	67466	69373	40054
17	75112	30485	62173	02132	14878	92879	22281	16783	86352	00077
18	80327	02671	98191	84342	90813	49268	95441	15496	20168	09271
19	60251	45548	02146	05597	48228	81366	34598	72856	66762	17002
20	57430	82270	10421	00540	43648	75888	66049	21511	47676	33444
21	73528	39559	34434	88596	54086	71693	43132	14414	79949	85193
22	25991	65959	70769	64721	86413	33475	42740	06175	82758	66248
23	78388	16638	09134	59980	63806	48472	39318	35434	24057	74739
24	12477	09965	96657	57994	59439	76330	24596	77515	09577	91871
25	83266	32883	42451	15579	38155	29793	40914	65990	16255	17777
26	76970	80876	10237	39515	79152	74798	39357	09054	73579	92359
27	37074	65198	44785	68624	98336	84481	97610	78735	46703	98265
28	83712	06514	30101	78295	54656	85417	43189	60048	72781	72606
29	20287	56862	69727	94443	64936	08366	27227	05158	50326	59566
30	74261	32592	86538	27041	65172	85532	07571	80609	39285	65340
31	64081	49863	08478	96001	18888	14810	70545	89755	59064	07210
32	05617	75818	47750	67814	29575	10526	66192	44464	27058	40467
33	26793	74951	95466	74307	13330	42664	85515	20632	05497	33625
34	65988	72850	48737	54719	52056	01596	03845	35067	03134	70322
35	27366	42271	44300	73399	21105	03280	73457	43093	05192	48657
36	56760	10909	98147	34736	33863	95256	12731	66598	50771	83665
37	72880	43338	93643	58904	59543	23943	11231	83268	65938	81581
38	77888	38100	03062	58103	47961	83841	25878	23746	55903	44115
39	28440	07819	21580	51459	47971	29882	13990	29226	23608	15873
40	63525	94441	77033	12147	51054	49955	58312	76923	96071	05813
41	47606	93410	16359	89033	89696	47231	64498	31776	05383	39902
42	52669	45030	96279	14709	52372	87832	02735	50803	72744	88208
43	16738	60159	07425	62369	07515	82721	37875	71153	21315	00132
44	59348	11695	45751	15865	74739	05572	32688	20271	65128	14551
45	12900	71775	29845	60774	94924	21810	38636	33717	67598	82521
46	75086	23537	49939	33595	13484	97588	28617	17979	70749	35234
47	99495	51434	29181	09993	38190	42553	68922	52125	91077	40197
48	26075	31671	45386	36583	93459	48599	52022	41330	60651	91321
49	13636	93596	23377	51133	95126	61496	42474	45141	46660	42338

TABLE A 1—*Continued*

	00–04	05–09	10–14	15–19	20–24	25–29	30–34	35–39	40–44	45–49
50	64249	63664	39652	40646	97306	31741	07294	84149	46797	82487
51	26538	44249	04050	48174	65570	44072	40192	51153	11397	58212
52	05845	00512	78630	55328	18116	69296	91705	86224	29503	57071
53	74897	68373	67359	51014	33510	83048	17056	72506	82949	54600
54	20872	54570	35017	88132	25730	22626	86723	91691	13191	77212
55	31432	96156	89177	75541	81355	24480	77243	76690	42507	84362
56	66890	61505	01240	00660	05873	13568	76082	79172	57913	93448
57	41894	57790	79970	33106	86904	48119	52503	24130	72824	21627
58	11303	87118	81471	52936	08555	28420	49416	44448	04269	27029
59	54374	57325	16947	45356	78371	10563	97191	53798	12693	27928
60	64852	34421	61046	90849	13966	39810	42699	21753	76192	10508
61	16309	20384	09491	91588	97720	89846	30376	76970	23063	35894
62	42587	37065	24526	72602	57589	98131	37292	05967	26002	51945
63	40177	98590	97161	41682	84533	67588	62036	49967	01990	72308
64	82309	76128	93965	26743	24141	04838	40254	26065	07938	76236
65	79788	68243	59732	04257	27084	14743	17520	95401	55811	76099
66	40538	79000	89559	25026	42274	23489	34502	75508	06059	86682
67	64016	73598	18609	73150	62463	33102	45205	87440	96767	67042
68	49767	12691	17903	93871	99721	79109	09425	26904	07419	76013
69	76974	55108	29795	08404	82684	00497	51126	79935	57450	55671
70	23854	08480	85983	96025	50117	64610	99425	62291	86943	21541
71	68973	70551	25098	78033	98573	79848	31778	29555	61446	23037
72	36444	93600	65350	14971	25325	00427	52073	64280	18847	24768
73	03003	87800	07391	11594	21196	00781	32550	57158	58887	73041
74	17540	26188	36647	78386	04558	61463	57842	90382	77019	24210
75	38916	55809	47982	41968	69760	79422	80154	91486	19180	15100
76	64288	19843	69122	42502	48508	28820	59933	72998	99942	10515
77	86809	51564	38040	39418	49915	19000	58050	16899	79952	57849
78	99800	99566	14742	05028	30033	94889	53381	23656	75787	59223
79	92345	31890	95712	08279	91794	94068	49337	88674	35355	12267
80	90363	65162	32245	82279	79256	80834	06088	99462	56705	06118
81	64437	32242	48431	04835	39070	59702	31508	60935	22390	52246
82	91714	53662	28373	34333	55791	74758	51144	18827	10704	76803
83	20902	17646	31391	31459	33315	03444	55743	74701	58851	27427
84	12217	86007	70371	52281	14510	76094	96579	54853	78339	20839
85	45177	02863	42307	53571	22532	74921	17735	42201	80540	54721
86	28325	90814	08804	52746	47913	54577	47525	77705	95330	21866
87	29019	28776	56116	54791	64604	08815	46049	71186	34650	14994
88	84979	81353	56219	67062	26146	82567	33122	14124	46240	92973
89	50371	26347	48513	63915	11158	25563	91915	18431	92978	11591
90	53422	06825	69711	67950	64716	18003	49581	45378	99878	61130
91	67453	35651	89316	41620	32048	70225	47597	33137	31443	51445
92	07294	85353	74819	23445	68237	07202	99515	62282	53809	26685
93	79544	00302	45338	16015	66613	88968	14595	63836	77716	79596
94	64144	85442	82060	46471	24162	39500	87351	36637	42833	71875
95	90919	11883	58318	00042	52402	28210	34075	33272	00840	73268
96	06670	57353	86275	92276	77591	46924	60839	55437	03183	13191
97	36634	93976	52062	83678	41256	60948	18685	48992	19462	96062
98	75101	72891	85745	67106	26010	62107	60885	37503	55461	71213
99	05112	71222	72654	51583	05228	62056	57390	42746	39272	96659

TABLE A 1—*Continued*

	50–54	55–59	60–64	65–69	70–74	75–79	80–84	85–89	90–94	95–99
50	32847	31282	03345	89593	69214	70381	78285	20054	91018	16742
51	16916	00041	30236	55023	14253	76582	12092	86533	92426	37655
52	66176	34037	21005	27137	03193	48970	64625	22394	39622	79085
53	46299	13335	12180	16861	38043	59292	62675	63631	37020	78195
54	22847	47839	45385	23289	47526	54098	45683	55849	51575	64689
55	41851	54160	92320	69936	34803	92479	33399	71160	64777	83378
56	28444	59497	91586	95917	68553	28639	06455	34174	11130	91994
57	47520	62378	98855	83174	13088	16561	68559	26679	06238	51254
58	34978	63271	13142	82681	05271	08822	06490	44984	49307	61717
59	37404	80416	69035	92980	49486	74378	75610	74976	70056	15478
60	32400	65482	52099	53676	74648	94148	65095	69597	52771	71551
61	89262	86332	51718	70663	11623	29834	79820	73002	84886	03591
62	86866	09127	98021	03871	27789	58444	44832	36505	40672	30180
63	90814	14833	08759	74645	05046	94056	99094	65091	32663	73040
64	19192	82756	20553	58446	55376	88914	75096	26119	83898	43816
65	77585	52593	56612	95766	10019	29531	73064	20953	53523	58136
66	23757	16364	05096	03192	62386	45389	85332	18877	55710	96459
67	45989	96257	23850	26216	23309	21526	07425	50254	19455	29315
68	92970	94243	07316	41467	64837	52406	25225	51553	31220	14032
69	74346	59596	40088	98176	17896	86900	20249	77753	19099	48885
70	87646	41309	27636	45153	29988	94770	07255	70908	05340	99751
71	50099	71038	45146	06146	55211	99429	43169	66259	97786	59180
72	10127	46900	64984	75348	04115	33624	68774	60013	35515	62556
73	67995	81977	18984	64091	02785	27762	42529	97144	80407	64524
74	26304	80217	84934	82657	69291	35397	98714	35104	08187	48109
75	81994	41070	56642	64091	31229	02595	13513	45148	78722	30144
76	59537	34662	79631	89403	65212	09975	06118	86197	58208	16162
77	51228	10937	62396	81460	47331	91403	95007	06047	16846	64809
78	31089	37995	29577	07828	42272	54016	21950	86192	99046	84864
79	38207	97938	93459	75174	79460	55436	57206	87644	21296	43393
80	88666	31142	09474	89712	63153	62333	42212	06140	42594	43671
81	53365	56134	67582	92557	89520	33452	05134	70628	27612	33738
82	89807	74530	38004	90102	11693	90257	05500	79920	62700	43325
83	18682	81038	85662	90915	91631	22223	91588	80774	07716	12548
84	63571	32579	63942	25371	09234	94592	98475	76884	37635	33608
85	68927	56492	67799	95398	77642	54913	91583	08421	81450	76229
86	56401	63186	39389	88798	31356	89235	97036	32341	33292	73757
87	24333	95603	02359	72942	46287	95382	08452	62862	97869	71775
88	17025	84202	95199	62272	06366	16175	97577	99304	41587	03686
89	02804	08253	52133	20224	68034	50865	57868	22343	55111	03607
90	08298	03879	20995	19850	73090	13191	18963	82244	78479	99121
91	59883	01785	82403	96062	03785	03488	12970	64896	38336	30030
92	46982	06682	62864	91837	74021	89094	39952	64158	79614	78235
93	31121	47266	07661	02051	67599	24471	69843	83696	71402	76287
94	97867	56641	63416	17577	30161	87320	37752	73276	48969	41915
95	57364	86746	08415	14621	49430	22311	15836	72492	49372	44103
96	09559	26263	69511	28064	75999	44540	13337	10918	79846	54809
97	53873	55571	00608	42661	91332	63956	74087	59008	47493	99581
98	35531	19162	86406	05299	77511	24311	57257	22826	77555	05941
99	28229	88629	25695	94932	30721	16197	78742	34974	97528	45447

TABLE A 2
ORDINATES OF THE NORMAL CURVE

Z	Second decimal place in Z									
	0.00	0.01	0.02	0.03	0.04	0.05	0.06	0.07	0.08	0.09
0.0	0.3989	0.3989	0.3989	0.3988	0.3986	0.3984	0.3982	0.3980	0.3977	0.3973
0.1	.3970	.3965	.3961	.3956	.3951	.3945	.3939	.3932	.3925	.3918
0.2	.3910	.3902	.3894	.3885	.3876	.3867	.3857	.3847	.3836	.3825
0.3	.3814	.3802	.3790	.3778	.3765	.3752	.3739	.3725	.3712	.3697
0.4	.3683	.3668	.3653	.3637	.3621	.3605	.3589	.3572	.3555	.3538
0.5	.3521	.3503	.3485	.3467	.3448	.3429	.3410	.3391	.3372	.3352
0.6	.3332	.3312	.3292	.3271	.3251	.3230	.3209	.3187	.3166	.3144
0.7	.3123	.3101	.3079	.3056	.3034	.3011	.2989	.2966	.2943	.2920
0.8	.2897	.2874	.2850	.2827	.2803	.2780	.2756	.2732	.2709	.2685
0.9	.2661	.2637	.2613	.2589	.2565	.2541	.2516	.2492	.2468	.2444
1.0	.2420	.2396	.2371	.2347	.2323	.2299	.2275	.2251	.2227	.2203
1.1	.2179	.2155	.2131	.2107	.2083	.2059	.2036	.2012	.1989	.1965
1.2	.1942	.1919	.1895	.1872	.1849	.1826	.1804	.1781	.1758	.1736
1.3	.1714	.1691	.1669	.1647	.1626	.1604	.1582	.1561	.1539	.1518
1.4	.1497	.1476	.1456	.1435	.1415	.1394	.1374	.1354	.1334	.1315
1.5	.1295	.1276	.1257	.1238	.1219	.1200	.1182	.1163	.1145	.1127
1.6	.1109	.1092	.1074	.1057	.1040	.1023	.1006	.0989	.0973	.0957
1.7	.0940	.0925	.0909	.0893	.0878	.0863	.0848	.0833	.0818	.0804
1.8	.0790	.0775	.0761	.0748	.0734	.0721	.0707	.0694	.0681	.0669
1.9	.0656	.0644	.0632	.0620	.0608	.0596	.0584	.0573	.0562	.0551
2.0	.0540	.0529	.0519	.0508	.0498	.0488	.0478	.0468	.0459	.0449
2.1	.0440	.0431	.0422	.0413	.0404	.0396	.0387	.0379	.0371	.0363
2.2	.0355	.0347	.0339	.0332	.0325	.0317	.0310	.0303	.0297	.0290
2.3	.0283	.0277	.0270	.0264	.0258	.0252	.0246	.0241	.0235	.0229
2.4	.0224	.0219	.0213	.0208	.0203	.0198	.0194	.0189	.0184	.0180
2.5	.0175	.0171	.0167	.0163	.0158	.0154	.0151	.0147	.0143	.0139
2.6	.0136	.0132	.0129	.0126	.0122	.0119	.0116	.0113	.0110	.0107
2.7	.0104	.0101	.0099	.0096	.0093	.0091	.0088	.0086	.0084	.0081
2.8	.0079	.0077	.0075	.0073	.0071	.0069	.0067	.0065	.0063	.0061
2.9	.0060	.0058	.0056	.0055	.0053	.0051	.0050	.0048	.0047	.0046

Z	First decimal place in Z									
	0.0	0.1	0.2	0.3	0.4	0.5	0.6	0.7	0.8	0.9
3	0.0044	0.0033	0.0024	0.0017	0.0012	0.0009	0.0006	0.0004	0.0003	0.0002
4	.0001	.0001	.0001	.0000	.0000	.0000	.0000	.0000	.0000	.0000

TABLE A 3
CUMULATIVE NORMAL FREQUENCY DISTRIBUTION
(area under standard normal curve from 0 to Z)

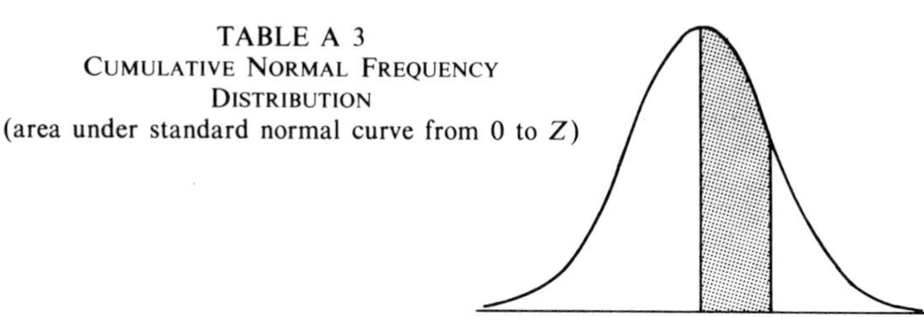

Z	0.00	0.01	0.02	0.03	0.04	0.05	0.06	0.07	0.08	0.09
0.0	0.0000	0.0040	0.0080	0.0120	0.0160	0.0199	0.0239	0.0279	0.0319	0.0359
0.1	.0398	.0438	.0478	.0517	.0557	.0596	.0636	.0675	.0714	.0753
0.2	.0793	.0832	.0871	.0910	.0948	.0987	.1026	.1064	.1103	.1141
0.3	.1179	.1217	.1255	.1293	.1331	.1368	.1406	.1443	.1480	.1517
0.4	.1554	.1591	.1628	.1664	.1700	.1736	.1772	.1808	.1844	.1879
0.5	.1915	.1950	.1985	.2019	.2054	.2088	.2123	.2157	.2190	.2224
0.6	.2257	.2291	.2324	.2357	.2389	.2422	.2454	.2486	.2517	.2549
0.7	.2580	.2611	.2642	.2673	.2704	.2734	.2764	.2794	.2823	.2852
0.8	.2881	.2910	.2939	.2967	.2995	.3023	.3051	.3078	.3106	.3133
0.9	.3159	.3186	.3212	.3238	.3264	.3289	.3315	.3340	.3365	.3389
1.0	.3413	.3438	.3461	.3485	.3508	.3531	.3554	.3577	.3599	.3621
1.1	.3643	.3665	.3686	.3708	.3729	.3749	.3770	.3790	.3810	.3830
1.2	.3849	.3869	.3888	.3907	.3925	.3944	.3962	.3980	.3997	.4015
1.3	.4032	.4049	.4066	.4082	.4099	.4115	.4131	.4147	.4162	.4177
1.4	.4192	.4207	.4222	.4236	.4251	.4265	.4279	.4292	.4306	.4319
1.5	.4332	.4345	.4357	.4370	.4382	.4394	.4406	.4418	.4429	.4441
1.6	.4452	.4463	.4474	.4484	.4495	.4505	.4515	.4525	.4535	.4545
1.7	.4554	.4564	.4573	.4582	.4591	.4599	.4608	.4616	.4625	.4633
1.8	.4641	.4649	.4656	.4664	.4671	.4678	.4686	.4693	.4699	.4706
1.9	.4713	.4719	.4726	.4732	.4738	.4744	.4750	.4756	.4761	.4767
2.0	.4772	.4778	.4783	.4788	.4793	.4798	.4803	.4808	.4812	.4817
2.1	.4821	.4826	.4830	.4834	.4838	.4842	.4846	.4850	.4854	.4857
2.2	.4861	.4864	.4868	.4871	.4875	.4878	.4881	.4884	.4887	.4890
2.3	.4893	.4896	.4898	.4901	.4904	.4906	.4909	.4911	.4913	.4916
2.4	.4918	.4920	.4922	.4925	.4927	.4929	.4931	.4932	.4934	.4936
2.5	.4938	.4940	.4941	.4943	.4945	.4946	.4948	.4949	.4951	.4952
2.6	.4953	.4955	.4956	.4957	.4959	.4960	.4961	.4962	.4963	.4964
2.7	.4965	.4966	.4967	.4968	.4969	.4970	.4971	.4972	.4973	.4974
2.8	.4974	.4975	.4976	.4977	.4977	.4978	.4979	.4979	.4980	.4981
2.9	.4981	.4982	.4982	.4983	.4984	.4984	.4985	.4985	.4986	.4986
3.0	.4987	.4987	.4987	.4988	.4988	.4989	.4989	.4989	.4990	.4990
3.1	.4990	.4991	.4991	.4991	.4992	.4992	.4992	.4992	.4993	.4993
3.2	.4993	.4993	.4994	.4994	.4994	.4994	.4994	.4995	.4995	.4995
3.3	.4995	.4995	.4995	.4996	.4996	.4996	.4996	.4996	.4996	.4997
3.4	.4997	.4997	.4997	.4997	.4997	.4997	.4997	.4997	.4997	.4998
3.6	.4998	.4998	.4999	.4999	.4999	.4999	.4999	.4999	.4999	.4999
3.9	.5000									

TABLE A 4
Distribution of *t* (Two-Tailed Tests)

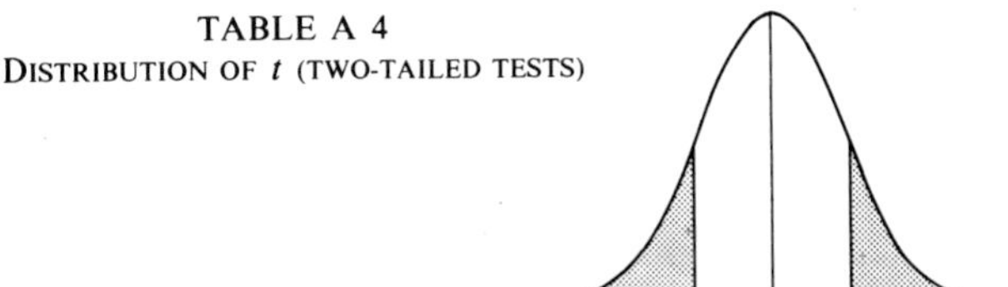

Degrees of Freedom	Probability of a Larger Value, Sign Ignored								
	0.500	0.400	0.200	0.100	0.050	0.025	0.010	0.005	0.001
1	1.000	1.376	3.078	6.314	12.706	25.452	63.657		
2	0.816	1.061	1.886	2.920	4.303	6.205	9.925	14.089	31.598
3	.765	0.978	1.638	2.353	3.182	4.176	5.841	7.453	12.941
4	.741	.941	1.533	2.132	2.776	3.495	4.604	5.598	8.610
5	.727	.920	1.476	2.015	2.571	3.163	4.032	4.773	6.859
6	.718	.906	1.440	1.943	2.447	2.969	3.707	4.317	5.959
7	.711	.896	1.415	1.895	2.365	2.841	3.499	4.029	5.405
8	.706	.889	1.397	1.860	2.306	2.752	3.355	3.832	5.041
9	.703	.883	1.383	1.833	2.262	2.685	3.250	3.690	4.781
10	.700	.879	1.372	1.812	2.228	2.634	3.169	3.581	4.587
11	.697	.876	1.363	1.796	2.201	2.593	3.106	3.497	4.437
12	.695	.873	1.356	1.782	2.179	2.560	3.055	3.428	4.318
13	.694	.870	1.350	1.771	2.160	2.533	3.012	3.372	4.221
14	.692	.868	1.345	1.761	2.145	2.510	2.977	3.326	4.140
15	.691	.866	1.341	1.753	2.131	2.490	2.947	3.286	4.073
16	.690	.865	1.337	1.746	2.120	2.473	2.921	3.252	4.015
17	.689	.863	1.333	1.740	2.110	2.458	2.898	3.222	3.965
18	.688	.862	1.330	1.734	2.101	2,445	2.878	3.197	3.922
19	.688	.861	1.328	1.729	2.093	2.433	2.861	3.174	3.883
20	.687	.860	1.325	1.725	2.086	2.423	2.845	3.153	3.850
21	.686	.859	1.323	1.721	2.080	2.414	2.831	3.135	3.819
22	.686	.858	1.321	1.717	2.074	2.406	2.819	3.119	3.792
23	.685	.858	1.319	1.714	2.069	2.398	2.807	3.104	3.767
24	.685	.857	1.318	1.711	2.064	2.391	2.797	3.090	3.745
25	.684	.856	1.316	1.708	2.060	2.385	2.787	3.078	3.725
26	.684	.856	1.315	1.706	2.056	2.379	2.779	3.067	3.707
27	.684	.855	1.314	1.703	2.052	2.373	2.771	3.056	3.690
28	.683	.855	1.313	1.701	2.048	2.368	2.763	3.047	3.674
29	.683	.854	1.311	1.699	2.045	2.364	2.756	3.038	3.659
30	.683	.854	1.310	1.697	2.042	2.360	2.750	3.030	3.646
35	.682	.852	1.306	1.690	2.030	2.342	2.724	2.996	3.591
40	.681	.851	1.303	1.684	2.021	2.329	2.704	2.971	3.551
45	.680	.850	1.301	1.680	2.014	2.319	2.690	2.952	3.520
50	.680	.849	1.299	1.676	2.008	2.310	2.678	2.937	3.496
55	.679	.849	1.297	1.673	2.004	2.304	2.669	2.925	3.476
60	.679	.848	1.296	1.671	2.000	2.299	2.660	2.915	3.460
70	.678	.847	1.294	1.667	1.994	2.290	2.648	2.899	3.435
80	.678	.847	1.293	1.665	1.989	2.284	2.638	2.887	3.416
90	.678	.846	1.291	1.662	1.986	2.279	2.631	2.878	3.402
100	.677	.846	1.290	1.661	1.982	2.276	2.625	2.871	3.390
120	.677	.845	1.289	1.658	1.980	2.270	2.617	2.860	3.373
∞	.6745	.8416	1.2816	1.6448	1.9600	2.2414	2.5758	2.8070	3.2905

Parts of this table are reprinted by permission from R. A. Fisher's *Statistical Methods for Research Workers,* published by Oliver & Boyd, Edinburgh (1925–1950); from Maxine Merrington's "Table of Percentage Points of the *t*-Distribution," *Biometrika* 32 (1942):300; and from Bernard Ostle's *Statistics in Research,* Iowa State Univ. Press (1954).

TABLE A 5
Cumulative Distribution of Chi-Square

Degrees of Freedom	Probability of a Greater Value												
	0.995	0.990	0.975	0.950	0.900	0.750	0.500	0.250	0.100	0.050	0.025	0.010	0.005
1	....	....	....	....	0.02	0.10	0.45	1.32	2.71	3.84	5.02	6.63	7.88
2	0.01	0.02	0.05	0.10	0.21	0.58	1.39	2.77	4.61	5.99	7.38	9.21	10.60
3	0.07	0.11	0.22	0.35	0.58	1.21	2.37	4.11	6.25	7.81	9.35	11.34	12.84
4	0.21	0.30	0.48	0.71	1.06	1.92	3.36	5.39	7.78	9.49	11.14	13.28	14.86
5	0.41	0.55	0.83	1.15	1.61	2.67	4.35	6.63	9.24	11.07	12.83	15.09	16.75
6	0.68	0.87	1.24	1.64	2.20	3.45	5.35	7.84	10.64	12.59	14.45	16.81	18.55
7	0.99	1.24	1.69	2.17	2.83	4.25	6.35	9.04	12.02	14.07	16.01	18.48	20.28
8	1.34	1.65	2.18	2.73	3.49	5.07	7.34	10.22	13.36	15.51	17.53	20.09	21.96
9	1.73	2.09	2.70	3.33	4.17	5.90	8.34	11.39	14.68	16.92	19.02	21.67	23.59
10	2.16	2.56	3.25	3.94	4.87	6.74	9.34	12.55	15.99	18.31	20.48	23.21	25.19
11	2.60	3.05	3.82	4.57	5.58	7.58	10.34	13.70	17.28	19.68	21.92	24.72	26.76
12	3.07	3.57	4.40	5.23	6.30	8.44	11.34	14.85	18.55	21.03	23.34	26.22	28.30
13	3.57	4.11	5.01	5.89	7.04	9.30	12.34	15.98	19.81	22.36	24.74	27.69	29.82
14	4.07	4.66	5.63	6.57	7.79	10.17	13.34	17.12	21.06	23.68	26.12	29.14	31.32
15	4.60	5.23	6.27	7.26	8.55	11.04	14.34	18.25	22.31	25.00	27.49	30.58	32.80
16	5.14	5.81	6.91	7.96	9.31	11.91	15.34	19.37	23.54	26.30	28.85	32.00	34.27
17	5.70	6.41	7.56	8.67	10.09	12.79	16.34	20.49	24.77	27.59	30.19	33.41	35.72
18	6.26	7.01	8.23	9.39	10.86	13.68	17.34	21.60	25.99	28.87	31.53	34.81	37.16
19	6.84	7.63	8.91	10.12	11.65	14.56	18.34	22.72	27.20	30.14	32.85	36.19	38.58
20	7.43	8.26	9.59	10.85	12.44	15.45	19.34	23.83	28.41	31.41	34.17	37.57	40.00

Condensed from table with 6 significant figures by Catherine M. Thompson, by permission of the editor of *Biometrika*.
Read the significance *P* from the normal table A 3. Use only one tail.

TABLE A 5—*Continued*

Degrees of Freedom	Probability of a Greater Value												
	0.995	0.990	0.975	0.950	0.900	0.750	0.500	0.250	0.100	0.050	0.025	0.010	0.005
21	8.03	8.90	10.28	11.59	13.24	16.34	20.34	24.93	29.62	32.67	35.48	38.93	41.40
22	8.64	9.54	10.98	12.34	14.04	17.24	21.34	26.04	30.81	33.92	36.78	40.29	42.80
23	9.26	10.20	11.69	13.09	14.85	18.14	22.34	27.14	32.01	35.17	38.08	41.64	44.18
24	9.89	10.86	12.40	13.85	15.66	19.04	23.34	28.24	33.20	36.42	39.36	42.98	45.56
25	10.52	11.52	13.12	14.61	16.47	19.94	24.34	29.34	34.38	37.65	40.65	44.31	46.93
26	11.16	12.20	13.84	15.38	17.29	20.84	25.34	30.43	35.56	38.89	41.92	45.64	48.29
27	11.81	12.88	14.57	16.15	18.11	21.75	26.34	31.53	36.74	40.11	43.19	46.96	49.64
28	12.46	13.56	15.31	16.93	18.94	22.66	27.34	32.62	37.92	41.34	44.46	48.28	50.99
29	13.12	14.26	16.05	17.71	19.77	23.57	28.34	33.71	39.09	42.56	45.72	49.59	52.34
30	13.79	14.95	16.79	18.49	20.60	24.48	29.34	34.80	40.26	43.77	46.98	50.89	53.67
40	20.71	22.16	24.43	26.51	29.05	33.66	39.34	45.62	51.80	55.76	59.34	63.69	66.77
50	27.99	29.71	32.36	34.76	37.69	42.94	49.33	56.33	63.17	67.50	71.42	76.15	79.49
60	35.53	37.48	40.48	43.19	46.46	52.29	59.33	66.98	74.40	79.08	83.30	88.38	91.95
70	43.28	45.44	48.76	51.74	55.33	61.70	69.33	77.58	85.53	90.53	95.02	100.42	104.22
80	51.17	53.54	57.15	60.39	64.28	71.14	79.33	88.13	96.58	101.88	106.63	112.33	116.32
90	59.20	61.75	65.65	69.13	73.29	80.62	89.33	98.64	107.56	113.14	118.14	124.12	128.30
100	67.33	70.06	74.22	77.93	82.36	90.13	99.33	109.14	118.50	124.34	129.56	135.81	140.17

For numbers of degrees of freedom greater than 100, calculate the approximate normal deviate $Z = \sqrt{2\chi^2} - \sqrt{2(\text{df}) - 1}$.

TABLE A 6

MULTIPLIER OF THE RANGE w THAT GIVES AN UNBIASED ESTIMATE OF σ, VARIANCE (V), AND RELATIVE EFFICIENCY OF THE ESTIMATE (NORMAL SAMPLES ASSUMED)

n	Multiplier	V/σ^2	Rel. Eff.	n	Multiplier	V/σ^2	Rel. Eff.
2	0.886	0.571	1.000	12	0.307	0.0571	0.815
3	.591	.275	0.992	13	.300	.0533	.797
4	.486	.183	0.975	14	.294	.0502	.781
5	.430	.138	0.955	15	.288	.0474	.766
6	.395	.112	0.933	16	.283	.0451	.751
7	.370	.095	0.912	17	.279	.0430	.738
8	.351	.083	0.890	18	.275	.0412	.726
9	.337	.074	0.869	19	.271	.0395	.712
10	.325	.067	0.850	20	0.268	0.0381	0.700
11	0.315	0.0616	0.831				

TABLE A 7

(*i*) SIGNIFICANCE LEVELS OF $t_w = (\overline{X} - \mu)/w$ IN NORMAL SAMPLES. (*ii*) SIGNIFICANCE LEVELS OF $(\overline{X}_1 - \overline{X}_2)/[(1/2)(w_1 + w_2)]$ FOR TWO NORMAL SAMPLES OF EQUAL SIZE. TWO-TAILED TESTS. DIVIDE P BY 2 FOR ONE-TAILED TEST

Size of Sample	Probability P			
	0.10	0.05	0.02	0.01
(*i*) 2	3.157	6.353	15.910	31.828
3	0.885	1.304	2.111	3.008
4	.529	0.717	1.023	1.316
5	.388	.507	0.685	0.843
6	.312	.399	.523	.628
7	.263	.333	.429	.507
8	.230	.288	.366	.429
9	.205	.255	.322	.374
10	.186	.230	.288	.333
11	.170	.210	.262	.302
12	.158	.194	.241	.277
13	.147	.181	.224	.256
14	.138	.170	.209	.239
15	.131	.160	.197	.224
16	.124	.151	.186	.212
17	.118	.144	.177	.201
18	.113	.137	.168	.191
19	.108	.131	.161	.182
20	.104	.126	.154	.175
(*ii*) 2	2.322	3.427	5.553	7.916
3	0.974	1.272	1.715	2.093
4	.644	0.813	1.047	1.237
5	.493	.613	0.772	0.896
6	.405	.499	.621	.714
7	.347	.426	.525	.600
8	.306	.373	.459	.521
9	.275	.334	.409	.464
10	.250	.304	.371	.419
11	.233	.280	.340	.384
12	.214	.260	.315	.355
13	.201	.243	.294	.331
14	.189	.228	.276	.311
15	.179	.216	.261	.293
16	.170	.205	.247	.278
17	.162	.195	.236	.264
18	.155	.187	.225	.252
19	.149	.179	.216	.242
20	.143	.172	.207	.232

From more extensive tables by permission of E. Lord and the editor of *Biometrika*.

TABLE A 8
SIGN TEST. SMALLER NUMBER OF LIKE SIGNS REQUIRED FOR SIGNIFICANCE, WITH TWO-TAILED PROBABILITIES

Number of Pairs	Significance Level			
	1%	2%	5%	10%
5	...	...	...	0(.062)
6	...	...	0(.031)	0(.031)
7	...	0(.016)	0(.016)	0(.016)
8	0(.008)	0(.008)	0(.008)	1(.070)
9	0(.004)	0(.004)	1(.039)	1(.039)
10	0(.002)	0(.002)	1(.021)	1(.021)
11	0(.001)	1(.012)	1(.012)	2(.065)
12	1(.006)	1(.006)	2(.039)	2(.039)
13	1(.003)	1(.003)	2(.022)	3(.092)
14	1(.002)	2(.013)	2(.013)	3(.057)
15	2(.007)	2(.007)	3(.035)	3(.035)
16	2(.004)	2(.004)	3(.021)	4(.077)
17	2(.002)	3(.013)	4(.049)	4(.049)
18	3(.008)	3(.008)	4(.031)	5(.096)
19	3(.004)	4(.019)	4(.019)	5(.063)
20	3(.003)	4(.012)	5(.041)	5(.041)

TABLE A 9
SIGNED-RANK TEST. SUM OF RANKS AT APPROXIMATE 10%, 5%, 2%, AND 1% LEVELS OF P† (these or smaller values indicate rejection; two-tailed test)

n	10%	5%	2%	1%
7	3(.078)	2(.047)	0(.016)	none
8	5(.078)	3(.039)	1(.016)	0(.008)
9	8(.098)	5(.039)	3(.020)	1(.008)
10	10(.098)	8(.049)	5(.020)	3(.010)
11	14(.102)	11(.054)	7(.019)	5(.010)
12	17(.092)	14(.052)	10(.021)	7(.009)
13	21(.094)	17(.047)	13(.021)	10(.010)
14	26(.104)	21(.049)	16(.020)	13(.011)
15	30(.095)	25(.048)	20(.022)	16(.010)
16	36(.105)	30(.051)	24(.021)	19(.009)
17	41(.098)	35(.051)	28(.020)	23(.009)
18	47(.099)	40(.048)	33(.021)	28(.010)
19	53(.096)	46(.049)	38(.020)	32(.009)
20	60(.097)	52(.048)	43(.019)	37(.009)

Reprinted from *Selected Tables in Mathematical Statistics,* vol. 1, by permission of the American Mathematical Society.

†The numbers in ( ) are the actual significance probabilities.

TABLE A 10
WILCOXON'S TWO-SAMPLE RANK TEST (MANN-WHITNEY TEST).
VALUES OF T AT TWO LEVELS
(these or smaller values cause rejection; two-tailed test; taken $n_1 \leq n$†)

0.05 Level of T

n_2 ↓ \ n_1 →	2	3	4	5	6	7	8	9	10	11	12	13	14	15
4			10											
5		6	11	17										
6		7	12	18	26									
7		7	13	20	27	36								
8	3	8	14	21	29	38	49							
9	3	8	15	22	31	40	51	63						
10	3	9	15	23	32	42	53	65	78					
11	4	9	16	24	34	44	55	68	81	96				
12	4	10	17	26	35	46	58	71	85	99	115			
13	4	10	18	27	37	48	60	73	88	103	119	137		
14	4	11	19	28	38	50	63	76	91	106	123	141	160	
15	4	11	20	29	40	52	65	79	94	110	127	145	164	185
16	4	12	21	31	42	54	67	82	97	114	131	150	169	
17	5	12	21	32	43	56	70	84	100	117	135	154		
18	5	13	22	33	45	58	72	87	103	121	139			
19	5	13	23	34	46	60	74	90	107	124				
20	5	14	24	35	48	62	77	93	110					
21	6	14	25	37	50	64	79	95						
22	6	15	26	38	51	66	82							
23	6	15	27	39	53	68								
24	6	16	28	40	55									
25	6	16	28	42										
26	7	17	29											
27	7	17												
28	7													

Table is reprinted from C. White, *Biometrics*, 8(1952): 37–38, who extended the method of Wilcoxon.

†The values of n_1 and n_2 are the numbers of cases in the two groups. If the groups are unequal in size, n_1 refers to the smaller.

TABLE A 11(*ii*)
TWO-TAILED SIGNIFICANCE LEVELS OF THE SPEARMAN RANK CORRELATION COEFFICIENT r_s; FIGURES IN PARENTHESES GIVE PROBABILITIES OF A VALUE $\geq |r_s|$ ON H_0
(since r_s is discretely distributed, these probabilities differ somewhat from nominal significance levels)

Sample size	Significance Level			
n	10%	5%	2%	1%
4	1.000 (.083)	none	none	none
5	0.900 (.083)	1.000 (.017)	1.000 (.017)	none
6	0.771 (.100)	0.886 (.033)	0.943 (.017)	1.00 (.003)
7	0.714 (.088)	0.786 (.051)	0.892 (.013)	0.929 (.006)
8	0.643 (.096)	0.738 (.046)	0.810 (.022)	0.857 (.010)
9	0.600 (.096)	0.683 (.050)	0.783 (.018)	0.817 (.010)
10	0.564 (.096)	0.648 (.048)	0.733 (.020)	0.781 (.010)
11 or more	Use table A 11 for r†			

†For sample size 11, use table A 11(*i*) for 9 df; for sample size $n > 11$, use table A 11(*i*) for $(n - 2)$ df.

TABLE A 12
TABLE OF $z = (1/2) \ln (1 + r)/(1 - r)$ TO TRANSFORM THE CORRELATION COEFFICIENT

r	0.00	0.01	0.02	0.03	0.04	0.05	0.06	0.07	0.08	0.09
.0	0.000	0.010	0.020	0.030	0.040	0.050	0.060	0.070	0.080	0.090
.1	.100	.110	.121	.131	.141	.151	.161	.172	.182	.192
.2	.203	.213	.224	.234	.245	.255	.266	.277	.288	.299
.3	.310	.321	.332	.343	.354	.365	.377	.388	.400	.412
.4	.424	.436	.448	.460	.472	.485	.497	.510	.523	.536
.5	.549	.563	.576	.590	.604	.618	.633	.648	.662	.678
.6	.693	.709	.725	.741	.758	.775	.793	.811	.829	.848
.7	.867	.887	.908	.929	.950	.973	.996	1.020	1.045	1.071
.8	1.099	1.127	1.157	1.188	1.221	1.256	1.293	1.333	1.376	1.422

r	0.000	0.001	0.002	0.003	0.004	0.005	0.006	0.007	0.008	0.009
.90	1.472	1.478	1.483	1.488	1.494	1.499	1.505	1.510	1.516	1.522
.91	1.528	1.533	1.539	1.545	1.551	1.557	1.564	1.570	1.576	1.583
.92	1.589	1.596	1.602	1.609	1.616	1.623	1.630	1.637	1.644	1.651
.93	1.658	1.666	1.673	1.681	1.689	1.697	1.705	1.713	1.721	1.730
.94	1.738	1.747	1.756	1.764	1.774	1.783	1.792	1.802	1.812	1.822
.95	1.832	1.842	1.853	1.863	1.874	1.886	1.897	1.909	1.921	1.933
.96	1.946	1.959	1.972	1.986	2.000	2.014	2.029	2.044	2.060	2.076
.97	2.092	2.109	2.127	2.146	2.165	2.185	2.205	2.227	2.249	2.273
.98	2.298	2.323	2.351	2.380	2.410	2.443	2.477	2.515	2.555	2.599
.99	2.646	2.700	2.759	2.826	2.903	2.994	3.106	3.250	3.453	3.800

TABLE A 13

TABLE OF r IN TERMS OF z

z	0.00	0.01	0.02	0.03	0.04	0.05	0.06	0.07	0.08	0.09
0.0	0.000	0.010	0.020	0.030	0.040	0.050	0.060	0.070	0.080	0.090
.1	.100	.110	.119	.129	.139	.149	.159	.168	.178	.187
.2	.197	.207	.216	.226	.236	.245	.254	.264	.273	.282
.3	.291	.300	.310	.319	.327	.336	.345	.354	.363	.371
.4	.380	.389	.397	.405	.414	.422	.430	.438	.446	.454
.5	.462	.470	.478	.485	.493	.500	.508	.515	.523	.530
.6	.537	.544	.551	.558	.565	.572	.578	.585	.592	.598
.7	.604	.611	.617	.623	.629	.635	.641	.647	.653	.658
.8	.664	.670	.675	.680	.686	.691	.696	.701	.706	.711
.9	.716	.721	.726	.731	.735	.740	.744	.749	.753	.757
1.0	.762	.766	.770	.774	.778	.782	.786	.790	.793	.797
1.1	.800	.804	.808	.811	.814	.818	.821	.824	.828	.831
1.2	.834	.837	.840	.843	.846	.848	.851	.854	.856	.859
1.3	.862	.864	.867	.869	.872	.874	.876	.879	.881	.883
1.4	.885	.888	.890	.892	.894	.896	.898	.900	.902	.903
1.5	.905	.907	.909	.910	.912	.914	.915	.917	.919	.920
1.6	.922	.923	.925	.926	.928	.929	.930	.932	.933	.934
1.7	.935	.937	.938	.939	.940	.941	.942	.944	.945	.946
1.8	.947	.948	.949	.950	.951	.952	.953	.954	.954	.955
1.9	.956	.957	.958	.959	.960	.960	.961	.962	.963	.963
2.0	.964	.965	.965	.966	.967	.967	.968	.969	.969	.970
2.1	.970	.971	.972	.972	.973	.973	.974	.974	.975	.975
2.2	.976	.976	.977	.977	.978	.978	.978	.979	.979	.980
2.3	.980	.980	.981	.981	.982	.982	.982	.983	.983	.983
2.4	.984	.984	.984	.985	.985	.985	.986	.986	.986	.986
2.5	.987	.987	.987	.987	.988	.988	.988	.988	.989	.989
2.6	.989	.989	.989	.990	.990	.990	.990	.990	.991	.991
2.7	.991	.991	.991	.992	.992	.992	.992	.992	.992	.992
2.8	.993	.993	.993	.993	.993	.993	.993	.994	.994	.994
2.9	.994	.994	.994	.994	.994	.995	.995	.995	.995	.995

$r = (e^{2z} - 1)/(e^{2z} + 1)$.

TABLE A 14, Part I
The 5% (roman type) and 1% (boldface type) Points for the Distribution of F

ν_2†	ν_1 df in Numerator																								ν_2†
	1	2	3	4	5	6	7	8	9	10	11	12	14	16	20	24	30	40	50	75	100	200	500	∞	
1	161	200	216	225	230	234	237	239	241	242	243	244	245	246	248	249	250	251	252	253	253	254	254	254	1
	4,052	**4,999**	**5,403**	**5,625**	**5,764**	**5,859**	**5,928**	**5,981**	**6,022**	**6,056**	**6,082**	**6,106**	**6,142**	**6,169**	**6,208**	**6,234**	**6,261**	**6,286**	**6,302**	**6,323**	**6,334**	**6,352**	**6,361**	**6,366**	
2	18.51	19.00	19.16	19.25	19.30	19.33	19.36	19.37	19.38	19.39	19.40	19.41	19.42	19.43	19.44	19.45	19.46	19.47	19.47	19.48	19.49	19.49	19.50	19.50	2
	98.49	**99.00**	**99.17**	**99.25**	**99.30**	**99.33**	**99.36**	**99.37**	**99.39**	**99.40**	**99.41**	**99.42**	**99.43**	**99.44**	**99.45**	**99.46**	**99.47**	**99.48**	**99.48**	**99.49**	**99.49**	**99.49**	**99.50**	**99.50**	
3	10.13	9.55	9.28	9.12	9.01	8.94	8.88	8.84	8.81	8.78	8.76	8.74	8.71	8.69	8.66	8.64	8.62	8.60	8.58	8.57	8.56	8.54	8.54	8.53	3
	34.12	**30.82**	**29.46**	**28.71**	**28.24**	**27.91**	**27.67**	**27.49**	**27.34**	**27.23**	**27.13**	**27.05**	**26.92**	**26.83**	**26.69**	**26.60**	**26.50**	**26.41**	**26.35**	**26.27**	**26.23**	**26.18**	**26.14**	**26.12**	
4	7.71	6.94	6.59	6.39	6.26	6.16	6.09	6.04	6.00	5.96	5.93	5.91	5.87	5.84	5.80	5.77	5.74	5.71	5.70	5.68	5.66	5.65	5.64	5.63	4
	21.20	**18.00**	**16.69**	**15.98**	**15.52**	**15.21**	**14.98**	**14.80**	**14.66**	**14.54**	**14.45**	**14.37**	**14.24**	**14.15**	**14.02**	**13.93**	**13.83**	**13.74**	**13.69**	**13.61**	**13.57**	**13.52**	**13.48**	**13.46**	
5	6.61	5.79	5.41	5.19	5.05	4.95	4.88	4.82	4.78	4.74	4.70	4.68	4.64	4.60	4.56	4.53	4.50	4.46	4.44	4.42	4.40	4.38	4.37	4.36	5
	16.26	**13.27**	**12.06**	**11.39**	**10.97**	**10.67**	**10.45**	**10.29**	**10.15**	**10.05**	**9.96**	**9.89**	**9.77**	**9.68**	**9.55**	**9.47**	**9.38**	**9.29**	**9.24**	**9.17**	**9.13**	**9.07**	**9.04**	**9.02**	
6	5.99	5.14	4.76	4.53	4.39	4.28	4.21	4.15	4.10	4.06	4.03	4.00	3.96	3.92	3.87	3.84	3.81	3.77	3.75	3.72	3.71	3.69	3.68	3.67	6
	13.74	**10.92**	**9.78**	**9.15**	**8.75**	**8.47**	**8.26**	**8.10**	**7.98**	**7.87**	**7.79**	**7.72**	**7.60**	**7.52**	**7.39**	**7.31**	**7.23**	**7.14**	**7.09**	**7.02**	**6.99**	**6.94**	**6.90**	**6.88**	
7	5.59	4.74	4.35	4.12	3.97	3.87	3.79	3.73	3.68	3.63	3.60	3.57	3.52	3.49	3.44	3.41	3.38	3.34	3.32	3.29	3.28	3.25	3.24	3.23	7
	12.25	**9.55**	**8.45**	**7.85**	**7.46**	**7.19**	**7.00**	**6.84**	**6.71**	**6.62**	**6.54**	**6.47**	**6.35**	**6.27**	**6.15**	**6.07**	**5.98**	**5.90**	**5.85**	**5.78**	**5.75**	**5.70**	**5.67**	**5.65**	
8	5.32	4.46	4.07	3.84	3.69	3.58	3.50	3.44	3.39	3.34	3.31	3.28	3.23	3.20	3.15	3.12	3.08	3.05	3.03	3.00	2.98	2.96	2.94	2.93	8
	11.26	**8.65**	**7.59**	**7.01**	**6.63**	**6.37**	**6.19**	**6.03**	**5.91**	**5.82**	**5.74**	**5.67**	**5.56**	**5.48**	**5.36**	**5.28**	**5.20**	**5.11**	**5.06**	**5.00**	**4.96**	**4.91**	**4.88**	**4.86**	
9	5.12	4.26	3.86	3.63	3.48	3.37	3.29	3.23	3.18	3.13	3.10	3.07	3.02	2.98	2.93	2.90	2.86	2.82	2.80	2.77	2.76	2.73	2.72	2.71	9
	10.56	**8.02**	**6.99**	**6.42**	**6.06**	**5.80**	**5.62**	**5.47**	**5.35**	**5.26**	**5.18**	**5.11**	**5.00**	**4.92**	**4.80**	**4.73**	**4.64**	**4.56**	**4.51**	**4.45**	**4.41**	**4.36**	**4.33**	**4.31**	
10	4.96	4.10	3.71	3.48	3.33	3.22	3.14	3.07	3.02	2.97	2.94	2.91	2.86	2.82	2.77	2.74	2.70	2.67	2.64	2.61	2.59	2.56	2.55	2.54	10
	10.04	**7.56**	**6.55**	**5.99**	**5.64**	**5.39**	**5.21**	**5.06**	**4.95**	**4.85**	**4.78**	**4.71**	**4.60**	**4.52**	**4.41**	**4.33**	**4.25**	**4.17**	**4.12**	**4.05**	**4.01**	**3.96**	**3.93**	**3.91**	
11	4.84	3.98	3.59	3.36	3.20	3.09	3.01	2.95	2.90	2.86	2.82	2.79	2.74	2.70	2.65	2.61	2.57	2.53	2.50	2.47	2.45	2.42	2.41	2.40	11
	9.65	**7.20**	**6.22**	**5.67**	**5.32**	**5.07**	**4.88**	**4.74**	**4.63**	**4.54**	**4.46**	**4.40**	**4.29**	**4.21**	**4.10**	**4.02**	**3.94**	**3.86**	**3.80**	**3.74**	**3.70**	**3.66**	**3.62**	**3.60**	
12	4.75	3.88	3.49	3.26	3.11	3.00	2.92	2.85	2.80	2.76	2.72	2.69	2.64	2.60	2.54	2.50	2.46	2.42	2.40	2.36	2.35	2.32	2.31	2.30	12
	9.33	**6.93**	**5.95**	**5.41**	**5.06**	**4.82**	**4.65**	**4.50**	**4.39**	**4.30**	**4.22**	**4.16**	**4.05**	**3.98**	**3.86**	**3.78**	**3.70**	**3.61**	**3.56**	**3.49**	**3.46**	**3.41**	**3.38**	**3.36**	
13	4.67	3.80	3.41	3.18	3.02	2.92	2.84	2.77	2.72	2.67	2.63	2.60	2.55	2.51	2.46	2.42	2.38	2.34	2.32	2.28	2.26	2.24	2.22	2.21	13
	9.07	**6.70**	**5.74**	**5.20**	**4.86**	**4.62**	**4.44**	**4.30**	**4.19**	**4.10**	**4.02**	**3.96**	**3.85**	**3.78**	**3.67**	**3.59**	**3.51**	**3.42**	**3.37**	**3.30**	**3.27**	**3.21**	**3.18**	**3.16**	

The function is computed in part from Fisher's table VI (7). Additional entries are by interpolation, mostly graphic.

†df in denominator.

TABLE A 14, PART I—*Continued*

ν_2†	ν_1 df in Numerator																								ν_2†
	1	2	3	4	5	6	7	8	9	10	11	12	14	16	20	24	30	40	50	75	100	200	500	∞	
14	4.60	3.74	3.34	3.11	2.96	2.85	2.77	2.70	2.65	2.60	2.56	2.53	2.48	2.44	2.39	2.35	2.31	2.27	2.24	2.21	2.19	2.16	2.14	2.13	14
	8.86	**6.51**	**5.56**	**5.03**	**4.69**	**4.46**	**4.28**	**4.14**	**4.03**	**3.94**	**3.86**	**3.80**	**3.70**	**3.62**	**3.51**	**3.43**	**3.34**	**3.26**	**3.21**	**3.14**	**3.11**	**3.06**	**3.02**	**3.00**	
15	4.54	3.68	3.29	3.06	2.90	2.79	2.70	2.64	2.59	2.55	2.51	2.48	2.43	2.39	2.33	2.29	2.25	2.21	2.18	2.15	2.12	2.10	2.08	2.07	15
	8.68	**6.36**	**5.42**	**4.89**	**4.56**	**4.32**	**4.14**	**4.00**	**3.89**	**3.80**	**3.73**	**3.67**	**3.56**	**3.48**	**3.36**	**3.29**	**3.20**	**3.12**	**3.07**	**3.00**	**2.97**	**2.92**	**2.89**	**2.87**	
16	4.49	3.63	3.24	3.01	2.85	2.74	2.66	2.59	2.54	2.49	2.45	2.42	2.37	2.33	2.28	2.24	2.20	2.16	2.13	2.09	2.07	2.04	2.02	2.01	16
	8.53	**6.23**	**5.29**	**4.77**	**4.44**	**4.20**	**4.03**	**3.89**	**3.78**	**3.69**	**3.61**	**3.55**	**3.45**	**3.37**	**3.25**	**3.18**	**3.10**	**3.01**	**2.96**	**2.98**	**2.86**	**2.80**	**2.77**	**2.75**	
17	4.45	3.59	3.20	2.96	2.81	2.70	2.62	2.55	2.50	2.45	2.41	2.38	2.33	2.29	2.23	2.19	2.15	2.11	2.08	2.04	2.02	1.99	1.97	1.96	17
	8.40	**6.11**	**5.18**	**4.67**	**4.34**	**4.10**	**3.93**	**3.79**	**3.68**	**3.59**	**3.52**	**3.45**	**3.35**	**3.27**	**3.16**	**3.08**	**3.00**	**2.92**	**2.86**	**2.79**	**2.76**	**2.70**	**2.67**	**2.65**	
18	4.41	3.55	3.16	2.93	2.77	3.66	2.58	2.51	2.46	2.41	2.37	2.34	2.29	2.25	2.19	2.15	2.11	2.07	2.04	2.00	1.98	1.95	1.93	1.92	18
	8.28	**6.01**	**5.09**	**4.58**	**4.25**	**4.01**	**3.85**	**3.71**	**3.60**	**3.51**	**3.44**	**3.37**	**3.27**	**3.19**	**3.07**	**3.00**	**2.91**	**2.83**	**2.78**	**2.71**	**2.68**	**2.62**	**2.59**	**2.57**	
19	4.38	3.52	3.13	2.90	2.74	2.63	2.55	2.48	2.43	2.38	2.34	2.31	2.26	2.21	2.15	2.11	2.07	2.02	2.00	1.96	1.94	1.91	1.90	1.88	19
	8.18	**5.93**	**5.01**	**4.50**	**4.17**	**3.94**	**3.77**	**3.63**	**3.52**	**3.43**	**3.36**	**3.30**	**3.19**	**3.12**	**3.00**	**2.92**	**2.84**	**2.76**	**2.70**	**2.63**	**2.60**	**2.54**	**2.51**	**2.49**	
20	4.35	3.49	3.10	2.87	2.71	2.60	2.52	2.45	2.40	2.35	2.31	2.28	2.23	2.18	2.12	2.08	2.04	1.99	1.96	1.92	1.90	1.87	1.85	1.84	20
	8.10	**5.85**	**4.94**	**4.43**	**4.10**	**3.87**	**3.71**	**3.56**	**3.45**	**3.37**	**3.30**	**3.23**	**3.13**	**3.05**	**2.94**	**2.86**	**2.77**	**2.69**	**2.63**	**2.56**	**2.53**	**2.47**	**2.44**	**2.42**	
21	4.32	3.47	3.07	2.84	2.68	2.57	2.49	2.42	2.37	2.32	2.28	2.25	2.20	2.15	2.09	2.05	2.00	1.96	1.93	1.89	1.87	1.84	1.82	1.81	21
	8.02	**5.78**	**4.87**	**4.37**	**4.04**	**3.81**	**3.65**	**3.51**	**3.40**	**3.31**	**3.24**	**3.17**	**3.07**	**2.99**	**2.88**	**2.80**	**2.72**	**2.63**	**2.58**	**2.51**	**2.47**	**2.42**	**2.38**	**2.36**	
22	4.30	3.44	3.05	2.82	2.66	2.55	2.47	2.40	2.35	2.30	2.26	2.23	2.18	2.13	2.07	2.03	1.98	1.93	1.91	1.87	1.84	1.81	1.80	1.78	22
	7.94	**5.72**	**4.82**	**4.31**	**3.99**	**3.76**	**3.59**	**3.45**	**3.35**	**3.26**	**3.18**	**3.12**	**3.02**	**2.94**	**2.83**	**2.75**	**2.67**	**2.58**	**2.53**	**2.46**	**2.42**	**2.37**	**2.33**	**2.31**	
23	4.28	3.42	3.03	2.80	2.64	2.53	2.45	2.38	2.32	2.28	2.24	2.20	2.14	2.10	2.04	2.00	1.96	1.91	1.88	1.84	1.82	1.79	1.77	1.76	23
	7.88	**5.66**	**4.76**	**4.26**	**3.94**	**3.71**	**3.54**	**3.41**	**3.30**	**3.21**	**3.14**	**3.07**	**2.97**	**2.89**	**2.78**	**2.70**	**2.62**	**2.53**	**2.48**	**2.41**	**2.37**	**2.32**	**2.28**	**2.26**	
24	4.26	3.40	3.01	2.78	2.62	2.51	2.43	2.36	2.30	2.26	2.22	2.18	2.13	2.09	2.02	1.98	1.94	1.89	1.86	1.82	1.80	1.76	1.74	1.73	24
	7.82	**5.61**	**4.72**	**4.22**	**3.90**	**3.67**	**3.50**	**3.36**	**3.25**	**3.17**	**3.09**	**3.03**	**2.93**	**2.85**	**2.74**	**2.66**	**2.58**	**2.49**	**2.44**	**2.36**	**2.33**	**2.27**	**2.23**	**2.21**	
25	4.24	3.38	2.99	2.76	2.60	2.49	2.41	2.34	2.28	2.24	2.20	2.16	2.11	2.06	2.00	1.96	1.92	1.87	1.84	1.80	1.77	1.74	1.72	1.71	25
	7.77	**5.57**	**4.68**	**4.18**	**3.86**	**3.63**	**3.46**	**3.32**	**3.21**	**3.13**	**3.05**	**2.99**	**2.89**	**2.81**	**2.70**	**2.62**	**2.54**	**2.45**	**2.40**	**2.32**	**2.29**	**2.23**	**2.19**	**2.17**	
26	4.22	3.37	2.98	2.74	2.59	2.47	2.39	2.32	2.27	2.22	2.18	2.15	2.10	2.05	1.99	1.95	1.90	1.85	1.82	1.78	1.76	1.72	1.70	1.69	26
	7.72	**5.53**	**4.64**	**4.14**	**3.82**	**3.59**	**3.42**	**3.29**	**3.17**	**3.09**	**3.02**	**2.96**	**2.86**	**2.77**	**2.66**	**2.58**	**2.50**	**2.41**	**2.36**	**2.28**	**2.25**	**2.19**	**2.15**	**2.13**	

†df in denominator.

TABLE A 14, PART I—*Continued*

ν_2†	ν_1 df in Numerator: 1	2	3	4	5	6	7	8	9	10	11	12	14	16	20	24	30	40	50	75	100	200	500	∞	ν_2†
27	4.21	3.35	2.96	2.73	2.57	2.46	2.37	2.30	2.25	2.20	2.16	2.13	2.08	2.03	1.97	1.93	1.88	1.84	1.80	1.76	1.74	1.71	1.68	1.67	27
	7.68	**5.49**	**4.60**	**4.11**	**3.79**	**3.56**	**3.39**	**3.26**	**3.14**	**3.06**	**2.98**	**2.93**	**2.83**	**2.74**	**2.63**	**2.55**	**2.47**	**2.38**	**2.33**	**2.25**	**2.21**	**2.16**	**2.12**	**2.10**	
28	4.20	3.34	2.95	2.71	2.56	2.44	2.36	2.29	2.24	2.19	2.15	2.12	2.06	2.02	1.96	1.91	1.87	1.81	1.78	1.75	1.72	1.69	1.67	1.65	28
	7.64	**5.45**	**4.57**	**4.07**	**3.76**	**3.53**	**3.36**	**3.23**	**3.11**	**3.03**	**2.95**	**2.90**	**2.80**	**2.71**	**2.60**	**2.52**	**2.44**	**2.35**	**2.30**	**2.22**	**2.18**	**2.13**	**2.09**	**2.06**	
29	4.18	3.33	2.93	2.70	2.54	2.43	2.35	2.28	2.22	2.18	2.14	2.10	2.05	2.00	1.94	1.90	1.85	1.80	1.77	1.73	1.71	1.68	1.65	1.64	29
	7.60	**5.42**	**4.54**	**4.04**	**3.73**	**3.50**	**3.33**	**3.20**	**3.08**	**3.00**	**2.92**	**2.87**	**2.77**	**2.68**	**2.57**	**2.49**	**2.41**	**2.32**	**2.27**	**2.19**	**2.15**	**2.10**	**2.06**	**2.03**	
30	4.17	3.32	2.92	2.69	2.53	2.42	2.34	2.27	2.21	2.16	2.12	2.09	2.04	1.99	1.93	1.89	1.84	1.79	1.76	1.72	1.69	1.66	1.64	1.62	30
	7.56	**5.39**	**4.51**	**4.02**	**3.70**	**3.47**	**3.30**	**3.17**	**3.06**	**2.98**	**2.90**	**2.84**	**2.74**	**2.66**	**2.55**	**2.47**	**2.38**	**2.29**	**2.24**	**2.16**	**2.13**	**2.07**	**2.03**	**2.01**	
32	4.15	3.30	2.90	2.67	2.51	2.40	2.32	2.25	2.19	2.14	2.10	2.07	2.02	1.97	1.91	1.86	1.82	1.76	1.74	1.69	1.67	1.64	1.61	1.59	32
	7.50	**5.34**	**4.46**	**3.97**	**3.66**	**3.42**	**3.25**	**3.12**	**3.01**	**2.94**	**2.86**	**2.80**	**2.70**	**2.62**	**2.51**	**2.42**	**2.34**	**2.25**	**2.20**	**2.12**	**2.08**	**2.02**	**1.98**	**1.96**	
34	4.13	3.28	2.88	2.65	2.49	2.38	2.30	2.23	2.17	2.12	2.08	2.05	2.00	1.95	1.89	1.84	1.80	1.74	1.71	1.67	1.64	1.61	1.59	1.57	34
	7.44	**5.29**	**4.42**	**3.93**	**3.61**	**3.38**	**3.21**	**3.08**	**2.97**	**2.89**	**2.82**	**2.76**	**2.66**	**2.58**	**2.47**	**2.38**	**2.30**	**2.21**	**2.15**	**2.08**	**2.04**	**1.98**	**1.94**	**1.91**	
36	4.11	3.26	2.86	2.63	2.48	2.36	2.28	2.21	2.15	2.10	2.06	2.03	1.98	1.93	1.87	1.82	1.78	1.72	1.69	1.65	1.62	1.59	1.56	1.55	36
	7.39	**5.25**	**4.38**	**3.89**	**3.58**	**3.35**	**3.18**	**3.04**	**2.94**	**2.86**	**2.78**	**2.72**	**2.62**	**2.54**	**2.43**	**2.35**	**2.26**	**2.17**	**2.12**	**2.04**	**2.00**	**1.94**	**1.90**	**1.87**	
38	4.10	3.25	2.85	2.62	2.46	2.35	2.26	2.19	2.14	2.09	2.05	2.02	1.96	1.92	1.85	1.80	1.76	1.71	1.67	1.63	1.60	1.57	1.54	1.53	38
	7.35	**5.21**	**4.34**	**3.86**	**3.54**	**3.32**	**3.15**	**3.02**	**2.91**	**2.82**	**2.75**	**2.69**	**2.59**	**2.51**	**2.40**	**2.32**	**2.22**	**2.14**	**2.08**	**2.00**	**1.97**	**1.90**	**1.86**	**1.84**	
40	4.08	3.23	2.84	2.61	2.45	2.34	2.25	2.18	2.12	2.07	2.04	2.00	1.95	1.90	1.84	1.79	1.74	1.69	1.66	1.61	1.59	1.55	1.53	1.51	40
	7.31	**5.18**	**4.31**	**3.83**	**3.51**	**3.29**	**3.12**	**2.99**	**2.88**	**2.80**	**2.73**	**2.66**	**2.56**	**2.49**	**2.37**	**2.29**	**2.20**	**2.11**	**2.05**	**1.97**	**1.94**	**1.88**	**1.84**	**1.81**	
42	4.07	3.22	2.83	2.59	2.44	2.32	2.24	2.17	2.11	2.06	2.02	1.99	1.94	1.89	1.82	1.78	1.73	1.68	1.64	1.60	1.57	1.54	1.51	1.49	42
	7.27	**5.15**	**4.29**	**3.80**	**3.49**	**3.26**	**3.10**	**2.96**	**2.86**	**2.77**	**2.70**	**2.64**	**2.54**	**2.46**	**2.35**	**2.26**	**2.17**	**2.08**	**2.02**	**1.94**	**1.91**	**1.85**	**1.80**	**1.78**	
44	4.06	3.21	2.82	2.58	2.43	2.31	2.23	2.16	2.10	2.05	2.01	1.98	1.92	1.88	1.81	1.76	1.72	1.66	1.63	1.58	1.56	1.52	1.50	1.48	44
	7.24	**5.12**	**4.26**	**3.78**	**3.46**	**3.24**	**3.07**	**2.94**	**2.84**	**2.75**	**2.68**	**2.62**	**2.52**	**2.44**	**2.32**	**2.24**	**2.15**	**2.06**	**2.00**	**1.92**	**1.88**	**1.82**	**1.78**	**1.75**	
46	4.05	3.20	2.81	2.57	2.42	2.30	2.22	2.14	2.09	2.04	2.00	1.97	1.91	1.87	1.80	1.75	1.71	1.65	1.62	1.57	1.54	1.51	1.48	1.46	46
	7.21	**5.10**	**4.24**	**3.76**	**3.44**	**3.22**	**3.05**	**2.92**	**2.82**	**2.73**	**2.66**	**2.60**	**2.50**	**2.42**	**2.30**	**2.22**	**2.13**	**2.04**	**1.98**	**1.90**	**1.86**	**1.80**	**1.76**	**1.72**	
48	4.04	3.19	2.80	2.56	2.41	2.30	2.21	2.14	2.08	2.03	1.99	1.96	1.90	1.86	1.79	1.74	1.70	1.64	1.61	1.56	1.53	1.50	1.47	1.45	48
	7.19	**5.08**	**4.22**	**3.74**	**3.42**	**3.20**	**3.04**	**2.90**	**2.80**	**2.71**	**2.64**	**2.58**	**2.48**	**2.40**	**2.28**	**2.20**	**2.11**	**2.02**	**1.96**	**1.88**	**1.84**	**1.78**	**1.73**	**1.70**	

†df in denominator.

TABLE A 14, PART I—*Continued*

ν_2†	ν_1 df in Numerator																								ν_2†
	1	2	3	4	5	6	7	8	9	10	11	12	14	16	20	24	30	40	50	75	100	200	500	∞	
50	4.03	3.18	2.79	2.56	2.40	2.29	2.20	2.13	2.07	2.02	1.98	1.95	1.90	1.85	1.78	1.74	1.69	1.63	1.60	1.55	1.52	1.48	1.46	1.44	50
	7.17	**5.06**	**4.20**	**3.72**	**3.41**	**3.18**	**3.02**	**2.88**	**2.78**	**2.70**	**2.62**	**2.56**	**2.46**	**2.39**	**2.26**	**2.18**	**2.10**	**2.00**	**1.94**	**1.86**	**1.82**	**1.76**	**1.71**	**1.68**	
55	4.02	3.17	2.78	2.54	2.38	2.27	2.18	2.11	2.05	2.00	1.97	1.93	1.88	1.83	1.76	1.72	1.67	1.61	1.58	1.52	1.50	1.46	1.43	1.41	55
	7.12	**5.01**	**4.16**	**3.68**	**3.37**	**3.15**	**2.98**	**2.85**	**2.75**	**2.66**	**2.59**	**2.53**	**2.43**	**2.35**	**2.23**	**2.15**	**2.06**	**1.96**	**1.90**	**1.82**	**1.78**	**1.71**	**1.66**	**1.64**	
60	4.00	3.15	2.76	2.52	2.37	2.25	2.17	2.10	2.04	1.99	1.95	1.92	1.86	1.81	1.75	1.70	1.65	1.59	1.56	1.50	1.48	1.44	1.41	1.39	60
	7.08	**4.98**	**4.13**	**3.65**	**3.34**	**3.12**	**2.95**	**2.82**	**2.72**	**2.63**	**2.56**	**2.50**	**2.40**	**2.32**	**2.20**	**2.12**	**2.03**	**1.93**	**1.87**	**1.79**	**1.74**	**1.68**	**1.63**	**1.60**	
65	3.99	3.14	2.75	2.51	2.36	2.24	2.15	2.08	2.02	1.98	1.94	1.90	1.85	1.80	1.73	1.68	1.63	1.57	1.54	1.49	1.46	1.42	1.39	1.37	65
	7.04	**4.95**	**4.10**	**3.62**	**3.31**	**3.09**	**2.93**	**2.79**	**2.70**	**2.61**	**2.54**	**2.47**	**2.37**	**2.30**	**2.18**	**2.09**	**2.00**	**1.90**	**1.84**	**1.76**	**1.71**	**1.64**	**1.60**	**1.56**	
70	3.98	3.13	2.74	2.50	2.35	2.23	2.14	2.07	2.01	1.97	1.93	1.89	1.84	1.79	1.72	1.67	1.62	1.56	1.53	1.47	1.45	1.40	1.37	1.35	70
	7.01	**4.92**	**4.08**	**3.60**	**3.29**	**3.07**	**2.91**	**2.77**	**2.67**	**2.59**	**2.51**	**2.45**	**2.35**	**2.28**	**2.15**	**2.07**	**1.98**	**1.88**	**1.82**	**1.74**	**1.69**	**1.62**	**1.56**	**1.53**	
80	3.96	3.11	2.72	2.48	2.33	2.21	2.12	2.05	1.99	1.95	1.91	1.88	1.82	1.77	1.70	1.65	1.60	1.54	1.51	1.45	1.42	1.38	1.35	1.32	80
	6.96	**4.88**	**4.04**	**3.56**	**3.25**	**3.04**	**2.87**	**2.74**	**2.64**	**2.55**	**2.48**	**2.41**	**2.32**	**2.24**	**2.11**	**2.03**	**1.94**	**1.84**	**1.78**	**1.70**	**1.65**	**1.57**	**1.52**	**1.49**	
100	3.94	3.09	2.70	2.46	2.30	2.19	2.10	2.03	1.97	1.92	1.88	1.85	1.79	1.75	1.68	1.63	1.57	1.51	1.48	1.42	1.39	1.34	1.30	1.28	100
	6.90	**4.82**	**3.98**	**3.51**	**3.20**	**2.99**	**2.82**	**2.69**	**2.59**	**2.51**	**2.43**	**2.36**	**2.26**	**2.19**	**2.06**	**1.98**	**1.89**	**1.79**	**1.73**	**1.64**	**1.59**	**1.51**	**1.46**	**1.43**	
125	3.92	3.07	2.68	2.44	2.29	2.17	2.08	2.01	1.95	1.90	1.86	1.83	1.77	1.72	1.65	1.60	1.55	1.49	1.45	1.39	1.36	1.31	1.27	1.25	125
	6.84	**4.78**	**3.94**	**3.47**	**3.17**	**2.95**	**2.79**	**2.65**	**2.56**	**2.47**	**2.40**	**2.33**	**2.23**	**2.15**	**2.03**	**1.94**	**1.85**	**1.75**	**1.68**	**1.59**	**1.54**	**1.46**	**1.40**	**1.37**	
150	3.91	3.06	2.67	2.43	2.27	2.16	2.07	2.00	1.94	1.89	1.85	1.82	1.76	1.71	1.64	1.59	1.54	1.47	1.44	1.37	1.34	1.29	1.25	1.22	150
	6.81	**4.75**	**3.91**	**3.44**	**3.14**	**2.92**	**2.76**	**2.62**	**2.53**	**2.44**	**2.37**	**2.30**	**2.20**	**2.12**	**2.00**	**1.91**	**1.83**	**1.72**	**1.66**	**1.56**	**1.51**	**1.43**	**1.37**	**1.33**	
200	3.89	3.04	2.65	2.41	2.26	2.14	2.05	1.98	1.92	1.87	1.83	1.80	1.74	1.69	1.62	1.57	1.52	1.45	1.42	1.35	1.32	1.26	1.22	1.19	200
	6.76	**4.71**	**3.88**	**3.41**	**3.11**	**2.90**	**2.73**	**2.60**	**2.50**	**2.41**	**2.34**	**2.28**	**2.17**	**2.09**	**1.97**	**1.88**	**1.79**	**1.69**	**1.62**	**1.53**	**1.48**	**1.39**	**1.33**	**1.28**	
400	3.86	3.02	2.62	2.39	2.23	2.12	2.03	1.96	1.90	1.85	1.81	1.78	1.72	1.67	1.60	1.54	1.49	1.42	1.38	1.32	1.28	1.22	1.16	1.13	400
	6.70	**4.66**	**3.83**	**3.36**	**3.06**	**2.85**	**2.69**	**2.55**	**2.46**	**2.37**	**2.29**	**2.23**	**2.12**	**2.04**	**1.92**	**1.84**	**1.74**	**1.64**	**1.57**	**1.47**	**1.42**	**1.32**	**1.24**	**1.19**	
1000	3.85	3.00	2.61	2.38	2.22	2.10	2.02	1.95	1.89	1.84	1.80	1.76	1.70	1.65	1.58	1.53	1.47	1.41	1.36	1.30	1.26	1.19	1.13	1.08	1000
	6.66	**4.62**	**3.80**	**3.34**	**3.04**	**2.82**	**2.66**	**2.53**	**2.43**	**2.34**	**2.26**	**2.20**	**2.09**	**2.01**	**1.89**	**1.81**	**1.71**	**1.61**	**1.54**	**1.44**	**1.38**	**1.28**	**1.19**	**1.11**	
∞	3.84	2.99	2.60	2.37	2.21	2.09	2.01	1.94	1.88	1.83	1.79	1.75	1.69	1.64	1.57	1.52	1.46	1.40	1.35	1.28	1.24	1.17	1.11	1.00	∞
	6.63	**4.60**	**3.78**	**3.32**	**3.02**	**2.80**	**2.64**	**2.51**	**2.41**	**2.32**	**2.24**	**2.18**	**2.07**	**1.99**	**1.87**	**1.79**	**1.69**	**1.59**	**1.52**	**1.41**	**1.36**	**1.25**	**1.15**	**1.00**	

†df in denominator.

TABLE A 14, PART II

THE 25%, 10%, 2.5%, AND 0.5% POINTS FOR THE DISTRIBUTION OF F

ν_2†	P	ν_1 df in Numerator 1	2	3	4	5	6	7	8	9	10	12	15	20	24	30	40	60	120	∞
1	0.250	5.83	7.50	8.20	8.58	8.82	8.98	9.10	9.19	9.26	9.32	9.41	9.49	9.58	9.63	9.67	9.71	9.76	9.80	9.85
	.100	39.86	49.50	53.59	55.83	57.24	58.20	58.91	59.44	59.86	60.20	60.70	61.22	61.74	62.00	62.26	62.53	62.79	63.06	63.33
	.025	648	800	864	900	922	937	948	957	963	969	977	985	993	997	1,001	1,006	1,010	1,014	1,018
	.005	16,211	20,000	21,615	22,500	23,056	23,437	23,715	23,925	24,091	24,224	24,426	24,630	24,836	24,940	25,044	25,148	25,253	25,359	25,465
2	.250	2.57	3.00	3.15	3.23	3.28	3.31	3.34	3.35	3.37	3.38	3.39	3.41	3.43	3.43	3.44	3.45	3.46	3.47	3.48
	.100	8.53	9.00	9.16	9.24	9.29	9.33	9.35	9.37	9.38	9.39	9.41	9.42	9.44	9.45	9.46	9.47	9.47	9.48	9.49
	.025	38.51	39.00	39.16	39.25	39.30	39.33	39.36	39.37	39.39	39.40	39.42	39.43	39.45	39.46	39.46	39.47	39.48	39.49	39.50
	.005	198	199	199	199	199	199	199	199	199	199	199	199	199	199	199	199	199	199	200
3	.250	2.02	2.28	2.36	2.39	2.41	2.42	2.43	2.44	2.44	2.44	2.45	2.46	2.46	2.46	2.46	2.47	2.47	2.47	2.47
	.100	5.54	5.46	5.39	5.34	5.31	5.28	5.27	5.25	5.24	5.23	5.22	5.20	5.18	5.18	5.17	5.16	5.15	5.14	5.13
	.025	17.44	16.04	15.44	15.10	14.88	14.74	14.62	14.54	14.47	14.42	14.34	14.25	14.17	14.12	14.08	14.04	13.99	13.95	13.90
	.005	55.55	49.80	47.47	46.20	45.39	44.84	44.43	44.13	43.88	43.69	43.39	43.08	42.78	42.62	42.47	42.31	42.15	41.99	41.83
4	.250	1.81	2.00	2.05	2.06	2.07	2.08	2.08	2.08	2.08	2.08	2.08	2.08	2.08	2.08	2.08	2.08	2.08	2.08	2.08
	.100	4.54	4.32	4.19	4.11	4.05	4.01	3.98	3.95	3.94	3.92	3.90	3.87	3.84	3.83	3.82	3.80	3.79	3.78	3.76
	.025	12.22	10.65	9.98	9.60	9.36	9.20	9.07	8.98	8.90	8.84	8.75	8.66	8.56	8.51	8.46	8.41	8.36	8.31	8.26
	.005	31.33	26.28	24.26	23.16	22.46	21.98	21.62	21.35	21.14	20.97	20.70	20.44	20.17	20.03	19.89	19.75	19.61	19.47	19.32
5	.250	1.69	1.85	1.88	1.89	1.89	1.89	1.89	1.89	1.89	1.89	1.89	1.89	1.88	1.88	1.88	1.88	1.87	1.87	1.87
	.100	4.06	3.78	3.62	3.52	3.45	3.40	3.37	3.34	3.32	3.30	3.27	3.24	3.21	3.19	3.17	3.16	3.14	3.12	3.10
	.025	10.01	8.43	7.76	7.39	7.15	6.98	6.85	6.76	6.68	6.62	6.52	6.43	6.33	6.28	6.23	6.18	6.12	6.07	6.02
	.005	22.78	18.31	16.53	15.56	14.94	14.51	14.20	13.96	13.77	13.62	13.38	13.15	12.90	12.78	12.66	12.53	12.40	12.27	12.14
6	.250	1.62	1.76	1.78	1.79	1.79	1.78	1.78	1.78	1.77	1.77	1.77	1.76	1.76	1.75	1.75	1.75	1.74	1.74	1.74
	.100	3.78	3.46	3.29	3.18	3.11	3.05	3.01	2.98	2.96	2.94	2.90	2.87	2.84	2.82	2.80	2.78	2.76	2.74	2.72
	.025	8.81	7.26	6.60	6.23	5.99	5.82	5.70	5.60	5.52	5.46	5.37	5.27	5.17	5.12	5.07	5.01	4.96	4.90	4.85
	.005	18.64	14.54	12.92	12.03	11.46	11.07	10.79	10.57	10.39	10.25	10.03	9.81	9.59	9.47	9.36	9.24	9.12	9.00	8.88
7	.250	1.57	1.70	1.72	1.72	1.71	1.71	1.70	1.70	1.69	1.69	1.68	1.68	1.67	1.67	1.66	1.66	1.65	1.65	1.65
	.100	3.59	3.26	3.07	2.96	2.88	2.83	2.78	2.75	2.72	2.70	2.67	2.63	2.59	2.58	2.56	2.54	2.51	2.49	2.47
	.025	8.07	6.54	5.89	5.52	5.29	5.12	4.99	4.90	4.82	4.76	4.67	5.47	4.47	4.42	5.36	5.31	4.25	4.20	4.14
	.005	16.24	12.40	10.88	10.05	9.52	9.16	8.89	8.68	8.51	8.38	8.18	7.97	7.75	7.64	7.53	7.42	7.31	7.19	7.08
8	.250	1.54	1.66	1.67	1.66	1.66	1.65	1.64	1.64	1.64	1.63	1.62	1.62	1.62	1.60	1.60	1.59	1.59	1.58	1.58
	.100	3.46	3.11	2.92	2.81	2.73	2.67	2.62	2.59	2.56	2.54	2.50	2.46	2.42	2.40	2.38	2.36	2.34	2.32	2.29
	.025	7.57	6.06	5.42	5.05	4.82	4.65	4.53	4.43	4.36	4.30	4.20	4.20	4.00	3.95	3.89	3.84	3.78	3.73	3.67
	.005	14.69	11.04	9.60	8.81	8.30	7.95	7.69	7.50	7.34	7.21	7.01	6.81	6.61	6.50	6.40	6.29	6.18	6.06	5.95

Reprinted from "Tables of percentage points of the inverted beta (F) distribution" by Maxine Merrington and Catherine M. Thompson, *Biometrika* 33 (1943):73 by permission of the authors and the editor.

†df in denominator.

TABLE A 14, PART II—*Continued*

ν_2†	P	1	2	3	4	5	6	7	8	9	10	12	15	20	24	30	40	60	120	∞
		ν_1 df in Numerator																		
9	0.250	1.51	1.62	1.63	1.63	1.62	1.61	1.60	1.60	1.59	1.59	1.58	1.57	1.56	1.56	1.55	1.54	1.54	1.53	1.53
	.100	3.36	3.01	2.81	2.69	2.61	2.55	2.51	2.47	2.44	2.42	2.38	2.34	2.30	2.28	2.25	2.23	2.21	2.18	2.16
	.025	7.21	5.71	5.08	4.72	4.48	4.32	4.20	4.10	4.03	3.96	3.87	3.77	3.67	3.61	3.56	3.51	3.45	3.39	3.33
	.005	13.61	10.11	8.72	7.96	7.47	7.13	6.88	6.69	6.54	6.42	6.23	6.03	5.83	5.73	5.62	5.52	5.41	5.30	5.19
10	.250	1.49	1.60	1.60	1.59	1.59	1.58	1.57	1.56	1.56	1.55	1.54	1.53	1.52	1.52	1.51	1.51	1.50	1.49	1.48
	.100	3.28	2.92	2.73	2.61	2.52	2.46	2.41	2.38	2.35	2.32	2.28	2.24	2.20	2.18	2.16	2.13	2.11	2.08	2.06
	.025	6.94	5.46	4.83	4.47	4.24	4.07	3.95	3.85	3.78	3.72	3.62	3.52	3.42	3.37	3.31	3.26	3.20	3.14	3.08
	.005	12.83	9.43	8.08	7.34	6.87	6.54	6.30	6.12	5.97	5.85	5.66	5.47	5.27	5.17	5.07	4.97	4.86	4.75	4.64
11	.250	1.47	1.58	1.58	1.57	1.56	1.55	1.54	1.53	1.53	1.52	1.51	1.50	1.49	1.49	1.48	1.47	1.47	1.46	1.45
	.100	3.23	2.86	2.66	2.54	2.45	2.39	2.34	2.30	2.27	2.25	2.21	2.17	2.12	2.10	2.08	2.05	2.03	2.00	1.97
	.025	6.72	5.26	4.63	4.28	4.04	3.88	3.76	3.66	3.59	3.53	3.43	3.33	3.23	3.17	3.12	3.06	3.00	2.94	2.83
	.005	12.23	8.91	7.60	6.88	6.42	6.10	5.86	5.68	5.54	5.42	5.24	5.05	4.86	4.76	4.65	4.55	4.44	4.34	4.23
12	.250	1.46	1.56	1.56	1.55	1.54	1.53	1.52	1.51	1.51	1.50	1.49	1.48	1.47	1.46	1.45	1.45	1.44	1.43	1.42
	.100	3.18	2.81	2.61	2.48	2.39	2.33	2.28	2.24	2.21	2.19	2.15	2.10	2.06	2.04	2.01	1.99	1.96	1.93	1.90
	.025	6.55	5.10	4.47	4.12	3.89	3.73	3.61	3.51	3.44	3.37	3.28	3.18	3.07	3.02	2.96	2.91	2.85	2.79	2.72
	.005	11.75	8.51	7.23	6.52	6.07	5.76	5.52	5.35	5.20	5.09	4.91	4.72	4.53	4.43	4.33	4.23	4.12	4.01	3.90
13	.250	1.45	1.55	1.55	1.53	1.52	1.51	1.50	1.49	1.49	1.48	1.47	1.46	1.45	1.44	1.43	1.42	1.42	1.41	1.40
	.100	3.14	2.76	2.56	2.43	2.35	2.28	2.23	2.20	2.16	2.14	2.10	2.05	2.01	1.98	1.96	1.93	1.90	1.88	1.85
	.025	6.41	4.97	4.35	4.00	3.77	3.60	3.48	3.39	3.31	3.25	3.15	3.05	2.95	2.89	2.84	2.78	2.72	2.66	2.60
	.005	11.37	8.19	6.93	6.23	5.79	5.48	5.25	5.08	4.94	4.82	4.64	4.46	4.27	4.17	4.07	3.97	3.87	3.76	3.65
14	.250	1.44	1.53	1.53	1.52	1.51	1.50	1.49	1.48	1.47	1.46	1.45	1.44	1.43	1.42	1.41	1.41	1.40	1.39	1.38
	.100	3.10	2.73	2.52	2.39	2.31	2.24	2.19	2.15	2.12	2.10	2.05	2.01	1.96	1.94	1.91	1.89	1.86	1.83	1.80
	.025	6.30	4.86	4.24	3.89	3.66	3.50	3.38	3.29	3.21	3.15	3.05	2.95	2.84	2.79	2.73	2.67	2.61	2.55	2.49
	.005	11.06	7.92	6.68	6.00	5.56	5.26	5.03	4.86	4.72	4.60	4.43	4.25	4.06	3.96	3.86	3.76	3.66	3.55	4.44
15	.250	1.43	1.52	1.52	1.51	1.49	1.48	1.47	1.46	1.46	1.45	1.44	1.43	1.41	1.41	1.40	1.39	1.38	1.37	1.36
	.100	3.07	2.70	2.49	2.36	2.27	2.21	2.16	2.12	2.09	2.06	2.02	1.97	1.92	1.90	1.87	1.85	1.82	1.79	1.76
	.025	6.20	4.76	4.15	3.80	3.58	3.41	3.29	3.20	3.12	3.06	2.96	2.86	2.76	2.70	2.64	2.58	2.52	2.46	2.40
	.005	10.80	7.70	6.48	5.80	5.37	5.07	4.85	4.67	4.54	4.42	4.25	4.07	3.88	3.79	3.69	3.58	3.48	3.37	3.26
16	.250	1.42	1.51	1.51	1.50	1.48	1.47	1.46	1.45	1.44	1.44	1.43	1.41	1.40	1.39	1.38	1.37	1.36	1.35	1.34
	.100	3.05	2.67	2.46	2.33	2.24	2.18	2.13	2.09	2.06	2.03	1.99	1.94	1.89	1.87	1.84	1.81	1.78	1.75	1.72
	.025	6.12	4.69	4.08	3.73	3.50	3.34	3.22	3.12	3.05	2.99	2.89	2.79	2.68	2.63	2.57	2.51	2.45	2.38	2.32
	.005	10.58	7.51	6.30	5.64	5.21	4.91	4.69	4.52	4.38	4.27	4.10	3.92	3.73	3.64	3.54	3.44	3.33	3.22	3.11
17	.250	1.42	1.51	1.50	1.49	1.47	1.46	1.45	1.44	1.43	1.43	1.41	1.40	1.39	1.38	1.37	1.36	1.35	1.34	1.33
	.100	3.03	2.64	2.44	2.31	2.22	2.15	2.10	2.06	2.03	2.00	1.96	1.91	1.86	1.84	1.81	1.78	1.75	1.72	1.69
	.025	6.04	4.62	4.01	3.66	3.44	3.28	3.16	3.06	2.98	2.92	2.82	2.72	2.62	2.56	2.50	2.44	2.38	2.32	2.25
	.005	10.38	7.35	6.16	5.50	5.07	4.78	4.56	4.39	4.25	4.14	3.97	3.79	3.61	3.51	3.41	3.31	3.21	3.10	2.98

†df in denominator.

TABLE A 14, PART II—*Continued*

ν_2†	P	ν_1 df in Numerator: 1	2	3	4	5	6	7	8	9	10	12	15	20	24	30	40	60	120	∞
18	0.250	1.41	1.50	1.49	1.48	1.46	1.45	1.44	1.43	1.42	1.42	1.40	1.39	1.38	1.37	1.36	1.35	1.34	1.33	1.32
	.100	3.01	2.62	2.42	2.29	2.20	2.13	2.08	2.04	2.00	1.98	1.93	1.89	1.84	1.81	1.78	1.75	1.72	1.69	1.66
	.025	5.98	4.56	3.95	3.61	3.38	3.22	3.10	3.01	2.93	2.87	2.77	2.67	2.56	2.50	2.44	2.38	2.32	2.26	2.19
	.005	10.22	7.21	6.03	5.37	4.96	4.66	4.44	4.28	4.14	4.03	3.86	3.68	3.50	3.40	3.30	3.20	3.10	2.99	2.87
19	.250	1.41	1.49	1.49	1.47	1.46	1.44	1.43	1.42	1.41	1.41	1.40	1.38	1.37	1.36	1.35	1.34	1.33	1.32	1.30
	.100	2.99	2.61	2.40	2.27	2.18	2.11	2.06	2.02	1.98	1.96	1.91	1.86	1.81	1.79	1.76	1.73	1.70	1.67	1.63
	.025	5.92	4.51	3.90	3.56	3.33	3.17	3.05	2.96	2.88	2.82	2.72	2.62	2.51	2.45	2.39	2.33	2.27	2.20	2.13
	.005	10.07	7.09	5.92	5.27	4.85	4.56	4.34	4.18	4.04	3.93	3.76	3.59	3.40	3.31	3.21	3.11	3.00	2.89	2.78
20	.250	1.40	1.49	1.48	1.47	1.45	1.44	1.43	1.42	1.41	1.40	1.39	1.37	1.36	1.35	1.34	1.33	1.32	1.31	1.29
	.100	2.97	2.59	2.38	2.25	2.16	2.09	2.04	2.00	1.96	1.94	1.89	1.84	1.79	1.77	1.74	1.71	1.68	1.64	1.61
	.025	5.87	4.46	3.86	3.51	3.29	3.13	3.01	2.91	2.84	2.77	2.68	2.57	2.46	2.41	2.35	2.29	2.22	2.16	2.09
	.005	9.94	6.99	5.82	5.17	4.76	4.47	4.26	4.09	3.96	3.85	3.68	3.50	3.32	3.22	3.12	3.02	2.92	2.81	2.69
21	.250	1.40	1.48	1.48	1.46	1.44	1.43	1.42	1.41	1.40	1.39	1.38	1.37	1.35	1.34	1.33	1.32	1.31	1.30	1.28
	.100	2.96	2.57	2.36	2.23	2.14	2.08	2.02	1.98	1.95	1.92	1.88	1.83	1.78	1.75	1.72	1.69	1.66	1.62	1.59
	.025	5.83	4.42	3.82	3.48	3.25	3.09	2.97	2.87	2.80	2.73	2.64	2.53	2.42	2.37	2.31	2.25	2.18	2.11	2.04
	.005	9.83	6.89	5.73	5.09	4.68	4.39	4.18	4.01	3.88	3.77	3.60	3.43	3.24	3.15	3.05	2.95	2.84	2.73	2.61
22	.250	1.40	1.48	1.47	1.45	1.44	1.42	1.41	1.40	1.39	1.39	1.37	1.36	1.34	1.33	1.32	1.31	1.30	1.29	1.28
	.100	2.95	2.56	2.35	2.22	2.13	2.06	2.01	1.97	1.93	1.90	1.86	1.81	1.76	1.73	1.70	1.67	1.64	1.60	1.57
	.025	5.79	4.38	3.78	3.44	3.22	3.05	2.93	2.84	2.76	2.70	2.60	2.50	2.39	2.33	2.27	2.21	2.14	2.08	2.00
	.005	9.73	6.81	5.65	5.02	4.61	4.32	4.11	3.94	3.81	3.70	3.54	3.36	3.18	3.08	2.98	2.88	2.77	2.66	2.55
23	.250	1.39	1.47	1.47	1.45	1.43	1.42	1.41	1.40	1.39	1.38	1.37	1.35	1.34	1.33	1.32	1.31	1.30	1.28	1.27
	.100	2.94	2.55	2.34	2.21	2.11	2.05	1.99	1.95	1.92	1.89	1.84	1.80	1.74	1.72	1.69	1.66	1.62	1.59	1.55
	.025	5.75	4.35	3.75	3.41	3.18	3.02	2.90	2.81	2.73	2.67	2.57	2.47	2.36	2.30	2.24	2.18	2.11	2.04	1.97
	.005	9.63	6.73	5.58	4.95	4.54	4.26	4.05	3.88	3.75	3.64	3.47	3.30	3.12	3.02	2.92	2.82	2.71	2.60	2.48
24	.250	1.39	1.47	1.46	1.44	1.43	1.41	1.40	1.39	1.38	1.38	1.36	1.35	1.33	1.32	1.31	1.30	1.29	1.28	1.26
	.100	2.93	2.54	2.33	2.19	2.10	2.04	1.98	1.94	1.91	1.88	1.83	1.78	1.73	1.70	1.67	1.64	1.61	1.57	1.53
	.025	5.72	4.32	3.72	3.38	3.15	2.99	2.87	2.78	2.70	2.64	2.54	2.44	2.33	2.27	2.21	2.15	2.08	2.01	1.94
	.005	9.55	6.66	5.52	4.89	4.49	4.20	3.99	3.83	3.69	3.59	3.42	3.25	3.06	2.97	2.87	2.77	2.66	2.55	2.43
25	.250	1.39	1.47	1.46	1.44	1.42	1.41	1.40	1.39	1.38	1.37	1.36	1.34	1.33	1.32	1.31	1.29	1.28	1.27	1.25
	.100	2.92	2.53	2.32	2.18	2.09	2.02	1.97	1.93	1.89	1.87	1.82	1.77	1.72	1.69	1.66	1.63	1.59	1.56	1.52
	.025	5.69	4.29	3.69	3.35	3.13	2.97	2.85	2.75	2.68	2.61	2.51	2.41	2.30	2.24	2.18	2.12	2.05	1.98	1.91
	.005	9.48	6.60	5.46	4.84	4.43	4.15	3.94	3.78	3.64	3.54	3.37	3.20	3.02	2.92	2.82	2.72	2.61	2.50	2.38
26	.250	1.38	1.46	1.45	1.44	1.42	1.41	1.39	1.38	1.37	1.37	1.35	1.34	1.32	1.31	1.30	1.29	1.28	1.26	1.25
	.100	2.91	2.52	2.31	2.17	2.08	2.01	1.96	1.92	1.88	1.86	1.81	1.76	1.71	1.68	1.65	1.61	1.58	1.54	1.50
	.025	5.66	4.27	3.67	3.33	3.10	2.94	2.82	2.73	2.65	2.59	2.49	2.39	2.28	2.22	2.16	2.09	2.03	1.95	1.83
	.005	9.41	6.54	5.41	4.79	4.38	4.10	3.89	3.73	3.60	3.49	3.33	3.15	2.97	2.87	2.77	2.67	2.56	2.45	2.33

†df in denominator.

TABLE A 14, PART II—*Continued*

ν_2†	P	ν_1 df in Numerator																		
		1	2	3	4	5	6	7	8	9	10	12	15	20	24	30	40	60	120	∞
27	0.250	1.38	1.46	1.45	1.43	1.42	1.40	1.39	1.38	1.37	1.36	1.35	1.33	1.32	1.31	1.30	1.28	1.27	1.26	1.24
	.100	2.90	2.51	2.30	2.17	2.07	2.00	1.95	1.91	1.87	1.85	1.80	1.75	1.70	1.67	1.64	1.60	1.57	1.53	1.49
	.025	5.63	4.24	3.65	3.31	3.08	2.92	2.80	2.71	2.63	2.57	2.47	2.36	2.25	2.19	2.13	2.07	2.00	1.93	1.85
	.005	9.34	6.49	5.36	4.74	4.34	4.06	3.85	3.69	3.56	3.45	3.28	3.11	2.93	2.83	2.73	2.63	2.52	2.41	2.29
28	.250	1.38	1.46	1.45	1.43	1.41	1.40	1.39	1.38	1.37	1.36	1.34	1.33	1.31	1.30	1.29	1.28	1.27	1.25	1.24
	.100	2.89	2.50	2.29	2.16	2.06	2.00	1.94	1.90	1.87	1.84	1.79	1.74	1.69	1.66	1.63	1.59	1.56	1.52	1.48
	.025	5.61	4.22	3.63	3.29	3.06	2.90	2.78	2.69	2.61	2.55	2.45	2.34	2.23	2.17	2.11	2.05	1.98	1.91	1.83
	.005	9.28	6.44	5.32	4.70	4.30	4.02	3.81	3.65	3.52	3.41	3.25	3.07	2.89	2.79	2.69	2.59	2.48	2.37	2.25
29	.250	1.38	1.45	1.45	1.43	1.41	1.40	1.38	1.37	1.36	1.35	1.34	1.32	1.31	1.30	1.29	1.27	1.26	1.25	1.23
	.100	2.89	2.50	2.28	2.15	2.06	1.99	1.93	1.89	1.86	1.83	1.78	1.73	1.68	1.65	1.62	1.58	1.55	1.51	1.47
	.025	5.59	4.20	3.61	3.27	3.04	2.88	2.76	2.67	2.59	2.53	2.43	2.32	2.21	2.15	2.09	2.03	1.96	1.89	1.81
	.005	9.23	6.40	5.28	4.66	4.26	3.98	3.77	3.61	3.48	3.38	3.21	3.04	2.86	2.76	2.66	2.56	2.45	2.33	2.21
30	.250	1.38	1.45	1.44	1.42	1.41	1.39	1.38	1.37	1.36	1.35	1.34	1.32	1.30	1.29	1.28	1.27	1.26	1.24	1.23
	.100	2.88	2.49	2.28	2.14	2.05	1.98	1.93	1.88	1.85	1.82	1.77	1.72	1.67	1.64	1.61	1.57	1.54	1.50	1.46
	.025	5.57	4.18	3.59	3.25	3.03	2.87	2.75	2.65	2.57	2.51	2.41	2.31	2.20	2.14	2.07	2.01	1.94	1.87	1.79
	.005	9.18	6.35	5.24	4.62	4.23	3.95	3.74	3.58	3.45	3.34	3.18	3.01	2.82	2.73	2.63	2.52	2.42	2.30	2.18
40	.250	1.36	1.44	1.42	1.40	1.39	1.37	1.36	1.35	1.34	1.33	1.31	1.30	1.28	1.26	1.25	1.24	1.22	1.21	1.19
	.100	2.84	2.44	2.23	2.09	2.00	1.93	1.87	1.83	1.79	1.76	1.71	1.66	1.61	1.57	1.54	1.51	1.47	1.42	1.38
	.025	5.42	4.05	3.46	3.13	2.90	2.74	2.62	2.53	2.45	2.39	2.29	2.18	2.07	2.01	1.94	1.88	1.80	1.72	1.64
	.005	8.83	6.07	4.98	4.37	3.99	3.71	3.51	3.35	3.22	3.12	2.95	2.78	2.60	2.50	2.40	2.30	2.18	2.06	1.93
60	.250	1.35	1.42	1.41	1.38	1.37	1.35	1.33	1.32	1.31	1.30	1.29	1.27	1.25	1.25	1.24	1.22	1.21	1.17	1.15
	.100	2.79	2.39	2.18	2.04	1.95	1.87	1.82	1.77	1.74	1.71	1.66	1.60	1.54	1.51	1.48	1.44	1.40	1.35	1.29
	.025	5.29	3.93	3.34	3.01	2.79	2.63	2.51	2.41	2.33	2.27	2.17	2.06	1.94	1.88	1.82	1.74	1.67	1.58	1.48
	.005	8.49	5.80	4.73	4.14	3.76	3.49	3.29	3.13	3.01	2.90	2.74	2.57	2.39	2.29	2.19	2.08	1.96	1.83	1.69
120	.250	1.34	1.40	1.39	1.37	1.35	1.33	1.31	1.30	1.29	1.28	1.26	1.24	1.22	1.21	1.19	1.18	1.16	1.13	1.10
	.100	2.75	2.35	2.13	1.99	1.90	1.82	1.77	1.72	1.68	1.65	1.60	1.54	1.48	1.45	1.41	1.37	1.32	1.26	1.19
	.025	5.15	3.80	3.23	2.89	2.67	2.52	2.39	2.30	2.22	2.16	2.05	1.94	1.82	1.76	1.69	1.61	1.63	1.43	1.31
	.005	8.18	5.54	4.50	3.92	3.55	3.28	3.09	2.93	2.81	2.71	2.54	2.37	2.19	2.09	1.98	1.87	1.75	1.61	1.43
∞	.250	1.32	1.39	1.37	1.35	1.33	1.31	1.29	1.28	1.27	1.25	1.24	1.22	1.19	1.18	1.16	1.14	1.12	1.08	1.00
	.100	2.71	2.30	2.08	1.94	1.85	1.77	1.72	1.67	1.63	1.60	1.55	1.49	1.42	1.38	1.34	1.30	1.24	1.17	1.00
	.025	5.02	3.69	3.12	2.79	2.57	2.41	2.29	2.19	2.11	2.05	1.94	1.83	1.71	1.64	1.57	1.48	1.39	1.27	1.00
	.005	7.88	5.30	4.28	3.72	3.35	3.09	2.90	2.74	2.62	2.52	2.36	2.19	2.00	1.90	1.79	1.67	1.53	1.36	1.00

†df in denominator.

TABLE A 15
Upper 5% Points, Q, in the Studentized Range

Degrees of Freedom, ν	Number of Treatments, a																		
	2	3	4	5	6	7	8	9	10	11	12	13	14	15	16	17	18	19	20
1	18.0	27.0	32.8	37.2	40.5	43.1	45.4	47.3	49.1	50.6	51.9	53.2	54.3	55.4	56.3	57.2	58.0	58.8	59.6
2	6.09	8.33	9.80	10.89	11.73	12.43	13.03	13.54	13.99	14.39	14.75	15.08	15.38	15.65	15.91	16.14	16.36	16.57	16.77
3	4.50	5.91	6.83	7.51	8.04	8.47	8.85	9.18	9.46	9.72	9.95	10.16	10.35	10.52	10.69	10.84	10.98	11.12	11.24
4	3.93	5.04	5.76	6.29	6.71	7.06	7.35	7.60	7.83	8.03	8.21	8.37	8.52	8.67	8.80	8.92	9.03	9.14	9.24
5	3.64	4.60	5.22	5.67	6.03	6.33	6.58	6.80	6.99	7.17	7.32	7.47	7.60	7.72	7.83	7.93	8.03	8.12	8.21
6	3.46	4.34	4.90	5.31	5.63	5.89	6.12	6.32	6.49	6.65	6.79	6.92	7.04	7.14	7.24	7.34	7.43	7.51	7.59
7	3.34	4.16	4.68	5.06	5.35	5.59	5.80	5.99	6.15	6.29	6.42	6.54	6.65	6.75	6.84	6.93	7.01	7.08	7.16
8	3.26	4.04	4.53	4.89	5.17	5.40	5.60	5.77	5.92	6.05	6.18	6.29	6.39	6.48	6.57	6.65	6.73	6.80	6.87
9	3.20	3.95	4.42	4.76	5.02	5.24	5.43	5.60	5.74	5.87	5.98	6.09	6.19	6.28	6.36	6.44	6.51	6.58	6.65
10	3.15	3.88	4.33	4.66	4.91	5.12	5.30	5.46	5.60	5.72	5.83	5.93	6.03	6.12	6.20	6.27	6.34	6.41	6.47
11	3.11	3.82	4.26	4.58	4.82	5.03	5.20	5.35	5.49	5.61	5.71	5.81	5.90	5.98	6.06	6.14	6.20	6.27	6.33
12	3.08	3.77	4.20	4.51	4.75	4.95	5.12	5.27	5.40	5.51	5.61	5.71	5.80	5.88	5.95	6.02	6.09	6.15	6.21
13	3.06	3.73	4.15	4.46	4.69	4.88	5.05	5.19	5.32	5.43	5.53	5.63	5.71	5.79	5.86	5.93	6.00	6.06	6.11
14	3.03	3.70	4.11	4.41	4.64	4.83	4.99	5.13	5.25	5.36	5.46	5.56	5.64	5.72	5.79	5.86	5.92	5.98	6.03
15	3.01	3.67	4.08	4.37	4.59	4.78	4.94	5.08	5.20	5.31	5.40	5.49	5.57	5.65	5.72	5.79	5.85	5.91	5.96
16	3.00	3.65	4.05	4.34	4.56	4.74	4.90	5.03	5.15	5.26	5.35	5.44	5.52	5.59	5.66	5.73	5.79	5.84	5.90
17	2.98	3.62	4.02	4.31	4.52	4.70	4.86	4.99	5.11	5.21	5.31	5.39	5.47	5.55	5.61	5.68	5.74	5.79	5.84
18	2.97	3.61	4.00	4.28	4.49	4.67	4.83	4.96	5.07	5.17	5.27	5.35	5.43	5.50	5.57	5.63	5.69	5.74	5.79
19	2.96	3.59	3.98	4.26	4.47	4.64	4.79	4.92	5.04	5.14	5.23	5.32	5.39	5.46	5.53	5.59	5.65	5.70	5.75
20	2.95	3.58	3.96	4.24	4.45	4.62	4.77	4.90	5.01	5.11	5.20	5.28	5.36	5.43	5.50	5.56	5.61	5.66	5.71
24	2.92	3.53	3.90	4.17	4.37	4.54	4.68	4.81	4.92	5.01	5.10	5.18	5.25	5.32	5.38	5.44	5.50	5.55	5.59
30	2.89	3.48	3.84	4.11	4.30	4.46	4.60	4.72	4.83	4.92	5.00	5.08	5.15	5.21	5.27	5.33	5.38	5.43	5.48
40	2.86	3.44	3.79	4.04	4.23	4.39	4.52	4.63	4.74	4.82	4.90	4.98	5.05	5.11	5.17	5.22	5.27	5.32	5.36
60	2.83	3.40	3.74	3.98	4.16	4.31	4.44	4.55	4.65	4.73	4.81	4.88	4.94	5.00	5.06	5.11	5.15	5.20	5.24
120	2.80	3.36	3.69	3.92	4.10	4.24	4.36	4.47	4.56	4.64	4.71	4.78	4.84	4.90	4.95	5.00	5.04	5.09	5.13
∞	2.77	3.32	3.63	3.86	4.03	4.17	4.29	4.39	4.47	4.55	4.62	4.68	4.74	4.80	4.84	4.89	4.93	4.97	5.01

Reprinted from *Biometrika* 39 (1952):192 by permission of the author, Joyce M. May, and the editor.

TABLE A 16

(*i*) Significance Levels of the Maximum Normed Residual max $|X - \overline{X}| / \sqrt{\Sigma(X - \overline{X})^2}$ for a Normal Sample. (*ii*) Significance Levels of max $|d| / \sqrt{\Sigma d^2}$, Where *d*s are Normal and Independent

Size of Sample n	5% level		1% level	
	(*i*) MNR	(*ii*) MNR	(*i*) MNR	(*ii*) MNR
5	0.858	0.917	0.882	0.963
6	.844	.884	.882	.940
7	.825	.853	.873	.915
8	.804	.824	.860	.891
9	.783	.799	.844	.869
10	.763	.776	.827	.847
11	.745	.755	.811	.827
12	.727	.736	.795	.808
13	.711	.718	.779	.790
14	.695	.701	.764	.774
15	.681	.686	.750	.758
16	.668	.672	.737	.744
17	.655	.659	.724	.730
18	.643	.647	.711	.717
19	.632	.635	.700	.704
20	.621	.624	.688	.693
22	.602	.604	.668	.671
24	.584	.586	.649	.652
26	.568	.570	.632	.634
28	.554	.555	.616	.618
30	0.540	.541	0.601	.603
35		.512		.564
40		.4.87		.542
60		.417		.464
120		0.316		0.350

Reproduced, with permission, from W. von Türk (formerly Stefansky), *Technometrics* 14 (1972):475–76.

TABLE A 17
SIGNIFICANCE LEVELS OF DIXON'S CRITERIA FOR TESTING EXTREME OBSERVATIONS IN A SINGLE SAMPLE

n	Criterion	Level 5%	Level 1%	n	Criterion	Level 5%	Level 1%
3	$\frac{X_n - X_{n-1}}{X_n - X_1}$ or $\frac{X_2 - X_1}{X_n - X_1}$	0.941	0.988	14	$\frac{X_n - X_{n-2}}{X_n - X_3}$ or $\frac{X_3 - X_1}{X_{n-2} - X_1}$	0.546	0.641
4		.765	.889	15		.525	.616
5		.642	.780	16		.507	.595
6		.560	.698	17		.490	.577
7		.507	.637	18		.475	.561
8	$\frac{X_n - X_{n-1}}{X_n - X_2}$ or $\frac{X_2 - X_1}{X_{n-1} - X_1}$	.554	.683	19		.462	.547
9		.512	.635	20		.450	.535
10		.477	.597	21		.440	.524
11	$\frac{X_n - X_{n-2}}{X_n - X_2}$ or $\frac{X_3 - X_1}{X_{n-1} - X_1}$	.576	.679	23		.421	.505
12		.546	.642	24		.413	.497
13		0.521	0.615	25		0.406	0.489

Reproduced, with permission, from W. J. Dixon, *Biometrics* 9 (1953):89.

TABLE A 18
SIGNIFICANCE LEVELS OF THE MNR = MAX $|\delta| / \sqrt{\Sigma\delta^2}$ IN A TWO-WAY TABLE WITH R ROWS AND C COLUMNS

C		R = 3	4	5	6	7	8	9
3	5%	0.648	0.645	0.624	0.600	0.577	0.555	0.535
	1%	0.660	.675	.664	.646	.626	.606	.587
4	5%		.621	5.90	.561	.535	.513	.493
	1%		0.665	.640	.613	.588	.565	.544
5	5%			.555	.525	.499	.477	.457
	1%			.608	.578	.551	.527	.506
6	5%				.495	.469	.447	.428
	1%				.546	.519	.495	.475
7	5%					.444	.423	.405
	1%					.492	.469	.449
8	5%						.402	.385
	1%						.446	.426
9	5%							.368
	1%							.407

Reproduced by permission of W. von Türk (formerly Stefansky), from table 6 in *Technometrics* 14 (1972):475–76.

TABLE A 19
ORTHOGONAL POLYNOMIALS

(At the foot of each column, the first number is the sum of squares of the polynomial values, the second is the multiplier λ_i needed to give integral values.)

n = 3		n = 4			n = 5				n = 6				
X_1	X_2	X_1	X_2	X_3	X_1	X_2	X_3	X_4	X_1	X_2	X_3	X_4	X_5
−1	+1	−3	+1	−1	−2	+2	−1	+1	−5	+5	−5	+1	− 1
0	−2	−1	−1	+3	−1	−1	+2	−4	−3	−1	+7	−3	+ 5
+1	+1	+1	−1	−3	0	−2	0	+6	−1	−4	+4	+2	−10
		+3	+1	+1	+1	−1	−2	−4	+1	−4	−4	+2	+10
					+2	+2	+1	+1	+3	−1	−7	−3	− 5
									+5	+5	+5	+1	+ 1
2	6	20	4	20	10	14	10	70	70	84	180	28	252
1	3	2	1	10/3	1	1	5/6	35/12	2	3/2	5/3	7/12	21/10

n = 7					n = 8					n = 9				
X_1	X_2	X_3	X_4	X_5	X_1	X_2	X_3	X_4	X_5	X_1	X_2	X_3	X_4	X_5
0	−4	0	+6	0	+1	−5	−3	+ 9	+15	0	−20	0	+18	0
+1	−3	−1	+1	+5	+3	−3	−7	− 3	+17	+1	−17	− 9	+ 9	+ 9
+2	0	−1	−7	−4	+5	+1	−5	−13	−23	+2	− 8	−13	−11	+ 4
+3	+5	+1	+3	+1	+7	+7	+7	+ 7	+ 7	+3	+ 7	− 7	−21	−11
										+4	+28	+14	+14	+ 4
28	84	6	154	84	168	168	264	616	2184	60	2772	990	2002	468
1	1	1/6	7/12	7/20	2	1	2/3	7/12	7/10	1	3	5/6	7/12	3/20

n = 10					n = 11					n = 12				
X_1	X_2	X_3	X_4	X_5	X_1	X_2	X_3	X_4	X_5	X_1	X_2	X_3	X_4	X_5
+1	−4	−12	+18	+ 6	0	−10	0	+6	0	+ 1	−35	− 7	+28	+20
+3	−3	−31	+ 3	+11	+1	− 9	−14	+4	+4	+ 3	−29	−19	+12	+44
+5	−1	−35	−17	+ 1	+2	− 6	−23	−1	+4	+ 5	−17	−25	−13	+29
+7	+2	−14	−22	−14	+3	− 1	−22	−6	−1	+ 7	+ 1	−21	−33	−21
+9	+6	+42	+18	+ 6	+4	+ 6	− 6	−6	−6	+ 9	+25	− 3	−27	−57
					+5	+15	+30	+6	+3	+11	+55	+33	+33	+33
330	132	8580	2860	780	110	858	4290	286	156	572	12012	5148	8008	15912
2	1/2	5/3	5/12	1/10	1	1	5/6	1/12	1/40	2	3	2/3	7/24	3/20

TABLE A 21—*Continued*

%	0	1	2	3	4	5	6	7	8	9
30	33.21	33.27	33.34	33.40	33.46	33.52	33.58	33.65	33.71	33.77
31	33.83	33.89	33.96	34.02	34.08	34.14	34.20	34.27	34.33	34.39
32	34.45	34.51	34.57	34.63	34.70	34.76	34.82	34.88	34.94	35.00
33	35.06	35.12	35.18	35.24	35.30	35.37	35.43	35.49	35.55	35.61
34	35.67	35.73	35.79	35.85	35.91	35.97	36.03	36.09	36.15	36.21
35	36.27	36.33	36.39	36.45	36.51	36.57	36.63	36.69	36.75	36.81
36	36.87	36.93	36.99	37.05	37.11	37.17	37.23	37.29	37.35	37.41
37	37.47	37.52	37.58	37.64	37.70	37.76	37.82	37.88	37.94	38.00
38	38.06	38.12	38.17	38.23	38.29	38.35	38.41	38.47	38.53	38.59
39	38.65	38.70	38.76	38.82	38.88	38.94	39.00	39.06	39.11	39.17
40	39.23	39.29	39.35	39.41	39.47	39.52	39.58	39.64	39.70	39.76
41	39.82	39.87	39.93	39.99	40.05	40.11	40.16	40.22	40.28	40.34
42	40.40	40.46	40.51	40.57	40.63	40.69	40.74	40.80	40.86	40.92
43	40.98	41.03	41.09	41.15	41.21	41.27	41.32	41.38	41.44	41.50
44	41.55	41.61	41.67	41.73	41.78	41.84	41.90	41.96	42.02	42.07
45	42.13	42.19	42.25	42.30	42.36	42.42	42.48	42.53	42.59	42.65
46	42.71	42.76	42.82	42.88	42.94	42.99	43.05	43.11	43.17	43.22
47	43.28	43.34	43.39	43.45	43.51	43.57	43.62	43.68	43.74	43.80
48	43.85	43.91	43.97	44.03	44.08	44.14	44.20	44.25	44.31	44.37
49	44.43	44.48	44.54	44.60	44.66	44.71	44.77	44.83	44.89	44.94
50	45.00	45.06	45.11	45.17	45.23	45.29	45.34	45.40	45.46	45.52
51	45.57	45.63	45.69	45.75	45.80	45.86	45.92	45.97	46.03	46.09
52	46.15	46.20	46.26	46.32	46.38	46.43	46.49	46.55	46.61	46.66
53	46.72	46.78	46.83	46.89	46.95	47.01	47.06	47.12	47.18	47.24
54	47.29	47.35	47.41	47.47	47.52	47.58	47.64	47.70	47.75	47.81
55	47.87	47.93	47.98	48.04	48.10	48.16	48.22	48.27	48.33	48.39
56	48.45	48.50	48.56	48.62	48.68	48.73	48.79	48.85	48.91	48.97
57	49.02	49.08	49.14	49.20	49.26	49.31	49.37	49.43	49.49	49.54
58	49.60	49.66	49.72	49.78	49.84	49.89	49.95	50.01	50.07	50.13
59	50.18	50.24	50.30	50.36	50.42	50.48	50.53	50.59	50.65	50.71
60	50.77	50.83	50.89	50.94	51.00	51.06	51.12	51.18	51.24	51.30
61	51.35	51.41	51.47	51.53	51.59	51.65	51.71	51.77	51.83	51.88
62	51.94	52.00	52.06	52.12	52.18	52.24	52.30	52.36	52.42	52.48
63	52.53	52.59	52.65	52.71	52.77	52.83	52.89	52.95	53.01	53.07
64	53.13	53.19	53.25	53.31	53.37	53.43	53.49	53.55	53.61	53.67
65	53.73	53.79	53.85	53.91	53.97	54.03	54.09	54.15	54.21	54.27
66	54.33	54.39	54.45	54.51	54.57	54.63	54.70	54.76	54.82	54.88
67	54.94	55.00	55.06	55.12	55.18	55.24	55.30	55.37	55.43	55.49
68	55.55	55.61	55.67	55.73	55.80	55.86	55.92	55.98	56.04	56.11
69	56.17	56.23	56.29	56.35	56.42	56.48	56.54	56.60	56.66	56.73

TABLE A 21—*Continued*

%	0	1	2	3	4	5	6	7	8	9
70	56.79	56.85	56.91	56.98	57.04	57.10	57.17	57.23	57.29	57.35
71	57.42	57.48	57.54	57.61	57.67	57.73	57.80	57.86	57.92	57.99
72	58.05	58.12	58.18	58.24	58.31	58.37	58.44	58.50	58.56	58.63
73	58.69	58.76	58.82	58.89	58.95	59.02	59.08	59.15	59.12	59.28
74	59.34	59.41	59.47	59.54	59.60	59.67	59.74	59.80	59.87	59.93
75	60.00	60.07	60.13	60.20	60.27	60.33	60.40	60.47	60.53	60.60
76	60.67	60.73	60.80	60.87	60.94	61.00	61.07	61.14	61.21	61.27
77	61.34	61.41	61.48	61.55	61.62	61.68	61.75	61.82	61.89	61.96
78	62.03	62.10	62.17	62.24	62.31	62.37	62.44	62.51	62.58	62.65
79	62.72	62.80	62.87	62.94	63.01	63.08	63.15	63.22	63.29	63.36
80	63.44	63.51	63.58	63.65	63.72	63.79	63.87	63.94	64.01	64.08
81	64.16	64.23	64.30	64.38	64.45	64.52	64.60	64.67	64.75	64.82
82	64.90	64.97	65.05	65.12	65.20	65.27	65.35	65.42	65.50	65.57
83	65.65	65.73	65.80	65.88	65.96	66.03	66.11	66.19	66.27	66.34
84	66.42	66.50	66.58	66.66	66.74	66.81	66.89	66.97	67.05	67.13
85	67.21	67.29	67.37	67.45	67.54	67.62	67.70	67.78	67.86	67.94
86	68.03	68.11	68.19	68.28	68.36	68.44	68.53	68.61	68.70	68.78
87	68.87	68.95	69.04	69.12	69.21	69.30	69.38	69.47	69.56	69.64
88	69.73	69.82	69.91	70.00	70.09	70.18	70.27	70.36	70.45	70.54
89	70.63	70.72	70.81	70.91	71.00	71.09	71.19	71.28	71.37	71.47
90	71.56	71.66	71.76	71.85	71.95	72.05	72.15	72.24	72.34	72.44
91	72.54	72.64	72.74	72.84	72.95	73.05	73.15	73.26	73.36	73.46
92	73.57	73.68	73.78	73.89	74.00	74.11	74.21	74.32	74.44	74.55
93	74.66	74.77	74.88	75.00	75.11	75.23	75.35	75.46	75.58	75.70
94	75.82	75.94	76.06	76.19	76.31	76.44	76.56	76.69	76.82	76.95
95	77.08	77.21	77.34	77.48	77.61	77.75	77.89	78.03	78.17	78.32
96	78.46	78.61	78.76	78.91	79.06	79.22	79.37	79.53	79.69	79.86
97	80.02	80.19	80.37	80.54	80.72	80.90	81.09	81.28	81.47	81.67
98	81.87	82.08	82.29	82.51	82.73	82.96	83.20	83.45	83.71	83.98
99.0	84.26	84.29	84.32	84.35	84.38	84.41	84.44	84.47	84.50	84.53
99.1	84.56	84.59	84.62	84.65	84.68	84.71	84.74	84.77	84.80	84.84
99.2	84.87	84.90	84.93	84.97	85.00	85.03	85.07	85.10	85.13	85.17
99.3	85.20	85.24	85.27	85.31	85.34	85.38	85.41	85.45	85.48	85.52
99.4	85.56	85.60	85.63	85.67	85.71	85.75	85.79	85.83	85.87	85.91
99.5	85.95	85.99	86.03	86.07	86.11	86.15	86.20	86.24	86.28	86.33
99.6	86.37	86.42	86.47	86.51	86.56	86.61	86.66	86.71	86.76	86.81
99.7	86.86	86.91	86.97	87.02	87.08	87.13	87.19	87.25	87.31	87.37
99.8	87.44	87.50	87.57	87.64	87.71	87.78	87.86	87.93	88.01	88.10
99.9	88.19	88.28	88.38	88.48	88.60	88.72	88.85	89.01	89.19	89.43
100.0	90.00	.....	.....	.....	.....	.....	.....	.....	.....	.....

TABLE A 22
TABLE OF SQUARE ROOTS

n	$\sqrt{n}$	$\sqrt{10n}$	n	$\sqrt{n}$	$\sqrt{10n}$	n	$\sqrt{n}$	$\sqrt{10n}$
1.00	1.00	3.16	2.00	1.41	4.47	3.00	1.73	5.48
1.02	1.01	3.19	2.02	1.42	4.49	3.02	1.74	5.50
1.04	1.02	3.22	2.04	1.43	4.52	3.04	1.74	5.51
1.06	1.03	3.26	2.06	1.44	4.54	3.06	1.75	5.53
1.08	1.04	3.29	2.08	1.44	4.56	3.08	1.76	5.55
1.10	1.05	3.32	2.10	1.45	4.58	3.10	1.76	5.57
1.12	1.06	3.35	2.12	1.46	4.60	3.12	1.77	5.59
1.14	1.07	3.38	2.14	1.46	4.63	3.14	1.77	5.60
1.16	1.08	3.41	2.16	1.47	4.65	3.16	1.78	5.62
1.18	1.09	3.44	2.18	1.48	4.67	3.18	1.78	5.64
1.20	1.10	3.46	2.20	1.48	4.69	3.20	1.79	5.66
1.22	1.10	3.49	2.22	1.49	4.71	3.22	1.79	5.67
1.24	1.11	3.52	2.24	1.50	4.73	3.24	1.80	5.69
1.26	1.12	3.55	2.26	1.50	4.75	3.26	1.81	5.71
1.28	1.13	3.58	2.28	1.51	4.77	3.28	1.81	5.73
1.30	1.14	3.61	2.30	1.52	4.80	3.30	1.82	5.74
1.32	1.15	3.63	2.32	1.52	4.82	3.32	1.82	5.76
1.34	1.16	3.66	2.34	1.53	4.84	3.34	1.83	5.78
1.36	1.17	3.69	2.36	1.54	4.86	3.36	1.83	5.80
1.38	1.17	3.71	2.38	1.54	4.88	3.38	1.84	5.81
1.40	1.18	3.74	2.40	1.55	4.90	3.40	1.84	5.83
1.42	1.19	3.77	2.42	1.56	4.92	3.42	1.85	5.85
1.44	1.20	3.79	2.44	1.56	4.94	3.44	1.85	5.87
1.46	1.21	3.82	2.46	1.57	4.96	3.46	1.86	5.88
1.48	1.22	3.85	2.48	1.57	4.98	3.48	1.87	5.90
1.50	1.22	3.87	2.50	1.58	5.00	3.50	1.87	5.92
1.52	1.23	3.90	2.52	1.59	5.02	3.52	1.88	5.93
1.54	1.24	3.92	2.54	1.59	5.04	3.54	1.88	5.95
1.56	1.25	3.95	2.56	1.60	5.06	3.56	1.89	5.97
1.58	1.26	3.97	2.58	1.61	5.08	3.58	1.89	5.98
1.60	1.26	4.00	2.60	1.61	5.10	3.60	1.90	6.00
1.62	1.27	4.02	2.62	1.62	5.12	3.62	1.90	6.02
1.64	1.28	4.05	2.64	1.62	5.14	3.64	1.91	6.03
1.66	1.29	4.07	2.66	1.63	5.16	3.66	1.91	6.05
1.68	1.30	4.10	2.68	1.64	5.18	3.68	1.92	6.07
1.70	1.30	4.12	2.70	1.64	5.20	3.70	1.92	6.08
1.72	1.31	4.15	2.72	1.65	5.22	3.72	1.93	6.10
1.74	1.32	4.17	2.74	1.66	5.23	3.74	1.93	6.12
1.76	1.33	4.20	2.76	1.66	5.25	3.76	1.94	6.13
1.78	1.33	4.22	2.78	1.67	5.27	3.78	1.94	6.15
1.80	1.34	4.24	2.80	1.67	5.29	3.80	1.95	6.16
1.82	1.35	4.27	2.82	1.68	5.31	3.82	1.95	6.18
1.84	1.36	4.29	2.84	1.69	5.33	3.84	1.96	6.20
1.86	1.36	4.31	2.86	1.69	5.35	3.86	1.96	6.21
1.88	1.37	4.34	2.88	1.70	5.37	3.88	1.97	6.23
1.90	1.38	4.36	2.90	1.70	5.39	3.90	1.97	6.25
1.92	1.39	4.38	2.92	1.71	5.40	3.92	1.98	6:26
1.94	1.39	4.40	2.94	1.71	5.42	3.94	1.98	6.28
1.96	1.40	4.43	2.96	1.72	5.44	3.96	1.99	6.29
1.98	1.41	4.45	2.98	1.73	5.46	3.98	1.99	6.31

TABLE A 22—*Continued*

n	$\sqrt{n}$	$\sqrt{10n}$	n	$\sqrt{n}$	$\sqrt{10n}$	n	$\sqrt{n}$	$\sqrt{10n}$
4.00	2.00	6.32	5.00	2.24	7.07	6.00	2.45	7.75
4.02	2.00	6.34	5.02	2.24	7.09	6.02	2.45	7.76
4.04	2.01	6.36	5.04	2.24	7.10	6.04	2.46	7.77
4.06	2.01	6.37	5.06	2.25	7.11	6.06	2.46	7.78
4.08	2.02	6.39	5.08	2.25	7.13	6.08	2.47	7.80
4.10	2.02	6.40	5.10	2.26	7.14	6.10	2.47	7.81
4.12	2.03	6.42	5.12	2.26	7.16	6.12	2.47	7.82
4.14	2.03	6.43	5.14	2.27	7.17	6.14	2.48	7.84
4.16	2.04	6.45	5.16	2.27	7.18	6.16	2.48	7.85
4.18	2.04	6.47	5.18	2.28	7.20	6.18	2.49	7.86
4.20	2.05	6.48	5.20	2.28	7.21	6.20	2.49	7.87
4.22	2.05	6.50	5.22	2.28	7.22	6.22	2.49	7.89
4.24	2.06	6.51	5.24	2.29	7.24	6.24	2.50	7.90
4.26	2.06	6.53	5.26	2.29	7.25	6.26	2.50	7.91
4.28	2.07	6.54	5.28	2.30	7.27	6.28	2.51	7.92
4.30	2.07	6.56	5.30	2.30	7.28	6.30	2.51	7.94
4.32	2.08	6.57	5.32	2.31	7.29	6.32	2.51	7.95
4.34	2.08	6.59	5.34	2.31	7.31	6.34	2.52	7.96
4.36	2.09	6.60	5.36	2.32	7.32	6.36	2.52	7.97
4.38	2.09	6.62	5.38	2.32	7.33	6.38	2.53	7.99
4.40	2.10	6.63	5.40	2.32	7.35	6.40	2.53	8.00
4.42	2.10	6.65	5.42	2.33	7.36	6.42	2.53	8.01
4.44	2.11	6.66	5.44	2.33	7.38	6.44	2.54	8.02
4.46	2.11	6.68	5.46	2.34	7.39	6.46	2.54	8.04
4.48	2.12	6.69	5.48	2.34	7.40	6.48	2.55	8.05
4.50	2.12	6.71	5.50	2.35	7.42	6.50	2.55	8.06
4.52	2.13	6.72	5.52	2.35	7.43	6.52	2.55	8.07
4.54	2.13	6.74	5.54	2.35	7.44	6.54	2.56	8.09
4.56	2.14	6.75	5.56	2.36	7.46	6.56	2.56	8.10
4.58	2.14	6.77	5.58	2.36	7.47	6.58	2.57	8.11
4.60	2.14	6.78	5.60	2.37	7.48	6.60	2.57	8.12
4.62	2.15	6.80	5.62	2.37	7.50	6.62	2.57	8.14
4.64	2.15	6.81	5.64	2.37	7.51	6.64	2.58	8.15
4.66	2.16	6.83	5.66	2.38	7.52	6.66	2.58	8.16
4.68	2.16	6.84	5.68	2.38	7.54	6.68	2.58	8.17
4.70	2.17	6.86	5.70	2.39	7.55	6.70	2.59	8.19
4.72	2.17	6.87	5.72	2.39	7.56	6.72	2.59	8.20
4.74	2.18	6.88	5.74	2.40	7.58	6.74	2.60	8.21
4.76	2.18	6.90	5.76	2.40	7.59	6.76	2.60	8.22
4.78	2.19	6.91	5.78	2.40	7.60	6.78	2.60	8.23
4.80	2.19	6.93	5.80	2.41	7.62	6.80	2.61	8.25
4.82	2.20	6.94	5.82	2.41	7.63	6.82	2.61	8.26
4.84	2.20	6.96	5.84	2.42	7.64	6.84	2.62	8.27
4.86	2.20	6.97	5.86	2.42	7.66	6.86	2.62	8.28
4.88	2.21	6.99	5.88	2.42	7.67	6.88	2.62	8.29
4.90	2.21	7.00	5.90	2.43	7.68	6.90	2.63	8.31
4.92	2.22	7.01	5.92	2.43	7.69	6.92	2.63	8.32
4.94	2.22	7.03	5.94	2.44	7.71	6.94	2.63	8.33
4.96	2.23	7.04	5.96	2.44	7.72	6.96	2.64	8.34
4.98	2.23	7.06	5.98	2.45	7.73	6.98	2.64	8.35

TABLE A 22—*Continued*

n	$\sqrt{n}$	$\sqrt{10n}$	n	$\sqrt{n}$	$\sqrt{10n}$	n	$\sqrt{n}$	$\sqrt{10n}$
7.00	2.65	8.37	8.00	2.83	8.94	9.00	3.00	9.49
7.02	2.65	8.38	8.02	2.83	8.96	9.02	3.00	9.50
7.04	2.65	8.39	8.04	2.84	8.97	9.04	3.01	9.51
7.06	2.66	8.40	8.06	2.84	8.98	9.06	3.01	9.52
7.08	2.66	8.41	8.08	2.84	8.99	9.08	3.01	9.53
7.10	2.66	8.43	8.10	2.85	9.00	9.10	3.02	9.54
7.12	2.67	8.44	8.12	2.85	9.01	9.12	3.02	9.55
7.14	2.67	8.45	8.14	2.85	9.02	9.14	3.02	9.56
7.16	2.68	8.46	8.16	2.86	9.03	9.16	3.03	9.57
7.18	2.68	8.47	8.18	2.86	9.04	9.18	3.03	9.58
7.20	2.68	8.49	8.20	2.86	9.06	9.20	3.03	9.59
7.22	2.69	8.50	8.22	2.87	9.07	9.22	3.04	9.60
7.24	2.69	8.51	8.24	2.87	9.08	9.24	3.04	9.61
7.26	2.69	8.52	8.26	2.87	9.09	9.26	3.04	9.62
7.28	2.70	8.53	8.28	2.88	9.10	9.28	3.05	9.63
7.30	2.70	8.54	8.30	2.88	9.11	9.30	3.05	9.64
7.32	2.71	8.56	8.32	2.88	9.12	9.32	3.05	9.65
7.34	2.71	8.57	8.34	2.89	9.13	9.34	3.06	9.66
7.36	2.71	8.58	8.36	2.89	9.14	9.36	3.06	9.67
7.38	2.72	8.59	8.38	2.89	9.15	9.38	3.06	9.68
7.40	2.72	8.60	8.40	2.90	9.17	9.40	3.07	9.70
7.42	2.72	8.61	8.42	2.90	9.18	9.42	3.07	9.71
7.44	2.73	8.63	8.44	2.91	9.19	9.44	3.07	9.72
7.46	2.73	8.64	8.46	2.91	9.20	9.46	3.08	9.73
7.48	2.73	8.65	8.48	2.91	9.21	9.48	3.08	9.74
7.50	2.74	8.66	8.50	2.92	9.22	9.50	3.08	9.75
7.52	2.74	8.67	8.52	2.92	9.23	9.52	3.09	9.76
7.54	2.75	8.68	8.54	2.92	9.24	9.54	3.09	9.77
7.56	2.75	8.69	8.56	2.93	9.25	9.56	3.09	9.78
7.58	2.75	8.71	8.58	2.93	9.26	9.58	3.10	9.79
7.60	2.76	8.72	8.60	2.93	9.27	9.60	3.10	9.80
7.62	2.76	8.73	8.62	2.94	9.28	9.62	3.10	9.81
7.64	2.76	8.74	8.64	2.94	9.30	9.64	3.10	9.82
7.66	2.77	8.75	8.66	2.94	9.31	9.66	3.11	9.83
7.68	2.77	8.76	8.68	2.95	9.32	9.68	3.11	9.84
7.70	2.77	8.77	8.70	2.95	9.33	9.70	3.11	9.85
7.72	2.78	8.79	8.72	2.95	9.34	9.72	3.12	9.86
7.74	2.78	8.80	8.74	2.96	9.35	9.74	3.12	9.87
7.76	2.79	8.81	8.76	2.96	9.36	9.76	3.12	9.88
7.78	2.79	8.82	8.78	2.96	9.37	9.78	3.13	9.89
7.80	2.79	8.83	8.80	2.97	9.38	9.80	3.13	9.90
7.82	2.80	8.84	8.82	2.97	9.39	9.82	3.13	9.91
7.84	2.80	8.85	8.84	2.97	9.40	9.84	3.14	9.92
7.86	2.80	8.86	8.86	2.98	9.41	9.86	3.14	9.93
7.88	2.81	8.87	8.88	2.98	9.42	9.88	3.14	9.94
7.90	2.81	8.89	8.90	2.98	9.43	9.90	3.15	9.95
7.92	2.81	8.90	8.92	2.99	9.44	9.92	3.15	9.96
7.94	2.82	8.91	8.94	2.99	9.46	9.94	3.15	9.97
7.96	2.82	8.92	8.96	2.99	9.47	9.96	3.16	9.98
7.98	2.82	8.93	8.98	3.00	9.48	9.98	3.16	9.99

Index

CURTAIN CALL

Previous Titles by Graham Hurley

The Faraday and Winter Series

TURNSTONE
THE TAKE
ANGELS PASSING
DEADLIGHT
CUT TO BLACK
BLOOD AND HONEY
ONE UNDER
THE PRICE OF DARKNESS
NO LOVELIER DEATH
BEYOND REACH
BORROWED LIGHT
HAPPY DAYS
BACKSTORY

The Jimmy Suttle Series

WESTERN APPROACHES
TOUCHING DISTANCE
SINS OF THE FATHER
THE ORDER OF THINGS

The Wars Within Series

FINISTERRE
AURORE
ESTOCADA

The Enora Andresson Series

CURTAIN CALL *

* *available from Severn House*

CURTAIN CALL

Graham Hurley

This first world edition published 2019
in Great Britain and the USA by
SEVERN HOUSE PUBLISHERS LTD of
Eardley House, 4 Uxbridge Street, London W8 7SY.
Trade paperback edition first published
in Great Britain and the USA 2019 by
SEVERN HOUSE PUBLISHERS LTD.

British Library Cataloguing in Publication Data
A CIP catalogue record for this title is available from the British Library.

ISBN-13: 978-0-7278-8861-7 (cased)
ISBN-13: 978-1-84751-989-4 (trade paper)
ISBN-13: 978-1-4483-0201-7 (e-book)

All Severn House titles are printed on acid-free paper.

Severn House Publishers support the Forest Stewardship Council™ [FSC™],
the leading international forest certification organisation.
All our titles that are printed on FSC certified paper carry the FSC logo.

Typeset by Palimpsest Book Production Ltd.,
Falkirk, Stirlingshire, Scotland.
Printed and bound in Great Britain by
TJ International, Padstow, Cornwall.

To
Lara, Harri, Enya, and Cormac

Tirez le rideau, la farce est jouée
Ring down the curtain, the farce is over

—the last words of François Rabelais

ONE

The neurosurgeon has a fondness for metaphor.

'The Reaper comes knocking at every door,' he says. 'I'm afraid yours might be one of them.'

I'm staring at him. Pale face. Pale eyes behind the rimless glasses. Pale everything. Half-dead already, he could be an apprentice ghost. Another metaphor.

'Should I lock the door? Hide? Pretend I'm not in?'

'Any of the above.' The eyes drift to the PC screen. 'Next of kin? A husband maybe?'

'He's in Stockholm.'

'On business?'

'He's about to re-marry. It might be the same thing.'

'I'm sorry.'

'So am I. The last thing the poor woman needs is Berndt.'

He reaches for his keyboard and taps a line I can't read. Is he making notes about some drug or other, some brave attempt to stay the Reaper at the corner of the street? Or is he having trouble spelling Berndt? I did once, so I wouldn't blame him.

With a tiny sigh he glances up, as if to check I'm still there. Then he half-turns to consult a calendar on the wall behind him. The calendar features a child's painting, stick figures in crayon, mainly red and yellow. There's a football, and birds, and a big whiskery sun. I rather like it.

'Do you have enough Percocet?'

'Yes.'

'Good. No more than one tablet every six hours and lay off the booze. Before we make any decisions, I'm afraid I'll need to see you again.' His finger has settled on the end of next week. 'Would Friday be convenient?'

'Friday would be perfect.' I manage a smile. 'My place or yours?'

* * *

Crying in public is something I try to avoid, in this case without success. This is a bar I've never been to before. It helps that everyone is a stranger. Moist-eyed, I order a large vodka and stare at my own image in the mirror behind the optics. In truth I feel undone, a parcel ripped apart by unseen hands inside me, but that's a complicated thought to share with anyone and thankfully no one seems very interested.

Less than two weeks ago I went to the doctor with a persistent headache and a problem with the vision in my left eye. Now, it seems, I ought to be thinking hard about a hospice. Private medical insurance definitely has its blessings but no one tells you how to cope with news this sudden and this final.

The neurosurgeon I've just left showed me the MRI scan they did on Thursday, tracing the outline of the tumour the way you might explain a new route home. I followed his thick forefinger as best I could, trying to imagine the cluttered spaces of my own throbbing head, but none of it made much sense. At the end, when I asked what next, he came up with the line about the Reaper. Now I leave my glass untouched and head for the street. Coping is something I've done all my life. Until now.

Home is a sixteen-pound cab ride to Holland Park. I live on the fourth floor of a 1930s block of flats with a sunny view south and the constant assurance from local estate agents that serious cash buyers are only a phone call away. The place is safe and beautifully maintained. I've spent the best years of my life here, even with Berndt, and until this morning it's never occurred to me that one day I might have to leave.

My immediate neighbour has lived here even longer than me. Her name is Evelyn. She's West Country, from a small village outside Okehampton. She's wise and kindly and Berndt always said she belonged in a homestead in frontier America with a rugged husband and an army of kids. Berndt was wrong about that because she's never married, probably never had a man, and maybe as a consequence she puts a great deal of thought into nurturing relationships she values. I flatter myself that I count as one of her friends.

Evelyn has sharp ears for the arrival of the lift but always waits until I've settled myself in before knocking lightly on the door. These calls are always for a purpose, another reason we get on so well. Since I've known her, she's worked as an editor for one of the smaller London publishing houses. People I know in the business tell me she's become a legend and I tend to believe them. She certainly knows that less is more, an editorial rule she applies unfailingly to her own life. She stands at the open door, a thick Jiffy bag in her arms. The pencil behind her ear is a signal that she's busy.

'I took this in,' she says. 'I think it's from your agent.'

She gives me the Jiffy bag and then pauses before stepping back into the hall.

'Is everything OK, my lovely?'

'No, if you want the truth.'

'Anything I can do?'

'No.' I force a smile. 'But thanks.'

She nods, says nothing. She'll be there if I need her, I know she will. But not now.

I put the kettle on and toy with making a cup of tea but give up, overwhelmed yet again by what's happened. I'm thirty-nine, shading forty. I'm in my prime. I jog round Kensington Gardens three times a week. My serious drinking days are long gone and I can't remember when I last had a cigarette. Only a week ago, a casting director swore he'd never seen me looking better. Now this.

Shit. *Shit.* Normally, I'm good at self-analysis. I can distinguish at once between a sulk and something more worthwhile, but this ability to read myself appears to have gone. Is this anger I'm feeling? Or bewilderment? Or, God help me, simple fear? The fact that I don't know only makes things worse. Helpless is a word I've never had much time for. It smacks of giving up, of surrender, of weakness. And yet that's as close as I can get. Helpless? Me?

I open the Jiffy bag. Evelyn's right, it's from my agent. It's a French-Canadian script and she rates it highly. The producers are still looking for finance, and despite everything it's good to know that my name attached might help them find the right kind of backer. So what do I do here? Do I lift the phone and tell my agent to hold all calls? Do I fess up and say I've joined the walking dead? Or do I just breeze on and hope that I can somehow make it through? In certain kinds of script we call that denial. Denial, under the current circumstances, sounds perfect. And so I pop another Percocet and curl up on the sofa.

My agent's right. Even with my mind still wandering up cul-de-sac after cul-de-sac, I give the script enough attention to know that it's very, very good. I play a French academic on a one-year sabbatical in Montréal. I fall in love with a handsome campus drunk who turns out to be already married. Life gets very difficult, and then impossible, but a clever plot gives me the chance to exact a little revenge in the final act.

The writing is serious and comic by turns, and the dialogue alone has won me over. By the time I've given the script a second read, I've already made that strange alchemical step into someone else's head. I *am* the woman on the page. I cut my bastard suitor far too much slack and I'm punished in subtle and inventive ways that bring a smile, albeit rueful, to my face. But Fate, thanks to the gods of the cinema, comes to my rescue. My jilted beau ends the movie on the very edge of Niagara Falls, contemplating a messy suicide, while I accept the applause of my peers for simply surviving. *If only*, I think.

Another knock on the door. It's Evelyn again, this time with a bottle in her hand. Very unusual.

'Good?' She's nodding at the script.

'Very good.'

'Tempted?'

'Oh, yes.'

She's brought whisky. I pour two measures, adding ice, aware of Evelyn monitoring my every move. On occasions, she can be very direct.

'So what's happened?' she asks.

I tell her everything. It doesn't take long. I'm sharing my brain space with a tumour. Soon it will kill me.

To my relief, she doesn't move. No arms around me, no whispered consolations, no invitation to share the pain. Just a simple question.

'And do you believe this man?'

'*Believe* him?'

'Yes.'

'You think he's kidding me? Some kind of joke? You think the guy's a sadist?'

'I'm just asking exactly what he said.'

I do my best to remember, word for word. Mention of the Reaper brings a scowl to her face.

'He said that?'

'He did.'

She nods. She clearly thinks it's unforgivably crass, even cruel, but she's also wondering whether he'd recognized me. I tell her my medical records are in my married name, Enora Andressen, but she doesn't think that's a factor. My last movie, a screen adaptation of a decent novel, has been doing good business in London art-house cinemas.

'Men can be funny around fame,' she says. 'Especially alpha males. I see it in the office sometimes. When we stoop to celebrity biogs, and the lady concerned pays us a visit, the Head of Sales always makes a fool of himself. It's primal behaviour. It belongs in the jungle. If I were you I'd ignore it.'

'That's a hard thing to do when he tells me I'm going to die.'

'He said *might*.'

'He did. You're right.'

'So hang on to that.'

A silence settles between us. It feels companionable. Warm. I think I love this woman. When things got really tough with Berndt and he started throwing the furniture around, she offered nothing but good sense. *Change the locks. Get yourself another man. Preferably someone bigger*. As it happens, I did neither but just now Evelyn is offering just a glimpse of something that might resemble hope.

'Did he talk about treatment at all?'

'No. I've got to see him again on Friday.'

'No mention of an operation?'

'No.'

'Radiotherapy? Chemo?'

'No.'

Another silence. I gulp the whisky, draining the glass. I haven't touched Scotch for years but I'm glad she's brought the bottle. The fierceness of the burn in my throat creeps slowly south. I'm alive. Everything's still working. Fuck the tumour.

'More?' She refills my glass, not waiting for an answer.

I nod. I'm gazing at her. My eyes are moist. I very badly don't want to cry. Not in front of Evelyn.

'And Malo?' she says softly. 'You think you ought to tell him?'

Malo.

I hold her gaze as long as I can and then I duck my head, holding myself tight, rocking on the sofa, letting the hot tears course down my face, howling with the pain of my grief.

Malo is my son. He's seventeen years old, impossibly handsome, impossibly difficult, and impossibly remote. I can forgive Berndt most of the stuff that went on between us, but not for stealing Malo. By the time he eventually left, far too late, I finally realized what he'd been doing with our son's affections. My ex-husband was always clever. He had a way with words. As a successful scriptwriter, he understood the magic of language. His move into direction taught him how to ramp up the pressure. His obsession with *noir* gave him the meanest of streaks. My poor Malo was putty in his hands. Given any kind of choice, what seventeen-year-old wouldn't opt for a penthouse apartment in Stockholm and the company of a blonde starlet with a huge Scandi fan base? Drunk, towards the end, Berndt had talked of trading me in, one washed-up actress for a younger model, but in my darker moments I wondered whether Malo hadn't shared the same thought. From what I can gather, Annaliese makes perfect cheesecake. Job done.

Enough. Evelyn is sitting beside me on the sofa. Practical as ever, she's found some tissues. I tell her that I've no intention of sharing my news with anyone, least of all Malo or my ex-bloody-husband.

'A secret, then? Just you and me? Until you're better?' Evelyn is smiling. I can tell she's pleased. She puts the cap on the bottle and suggests I go to bed. Anytime, day or night, all I have to do is lift the phone. I do my best to thank her, to apologize for the tears, but already she's on her feet.

At the door, struck by a sudden thought, she turns back.

'I forgot,' she says. 'You had another caller this afternoon. He knocked on my door as well. Mitch, he said. Mitch Culligan. Ring any bells?'

I shake my head. The name means nothing.

'He said he'll give you a ring. He must have your number.' She nods towards the bedroom door. 'Sweet dreams, my lovely.'

Sweet dreams? It's eighteen years ago. For the second time running, a picture of mine is up for a major award. My mum and stepdad have trained it down from Brittany in case my movie makes the Palme d'Or. Expecting me to meet them at the station in Cannes, they take a taxi to the Carlton in time to catch me deep in conversation with Berndt Andressen. Berndt is hot just now. He's just penned a script which will – in time – open the floodgates to a torrent of Scandi crime *noir* and he's in town to court some of the international finance people who might make his script happen.

Like everyone else in Cannes, I've read about Berndt in the weightier trade magazines, but in person he comes as a bit of a surprise. He's a decade older than me but it hardly shows. He's slim, quiet, and decidedly opaque. He has a thatch of blond hair and the good taste to wear a carefully rumpled suit instead of the designer jeans and collarless linen shirts that have become standard combat issue in certain corners of the media world. We've been talking non-stop for hours by the time my parents show up, which is a weak excuse for not meeting their train. I've never been able to fool my mum, no matter how hard I try. Berndt is courteous and attentive to them both, and insists on buying a bottle of champagne to toast their arrival. 'You'll marry that man,' my mum tells me later when Berndt has left for yet another interview. And she's right.

We slept together that night. My movie didn't win the Palme d'Or but I was way past caring. If Berndt's *noir* script measured up to his talents in the sack then he was heading for stardom. The second time we made love, in the way that only a woman can sense, I knew that Berndt and I had made a baby. As it happens, I was wrong but that – as they say in La La Land – doesn't play well on the page.

The rest of the festival we talked, drank, swam a little, and compared endless notes. That summer I'd been along the Côte d'Azur for a month already, holed up in Antibes waiting for a French production team to finish a rewrite before shooting extra scenes for a gangster movie in which I'd won a smallish but important role. The script work went far too slowly and at the very end of the shoot some of us had fallen into bad company aboard one of the grosser cruisers docked in the marina. This was nothing I especially regretted – in those days I could put anything down to research – but at Cannes a day later it was wonderful to be in the company of someone thoughtful, someone who knew how to listen, someone whose interest in yours truly extended beyond the taking of yet another scalp. Berndt had the kind of attentive curiosity I've yet to meet in any other man. It took me years to realize how predatory that can be, but by the time the festival came to an end I knew I was in love.

Berndt and I said our goodbyes at the airport. His flight to Stockholm was the first to leave. I remember standing in the hot sunshine on the balcony at the airport, watching his plane climb away into the blue and wondering whether I'd ever see him again. Eight days later, on the phone from Copenhagen, he proposed. We were married in London a week before Christmas. By then, I was well and truly pregnant.

My alarm is always set for 06.30, a habit I picked up on countless locations. For most of the night I've been dreaming about my father. He's been dead for a long time now, a victim of throat cancer, but when I was a child he used to entrance me with puppet shadows on the wall. He'd make shapes with his hands, sometimes one, occasionally two. The shapes would be a barking dog, or an owl with flappy wings, or some nameless beast with three heads, and my world was all the richer for these sleights of hand. At the time he called them *trompe l'oeils* but it was a while before I realized they were simply optical illusions.

Now, a thin grey daylight washes across the big double bed. For the first time in weeks, I realize that I haven't got a headache. I silence the alarm clock and think hard about the stillness inside my skull. Has the tumour taken fright and left me for someone else? Do I owe my life to Johnnie Walker Black Label? I get up, moving very carefully, the way you might carry an overloaded tray. Last night's glasses are still where I left them in the sink. Gratitude smells of stale Scotch.

The phone rings at one minute past nine. I'm on my third cup of tea, still pain-free, still marvelling at this small moment of release. The voice on my mobile phone belongs to a stranger: northern accent, bit of a cold.

'Enora Andressen?'

'Who is this?'

'My name's Mitch. Mitch Culligan. You're OK to talk?'

Culligan. The name rings a bell but I can't think why. 'How did you get my number?'

'Friend of a friend.'

'Like who?'

'Can't say. Sorry.'

'So why should I talk to you?'

'No reason at all. I'm in a car outside your block. Red Fiat. Seen better days. If this sounds creepy, it isn't. Fancy breakfast?'

I have to take the phone into the spare bedroom to check the street. A tallish guy, visibly overweight, gives me a wave. Grey anorak. Battered day sack. Baggy jeans. Terrible hair. I'm staring hard at his face. I saw him once on *Newsnight.* He's a journalist. And he once did a couple of decent articles on the land mines issue.

'Are you the guy who called on my neighbour yesterday?'

'Yes.'

'So what do you want?'

He won't say, not unless we have a proper conversation. In my job, exchanges like these come with the territory. Normally I'd bring the exchange to an end rapido but the land mines issue was important to me, still is, and in any case yesterday has done something to my normal sense of caution. Time, for one thing, has become a commodity I can no longer take for granted. What the hell.

'Breakfast,' I tell him. 'On you.'

We go to a wholefood cafe off the Bayswater Road where they know me. At my insistence, we walk. He's much taller than I am. He has a strange gait, lumbering, flat-footed, body thrust forward, hands dug deep in the pockets of his anorak, as if he's heading into a stiff wind. When I mention land mines he nods. Angola. The Balkans. Afghanistan. Any country touched by conflict has been left with a legacy of buried mines and a generation of kids who've stepped on them.

'You know about this stuff?' He seems surprised.

'I do. Not first-hand but through someone close to me. There are charities who work in the field. I've always done my best to help.'

He nods in approval. We're definitely on the same page here. Nice to know.

The cafe is comfortably full but there are generous spaces between the tables. Wealth has its own smell, in this case freshly brewed Java Pure.

My new friend studies the chalkboard in disbelief.

'They do bacon?'

'No.'

'Sausages?'

'Only soya.'

'Anything edible?'

We order scrambled eggs on five-grain wholemeal toast. En route to a table at the back, I accept an air kiss from an Iranian art dealer who owns the gallery on the corner. Word on the street suggests that nothing costing less than $10,000 gets out of his door.

We settle at the table. Mitch, his day sack tucked under the table, is taking a lively interest in the cafe's clientele. I'm beginning to be intrigued by this man. In my business you spend half your life pretending to be someone else and unconsciously or otherwise you're forever on patrol, watching other people, making mental notes, tucking away their tiny idiosyncrasies – little give-away tics – in case they might prove useful later.

Mitch, unusually, offers few clues. His lumberjack shirt, which is missing a button, could do with an iron. He badly needs a proper shave. His lace-up boots are caked with mud. But this air of neglect doesn't appear to trouble him in the least. On the contrary, he seems to be a man thoroughly at ease with himself – not as common as you might think.

He wants to know what I thought about the recent election sprung on us by Theresa May. The question takes me by surprise.

'Nothing,' I tell him. 'I was on location in the States.'

'Trump?'

'A buffoon. And allegedly a serial groper.'

'You ever get to see him in the flesh? Meet him, maybe?'

'Christ, no. That man puts his smell on people. He's a dog. I'd be washing him off for a week.'

He grins at me, his big face suddenly younger. He says he spent the three weeks before the election touring parts of the UK, taking the pulse of the place, looking for clues.

'And?'

'This country's a crime scene. Make that a plural – crime *scenes*. We've been screwed by neglect, by clever lies, and by a long list of politicians who should have been paying more attention. Behind them is an even longer list of names you've never heard of and they're the ones you really need to watch.'

'You write about all this stuff?'

'Of course.'

'So should I have recognized your name?'

'Depends what you read.'

'Very little, I'm afraid. I gave up on the press years ago. If a girl wants fiction she should stick to novels.'

He shoots me a look. I think I've offended him. Then he tells me he's recently gone freelance after years with a major broadsheet. Facebook and the rest are killing the print business but there's still money to be made.

'Is that what this is about?' I nod at the space between us. 'You're after some kind of exclusive?'

This time I know I've hurt him. Worse than that, he's got me down as a spoiled celeb, tucked away in Holland Park, protected by a thicket of agents, publicity machines, and very big ideas about myself. He's talking about the recent election again, how three weeks on the road talking to people about the kind of lives they lead should be compulsory for every politician.

'But it is. That's what elections are about. No?'

'No.' He shakes his head. 'It's a comedy show. These people are looking for votes but they have no time. Here's a sticker. Nice dog you've got there. Lovely baby. Remember my name.' He scowls, leaning forward over the table. 'This country is dead on its feet. The ones with money might vote. The rest have nothing to protect. On some levels, believe me, it's scary. And you know why? Because the system doesn't work any more. Because the system is fucked. I could spend half a day in some run-down shopping centre and not meet anyone who had a clue what to do with his vote. Either that or they couldn't be arsed. This country has become world class at giving up. Not here so much. Not in London. But up there. You ever been to Burnley?'

'Once. I was playing in a Rattigan at Blackburn. You?'

'Born and bred. My dad was a vicar. Can you believe that? Burnley was a proper place. Once.'

He makes space on the table and sighs while the waitress delivers the food. He watches her return with a carafe of mango juice, on the house. My new friend is impressed.

'Are they always this nice to you?'

'Always.'

'Why?'

'They like my films.'

'Good. I've been meaning to tell you. That scene at the beginning of *Arpeggio* – you seriously underplayed it. The rest of the movie? Excellent. But you were nuts to kick it off that way.'

I take this as a compliment, partly because I'm warming to his bluntness but mainly because he's right.

'That was the director,' I tell him. 'I wanted to play it full-on. He thought we'd lose the audience.'

'You nearly did. And I'm a fan.'

We talk movies while he demolishes the scrambled eggs. To my slight surprise, his knowledge of films is huge, his taste impeccable. Early Chris Nolan. Anything by Almodóvar. Sean Penn in *21 Grams*. Perfect.

'Not hungry?' He's looking at my brimming plate. I've barely touched it. His interest is obvious.

'Help yourself.' I reach for the mango juice. 'And while you're at it you might tell me why we're here.'

He forks scrambled eggs on to his plate and then bends to the day sack. Moments later I'm looking at a thickish file. Handwritten on the front is a single word: Cassini.

Already I'm intrigued. I've come across the word recently but I can't remember where. *Cassini*?

Mitch wolfs several mouthfuls of egg and then opens the file. Sheets of text hide a pile of photos. He's about to show me one of them but then his hand pauses. Long fingers. Cared-for nails. No rings.

'This is a long story,' he says. 'Which is rather the point.'

'I like stories. But what's this got to do with me?'

'That's my question, I'm afraid.' He pauses, looking up, catching something in my voice. 'You OK?'

I'm not. I have a sudden, blinding pain behind my left eye. I can see two versions of Mitch, both of them shading into grey, and the tables around us are blurred beyond recognition. Mitch is already on his feet. I'm clinging on to the edge of the table now and I'm dimly aware of throwing up on to my plate. Somebody – Mitch? – is holding me from behind. I try to lift my head. The pain is unbearable. Then everything goes black.

TWO

The next few days are lost to me. Time expands, condenses, wriggles around, expands again, and then snaps like some worn-out elastic band. My first moment of consciousness comes in the back of an ambulance. After that, I'm adrift again until a face swims into focus at my bedside. Only slowly do I associate the rimless glasses and the paleness of the face with my neurosurgeon. He's wearing blue surgical scrubs and a white face mask dangles beneath his chin. Costume drama, I think dimly. Death-bed bye-bye scene. Not good.

The face fades, only to return later. It might have been a couple of minutes later, it might have been the following day, I've no idea, but this time we have a conversation.

'It went well.' He's sitting on the bed. 'Much better than we expected.'

I try to speak. Frame an answer. Make some kind of comment. Nothing. I start to panic. An exposed nerve in my poor failing brain? A parting *billet-doux* from the tumour? A slip of the scalpel cutting whatever goes to my vocal chords? I try to swallow. Find it nearly impossible. What kind of future awaits an actress who can't speak? Can't even breathe properly? Then comes a soft pressure on my lower arm and it takes me a moment to realize what it is. Reassurance. Comfort. Kindliness.

'You've still got tubes down your throat,' the face says. 'Don't try and talk. Just nod or shake your head.' The face smiles. 'Any pain?'

I shake my head.

'Nothing? No discomfort?'

A tiny shake this time.

'Just a bit?'

I nod.

'Good girl.' The hand on my arm, again. 'Take it easy. No dancing. I'll be back tomorrow.'

* * *

Bless him. Bless them all. My first visitor is Mitch. He lumbers into the High Dependency Unit, his big face invisible behind an explosion of blooms.

'Lilies,' he says. 'My dad used to swear by them. Dark properties. Keeps the devil in his place.'

He disappears to find a nurse and a vase, and returns within moments. No anorak today. Apparently an Indian summer has descended on West London.

'OK?' His eyes are mapping the tangle of leads that keep my vital signs on track.

'Fine.' I can talk now. The tube has gone.

'And?' He wants an update, a prognosis. No messing.

'They say I might get better.'

'Might? That sounds a bit provisional.'

'Might,' I confirm. 'All these conversations depend on where you start. They wrote me off the other day so I can definitely handle "might". In fact "might" could become my very best friend.' I lie back and close my eyes. My little speech has exhausted me.

When I come to again, Mitch is very close. He's found a chair. He's been eating garlic.

'This has been going on a while? Only I feel a bit guilty.'

'You mean the cafe?'

'Yes. Dragging you out. Bothering you.'

'It was good you were there. It would have happened anyway.'

'How do you know?'

'I don't. I'm just being nice.'

'Appreciate it.' His hand lingers briefly on my shoulder. Then he's on his feet again. 'Anything I can bring? Next time?'

My hand goes to the bareness of my scalp where they shaved me before the operation.

'A beret?' I suggest. 'Maybe red?'

After Mitch, a succession of friends and relatives arrive. Evelyn is first, bearing a box of Belgian chocolates she knows I'll kill for. Then my agent and the actor who played opposite me in *Arpeggio*. The news that our movie has been short-listed for a BAFTA sparks a big smile from the nurse who's sorting my tablets. Next day, my parents arrive from Brittany. Evelyn has given them the key to my flat and my mum will be here for as long as I need her. She stays all afternoon and she can't wait to tell me the news from Stockholm. Malo has been on the phone. He's had a big row with his father. His love affair with Sweden is over. This is very good news indeed.

'He's coming back to London?'

'He didn't say that.'

'What, then?'

'He wants to travel a little.' She's frowning. 'I told him he should still be at school.'

School has been a festering sore between the three of us – Berndt, myself and Malo – for some time. Malo bailed out of the local comprehensive last year. My marriage was in ruins and I admit to taking my eye off the ball, but it was months before either Berndt or I realized that on many days he simply wasn't turning up. When I confronted him, he said that the teaching was crap and he was wasting his time, but I think the real reason was that he just couldn't cope. Under the circumstances, our circumstances, I couldn't really blame him. Then he went to Sweden and what was left of his education became Berndt's problem.

'So what did he say on the phone?'

'He wants to borrow five thousand euro. I told him to ask his father.'

My smile is unforced. I've learned recently from my stepdad that my Christian name was my mum's choice. Enora is Breton in origin, homage to a saint of the same name. St Enora is celebrated for entering a convent on the day she got married. Maybe I should have listened harder and picked up the hint.

'So where does Malo want to go?'

'Polynesia. He says it will improve his French.'

'His French is fine already.'

'*Bien sur.* But he says that would make New Caledonia a kind of university. *Pas stupide, notre Malo.*'

She's right. Despite his academic record, Malo has always been bright, as well as manipulative and feckless, but given the chance I'd write a cheque here and now for both of us. Two tickets for New Caledonia? Two glorious months in the sunshine with my lovely boy? Ten thousand euro for the pair of us? A steal.

Mum has had a word with the neurosurgeon.

'He says you're doing OK. Maybe you should come back to France with us.'

I do my best to smile up at her. There are worse prospects just now than an autumn in their manor house overlooking the beach at Perros-Guirec. My mum cooks like a French mother should and I've never had a problem with my stepdad. He used to be a consultant engineer in the oil business, which is where all the money came from. His career made him a perfect role model for my son but his occasional attempts to talk Malo into doing something useful with his life have so far come to nothing.

I say I'm grateful for the offer.

'You mean no?'

I nod. For now, I tell her, I need to be in London.

'For the doctors?' Her eyes flick towards the nursing station. 'Just in case?'

'Of course. And maybe one or two other things.'

Mitch is back a couple of days later. He's knocked at Evelyn's door again and on her instructions he's arrived with my iPad, plus a beret she's found in my wardrobe. I seize the iPad greedily. I want to know where to find some of the stuff he's been writing.

'You're serious?'

'Always.'

He directs me to a handful of websites. One of them is the *Guardian*, another the *New Statesman.* A third, the most interesting, is a web-based digest of investigative journalism that calls itself Finisterra.

'It's Latin,' he explains. 'Posh. It means the end of the world.'

'Light reading? Something to keep the kids amused?'

He smiles. We've met just three times, the first a near-death experience, but we seem to have already developed that unvoiced kinship that can happen between two strangers who share the same sense of humour. In my drowsier moments I wonder whether it bears any comparison to my early days with Berndt, but I know already that this is something utterly – and thankfully – different. We're older. We've knocked around a bit. We've left footprints in the sand. Not that I know the first thing about this man, except that he makes me laugh.

'Tell me where you live.' It's early evening, rain streaming down the window, the hiss of rush-hour traffic from five floors below.

'Hither Green,' he says.

'Where's that?'

'South of the river. A Tesco Express and a couple of decent pubs. Old working-class London plus young couples on the make. You're getting the picture?'

'Alone? Married? Kids? The whole shtick?'

'Once.'

'And now?'

'Just me and a Syrian guy from Aleppo. He's a refugee but he's got leave to remain.'

I nod. I badly want to dip into the websites this man's tallied for me but I like his company, too. Tough call.

'You want me to take a walk? Find a coffee?' He's nodding at the iPad. 'Then come back?'

I gaze up at him and nod. This man reads my mind. Remarkable.

I start with the *Guardian*. Mitch Culligan turns out to be an occasional contributor to the Opinion page. A couple of his pieces deal with the aftermath of the election. They're punchy, beautifully written, closely argued, full of attitude. His pre-election tour of the kingdom, as I already know, filled him with a bewilderment verging on despair. Now, to everyone's astonishment, Corbyn has pricked the Tory bubble and the electoral future, just for a month or two, looks a great deal sunnier. Brexit, in Mitch's book, remains a huge own goal but the kids are at last on the march and Jezza has given them some sense of destination.

I find myself nodding. I met Corbyn once at a college in North London where I was doing an acting workshop. He struck me as sincere and companionable in ways I've never associated with politicians. He was also, like Mitch, a bit of a film buff.

I scroll through the rest of the *Guardian* pieces, then sample an offering or two from the *New Statesman.* By now, Mitch's preoccupations are very clear. He detests, in no particular order, Donald Trump, Steve Bannon, Margaret Thatcher, Liam Fox, fiscal favours to public schools, the rotting corpse of UKIP, and the whole swamp of corporate greed. He suspects Theresa May is no great brain and he'd very much like to ride to work on a re-nationalized railway. He also thinks, joy of joys, that the Tory party is about to implode, and that a judicious poke or two might reveal the blackness of its soul. 'Expose these people for who they really are,' he writes, 'and they'll be out of power for a generation.'

I've found the investigative website by the time he returns. He must have been gone at least an hour. Some coffee.

'I had to make a couple of calls,' he grunts. 'Most days it would have been twice as long.'

I take this as a compliment. Then I mention Finisterra. This website, as I understand it, is Mitch's little allotment, the place where he can plant whatever thoughts he wants and share them with a wider world. Fair?

'Completely. Allotment is right. I take all this produce to market and wait for takers. That's the magic of the internet. You can kid yourself you've put the mainstream media out of business until you take a proper look at the traffic figures.'

'So why the Black Death?'

'You mean the piece on Boccaccio? The *Decameron?* The week the plague came to Florence?'

'Yes.'

'Because no one else would take it.'

'I understand that. I'm asking you why you wrote it in the first place.'

'Easy. It's the stats. One in two of the population died. Not just Florence, everywhere.'

'Here?'

'Everywhere. Do the sums. The Black Death gave you a fifty per cent chance of making it through to the weekend.'

These are odds I can relate to, thanks to my neurosurgeon. His playful introduction to the Grim Reaper still rankles, though more recently he appears to have saved my life.

'You've studied this stuff?'

'I have. I know it makes me a sad old git but maybe you should put it down to perspective. Grief's an easy sell these days. Why? Because there's serious money in other people's pain. Some lunatic goes bonkers on Westminster Bridge and the whole country – the whole fucking planet – wrings its hands. This is heavy stuff. Wrong time, wrong place. But we need to get a grip because these are buttons the bad guys know exactly how to press.'

'You think we should just ignore it?'

'I think you should log on again,' he nods at the iPad, 'and read about the Black fucking Death.'

This conversation is getting nowhere but I have the sneakiest feeling that Mitch is right. Living with Berndt's fevered imagination bothered me for years. He wrote like an angel and knew exactly how to tighten a plot until the audience was gasping for air. These are genuine talents but like everyone else in the marketplace he was aware that movies and books and everything else we call entertainment had to get blacker and blacker. Cannibalism. Animal rape. Multiple sex with Siamese twins. Anything to keep the cheques rolling in. I shudder to think about the impact on kids, and on Malo in particular. Serious money in other people's pain? Too right.

Mitch is delving in his day sack. Out comes the sweetest umbrella, powder blue, plus the cable to charge my iPad. Apparently the rain's in for the evening and Mitch has to make an event in deepest Camden. The four-way debate will be exploring the proposition of something called the Deep State. When I ask who's on the panel, he gives me names. One's a Labour MP. One's a rebel cleric. Another guy is a Venezuelan academic I've never heard of.

'That's three,' I say. 'Who else?'

'Me.'

Ouch.

Instinctively I reach out for him but he backs away. Then, just in time to delay his departure a little longer, I remember the name on the file in the cafe.

'What's Cassini?'

He looks at me a moment, then checks his watch before nodding down at the iPad.

'Google *Death Dive*,' he says. 'And wish me luck.'

Cassini turns out to be an inter-planetary space probe launched to investigate the rings of Saturn. It's done most of its business and just now it's barely weeks away from a final plunge towards the planet's surface. With the batteries all but dead this last act may reveal just a little more about Saturn's secrets before its atmosphere burns it alive. Hence the term 'Death Dive'.

Death dive? I'm trying to connect the dots here. Mitch is a campaigning journalist with a national reputation. He has talent, motivation, a squillion media contacts and thus a sizeable audience. People read his stuff, probably lots of people, and so they should. But what has any of this got to do with little me? I circle the question all evening, a bit like the poor space probe, getting no closer to an answer. Mitch has made no promises to come back to the hospital and in any case I may be discharged within days. Is this the last I'll see of him? Or might there one day be a chance to get a proper look at his mysterious Cassini file?

Next morning they get me into a wheelchair and take me several floors down to see my neurosurgeon again. This turns out to be the office where we first met, but by this time we've become old friends. He takes me through the operation and shows me scans and surgical images the way you might share holiday snaps. The tumour, it seems, was a glioblastoma, malignant, aggressive, difficult to get at. Pressing on my optic nerve, it was responsible for my headaches and my wonky eye. Untreated, it would undoubtedly – in time – have killed me.

'You got it all out?'

'We think so. We can't be absolutely certain but the chances are good.'

'So it won't come back?'

'It may. We never say "never". We'll be scanning you every three months to check.'

'And if it does? Come back?'

'Then we may have to pay it another visit.' A thin smile. 'The Reaper shouldn't have it all his own way.'

Under the circumstances, I later reflect, the comment is as close as I'm going to get to an apology. When I told the Reaper story to my mum she was horrified. It was my stepfather who had a private word with the hospital management and I imagine they did the rest. Either way, I'm genuinely grateful for everything this man and his team have been able to do for me.

'You saved my life,' I tell him. 'At least for now.'

Back home, I find three more Jiffy bags awaiting me. My stepfather has gone back to Brittany but my mum has truly made herself at home. My appetite for her very distinctive take on onion soup has returned with a vengeance and I'm even able to help her out with a decent bottle of Sauvignon.

Two of the Jiffy bags have come from my agent. Both contain scripts and both are duds. Of the French-Canadian project there's so far no news though these things take forever to develop so I'm far from disappointed. In any case, the last couple of weeks seem to have loosened my career's grip on me. Things I once viewed as important – an evil review, for instance – have joined the rest of the background static in my life. No longer will my precious future be threatened by a critic's crap judgement or a casting director's inexplicable failure to see me as the obvious choice for the female lead. No, from now on – just like everyone else on the planet – I'm simply a billion cells in loose formation, any one of which could turn rogue at any moment.

This realization, to my surprise, comes as a bit of a relief. All told, I've done OK. I've been in some quality films. I've won some devoted fans. I've made more money than I ever dreamed possible. I might even get to live with my precious boy again. Not bad. Not bad at all.

One evening, late-ish, I try to explain some of this to my mum but, like most French women of her generation, she's fiercely proud of what I've achieved and refuses to accept it might be over. In time, she tells me, I'll get back to my old self, maybe put on a bit of weight, start looking for new roles to conquer. I nod, and agree, mainly to keep her sweet but later, after she's finally retired to the spare bedroom, I stand by the window in the lounge, staring up at the night sky.

Thanks to the street lights and all the other pollution there's very little to see but I know that out there, way beyond the creamy blob of the moon, a tiny little morsel of high-tech is readying itself for its final curtain call. I've read the websites. I'm word perfect on the script. One last fizz of data transmission. One last fuel burn. And then the plunge into oblivion.

I hang on to the thought for a moment and then remember the third Jiffy bag. I've a feeling it might have come from Mitch and the moment I open it I know I'm right. He's sent me a framed shot of Saturn from the Hubble telescope. The image is truly beautiful, the huge blue planet girdled in mystery. I hold it at arm's length, and then return to the window.

Poor Cassini.

THREE

Two days later, it's Mum who takes the call.

'Malo,' she mouths, bending to the phone.

My angel child is at Stockholm airport. His flight is due to take off in half an hour. By mid-afternoon, he should be in Heathrow. Any chance of a lift?

'Ask him which terminal.'

My mother gives me the phone. I put the question myself. Malo says he hasn't a clue. He sounds sleepy.

'Give me the flight number,' I tell him. 'It'll be on the ticket.'

There's a long silence. My pulse quickens. For a moment or two I think I've lost him. Finally he gives me the flight number. Then he's gone.

The last time I saw Malo was more than a year ago. Berndt had moved out after a particularly vicious row. He'd come at me with an empty bottle of vodka he'd smashed on the edge of a kitchen work surface and after I'd locked myself in the loo, I called the police. Malo stayed in contact with his father after the exclusion order and disappeared within the month. No farewell hug. No note of explanation. No forwarding address. Not even a hint that Berndt's credit card had bought him a ticket to Sweden. At the time, it had been hard to forgive either of them. Now, though, I'm telling myself that Malo is my best chance to rescue something from the ashes of the past couple of years.

At the airport, my mum is the first to spot him among the tide of incoming passengers.

'Malo!' she shouts in her throaty Breton twang. I follow her pointing finger, her madly waving hand. The same mop of black curly hair. The same adolescent slouch. The same ripped jeans. But something has happened to his face. The hollows are deeper, more sculpted. He's started shaving in earnest. To my slight surprise, I'm looking at a man.

My mum, like me, has always been a sucker for the sudden warmth of Malo's smile.

'Tout va bien, mon petit?'

Malo accepts her hug and assures her he's fine. His voice is deeper, his French perfect. So far, he hasn't even looked at me.

The stream of incoming passengers is thinning. Malo stoops to retrieve his bag. Soft red leather. Not cheap. Like his father, my son has always travelled light.

'Why Einstein?' I'm gazing at the image on his T-shirt.

Malo ignores the question. He can't take his eyes off the neat line of stitches across my shaved head.

'Some kind of accident?'

'In a way, yes.'

'You OK?'

'I'm fine.'

I resist the urge to kiss him, to tell him how wonderful and unexpected it is to see him again, how the best news always takes you by surprise. Instead, I tell myself that this is a scene best played long. Life has moved on, his and mine. We need to take our time.

On the way back into London I insist that Malo sits up front with my mum. She's always been a natural with Malo, ignoring his reticence, filling his silences with a blizzard of news from home. This is a knack I've never been able to master, not least because to me gabbling is always a sign of nerves, but by the time we're back home I realize that nothing has changed. My son is as gorgeous, and opaque, and impenetrable as ever. Except when there's something he needs in a hurry.

It's late evening. Before retiring, my mum has broken the news that she's off back to Brittany the day after next. I've told Malo about the operation, and the various tests to come, and it's clearly Mum's expectation that her grandson will take over as my guardian and nurse. This responsibility Malo appears to be ready to shoulder. With one proviso.

'I've got a friend, Mum. From Stockholm.'

'He? She?'

'She. We could share the spare bedroom after Gran's gone, yeah?'

Her name, he tells me, is Eva. She's just graduated from university with a degree in psychiatry. She speaks perfect

English. Better still, she has a driving licence. Perfect for running me around.

He presents the proposition with a yawn, and wanders into the kitchen to make himself a coffee. My agreement is evidently a given, a technique I suspect he's picked up from his father. Irked, I call him back.

'How old is Eva?'

'Twenty-six.'

'And how long might she be here?'

'Dunno.'

'Weeks? Longer?'

'Depends.'

'On what?'

'Loads of stuff. You'll like her. She's cute.'

'Cute' is a word I don't recall Malo using before. New, as well, is his invitation to join him in a spliff or two. The coffee abandoned, he produces a bag of loose tobacco, a box of papers, and a cube of resin untidily wrapped in silver foil.

'You brought that through customs?'

'Yeah.'

'You're crazy. These days they lock people up for less.'

He shrugs, doesn't react. I watch him rolling the joint. He does it quickly, deftly. For the first time in days my head is beginning to throb.

'If you really want to smoke you'll have to do it outside, preferably on the pavement.' I nod towards the door.

'It's raining.'

'I don't care. The choice is yours. I hate smoking. You know that.'

He studies me for a long moment and I sense something in his eyes that I've never seen before. My precious son is full of rage.

'Tell me about this thing with your father.' I reach for the spliff and put it carefully to one side.

'What do you want to know?'

'Everything. What happened between you. Why you decided to leave. What drove you out.'

He gives me that same look but this time the inference is clear. This is none of my business. I try again, same questions, different phrasing, and this time – briefly – I spark a response.

'The man's a dickhead,' he says. 'But I guess you'd know that already.'

I want to know more about the dickhead, about Eva, about his year and a bit in Stockholm, but Malo is already on his feet. He picks up the spliff and heads for the door. Moments later, he's gone.

It must be twenty minutes before the phone rings. Thinking it might be Malo locked out in the rain, I'm already heading for the panel in the hall that houses the entry phone.

'Enora? That you?'

It's Mitch Culligan. Before I get a chance to say anything he's apologizing for the lateness of the hour. He's been tied up with a bunch of Marxists who don't know how to end a sentence. He'd have phoned earlier but he couldn't find a way of putting the conversation out of its misery.

'That bad?' Despite myself, I'm laughing.

'Worse. Two of them belonged in a museum. The other was borderline deceased. Marx never knew what he started. How are you?'

'Terrible, if you want the truth.'

'It's come back? The headache? All the rest of it?'

'No. Yes. Oh, shit . . .' I'm staring at the video camera. There's no one at the main entry door but where the gardens lap against the exterior wall I can make out a slim figure, bent over a cigarette, sheltering under a tree. It's an image that makes my blood run cold. To me it speaks of something beyond pathos. It has to be Malo.

'Enora? What's the matter?'

'Nothing.'

'You're crying.'

'How do you know?'

'I can tell. It's a man thing. A night with a bunch of Marxists tells you everything you want to know about grief. That's a joke, by the way. You want me to come round?'

I know the answer is yes. Just now, I'd like nothing better. But an evening with my only child, the only human being I'm ever likely to bring into the world, has taught me what a crap mother I must be.

'No,' I say as firmly as I can. 'Give me a ring tomorrow.'

* * *

Malo, when he finally returns, is soaking wet. I let him in at the main door downstairs and fetch a towel while he walks up to the fourth floor. His curls are plastered against the whiteness of his scalp, and his T-shirt and jeans are wet through, but he's calmer, even talkative. He seems to have realized that I hold the keys to the flat, that I own the place, that I might conceivably have ideas of my own, and this – at last – prompts the beginnings of a conversation.

'It must have been scary,' he says. 'Nearly dying like that.'

He's standing in the middle of the living room, naked except for his boxer shorts. I want to towel every inch of him dry. I want to make him hot chocolate and French toast just the way he used to like it. I want to have the two of us back together again and maybe have another go at getting it right. Instead I ask him to tell me more about Eva.

'I met her through Dad. He says he went to her for advice on some character or other but that's bullshit. When I turned up he was giving her this big line about a part he had in mind. You know the way he can be? Close but not too close? Hanging back? Pretending he isn't interested?'

I nod. I know exactly how Berndt plays scenes like these, artful, canny, seemingly disengaged. He once boasted to me that he could talk any woman into bed without laying a finger on her.

'This was in the apartment? The penthouse?'

'Yeah.'

'So where was Annaliese?'

'She's been on location in Russia. Before that, she was doing some movie or other in Hamburg.'

'So she's away a lot?'

'Most of the time.'

'And is that a problem?'

'Not for her. She's got a brain in her head. A couple of weeks living with Dad and she couldn't wait to get on the road again.'

'But they're going to get married. Have I got that right?'

'No. That was the plan. They'd made the big announcement. Then Annaliese told Dad she'd prefer to wait a bit.'

'Why?'

'She never said. Not to me.'

'So will they ever get married?'

'No chance.'

'And your dad?'

'He never mentions it. Eva calls it denial. You mind if I have a shower?'

He disappears towards the bathroom without waiting for an answer, leaving me alone in the living room, gazing at the wet footprints on the carpet. I'm grinning. The news from Stockholm is excellent. Not just me, I think. Her, too.

In the kitchen, I put the kettle on. The last face-to-face conversation I had with Berndt took place at a pub we occasionally used in Soho. He was days away from moving back to Stockholm and he wanted to settle what he quaintly called 'our private affairs'. In terms of money and possessions this boiled down to a simple split. He'd keep his place in Stockholm, everything he owned personally, and the contents of his several bank accounts, while I would end up with Holland Park, my accumulated savings and investments, my three-year-old Peugeot, and anything else with my name on it.

At the time, I didn't believe him. I'm worth a great deal more than Berndt and I thought that this was a big, splashy gesture calculated to belittle me. In the world of Berndt Andressen, he seemed to be saying, money counted for nothing because only his work, that sequence of movies stretching from here to eternity, really mattered. That was why he was so glad to be shot of me. That was why he couldn't wait for us to get formally divorced.

Thinking about it that night, back in Holland Park, I tried to imagine what would happen next. Berndt would talk to his lawyers. They'd point out I was worth a bit and that a bit of pressure might entitle him to half the joint estate, maybe with some kind of adjustment for Malo. At that point he'd nod and agree, and we'd then settle into our respective trenches, hunker down, and await the opening bombardment.

My lawyer said much the same thing. He seemed to have been dealing with divorces for most of his professional life and in his view my soon-to-be ex-husband was suffering from something my lawyer called 'shaggers' remorse'. This boils down to guilt at the break-up and there are some men who believe that a generous settlement will wash all that nastiness away. To be frank, I never thought Berndt was one of them. Deep down, as his darker plots might suggest, he's a selfish, narcissistic, ruthless bastard, and as the weeks passed without any word from his lawyers, I began to wonder if I was fated to spend the rest of my life in this matrimonial twilight. When I discussed this with my own lawyer, he counselled patience. Early days, he said. Better to wait and see.

Nothing happened. Neither then nor now. There's still been chatter in the trade press about Berndt's forthcoming nuptials but unless he's planning on becoming a bigamist I can only believe that the journalists have got it wrong. Now, thanks to Malo, I know that yet another relationship in his busy, busy life has crashed and burned. Annaliese, bless her, has seen through the man. Soon enough, in her own interests, she'll probably be gone. Either way, I'm very glad he's out of my life. The only thing he had that mattered to me is my precious son and just now he's within touching distance. Berndt, to be blunt, is history.

Malo joins me in the kitchen. He's wearing my dressing gown he must have found in the bathroom. The tea's nearly brewed.

'So what happened with Eva?' I ask. 'When your dad wanted this advice?'

'She turned him down. She told him she didn't believe in the character and neither did he. Dad's not used to women answering back. Not these days. Not with his reputation.'

I pour the tea. I can imagine this conversation in every detail. Malo's right. Berndt always had the upper hand.

'You think I was like that?'

'Like what?'

'Compliant?'

'Never. You fought back. If you want the truth it was horrible being around you both.'

'But why did you go with him?'

'Because he did the same number on me. And I was stupid enough to believe him. Talk to Eva. She's the one who's seen through all the bullshit.'

I nod. Grin. I want to give my wayward boy a hug. By some miracle I don't properly understand we seem to have become comrades in arms, refugees from the same dark presence I should never have let into my life. Looking for the sugar bowl I wonder whether I owe my son some kind of apology for the madness of the last couple of years, but when I begin to explain just how things got so bad so quickly he shakes his head and turns away.

'Mum . . .' he mutters.

I feel a sudden blush of embarrassment. Playing mother is a tough gig. I've gone too far, too fast. I ask again about Eva. Does he have a photo? He fetches his phone from his sodden jeans next door. I find myself looking at an urchin face caught in bright sunshine as if someone's just called her name. Wide smile. Perfect teeth. Punk hair. There's a tram in the background, and a priest on a bicycle.

'Twenty-six, you say?'

'I know. I couldn't believe it either. Sometimes I think she could be my kid sister.'

'And your father?'

'He couldn't handle her. Or us, either. Once he knew we were screwing it got really ugly.'

'Is that why you're here?'

'Yes.'

'Because you've got nowhere else to go?'

'I didn't say that.'

'But is it true?'

'No.' He blinks and looks away, embarrassed by this small admission. I want more. I want him to tell me he's glad to be away from his father. I want to know he's going to be around for a while. I want him to call this cave of mine home.

At last, I pour the tea. Malo always had a special mug, a souvenir from a muddy weekend the two of us shared at Glastonbury, but he barely touches it. Instead, he begins to yawn. I do my best to reignite the conversation but it's obvious he's no longer in the mood.

He's looking round the kitchen as if he's never lived here, never even seen it before, and then he notices one of the photos I've pinned to the corkboard. It comes from way back and it shows the pair of us, Malo and me, on the South Bank. He'd accompanied me to a rehearsal at the Old Vic and afterwards I'd taken him for coffee at the Festival Hall. These little get-togethers were much rarer than you might imagine, one of the reasons I've found space on the corkboard, and Malo seems fascinated by it.

He was still a boy then, pre-adolescent, vocal, curious, prone to mischief and wonderment in equal parts, and I still treasure the memory of that afternoon. He'd sat in the stalls in the theatre while we went through a sequence of key scenes in the last act and he'd been fascinated by the play of light on the set when the director was experimenting with a couple of effects. I suspect this was the first time he'd properly grasped the idea of theatre and there was pride as well as curiosity in the questions he fired at me over the strudel. Who was Ibsen? How come I'd got the part of someone who kills herself? Was I really crying when I was crying?

We were with another actress friend of mine at the time and she thought Malo was really sweet. She was older than me, with two grown-up sons, and when she produced her camera to take a shot of the pair of us she told Malo to cuddle close to me. Good as gold, he did exactly as he was told and next day, back at the theatre, she told me how envious she'd been. Seven's a magic age, she said. You'll spend the rest of your life trying to get back to that moment.

At the time, I'd thought nothing of it. Malo had always been trusting, as well as tactile, and I'd somehow thought that would last forever. Now, in the kitchen, I ask him whether he remembers that afternoon in the empty stalls at the Old Vic.

He's still gazing at the photo. I think I catch the faintest nod though he won't answer my question. Finally he steps back from the corkboard. For whatever reason, he won't meet my eyes.

'I'll crash on the sofa.' He turns towards the door. 'Have you got a couple of blankets?'

FOUR

I phone Mitch Culligan two days later. After a second night of broken sleep, Malo has swapped the sofa for the privacy of my bedroom during the day and Mum is out scouring the local shops for goodies for my stepdad. Life without certain foodstuffs is a hanging offence in their house and she's dangerously low on Marmite and decent marmalade.

Mitch is on a bus, stuck in traffic somewhere near the Elephant and Castle.

He tells me we need to meet.

'Why?'

He won't say. When I mention Cassini, he just grunts. Whether the doomed little explorer has become code for something else I've no idea but the realization that Malo may be around for a while has given me a whole new take on my immediate future. I have responsibilities here. One of them is to get better. Another is to my wayward son. And the third is to find out exactly what Mitch Culligan might be after.

We agree to meet. He has an appointment with his agent to discuss a first-draft submission. This is the first time I become aware that he writes books, as well as tub-thumping, bravura pieces for the left-wing press, and I'm back on the internet within seconds of ending the call. Two crime novels and a biography of an ex-Navy diver I've never heard of, a man called Joe Cassidy. A visit to the Amazon site reveals a four-star average reader score for the latter with sixty-three posted reviews. Impressive.

We meet, as planned, at the office of his agent in Putney. I'm curious about the biography.

'Why Joe Cassidy?'

'I did an interview with him back in the day. He was involved in all kinds of mischief over the years and he fascinated me. I never believed in greatness until I met that man. He was still middle-aged but he'd been through a lot of stuff in the Falklands War and he was so *wise*. Fame had passed him by, not that he seemed to care a toss, and I thought that was a shame. When I suggested a book he wasn't keen, but if you ask the same question often enough you'll end up with the answer you want. I think he said yes just to get me off the phone. After that everything went like a dream.'

'He didn't disappoint?'

'Not at all. And he even seemed to like what I made of him.'

At this, he gets up and leaves the borrowed office. Seconds later he's back with a copy of the book I recognize from Amazon. It's slimmer than I expected and the photos include a shot of a cadaverous man in baggy shorts sitting on the edge of a harbour wall.

'He's only got one leg,' I point out.

'Exactly. Yet he still swam year round, every morning, without fail. He ran a dive business, too. Still does, as far as I know. Amazing bloke.'

'A kind of homage, then?' I'm still looking at the book.

'Definitely. A privilege to meet the guy. He became a kind of friend.'

I nod. Hand the book back.

'You'd rather not?'

'I'd like you to sign it. Do you mind?'

'Not at all.' He's looking pleased. He finds a pen and pauses a moment before scribbling a line and then adding his signature. When I get the book back, his handwriting is indecipherable.

'What does it say?'

'It's a quote from Joe. *Never let the bastards have it all their own way*. I think that's what kept him going. That and Scotch.'

'Bastards?'

'Bastards.'

'Anyone in particular?'

'Plenty. He was a good hater, Joe. He liked to recruit an army of enemies and see them all off. It was a while before I realized he was a blunter version of what kept this country going. Stubborn as you like. Brave. Stoic. Difficult. Unforgiving. Brilliant man.'

I nod, not altogether convinced.

'Last of a breed?'

Mitch nods, smiles, says nothing. I thank him for the Saturn photo.

'You like it?'

'I love it. September fifteenth? The final plunge? The death dive? Am I right?' I want him to get the file out, the one with *Cassini* on the front. Once again, he seems to read my mind, stooping to his day bag.

Then he pauses. I know exactly what's coming next.

'Last time we did this wasn't good,' he says carefully. 'Are you sure you can handle it?'

As a question, this has its drawbacks. Handle what? Mitch acknowledges the question with a frown. His fingers have already slipped inside the file. Moments later he produces a sheaf of photos.

I'm looking at a shot of a middle-aged man in an expensive suit. He's posed against a group of younger women who look like models. The setting, maybe a hotel, could be some kind of fashion event. He's on the short side and he needs to cut down on the lunches but he has the look of someone unbothered by the opinions of others. He holds himself like a boxer, squat, nicely balanced, ready for anything. The grin is wide. There isn't a thread of grey in his curly black hair and the three-day stubble is camera-perfect. He has his arm around the nearest of the women and the fact that she must be at least a couple of inches taller doesn't bother him in the least. Confidence, I think. And maybe serious wealth.

'Do you recognize him?' Mitch asks.

'No. Should I?'

Mitch lets the question ride. The guy's name, he says, is Hayden Prentice. He pauses, looking at me.

'No?'

'No.'

'His mates in the early days used to call him "Saucy". Some still do.' Another pause. 'HP? Saucy?'

'I'm sorry.' I shake my head, still staring at the photo. 'He's in the rag trade? Some kind of businessman?'

'Only as an investor. The guy's a moving target. Think fingers. Think pies. He made his first fortune in the Alps. Affordable ski holidays for the nearly rich. Some say he got lucky with the exchange rate and perfect snow but if that's true then he's been lucky ever since. Just now he has a big fat portfolio of businesses.

They range from a stake in a major entertainment complex to a couple of niche insurance companies, and every one of them makes him money. Luck, I'm afraid, doesn't cover it.'

I'm looking harder at the face in the photograph. Something is stirring deep in my poor brain. Maybe a long-ago conversation. Maybe more than that.

'Alps, you say?'

'Yeah.'

'The French Alps?'

'Courchevel. Chamonix. Les Houches. Combloux. Draw a circle round a decent airport. Talk to the locals about weather patterns. Get the chalet offer right. Suss the nightlife. Nothing's fucking rocket science, not if you know what you're about.' He pauses, looking for another photo. 'That's him talking by the way, not me.'

This time Prentice has been snapped on the deck of what looks like a sizeable yacht. It might be the south of France, Sardinia, Marbella, anywhere with that overwhelming blue of sky and sea. Prentice is squatting on the deck with a handful of sun cream. Lying full length on a turquoise mattress is a naked blonde with huge Guccis, and an all-over tan. Whoever took the photograph seems to have taken Prentice by surprise. He's looking up at the camera, visibly annoyed. His own tan matches his companion's and I'm right about the boxer bit. He's carrying a couple of pounds he doesn't need around his belly but his shoulders are broad, his chest lightly matted with black hair. This is someone who plainly works out, someone you'd be wiser not to upset.

'Well?' A tiny edge of impatience has crept into Mitch's question.

I'm still looking at the photo. Especially the chest.

'He speaks French? Prentice?'

'Not a word. Wouldn't hear of it. Mark of weakness in his book.'

I nod. Sit back. Close my eyes. It's all coming back, every single detail. *Saucy*, I think. Fuck.

FIVE

It's the summer of my first appearance at Cannes, my first glimpse of Berndt Andressen, the summer when I prefaced my film festival moments with a lazy, paid-for month shooting a movie at Antibes. With the budget on its knees, the French production company put us up in a fleapit *pension* in a back street near the station. According to the director, an ageing *faux philosophe* still looking for his first major breakthrough, the experience was meant to sharpen our appreciation for the class of criminal around whom his sagging scenario was constructed. We all knew this was an excuse to save a few more precious francs and when the invitation arrived for drinks aboard a newly docked vessel in the marina, we gladly decamped from the cafe-bar which we'd made home.

The boat was clinching proof that no one with any money has any taste. It was big, sleek, and very white. It had the kind of lines a child might have drawn if he wanted to turn a series of waves into a boat. You boarded via a gangplank that looked as box-new as everything else. A uniformed lackey who probably doubled as a bodyguard met you at the other end and when you stepped down into the salon, along with a blast of air-con came gusts of an overpowering perfume I can only describe as emetic. The name of the boat was everywhere: on the bow, on the stern, on the lackey, on the lifebelts, embossed in brass on the bridge controls, even appliquéd on to the thick courtesy towels in the closet. *Agincourt.* Subtle.

Smirking in French is easy. You put on your brightest smile, shoot a look at your buddies, and murmur that you think the whole thing is *impressionnant.* In this context *impressionnant* means *merde*.

Our host was a man called Dennis who apparently owned a football club along with some huge construction company. He was tall, thin, and rarely smiled. Unlike everyone else, he'd forsworn shorts for a suit. The pastiness of his face was untroubled by the sun and within minutes of our arrival, having muttered instructions to the lackey about the champagne in one of the huge fridges, he made his excuses and left for the airport at Nice.

At this point we were beginning to wonder quite why we'd ever been invited but minutes later the rest of the party arrived. They'd spent most of the afternoon in a harbourside bar and it showed. They were all English, mainly men. They tramped aboard and there was a blizzard of oaths when one of them lost his footing on the deck. This turned out to be Saucy. My first impression was slightly marred by the blood pouring out of a gash on his cheek. In another life I happened to have done a first-aid course.

A woman we hadn't seen before was already on hand with a towel and a handful of plasters but Saucy ignored her. He liked what I'd had to say about ice cubes. The woman fetched a bucketful from down below while Saucy sprawled on one of the hand-stitched leather banquettes. I had the towel pressed hard against the side of his face. He couldn't take his eyes off me.

'I've seen you before, ain't I? Give us a name.'

'Enora.'

'I meant the film. The movie. Starts in London. Two dykes in bed. Not you, love.'

'*Kalendar*.'

'That's it. You were good. You were great. Easy with that ice, eh?'

The bleeding stopped within minutes. I closed the wound, which wasn't deep, and applied the biggest of the plasters. All the while, Saucy was telling me about the movie, about what happened, as if I'd never heard of it before. Finally he swung himself up into a sitting position, fingering the plaster. By now everyone was up on deck. Drinks had been served and there was a great deal of laughter. When I suggested we join them, Saucy shook his head.

'You got a bloke with you? One of that lot?'

'No.'

'Good. A gal needs looking after. The least I can do.'

Cold on the page this sounds less than enticing, but there was something about him, a cheekiness or maybe a hint of vulnerability that I liked. He was bouncy, like a dog you'd be happy to have around, and when he said he'd fix us a proper drink, not poncey Krug, I said *oui.*

'You speak the lingo?'

'*Oui.*'

'Wash your mouth out. Now tell me you like a decent margarita.'

'*Oui.*'

'Yeah? We like that. Stay there. I'll be back.'

I did what I was told, trying to decide which of the framed sunset shots I loathed most. The margarita, when Saucy finally returned from below, was excellent.

'So tell me about you,' I said. 'You're a friend of the guy in the suit?'

'Yeah. Like brothers on a good day, me and Den. Other times it ain't so sweet, but who's keeping score?'

'And this is his boat?'

'Every nut. Every bolt. Every gold fucking door knob. Hideous, innit?' Laughter suited his face, even when blood began to seep beneath the plaster. He told me about Den, about the money he'd made, about the way he couldn't keep his hands off all the gold clobber. He made this obsession with self-advertisement sound like an addiction, which in a way I suppose it was. Finally we got round to the question I really wanted to ask.

'And what about *your* football club?'

'Pompey. Till I die.'

'Pompey?'

'Portsmouth.'

'And you own it? Like Den?'

'*Own* it?' More laughter. 'Fuck me, I'm just a punter. Tell you the truth, it ain't going too well just now but disappointment never hurt anyone. We miss Ballie big time though he lost it towards the end.'

Ballie? I hadn't a clue what he was talking about but it didn't seem to matter. I'd finished the margarita and my new barman was down below fixing another. By the time he came back, to my slight regret, we had company. The new arrival, a refugee from the drinks party on deck, had yet to introduce himself. Saucy spared me asking.

'This is Terry,' he said, 'And he's about to fuck off back upstairs. Right, Tel?'

Terry spared me a nod, told me to watch my arse, and did what he'd been told. Saucy and I were back on the banquette. Already the first margarita had gone to my head. When I asked Saucy about his last port of call he described a city that could only have been Nice. A single night tied up alongside a boat full of partying Russians. Total oblivion.

'You never went ashore?'

'No point. Vodka's fucking vodka. Especially when it's free.'

I couldn't think of an answer to that but Saucy didn't want one. Instead, he asked me what it was like to be an actress. This is a question we thesps get less often than you might imagine. Strangers often want to know about a particular performance, or a favourite fellow actor, but what they're really after is gossip. Saucy wanted something far more interesting.

I remember doing my best. I told him I'd always liked make-believe: reading it, watching it, doing it. Sit in the darkness in the cinema, put yourself into the hands of a good director, and you can find yourself mapping a whole new world. Not just outside but inside.

'What the fuck does that mean?'

'It means you can become someone else. Which is what an actress does.'

'That's good.' Saucy appeared to be thinking hard. 'I never thought of it that way.'

He seemed genuinely impressed. He said he liked movies, make-believe, all that shit, but that nothing could beat real life.

'Where I come from,' he said, 'you get one chance. Either you take it or you spend the rest of your fucking life sat on your arse.'

'And you?'

'I took it. Big time.'

'So what did you do?'

'Back then I was an accountant. Don't laugh. It's not what you're thinking. In life you follow the money and the guys who know the best route up the mountain are people like me. All you need is a bit of nerve, a head for heights, a bit of bottle.'

This wasn't what I'd been expecting. Saucy came at you with the force of an express train. It wasn't just his physique, the way he held himself, but there was something in his eyes, a hint of apartness, that I've only ever associated with people who've made the big time, either through wealth or fame or – more usually – both. Marquee names in showbiz have it. A top French politician who used to fancy my mum had it. These are people who trust their own judgement, who always know which gambles pay off, and who never entertain a moment's self-doubt.

I don't know why but I didn't want this conversation to end. Saucy had swallowed the last of his second margarita.

'Are you rich?' I asked him.

'Yes.'

'But accountants don't get rich. Not properly rich. Not drifting-round-the-Côte-d'Azur-getting-pissed rich.'

'This one does.' He put the glass to one side and kissed me softly on the lips. Then he stood up and tugged at his shorts. 'Later,' he said. 'Then we'll have a proper chat.'

We joined the revellers on deck. More bottles of champagne. More stories traded back and forth. More extravagant toasts to passing strangers on the marina pontoon. Then, as the sun set, the arrival of a mountain of take-out from a *resto* on the quayside. By now, we were all royally pissed. Thick wedges of anchovy, olive, and tomato pizza made little difference. By this time, Saucy had abandoned an Italian actress who was part of our little *equipe* and was back at my side. Some of my buddies were trying to negotiate the gangplank. In the warm darkness, totally blitzed, they looked like battlefield casualties and when Saucy suggested I stay the night and spare myself the embarrassment of following them it seemed like a totally reasonable proposition.

He was occupying a stateroom next to his mate Dennis. I remember giggling when Saucy confirmed that the sheets were silk and I think there may have been mirrors on the ceiling overhead. Either way, Saucy was a considerate lover and didn't seem to mind when I went to sleep on him. By the time I awoke, nursing the grandmother of all hangovers, it was daylight again and he'd gone.

SIX

Mitch is impressed by my recall.

'Word perfect,' he says. 'So why didn't you recognize him?'

There are more photos on my lap. Saucy snapped on a busy street of what looks like London. Saucy at the wheel of a rally car. Saucy blowing out the candles on a huge cake. I count them, still thinking about Mitch's question.

'He's really fifty?'

'Yes. Last month. He was abroad at the time so the big party's still to come.'

'He lives in London?'

'Dorset. He bought a pile a couple of years back and spent a fortune doing it up.'

'You've been there? Seen it?'

'Not personally but someone I trust got an invite down. She says he's done it beautifully. Bought the right interior design talent. Got the décor spot-on.'

Mitch produces another photo, this time scissored from a copy of *Country Life*. The house – Flixcombe Manor – looks comfortably Georgian. Seven bedrooms. Two reception rooms. Sauna. Inner courtyard. Heated indoor swimming pool in adjacent stable block. Self-contained guest cottage. Timber-framed barn. Sherborne six miles.

I look at the house again, trying to imagine it at the hands of the man I met so briefly all those years ago in Antibes. Mitch, I suspect, is wrong. Saucy would have masterminded the job himself.

'It was eighteen years ago,' I tell Mitch. 'We were different people, both of us. That's why I needed the name.'

'I get the impression you liked him.'

'I did. You're right. He was real. In those kinds of circles that's rare.'

'You saw him again?'

'No.'

'Never tried to make contact? Never went back to the boat?'

'There was no time. They'd finished the re-write on the script and I was back to work. The next day I went along the coast to Cannes. By that time, if you want the truth, I'd forgotten about Saucy. It was an interlude. An amusement. I met someone else at Cannes and that was completely different.'

Mitch nods. So far I've never mentioned my ex-husband and I sense this isn't the place to start. Instead, I ask how Mitch had made the link between me and Saucy.

'I'm writing a book about him,' he says, 'and I've been talking to loads of people. Some of them were on that boat and a couple remembered you.'

'There's a book in Saucy?' I'm staring at him.

'Yes. Saucy and one or two friends of his. Like-minded folk.'

'How come?'

He won't answer the question, not yet. My phone is ringing. I fetch it out and check the caller.

'You want to take that?'

It's my agent, Rosa. She checks I'm OK – a question that has suddenly become more than a courtesy – and then says I'm up for a BBC radio play. She names a writer she knows I admire. It's a one-off, an hour long. Rehearsals are scheduled for a couple of weeks' time. Might I be interested? I say yes without really thinking. She sounds pleased. She'll put the script in the post.

'Head OK, sweetness?'

'Head's perfect.'

Rosa hangs up and I'm back with Mitch. He's found yet another image. This time it's on his iPad. He passes it across. I blink, look harder. This has nothing to do with Saucy or two-million-pound properties in rural Dorset. The segment of a black orb, delicately edged in silver, pushes in from the top left-hand corner. It looks both sinister and beautiful. Beyond it, the curve flatter, is a biscuit-coloured girdle that seems to act as a shield or a fence against the infinite blue-black background. Look harder, and this yawning void is faintly pricked with tiny points of light.

'Saturn.' I look up. 'Has to be.'

'You're right.'

'Taken from Cassini?'

'Right again. But that's not the point.' He gets up and crouches beside me. His finger finds one of the bigger points of light. 'That's us,' he says. 'That's Earth.'

For a second or two I don't believe him. Then I remember the stats I've picked up from the various websites: twenty years in ever-deeper space, a journey of three and a half billion kilometres. No wonder we've reduced ourselves to just another twinkle, one among trillions.

I shake my head. Little Cassini has given us a truth we've all found so hard to grasp, or even imagine. Our own planet, our own teeming world, is – in truth – nothing more than a speck of bluish light: small, frail, lonely, steeped in darkness. For a split second, madly, I'm tempted to relate all this to the cellular vastness of my own brain, equally frail, equally vulnerable. In both dimensions, inner and outer space, we never get the real picture. We kid ourselves we're unique. We tell ourselves we're important. We even pretend we're somehow in charge. When all the time we're the merest smidge from oblivion.

Just now, just here in this small moment of wonderment and revelation, I owe little Cassini everything.

'When are they putting her to death?' I ask.

'Next week.'

'We should hold a wake. Raise a glass or two. Thank her from the bottom of our feeble hearts.'

There's alarm in Mitch's face, as well as puzzlement. He asks my agent's question.

'You OK?'

'I'm good.' I nod at the iPad. 'And I'm serious about the wake.'

It's at this point that I realize how seriously Mitch is taking my welfare. He's collecting the shots of Saucy and returning them to the file. When I ask why, he says my getting better is way more important than anything he might have in mind. When I realize he means it, that he's acting as a kind of guardian, this becomes a real piss-off. I'm hooked on Saucy. I want to know more. I want to know what he's done to merit an entire book and I want to know what role Mitch might have in mind for yours truly in whatever happens next.

Mitch has ignored every one of my questions. He's busy with

his iPad. He lifts his head and checks my email address. Then he squints at the screen and hits Send.

'Take a proper look when you get home,' he says. 'And have the rest of the day off.'

I don't move. If this is my cue to leave it has to be on my own terms.

'Just tell me one thing,' I say. 'Tell me why you've put the word "Cassini" on the file.'

Mitch nods, like he's been half expecting the question. Those long fingers return the image to the screen. He seems to brood for a while. Then he smiles.

'Some said the mission was impossible,' he murmurs. 'Saturn up close and personal? A robot parked on one of her moons? Clues to what made the universe in the first place? Turns out they were wrong.'

SEVEN

I show Malo the Cassini image when I get home. My mum is packed and ready for the airport. Evelyn, who'd volunteered to drive us, has been called to some kind of emergency meeting and so it's going to be me behind the wheel. This is exactly what I shouldn't be doing, not if I'm paying attention to my neurosurgeon, but I haven't shared his sensible advice about not driving with either my mother or my son, and in any case Cassini's take on little us has revised my view of more or less everything.

'Neat.' Malo has spared the image a glance on my iPad. 'Cool.'

My mum, in her practical French way, finds the stats difficult to grasp. She's also a devout Catholic. God created the universe and she believes we unpick its mysteries at our peril. She's also worried about missing the flight.

'You don't think that's incredible?' I'm still looking at my iPad.

'I don't think anything, *cherie.* It's Friday night. The traffic will be *penible.*'

She's right. Her flight leaves at 18.35. It's already nearly four. At the airport, thanks to the latest terrorist attack, we should add an extra hour for the queues at security but a check on my iPad tells me that traffic on the M4 is moving sweetly west.

We join the motorway at Chiswick. My iPad lies. All I can see are three lines of traffic inching forward. It's too late to look for a diversion. We crawl up the feeder lane and barge our way into stalling traffic. My mum, who has an acute instinct for impending disaster, is already convinced that we've missed the plane.

'The Tube,' she says. 'Drop me at a station.'

Easier said than done. My knowledge of exits on the M4 is less than perfect but I've a feeling I can get off at the next junction and try and find the Piccadilly line at South Ealing. Mum is only too game to take on Heathrow alone but I use the airport a lot and I know the kind of shortcuts that might still get her on to the plane.

The traffic has come to a complete halt. Tension has a way of killing conversation and this is the perfect example. I try and find something on the radio. Sarah Dunant interviewing Rosamunde Pilcher. Steve Wright in the Afternoon. Simon Rattle conducting Poulenc. Anything. My mother isn't fooled.

'We should have gone straight to the nearest Tube. This is crazy.'

I can only agree. Mile after mile of traffic seized solid? Mad. All that filth pumping into God's good air? *Fou*. The most expensive city on earth moving at less than the speed of a horse and cart? *Incroyable.* I tell myself to relax, to ignore the growing pressure in my head, to concentrate on that lonely speck of brilliance in the deepness of space. Cassini, I say softly. Cassini.

Like a one-word prayer, it seems to work. Very slowly, stop-start, the traffic is on the move again. By now, my mother has become fixated by the digital clock. She's winding backwards from the moment when they close the departure gate. She wants to know how often the trains run on the Piccadilly line.

'All the time,' I say. 'One every minute.'

She knows I'm lying but she also knows I'm doing my best. And more than that, she's begun to blame herself for putting me in this situation. Her hand settles briefly on my arm.

'I'm sorry,' she says simply.

Minutes later, we're off the motorway and heading north. South Ealing station is less than a mile but there's no parking.

'How's the time?' I ask her.

'*Pas de problème.*'

'I don't believe you.' I steal a glance at the clock. Nearly quarter to five already. This is going to be very tight.

I wait for a break in the oncoming traffic and haul my little car into a U-turn. I can't see a kerb without double yellow lines but this isn't the moment to lose one's nerve. I park outside a florist, grab my mother's heavier bag, and hurry her across the road. The driver of a Polish truck brakes in time to avoid killing us both. Glancing back, I catch him crossing himself.

'That was God's doing,' I tell my mum. 'Three more Hail Marys and we might make it.'

We do. Just. At Terminal Two, I have a second and a half to plant a kiss on my mum's cheek and tell her I love her before she's swallowed by the queue for Departures.

Exhausted, I find a bar. I haven't popped a Percocet for a couple of days now and I tell myself I've earned a drink. My head is thumping but I put that down to stress. When the barman slides into view I wonder what to order. Memories of Saucy tempt me to ask for a margarita but there's something slightly shameful about a lone woman drinking cocktails at half past five so I settle for a glass of San Miguel and find myself a table with a view out across the concourse. Moments later, I feel a pressure on my shoulder and look up to find an actress friend of mine called Fern. Her flight to Rome has been delayed by an hour and a half. Would I mind if she joined me?

We drink for the best part of an hour. Fern knows nothing about my tumour, about my operation, or even about Cassini. I'm wearing a beret that hides the stitches and I realize at once that I can pretend nothing's happened. This, believe it or not, is enormously liberating. We talk about a movie part she's just landed. She tells me a very rude story about a gay director we both adore. And when she enquires what's next in my busy, busy life I'm more than happy to talk about the radio script.

'Brilliant,' she says. 'No lines to learn. No hanging around. Bosh-bosh. Done.' She glances at her watch. 'Shit.'

She takes a quick peek at her make-up mirror and seconds later she's gone.

'*Addio*,' I whisper, turning to watch her running for the departure gate.

By now, I'm pissed. Under normal circumstances a pint and a half of Spanish lager would impart nothing but a cosy glow but something troublesome seems to have happened to my balance. Finding my way back to the Piccadilly line is a lot more difficult than it should be. On the train I pick up a discarded *Metro* and feign interest in a story about some Brexit deadline. By South Ealing, I'm trying very hard not to throw up. I emerge from the station and look in vain for my precious Peugeot.

Hopeless. It's gone.

The flower shop, as it happens, is still open. I join the queue of customers selecting blooms for their Friday date. Every transaction takes an age – scissors to trim the stalks, showboat ribbons for the paper, those little sachets of plant food to keep the dream alive after the weekend – and by the time I get to the head of the queue, I'm nearly as exhausted as the owner.

He's late middle-aged, thickset, heavy foreign accent. When I ask whether he noticed a red Peugeot parked outside his shop, he nods.

'I phoned the council,' he says. 'The guys with the truck took it away an hour ago.'

I find a lavatory in a pub up the road. After throwing up twice I feel a little better. I have neither the energy nor much interest in reclaiming my car, but nor do I trust myself on public transport. There are all kinds of ways of admitting defeat but the truth is I need help. Malo, bless him, would be useless. Instead I call Mitch.

He says he's at home in Hither Green. I have a feeling he's probably very busy but the minute he hears my voice – the hint of panic, he says later, not what I'm trying to say – he tells me to stay put. He'll be there as soon as he can. I give him the name of the pub and mention the Tube station. An Uber brings him to the door within half an hour.

I'm sitting at a table where I know he'll spot me. He comes across. My glass of Perrier is untouched but I'm feeling better enough to apologize at once for calling him out.

'It's not a problem.' He's looking around. 'Do they do food here? And if they do, could you cope with me having something to eat?'

I say yes to both. It turns out he's eaten virtually nothing all day, the result of an empty fridge and an impossible deadline. He scans the food board and then goes to the bar to order a plate of pasta. By the time he comes back, Donald Trump is on all three TV screens threatening Pyongyang with fire and fury.

Mitch watches for a moment or two. It turns out he was in America this time last year to file copy on the hustings, a couple of weeks he was careful to spend with the less reported parts of the electorate.

'You had to get away from the cities. Places the media discount. West Virginia. Coal country. South Carolina. The Georgia boondocks. An hour or so off the freeway you couldn't move for Trump posters. It wasn't subtle, any of this stuff. People were pissed off, left out, and he scooped them up by the millions.'

None of this is new. I'm no politics junkie but I must have heard the same message dozens of times.

'You thought he'd win?'

'Never. But I thought he'd come close . . . much closer than the East Coast ever gave him credit for.'

'So how did you feel? When it happened?'

Mitch gives the question some thought. He'd stayed up most of the night, he says, watching the live coverage, bouncing from CNN to Fox to NBC and back again. After AP called Pennsylvania for Trump everyone knew the game was up. Clinton was slow to lift the phone and concede but she and her buddies were in shock.

'I went to sleep around dawn. By the time I was up again it was gone nine. The paper wanted a day-after piece, America waking up to President Trump, and so I left the hotel and just walked a block or two, looking for a diner. I was in New York. This is a city I know well, really well, yet it looked completely different. There's a guy called Sebastian Haffner. He was a German lawyer, wrote a wonderful book about trying to survive in the Third Reich. The day after Hitler came to power, back in 1933, he felt something similar. He happened to be in Berlin but he said exactly the same thing. It's like you're walking through some place you've never been before. Everything looks different, feels different. The people. The traffic. The cops. The bums on the sidewalk. Even the weather, even the clouds. You've shut your eyes, opened them again, and you're in the world of make-believe. A dream? No. A nightmare? Possibly. But so, so weird.'

I nod. *Some place you've never been before.* I know exactly what he means. This has nothing to do with American politics nor New York but trying to cope with a stranger, a presence, in your own brain takes you to some very strange places. I wait for Trump to leave us in peace and then I try and explain.

'You start to doubt your own body,' I tell him. 'You think you know it and you don't because it's let you down, big time. Is that treason? Betrayal? Have you insulted it? Too many margaritas? Too much sunshine? That may be true. You have to admit it. But in that case, what *is* your body? And what is the you that is you? And where, exactly, is the line between them? I can think of no good reason why I've bred a tumour, given it house room, but there it was, making itself at home, and I'd be mad to assume it might not be back. And that changes everything, believe me.'

Mitch has lost interest in the news feed. 'You sound angry,' he says.

'I am. Take now. How come I'm throwing up? How come I can't trust myself to make it back home? How come I have to drag you halfway across London to sort me out?'

'You've had major brain surgery. That makes a difference.'

'Of course it does. I should never have left home in the first place. I understand that.'

Briefly I explain about taking my mum to Heathrow. Given the circumstances, I'm still telling myself I had no choice. A couple of drinks afterwards was probably foolish but it felt very all right at the time. When I mention the car, Mitch says he'll sort it tomorrow. All he'll need is my driving licence and registration documents.

'You don't have to do that.'

'You're right. But I want to.'

'Why?'

This turns out to be a question he's not prepared to answer. Instead, as the barman signals that his meal is ready, he gets to his feet.

'Life's always on loan,' he says. 'It's just that none of us know it.'

I watch him eat. I'm starting to wonder whether he's had some kind of encounter with cancer himself. He seems so empathetic, so prescient. Then I tell myself that these are talents that any good journalist needs. They have to listen, to understand, to be able to clamber into someone else's heart, or – God help them – head. And I know this because acting, if you want to do it right, is very similar.

No harm in asking the question, though. Cancer? Mitch looks surprised.

'No way,' he says.

'You think you're immune?'

'I think I'm lucky.' He stabs at a curl of calamari. 'You told me no radiotherapy in hospital. And no chemo.'

'That's right.'

'And that's still the case?'

'As far as I know.'

'So your hair . . .?' He's looking at my beret.

'It should grow back pretty quickly. I've been bald before. I did a film in Syria before it all kicked off out there. It was set in Palmyra, way out in the desert. The place is truly ancient. Ruins everywhere. Me included.'

Mitch wants to know more and while he finishes the pasta I tell him about the script. I played a fast-rising French business executive caught at a key moment of her convalescence. One of my buddies to make the trip from Paris to offer spiritual nourishment was an ageing leftie with whom I had unfinished business. Scenes towards the end of the movie called for us to chew the philosophical cud in the ever-lengthening shadows of an ancient Greek temple.

The film bombed and looking back I can understand why. Too wordy. Too ambitious. And too bloody long. But this was the first time I'd been to the Middle East and I had a soft spot for the desert. It was winter, freezing at night but delicious for most of the day, and I came away convinced that three weeks among the ruins of an ancient civilization were the perfect antidote to most of society's ills.

'Convalescence from what?'

'Cancer.' I touch my left breast. 'Aggressive and sadly inoperable. Hence the heavy chemo and the loss of hair. I died a beautiful death in a fifth-floor flat on the Beirut corniche. By which time most of the audience had wisely left.'

'Was the part useful? Looking back?'

'Sadly not. It sounds harsh but I never really took cancer seriously at the time. I was twenty-six. At that age you know you're immortal.'

The waitress arrives to take Mitch's plate away. We stare at each other for a long moment and then – as if by some strange biochemical accident – we both laugh.

'Well?' I say. 'Are you going to tell me?'

'About Prentice?'

'About Saucy, yes.'

'Only if you're sure I should.'

'I don't know what that means.'

'It means we both have responsibilities here. And the first is to you getting better.'

'Isn't that my business?'

'Of course. But I'd like to think I'm here to help you.'

'You are. So tell me more about Saucy.'

He holds my gaze and then shrugs. In a way I suspect this is a gesture of surrender. I get the feeling he'd spend the rest of the evening quite happily talking about something else but time, as I keep telling him, may no longer be on my side.

'Think of this as a movie,' I say. 'So far you've given me the trailer. It's a good trailer. I'm intrigued. I'm hooked. So what's all this really about?'

Another shrug. He plaits his fingers, leans forward, changes his mind, shifts in his seat, tries to relax.

'Back in the day at Antibes, you say he told you about Pompey, about Portsmouth. Am I right?'

'Yes.'

'OK. So we have the young Prentice. He's a grammar-school boy, believe it or not, but he's a tearaway as well. No patience. Bright as you like but a rebel. The last place he ever wants to go is university, which is a bit of a shame from his parents' point of view. No one in the family has ever got further than school-leaving age. Sending young Hayden to the Pompey Grammar has cost them a lot of money. According to his year tutor, university is well within his grasp, even Oxford or Cambridge, yet here he is, running with the wrong crowd, pissing it all away.'

I press for more details. The wrong crowd turns out to be a bunch of football hooligans who call themselves the 6.57 crew, a witty homage to the train they take every other Saturday morning to follow their beloved club to the four corners of the kingdom.

'He's mad about football?'

'He's mad about violence. Him and his mates go for the ruck, not the match. Before, during, afterwards, it makes no difference. The fights are pre-arranged. Pompey are playing away at Liverpool, right? That puts them on the concourse at Waterloo around quarter past nine. A bunch of thugs from Millwall are waiting for them. Before Prentice and his mates get the Tube to Euston there's debts to settle. It sounds medieval and in a way it is. This is the language of ambush, of facing impossible odds, of mateship and of never losing your bottle. Think the Hundred Years War. Think the Alamo. This is serious stuff. Chains, belts, hammers, knuckledusters. The chances of never making it to Anfield aren't low.'

'Anfield?'

'Liverpool. It's the football ground. Their destination.'

'And what then?'

'Another ruck if you're lucky, and maybe something tasty on the way home, too. Like I say, his parents were at their wits' end. This little journey was supposed to end at Oxford or Cambridge, not the magistrates' court.'

I nod, trying to associate this fierce little pageant with the man I'd so briefly shared a bed with.

'This went on for a while?'

'Yes and no. Prentice turned out to have a keen sense of where his best interests really lay. No one ever questioned his bottle. This wasn't the biggest guy you've ever met but he was mixing it with some real animals. At the same time he wasn't stupid. He knew exactly what was really going down and he was determined to be part of it.'

'Part of what?'

'Drugs, chiefly dealing. It was ecstasy to begin with but the real money was in cocaine. The six-five-seven was a gift for someone like Prentice. Put all those crews together, hooligans or otherwise, and you've got a ready-made distribution network. They fought most of the time but they still got to know each other. All it took was for some of them to get properly organized, and they did.'

'And Saucy was part of that?'

'Big time. Not the number one. Not even the number two. But Prentice took it to another level because Prentice was ahead of the game and had the foresight to ask the key question. You sort out the wholesale side. You lay your hands on industrial quantities of cocaine. This stuff was shipped out from the Dutch Antilles, Aruba in particular. You build your supply chain, your client base, and – bingo – the money starts rolling in. We're not talking thousands or even tens of thousands. Before very long, couple of years max, we're talking millions. So what happens to all that dosh? You can't drink it all. You can only drive one Bentley at a time. No, somebody has to play the magician. Someone has to wave the wand and convert all that loot into legitimate profits.

On one level it's called money-laundering. The likes of Prentice would call it investment.'

'Saucy became a businessman?'

'An accountant to begin with. He did the training, passed the exams, got himself articled, sussed the way it all worked, but all the time he was looking for likely propositions, ways he could wash all the drugs money and turn yet more profit. We're talking legitimate investments here. Taxi firms. Nail salons. Cafe-bars. Nursing homes. Even a security consultancy.'

I nod. Saucy had told me all this before, or at least the bit about becoming an accountant, but only now does it make sense. The world of business, Berndt used to say, is one huge safe. Crack the combination, get inside, and you can't fail to make serious money.

'And the police?' I ask Mitch.

'The drugs operation drove them nuts. They knew exactly what was happening but Prentice and his mates were always three steps ahead. Pay attention to the small print, hire the right talent, play the establishment at its own game, and they can't touch you. Not if you watch your back. Not if you're clever. And not if you keep your own house in order.'

'What does that mean?'

Mitch leans back a moment and takes a quick look round. It's Tuesday evening. The pub is half-empty.

'These guys were no strangers to violence. That's how it all kicked off. Literally. That builds a loyalty, a family bond. If you've stuck it to the Millwall against impossible odds then you're mates for life. No one grasses you up. And if they ever do, or if some other twat arrives who fancies his chances, then you wouldn't want to be around to watch the consequences.'

'Like?'

'It doesn't matter. People disappeared. And they were the lucky ones.'

I'm starting to find some of this stuff problematic. Has Saucy killed someone?

'Quite possibly. He's certainly hurt people.'

'And you can prove that?'

'Yes.'

'So why does it matter? Why does *any* of this stuff matter?'

At this point Mitch gets out his phone and summons a cab. His place, not mine. We sit in the back of a new-smelling Uber, not saying a word. The driver, who looks like a refugee from the Sixties, has recognized me. He turns out to be a big film buff. He's even watched a pirated version of my Palmyra movie on the internet.

'You speak French?' To my knowledge no subtitled versions are available.

'No way. But I'm not sure the dialogue mattered that much judging by a review I once read.'

Whether he knows it or not, this is a very shrewd judgement.

Beyond New Cross, Mitch gives the driver directions. The house, to my surprise, is detached, brick and stucco, with newly painted woodwork and a line of pot plants beneath the single bay window. All it needs for the full suburban dream is maybe a Neighbourhood Watch sticker and a poster for the church bring-and-buy sale. Mitch unlocks the front door and stands aside to let me in. He says his Syrian housemate is at work.

'What does he do?'

'Care home round the corner, cash in hand. He double shifts when he's got the energy. The guy that owns it thinks it's Christmas.'

'Why?'

'Sayid was a hospital consultant back in Aleppo, specialized in geriatric medicine. Now he's wiping arses down the road. Someone might explain the logic someday but I'm not holding my breath.'

Mitch shows me into the living room at the back of the property. The house feels warm: warm colours, warm furnishings, an essay in making a stranger feel comfortable. Nothing looks new, not the rugs on the carefully stained floorboards, not the G-plan sofa flanking the open fireplace, not the old Dansette Major record player with its stack of LPs beside it.

There's a big fish tank in a corner of the room, cleverly backlit. I stare at the bubbles, at the handful of exotic fish. They look like scraps of multi-coloured paper carried lightly on a whisper of wind.

'Lovely,' I murmur.

Mitch brews fresh coffee. Just the smell takes me back to the wholefood cafe where we first had the beginnings of a conversation. I join him in the kitchen. He wants to know how I feel.

'Better,' I tell him.

'Seriously?'

'Yes.'

I study his whiteboard beside the fridge while he attends to the coffee. A calendar hangs beside the board and September is brimful of names and times and phone numbers. One name in particular keeps cropping up.

'Who's Jennie?'

'My editor.'

'On the paper? Some magazine?'

'At the publishing house.'

I nod, realizing I'm pleased she doesn't mean more to him, trying hard not to let it show. I badly want to know whether there's another woman in his life but short of asking, I'm not quite sure how to find out. Once again, as psychic as ever, he spares me the effort.

'No girlfriend,' he says. 'In case you were wondering.'

'Any particular reason?'

'Sure. These last three years I've been more comfortable with a man.'

'Your lodger?'

'My partner. Sayid has the gift of happiness. One day you might meet him.'

'Love to.'

'Then I'll arrange it.'

The gift of happiness. I'm still looking at the calendar, fighting a tremor of disappointment at the news that Mitch is gay. My finger finds the fifteenth. It's circled in red.

'Cassini Day?' I ask.

'Exactly.'

'And we're still on for the wake?'

'Of course.'

'Then maybe Sayid should be there, too.'

We have the coffee back in the living room. I'm about to bring Saucy up again but Mitch needs no prompting.

'Prentice got very rich very quickly. They all did but the first real sign that he'd made it was the house he bought.'

'Big? Flash?'

'Not at all. That was the point. It was in Portsmouth, up on the hill on the mainland. This is an address that matters in Pompey. The views are sensational. You can see right over the Solent to the Isle of Wight but the houses themselves are pretty suburban, a bit like this place. Thirties stuff. Double bay windows. Lots of pebbledash. Nice gardens. Maybe a gnome or two. But not much else.'

'So why did he want to move there?'

'He didn't. The place was for his mum and dad. It was Prentice's way of saying there was more to life than Oxford and Cambridge. From what I can gather his mum was delighted. They've both moved to Spain now but that's another story.'

I nod. Sweet, I think. And clever. No need to brag or boast. Just hand them the keys. I'm starting to wonder whether I should ever expect something similar from Malo but I know it's unlikely. He knows I already have more money than I need. What would be the point?

I want to know more about the darker moments of Saucy's giddy rise to fame and fortune. Did these threats to his empire really disappear?

'Yes. And what's more, everyone knew it. In every case you're talking low-life chancers who never realized what they were getting into. Two of them just vanished from sight. A body part from the other guy reappeared on the beach at low tide.'

'Which body part?'

'You don't want to know. But if you're posting a message it doesn't get more graphic. Mess with us and this is what happens. End of. Take a proper look at these guys and it's hard not to be impressed. You've met Prentice. He can be a charmer when it suits him.'

Indeed. Mitch's use of the surname, Prentice, is beginning to grate.

'Why don't you call him Saucy?'

'Because that makes him what he isn't. He's not Jack the Lad. He's not one of life's treasures. He's not Robin Hood. He's a gangster with the brains and imagination to have made himself very rich. He may well have killed people. He's ruthless, and bent, and refuses to take shit from anyone. All of which puts him in very interesting company. Prentice has a very short attention span. He's like the shark. He needs to keep on the move.'

'You're telling me he's bored with being rich?'

'I'm telling you he's ambitious. He's done money. Money is easy. Money was yesterday's gig. Just now he has more of the stuff than any reasonable man can possibly want. What lies beyond money is influence and maybe recognition. Prentice wants his little niche in history, his own little place on the bookshelf. Now is a good time. Now is when a bunch of guys have burst in and kicked over all the furniture. The UK's in pieces. So is the government. If you're clever, and if you know how to keep your nerve, that's a very rare opportunity and Prentice knows it. Money, if you've got enough of it, will buy you anything in this country.'

I nod. I'm beginning to sense where all this might be leading.

'We're talking politics?'

'We are.'

'Brexit?'

'Of course.'

'New faces around Saucy's dinner table?'

'Yes.'

'Anyone I might have heard of?'

Mitch won't say. Brexit, he tells me, has opened countless doors in the nether reaches of the British establishment. The people, by a laughable majority, have spoken and after the first brief moments of disbelief it's dawned on some of the major players that everything is up for grabs. There's no going back. There's no surrender to economic logic or buyer's remorse. Just a big fat chance to barge through some of these newly opened doors and help yourself to a big fat chunk of a highly uncertain future.

'That's a gift for the likes of Prentice. That's what he sussed from the start, way before the referendum. And that's why he's started putting his money where other people's mouths were. He's always had an eye for an investment. That's his genius. That's his special gift. He's nerveless and he's quick on his feet.

Brexit, before the vote, was just another business opportunity. Now, it's very different.'

'Why?'

'Because people owe him. Powerful people. People who may soon be running the country. So Prentice needs to call in all those debts. He's fifty. Just. Fifty's nothing.'

'So what does he want?'

'Power. Influence. He's spent half his life shafting the establishment. In a couple of years, unchecked, he could *become* the establishment. Him and a bunch of like-minded mates.'

'You said unchecked.'

'I did.'

'So what does that mean?'

'It means we get to know the real story.'

'About Saucy?'

'About Prentice. And the company he's been keeping.'

'You mean someone grasses him up?'

'Yes.'

'Someone like you?'

'Yes.'

'By writing a book? Getting it all out there?'

'Yes.'

'To what end?'

Mitch just looks at me. I suspect we both know that this is the key question, the fast-approaching bend in the road. In some ways, like trillions of others in this country, I'm a babe in the wood when it comes to politics. But in others, when we're talking plot and motivation, I'm a whole lot sharper.

'You want to cry foul,' I suggest. 'You want another referendum. You want this Brexit thing gone.'

EIGHT

I sleep late next morning. It happens to be Sunday. Malo stumbles into the bedroom with a mug of tea and a letter he must have found downstairs on the mat.

'Where were you last night?' I ask him.

'Out.'

'Where?'

'With mates.'

'So when did you come back?'

'Just now.' He drops the letter on the duvet and backs cautiously out of the bedroom.

I want to know more. I want to know who these mates are. And I want to know what's happened to the little stash of ready money I use to pay my bills. Late last night I remembered to leave a tenner for the milkman in the lobby downstairs but the jam jar I keep in the kitchen was empty. Last time I counted, I was looking at more than forty pounds. All gone.

I catch up with Malo in the kitchen and pour the tea down the sink. I gave up sugar last year but he wouldn't have known that. He doesn't seem to take this gesture personally. Indeed, he seems to have forgotten he made the tea in the first place. His eyes are everywhere and nowhere. He seems to have slipped his moorings.

'Are you OK?' I'm trying my best to sound mumsy.

'I'm good.'

'So where did you go?'

He yawns and tries to shrug the question off. I'm barring the way to the door and I'm not about to move. My empty jam jar is still on the side. Challenged, Malo says he'll replace the money as soon as he can.

'Like when? How?'

'Dunno.' Another shrug.

'So what did you use it for? What did you buy?'

'Drink. Stuff.'

'What stuff?'

'Gear.'

'You mean drugs?'

'Yeah, drugs.'

'Anything in particular?'

'Cocaine.'

I'm angry, and shocked. I also feel taken for granted. Forty quid's worth of coke up your nose? No money? Just help yourself and see if she notices? All of this is pretty dire but deep in my soul I also feel a flicker of respect. Malo hasn't denied taking the money. And when I asked him a straight question he gave me the straightest of answers.

'How long has this been going on?'

'What?'

'The cocaine.'

'A bit.'

'What's a bit?'

'A bit of a while.'

'Does your father know?'

'No. He knows nothing. Nothing about me. Nothing about anything.'

'So where does the money come from?'

For the first time he falters. He doesn't want to tell me.

'Does he give you an allowance? Your dad?'

'Allowance?'

'Money. Regular money.'

'Sort of.'

'What does that mean?'

'It means I ask him for money and mostly he says yes. I've also got a couple of his credit card numbers.'

'And you can buy cocaine that way? And cannabis? And whatever else?'

'Yes.'

'How?'

He won't say. I get the impression the conversation is over. I'm looking at the empty jar.

'You know what gets to me most? The fact that you just helped yourself. I want this to be your home but it's mine, too. That means we have to trust each other. Am I making sense here?

Don't go helping yourself again. Ask me first.' I force a smile. 'Deal?'

I leave the kitchen without waiting for a reply. I don't know it yet but a crap morning is about to get a whole lot worse.

Back in the privacy of my bedroom, I remember the letter. It turns out to be from my solicitor. As we'd agreed, divorcing Berndt isn't going to be as simple as he'd first suggested. My husband's Swedish solicitor is back in the game with a vengeance. His client, Berndt Andressen, is in some financial difficulty. A movie in which he'd invested heavily has gone belly-up. As things stand, he's in some danger of losing everything.

I skip the small print – three dense paragraphs listing Berndt's exposure on various projects – and go straight to the meat of the letter. Your ex-husband, writes my solicitor, now wants to lay claim to half of your joint estate. That effectively means half of everything you're worth. These are still early days. We can fight fire with fire. But I'd be remiss if I didn't keep you abreast of events.

Merde. I fold the letter and slip it back into the envelope, wondering whether this is the real reason for Malo bailing out of his father's life. His prospective stepmother gone. His father's money gone. The penthouse apartment probably up for sale. What's left to keep my wayward son with a father he appears to despise?

But I know this is the least of my problems. The conversation with Malo has left me seriously worried. No seventeen-year-old should be out all night hunting for the white powder. Where did he get it? Where did he use it? Was he in company? Mates? Someone he's met more recently? Or is he the loner I've always imagined him to be? The slight figure bent over a spliff I glimpsed in the video phone only a few nights ago?

The latter thought makes me shudder. This is where the real problems start, I tell myself. This is where you fall out of love with yourself and get into seriously deep water. I know it happened to Berndt once, because he told me so. That kind of desperate cut-off-ness has also lurked on the edges of my own life from time to time. Poor bloody Malo, I think. With parents like us, the child never had a chance.

I find him watching TV. *Peppa Pig*, at half past eight in the morning, is probably worse than cocaine. I sit down beside him on the sofa and it's a moment before I realize that he's gone to sleep. His breathing is light. He jerks from time to time, tiny spasms of reflex movement. A thin dribble of saliva tracks down his chin.

I study him for a moment, knowing that something is badly wrong, then I whisper his name. Nothing. No reaction at all. I watch him a moment longer, then leave the sitting room. The spare bedroom, after my mum's careful stewardship, is a mess. Clothes everywhere, trainers that don't match, and a copy of *The Sun* open at one of the sports pages. I'm looking for Malo's phone. I find it under the duvet. Eva is listed in the directory. I copy down her number. It seems to include the prefix for Sweden.

Back in the living room, Malo is still asleep. I turn the volume up on the TV a notch or two and retire to my bedroom. I use my own phone to dial Eva. The number rings and rings before an answering machine kicks in. I'm listening to a recorded message in Swedish. It's a woman's voice, deeper and more throaty than I'd expected. It may or may not be Eva. I wonder about leaving a message of my own and then decide against it and hang up. Better to keep phoning until we can talk properly.

After I've turned off my phone, I find myself looking at the letter from my solicitor. I read it again, more carefully this time, and wonder whether to call him or not. Berndt's world appears to have imploded, a perfect storm which has robbed him of his live-in fiancée, his trophy penthouse, and now his son. His talent for raising production money seems to have deserted him and when I Google the reviews for his latest TV series I even begin to feel sorry for him. 'Vapid', 'repetitive' and 'empty' are the kinder judgements. A couple of years ago I'd have still been part of his life, telling him what a genius he is. Now, at best, I feel mildly curious. How come a stellar career like his can end so suddenly?

Because it can. Because stuff happens that way. Cassini, I think. The beckoning rings of Saturn. And next week, the final dive into oblivion. Before I left Mitch last night, at my request, he gave me contact details for Hayden Prentice. He did it with a show of reluctance which I took at the time to be phoney.

Waiting for the Uber, I asked him to be straight with me. He knew I'd once slept with Saucy. His precious book, as far as I can gather, badly needs an inside source a great deal closer to its subject. Everything he's told me about the man suggests his mates are family: tight-lipped, suspicious of outsiders, paranoid about journalists. Hence Mitch's bid to tune in on a different frequency. This happy phrase was his. Whether he admits it or not, he wants me to try and rekindle whatever brief spark existed in the first place. And after that he wants me to listen, and ask a playful question or two, and report back. That's what last night, and all our other encounters, have really been about.

Mitch hadn't foreseen my tussle with cancer, and neither had I, but sitting on the bed I begin to wonder what difference it makes. Back in my old life, when I was well, I'd have dismissed the proposition out of hand. I was an actress, moderately successful, not a bloody spy. My life was full of opportunities and the only cloud in the blueness of my sky was Malo. His absence from my life was a running sore. On screen I could more than hold my own but as a mum I was rubbish. That hurt badly.

Now, the tide appears to have turned. Washed up on the beach is my wayward son and his bankrupt father plus a journalist I happen to respect as well as like. As long as I can remember I've always believed that things never happen by accident. Stare hard enough at the bottom of any cup and the tea leaves will send you a message. Thanks to my neurosurgeon and his team, I seem to have been granted a second chance at life. So what now?

First things first. I try Eva again on the off-chance that she may have been on another call. This time she answers. It takes several attempts on my part before she realizes who she's talking to. Her English is near perfect.

'You mean you're Malo's mother?'

'Yes.'

'Then how can I help you?'

'I understand you're coming to London. I'm just phoning to say you're more than welcome to stay.'

'London?' She sounds surprised.

I repeat the invitation. My son has moved in. She's welcome to join him. She begins to laugh, then apologizes.

'Malo told you this? About me coming to London?'

'Yes.'

'It's not true. Why would he say such a thing?'

I stare at the phone. I'm beginning to feel foolish. Maybe we should start this conversation again.

'Malo tells me that you and he are . . .'

'Friends?'

'Yes. And maybe more than friends. Have I got that right?'

'No. We have coffee together sometimes. He likes to talk. Once we went to a party together because he had no car. But no, nothing more.'

'Then I'm sorry to have bothered you.'

I'm about to hang up but she hasn't finished.

'Wait,' she says. 'How is he? Your son?'

'Why do you ask?'

'Because I know he has problems. He tells me about them. For a boy his age he's very open, very honest. Maybe too much honest.'

A boy his age.

I swallow whatever pride I have left and ask her to be more precise. What kind of problems exactly?

'I know he found it hard. The break-up.'

'And?'

'Kids can do silly things. Go looking in the wrong places for help.'

'You mean drugs?'

'Yes.'

'Cocaine?'

'No. Cocaine is expensive here. He couldn't afford it.'

'Then what?'

There's a long silence. I'm praying that we're not talking heroin. Then she's back on the phone.

'This is difficult,' she says.

I shut my eyes, trying hard not to lose my temper. This woman is playing the professional counsellor, hiding behind client confidentiality. Grotesque.

'He's your friend, not your patient,' I point out. 'And he also happens to be my son. Of course he's had a hard time. We all have. But he's arrived with a cannabis habit and he's taken to staying out all night. He says he bought cocaine. That may or may not be true. A couple of days ago I thought I knew him. Now I'm not so sure. As we speak, he's unconscious on the sofa. He's not well, Eva. Even I know that. All I want to know is why.'

Another silence. I'm trying to be patient, trying somehow to bond with this image on Malo's phone. Finally she seems to have made a decision.

'There's a form of cannabis around here we call Zombie. I think you call it Spice. It's very strong, very dangerous. It's synthetic, not from a plant. Malo uses it all the time. Say hi from me, *ja*?'

Malo uses it all the time. The drooling figure on the sofa. Zombie. The perfect description. I blink, fighting the tears. I try hard to focus on the window, on the door, on anything that will clear my vision. I have a thousand questions to ask, a thousand reasons to keep this woman on the line, but then I realize that the phone has gone dead and it's far too late.

Malo. Zombie. Spice.

I curl up in a ball, bury myself under the duvet, and try – just for a moment – to shut out the rest of the world. Hopeless. My mobile is ringing. I surface to find a text from Mitch: *Ring me soonest. Urgent.*

He answers on the second ring and senses at once that something's wrong. In this respect, as in many others, his instincts never let him down. He doesn't need to be face-to-face. Just something in my voice prompts the inevitable question.

'What's happened?'

'Nothing.'

'I don't believe you.'

'You should.' I tug at the duvet. 'So what's so urgent?'

There's a silence. He's evidently in two minds whether to continue this conversation. What little I know about investigative reporters suggests they have no conscience. In any kind of ethical toss-up, the demands of the story win every time. Wrong.

'Just tell me,' I say. 'Maybe I can help. Maybe I can't. It's about Saucy?'

'Prentice, yes.'

'So what's happened?'

'He's had an accident. He's in hospital in Dorchester. It's not life-threatening and he's not going to die but he'll be in there for at least a week. Dislocated shoulder. Suspected damage in the neck area. Broken ribs. Suspected fracture in the pelvis.'

He fills in the details. It seems that Saucy is a big supporter of Front Line, a charity that looks after injured or traumatized ex-servicemen. Yesterday he'd organized a motocross event at his Dorset estate and charged punters ten quid a head to watch. He'd called on mates and a bunch of minor celebs to turn up for the afternoon. Saucy had once ridden bikes and put himself down for the last race of the day. Cocky as ever, according to Mitch, he'd borrowed something called a Racing Warrior and come to grief on the third lap. For full coverage, I'm to go to YouTube and tap his name in.

'He's got a private room on the second floor.' Mitch is telling me about the hospital. 'The nearest ward is called Little Bredy.'

'So what do you want me to do?'

'I thought you might pay him a visit.'

'Out of the blue? Just like that?'

'You could say you've picked up the story somewhere, seen the YouTube clip. Old mates. Bunch of flowers. Whatever.'

'It's eighteen years ago,' I point out. 'He won't have a clue who I am.'

'I doubt that.'

Something in his voice prompts me to pause. He knows more than he's letting on.

'Why do you doubt that?'

'Because he's never forgotten you. And that's according to people who know him very well. Your name comes up more than you might imagine.'

'Trophy fuck?'

'I doubt it. Apparently he's seen all your movies. That makes him more sensitive than we might imagine.'

Despite everything, I'm beginning to feel better. The thought of Saucy sitting through some of my more obscure films makes me smile.

'And supposing I go down there, supposing I get in, what do I say?'

'You empathize. You make friends again. You cosy up.'

'But what do we have in common?'

'Hospitals.'

Coming from Mitch, this is uncharacteristically blunt. Maybe, after all, he's just like every other journalist. Maybe he's as hard-boiled as the rest of them. Oddly enough, I don't protest, don't end the call. Instead I enquire what might lie further down the line.

'That's what you asked me last night,' he says. 'and to be frank I've no idea. Which is rather the point. I know a great deal about this guy. I've gone as far as the cuttings and conversations with some of his rivals will take me. What I don't know is what he's really like because only his mates and his ex-missus would know and they're not telling.'

'Which leaves me.'

'Exactly.'

'Your precious source.'

'Yes.' Another silence. 'So what do you think?'

I keep a pad and a pencil on the carpet beside the bed. I roll over, the phone still to my ear.

'Give me the name of the hospital again,' I say.

'You'll go?'

'I might.'

I spend the rest of the morning keeping an eye on Malo and having a think. Once he's woken up and I've put something legal inside him, we watch the YouTube clip together. The footage must have been posted by one of Saucy's army of punters. The opening shots show a decent crowd, hundreds certainly, massed at the start. Most of them are men and many of them are wearing an assortment of what I can only describe as combat gear: camouflage smocks, heavy lace-up boots, the whole military shtick. It's raining and the up-and-down race track is plainly a nightmare but the first couple of races come and go without a major incident. Faceless riders in huge helmets lose control and fall off but no one seems seriously hurt.

Then the guy with the camera changes position and we find ourselves looking down on a sharp right-hand bend at the end of a long descent. It's obvious from the skid marks that this bit of the course has seen a good deal of action already but nothing prepares us for the compact little figure in black leathers who comes hurtling down the slope at least ten metres ahead of the rest of the field.

In retrospect it's obvious that he's lost control already but watching the footage for the first time is like watching a movie. You know that something ugly and probably very painful is about to happen because the laws of physics tell you so, but at the same time you assume that the god of good intentions will somehow intervene. Alas, he doesn't. The figure in black goes sideways into the corner, hits the raised bank, and parts company with the bike. The riders behind him desperately try to take avoiding action but it's far, far too late. Bike after bike piles in and a second or two later we're looking at a major incident. The soundtrack alone is truly alarming, impact after impact, but in the silence that follows there comes another sound and it takes me a moment to realize what it is. Laughter.

The footage cuts to a final sequence. We're down among the wreckage. Wherever you look there are men in muddy leathers limping around, nursing their wounds. In the distance, I can hear the approaching clatter of a helicopter. Then the camera pans down to a lone figure sprawled on his back on the wet grass. His helmet lies beside him. His left hand is clutching his right wrist and there's something badly wrong with the set of his shoulder. There's blood on his face and he's obviously in pain but he's summoned a grin for the punters. Even eighteen years can't erase the memory of that grin.

Malo wants to know who he is.

'His name's Hayden. Hayden Prentice. His friends call him Saucy.'

'Why?'

'It's to do with his initials.'

'HP? The sauce?'

'Exactly.' I nod at the screen. 'This whole place belongs to him. He owns it.'

Malo is impressed. Both by what we've seen of the estate and by the little man's courage. I explain about the fund-raising, about Front Line. Malo has rewound the footage for another look at the crash and has now freeze-framed the face on the wet grass.

'You know this Saucy?'

'Yes.'

'Well?'

'No. I met him once a while back. Before you were born.'

'What was he like?'

'He was OK.'

'Rich? Then?'

'Yes.'

'Very rich?'

'Probably, yes. I don't know. Does that matter?'

Malo doesn't answer. He seems fascinated by Saucy. He can't take his eyes off him.

'He's mad,' he says at last. 'Coming down that hill at that speed was crazy.'

'You're right. But that's the way he's led his whole life.'

'How do you know?'

The question is sharp, proof that my poor damaged Zombie isn't quite as lost as I'd thought. I bluff about keeping up with the news but I can tell he isn't convinced. Then an idea begins to shape itself deep in my head.

'Would you like to meet him?' I ask. 'Saucy?'

'Yeah, I would.' Malo lifts his head at last and glances round at me. 'Like when?'

NINE

The next day we take a mid-morning train from Waterloo. Malo has spent the rest of Sunday searching the internet for more clues about Saucy, something I should have done myself. Among his findings are shots of his ex-wife, who happened to have been an actress. I've never heard of her but the eye-popping cleavage and split skirt are all the clues you'd need. Her name was Amanda and even without heels she must have been three inches taller than her husband, but I can't imagine physical odds ever putting Saucy off. He'd have treated his new bride exactly the way he treated Saturday's race: full throttle and fuck the consequences.

Other gems from Malo's trawl include a series of contributions from fellow tycoons he'd bested in various deals. All of them were rich and all of them were cross and when you boiled down what they had to say you ended up with the same basic complaint: that Hayden Prentice was a rogue operator who'd never encountered a rule or made a promise he didn't break. The guy was a force of nature. The guy belonged in a zoo. The guy, to be frank, was impossible.

Beyond Woking, Malo wants to know how I'm going to explain our presence at Saucy's bedside.

'I'll tell him you're my son,' I say.

'That's not what I meant.'

'I know it isn't.'

We're sitting beside the window with a table between us. I put my hands briefly over Malo's. He doesn't flinch. Last night, to my immense relief, he seemed happy to spend the evening with me in front of the TV. We even had a chat about the nonsense we were watching and when I deadlocked the front door and kept the key he didn't say a word.

Now he wants to know how I'll be handling Saucy. It's a good question, one I asked Mitch. Mitch was worse than vague but overnight I've had a thought or two.

'Your father spent some time in Angola a while back,' I say lightly. 'He may have mentioned it.'

'No.' Malo shakes his head. 'Angola?'

'It's in Africa. There was a civil war in the Nineties and the place was full of minefields. There are lots of kids in Angola and some of them got badly injured. Your dad wanted to set a film there. He spent time in the minefields to research a script.'

'It got made? This movie?'

'It did. It was called *Campos dos Sonhos*. That's *Fields of Dreams* in Portuguese. It's ironic. It's what the kids called the fields where the soldiers buried the mines.'

I made a point of finding a copy of *Campos* after Berndt and I first met. It was one of the reasons I married him. The film, shot in Angola with a Swedish crew, was a revelation. Powerful was too small a word. Kids with no arms, no legs. Primitive hospitals. Tens of thousands of crippled street orphans in Luanda with nowhere to sleep. Watch a movie like that and you despaired of the human race.

Afterwards, at Berndt's prompting, I got involved with a charity set up by a bunch of ex-squaddies who'd worked in mine demolition all over the world. They wanted profile and I was happy to add what little I could. The charity is still out there, still effective, and I still give them whatever help I can. In the earliest days of our relationship, Berndt and I had toyed with adopting a couple of these kids but then Malo had turned up, and life had bustled on, and we'd never even completed the first set of forms.

'What's the name of this charity?'

'MAG. Mines Advisory Group.'

'And Saucy?'

'Wait and see.'

We take a cab from Dorchester station. Stepping into a hospital again is something that's been bothering me for the last hour or two – the same bustle of nurses with too many patients and not enough time; the same brief glimpse of impossibly young doctors; the same drawn faces of anxious relatives waiting for news – but the experience is oddly comforting. I owe my life to a place like this. Why should I feel anything but gratitude?

It's Malo who steps out of the lift on the second floor and

leads the way to the nursing station. The sister in charge is sipping a mug of tea while she scrolls through endless emails on her PC. She barely looks up when Malo enquires about Mr Prentice.

'Does he know you?'

'He knows my mum. They were big friends once.'

'Really?' This time she spares me a proper look.

I nod. Smile. Me and Mr Prentice? Best buddies. Malo gives her our names. She gets up and disappears into a side room at the other end of the open-plan ward.

A minute or so later, she's back.

'He's in a bit of a state at the moment. That's his description not mine. If you've got a sense of humour, he says you're very welcome. Otherwise he won't be offended if you bugger off.'

This raises a grin from Malo. I'm still looking at the sister.

'So how is he?'

'Lucky. Lucky to be in one piece. Lucky to be alive. You've seen the footage? Everyone else has. An incident like that we'd normally be counting the bodies. Remarkable.' She holds my gaze. 'He says you're famous. A film actress? Is that true?'

'No, not famous. Just . . .' I smile. 'Lucky.'

We head for Saucy's private room. Malo, once again, is in the lead. He knocks lightly at the door and then stands aside to let me in. The room is smaller than I've been expecting. Saucy is lying prone, his head in a neck brace, one plastered arm secured to an overhead pulley. A vase brims with flowers on his bedside table and there are more blooms in a bucket on the floor. The TV at the foot of the bed is showing some panel game though I'm not certain that he's in any position to watch. A plastic tube trails to a bag beneath the bed. The bag is yellow with urine.

Malo has spotted a line of teddy bears on the windowsill. Some of them have Union Jack waistcoats. The one beside Saucy's pillow is wearing camouflage.

Saucy is aware of our presence but can't move his head. There are cuts and abrasions on his face, and swelling has nearly closed one eye. The other inspects me as I step into his view.

'Antibes,' he says at once.

'That's right.'

'Great night. One of the best. You?'

'I'm fine.'

'That wasn't my question. Enjoy it? Remember it?' He frowns. 'Who's that?'

He's caught sight of Malo. I do the introductions.

'Come here, son. Closer. The name again?'

'Malo.'

'Malo?' Saucy's eye is back on me. 'Yours, you say?'

'Mine,' I confirm.

'That makes you lucky, son. Luck is everything. And I'm the living proof.'

Malo gazes down at him. 'That's what the nurse said. We saw the clip on YouTube. Why so fast down that hill? You mind me asking?'

I've never seen Malo like this, so respectful, so coherent, so *polite.* Saucy is still staring up at him. His good eye seems to be watering. He starts to blink. Then he tries to move, wincing with the pain.

'Kleenex, son. Fuck knows where. Left eye. Quick as you like.'

Malo finds the box of tissues and extracts a couple. He has the touch of a born carer, deft, firm, unembarrassed. I can't be more proud of him. And more surprised.

'Well done, son. How old are you?'

'Seventeen.'

'Good age. *Brilliant* age.' He winces, trying to make himself more comfortable. 'Live with your mum, do you?'

'Yes.'

'Dad?'

'No.'

'Gone?'

'Yes.'

'Miss him?'

Malo doesn't answer. Saucy's good hand crabs sideways across the whiteness of the sheet, looking for Malo.

'Proud of your mum, are you? All the stuff she's done? All them movies?'

Malo nods, says nothing, but he's beaming with a naturalness I haven't seen for years. I remember exactly the way Saucy made me feel at home on the boat in Antibes and he's doing it again, in front of my eyes, as if he and Malo have been best mates all their lives. I came expecting, at the very least, awkward silences between us all. Instead I've become part of a love-in.

Malo still wants to know about the accident. How come the death dive into that final corner? Forty miles an hour. Had to be.

'Dunno, son. You want the truth? I don't remember any of it. First I knew was getting me into the chopper. Mustard, those blokes were. Thought I'd broken my neck.' This little speech comes out in brief instalments with pauses for breath in between. Berndt broke a rib once, and spoke exactly the same way, passing a message between jolts of pain.

'You don't remember a bloke with a camera?' Malo asks. 'When you were flat out on the grass?'

'No.'

'But you were grinning. You were telling him you were OK.'

'OK?' Saucy tries to suppress a cackle of laughter. 'You call this OK?'

'No, but you did. It's there. On YouTube. I saw it.'

Saucy nods and closes his eyes. Time, I sense, is strictly limited. I move Malo gently out of the way and take his place beside the bed. When Saucy opens his eye again he's looking at me.

'Enora, yeah?' His hand finds mine. It's warm.

'Enora,' I confirm.

'Pretty name. Pretty lady.' He gives my hand a squeeze. 'Nothing changes, eh? Me, the mad one. You, class act. In life, you get what you deserve. God bless you both.'

He closes his eyes again. He's obviously exhausted but when I try and free my hand and mutter something about letting him get some sleep he won't let me go.

'Why did you come?'

'We came to see how you were.'

'And what do you think?'

'I think you need to rest. It may take a while but I'm sure everything will be OK in the end.'

'That's what they all say. So why did you really come?'

'It's complicated. And cheeky.'

'I like cheeky.'

I explain about the mine charity, about the killing fields in Angola, about the film Berndt made, and finally about all Saucy's work for Front Line. This suggests he's into good causes.

'You want money?'

'Yes.'

'How much?'

'A hundred thousand.'

'That's a lot.'

'I know.'

He nods. He wants to see Malo again. I step aside. Malo takes his hand uninvited, asks him whether he needs more tissues. Saucy says no and then mumbles something neither of us catch. Malo squats beside the bed, gets as close as he can, his ear inches from Saucy's mouth. Another mumble. Malo is staring at him. Then he gets up. He wants a pen and something to write on. I find an envelope in my bag, and a biro. Malo scribbles something and tucks it beneath Saucy's pillow.

'Where's your mum?' Saucy whispers.

I return to his bedside. Saucy wants me to kiss him goodbye and then fuck off. His lips are dry. He hangs on to me for a moment with his good hand. He tells me to come back any time and bring the boy. Then he waves us both away.

Only in the lift returning to the ground floor do I ask what Malo had written on the envelope.

'My birthday,' he says. 'He wanted to know when I was born.'

TEN

Malo and I take the train back to London. Our fifteen minutes at Saucy's bedside have shaken me. Not because of his physical state, which was worse than I'd anticipated, nor because he remembered so much of our previous encounter, but because of his reaction to Malo. Watching them both in that airless private room, I began to wonder for the first time about my son's real paternity. The black curly hair. The broad set of shoulders. The waywardness. The refusal to listen to anyone else. And now the unspoken kinship between them. They seemed to bond in an instant in ways that only shared genes can explain. Father and son? Christ.

Malo can't possibly be having the same thoughts because he knows nothing about what happened at Antibes but nonetheless he can't stop talking about this new man in his life. How brave he is. How gutsy. How easy to talk to. How real. All of this is true and it puzzles my son because I realize how much he mistrusts wealth.

As a family, we always took money for granted. Both Berndt and I were earning loads. Yet the more comfortable our lives became the faster everything began to fall apart. I've often thought about this in the small hours – how our separate careers spun us into deepest space, leaving Malo alone in the family home. Success, I told myself, did that because success gave us money, lots of money, and money – in turn – bought us a life of infinite choices that turned out – for my poor son – to be no life at all. His mum packing her bags for yet another five weeks on some far-flung location. His father beyond reach, caged by post-production problems, by funding issues, by scripts in development. No wonder the poor soul turned to drugs.

Drugs. We've left Winchester and the train is suddenly roaring through a tunnel. In the half-darkness, I'm watching my son. If I'm right about Saucy, if he really is Malo's dad, then I'm looking at a situation rich in irony. Hayden Prentice, tearaway and apprentice drug baron, the Pompey hooligan who would learn the dark crafts of accountancy and turn bucketloads of cocaine into a legitimate fortune. Malo Andressen, conceived one drunken night in Antibes, marooned in late adolescence with no one to count on but his neighbourhood drug dealer. If Berndt were here, I think, he'd be reaching for his note pad already. Brilliant idea. Thick with possibilities.

We're out of the tunnel. Fitful sunshine. Fields of sheep. And a blur of faces as we race through one of the smaller stations. I'm trying very hard to remember exactly what happened that night in Antibes. Saucy and I definitely had sex because I could smell it the next morning. And it's true, too, that I'd abandoned birth control after some bad experiences with both the pill and a sequence of coils. Busy and distracted as ever, I'd doubtless planned to sort myself out once life gave me the chance. Then Berndt happened and I was only too glad to leave the consulting room with the test results and take the glad tidings home. Poor Malo, I think. Poor bloody child. Not Berndt's fault at all. But mine.

The next couple of days, I'm not well. It starts with a headache, a distant growl of thunder to begin with but quickly building into something that feels like a full-blown migraine. Nauseous and irritable, I give up eating and take to my bed. I throw up a number of times, a process that becomes ever more painful as my stomach empties, and I do my best to sleep. Dark thoughts lurk beyond the edges of the throbbing pain. Might this signal the return of the tumour?

Malo, bless him, does his best to comfort me. He brings tea and a slice or two of dry toast. Both help a little, as does a conversation on the subject of his smoking habits. I feel too horrible to bother dancing round the point and simply propose a deal. I'll tolerate him having the odd spliff but it has to be within our four walls, and it has to be ordinary cannabis. Thanks to the internet, I'm now one of Holland Park's experts on the evils of Spice. I know it's powerful, quickly addictive, relatively cheap, and easy to get hold of. Four reasons, I tell him, to never use it again.

'What do you want?' He's looking at my upturned palm on the duvet.

'I want you to give me whatever you've got.'

'You mean the Spice?'

'Yes, and I want you to promise me you won't use it again.'

He nods. He understands. He says it's going to be difficult.

'Of course it's going to be difficult. That's why they invented this stuff in the first place. The choice is simple. You can put yourself at their mercy or not. If it's the former, you can go back to Sweden and waste your life there. If it's the latter, we can work something out.'

'Like what?'

'Like staying here. Like getting some kind of life together. It's called looking out for each other. You could phone your dad for advice but I'm not sure he'd know what you were talking about.'

The dig at Berndt raises a smile. Malo ducks his head and picks at a loose thread in the stitching around the duvet.

'I've been thinking,' he says at last.

'Good sign. Tell me more.'

'Did you fuck him?'

'Your dad? Of course I fucked him.'

'I meant Saucy.'

'Saucy?'

'Saucy. I was watching you both in the hospital. He acted like he was used to fucking you.'

'What does that look like?'

'Like he owned you. Like you were his.'

'Then the answer's no.'

'Never?'

'I didn't say that.'

'When then?' He's abandoned the duvet. His eyes meet mine.

'A while back.'

'Like when?'

I'm reluctant to go much further but then it occurs to me that I might never get this opportunity again. My bursting head may take me back to hospital, back to the operating theatre, and to whatever may – or may not – lie beyond it. So far, Malo hasn't said a word about kicking his drug habit. One false move, one tiny miscalculation, and I might never see him again. For a brief moment I'm back beside the video phone in the hall, watching my ghost of a son batter himself senseless with another lungful of Spice.

'You know it's a prison drug, don't you?'

'You're changing the subject.'

'I don't care. That's where it comes from. It feeds off a captive market. It feeds off boredom and despair. You really think you're not better than that?'

Malo shakes his head, refuses to answer. When I finally manage to coax a reply he says that boredom and despair pretty much covers it. He's been banged up for years. With us.

My head hits the pillow. My hand claws wildly over the edge of the bed.

'You want the bucket again?'

'Please.'

Malo has it there in seconds. It smells lemony from the disinfectant he used last time he cleaned it. He helps me pull myself upright in the bed and then lean sideways as I retch a thin stream of yellow bile. It tastes foul. I'm gasping for breath. I've rarely felt worse.

'You want me to phone someone?' He sounds on the edge of panic. 'You need help?'

'You,' I mutter. 'I need you.'

He cradles my head, asks me whether I want to be sick again. When I say no he goes to the bathroom for a flannel and a towel. If my son does nothing else with his life he's going to make a fine carer. A second trip to the bathroom produces a toothbrush and some mouthwash.

'Use the bucket to spit in,' he says.

I do his bidding. Slowly the bitter taste of bile goes away. My face washed, my mouth rinsed, I feel a whole lot better. I pat his thin arm, a gesture of appreciation.

'I fucked him once, if you really want to know.'

'Saucy?'

'Yes.'

'When?'

'The night before I met your father.'

'And him? You fucked him too?'

'I did. Lots and lots of times. We got married soon after.'

'And Saucy?'

'I never saw him again. Until just now.'

'He never tried to get in touch?'

'Never.'

'And you?'

'I was in love.'

Malo shifts his weight on the bed. I sense he finds the phrase uncomfortable, or maybe it's the very idea. So far it hasn't served him well.

'Does Dad know about Saucy?'

'Christ, no.'

'So what does he mean to you? Saucy?'

It's a very good question. Only days ago, before we made the trip down to Dorchester, I'd have said nothing, zilch, *nada*. A night on the piss. An exchange of body fluids. Barely even a memory. Now, though, I'm not so sure. Seeing Saucy and my son together was a revelation. They were a perfect fit.

Malo spares me pursuing this line of thought any further. His face is very close. He thinks that Hayden Prentice could be his real dad. Might that be possible?

'Yes,' I say. 'It might. And what's more, I think Saucy knows it. That's why he asked you about your birthday. He's good with figures, Saucy. The night we had sex we were both on a boat in Antibes. He might remember the date. All he has to do then is add nine months.'

'Seventeenth of February?'

'Exactly. I was in Cannes the third week in the previous May. The dates work perfectly.'

'So it could be true? Is that what you're saying?'

'Yes.'

Malo doesn't say a word. He's looking thoughtful. Then he grins and gets off the bed.

'Where are you going?'

He doesn't reply. I hear his footsteps outside in the hall before he disappears into his bedroom. When he returns he presents me with a smallish cube wrapped in silver foil and a sealed plastic sachet containing fine shreds the colour of dry sand. This is a fairy tale, I tell myself. I want to believe it but I dread disappointment.

'Spice?' I'm looking at the sachet. Malo nods. 'And this stuff?'

'Resin. That's the lot. That's all I've got. And now it's yours.'

I gaze at him, shaking my head in wonderment. Then I open the sachet and empty the contents into the bucket. The crumbled resin comes next, the residue sticky on my fingertips. I spit in the bucket for good measure and ask Malo to do the same. He complies without protest, and then watches me give the bucket a final shake. Then I nod at the door.

'Lots of water and down the loo,' I tell him. 'No cheating.'

I listen to him doing what he's been told. The flush of the lavatory never sounded sweeter. Then he's back in the bedroom, looking down at me.

'What now?' he says.

Good question.

Mitch phones later, around six. Much better now, I thank him for leaving us alone.

'Us?'

'Me and my son. His name's Malo. I may have mentioned him.'

'You're OK?'

'I'm good.'

'And Prentice? You're going to pay him a visit?'

'We saw him a couple of days ago. Things have been a bit tricky since.'

'So how was he?'

I tell Mitch what we found down in Dorchester. I keep it brief. Saucy, in the view of people who know about such things, is lucky to be alive.

'He recognized you?'

'Yes.'

'Made you welcome?'

'Yes.'

'And?'

'And nothing. He's in a bad place. He's a mess.'

I don't tell him about my pitch for money for the mine charity. Nor do I describe his fascination with Malo. For once I'm bossing this conversation, no longer at the mercy of events, and the feeling is indescribably wonderful. Then, all too typically, I make a major mistake.

'Tell me something,' I say casually. 'If you want a sample of someone's DNA, how do you go about it?'

Mitch wants to know why I'm asking. I tell him it's a favour for a girlfriend. She wants to be sure of the link between her newborn baby and the three men in her life.

'I told her I hadn't a clue,' I say. 'But I'd make some enquiries.'

'You mean talk to me.'

'Yes.'

'Right.' There's a longish silence. I picture him thinking, trying to join the dots, trying to separate the thinness of the lie from the real reason for me asking. Silly bitch. Silly, silly bitch.

'You'll need saliva, or a nail clipping, or a hair with follicles, just a trace,' he says at last. 'And that's from the pair of them.'

'You mean the four of them. You mean *she*'ll need.'

'Of course. The technology's amazing these days. LCN. Low Copy Number. Just a used toothbrush will be enough. Extraordinary when you think about it.'

I thank him for the advice. My girlfriend, I say, might want to buy him a bottle or two of wine. What should she be looking for?

'A 2011 San Vincente Rioja.' He's laughing. 'But you knew that already.'

Malo and I are back on the train within the week. I've made a precautionary phone call to the hospital in Dorchester and have been assured that Mr Prentice is still in residence. When I asked about a possible discharge date the woman at the nursing station said the doctors couldn't be sure. At least a week, she thought. Probably longer.

This time we take a later train, arriving at the hospital in the early afternoon. Saucy's already busy with a visitor and we find a perch in the waiting room outside the ward. Malo eyes a nest of soft toys in the corner while I thumb through a battered copy of *Hello!* magazine.

So far, thank God, my son is showing no signs of the withdrawal symptoms I've been dreading – no sweats, no sleeplessness, no desperate desire to hit the streets again. On the contrary, he seems determined to make friends with me. Is this Saucy's doing? Is Malo that glad to have turned his back on Berndt? In truth I've no idea but each day seems just a little better – sunnier – than the last, and for that I'm deeply grateful.

I've just finished a breathless account of Catherine Zeta-Jones' first date with Michael Douglas when the door opens to admit a man in his early fifties. He's overweight, flushed, grey suit, purple tie, expensive-looking shoes. He spares Malo a nod and then asks me whether I'm here to give young Hayden a shake or two. When I say yes, he apologizes for keeping me waiting. Northern accent.

'He's free now,' he says. 'What he really needs is a bloody secretary.'

He laughs at his own joke and steps back outside. I've seen his face before, I know I have, but I can't think where. When I ask Malo, he shakes his head. No idea.

Saucy is pleased to see us. With the exception of the flowers, which have disappeared, not much in the room has changed. The patient is still in a neck brace, his arm is still pointing skywards, and he's still on a catheter, but the moment we step into the room I know I'm looking at a different man. He seems younger. Much of the pain that was so evident last time appears to have gone. His voice is stronger too, and he's mastered the art of speaking in whole sentences without having to stop and draw breath.

He pats the sheet for me to take a seat and pouts his lips for a kiss. I oblige on both counts. The swelling around his left eye has settled down. He's looking hard at Malo.

'Sorry about last time, son. Not at my best.'

Malo shrugs the apology away, says it's not a problem. Saucy manages a grin.

'Fucking bikes. That's over. In this life, son, you get to know your limits. Once is enough.'

Saucy asks me to pass him the cup on his bedside table. It's plastic, sealed top with a bendy straw, the kind you give to young kids. He sucks it greedily while I hold the cup. Afterwards he's dribbling what looks like orange juice but Malo is on hand again with the tissues. What a team.

Saucy seems pleased. He gives my hand a squeeze and settles back against the pillows.

'Know how many hits that piece on YouTube got?' he says brightly. 'Over a million. A *million.* In this world it pays to make enemies but a million is way over the top. Should I be flattered? Is that a compliment?'

I tell him it's a sign of the times. People love seeing others come to grief. He ignores the comment. My red beret seems to intrigue him.

'You cold or something? Why don't you take it off?'

I remove the beret. My hair has begun to grow again, a soft brown fuzz under my fingertips. Saucy is gazing at the line of stitches.

'You, too?' he says at last. 'Too quick into the corner?'

Malo laughs. I shake my head. Saucy wants to know more – demands to know more. I explain about the tumour, the operation, and coming to in a room much like this.

'You mean cancer? In the brain?' He seems to be taking the news personally.

'Yes.'

'Shit. I never knew.'

'I never told you.'

'Yeah, but . . . your *brain*? You had a headache? You couldn't think straight? Then everything went black? Is that what happens?'

'More or less.'

'And now?'

'Now's different. You take things for granted before. Plus all the wrong things matter far too much. Work. Disappointments. Crap reviews. That's all gone. Thanks to my little visitor.'

'That stuff used to get to you? Are you serious?'

'Of course. And it gets to everyone, no matter how much smoke they blow up your backside.'

He rolls his eyes. There's wonderment in his face, as well as surprise.

'But you were beautiful,' he says. 'You were so beautiful. You still are. Even fucking bald you're beautiful. Can I say that? Do you mind?'

'Not at all.' I gaze down at him. He looks like a child. He might be Malo's age when the marriage first went seriously wrong, maybe even younger. What I'm seeing, what I'm hearing, is hard to associate with the darker bits of Mitch's description. This man has killed people? Hurt people? Broken every rule in the book? Really?

'I've brought a hairbrush,' I tell him. 'It's nothing personal but maybe you need a bit of TLC.'

'Be my guest.' He sounds delighted.

Malo looks on, surprised, as I bend over the bed and begin to attack the tangle of greying curls. I work slowly, exactly the way I used to with my son, using the brush to tease out knot after knot, trying not to hurt him. His scalp is bone-white beneath the matted hair. I've nearly finished when an orderly appears at the door. He wants to know whether we'd like tea or coffee. I say tea, Malo coffee.

'Top work.' The orderly nods down at Saucy and then shoots me a grin. 'You want a job?'

Moments later, I'm finished. At least half a dozen of Saucy's hairs have ended up on the brush. I return it carefully to my bag and then perch on the side of the bed. I want to know what Saucy's been up to all these years our paths haven't crossed. I want to know where he lives, who he shares his life with, how he keeps himself amused. In short, I want to know everything.

'Why?'

It's a reasonable question. I'm playing this scene way too fast. I need to take my time.

'Because I'm nosy,' I say. 'And because I've never forgotten you.'

'No bullshit?'

'No bullshit.'

'And the boy?' He manages a tiny shift of his head in Malo's direction.

'He knows.'

'Everything?'

'*Everything?*' My laughter is totally unforced. 'You seduced me. You got me very drunk. And then you fucked me. That's it. There's nothing else to know. Except you made me laugh.'

Saucy is frowning. This isn't the answer he wanted. He's looking a little bruised. I lean over and kiss him again, this time on the forehead. 'No offence,' I tell him, 'but I'm still wondering about that hundred grand.'

'For the mine people?'

'Of course.'

He closes his eyes for a moment or two, says nothing.

'And there's me thinking I had you for free,' he mutters at last. 'Talk to my accountant. I'll give you the number. He'll put a cheque in the post.' The eyes open again. 'So what else does it take, apart from money?'

'To do what?'

'To stay in touch.'

I have a late supper with Mitch that night, at an Italian place in Hither Green. I called him from the train while Malo was using the toilet and described the visitor we'd so briefly met at the hospital. Thinning hair. Rimless glasses. Nice shoes. Purple tie. Purple watch strap. Now, over seafood pasta and a bottle of Soave, Mitch spreads a handful of photos beside my plate.

'Him,' I say without hesitation. I'm pointing to a suited figure on a raised stage. He has his arm around a tall, blonde woman in late middle age. She's wearing a purple rosette. They're both waving to what I assume must be an audience.

His name, according to Mitch, is Erik Dubarry, Erik with a 'k', though his real name is Derrick Johnson. He's been one of the more anonymous figures in the wings of the pantomime that is UKIP and Mitch, as ever, seems to know everything about him.

'Long-time mate of Prentice. Businessman in his own right. Yorkshire born and bred. Brewery interests in the North East and a huge property portfolio, mainly in seaside towns. This is the guy who buys all those rundown hotels that no one wants and converts them into DSS. Our masters wouldn't thank us for saying so but without the likes of Dubarry, these people would be sleeping in the streets.

'And the UKIP link?'

'He was early on the bus. I'm not sure there's a political bone in his body but he's like so many of the rest of them. Boil the offer down. Press the right buttons. *England for the English. Take back control. Don't let Brussels steal our money. Immigrants back where they belong*. You can write this stuff in your sleep and people like Dubarry didn't take much notice until UKIP started winning votes. At that point they realized that they could really get under the Establishment's skin and that's when the cheque books came out.'

'And you're telling me Saucy's part of this?'

'Prentice is part of anything that threatens to get out of hand. He can't resist a ruck, even now. That's been the appeal of UKIP. They promised to get stuck in and that's exactly what they've done. The Tories never saw them coming. Not really. They dismissed them as a bunch of fruitcakes and it never occurred to them for a millisecond that they might lose the referendum. They thought people would vote with their wallets. The one thing they never took on board was that most of those wallets were empty. That was the genius of UKIP. People had literally nothing to lose so why not stick it to the rich bastards? Prentice understood that. He got rich by screwing the poor guys. He knows the way the trick works because he understands the kind of lives that quite a lot of people lead. We're not talking drugs here. We're not even talking unemployment. You can get a job these days, even two jobs, and still have trouble paying the rent. Prentice and his mates know that. They're pushing at an open door. Show them a glimpse of the Promised Land, wind them up about the fat cats, about the London political set, about the blokes in suits that make all the decisions, and they'll vote in their millions. That's music to Prentice's ears. Talk to him about the issues, about the small print, about trade treaties and preferential tariffs and all the rest of it and he'll go to sleep on you. Show him the way to give the suits a good kicking and he's your friend for life. These people have money, lots of money, and that puts them in a very special place.'

Gradually it's dawned on me that Mitch is seriously angry. He hates people like Prentice because he hates what they're doing to the country. This is mayhem for mayhem's sake, he seems to be saying. This is politics waged like the Pompey away-games where he first tasted blood. I also get the feeling that Mitch doesn't entirely trust me any more. He senses that Saucy may have turned my head again and he's exceptionally keen to make a course correction or two. He wants me to know that Prentice and his mates are out of control. They're rich. They don't care who they upset. And they have absolutely no interest in the real-life consequences of what they're up to. We're a year or two away from catastrophe, he says. And no one knows what to do.

Catastrophe is a very big word. I ask him what, exactly, it means.

'Brexit,' he says simply. 'Telling Europe to fuck off. Turning our backs on what makes life sweet. Assuming we can pig out at someone else's trough. You really think Trump won't screw us? You really believe the Indians will do a deal without a trillion work visas? This is madness. Dubarry knows it. Prentice knows it. And that makes them very happy.'

'Because . . .?'

'Because there'll be rich pickings in the wreckage and they all know it. In five years, maybe ten, there won't be a National Health Service worth the name. All gone. All privatized. All in the hands of big business. The same with education. With care of the elderly. With parks and pot holes and rubbish collection and libraries and the rest of it. The world we take for granted will be gone and all you'll hear from the likes of Prentice will be laughter. You think I'm joking?' He suddenly looks up, mid-sentence, and gets to his feet. 'This is Sayid,' he says. 'If you want to know about consequences, you're looking at an expert.'

So I finally get to meet Sayid, Mitch's housemate and partner. He's tall. He's wearing a pair of grey trackie bottoms and a loose crew-neck top. His head is shaved and his skin is the lightest shade of olive. He moves with the grace and poise of a natural athlete and he also has what a publicist of mine once called 'a melting smile'. One could do a lot worse than fall in love with a vision like this, I think. Assuming you had any choice in the matter.

He apologizes for interrupting our meal. He's been running on Blackheath and has somehow misplaced his keys. I'm listening hard. The clue is in the language. Not lost. Misplaced. Mitch dismisses his apologies and insists he join us. The waiter brings an orange juice and a menu. Sayid says no to food but sips the juice. After Mitch's little diatribe, the passion has gone out of the conversation but we have a quietly civilized exchange of views on the iniquities of Home Office harassment of asylum seekers.

The restaurant is barely a quarter full but Sayid is clearly ill at ease sitting here in his running gear. He keeps glancing over his shoulders at other diners, at the door, at the street outside, and as soon as he's finished the juice I push my plate to one side and tell Mitch I have to go.

'You mean home?'

'Yes.'

'You can't.'

'Why on earth not?'

'It's the fifteenth,' he says. 'Cassini left us this afternoon.'

We take an Uber back to Mitch's place. The driver seems to think the company might be in trouble in London but Mitch won't have it. Too many drivers. Too many customers. A tweak to the terms of employment and everything will be fine. Sayid says nothing during the journey. He's like an unpresence in the back of the cab, a shadow of a man with nothing to say. Even his sweat smells divine.

In Mitch's kitchen, a major surprise. Sayid turns out to be a gifted cake-maker, and he's presented his landlord and lover with a very personal take on Saturn. Strictly speaking it's only half a planet, cut through at the equator to sit neatly on the presentation plate, but the delicate shades of ochre and blue are perfectly rendered and he's even added two rings carefully sculpted from the icing he had left. Best of all is a tiny Cassini lookalike he's whittled from the remains of a discarded window frame he says he found in Mitch's shed. I stand in the kitchen, lost for words. Is there no end to this man's talents?

Mitch, I know, is touched. He's been out all day on various assignments and he'd no idea of the surprise in store. He wants to know how Sayid could possibly have kept a secret like this when their lives are so closely interlinked. Sayid smiles. He's sparing when it comes to conversation but his eyes – the deepest brown – fill in the gaps between words.

'It's a trick you have to learn,' he murmurs, 'in this country.'

'Cake making?' I ask.

'Discretion. London is a big city. I'm good at keeping secrets.'

Mitch produces a bottle of Moët. Sayid sticks to juice from the fridge. When he finds a knife for the cake he insists that I do the honours. Cutting into the perfect half-sphere, I marvel at the skill that went into its creation. This is Big Bang all over again, I tell him. A second go at confecting Saturn from the dust and debris of deep space.

'I found a mould at the care home,' he says. 'After that it was simple.'

The cake is delicious. Mitch pulls the curtains tight against passers-by before we troop into the sitting room. His anger has gone now. He's softer, gentler, something I put down to the presence of Sayid. I've acquired a lot of gay friends in my career and I've always noticed how comfortable they are around their partners. Mitch and Sayid are the same. Their obvious affection is unforced. They touch each other a lot – nothing heavy, nothing sending a message, simply a hand lingering briefly on an arm or a thigh. They sit together on the battered old sofa, companionable, enjoying the cake, sharing each other's days. Being part of this, being so wholly accepted, feels like a privilege as well as a pleasure. Nice people. The best.

I ask about Aleppo. I spent some time there before the war when we were shooting the movie in Palmyra. I remember the souk and the dust of early evening that turned the sunsets into magical shades of gold. It was a lovely city, proof that ancient civilizations can teach us how to live better lives.

'You think so?'

I can't make out what lies behind Sayid's smile. I don't want to trespass on this man's past, I don't want to ask the obvious questions about what it must be like to lose everything you count as precious, but at the same time I'm becoming horribly aware that nothing is forever. We need to ask ourselves the harder questions, I tell myself. We need to be braver, more candid. That's why I forced Malo to a decision about his young life. Be bold. Make friends with the truth. Because everything else, in a phrase that might have dropped from Saucy's lips, is bollocks.

'And Aleppo? The state of the place? Now?' I ask.

'Gone.' Sayid makes a soft, sideways motion with his hand. 'You never think it possible until it happens.'

He'd fled with his cousin and her family. I get the impression it had cost them all a great deal of money, probably his money, but he's far too classy to go into the details. At the coast they took a boat to one of the Greek islands. Waited weeks to find a place on the ferry. Walked and trained it north through the Balkans. A life of queues and freezing rain and shitting in the fields and untold misery. When we get to Calais the story stops.

I assume he jumped a lorry but he isn't saying. Instead he tips his head back and stares up at the ceiling.

Mitch pours me more champagne and raises his glass.

'Cassini,' he says.

I'm home shortly after eleven. I let myself in very quietly, not wanting to disturb my sleeping son, but when I peek round his bedroom door the room is empty. Quickly, I check the rest of the flat but there's no one there. Malo has disappeared.

ELEVEN

I collapse on the sofa, close to tears, regretting the wine and the champagne. I can't think straight. All I can imagine, all too clearly, is the moment when the emptiness of the flat and his own inner demons crushed my poor benighted boy and sent him back to the streets. He'll have got money from somewhere. He'll have bought papers and tobacco and scored another sachet of Spice. He'll have rolled himself a big fat spliff, liberally dusted with the Zombie powder, and stumbled back into his cave. I should never have left him, I keep telling myself. Either that, or he should have come with me. My fault. My fault again. So much for motherhood.

I reach for my phone, checking in case he's left a message. My agent's been in touch about the radio play. Heal's have an autumn sale. Flights to Alicante and Murcia are mine at a knock-down price. No word from Malo. I shut my eyes, pull my knees up to my chin, rock softly to and fro. Should I get out there? Start looking for him? Start asking strangers whether they might have seen a thin seventeen-year-old with black curly hair and trouble walking straight? Or should I cut to the chase? Go to the police? Tell them I've tried and I've tried but that nothing seems to keep my precious son out of harm's way? I shake my head. None of the above, I think. Not if I'm to hang on to a shred of self-respect.

Moments later, I think I catch a noise at the front door, a strange scuffing sound. Then comes a knock, soft, then another. I get to my feet, checking the time. 23.19. Malo, I tell myself. He knows the combination to the main door downstairs but he hasn't got the key to the apartment. Another knock. I'm at the door now. I pull it open. I'm right. It's Malo.

I stare at him, not quite knowing how to handle this scene. Part of me is swamped with relief. But the rest of me is very, very angry. He's wearing jeans and the usual Einstein T-shirt. His eyes are moist and he's swaying a little. But what draws my attention is the state of his right hand. It's wrapped in a loose crepe bandage and blood is seeping through.

'You've been in some kind of fight?'

'Fight?'

I don't bother answering. The possibilities are endless. A quarrel over the deal. A fall in the street once he'd scored. A hand through a window. Any fucking thing.

I take him into the bathroom and remove the bandage. There are cuts on his palm and two of his fingers, nothing deep, nothing serious. He watches me sponge the blood away. He shows no traces of guilt. Neither, I realize, does he smell of tobacco.

'So where did you go?'

'Next door.'

'You mean Evelyn's?' I'm staring at him.

'Yes.'

'Why?'

'She came round, knocked on the door. She needed help.'

He begins to explain about her oven. It's electric and it has a special bulb that lights up when she needs to check on what's happening. The bulb goes *phut* from time to time and it's tricky to get out. I know this because Evelyn has arthritis in both hands and it's normally me who has to unscrew the thing.

'So what happened?'

'I got it out but I broke it in the process.' He's looking at his hand. 'I put another one in but there was glass everywhere so I ended up cleaning the whole oven.'

'*You* did that?'

'Me. Yeah. She wanted to give me money afterwards but I said no. She had some wine. We split a bottle.' He smiles at me. 'Maybe two.'

I gaze at him a moment and then give him a hug. My junkie son cleaning my precious neighbour's oven. When will I ever learn to trust him? When will I ever learn to trust anyone?

'Why are you crying, Mum?'

'Because I'm useless. Because I get everything wrong. But you're back. And that's all that matters.'

After I finish with his hand we celebrate with the remains of a bottle of Crianza I've got lurking in the kitchen. Malo is normally quick on the uptake but it takes him a while to realize that I'd jumped to the wrong conclusions.

'You really think I'd gone to score again?'

'Yes.'

'Why?'

'Because . . .' I shrug hopelessly. 'Just because.'

'Because you don't trust me?'

'Because I know how evil this stuff is.'

'And you think I don't know that?'

'Of course you know that. But knowing isn't enough. Not at your age. Not at any age. It doesn't matter. Come here.'

I give him another hug, clumsy this time, and spill red wine down his T-shirt. It looks bloodier than his poor hand. Poor Einstein.

'Crime scene.' I'm laughing now. 'I'll buy you another one.'

Malo wriggles free and holds me at arm's length. For the first time in years, looking at the expression on his face, I get the feeling that he might like me. Not as his mother. Not as his scold. But as a friend. He seems to be taking responsibility here, not for his own life but for mine.

'You're pissed,' he says gently. 'You need to look after yourself.'

We both surface late next day. Still in bed, I spend half an hour on my iPad trying to get to grips with DNA tests. If I want to be sure about Malo's real father then I need to submit two samples to a specialist laboratory for testing. For a very reasonable £59, an outfit called Paterfamilias will look for a match. I already have strands of Saucy's hair, all with follicles. What I need now is something from Malo. He's just got up. I can hear him in the bathroom. I wonder about stealing his toothbrush after he's used it but then I remember last night's encounter. You need to be bigger than this, I tell myself. This needs to be a team effort.

I table the proposition over a late breakfast. Malo loves the idea of Paterfamilias. He hasn't trimmed his nails for a while. Might I like to help myself?

Malo has shortish fingers, Saucy's fingers, and I use my nail scissors for the trim. I store the cuttings in an envelope and then return to the Paterfamilias website to order a DNA kit. I'm about to supply my credit card details when I notice that an extra £99 buys you a premium service: collection within two hours, a three-day turnaround, and courier delivery of the results. I hesitate for perhaps a second. £156? For acquiring a new dad for Malo? A bargain.

Back in the kitchen, Malo is looking troubled. 'What if it doesn't work?'

'You mean a negative result?'

'Yes.'

'Then you're lumbered with Berndt.'

'Ah . . .' He's frowning. 'You didn't fuck anyone else around that time?'

'I'm afraid not.'

'You're sure?'

'Absolutely.' I put a hand on his arm. 'Sorry to disappoint you.'

The courier with the DNA kit turns up within the hour. She's outside in the sunshine at the front door and Malo, after a double take, buzzes her in. Thirty seconds later she's up on the fourth floor. Malo already has the door open and stands aside to let her in. She's on the short side. She's wearing a tight jacket in dark red leather and trousers to match. Her short cropped hair is jet black, similar to Malo's, and there are hints of Aztec Indian in her face. A big helmet dangles from one hand and she has the sample kit in the other. Saucy, I know, would eat her alive.

She's given me the DNA kit. I'm in the kitchen with the door open while Malo talks to our visitor. I'm in the process of choosing a couple of his nail cuttings for one of the sample envelopes when it occurs to me he's chatting her up. He's asking her about the helmet. She says it's state of the art. Her English is perfect but heavily accented. It seems her father has connections in the world of Formula One and the helmet has come from one of the race teams. She's unzipped the jacket and she's standing in front of him, showing him how the helmet works. You input an address and it gives you audio directions via Bluetooth plus a head-up display on the visor.

Malo is impressed. He wants to put it on. She reaches out, making adjustments, telling him how to lower the visor. From where I'm standing it's hard to judge her age – early twenties, maybe younger – but there's a gleam in her eyes that tells me my son has touched a nerve. She's laughing at something he's just said. Good sign.

I'm finished with the samples. I pack them carefully into the kit, add my personal details, and seal the Jiffy bag. Next door, to Malo's visible disappointment, our courier is getting ready to leave. She takes the Jiffy bag and then checks a document. The results, she promises, will be back in three days, maybe even earlier. She does jobs for these people all the time. They're very good.

Malo wants to know whether she'll be delivering personally.

'Would you like that?'

'I would.'

'Then maybe yes.'

They lock eyes for a moment, then she's gone.

Malo and I spend the day together in town. We go to Fat Face, and one or two other stores, and I buy my son two new pairs of 501s, a rather stylish North Face jacket, a selection of T-shirts, and a pair of heavily discounted Nike trainers. In every store he seems reluctant to see me parting with money on his behalf, a trait I find oddly endearing, and when we pause for a late lunch at a Nandos he goes for the cheapest option on the menu. When I ask him why he needs to be so mean with himself he says he doesn't know. One day, he tells me, we'll go somewhere really posh, really expensive, and he'll pick up the bill. I smile. I can hear Saucy in his voice. Who needs DNA tests?

Late afternoon, I have an appointment with my agent. Malo is reluctant to come with me but I insist. She's often heard me talking about you, I tell him, and now she can meet the real thing. 'The real thing' is a phrase that seems to amuse him. We take a cab to Camden Town.

Rosa has been my agent for longer than I care to remember. She's nearing retirement now, a prospect which most people in the industry dismiss out of hand. A big woman, brisk, funny, fearless, she began life as a continuity girl out at Shepperton Studios. In time she worked on some huge movies under directors with brutal reputations but survived intact. For a while after that she ran an agency for camera crews, chiefly working for the big network TV companies, but she missed what she calls 'the talent' and started to represent us thesps. I've known her all my working life and trust her implicitly. She's done me some eye-watering deals but most of all I treasure her contempt for the vainer side of our profession. Like Saucy, she has no time for people who take themselves too seriously. And, like Saucy, she makes me laugh.

This is the first time she's seen me since I left hospital. She gives me a big hug, and then looks Malo up and down.

'He's looking after you? Your boy?'

'He is.'

'Glad to hear it. You're either lying or you're lucky. Either way it's five o'clock. Vino, or something more interesting?'

I settle for a gin and tonic. Malo asks for water. Rosa shoots him a look.

'Something the matter? You're ill?'

'Thirsty.'

'Lager, then? Peroni?'

'Water, please.'

She shrugs and sorts out the drinks. We talk briefly about the Montréal project and agree that the script looks deeply promising. When she asks me how soon I could be ready for the rigours of location work I tell her there's nothing I couldn't face with a decent wig. She's been eyeing my beret for some time.

'Do you sleep in it?'

'I sleep alone. The beret lives on the hook on the back of the door.'

'No plans in that direction?'

'None at all. I like my own company. And I have Malo.'

'Very sane. You see Berndt at all these days?'

Rosa knows Berndt well. When we were a couple we'd all run into each other at various industry get-togethers and as the years went by I had the tiniest suspicion that she quite fancied him. Something about the way she'd trap him in a corner and make sure his glass was always full. My husband, she once told me, was every thinking woman's fantasy.

'He's hit a rough patch,' I tell her. 'Shit happens. Even in Sweden.'

'Shame. I'm hearing the same stories. In his part of the wood creative talent isn't enough. You have to be clever with money, too.'

'And he isn't?' Malo is looking at Rosa.

'No, my precious, he most definitely isn't. I shouldn't be the one telling you but you've probably worked it out already. You should be proud of his work, though. He's made some fine films.'

Malo nods and then studies what's left of his fingernails. On the subject of Berndt, he has nothing more to say.

Rosa wants to talk about the BBC radio commission. The script has been adapted from a novel. It's called *Going Solo* and I get to play the wife of a pilot, widowed after he disappears mid-Channel while flying his light aircraft to France. The husband owns a couple of classic World War Two fighters, a Mustang and a Spitfire, both two-seater conversions. Between them, they've run a business offering flights of a lifetime, mainly to WW2 veterans, and now she determines to step into her husband's shoes and learn to fly.

All this, to be frank, is a bit techie for my taste, but the script brightens when she begins to suspect that her husband's death is no accident. The more questions she asks, the more troubled she feels, and as she begins to master the Spitfire she realizes that her fate rests entirely in her own hands. The play ends – improbably – with a winner-takes-all dog fight over a major air show, but by this time she's found a kind of redemption. Given the events of the past few weeks, this is something I can relate to. As one of Berndt's characters once said, we are all – in the end – totally alone.

Rosa wants to know what I think of the script. The writer, Pavel Sieger, has a growing reputation.

'I like it. I like it a lot. We're not talking Pinter or something super-edgy but just now that's a bit of a bonus. It feels quite old-fashioned but in a good way. Boys' Own with a splash of feminism. Plus I don't have to learn a word.'

This, of course, is the crux. Given my recent medical history, and all the stress of location work, it's been clever of Rosa to think of radio. She reminds me of the rehearsal dates, and the two days set aside for the recording. I assure her they're all in the diary.

'Anything else out there?'

'Not so far, sweetness. You'll be the first to know. Take it easy for a while. Watch some of those Battle of Britain documentaries. Get yourself in the mood.'

We say our goodbyes after the second gin. Malo, who's known nothing about *Going Solo*, is full of questions but I can only disappoint him. 'This is radio,' I keep insisting. 'If it was a movie you could have the time of your life on location. All those planes. All that hardware. But a radio studio is a shed with thicker walls and the closest you'll ever get to the real thing is a whole bunch of sound effects. It's all about language, dialogue, timing. Come and watch by all means. But prepare to be bored.'

We're back in Holland Park as the daylight begins to fade. I stand in the living room for a long moment, watching the big jets hanging in the dusk as they descend into Heathrow, thinking of nothing in particular. I'm still basking in the soft glow of that second gin. I'm still telling myself how lucky I am that Malo is still around, still clean, still interested. When the buzzer goes on the entry video I barely hear it. Malo leaves the room to investigate. He's back within seconds.

'It's Dad,' he says. 'What shall I do?'

TWELVE

Berndt says he's come straight from the airport. It's raining now and his hair is plastered against his skull. He says he's spent his last five hundred krona on the taxi fare and a bunch of flowers from the shop on the Bayswater Road. Malo has put the flowers in the washing-up bowl in the sink. He seems in no rush to find a vase.

Berndt looks wrecked. Worse still, he looks plaintive. This isn't the man I fell in love with, the man I was crazy about, the man I married. His spirit has gone and so has the faint air of mystery that made him so irresistible. It took me years to realize just how hard he'd worked to perfect this look but even towards the end – with a half-smile or a duck of his head – he could still take me by surprise. No longer. The man sitting alone on the sofa belongs in one of Berndt's early *noir* scripts. He's become one of life's losers, deflated, defeated, awaiting the knock on the door that will finish him off. I go to the bathroom and get him a towel.

'So what do you want?' I ask for the second time.

'Help. Money. Time.' He's knotting the towel between his hands. 'Maybe a bed?'

I tell him there's no chance he can sleep here. Both bedrooms are occupied.

'The sofa?'

'No.'

'Why not?'

'Because we don't want you here.'

The 'we' hurts him. I can see it in his face. Is he acting? I can never be sure. I steal a glance at Malo. He's watching this little drama carefully, his face impassive. Dialogue, delivery, timing, I think. My boy is learning fast.

'So where shall I go?' Berndt asks.

'I've no idea. It's often best to have a plan before you make a start on the first page. That's what you always told me. So what's happened to the plan?'

'There is no plan. You want letters from the debt collectors? I can give you dozens. You want emails from my bank? My pleasure. You want a look at the message I had yesterday from Scantrax? About pulling out of the distribution deal on the Svalbard series? No problem. All of that I have in my little briefcase. But no plan.'

This is pathetic, I tell myself. He knows me too well. He knows I'm prone to lending an ear, to listening hard, to sympathizing, to nodding in agreement, to understanding, to trying to find a way through other people's problems. But what he doesn't know is that things are different. Because I have Malo now and that has changed everything.

'So why did you come?' I ask him.

'Because there's so much to discuss.'

'There's nothing to discuss. We're getting divorced. They're called solicitors, or they are in this country. They talk to each other. Sometimes they lift the phone and talk to me, and maybe you. That's what our money buys. That's what's so sweet about the deal. Arm's length. Job done. Back to square one.'

'I have no money.'

'I'm not sure I believe you.'

'Is that your solicitor speaking or you?'

'It's me,' I say. 'I'm not sure I believe you because I know nothing about the small print. My accountant tells me life is a dry river bed. Under every stone is a hidden cache of money. It's like that everywhere in the world. Business deals, tax evasion, divorces, everywhere you look. You tell me you're broke, bankrupt, whatever. How do I know you haven't hidden money away? How do I know you haven't been down to the river bed? That's why I need my accountant. Because he'll know.'

This has the makings of a speech. Malo's visibly impressed. Soon I might expect applause, but for now all I can see is Berndt shaking his head.

'More money,' he says, 'wasted on accountants. These people cost a fortune and you know something else? When the end comes, and you're through, their fees are even bigger. I hate accountants. I loathe accountants. The only good accountant is a dead accountant.'

'That bad?'

'Worse. You know something? I had a meeting yesterday in Stockholm. My accountant was there, in the flesh. I'd asked him along to vouch for me, to fend off a particular creditor, but he did exactly the opposite. He said I'd *exhausted my assets*. That was his exact phrase. He came into that room as an accountant. He left it as an undertaker. Turns out he'd come to bury me. So here I am.'

This is better, I think. This is near-vintage Berndt. There's life in the old bastard yet.

'You have to go,' I say, 'because I'm not having any kind of discussion. You want to do that, then talk to my solicitor. He charges three hundred pounds an hour so you might have to keep it brief.'

I'm tempted to offer him a drink, purely out of charity, but Malo beats me to it. He offers Berndt one of the Kronenbourgs in the fridge. Watching him fetch a glass from the kitchen I feel a tiny tremor of apprehension.

Berndt opens the can and takes care to tilt the glass at exactly the right angle. This is a trick he taught me years back and I've used it ever since. Perfect head. Perfect presentation.

'*Skål*.' Berndt raises his glass to Malo. Ignores me.

Malo wants to know about Annaliese.

'There's nothing to know. The last time we talked she was on location in Copenhagen. I gather she's found herself a Danish boyfriend.'

'She's not coming back?'

'To Sweden, yes. But not to me.'

'And the apartment?'

'Gone. Repossessed.'

'So where have you been living?'

'With my sister in Norrköping. I stayed three days. By the end I couldn't work out which of us was the more depressed. It kills you, Malo.'

'What does?'

'Staking everything. Losing everything. Failing. I used to write movies about situations like these. Not for a second did I ever . . .' He turns away, burying his head in his hands. His shoulders are heaving. Small animal noises are coming from somewhere deep inside him. Impressive, I think.

Malo's face is a mask. He's totally impassive, totally unmoved. Like mother, like son.

The broken figure on the sofa seems to have regained some kind of control. He has the grace to apologize. I say that's unnecessary. Malo wants to know why he hasn't arrived with more baggage.

'Light,' he says. 'I travel light.'

'Just the one bag?'

'Yes.'

'No change of clothes? No spare jeans?'

'No.'

Malo and I exchange glances. Then Malo goes to his bedroom and reappears with one of the pairs of 501s I'd bought him on our shopping trip. Berndt looks up at him, uncomprehending. Malo asks him to stand up.

'Why?'

'I just need to check something.'

'OK.'

Berndt stands up. The carpet is wet where his feet have been. Malo holds the jeans against Berndt's still-rangy frame. They're way too short.

'Shame,' he says.

I stare at him. This is beyond subtle, beyond cruel. Whether Berndt realizes it or not, his alleged son is making a point. Different build. Different genes. Different everything. What I'd assumed was an act of charity is anything but. Just faintly, I'm beginning to feel sorry for the man I'd once loved.

'So what will you do?' I ask him. 'Now? Tonight?'

'I've no idea.'

'You've really no money?'

'None.' He fetches out his wallet, tosses it across to me. 'Check, if you want.'

'Credit cards?'

'Insufficient funds.'

'An account with some hotel chain?'

'Those days have gone.'

I nod, think, make a calculation or two. I have a personal account with two hotel chains. I use them when I have downtime on location and I can get away for a couple of days. One of them, the Holiday Inn, has an outpost in Kensington. It's not cheap but just now it might ease my conscience.

I give Berndt the address and order an Uber. The Uber, I tell him, will be on my account. Also the hotel. Three nights, absolute max.

'And after that?'

'I suggest you go home. You have a return ticket?' With some reluctance he nods. 'Then use it.'

He looks up at me and for a moment I think we're in for tears again but then he seems to pull himself together. He's come to try and negotiate some kind of settlement and he's failed completely. Neither I nor Malo want anything to do with him. This is public failure compounded by private humiliation. This is what happens when you belittle and threaten and bad-mouth someone you're supposed to love. This is what happens when, for the briefest moment, you start to believe in your own legend. I've seen it happen dozens of times but never like this, so close, so intimate, so personal.

Malo offers to walk him down to wait for the Uber. Berndt gets to his feet and reaches for his bag. He has one question left and I spare him the indignity of having to ask it.

My bag is in my bedroom. I count £270 in notes, all I can find. I extract twenty pounds and give the rest to Berndt. He looks at it for a moment and then folds it into a pocket in his jeans.

'Don't eat it all at once,' I suggest. 'London isn't cheap.'

He looks me in the eye and then turns away.

'*Tack*,' he mutters, heading for the door.

THIRTEEN

That night I can't sleep. My head's full of DNA kits, of couriers in red leather, of Berndt, of Saucy. I try and visualize the passage of Saucy's hair, of Malo's tiny crescents of fingernail, through all the scientific processes, faceless boffins with latex gloves and glass pipettes, whirling machines, multi-digit read-outs, the whole mad sci-fi world which will, by Friday, announce a winner in this bizarre race to claim authorship of my son.

In my own mind, it's already a foregone conclusion. It has to be Saucy. Malo's build, Malo's temperament, Malo's fitful charm, even Malo's hair. Why did I ever suspect otherwise? Why – in seventeen long years – did it never occur to me that Saucy's sperm might have crossed the line before Berndt's? In the face of so many clues, so much evidence, why didn't I take a tiny step back and acknowledge at least the possibility that I was sharing my life with a child called Malo Prentice?

The answer, of course, was simple. I didn't want to. I was in love with a man who was taking me places I'd never even dreamed about. I was making my name as an actress. I was building a career. Only when our marriage began to falter did I start looking at Malo in a new light – as someone I had to hang on to – and even then it never occurred to me that his loathsome father might be a genetic impostor. That my precious Malo might belong to someone else.

Saucy. I can see him in hospital. I can see him on the boat in Antibes all those years ago. I think I know the kind of man he is. And this worries me greatly. Why? Because I like him.

I roll over. It's 03.34. I phone Mitch. I like him, too, and I respect him greatly, but this thing is beginning to get out of control. I want a simple answer, a couple of sentences, no more.

He answers quicker than you might imagine. He's seen my name on caller ID.

'What's the matter? You want me to come round?'

'Sweet thought, but no. Sorry to call at this time of night but just tell me one thing. What do you really want from me?'

'I don't understand the question.'

'It's about Saucy. What do you want me to find out?'

'I want to know who his mates are. I want to know who he's talking to politically. I want to know where he's putting his money. And I want to know what went down in Pompey.'

'When?'

'When he was running with the 6.57 crew. When he started making money. Friends and enemies. People who might still be around.'

'People he hurt?'

'Yes.'

'People he might have killed?'

'Yes.'

'You really believe all that stuff? Or is that where I come in?'

'Yes.'

I hear another sleepy voice in the background and then the line goes dead. I'm staring at the clock. 03.35.

I wake at nine. This time it's Berndt on my mind. It was his decision to come to London, his decision to knock on our door, but I can't rid myself of the image of my nearly ex-husband on the sofa, sobbing and humiliated, his head in his hands. Neither can I rid myself of a lingering sense of guilt for pushing him out into the night. He was never great with money. He loved ideas, scripts, narrative arcs, mould-breaking *noir* drama. He could conjure magic on the set and on location. He could, by his own account, talk any woman into bed. But when it came to making any kind of sane investment in the future his attention would wander. I learned this early. Hence my decision to look after my own financial affairs within the marriage. Alone now, he's adrift into a script he doesn't begin to understand. Maybe I should help him get his bearings.

I phone my solicitor. He's the same vintage as Rosa – sturdy, nerveless, a matchless professional happy in his own skin. His real name is Charles but I've called him Carlos since we met. He's retained the loyalty of a smallish clientele and whenever I've needed advice in a hurry he's always found the time for us to meet.

This morning is no exception. I take a bus to the Strand and plunge into the maze of streets leading down to the river. He has an office on the third floor of a building that may well have survived the Great Fire of London. From his window I can see the top of the London Eye and clouds of seagulls drifting in the wind.

Carlos is tall, slightly bent, with a big pale face framed by an explosion of grey curls. His hair always reminds me of certain pictures of Beethoven towards the end of his life, which is entirely apposite because Carlos is a gifted pianist. On boisterous social occasions, if he's in the mood, he will commandeer a piano and belt out a tune or two. When the company is more intimate, his rendition of Schubert's *Impromptu Opus 90* has made me cry.

'Well?'

I tell him about Berndt suddenly appearing at our apartment. I try my best to capture just how helpless he's become.

'You gave him money?'

'Of course I gave him money. He's in a decent hotel for a couple of nights at my expense. But that's just for now. I don't want him at my door again. I want him gone. But I want him happy, too.' I offer Carlos a bleak smile. 'So what do I do?'

Carlos briefly reviews the options. I'm filing for divorce. So far, Berndt hasn't responded. Whether or not he even has a solicitor at this point is far from certain. We can press the issue, insist on full financial disclosure, and arrive at some kind of settlement. Or we can wait.

'For what?'

'For matters to resolve themselves.'

'And if they don't? Ever?'

'Then we're all dead. Full financial disclosure will get things moving. There may be merit in that.'

'But he has no money.'

'Who says?'

I frown. This is exactly the response I expected. My accountant would say the same thing. Everyone, in the end, is unknowable. Especially someone as opaque as Berndt.

'So what do I do?'

Carlos is rolling himself a cigarette. I think he took up smoking as a protest against political correctness. With his earnings he could afford any cigar on the market. Typically, he seems to prefer shag tobacco.

'Is he still a good listener?'

'If it's in his interests, yes.'

'And he's here for how long?'

'Three days, max.'

'Then get hold of him and give him one of these.' He sorts out a card from the chaos of his desk drawer and flips it across. 'Tell him the first half hour's free. After that he's on the meter like everyone else.'

'And what will you tell him?'

'That's client confidentiality. I can't possibly say.'

'Are you serious?'

'Always.'

He gets to his feet and extends a bony hand. I thank him for his time, and for his gesture to Berndt.

'My pleasure. He made some decent films in his time.'

I take another bus, heading west this time. The Kensington Holiday Inn is on the Cromwell Road. I've tried to raise Berndt on both his mobile numbers but neither seems to work. My only option, therefore, is another face-to-face.

It's lunchtime and Kensington is buzzing. I wait for a break in the traffic to get across to the hotel. The estate agents around here don't even bother with property details in English. You need Arabic to buy anything over three million, though the photos speak for themselves. Why anyone would need a Jacuzzi trimmed in purple ermine is beyond me.

I ask for Berndt Andressen at reception. The girl behind the counter recognizes the name at once. She tells me he's at lunch just now.

'You happen to know where?'

'Of course.' She nods in the direction of the hotel restaurant. 'He phoned to book a table this morning.'

I lurk in the lobby for a moment or two, wondering if this is such a good idea, but the restaurant is busy, a sea of faces, and I tell myself there's comfort in numbers.

I step inside, trying to spot Berndt. After a moment or two I locate him at a table in the far corner beside the long plate-glass window that looks out over the street. In keeping with last night's encounter, I'm expecting him to be eating alone, picking at something soulful, perhaps deep in a book. Instead, he has company. She's facing me, unlike Berndt. She's youngish, probably early thirties, certainly younger than me. Her auburn hair is swept tightly back, exposing her face. It's an extraordinary face, perfect bone structure, wide eyes, full mouth, a face made for the camera. For a moment I wonder whether it's Annaliese but then I remember that she's always blonde.

'Madame?' It's the maître d'. He has a list of reservations on a display board. He wants to know whether one of those names is mine.

I ignore him. I can't take my eyes off the table in the corner. I've played this scene so often myself, been the willing target, the listening ear, the rapt companion, while Berndt casts his spell. I know exactly what he's at, what pitch he's making. I could even voice the dialogue, word for word. His hand reaches for hers. He's leaning into the conversation. He's pushing for some kind of commitment. Bed. Money. A role he needs playing. Anything.

'Madame?' The maître d' again, insistent this time.

At last I drag my attention elsewhere, spare him a glance, make my excuses, turn to leave.

'I have a table for one,' he says. 'Any use?'

'None at all.' I manage a smile. 'But thanks for the thought.'

Bastard. *Bastard.* I'm sitting on the bus again, inching slowly home. The sight of Berndt doing his trust-me number has brought tears to my eyes. Not because I'm jealous. Not because I'm ever going to miss him again. But because I've been so easily duped. I've no idea who this woman is. She might be someone he's known forever. She might be an actress or a model, someone he's been around lately and added to his must-have list. She might have been in the bar last night, a bird of passage, moving through. A tall stranger suddenly appears on the bar stool beside her. There comes a casual word of introduction, a name left dangling lightly between them. And then, if the prospects look promising, he'll ease into a proper conversation.

Berndt has always been brilliant at this. I must have watched him playing strangers, mainly women, a million times. The fluency of his English, spiced with a Scandi accent. The hint of his standing in the film business. The mention of a particular script or a special actor, names that will resonate and intrigue. And the whole shtick, cleverly parcelled in bite-sized morsels, complete with those sudden trademark silences that confirm what a thoughtful and reflective guy he really is. Bastard, I think again, and then, as the bus begins to pick up speed, I feel abruptly better. In real life, I tell myself, she's probably a hooker. What's left of my £250 won't buy what he's after but she's willing to join him for lunch and put up with all the arty bullshit. My poor Berndt. Reduced to a mere punter.

Thus consoled, I park my fantasies, get off the bus, walk home, and climb the stairs to the apartment. I find Malo in the living room, staring at his mobile, a huge grin on his face. Impatient as ever, he's been busy since I left. He'd made a note of the courier agency, got their number from the internet, and given them a call. The lady in red leather who delivered to Holland Park yesterday? She left her silk snood. Might they give him her number so he can arrange to get the thing back to her? The agency, of course, wouldn't release her number but promised to get her to call. She's just phoned back. Her name is Clemenza. They're meeting this evening for a drink.

I nod my head in approval. This is very direct, very Saucy. No bar stool. No cosy little restaurant table. No creepy come-ons. Just a surprise exchange on the phone and a full-frontal invite out.

'You'll need money,' I tell him. 'My pleasure.'

I spend the afternoon working on the radio script for *Going Solo* and towards six o'clock I knock on Evelyn's door. Happily, she's finished work for the day and offers me a drink. I settle in and thank her for keeping an eye on my son the other night.

'I gather you thought he'd gone AWOL,' she says.

'Who told you that?'

'Malo. He thought it was funny. He's a nice lad. He's got a way with him. He knows how to listen. That's rare in my experience.'

This, from Evelyn, is praise indeed. She's merciless in most of the judgements she makes, especially where men are concerned. For the briefest moment I wonder whether to tell her about Paterfamilias, and the likelihood that Malo might belong to a different father, but she beats me to it.

'Malo told me about the DNA test. He says his real name's Prentice.'

'He's decided? Already?'

'I think it was the wine. He said he'd met his real dad twice. And then he showed me the pictures.'

'The YouTube stuff?'

'The motorbikes. The crash. He seems to think his dad was lucky to survive.'

'Malo's right. He is.'

Evelyn fetches a tube of Pringles from the kitchen. The editor in her is greedy for more detail. Work in publishing and you want the whole story.

I tell her what little I can remember. Antibes. Far too much to drink. And waking up next morning.

'You can't remember?'

'Not really.'

'How disappointing.'

Disappointing? Words are Evelyn's business, mine too, and I'm thinking particularly hard about this one.

'Disappointing's wrong,' I say at last. 'I remember feeling relaxed with him. I remember the way he made me laugh. Berndt always told me that conversation was the great aphrodisiac, getting close to someone, getting them to open up, getting them to surprise themselves. But he's wrong. What got me into bed that night was laughter. Laughter and a million margaritas.'

'Do you regret it?'

'Not at all. Not then and certainly not now.'

'What's the difference?'

'Then I knew nothing. I was a child. A naïf. An innocent. Now I sometimes think I know too much. Berndt taught me that, oddly enough. Not deliberately, not on purpose. It was just something I learned by being around him, by watching and listening. The Berndts of this world are predators. Most of us wander around looking at nothing in particular. Berndt hunts. All the time. Because he can never resist the next meal.'

'That says greed to me.'

'Greed's close but it's a possession thing. Think dogs. Think lamp posts. People like Berndt always need to mark their territory, to leave their smell. Then, like any dog, they move on.'

I was suddenly back in Kensington, back in the restaurant, letting Berndt's latest little tableau sink deep into my subconscious. I know I'll live with that image for a very long time and in a way I'm grateful. Better that moment of luminous clarity than some splashy gesture on my part to try and ease his financial agonies. Trust what you see, what you know. Because the rest can only hurt you.

Evelyn enquires whether my son enjoyed helping her out with the oven.

'He did. He thought you were great. Really easy to talk to.'

'I'm flattered. So what next?'

I tell her I don't know. It happens to be the truth. At last I have a clear brief from Mitch but I have no intention of discussing that with Evelyn. As far as she's concerned, Mr Hayden Prentice is a work in progress.

'You'll see him again?'

'Definitely.'

'He knows about Malo?'

'He can't. None of us can. Not yet. Not for sure. I'm sure he's got his suspicions. Maybe he's got his fingers crossed. But that's not the same as knowing.'

'And Malo?'

'Malo approves. Malo likes him. Saucy calls everyone "son" but in Malo's case it seems to fit perfectly. After you've had Berndt as a father all your life, that matters, believe me.'

Evelyn nods, raises her glass. 'To Malo,' she says. 'Tell him the oven light works perfectly.'

Malo is back earlier than I expected, just after nine. He's toting a motorcycle helmet and a big fat smile.

'Clemenza's spare,' he tells me. 'I rode pillion. We went down to a bar she knows in Putney. She drives that bike like you wouldn't believe. People just get out of the way. How cool is that?'

Clemenza, it turns out, is the daughter of a leading Columbian businessman currently living in London. At the mention of

Colombia, my heart sinks. She'll have limitless access to cocaine, probably by the kilo. Malo, when I risk a cautionary word or two on the subject, does his best to ease my anxieties. Clemenza, he tells me, is straighter than straight. Goes to Mass every Sunday. Doesn't smoke. Doesn't do drugs. Never touches alcohol. Because biking is all. And she's brilliant at it.

'You've met an angel,' I tell him. 'If I were you I'd sell tickets. Charge by the minute. You're telling me she's Catholic?'

'She is. How did you know?'

'Mass is the clue. Is she super-bright? As well as virtuous?'

'You're taking the piss. You should meet her.'

'I did. And I'm not surprised you tracked her down.'

'I mean properly meet her. Talk to her. She's lovely.'

'Good. I'm glad. So what next?'

'Tomorrow, same time. She's going to pick me up.'

He gives the helmet a little swing, holding it by the chin strap, and then retires to his bedroom. For the first time ever, I can hear him singing. Later I discover it's a track from an xx album. He doesn't know the lyrics and his voice is beyond flat but none of that matters. Thanks to my years with Berndt, I've learned the difference between an act and the real thing. My precious boy is very happy and that – as miracles go – is up there with the loaves and the fishes.

I remember a line from a hymn we used to sing at school a trillion years ago.

Our God, our help in ages past, I murmur, *our hope for years to come.*

Yes, I think. Yes, yes.

FOURTEEN

On Friday, Clemenza arrives with the DNA results. Malo has obviously shared every last detail of the story so far but she's discreet and well-bred enough to hand over the envelope and head for the door. Malo calls her back.

'Wait,' he says. 'Let's see what it says.'

The envelope is marked *Private and Confidential* and is addressed to me. He rips it open, pulls out the single sheet of paper, and gives it a quick read. Then he shows it not to me but to Clemenza. The girl, embarrassed, takes the notification and gives it to me. I'm looking at the news we've both been expecting for the last three days. Hayden Prentice, aka Saucy, is Malo's dad.

Malo is giving Clemenza a hug. Then it's my turn. Somewhere in all of this I catch the words 'thank you'. For a moment I think the poor boy's on the verge of tears. When he looks up and steps back, Clemenza has gone.

'What do you expect?' I ask. 'Making a scene like that?'

We go back to the notification. These agencies are always careful to include a health warning but we should be 98.3 per cent certain of the DNA match between Malo Andressen and Hayden Prentice. This, it seems, is a document we can use in any court of law, any passport application, any setting that requires proof of paternity. I think briefly about Berndt, about where he might be at this moment, but then decide he's not worth the effort. One day I'll have to tell him, unless Malo gets there first, but the time and opportunity will be of my choosing. For now, we have something else on our minds.

'Saucy,' Malo says. 'Dad. He has to know. We have to tell him.'

I agree. I've been on to the hospital and the news is good. There's nothing broken. His dislocated shoulder has reset nicely and the bones in his neck are stable. The hairline fractures in his ribs are on the mend, his pelvis has turned out to be undamaged, and he doesn't need a catheter any more. For weeks to come he'll be wise to take it easy but he no longer needs hospital care.

'Who told you all this?' Malo asks.

'Saucy. They're taking him home by ambulance this afternoon. He says he's over the moon. He wants us to pay him a visit.'

'When?'

'Tomorrow. We'll train it to Dorchester again. He'll have someone pick us up at the station. Take us out there.'

'You mean to his house? The big one? The estate?'

'I assume so.'

The prospect of visiting the family acres fills Malo with something I can only describe as glee. For the second time in half an hour he's close to euphoria. Less than a week ago he was sharing his life with Peppa Pig and a sachet of Spice. Now he has the world at his feet.

Next day, as instructed, we head down to Dorchester on the train. A woman my age is waiting at the ticket barrier to pick us up. For a second or two, looking at her, I'm furious with myself for not asking Saucy the obvious question. Is this his wife? The new Amanda in his life? Is this Mrs Saucy going to take kindly to a couple of surprise house guests, one of whom is the mother of his long-lost son?

I needn't have worried. Her name is Jessie and it turns out she's partnered with the guy who runs the estate. They live in a bungalow in the grounds of the main property and she and Andy have known 'H', as she calls him, most of their lives. This explains her accent, gruff, almost cockney, exactly the same as Saucy's.

We're in the back of a new-looking Range Rover, leaving the outskirts of Dorchester. I'm doing my best to pump her for more information but blood, as ever, is thicker than water. No, H isn't currently married. Yes, he leads a very busy life. Yes, he occasionally manages an old-times visit to Pompey. And yes, he's very much looking forward to seeing us both.

Malo is beaming. We've turned off the main road to Yeovil and he's glimpsed a handsome Georgian pile through the trees. In his jacket pocket is the good news from Paterfamilias. Life could scarcely be sweeter.

The drive sweeps up through the trees. Off to the right, at

Malo's prompting, I recognize the motocross course where Saucy came to grief, a zig-zag of deep tyre ruts disappearing over a rise. Ahead is the building we've just glimpsed.

Flixcombe Manor nestles squarely in the greenness of the landscape. Golden points of sunlight reflect from a dozen windows. Smoke curls from one of the chimneys. The entrance to the place is pillared, imposing, a brilliant shade of white that draws the eye. This is a million miles from Pompey. No wonder Saucy couldn't wait to get home.

Malo has spotted Saucy emerging from the front door. He shuffles carefully forward with the aid of a pair of crutches. Game as ever, he starts to edge himself sideways down the steps. The proper place to meet guests is on the circle of gravel at the front of the house.

Face-to-face, once we're out of the car, he looks more frail than I'd been expecting. He has a kiss for me and a hug for Malo but the slightest misstep registers on his face. It happens again as he leads us back towards the house, making him wince with pain.

'Don't push it,' I tell him. 'Take it easy.'

'Easy? Bollocks,' he mutters darkly. 'Good to see you both.'

I help him on the way up to the house, Malo on his other side. For the last couple of steps we're virtually lifting him.

'Tougher than you thought, eh?' He's looking pleased.

Inside, the place is beautifully furnished, antique pieces everywhere. The smell of freshly cut flowers hangs in the air and I'm still doing a mental audit on the furniture when Saucy spares me the effort.

'Seventy-eight grand the lot,' he says, 'including the pictures. I bought it all with the house. Bloke who owned the place knew what he was doing. Would have cost me a fortune to replace this lot.'

'What's happened to him? Where's he gone?'

'He died. Drank himself to death. I bought the estate off his eldest boy. He and his missus live out in Dubai. Can't stand the fucking rain.'

Jessie has laid tea out in one of the drawing rooms but Saucy takes us into the kitchen. He says it's the one room in the house where he feels really comfortable and looking round I know exactly what he means: wooden beams, hanging pots and pans, plenty of space, stone floor, a big range with a wood-fired Aga, and a huge old kitchen table most set designers in the movie business would kill for. Big windows look out on to a vegetable garden to the rear of the house and when I get a proper look I count half a dozen chickens peck-pecking among the onion sets.

'All you need is a dog,' I tell him. 'Who does the cooking?'

'Me.'

'*You?*'

'Yeah. I'm not too handy just now so Jessie helps with the heavy stuff. I just did a load of flapjacks, though. In your honour.'

The flapjacks are delicious, moist as well as crunchy, a difficult trick to pull off. Malo sits beside us at the kitchen table. When his hand goes to his jacket pocket I tell him to wait until the three of us are alone.

At last Jessie leaves us. This is Malo's cue. Saucy has near-animal instincts. He knows something's going on. He can smell it.

Malo is flattening the Paterfamilias results against the grain of the table. Saucy squints at it, then tells Malo to fetch his glasses. Hall table. Next to the door.

'OK, Dad.' The word doesn't seem to register. Malo waits a moment longer for a reaction and then heads for the hall.

'What's all this about?' Saucy is looking at me.

'I'd prefer if Malo tells you.'

'Tells me what?' There's an edge in Saucy's voice I haven't heard before. He doesn't like surprises.

Malo is back with his glasses. They're nice enough – rimless, inoffensive – but when he puts them on Saucy looks ten years older. He reaches for the Paterfamilias document, scans it quickly, then reads it again.

'Fuck me,' he says softly. 'Who'd have thought?'

It's at this point that it occurs to me that Saucy probably doesn't have any kids of his own. Another unasked question.

'All right, Dad?' It's Malo again. He loves the word.

Saucy is still studying the document, reading the small print, exactly the way a businessman might. He wants to make sure he's got this thing right. Finally his head comes up and he takes his glasses off. He's looking at me, not Malo.

'Give us a kiss,' he says.

I get up and circle the table, settling beside his chair and putting my arms round him. He feels softer and perhaps more frail than I'd been anticipating. I kiss him on the lips. His eyes are closed. He's smiling.

'How long have you known?' he asks.

'Since yesterday.' I nod at the document.

'But really? In here?' His head sinks to my chest.

'A couple of weeks. It started when I saw the pair of you together in hospital. Peas in a pod. You couldn't miss it.'

'Not before?' He sounds disappointed.

'No. Things have been difficult recently. In fact things have been bloody awful for quite a while.'

'Anything I should know about? Anything you want to tell me?'

'No. Maybe later. Maybe one day. But not now.'

I get to my feet again and take a tiny step backwards. This should be Malo's day, not mine. Saucy reaches out to him, then circles my waist with his other hand.

'Family,' he says softly. 'Right?'

We both nod. Malo wants to know what he should call him.

'Dad,' he says. 'What else?'

Malo looks slightly crestfallen. I'm not sure what kind of reaction he's been expecting but it certainly wasn't this. There's almost an air of entitlement about Saucy, as if he'd been expecting the news.

'Tell me you're surprised,' I say. 'Tell me it's made your day.'

Saucy says nothing. He glances at Malo, then chuckles.

'Who's the lucky one, son? You or me?'

This, thank God, breaks the ice. Saucy calls for champagne. Jessie appears. When Saucy shares the glad tidings she looks amazed.

'You old dog,' she says. 'You never said.'

'I never fucking knew.'

'But it's true?' She's putting the question to me. I nod at the Paterfamilias confirmation, say nothing.

Jessie picks it up.

'D'you mind?'

'Not at all,' I say. 'Help yourself.'

The news puts a big smile on her face. She wants to tell Andy. She wants to start hitting the phones. She wants the world to know.

'Steady, gal.' Saucy's looking alarmed.

'Steady? You have to be joking. Party time. We'll get some people up. Sort out some invitations. Your fiftieth and now this. Brilliant.' She departs to lay hands on a couple of bottles of Moët. My gaze returns to Saucy. Content isn't a word I'd ever associate with my son's new dad but just now it comes very close.

'Get a few people up from where?' I ask him.

'Pompey.' he says. 'Where else?'

'And what should I call you?'

'H.' He blows me a kiss. 'Like everyone else who ever fucking mattered.'

And so H he becomes. In a way I'm sorry because I liked 'Saucy'. It chimed exactly with the personable chancer I met down in Antibes, and the more Mitch told me about the mad days of his youth, the more the name seemed to fit. Now, though, his circumstances call for a different persona. Not because he's rich. Not because his eye for a deal has won him this glorious house and countless acres of prime real estate. But because new friends and a bigger address book call for something a little more dignified. The Hayden Prentice that Mitch suspects of keeping interesting political company will never turn his back on the mayhem that was his past. But I'm beginning to understand that H – clipped, intimate, gently assertive – is a whole lot closer to the essence of the man than 'Saucy'.

H, I think. Close to perfect.

We agree to stay for a couple of days. H spends a great deal of time on the phone in the small upstairs room he uses as an office. He invites me up there on a couple of occasions for a chat about Malo, and how things have been for both of us, and I'm surprised by how bare the room is. The only items worth a second glance are the view, which is sensational, and the desk. The desk is far from antique but I fall in love with it at first sight. It's huge and purposeful, deep drawers on either side, and a spread of dark green leather on the top. I run my fingers over the leather, tracing a path around the big PC screen.

H is delighted by my interest. He tells me he lifted it from Pompey dockyard back in the day when no one was looking. It belonged in the big quayside building where the Naval Writers worked. He fancied it because his dad had been in the Navy and was looking for something half decent for the new house on Portsdown Hill. With a mate he'd man-handled it down three flights of stairs and smuggled it out of the dockyard in the back of a plasterer's van. His dad had used it to keep track of his paperwork and once he'd passed on, H had reclaimed it.

'So when did he die? Your dad?'

'Years back. Asbestosis. These days they give the family a whack of money in a case like that, but back then you'd bury the poor bugger and try and make the best of it. It was my mum I felt sorry for. She was lost without him.'

'And she's still alive?'

'No. She turned her face to the wall. Once he'd gone that was it. Finito. End of. Sad as you like. Here . . .'

He stabs at the keyboard, opening an icon on his screensaver, and moments later I'm looking at a black and white photo of a couple in what I take to be their early sixties. These are faces from a different era – work-seamed, shy, peering uncertainly at the camera. H's dad looks taller than his son but his mum is tiny, a little pudding of a woman.

'Their names?'

'Arthur and Gwen. Nicest people you'd ever meet. Never deserved a son like me. I gave them nothing but grief. At least in the early days.'

I smile, returning to the photo. Arthur has his bony arm around his wife, at once proud and protective. It isn't hard to imagine their bewilderment when their young scrapper of a boy began to come home with serious money.

'What about you?' he asks. 'Your folks?'

I tell him about my mum. When he realizes she was French, and that I spent part of my youth in Brittany, he seems pleased.

'I never knew that,' he says. 'When we were on the boat I had you down as a toff, private education, posh school, university, whatever. That's where I figured you learned French, that fancy accent of yours.'

'And now you know different?'

'Too right. You lived there. That's allowed. You fucking *are* half-French. That turns you into a human being.'

'So you fucked me because you thought I was posh? You're telling me this was class war?'

'I fucked you because you were pretty. Not pretty, beautiful. Then I started watching some of those movies of yours and I twigged you were fucking bright as well, really talented. Thick old me, eh?'

I half expect a hand on my arm, or maybe somewhere else, but to my relief he's only interested in conversation. He wants to know about Malo, about the lives we've made for ourselves over the last decade and a half, about what kind of boy he really is.

'Troubled,' I say at once. 'But these days that comes with the turf. Every mother says the same thing. Which means it's probably our fault.'

'Troubled how?'

I tell him about the lives that Berndt and I used to lead – busy, pre-occupied, neglectful.

'We both made a point of trying to involve him in the showbizzy bits, take him on location, show him around the set, introduce him to people he could boast about later. We thought that was enough. In fact we thought he was lucky, spoiled even, because that was way more than any other kid ever got. What he really wanted was a new skateboard and mum and dad somewhere close, somewhere he could rely on, somewhere he could call home. We never sussed that.'

'Because?'

'Because we were stupid. Because we were selfish. And because we never paid him the right kind of attention. Had we listened hard enough all the clues were there but somehow we never had the time.'

'*Made* the time.'

'Exactly. Does that make you feel guilty? Of course it does. But then it didn't seem to matter.'

'And this Berndt?'

'He was the same. He does a wonderful guilt trip but that's because he's a much better actor than I am.'

'He's gone now? Fucked off?'

'Yes.'

I tell him briefly about the depths we hit once the marriage had collapsed. I don't mention Berndt's recent visit.

'You hate him?'

'I used to feel sorry for him. Now I'm indifferent. Indifference hurts someone like Berndt more than anything else so maybe he's getting what he deserves. This is a man who needs to be noticed.'

'And the boy?'

'Berndt stole him from under my nose. Made promises he never kept. Malo's brighter than me. It didn't take him long to see through all that bullshit. Kids can be unforgiving. After Malo bailed out, Berndt was history. That hurts him too. He only thinks in the present tense.'

'Nice.'

'Nice what?

'Nice phrase. You really do hate him, don't you?'

We hold each other's gaze for a long moment. Then I nod.

'I do,' I say. 'You're right. Which is why this paternity thing really matters. I don't know whether you know it but Malo has been like a kid at Christmas. Life hasn't done him too many favours recently but suddenly he's looking at the present of his dreams.'

'You mean me?'

'I do.'

'Shit.' He spares Mum and Dad a glance on the screen. 'No pressure, then.'

'You're disappointed?'

'Not at all. How could I be?'

'You're worried? Intimidated?'

'*Intimidated?* Fuck, no. Never.'

'You think he might be a bit of a burden? A bit surplus to requirements?'

'No. If you want the truth it's not him I worry about, it's me.'

'I don't believe you.'

'Try. I have no kids of my own. I'm clueless. I don't know where to start.'

'You could start by talking to him, by getting to know him. Don't tell me you're shy.'

'Of course I'm not shy. I can talk to anyone. Just ask Jessie, Andy. Ask any of the geezers we're gonna get up from Pompey. Ask the guy who's coming to dinner next week. He's like you. You'll see his name in the paper. No, it's not shyness. It's something else.'

'Like what?'

'Like what happens if I get it wrong? You say he's troubled. You say he's damaged already. What if I can't put that right?'

I wonder whether this is the moment to tell Malo's new dad about his son's dalliance with Spice, but decide to adopt a slightly harder tone.

'You make him sound shop-soiled,' I say. 'He's not. Underneath, he's just a kid. He's confused. He wants a middle to his life.'

'You mean us.'

'I mean you. I'm his mother. I'm also recovering from a brain tumour. That, believe it or not, is a bit of a wake-up call. He has my full attention. It might be a bit late but he knows I'm there for him.'

'But you'll help me, yeah?' He sounds almost plaintive. 'You'll be around when I fuck up? You'll be around when I'm getting it all wrong and he thinks I'm just another cunt? You'll do that? For me?'

'For us.'

'Promise?'

'Of course. As long as you make an effort.'

He nods, thoughtful. Then he looks at the screen again. He looks, if anything, a bit lost. I put my arms round him, kiss his temple, hold him close. What I want is a commitment. Not to me but to Malo.

'It's just parenthood,' I murmur. 'Piece of piss.'

'You really think so?'

'I do.'

'OK.' He offers me a rueful grin. 'I'll give it a go.'

FIFTEEN

He's as good as his word. So far, we've spent three days at Flixcombe Manor. On the morning of the fourth, as I'm packing my bag to return to London, I get a call from my agent. Rosa tells me the BBC have had to rejig the rehearsal and recording schedule for *Going Solo* because the male lead has gone down with whooping cough. I won't be required for at least a week, probably a lot longer. This is very good news indeed because H and his new son are beginning to bond and it gives me a great deal of pleasure to be on hand to watch it happen.

It begins, fittingly enough, with one of H's trail bikes. H doesn't want Malo to end up in hospital like his dad and so he tasks Andy, Jessie's partner, to teach Malo how to ride the thing properly. Andy, it turns out, is a genius when it comes to dealing with kids Malo's age. He has the looks – tall, stylish, fabulously unshaven – and his patience is seemingly limitless. He has the kind of dry wit, often cruel, that kids adore and he knows how to offer praise when it will make a difference. He's also a natural on two wheels, gunning the mud-caked machine around the estate course while the three of us – Malo and his mum and dad – look on.

Towards the end of the first lap, where H came to grief, Andy brakes late, slides the machine into a perfect turn, and then accelerates hard for the climb out. The bike snakes left and right before the rear tyre bites and then he's gone. Malo is open mouthed. Me too. While H, remembering the price of getting it wrong, just shakes his head.

After this display, it's Malo's turn in the saddle. Andy has taken him to the flattest part of the estate where he can fall off to his heart's content. Watching him clamber on to the machine and listen to Andy carefully running through each of the controls, I wonder how much of him is thinking about Clemenza back in town. She, he's told me, is an artist on two wheels and now he's about to find out how tough it can be.

The briefing over, Malo puts on his borrowed helmet and lets Andy make the final adjustments. Then, more carefully than I ever expected, he revs up and heads for the figure-of-eight course that H has already coned out. For the rest of the morning, with a pit-stop to refuel, he slowly masters both the bike and the trickier corners. At noon, H calls a halt. Jessie has done lunch. Over egg and chips around the kitchen table, H and Andy take it in turns to analyse where Malo might be having problems. Both of them are careful to let him have his say when he doesn't understand something or even disagrees and by the time we're back outside Malo seems to have found a real confidence. They're treating him like a grown-up, like an adult, and he loves it. For the first time I wonder why Berndt never tried something similar.

That night we all watch *The Dam Busters* in the little cinema H has had installed. This turns out to be H's all-time favourite movie. To my shame I've never seen it before but I've had dealings with one or two of the actors, now in their dotage, and a story of mine about the tail gunner in one of the bombers that doesn't make it reduces H to fits of laughter. As the second dam begins to crack and the music on the soundtrack swells he takes my hand in the half-darkness. Malo is aware of this and when I let it happen I catch a tiny nod of approval. Malo wants us all to be friends and just now I know he's right. Aside from anything else, my poor boy deserves a little affection in his life.

The following evening, we have company at dinner. H has already mentioned a special guest and I've been tempted to ask for more details. Visitors like this, of course, are exactly what Mitch needs for his precious book – a taste of the milieu H is trying to make his own – but already I'm starting to understand something that Mitch, in one of his *Guardian* articles, called Stockholm Syndrome. This has nothing to do with either Berndt or Malo but refers to that strange about-turn that often happens to kidnap victims. You get taken at gun point. You get banged up by your captors. You have no doubts that they're thoroughly wicked. But somehow you start to see it their way.

This, I know on my nerve ends, is beginning to happen to me. I'm not a prisoner in this house. I'm not here against my will.

But H has been on the receiving end of a thoroughly bad press and somehow, in my heart, I'm not altogether sure he deserves it. He's also the father of my only son and that matters more than anything.

Our guest for dinner turns out to be a UKIP politician. His name is Spencer Willoughby. As H has anticipated, I dimly recognize his face from some newspaper or other, or maybe the telly. He's tall, quite articulate in a shouty, superficial kind of way, nearly good-looking, and drinks far too much of H's burgundy for his own good.

Jessie serves fillet steak with veggies from the garden. We know by now, thanks to Willoughby, that H has been more than generous with his contributions over the past few years. A cheque for a hefty sum ahead of the 2015 election. Another to help fund the push to win the Brexit vote. This draws nothing more than a curt nod from H. The relationship between the two of them has been obvious from the start, with H emphatically in control, and when Willoughby starts naming specific sums of money H does nothing to shut him up.

This collusion on H's part is, I suspect, for our benefit. He wants to show me, and perhaps Malo as well, that he bankrolls some of their efforts, that he puts his money where his mouth is, and that he's therefore become something of a player in a high-stakes political game. Willoughby's mission this evening is clearly to leave with another cheque. UKIP are about to name a new leader and they want to claw back territory and profile they lost in the last election. Nothing, it seems, comes cheap in the world of politics.

By the time we get to H's contribution to the meal, a workmanlike baked Alaska, Willoughby is drunk. This doesn't please H at all. He didn't invite this loudmouth to dinner to make a fool of himself and when Willoughby leans across the table to suggest that half a million from his host might make all the difference under the new leadership, H starts to lose it.

'So what the fuck will you do with all this dosh? Supposing it ever happens?'

Willoughby is having trouble keeping anything in focus. I think he knows H has a short fuse but he adores the sound of his own voice.

'Home truths,' he roars.

'What the fuck does that mean?'

'Blacks. Browns.' He leans forward over the table. 'Even the fucking French.'

'What about them?'

'We have to be honest. We have to listen to the people. We have to chuck these leeches out. There's serious votes in immigration. That's how this whole thing started.'

This is a big, big mistake. H isn't racist and – given my presence at the table – he definitely likes the French.

'You need to get a grip, my friend,' H growls. 'Since when did the frogs do us any harm?'

'That's not the point. They've got their own country. They're different to us. That's what got us over the line last year. That's what won us the referendum. We need to build on that. British jobs. For British workers. British laws. British courts. Taking back control means what it says. We need to look after our own. When did you last sit on a bus and hear a conversation in proper English?'

I doubt whether Spencer Willoughby has been on a bus in years but that's not the point. He's thinks he's addressing some meeting or other and he's not going to stop. He's also after H's half million quid.

'You want to know what we're really about? How we're going to pull it out again? This is just you and me, Hayden. Inside track stuff. Strictly embargoed. *Comprende?*'

H gestures for him to go on. He obviously thinks that money on this scale should buy him the odd party secret which seems, to me, entirely reasonable.

Willoughby reaches for what's left of the burgundy and refills his glass. The Tories, he tells us, are in deep shit. Theresa May, surrounded by headbangers, is a hostage in Downing Street. A lost vote in parliament, more treachery from the likes of Boris Johnson, no movement in the Brussels negotiations, and we could suddenly be facing another election. When that happens, Labour will be after loads of constituencies in the north. That's where the election will be won or lost. These are people who voted for Brexit. People who've had enough. The last thing they should be doing is putting a fucking cross against the Labour candidate.

'So you take them on again? Is that what you're telling me?'

'Not at all. We're having quiet conversations with all sorts of people, top Tories all over the country. We have lots in common. When it comes to Brexit we're there to keep them honest, to hold their fucking feet to the fire, to make sure they deliver. There are places up there where we'd still poll well, have lots of support. The Tories could use those votes because the punters have got nowhere else to go. Labour want to stay in the EU. They won't admit it but that's the truth. So when the election happens, if the deal is right, there are places we'll simply stand aside.'

'You mean a deal?'

'I mean no UKIP candidate in key selected constituencies.'

'You're telling me you'll throw in the towel?'

'Exactly. As long as the Tories do what they're told.'

H is toying with the remains of his baked Alaska. He's trying to make sense of all this. Finally he looks across at Willoughby again.

'So I'm bunging money to the Tories, is that what you're saying?'

'You're bunging money to us. To make sure that Brexit happens the way we want it.'

'But the Tories spend it? To help win all these constituencies?'

'In effect, yes. It's nothing we'd ever own up to, and they won't either, but that's the way these things sometimes work. Means and ends, Hayden. We keep the Tories in Downing Street and we make sure we get out of Europe on our terms. That's what it's about. Just like always.'

H nods, says nothing. Even I understand the dark logic of the plan. I'm thinking of Mitch. He should be here at the table. He should be listening to all this.

I steal a glance at Malo. He's trying to stifle a yawn. Just for a moment, Willoughby appears to have stopped banging on. I take advantage of this brief silence to make my apologies for beating a retreat. Malo's had a heavy day. I'm exhausted. Much better that you two have the rest of the conversation in peace.

H looks up at me. His wine is virtually untouched.

'Fucking good idea,' he says. 'Leave this clown to me.'

SIXTEEN

I come down early next morning after a sleepless night. I find Jessie in the kitchen. She's filled the dishwasher with last night's dirty plates and now she's drawing up a tally of names on a big foolscap pad. I suspect this is the guest list for the celebration bash that H is determined to throw to mark his fiftieth birthday, and – much more importantly – the surprise acquisition of a son. It turns out I'm right.

'How many?' I enquire.

'Dozens. Hundreds. Thousands. H isn't short of mates in Pompey.'

'Seriously . . .'

She shoots me a look. I sense we've become friends but I'm learning fast that being part of H's teeming ménage is something you'd never take for granted. There are unspoken rules, lines you should never cross, and one of them is the need to keep your curiosity under control. There might be lots of questions you want to ask but you'll never get the answer until H is good and ready.

This I'm beginning to understand. What comes as a surprise is Jessie's own enquiry.

'Did you really only meet him that one time?'

'Yes. How did you know?'

'H told me. France, wasn't it?'

'Antibes. Celeb land, even then. Not us. We were cheap hires on a movie, ten a penny if you want the truth.'

'So what did you make of him?'

'I thought he was funny. And bold.'

'That's H. Still is. Though he doesn't laugh as much as he used to. Nice man underneath it all. Heart of gold.'

I nod, tell her I'm glad to hear it. She reaches for her pen again, remembering yet another name. Then she looks up.

'You will be here, won't you? For the party? H likes to pretend it's all about being fifty and about finding Malo but I'm not sure that's true.'

'I'm flattered.'

'You should be. H isn't an easy man to get close to.'

That morning, Malo takes a break from learning to ride a motorbike. Instead, Andy takes him shooting in the woodland that fringes the estate to the west. All morning, sitting by an open window in the big library downstairs, I hear distant gunshots and by the time the two of them return for lunch, Malo has bagged his first pigeon. With some pride he shows it to his dad.

'Top job, son. Spot of red wine gravy? Game chips? Lovely.'

I later find the poor dead thing on the compost heap beyond the vegetable garden but that doesn't matter. What's infinitely more important is the ease with which my son appears to be settling into his new life. Neither of us know how long we'll be staying. The subject is never broached and in any case, apart from the BBC radio play, neither of us have other commitments. And so the days roll by, the weather holds, and Jessie and H between them conjure delicious meals. In the evenings, when we're not watching yet more movies, we play Monopoly, or backgammon, or maybe a card game. As I say to H at the end of the first week, it feels like an endless Christmas.

'Is that a compliment?'

'Definitely. I've always loved Christmas. Malo does, too. You're spoiling us rotten. We like that.'

H nods. That slight defensiveness I'd noticed earlier has gone completely.

'It's convalescence,' he says, 'For all of us.'

On the following Monday I take a call from Mitch. I happen to be out on my morning walk, something I've adopted to try and keep my weight down in the face of so many meals. From where I'm standing, the house is a grey dot in the distance.

'Are you OK?' I ask him.

'I'm fine. Busy but fine. Where are you?'

'H's place.'

'Who's "H"?'

I laugh. This is the measure of just how far Malo and I have come. I explain about Saucy's new name. I can tell from Mitch's voice that this isn't altogether good news. I sense I'm supposed to keep him at arm's length. Not succumb yet again to the gruff Prentice charm.

'You're OK with him?'

'I'm a house guest, Mitch. And so is my son. If you want the truth it's a bit of a break for both of us. Think Jane Austen. I'm in temporary residence. I spend most of the day with my nose in a book. I might even take up crochet-work. Aside from that I do what a good spy's supposed to. I watch and I listen. That's when I'm not bloody eating.'

'And?'

'I'm getting fat.'

'That's not what I meant.'

'Of course it isn't.'

'So? You've got anything for me?'

I don't answer, not immediately. First I want to know why he's phoned me.

'We have to meet,' he says. 'I've been getting some grief from my publisher. He needs to know when I can deliver.'

'Which is what you're asking me.'

'Indeed.'

I tell him it might be difficult. 'We're down here for a while.'

'But I thought you had something to do for the Beeb?'

'I have. It's been postponed.'

'I see.' There's a longish silence. Then he's back with a proposition. He has to go to Salisbury tomorrow for a meeting with a source. We could have lunch afterwards.

I tell him that might be possible. I've no idea how far Salisbury is and I've even less idea how I might get there but H is in the business of making things happen so I say yes.

'I'll phone to confirm,' I tell him. 'And you can tell me where to meet.'

Back at the house, an hour or so later, I broach the idea to H. I have a friend I need to see. We want to meet in Salisbury. Any ideas? As it happens, Jessie is going to an interior design agency in Romsey tomorrow to discuss some plans for one of the rooms upstairs. Romsey is near Southampton. She could drop me in Salisbury and pick me up on the way back home.

'Sorted?'

'Sorted.'

That night we're back in the cinema. At Malo's request, we're watching a movie called *Gravity*. Sandra Bullock and George Clooney, in fetching space suits, cavort in orbit while the gods of weightlessness do their best to wreck the party. The plot line is non-existent and the dialogue is beyond parody and even the special effects aren't enough to keep me awake. The movie over, H gently shakes me back into something approaching consciousness.

'Who won?' I mutter.

'The Germans. On penalties. Same old fucking story.' H peers down at me. 'You want me to come? Tomorrow?'

'Not really. It'll be fine.'

'You're sure?'

'Yes. But thanks for the thought.'

He studies me a moment longer, half-expecting me to change my mind, then gives my elbow a little squeeze.

'Take care out there, yeah?'

Late next morning, around twelve, Jessie drops me outside the hotel in Salisbury where I'm due to meet Mitch. We've chatted for most of the journey from Flixcombe Manor. H, she says, gives her all the freedom she wants when it comes to decoration. She loves muddy colours, dark greens and browns, and H indulges her passion for wood panelling, of which the house has a great deal. Budget-wise, unless she gets really silly, he simply signs the cheques.

Towards the end of the journey, with the city's cathedral spire in sight, I'm starting to wonder whether she thinks I'm going to be some kind of threat to her monopoly on interior decoration and before she drops me outside the hotel I try to offer a little reassurance.

'I'm Malo's mum,' I tell her. 'Nothing else.'

The hotel is called the White Hart. There's a bit too much red velvet and polished brass for my taste but I'm not here for the decor. I find Mitch at a table in the lounge. He makes his excuses about not being able to stretch to the restaurant down the hall but I tell him that the bar menu will be fine. Something light with salad? Perfect.

He fetches me a gin and tonic. I notice he's drinking Perrier.

'Liver packed up? Or are you driving?'

He says he took the train down, working non-stop on his laptop.

'The woman opposite me was raving about the countryside. Said Wiltshire in autumn was close to perfect. I had to take her word for it.'

'Poor you. Sayid?'

'Sends his best. He's had a bit of a cold these last few days. In fact we both have.'

I nod, sympathize. I've been wondering why he looks so peaky. He wants to know what Malo is doing with me down at Chateau Prentice.

'He came for autumn,' I say lightly. 'Like your lady on the train.'

'You're kidding.'

'You're right, I am. He's keeping his mum company. It happens sometimes. It's a parent thing.'

'But he's a city kid, no?'

'Of course.'

'So isn't he bored?'

'Far from it.'

I've been wondering whether to tell Mitch the whole story about Malo and this would be the perfect opportunity, but it's still something I want to keep to myself. On the other hand he already knows about Antibes and he might have drawn one or two conclusions of his own. Either way I'm keen to move the conversation on.

'You mentioned your publisher on the phone. Is there some kind of problem?'

'Yes.'

'Like?'

Mitch checks left and right before answering. I find this unsettling. Life in the country has been so simple.

'Why do you want to know?' he asks.

'Because there are ways I might be able to help.'

'With the publisher?'

'With the book. You've told me it's about Prentice and some friends of his. You called them "like-minded folk". It was an interesting phrase. I remembered it. What, exactly, does it mean?'

'It means people who think the way he does.'

'Politically?'

'Yes.'

'Coming from the same kind of background?'

'The same pedigree, yes. Not the same city, even the same area. But the same mindset.'

'We're talking rich people?' I ask.

'By definition, yes. Without money, serious money, none of this would matter.'

'None of what?'

'Influence. Leverage. Putting your money to work.'

'In the interests of what?'

'Of yourself, in the end. But en route you kid yourself it's for the good of the nation.'

'You mean Brexit?'

'Of course. But we've talked about that.'

I nod. He's right. We have. I remember the pub in Ealing the day when my poor Peugeot had been towed away. On that occasion Mitch was nice enough to get it back for me, something I shouldn't forget.

'H had a politician to dinner the other night,' I tell him. 'His name was Spencer.'

'Spencer Willoughby? UKIP?'

'Yes. Is he big? Important? Should I have known about him before?'

'He's middling. Probably less important than he thinks. But he's certainly ambitious.' Mitch stoops to the bag beside his chair and produces a pad. 'So what happened?'

I tell him about H's contributions to the UKIP fighting fund over the last couple of years. Mitch appears to know about this already but when I mention Willoughby's fresh bid for yet another cheque, Mitch scribbles himself a note.

'Did he say how much?'

'He wanted half a million.'

'And Prentice?'

'He didn't say anything. Not while I was there.'

'So it might have been a yes?'

'Might have been. There was something else though, nothing to do with money.'

I tell Mitch about Willoughby's plans for fighting the next election. How UKIP planned to leave the Tories a clear run in constituencies they might not otherwise win. And how H's money might help make that happen.

'You mean the money would go straight to the Tories?' His pen is poised over the pad.

'That wasn't clear. You'd have to ask him. This is Willoughby talking. That's all I can tell you.'

'Did he mention any constituencies in particular?'

'No. Except that most of them were in the north.'

Mitch nods. He seems to think that Willoughby might be out of his depth. The word he uses is 'flaky'. UKIP, he says, were virtually wiped out at the last election and it isn't clear why the Tories would ever be listening to them.

I'm desperately trying to remember some of the other things Willoughby said that night at H's table.

'He thinks they still have a direct line to the grass roots. He told H they're still plugged in. The way he saw it, only UKIP could guarantee a proper Brexit. Does that make sense?'

'It does. UKIP always claim they were there at the birth, and they're right. They also did most of the heavy lifting when it came to the referendum. That's where donors like Prentice were so important. Now, the Tories don't need UKIP marking their homework. They've got enough headbangers of their own.'

'So Willoughby . . .?' I shrug. 'Am I wasting your time here?'

'Not at all. Especially if you're right about Prentice's money ending up with the Tories. You think it did? You think that's true?'

'I've no idea.'

'Can you find out?'

The question couldn't be more direct and it puts me on the spot. Do I steal a look at his cheque book? Access his online bank account? Wait until his office is free and root through the drawers in that wonderful desk?

'I don't know,' I said. 'I suppose I could try.'

'That would be helpful. Especially if you took a photo or two.'

'Of the cheque stub? If there is one?'

'Yes. And maybe some shots of the house, too. We call it "colour" in the trade. It sets the mood. Shows the reader just how far our Pompey boy's come.'

Mitch gives me a shopping list: photos showing the house in the landscape, the sheer size of the estate, the motocross circuit where H came to grief. He also wants as much as I can get from inside: the baronial hall, the main reception room, the library, the newly installed cinema, the kitchen, even a taste of the bedrooms upstairs. This, of course, would be much easier for yours truly but I'm not at all sure I want to do it.

Mitch offers me a sheet of paper from his pad and a pen. He thinks I should be making a list.

'I've got a perfectly good memory,' I tell him. 'I think I know what you want.'

'Am I pissing you off?'

'Yes, you are.'

'So what's changed? I was pushing at an open door when you were in London. Or at least I thought I was.'

'It's just . . .' I'm trying to marshal my thoughts. 'None of this is easy . . .'

'You're bloody right it isn't. Try being on the sharp end. Try getting anywhere even half close to these people.'

'These people?'

'People like Prentice. Self-made. Powerful. Minds of their own. They know exactly what they're after and they don't care who they hurt on their way. They might corrupt the political process in getting there and you might not think that matters but there are still some people who do. Forgive me, but I assumed you were one of them. Which just goes to show how wrong us fucking scribes can be.'

Mitch very rarely swears. He's as angry as I am. I get to my feet. I've had enough of this. It wasn't me who made the phone call in the first place, who stood outside my flat and suggested breakfast. It wasn't me who's been on my case ever since, pushing me gently towards H and his mates.

'You've been very kind to me,' I tell him. 'You've been there when it mattered and I really appreciate that. You've probably got me down as a spoiled celeb with more money than sense. For the record I do care about democracy and the health service and pot holes and libraries and all the rest of it, but where we're heading at the moment is, to be frank, just a wee bit personal. So you'll forgive me for skipping lunch.'

'And this?' Mitch nods at his pad. 'Willoughby? The cheque? All that?'

'We'll have to see.'

'That might not be an option. My publisher's anticipating an election within the next six months. We have to get our ducks in order. Otherwise I've just wasted a year's work, probably more.'

This is emotional blackmail and both of us know it. I'm still lingering by the table, wishing I'd hadn't heard that last little speech.

The stand-off is becoming awkward. The barman is eyeing us with some interest. I think he may have recognized me but I can't be sure. Finally, it's Mitch who breaks the silence.

'This is about Malo, right?' He gestures at the chair I've just abandoned. 'Do you want to talk about it?'

I stare down at him. Then I shake my head.

'No,' I say. 'I don't.'

SEVENTEEN

I'm back at Flixcombe Manor by mid-afternoon. H has spent most of the day in his office conferring with his accountant on a business plan for new insurance products for the over-sixty-fives. Two recent developments, he tells me, have told him there's squillions of quid to be made. One of them is the fact that older souls – both men and women – are a great deal fitter than they used to be. They still want to push themselves. They're still hungry for physical risk. Yet older bodies break more easily, hence the call for extreme sports insurance cover to put them back together again without, he says, costing an arm and a leg. Ho ho.

'Did you dream that up in hospital?'

'Yeah, as a matter of fact I did. If you hurt yourself in this country it's not too bad. If it happens anywhere else, you're stuffed. I might call it Easy-Mend. What d'you think?'

I laugh dutifully and enquire about his other wheeze. This also has a great deal to do with abroad. The day after we leave the European Union, millions of ex-pats are going to find themselves suddenly stripped of medical care. For the elderly, he says, this is going to be a real problem. Accessing the local clinic in Provence or Malaga won't be cheap but H and his accountant are busy cooking up ways these orphans can avoid selling up and coming home.

'I've been in this business years,' he says. 'I've got the contacts. I know what I'm about. Brexit is a big fat pay day if I get it right.'

A big fat pay day if I get it right.

I'm suddenly thinking of Mitch. For weeks now he's been telling me what a disaster awaits millions of Brits both here and abroad. How bits of the country will implode post-Brexit. And how the likes of Hayden Prentice are lurking in the shadows, waiting for the dust to settle, making their plans. Now I'm hearing it from the horse's mouth. *Easy-Mend*. Very funny.

H is watching me carefully. He's abandoned the paperwork on his desk.

'Go well, did it? That little lunch of yours?'

'It was lovely. Girlfriend I haven't seen for years. Bit of a catch-up really.'

'Great.' His hand lingers briefly on mine. 'Jessie's finished that list of hers for the weekend. You want me to talk you through it?'

Jessie has typed the list out and sent it to H. A couple of keystrokes and it hangs on his PC. Nearly sixty names, not one of which I recognize. H is scrolling slowly through them, a smile on his face, and it occurs to me that in some respects this little gathering – assuming they all come – will be like one of those regimental reunions.

My stepfather's dad was in the army and saw service in Korea. His regiment had a hard time in a particular battle at some river or other. I remember him telling me they were holding a hill against the Chinese. A third of them were killed or wounded and the rest of them spent the next two years behind the wire. At the war's end they made it back to the UK, all damaged to various degrees, and every year that followed – on St George's Day – they got together to pay their respects to the dead and get roaring drunk.

Is H doing something similar? Am I in for a night of war stories and a monster hangover?

H has settled on a particular name. Mick Pain.

'Top bloke,' he says fondly. 'Hard as you like. I dug him out of the shit the morning we got ambushed by the Millwall. Waterloo fucking station. Just down that little alley behind the arches. Cunts were waiting for us. Hundreds of them. Mick was out front. Chinned one of their generals. Huge bloke. Went down like a sack of cement. Bad move on Mick's part because the Millwall never liked us taking the piss. There were some nasty bastards there but we still gave them a hiding.' He shakes his head. 'Brilliant.'

Another name. Tony Morse.

'Classy guy. Still is. You'll love him. Clever as fuck and likes you to know it. Once we started making serious moolah we couldn't do without him. Wised us up. Taught me stuff I'd never have known otherwise.'

'Another hooligan?'

'Fuck, no. Bloke was a solicitor. Retired now, thanks to us. Lovely guy but silly around women. Three wives? That's fucking careless.'

The list goes on. A woman called Leanne who rose to the challenge of running a string of themed wine bars, all funded by cocaine loot. A planner, Alan Ransome, who knew every short cut in the book when it came to making development applications. And a streetwise bodybuilder, Wesley Kane, who H still calls The Prefect.

'Wes kept things nice and tidy,' H tells me. 'Any problems, Wes would sort them out. We bought him a house in Fratton for the day job. The sound insulation cost us a fortune but an hour with Wes in that back room and you wouldn't fuck about any more.'

I edge my chair back an inch or two. I'm beginning to get the picture here: a bunch of football thugs with a limitless appetite for violence discover the joys of drug dealing and hit the mother lode. The money rolls in and sensibly they use some of it to buy proper advice. Solicitors. Accountants. Planners. All helping themselves.

'And the police?' I wonder aloud.

H shoots me a look and reaches for the mouse again. Scrolling backwards he settles on another name. Dave Munroe.

'Fat Dave,' he says. 'DC on the drugs squad. Kept us abreast of events. Most of the time the Filth had their heads up their arse, totally useless, but times when they became a nuisance Dave would mark our card.'

'And you paid him?'

'We did. Money mostly, because Dave had a bit of a gambling habit, but women too. Dave loved big black women. We could help him out there. Jamaican called Gloria who ran a couple of knocking shops.' He laughed. 'We'd bought one of those big seafront hotels by then. Dave had screwing rights in a room on the top floor. She used to eat him alive.'

'And the police never realized?'

'About Dave? Never. In the end he retired on full pay. These days he's on the Isle of fucking Wight drinking himself to death but once he knows our booze is free he's bound to turn up. Here's someone else you might meet . . .'

Marie Mackenzie. H is staring at the name. His mood has darkened.

'Lovely woman,' he says softly. 'Real class.'

'So how come she liked football?'

'She didn't. She married into it. Worst decision she ever made except she loved him.'

'Who?'

'Bazza. Bazza was our top guy. Bazza was King. Bazza was someone you'd never fuck with. Mad cunt but clever. The Filth spent years trying to nail him but never got close. The way these things work, you need a leader, someone with a brain in his head, someone who knew the difference between taking a risk and pissing it all away because you've made the wrong fucking call. Bazza was on top of that, all day, every day. He was the one who took us all legit. He was brilliant in business, brilliant in everything. Fucking sad.'

'What happened?'

'He made one mistake. Just one. Bloke called Winter. Paul Winter. He was Filth to begin with, drugs squad again, and a real pain in the arse because he was the only cunt who knew what he was doing. They stuck him undercover and then got him blown because they were so fucking useless, so he had a meet with Bazza and told his bosses to fuck off. Bazza thought that was funny. So funny, he offered him a job.'

'With you lot?'

'Yeah. Some of us thought Bazza had lost it. Big mistake. And in the end we were proved right.'

'He was still undercover?'

'Not at all. He did the biz for us for years. He was good. And he was funny, too. Then Bazza got ideas above his station, stood for parliament. Daft move. Absolutely insane. He was king of the city already. Why become a fucking politician?'

'He went through with this? Campaigned?'

'Yeah. And didn't do bad, neither. But that wasn't the point. Him and Winter fell out, big time, and the next thing Bazza knows Winter's grassed him to the Filth. That was the end of it. Even Bazza couldn't survive that. Served him right in a way. He should have listened to us.'

'He's still around? Bazza?'

'No.'

I want to know more. I want to know exactly what brought this stellar career to an end, but H won't tell me. All I need to know, he says, is what I see around me. The house. The estate. The cars. The bikes. All the other toys that make life so sweet. Not just for him, not just for Hayden Prentice, but for so many of the names on Jessie's invitation list. All this, he seems to be saying, is down to Bazza and there isn't anyone on the planet that can touch any of it. That's how fucking good he was.

He's staring at the PC screen. Marie Mackenzie.

'You think she'll come?'

'I doubt it. She hates us now.' He falls silent, shakes his head, rubs his face. 'We'll have a minute's silence on Saturday night,' he says. 'Before we all get too pissed.'

While H tidies the last of his paperwork, I go for a walk. Through no fault of my own, unless you count Antibes, I appear to have ended up in a weird cult of ex-gangsters who've made their money but still can't quite believe it. I know very little about drug dealing and even less about putting all that loot through what H quaintly calls 'the laundromat' but I can only take H at his word.

We live in a free market. You source something that people need badly. You charge the earth, watch your back, and make a fortune. Everything you do, everything you touch, is illegal but if you hire the right advice and pay the right money you're home free. This is the kind of conjuring trick that makes good guys like Mitch Culligan froth at the mouth – and I don't blame him because even to me the implications are troublesome.

It can't be just Portsmouth. This must be happening all over the country. Name any city and you'll find a Bazza, and an H, and a bent solicitor or two, and rogue cops, and corrupt accountants, and a bunch of crazies with far too much money. That's how the trick works. That's why capitalism is such a lovely word.

Should I phone Mitch? Should I apologize for walking out on him at lunchtime? Should I tell him he's right? That Brexit is a disaster in the making? That the entire country has suddenly found itself at the mercy of Hayden Prentice and his brethren? Or should I wait until Saturday, and put on my best dress, and settle in for the Command Performance? Make a few notes, keep my phone on Record, play the spy for real?

In truth, I'm out of my depth. This is a role I've never been offered before, and don't much want, but I am where I am and there's something shaping in the wreckage of my poor brain that feels suspiciously like duty. I owe it to a number of people to try and figure out what's really going on here. One of them is obviously Mitch. Another, far more important, is Malo. I pause to phone him. I need to get my bearings. Or, more precisely, his.

When he finally answers, he tells me he's busy. A local farmer has lost his stockman to an attack of flu and after a call to Andy, Malo has been drafted in to help get his cows to the dairy.

'You're OK with that?' I ask.

'It's fine. Cool. Cows? You've no idea how big they are. Amazing.'

I can hear cattle lowing in the background and the slow clop-clop of hooves on a tarmac road. An act of simple neighbourliness, I tell myself, has opened yet another door in Malo's young life and who am I to get in his way?

Standing in the middle of this huge estate, listening to my son trying to get some action out of a cow, it's hard not to remember that moment on the video phone that still haunts me, his thin body bent over a Spice-filled spliff. Down here he has a million distractions, a trillion million better things to do, and that very definitely matters. On the other hand, although he doesn't know it, we owe this sudden whiff of redemption to the narco biz, to drugs money.

Impossible, I tell myself. Just what does a girl do?

I allow myself the beginnings of a smile. This little irony would be music in my ex-husband's ears. Berndt would be on it in seconds. He'd block out the story, add a cast of low-life characters, write a trial scene or two, hone the dialogue, and doubtless find a backer. And within less than a year, if all went well, I'd find the resolution to my little dilemma on screen.

I shield my eyes against the low sun. I'm probably kidding myself but several fields away I think I can hear cattle.

'Mum?' It's Malo. 'What's going on?'

'Nothing,' I tell him. 'As long as you're OK.'

* * *

That evening, H gets Jessie to drive us to a gastropub he likes in a village down the road. This is excellent news, partly because there are questions I need to ask, but mainly because I'm starving hungry.

The pub is full but H has made a reservation. The woman in charge of the restaurant gives him a careful hug and asks how he's getting on. H's accident has evidently been the talk of the village and she tells me to take great care of him.

'Boys will be boys,' she says. 'But there are limits.'

Quite. Our table is beside the fire, which is mercifully unlit. The nearest couple to us are locked in a fierce debate about the Tory party conference which takes place next week. The husband – tweeds, brogues, fine-spun hairs on the backs of his hands – thinks Theresa May's useless. His wife predictably disagrees. Our PM, she says, has the measure of the cabinet rebels. She'll see them off without any great fuss, exactly the way she glided into Downing Street in the first place.

'Stole,' the husband says. 'She stole into Downing Street like a thief in the night. I give the bloody woman until Christmas. And that's if she's lucky.'

H is rolling his eyes. When he reaches for the menu I ask him just how much of this stuff really matters to him.

'You mean politics?'

'Yes.'

'Most of it's bollocks. Most politicians are on the make except for the Labour lot and they're fucking lunatics. The steak here is brilliant. Fill your boots.'

I order fish with hand-cut chips and a dish of veggies. H sorts out a bottle of Chablis largely, I suspect, for my benefit. While we wait for the wine, I go back to the matter in hand. The couple next door, mercifully, have paid up and gone.

'You told me this Bazza stood for parliament. You give loads of money to UKIP. If most of it's bollocks, how does that work?'

He's dismissive. 'It's a game. Sometimes you like seeing these people dancing for your money. Toss them a quid or two and they'll do anything.'

'What about Bazza? Standing for parliament takes some doing.'

'He was taking the piss. Seeing how far he could get.'

'I don't believe you.'

'You don't?' I've caught his interest at last. 'So what's your take, then? You don't think I've got better things to do than ponce around in suits all day?'

'I think you want power.'

'I've got power. It's called money.'

'But political power. Calling the shots. Making the odd law.'

'Law?' He's laughing now. 'Laws are there to be broken. I've built a whole career on that. You might have noticed.'

'So why UKIP?'

'Because they were the underdogs. Because they were a load of old dossers in ties and blazers with lots to get off their chests. If my dad had ever been into politics he'd have been a Kipper. I know these people. I know the way they think. I know what gets under their skins. They're not toffs like the Tories and we're not talking Nazis like the English fucking Defence League. They just wandered into politics like you or me might go shopping and they decided to make a noise. They don't much like abroad, and they definitely don't like Pakis nicking their seat on the bus, and given half a chance they'd shut their doors and spend the rest of their lives listening to the fucking Archers. Do I have a problem with any of that? No. Do I think the country's going to the dogs? Yeah. And the sooner the better. Because then we can have a proper sort-out.'

I nod. I've heard Mitch say something similar, especially the last bit. He called it anarchy.

'So did you give them the half million in the end? UKIP?'

'No way. I wrote them a cheque for five hundred. That's bye-bye money for old times' sake. Spencer Willoughby is a twat. Even the Kippers can do better than that.'

The waiter arrives with the wine. After he's gone, H raises his glass.

'Let's talk about something interesting,' he says. 'Here's to the boy.'

'To Malo.' We touch glasses.

H leans back in his chair, trying to ease his shoulder. He's been with his son a lot these last few days and he likes what he's seen.

'He loves it down here,' he says. 'He really does. It's opened his bloody eyes. He's even put a bit of weight on.'

'I know. I've noticed.'

'You're telling me that's a bad thing?'

'Not at all. Far from it. He hasn't been in a good place recently.'

'Really? You want to tell me more?'

I explain about the fall-out with Berndt, about the phantom affair with Eva, about turning up in London with a stash of cannabis and a Spice habit.

'*Spice?*' H looks horrified. 'That's low-life. That's B fucking wing. You're serious?'

'I am. He could have been one of your customers back in the day. How does that make you feel?'

This sounds more aggressive than I intended but it was hard to resist the temptation. Live by the sword, die by the sword.

'Nice one. It wasn't around at the time but we'd never have touched Spice. Nor smack. Nor a whole load of other rubbish. We were there to spread a little happiness. We wanted people to love each other up. That was always Bazza's line.'

H is grinning at me. Uppers, not downers, he says. Loads of ecstasy and as much of the laughing powder as Pompey could handle. I smile back. He rides the punches well. But he still wants the full story.

'About what?' I enquire.

'Malo.'

I tell him as much as I know: my son slipping his moorings, my son disappearing at odd hours, my Zombie boy back at the door, totally wasted.

'Wasted is fucking right. That stuff's evil. Which is why he's best off down here.'

'Where you can keep an eye on him?'

'Exactly.'

To be fair, this is a proposition I've been half-expecting but now he's spelled it out it makes me anxious. Stockholm Syndrome, I think. My son held hostage for a second time, first by Berndt, and now by his real father. Is this really what I want?

'What about Malo?' I say. 'What does he think?'

'He's up for it. Room of his own? Learn how to drive a car as well as a bike? Money in his pocket? The boy can't wait.'

'You're buying him off. That's too easy. There has to be more.'

'Of course there fucking does. He needs a project. Something he can run with. A bit of responsibility. If I make it sound a doss, it won't be. No one gets a free pass. Not here. Not anywhere.'

'So what will he do?'

'I dunno yet. I've got a couple of ideas. We'll see.'

'Do I get any clues?'

'No.' The glass extended again. 'Happy days, eh?'

Happy days. I'm starting to get a feel for life in H's boiling wake. He hasn't recovered yet, far from it, but the pace he sets is still blistering. He's up by half past seven, a cup of coffee and two slices of white toast. He summons a series of business associates and sundry others for brainstorming sessions in the downstairs reception room he's reserved for get-togethers like these. On more than one occasion I watch the faces of these people as they leave. They look dazed. They look as if they've been under sustained bombardment. And they probably have.

By now I've realized that H isn't one of life's democrats. He mistrusts discussion and he has no time for the word 'no'. So far, I've only seen the softer side of him, and it's Andy, of all people, who marks my card.

'He wants to get you down here,' he says. 'As well as Malo.'

'And he thinks that's possible?'

'He thinks anything's possible.'

'So what would you suggest?'

The question takes him by surprise. I know he likes me, maybe more than that. I've seen it in his eyes, and in the odd aside from Jessie. I suppose I'm a bit exotic. I come from a different world, the world of feature films and the theatre and the odd TV series. That might be interesting in its own right but more importantly it keeps me at arm's length from H. I owe him nothing, at least not yet, a situation I might – in Andy's phrase – be wise to hang on to.

'H likes to own people,' he says. 'Half the time he doesn't even know he's doing it. It's just part of his nature. He wants you on the journey. He wants Malo, too. He wants all of us. With H, it's black and white, all or nothing. This is a guy with no gearbox. It's stop or flat out. And stop bores him stupid.'

That's nicely put. I make a mental note. What other tips does he have?

'Don't let him shag you. He wants to. Badly. Do I blame him? Of course I don't. But he knows he's done it once and he sees no reason why it shouldn't happen again.'

'What about me? Do I get any say in this?'

'Of course you do. But you asked me a question. H is a creature of habit. He's also impatient, and headstrong, and sees no point in fannying around. What you see is what you get.'

I nod, and thank him for his time. This is sensible advice. We're talking in the meadow beyond the kitchen garden where Andy has been tipping poison down badger holes. This spot, as far as I can judge, is invisible from the house.

Andy hasn't moved.

'Does he ever let you out at night?' he asks.

'Never. I'm gagged and bound. That's the way he likes me best.'

'Fancy a drink at all? Whenever it suits? There are some neat pubs I know. Yeovil isn't far.'

I study him for a long moment. Good face. Good eyes. And the kind of smile any girl would enjoy waking up to. I beckon him closer. This is confessional stuff.

'You know what that makes you?' I ask. 'Apart from brave and reckless and quite attractive?'

'Tell me.'

'It makes you crazy. I've spent all morning trying to imagine the wrath of Hayden Prentice and it does nothing for my peace of mind. I suspect you've known him forever. Otherwise you wouldn't be here. So why are you coming on to the mother of his long-lost son?'

'Because I can.' He grins. 'And because we'd get away with it.'

'That's supposition. You're delusional. And you know what?' I lean forward and kiss him lightly on the lips. 'That makes you just like H.'

The following day I have to stage a brief return to London for a conference with my agent. Jessie drives me to Dorchester but my attempts to spark a conversation come to nothing. At the station I wave goodbye but she doesn't spare me a backward glance. She's been watching Andy, I conclude. Very wise.

Rosa, when we go to lunch, shares the latest news from Montréal. The movie to which I've attached my name is definitely going ahead, but before I sign any contract Rosa has to be sure about my medical status.

'I need more tests,' I say at once. 'Do you want me to ask for a schedule?'

'It might be more complicated than that. We have to do the usual with the insurance forms. I've no objection to lying but if you collapse on set it might get awkward. The Montréal people are thrilled to have you but the word seems to have spread. They need to be certain you'll last the course.'

Last the course. After my break in the country I'm suddenly back in the world of movie budgets, and deadlines, and very probably another session with my neurosurgeon.

'When do they want to start?'

'After Christmas. Montréal? Wrap up warm.'

Christmas is nothing. Less than three months away. How will I feel? Where will we all be?

I call my neurosurgeon for advice. His secretary says he's busy at another hospital. I tell her I'll call back later and she says early next week would be best because he's up to his ears. Rosa has been half-listening to the conversation.

'You'll be around tomorrow?' she asks. 'It'd be nice to get this thing wrapped up.'

Tomorrow is Saturday. I shake my head and tell her I have a pressing social engagement in deepest Dorset.

'Party time?'

'In spades.' I nod. 'Wish me luck, eh?'

It's Jessie who picks me up at Dorchester station but this time she has Malo in the car. It's raining hard and his Wellington boots have left mud all over the Range Rover's rubber matting. He talks non-stop on the drive back to Flixcombe Manor, mainly about tomorrow's party. He and Andy have taken the trailer to the big Tesco Extra in Yeovil. He's a bit woolly about the details but between them they've plundered shelf after shelf of canapés, sausages, crisps, cold chicken, plus five packs of plastic glasses and a mountain of booze.

'Mainly Stella,' Jessie says drily. 'And bourbon for afters.'

I press Malo for more news. 'How many might be coming?'

'Forty-eight so far,' Jessie confirms. 'And counting.'

Forty-eight guests is a small crowd. Or maybe a big riot. Forty-eight people will need an awful lot of looking after.

'Where will they all sleep?' I enquire.

'Where they drop.' Still Jessie. 'I don't think sleeping's on the agenda. If push comes to shove I've had Andy put straw down in the barn.' She shoots me a look. 'He may have mentioned it.'

'Sadly not,' I say. 'But it sounds very practical.'

By now we've taken the turn from the main road and Jessie floors the accelerator as the back of the Range Rover shivers on the wet mud. The big grey house appears through the trees. Malo has glimpsed it already.

'Home,' he says absently. 'That was quick.'

We spend the evening brightening the house under H's direction. Between them, Jessie and H have dug out countless photos and mementos from the glory days in Pompey. A pile of football programmes will greet guests on their arrival. In one of them is a travel voucher for a month in Mauritius, H's idea for warming up the party from the off. I flick through a couple of the programmes, trying to grasp the magic appeal of Fratton Park, the legendary stadium where H's journey began. Andy helps me out.

'Fortress Fratton,' he says. 'Think Alcatraz on a wet day. Back to the Eighties. Back to the Seventies. Rule one: the old days were even better than you thought. Real football. Blood, sweat and shit meat pies. Ring any bells?'

'None,' I say. 'Roland Garros was about my limit.'

'Roland what?'

'Garros. It's in Paris. It's French for Wimbledon.'

Andy loves banter like this. Twice this evening he's confessed to being a closet thesp, deeply frustrated, and I'm beginning to believe him. Of Jessie, for the moment, there's no sign.

Andy has found a cutting from a French newspaper. He can't resist asking for a translation.

I take a look. The cutting comes from *Le Havre Libre*. The date is badly smudged. Nineteen eighty-something. At the top of the story is a black and white photo of a cafe. My first impression is that the Vikings have paid a visit. The place has been sacked: tables on the terrace overturned, windows smashed, glass everywhere. Andy is standing beside me. I can feel his presence, the heat of his body. His finger rests briefly on the top of the photo.

'The clue's in the name,' he says.

Café de Southampton? I'm none the wiser.

'Southampton's up the coast from Pompey. The two cities hate each other. Anyone from Southampton is a Scummer. Anyone from Le Havre is even worse. The local frogs are twinned with Scummerdom. Something like that, you don't need an excuse.'

'For what?'

Andy describes the cross-Channel expedition. Bazza and H and the rest of them jumped the ferry in Pompey and made the crossing to Le Havre. They attended a pre-season friendly, Pompey against the French lads, and then descended on the cafe. The rest, he says, is history.

'You were there?' I'm still looking at the photo.

'No. I was late to the party. Nineties boy.' He picks up the cutting. 'This became a battle honour. Bazza couldn't stop talking about it, H too. Run with these guys and you realized anything was possible.'

'Even a place like this?' I gesture round the wood-panelled walls.

He laughs, looks at me. We're alone in the big reception room, surrounded by memorabilia. The door is shut. Very politely, he asks for a kiss. I tell him no.

More photos. More faces. More tattoos. Episode by episode the story H wants to celebrate unfolds. Some of these young faces, mostly pissed, will reappear tomorrow night. They'll be older, maybe wiser. Some of them might even be semi-respectable, investing all that narco-loot in a wife and a family and a decent address.

Slowly piecing together this pageant, I begin to wonder about all our lives. Berndt and I never did drugs, not seriously, but showbiz and the whole creative shtick is just as narcotic. Applause, like a good review, can become addictive. You start to yearn for the approbation of your friends, for that arty three-page profile in *Cahiers* that says you've arrived, for the whisper at boozy gallery openings that you've been down to Cannes not once, not twice, but umpteen times. H and his army of marauding cavaliers were chasing their dream as hard as we were chasing ours. What right do I have to raise an eyebrow or wag a finger? To suggest that we were perfect parents? Proper human beings?

It's at this stage that Malo steps into the room. Andy, mercifully, is on his knees on the polished wood floor, arranging a grid of photos. I'm aware that this is priceless stuff from Mitch's point of view but I can't bring myself to get my phone out and take a shot or two.

Malo says that H wants to call a halt. He thinks we've done well. He wants us to get in the mood for tomorrow night. The parallel is obvious. I can't resist it.

'You mean some kind of rehearsal?'

Yes. And yes again.

We drink vintage Krug in the kitchen. Jessie has done two big bowls brimming with chips. We have a choice of sauces, red or brown. By ten o'clock, as H has doubtless anticipated, we're blind drunk.

After a final round of Jim Beam, Jessie steers Andy towards the door. Malo has already beaten a retreat, coshed by the champagne. That leaves me and H. Dimly I'm aware that this is a repeat of Antibes without the bougainvillea and the heat.

I'm sitting on H's lap. It's his idea and I'm trying very hard not to hurt him. He's telling me that he thinks he might be in love with me, which is definitely something I don't remember from the last time, but there's a cautionary, slightly provisional note in his voice.

'You are or you're not?' I let him suck my finger for a moment.

'Yes,' he says.

'Yes what?'

'Yes, please.' He starts to laugh. The laugh has an innocence I find wildly attractive because I've never heard it before. He's laughing like a child. He's laughing like Malo used to.

'You're pissed,' I tell him gently. 'It's nice.'

'You want to fuck me?'

'I don't know. Tell me why I should.'

The question seems to floor him. For the first time ever he doesn't know what to say.

'You weren't this way before.'

'You never told me you loved me before.'

'Might. I said might.'

'Yeah. And that's what I said back.'

'So what does it take?'

He's trying to juice me through my knickers. I remove his hand and give it a little squeeze.

'Is this some kind of negotiation?' I ask him. 'Some kind of business thing?'

'Not at all. I wish it fucking was.'

'Why?'

'Because you've got me. You've nailed me. That's not supposed to happen. Not to me. Not to H.'

Our faces are very close. He has a tiny smudge of brown around his mouth. I moisten a fingertip and remove it. Saucy, I think. HP.

'It's the Krug,' I tell him. 'You'll get over it.'

'I don't want to get over it. I want it to go on and fucking on. You don't have to do it. You don't have to fuck me. You don't even have to sleep with me. I just want you to be honest. I just want to know where we are.'

'We're here,' I tell him. 'You're Malo's dad and we're the ones who made it happen. Isn't that enough?'

'Yeah.' He pauses for a think. 'Yeah, it is.' He looks up at me. There's a lostness in his eyes. The last time I saw anything similar was with Malo, very recently, back in London. 'You're good with me, right?'

'I am.'

'You don't think I'm a twat?'

'Never.'

'You mean that?'

'I do.'

'And you'll stay here? Move in?'

'No.'

'Never?'

'I never say never.'

'One day then?'

'It's possible.' I kiss him properly this time. 'You're a good man. All you have to do is believe it.'

EIGHTEEN

No one gets up next morning until way past ten. I prowl around the kitchen trying to remember where Jessie keeps the Sumatra strength five coffee. Footsteps on the stairs falter, resume, then come to a halt. I find Malo in his boxers holding his head. He thinks he might be dying. He asks me to check. I know exactly how he's feeling. It's probably a genetic thing.

By mid-afternoon, with the exception of Malo, we're combat-ready again. The food is laid out under cling film, including the stuff that has to go into the oven, and H has liberated yet more bottles of Krug. Souvenirs from the old days have penetrated every corner of the house until it's started to resemble a museum. This seems to please H no end. All his life, I suspect, he's hankered after an event like this, an opportunity not just to showcase his precious house but to prove that he can father a son. He even appears to have forgiven me for not sleeping with him. Either that, or it's slipped his mind.

Our first guest arrives early. To H's surprise, it's Marie Mackenzie. She must be in her fifties but she's kept her looks. She's blonde and slightly gaunt, testimony to a gym addiction. She wears a Balenciaga three-quarter-length dress, in the subtlest shade of a burnt autumnal red, a woman with taste as well as poise, and I catch a whiff of Coco Mademoiselle as she offers H a token hug. By now I know that Bazza is dead, because H – very drunk – has told me so. At the first opportunity, once I've been introduced, I offer her a tour of the house.

It turns out that this woman has recognized me. She's a big fan of the kind of movies I've been involved in all my professional life, which – given what I know about Bazza's social life – comes as a slight surprise.

'You speak French?' I ask her.

'*Oui.*'

'And you liked the films?'

'Very much.' We happen to be in H's bedroom. She looks me in the eye. 'So what are you doing here?'

'I have a son by Hayden.' I see no point in lying.

'He's the boy downstairs?'

'Yes. A one-night stand. Eighteen years ago.'

She nods. Some of the iciness has gone. I sense she's beginning to warm to me.

'It's hard to miss the resemblance,' she says. 'Poor boy.'

'You think Hayden needs a makeover?'

'Not at all. It's the inside I always worry about.'

I blink. This woman was married to the legendary Bazza, who was by all accounts H on steroids. What right does she have to badmouth Malo's dad?

She's looking at a display of trophies neatly arranged on H's bedside table. They include a brown chunk of turf ripped from the Wembley pitch when Pompey won the FA Cup, and a beer glass liberated from the Café de Southampton.

'They were all crazy,' she murmurs. 'You know that, don't you?'

I nod, say nothing. She's looking at the bed now. Before she draws any conclusions I assure her that I sleep in a room on the next floor.

'Very wise. These men like to take scalps. There comes a day when you realize they're all dogs. They still follow their dicks, believe it or not, and you're talking to someone who knows.'

'You're still fighting them off?'

'I am. The professionals are the worst. Give a man a law degree and he thinks you can't resist him. There'll be some of them here tonight so don't say you weren't warned.'

'You still live in Portsmouth?'

'Of course.'

'So why don't you move, if it's so awful?'

'Why should I? It's my home. My kids are there. And their kids. Maybe it's become a habit. Maybe I've become too idle. But I happen to like it.'

'And now? Here? Why make an appearance?'

'Good question.' She walks across to the big window where H probably stands every morning, surveying his estate. She looks out at the view, silent, reflective. Then she turns back to me.

'My husband would have wanted me to be here,' she says. 'This is for his sake, not mine.'

I nod, say nothing, trying to weigh whether or not to put the obvious question. In the end I realize I have no choice. I may never get the chance again.

'So what happened?' I ask. 'To your husband?'

She looks me in the eye. Not a flicker of emotion.

'The Filth killed him,' she says.

'Filth?' I've heard H use this word, but somehow I don't expect it from Marie.

'The police. Down in Portsmouth. Bazza should have paid a little more attention. He could be a lovely man sometimes but he never listened. He was hopelessly outnumbered. He could have surrendered, handed himself in, but that was never his style. I understand that one bullet was all it took. Not that anyone ever tells you the truth.'

She shoots me an icy smile and moments later she's gone. I hear her footsteps on the staircase. Then comes a murmur of voices in the hall before the front door closes behind her. H struggles up the stairs and appears at the open door.

'What was that about?' he asks me. 'Why won't she stay?'

I wonder whether to be honest, to tell him that her take on the last thirty years might be at odds with his, but then I decide against it. Tonight is, after all, his party.

'Previous engagement,' I say lightly. 'Early drinks with friends.'

By mid-evening, the party is beginning to rev up. Malo, chastened by last night's Krug, is taking it easy on Pepsi with a splash of bourbon. I'm doing something very similar, except I won't let anything alcoholic near my lime juice. I drift from room to room with H in attendance, trying to put faces to the names on Jessie's list. One of the earliest guests to arrive has been Dave Munroe, the bent cop. He's easy to spot because he's in a wheelchair, testament – I imagine – to his failing liver. He's a balloon of a man, tiny head perched on a huge body, affable, garrulous, with a winning smile and a habit of patting everyone's extended hand. Once captured, it's difficult to get free but I squat beside his chair for a minute or two while he quizzes me about exactly what I've seen in H.

'He's the man of my dreams,' I tell him. 'Everything a girl could ever want.'

I get a special pat for that and a big wink. This is someone, I suspect, who loves complicity and can spot a lie a mile off. I fetch him a couple of sausage rolls and we agree to have a proper chat later.

By now, the big rooms downstairs are filling up as more guests pile in. Malo is manning the bar in the entrance hall. So far the gender balance is heavily skewed, men everywhere, middle-aged, heavy-set, baggy jeans and a variety of eighties T-shirts, and most of them go for the Stella. None of them seem to have wives or partners but according to H this is deliberate. They've left their women at home. Tonight, like the old days, is strictly for the blokes.

H introduces me to Mick Pain, his comrade in arms when they battled rival hooligans the length and breadth of the kingdom. Mick has a can of Stella in either hand and looks truly scary. He has a deep scar along the line of one cheekbone and far fewer teeth than the usual complement but in conversation he turns out to be a real softie. On weekends he works as a volunteer, helping to maintain an old fishing boat moored in a Pompey dock. He also has a passion for bird watching and he's telling me about a nature reserve on the Isle of Wight which is dreamland if you happen to like little egrets.

It's at this point that someone cranks up the music. Mick taps his ears and shakes his head, a clue that further conversation is impossible. Other guests are having the same problem so I set out to look for Andy, who spent most of the afternoon setting up the audio.

I find him DJ-ing in a corner of the hall towards the back of the house. He lifts one earphone and asks me what I think of his playlist. To be honest I'm no expert when it comes to Eighties indie music, headbanging or otherwise. I've never heard of The Beat nor The Pixies. Neither can I hear what Andy's trying to tell me unless I get very close indeed. I'm pressed up against him, trying to suggest that a little less volume might help the party along, when I catch sight of a face through a mill of guests. It's Jessie with a plate of sausage rolls. She's moving from group to group, trying to exchange banter, share gossip, catch up, but every now and then she shoots us the kind of look that leaves little room for ambiguity. She plainly doesn't trust Andy. And neither, after a promising start to our relationship, does she much like me.

A minute or so later, the sausage rolls gone, she's at our side.

'Hate to break this up,' she yells, then whispers something to Andy and disappears. This turns out to be a cue for H. Andy at last winds down the volume until the music dies under the buzz of conversation. Then comes a double blast on an air horn and H mounts an antique dining chair in the middle of the hall. He's calling the party to order. First he wants to introduce someone very special. I'm trying to attract Malo's attention, to warn him to take a bow, but it turns out that H means me. This is something I haven't expected. Whistles and cheers greet my name. I imagine that most of the people in this room haven't got a clue who I am but it doesn't seem to matter. Listen to H, and you might think I'm the toast of Hollywood.

'She's come to raise the tone,' he says. 'And about fucking time.'

At this point the main door opens and Gloria appears. This is the woman who's been making Dave Munroe so happy all these years. She's very black, with a huge chest and a Caribbean dress the colour of sunset that's probably visible from the moon. In tow, behind her, come a bevy of girls, mainly Asian, laid on as a present for H. The fact that they're all dressed in Pompey strip – red, white and blue – raises another storm of applause. Gloria blows H a kiss and says she's sorry for being late. Borrowed minibus. Puncture on the motorway.

The girls begin to circulate while H gets on with his speech. This time it's Malo's turn. H doesn't bother to gild the lily.

'Nearly twenty years ago,' he says, 'I had the pleasure of sharing an evening with this lad's mum. Fuck knows how but he turns out to be the junior version of me. Malo, son, you're very welcome in our house. And you know why? Because it's your house as well. And these people here tonight will be your mates forever. Because they'll make it their business to look out for us both.'

This might not play well in cold print but tonight, in this company, H's little announcement truly touches a collective nerve. His mates aren't used to seeing H so natural, so sincere. There's a kind of humility, as well as pride, in the way he doffs his cap to our son and I know from the expression on his face that it means the world to Malo. At the news that H has become a dad, there's a stir among the watching faces, first astonishment, then delight. Dave Munroe calls for three cheers for Malo. The roar that follows fills the house.

But H hasn't finished. He has one last thing to say.

'Not everyone has turned up.' He's looking round. 'And that's their fucking loss. But there's one guy that should be here and will never make it. So a minute's silence, boys and girls, for the one man among us that we'll never forget.'

'Bazza.' This from Dave Munroe.

Heads bow. Feet shuffle. The Asian girls stop giggling. Then there's total silence. I'm still thinking about Marie, and her errant husband's private Alamo, when H's head comes up. This is a signal to Andy. He hits a button on his master board, pushes the volume paddle to its limits, and suddenly we're listening to Tina Turner.

I nod. I never met Bazza but it's hard not to choke up. *Simply the best.* What could be more perfect?

People are singing now. They're world class on the lyrics. They belt it out, line after line, linking arms, swaying with the music, punching the air as the song explodes towards the big finish, and then they throw their arms around each other as the music dies.

'Fucking wonderful,' H says, as he dismounts from the chair. And he's right.

Much later, past midnight, I find myself alone in the barn with Tony Morse. Tony is the gang's solicitor, the sleek legal eagle charged with transforming decades of Bazza's narco-loot into money he can safely spend without courting arrest. In his prime, I suspect, Tony was a real looker. He's still attractive – tall, nearly slim, with a mane of greying hair and greeny-blue eyes he uses to some effect. He's also wearing a beautifully cut tuxedo with a red velvet bow tie, which certainly marks him out from the sea of denim we've left behind in the house. I especially like his cuff links, solid gold, weighty, yet somehow discreet.

'Why the dolphin motif?'

'They were a present from Bazza and Marie. Marie was the one with the taste. Bazza wanted her to find something that would appeal to me and this is what she came up with.'

'Was she right?'

'She was. I like company, just like dolphins. I'm also partial to crowds. If you're practising as a criminal lawyer you have to be a bit of a performer. You're in court. Whether it's the magistrates or a jury you have to make an impact, you have to get people onside. It's persuasion. It's hearts and minds. It's what dolphins do. Am I ringing any bells here?'

'Of course you are. It's exactly the same in my business. Except no one's life hangs in the balance.'

Like me, this man is stone-cold sober. It was his idea to get out of the house, mine to wander into the barn. Buy enough booze, play the right music, invite like-minded people, and there always comes a moment in any party when things start to get out of hand. Judging by the noise we can still hear through the open windows, we've made exactly the right decision.

'You're driving back tonight?' I ask.

'Sort of.'

'What does that mean?'

'I've got a room booked in a hotel in Shaftesbury. You're more than welcome if it appeals.'

It's a civilized invitation and I appreciate the absence of pressure. Stepping away from the house I'd briefly watched two of the Asian girls servicing a lone partygoer in the back of Gloria's minibus. I suspect that H would have preferred them to use one of his many bedrooms but old Pompey habits obviously die hard.

Now I ask my new friend about the wilder extremes of Bazzaland. People, I say carefully, talk about serious violence. Are they making it up? Or did the likes of Wesley Kane really play the Enforcer?

The question seems to amuse him. He lights a cigar, looks for somewhere to lose the match among the loose hay without setting the barn on fire.

'Wesley was an animal,' he says. 'Still is. Did they tell you about the house they bought him in Fratton?'

'They did.'

'He's still got it. He's off the reservation now, freelancing for anyone with debts to settle. Last time I heard he was into kettling.'

Kettling Pompey-style, it turns out, has nothing to do with riot control. It needs nothing more than an electric kettle, a chair, and a length of rope. First you do a spot of tying up. Then you boil the water. Then you ask a question or two. If the answers aren't to your liking, the water goes over the guy's lap.

'It's very effective. Especially when you drive him to hospital and dump him outside A&E. By this time, if you've got it right, the skin is off his todger. People who know Wes tell me he loves punch lines. He thinks that's one of the best.'

'I think it's horrible.'

'That puts you in a minority of just two.'

'Me and you?'

'Indeed.'

I gaze at him for a moment and then ask about what exactly happened to Bazza Mackenzie.

'He died. Took a bullet for his sins. There were cops in Pompey still celebrating a week later.'

'But how? Where?'

'There was a bent cop called Paul Winter who'd gone over to the dark side and decided to work for Mackenzie. In the end they fell out. Big time.'

'Why?'

'Mackenzie thought Winter was screwing his wife.'

'You mean Marie?'

'Yes.'

'And was he?'

'No. But Winter was clever. It was his job to light Mackenzie's fuse and that's exactly what he did. They ended up in the middle of the night in a shop that sold reptiles, mainly snakes. Deeply appropriate.'

'And?'

'It was all a sting. The cops had wired Winter. By the time they got there, Mackenzie had Winter all trussed up. He'd lost it completely. As I understand it, he was about to fill Winter's mouth with expanding foam, the stuff you use to seal double glazing. One verbal warning to Mackenzie, and then they shot him dead. Bang. End of.'

'How do you know all this?'

'The officer who pulled the trigger told me later. Over a drink.'

'You always talk to both sides? Play both ends against the middle?'

'Of course.' He's smiling. 'I'm a lawyer. That's what lawyers do.'

I'm still staring at him, totally hooked. First kettling. Now death by expanding foam. Tony takes my arm and walks me towards the big open doors and the fresh air. Further away from the house, we pause in the shadow of a huge elm tree. The wind's got up and tiny lanterns are dancing against the blackness of the sky. Tony's head is back. He draws on the cigar and expels a plume of blue smoke. I've always loved the smell of cigars.

'One more question,' I say softly. 'Then I'll let you go.'

'That would be a shame.'

'It's about H.'

'Surprise me.'

'Did he ever kill anyone?'

There's a long silence. My arm is still linked through his. From outside the house comes the clunk of the minibus door sliding open. I'm still waiting for an answer.

'No,' he says at last. 'Never. Not to my knowledge.'

'Did he ever come close?'

'That's a different question. Why do you want to know?'

'Because H intrigues me. And because he's the father of my son.'

'You've got your doubts about young Malo? You think he might fall into bad company? Take to killing people?'

'Not at all. Doubts? No. Hopes? Plenty.' I put the gentlest pressure on his arm. 'H is a good man. When people tell me otherwise I need to know why.'

'Maybe they have their own debts to settle. Theirs was a rough old world. They broke every rule in the book and they certainly hurt people. That came with the territory. What I'm not going to discuss are individuals. To work for Bazza was to pay your dues and button your lip. If you think it's a Mafia thing you wouldn't be far wrong. *Omerta, capisci?*'

Omerta. The code of silence. I nod. I understand.

I hesitate a moment. Part of me says I owe Mitch nothing. But one more question wouldn't hurt.

'So you're not going to tell me? About H?'

'Of course not.' He smiles down at me. 'And that's for your sake, not mine.'

NINETEEN

The morning after is predictably glum. I come downstairs to find bodies everywhere. In a movie this would be a crime scene, lots of scary music and lingering close-ups. Bodies curled in corners. Bodies hogging sofas and armchairs. Even a body glimpsed through the kitchen window, gently swinging in H's favourite hammock. Of Gloria and the Asian lovelies there's no sign, and when I check outside the minibus has gone.

Back in the house no one has stirred. Then Malo joins me. Between us we lay hands on a couple of big shopping bags and start the clean-up. Thank God Andy invested in all those plastic glasses. With no need to trouble the dishwasher, we move from room to room, picking up cans, smeary paper plates, olive stones, plastic knives and forks. There's been a major crisp spillage in a corner of the library. I despatch Malo to look for a dustpan and brush while I take a wet cloth to a mysterious stain on one of H's fluffy white rugs. When I sniff my fingertips it turns out to be lime pickle.

H himself appears at exactly the moment when we're starting to win. He must have had some kind of accident last night, or maybe took a fall, because he's walking with a limp. He's wearing a pair of black pyjamas. He gives the nearest body a poke with a bare foot. The body grunts, rolls over, goes to sleep again.

'Scrambled eggs,' H grunts. 'I'll sort these tossers out.'

Malo and I retreat to the kitchen. Someone, probably Jessie, has had the foresight to invest in three big trays of eggs. Malo finds half a dozen loaves of sliced white in the pantry while I raid the fridge for butter and marge. In minutes, we're ready to feed the small army of casualties next door. Waiting for our first customers to appear I tell my son we could make a living out of this.

'Disaster relief,' I say. 'All we need is a big tent and a couple of film crews.'

Malo shoots me a look. He isn't convinced.

The first faces appear, pale, largely silent. I do the eggs while Malo butters the bread. Soon the half dozen chairs around the kitchen table are all occupied and more and more people are piling in. Most of them want nothing but coffee. H takes it upon himself to spoon instant coffee into the plastic glasses we never managed to use last night and then gets the kettle on. Watching him pour the boiling water it's hard not to think about Wesley Kane.

With the kitchen full, H takes the opportunity to make another little speech. Last night was great, he says, but nothing comes for free. Later today he'll be circulating details of another little wheeze he's dreamed up to raise some moolah for Front Line. This time he has no plans to end up in hospital. The joke raises a chuckle from Dave Munroe but no one else has a clue what H is talking about. Pressed for details, he says he just wants people to pile in and help.

'I'm after six volunteers,' he says. 'At five grand a pop. If you can't manage that kind of money then find someone who can.'

Five grand for what? He won't say. Just wait for the email, he says. Just be fucking patient for a change.

Later in the day I get a chance to corner him. Everyone but Dave Munroe has pushed off. Dave's waiting for his carer to arrive and has holed himself up in the library with one of H's true-crime books. I fetch him a coffee and we have a chat. H has told me he lives on the Isle of Wight. How come?

'No choice, really. My lovely colleagues always suspected I was on the take but no one could prove it so once I retired I didn't want to hang about in Pompey. I had a thing going with this woman down in Ventnor. She ran a B&B and the invite was there to move in. So I did. Best move ever. She looks after me a treat. Heart of gold. Lucky me, eh?'

When he asks for sugar, I return to the kitchen and find H sitting at the table. He says he wants a word or two about Malo. I take the sugar to Dave and rejoin H.

'It's about this project,' he says. 'You remember in the pub the other night we agreed he needs something to get stuck into? Something to put his smell on? I think I've got just the thing.'

Put his smell on. Very H.

'And?'

'I was talking to Mick Pain. Turns out he works on an old boat in the Camber Dock. Makes sense in a way. He's a decent chippie.'

The boat, H says, is a Brixham trawler, old as you like, lovely thing according to Mick. It's owned by some charitable trust or other and they're always looking for ways of trying to raise money to cover the upkeep.

'I thought you said Front Line?'

'I did. Here's the deal. They can take six paying guests. They charge a grand each for a long weekend. We find six punters, five grand each, and off we go. We pay the trust people their whack and the rest goes to Front Line. That's more than twenty grand if I've got the sums right. Not bad for a weekend's work.'

'Go where?'

'France. Normandy. This is November. We push off on the tenth. The eleventh is Remembrance Day. We take lots of flowers. We look smart. We find a war memorial on one of those invasion beaches, and pay our respects. Then we have a beer or two and come home.'

I nod. I see the logic. Clever.

'It's mustard,' he says. 'And the boy will be doing the work, keeping it together, keeping the show on the road. I've already mentioned it. He loves the idea. Big fucking smile on his face.'

'Five thousand pounds is a lot. What makes you think people will pay that kind of money?'

'You.'

'Me?'

'You're coming, too. Star of the show. Three days with a movie queen? Cheap at the price. Can't fail.'

I'm speechless. I've never much liked boats, and I detest being the centre of that kind of attention. Neither reservation makes any difference. H will be booking the boat tomorrow. Everything else falls to Malo.

'Do it for him,' H urges me. 'You can't let the boy down.'

I shake my head. Try to think of other excuses. But H is right. If Malo is to spend his days doing something useful then this sounds near-perfect.

'This boat's got a name?'

'Yeah. *Persephone*.'

'*Persephone*?' I start to laugh. I did a Greek tragedy once on a bare clifftop stage near Land's End.

H wants to know what's so funny.

'You know about Persephone? You know who she was?'

'Tell me.'

'Queen of the Underworld.'

H is staring at me. Then he shakes his head as if he can't quite believe it.

'Brilliant,' he says. 'You couldn't make that up.'

For the rest of Sunday we're in a bit of a trance. H mutters about Post Traumatic Stress Disorder and Dave Munroe departs when his specially adapted van arrives to collect him. That evening, deep into a showing of a Clint Eastwood movie, I get an apologetic email from Rosa, my agent. The actor playing against me in *Going Solo* has recovered. The producer wants to start rehearsals on Tuesday. Can I manage that? I text her a yes and mention it to H once the film's over. Either he doesn't hear me properly or he doesn't care.

'Whatever,' he says.

On Monday, Jessie once again runs me to the station in Dorchester. I try and dream up ways of convincing her that Andy hasn't got into my knickers but I sense she's in no mood to listen. Only when I mention that I might be away for a while does she brighten up.

'Take your time,' she says. 'We'll look after Malo.'

Back in London I phone my neuro-consultant and explain the issue over the insurance for the Montréal movie. He's sympathetic but can do nothing until I've had the three-month scan. Only then can he be sure of my prognosis. I relay this news to Rosa, who isn't best pleased. I suggest that maybe we can skate round it somehow. She says she'll try.

I spend the evening going through the BBC script. Rehearsals start the following morning in a basement studio in Broadcasting House. I've never done radio before and I like it a great deal. It feels very intimate and I love playing from a script I can see in front of me. The producer is young, fresh out of Cambridge, and he reminds me a little of Berndt in his earlier days. He has a delicate way of suggesting changes of nuance and tone in delivering particular lines and I quickly learn to trust him. By lunchtime we've made a great deal of progress and the afternoon goes so well that he thinks we might make a start on the recording a little earlier than planned.

That night I take Evelyn to a play I know she wants to see at the Harold Pinter theatre. In a way it's a thank-you for all the support she's given me over the past few weeks and I know she's touched that I've managed to lay hands on a couple of tickets. The play – *Oslo* – is excellent, richly deserving the rave reviews in every corner of the press, and she insists on treating me to a late supper in an Italian place she knows in Covent Garden. The place is buzzing and we linger over coffee and a couple of brandies after the meal. I tell her about Malo, and a little about the party at the weekend. A family reunion doesn't really do it justice but that's the story I'm sticking to. Evelyn might be incredibly well-read but deep down I suspect she's a woman of tender sensibilities. I don't want to shock her.

'So what's this Hayden person like?'

It's hard to do justice to H. In a way he belongs among the pages of a novel.

'He's larger than life,' I tell her. 'He's bold and he's loud and what you see is pretty much what you get. But Malo seems to think the world of him and that's good enough for me.'

'Malo's happy down there?'

'Very.'

'That's good. I'm glad.'

She asks about my journalist friend. She's met Mitch on a number of occasions and I think she approves. I tell her we met recently for lunch but again I spare her the details.

'You don't think you two might . . .?'

'No.'

'Shame. I just wondered . . .'

'I'm afraid not. Just now I'm trying very hard just to stay in one piece. You've no idea how wearying that can be.'

'And the radio thing?'

I tell her about today's rehearsal. The sound effects, in particular, I find astonishing. Half close your eyes during various flashback sequences and you could have been living in the Blitz. This makes Evelyn laugh. She settles the bill and I splash out on a cab back to Holland Park. We say our goodbyes outside her door and I give her a hug.

'Thank you,' I say, 'for everything.'

When I let myself in my answerphone is flashing red in the darkness. I turn on the lights and press the play button to access the message. It's Mitch. He's been trying my phone all evening but I've obviously got the thing switched off. We need to talk. It sounds urgent.

I reach for the phone, then have second thoughts. A check on my mobile reveals two more messages, both from Mitch, but he's right, I've been unobtainable. I fetch a glass of water from the kitchen and make myself comfortable on the sofa. I know I should phone him. I know it's what he wants. But there's something inside me that is weary of playing the agent in enemy territory.

I'm sure that H has been up to all kinds of mischief in his life, but as far as Malo and I are concerned he's been nothing but generous. Is it really my job to hazard all that? No matter what Mitch tells me about UKIP, and the perils of Brexit, and all the rest of the doomster nonsense? Shouldn't I simply be grateful that fate has given my son a father who really cares about him?

I think of the trawler down in Portsmouth and my adolescent boy suddenly handed responsibilities he could never have dreamed of. A couple of weeks ago he was slumped on this same sofa, totally Zombied. Now he's tasked with extracting a great deal of money from people he barely knows. That's a steep learning curve, something Berndt would never have dreamed up, and when I've drained the glass I decide to call it a night. Mitch will have to wait. Maybe I'll phone him first thing tomorrow. Maybe.

I don't. For whatever reason I fail to hear the alarm clock and by the time I'm properly awake I have barely an hour to make it to Central London. I swallow a bowl of cereal and skip my make-up. By the time I make it to the basement studio, the rest of the cast and production crew are already eyeballing the day's first scene. I mutter my apologies and slip into my chair in the recording booth.

My script is still on the music stand where I left it. I find my place and we kick off. Mercifully the male lead has a protracted coughing fit which gets us all wondering whether he's really better. The producer fetches water from the nearby washroom. It doesn't do the trick and so the producer calls a break while the male lead pops out to the nearest chemist for throat lozenges. With nothing better to do I ask the recording engineer if I can take a peek at his paper. It happens to be the *Guardian.*

I briefly scan the first couple of pages and I'm about to turn to the arts section at the back when I take a second look at the photo beneath the main story on page four. It shows a UKIP politician snapped on his own front doorstep. The last time I saw Spencer Willoughby he was roaring drunk at H's dinner table. Now he's opened the door to a press photographer and he looks less than pleased.

The accompanying story has been lifted from Mitch's Finisterra website. It reveals that UKIP may be standing aside to give Tory candidates a clear run in the next general election. Not only that but sizeable donations to the Kippers are alleged to be ending up in Conservative Central Office. The most recent donation is alleged to be half a million pounds. More details on the website.

My heart is sinking. I use my smart phone to access the full story from Finisterra. An old photo of H shows him inspecting a vast gin palace at the Southampton Boat Show. I'm not aware of any plans on H's part to acquire this trophy motor cruiser but that's far from the point. This single image has done untold damage to the Tory party, to UKIP, to Spencer Willoughby and to H himself. Anyone who suspects that democracy has been handed to the fat cats need look no further. My eye drifts to the Finisterra byline. The name of the journalist responsible for this damning exclusive? Mitch Culligan.

Rehearsals begin again with the return of the male lead. Mid-morning, the producer calls a break. I step out of the studio and call Mitch. I've read the story again and I'm beyond angry.

I ask him what the fuck's going on.

'I've been trying to contact you,' he says.

'I know.'

'Since yesterday.'

'Yes. So what did you want? Were you trying to warn me?'

'I needed to let you know.'

'And if I'd objected? Asked you to think again? Asked you not to publish?'

'That might have been difficult.'

'*Difficult?*' I'm ready to explode. 'I gave you that information on trust. You told me it was for your book. That's what I believed. That's what you said. A book we can sit down and talk about. In a book you can put all this stuff at arm's length. Now I find it all over the fucking internet. And it's wrong, by the way.'

'What's wrong?'

'The figure you quoted. The half million pounds.'

'That's what you told me.'

'I said "may". I said "might". I said I couldn't be sure. If you want the real figure it's five hundred quid. That might keep UKIP in beers for an evening or two but it's not half a million. You should have checked.'

The phone goes dead. He's cut me off. Bastard. Then I start to wonder about H. Did he really mean five hundred? Or has he been stringing me along as well? These questions are impossible to answer, especially now. The studio manager is beckoning me inside. My day job awaits. Shit, I think. Shit. Shit. Shit.

We break for lunch at one. So far cast and crew have shared adjoining tables at the Broadcasting House canteen. On this occasion I skip the Beeb salad and head out to the street where I can try and get my teeming thoughts in order. When I check my phone this time there are a couple of texts, both of them from people I've never heard of. I scroll slowly through them. They're both demanding a call back, I've no idea why, then I pause on a name I recognize. Spencer Willoughby. The message is blunt. I've grassed him up. He thought he was in discreet company at H's table and it turns out he wasn't. Very shortly I may be hearing from his solicitor. Over and out.

Worse and worse, I think. Little me in the middle of a media shit storm with nowhere to hide. I phone Mitch again. He apologizes at once for hanging up on me.

'Someone more important to talk to?'

'Something like that.'

I ask him again why he chose to publish now and not later. This time he's blaming his publisher.

'They wanted advance publicity,' he says, 'for the book. It's Theresa May's big speech tomorrow, up in Manchester. We needed to get in her face.'

Part of me marvels at the little party tricks these people get up to. Shafting the Tories. Winding up the Maybot. The rest of me is still very angry. Angry and betrayed.

'Betrayed? How does that work?'

'I trusted you.'

'Trusted me how?'

'To protect our interests. Mine and my son's. I won't bother you with the details but I dread to think where this thing is going to end.'

'I don't understand.'

'Yes, you do. You asked me to keep my eyes open. To look around. To report back. And that's exactly what I did because I assumed you'd act like any decent human being when it came to using any of this stuff. Instead, I find myself all over a national paper.'

'Nonsense. You're not mentioned anywhere.'

'And you really think that matters? You really think there aren't people out there who might draw the odd conclusion?'

I tell him about the text from Willoughby.

'So what's he going to do about it?'

'I've no idea. Until his solicitor tells me.'

'He's brought in the lawyers? Already?'

'So he says.'

'He's bluffing. They all do that. Forget it.'

Forget it? I explode on the phone. Tell Mitch what a snake he is. All the time I thought he had my best interests at heart. Now this.

There's a long silence. I'm watching a couple snatching a kiss across the road and for a moment I wonder whether Mitch has hung up on me again. Then he's back on the line.

'This is about Prentice, isn't it? And about Malo?'

'I don't know what you mean.'

'Is he your son's father? All I need is a yes or a no.'

'Why? What would you do with either?'

'That's immaterial. I just want to get the story straight. Do I hear a yes?'

'You hear nothing.' It's my turn to hang up.

I feel like crying. I feel like setting off west and taking the long way home. I feel like spending the entire afternoon blanking myself off from my phone, from my army of new friends, and from anything else that can hurt me. Instead, as I must, I go back to the studio.

By early evening, we've made a decent start on the recording. Another day and a half should see it finished. The male lead suggests a drink in a pub he knows nearby. Exhausted, I accept.

He buys me a large gin and tonic. He settles into the chair across the table. He's drinking Guinness.

'You look terrible,' he says.

'Thanks.'

'What's the matter?'

I shake my head. Everything is blurring. Everything is slipping out of focus. Unusually for a man, he's carrying a handkerchief.

'Here.' He's trying to be gentle. 'Blow your nose.'

I try and dry my tears. It's hopeless. He slips into the chair beside me and I feel myself collapsing against him. Pathetic, I tell myself. Pathetic and unnecessary. He puts one arm round me, offering comfort. Mercifully, he doesn't ask me what's wrong.

The great thing about London is that nobody else seems to care. I'm not talking about the male lead. I'm talking about everyone else in the pub. I'm sobbing my heart out and no one even looks up from their smart phones.

When I've finally stopped crying and got some kind of grip I bring this to the attention of my new companion.

'Don't you think that's remarkable?' I ask him.

He smiles and asks me whether I'd prefer otherwise. Anonymity is bred deep in this city's bones. Speaking personally, he thinks that's rather wonderful.

'But you're an actor,' I say. 'And we're all show-offs. It's in the job description. The last place we need to be is off the fucking radar.'

The fucking radar is a phrase he says he'll cherish. Praise is a currency actors tend not to waste on each other and for the first time I start taking a proper look at this man. In the studio, as scripted, he's pulled off an American accent with an underlying hint of German and done it with some style. He's maybe five years older than me but he laughs a lot and has a fund of good stories about his days in Australian soaps. Now he's wondering about another drink.

'Yes, please.' I fumble for my bag. 'And make mine a double.'

We stay in the pub most of the evening. Supper is three bags of crisps, one of them shared. When he asks me where I live he says his flat is a lot closer. I've told him nothing about Malo, about H, about Flixcombe Manor, about Mitch, about Spencer fucking Willoughby, about Brexit, and I feel a great deal better. Drunk but better. His flat turns out to be a ten-minute walk away. It's small, cosy and very male. As we strip, I establish the rules of engagement. We'll fuck until we go to sleep and at some point after that I'll get up and go home. No commitments. No regrets. Not a whisper of this to anyone else.

'Perfect.' He's on his knees, easing my thighs apart. He has a busy tongue and he knows exactly what he's doing. Hours later, dressed again and waiting for the Uber, I wake him up.

'How come you can afford an address like this?'

He peers up at me, still groggy with the Guinness.

'Australian soaps,' he says. 'You won't believe the residuals.'

Next day, in the parlance, we put *Going Solo* to bed. Neither I nor my male lead betray the slightest evidence of our night together and after a second take on the final scene we all share a couple of bottles of Prosecco brought in by the producer. We toast the health of the production, offer effusive thanks to the producer for his hospitality, have a hug or two, and make for the street. It's only on the bus home that I realize that in all likelihood I'll never see my one-night stand again. I know his name, Desmond Reilly, I dimly remember where his flat is, but I have no contact details. Nice man. Gifted lover. Amusing stories. End of. Of this, believe it or not, I'm strangely proud and it's only when I'm walking home from the bus stop that I realize the real irony. Eighteen years ago something similar happened in Antibes. And look where that took me.

Back in the apartment, I call Malo. I want all his news, especially how he's getting on with Project *Persephone*, but as well I'm keen to know about H.

'He's fine. Just like always.'

'Not upset in any way?'

'Not at all. Why should he be?'

I know I can't take this line of enquiry very much further, not without betraying my hand. Malo, it turns out, is in Portsmouth with Andy and H. They've spent most of the day aboard the old fishing boat, talking to the people from the charity. H has secured a handy discount on the hire fee, releasing more money for Front Line, and Malo has lots of photographs he can use for his presentation to would-be passengers. I'm about to ask him what happens next when he puts me on to H.

'You OK?'

I tell him I'm fine. Malo's right. He sounds as blunt and bouncy as ever. Maybe he never reads the *Guardian* or Mitch's fucking website. Maybe he hasn't had an earful from Spencer Willoughby. Maybe the media shit storm has passed him by.

He wants to know how the BBC thing went. I tell him it's in the can.

'They're pleased?'

'I think so.'

'Anything else you have to do up there?'

'Not much.'

'So when are you coming home?'

'Home' is the key proposition here. He's used the word before, glancing references *en passant*, but this is the first time he's put it in lights. Home. Unadorned. Unqualified. Unambiguous. The place where you hang your hat. The place where your heart is. Home.

'I'm at home now,' I say lightly. 'You must come up some time, pay me a visit. It's not Flixcombe Manor but I'm very fond of it.'

'We miss you,' H says. 'Especially the boy. Exciting times, this *Persephone* thing. We need to get our shit together.'

I agree. We certainly do. There's a hiatus, a moment or two of silence, slightly uncomfortable. Then he's back.

'That's a deal, then. Friday? End of the week? Just the three of us? No Tina fucking Turner? Give me a bell. All I need is a train time.'

He's gone. I'm back with my son. It seems he's very busy. He has to find out about the weather in mid-November and the shipping lanes off the French coast. The people he's talking to are very nice and – just for the record – he's loving what he's seen of Pompey.

Pompey? He uses the term as if he's lived there all his life. He uses it the way you might refer to a much-loved relative. Is this genetic, too? Something he's inherited from his dad?

'Take care,' I say. 'See you soon.' But already he's gone.

As it turns out I'm on the train again, as ordered, on Friday. Even in the early afternoon London is emptying ahead of the weekend and every carriage is packed. At Dorchester station, to my relief, there's no sign of Jessie. H is alone at the wheel.

'Your doctor's signed you off?' I ask him. 'No issues with driving?'

'Issues? You want to take the bus?'

We leave Dorchester. He drives fast but the deeper we get into the countryside, the narrower the roads, the more carefully he gauges every corner. This is a man, I remind myself, who's successfully eliminated every possible risk in his life. Not from an excess of caution but by being consistently smarter than the enemy. His luck may have run out on the trail bike but when it comes to dealing with threats on two legs he appears to be beyond reach.

We chat about Malo, about *Persephone*, about the job our son's doing on the task H has handed him. H has located a graphics consultancy in Yeovil, people with a real flair for marrying images and text. Malo spent most of yesterday with them and came back with some impressive ideas.

'They've sourced a load of pics of your lovely self,' H says without a hint of irony. 'We thought we'd leave the final choice to you.'

I nod, say nothing. I want, very badly, to find out whether he saw Tuesday's *Guardian* but I can't think how to ask. Minutes before we hit the turn-off for Flixcombe Manor I enquire whether he saw anything of May's big speech at the Tory conference. A joker in the front row presented her with a P45 when she was in mid-flow, a clip that made countless news feeds within seconds.

H saw it. Thought it was funny. Along with the cough sweet and bits of the backdrop falling apart. He especially liked the man with the P45.

'Same bloke presented Trump with a bunch of golf balls when he was over here on business. Each one had a swastika on it. That's my kind of politics. Get under their skins. Give the bastards a poke. Everything goes tits-up when you take these clowns seriously.'

'And does the same go for Willoughby?'

'Willoughby is a prat. The field's pretty thin just now but he can't even make it as a politician. That man's out of his depth in a fucking puddle.'

And that's it. No more mention of politics, of *Guardian* exposés, of fictitious six-figure donation to UKIP funds. Case closed.

We spend a very civilized weekend helping Malo shape the pitch he wants to make to lure paying passengers aboard *Persephone.* As H promised, the graphics people in Yeovil have found umpteen photos of yours truly and I spend one of the nicest evenings of my life in the downstairs library with my son as he goes through them. What were you doing in Rome with Liam Neeson? How old were you when they shot the adaptation of *The Reichenbach Falls*? What does it feel like to walk that red carpet into the Leicester Square Odeon? I answer each of these questions as best I can, aware of H cocking an ear in the background. Many of these locations I've virtually forgotten, lost beneath the drift of the years, but when we get to a shot towards the bottom of the pile, the memories put a smile on my face.

We were shooting, in all places, on the Isle of Wight. It was a biopic about a famous Victorian photographer, Julia Margaret Cameron, and I was playing one of the younger models she used to perfect various lighting techniques. It was a lovely time of year, late spring, and we were staying in a hotel – the Farringford – which had once belonged to Tennyson.

My role was undemanding and I treated the location a bit like a holiday. I'd get up every morning shortly after dawn and take the path down through the fields and up on to the long fold of chalky turf that stretched west towards The Needles. It was called, appropriately enough, Tennyson Down and I'd spend a glorious hour or two in the company of rabbits, marauding seagulls, and the odd hare without once laying eyes on another human being. Those were the weeks, I tell Malo, when I truly fell in love. Not with someone else but with that sense of kinship, of belonging, of being mysteriously entwined with something infinitely bigger than I'd ever be.

I'm not sure Malo really understands what I'm trying to say. Neither am I making an especially good job of describing what, in the end, became moments of transcendence. But after Malo discards the shot of me dressed up as a nineteenth-century photographer's model, and moves briskly through the photos that remain, I catch sight of H sprawled in his armchair. He offers me a nod of what I suspect might be approval. Praise indeed, I think.

On the Sunday, H takes us all out for lunch. It's a different pub this time, more down-to-earth, and Jessie and Andy come too. Jessie, for some reason, seems to have become more comfortable in my company. I can't say she's forgiven me because there's nothing to forgive, but we manage a half-decent conversation when we're waiting for the mains. The men are talking football. Pompey, it seems, are struggling a bit but Manchester City, according to Andy, are a joy to watch.

Jessie rolls her eyes. She's never had much time for football, which must be a problem if you're living with Andy, but she tells me about another expedition she made only days ago. The Romsey interior design place she visited when I was last down turned out to be a real disappointment but she's found an interior design outlet in Salisbury that has exactly what she's been after. I listen to her rhapsodizing about granite worktops, and soft-close drawers, and clever solutions to crockery storage, and all the other bits of kitchen porn, and agree that our parents' generation were lucky to survive without a walk-in freezer. Then, as the men's attention turns to the approaching waitress, I feel the lightest touch on my thigh.

'Watch out for yourself,' Jessie says softly.

Watch out for myself? I fork my way through a plateful of slightly overcooked scampi and wonder what on earth she means.

Andy and H have done football by now and we're all discussing last weekend. A torrent of emails, most of them obscene, have been flooding in after the party and as far as H can judge it seems to have been a roaring success.

'Only you would know,' I suggest.

'Yeah?' He turns to eyeball me. 'And what's that supposed to mean?'

I shrug. I've meant no offence. All I'm saying is the obvious. That these are people H has known most of his life. They're friends, buddies, mates, brothers-in-arms. They must have been through a lot together and a closeness like that can only breed loyalty. No wonder the weekend went so well.

H frowns. He seems to be exploring the notion in his head, putting it to the test. Finally he picks up the remains of a lamb chop and gives it a chew. Then he licks his fingers and gives his mouth a wipe on a paper napkin.

'You're right,' he grunts. 'Loyalty? Nothing fucking like it.'

I'm back in London for the start of the new week. Jessie's warning still bothers me but I'm doing my best to ignore it. I spend a lovely afternoon drifting around the Singer Sargent exhibition at the Dulwich Gallery and return with a bag of crumpets from the Londis round the corner. I make tea, toast the crumpets, watch the evening news, try calling Malo. He's on divert so I leave a message asking him to phone, nothing urgent, then settle in to let the evening drift by. By ten I've decided I deserve an early night. By half past I'm asleep.

It's five in the morning when the phone rings. Thinking it might be Malo, I fumble for the mobile. It's Mitch. He sounds terrible.

'What's happened?'

'We have to talk.'

'Where?'

'My place. Soon as you can. Please.'

TWENTY

I'm in Hither Green within the hour. The Uber drops me outside Mitch's house. Dawn is breaking over the rooftops across the road. Mitch has the front door open before I've even made it through the gate. He's fully dressed and he has the look of someone who's been up all night. Exhausted. And fearful.

'What's happened?'

He gets me into the house and checks the street left and right before shutting the door. I hear the slide of two bolts. Then we're in the kitchen at the back and I'm looking at what appears to be a business card on the work top. I recognize the Met Police logo. Not a good sign.

'Sayid,' he says.

I put my arms round him. I think he needs to cut loose, let everything out. I tell him to take his time. Deep breaths.

He nods, tips his head back, gulps. I suggest we both sit down. Does he need a drink?

He shakes his head. No drink.

'I got home around ten,' he says. 'I couldn't understand why the lights weren't on. Sayid should have been home. He never runs on a Monday. Never.'

He let himself in and looked for a note. Nothing. He waited ten minutes or so, thinking Sayid might have gone out for milk or whatever, and then he phoned the care home. The manager on duty said he'd left as usual around seven. As far as he was aware he was heading straight back home.

It was now half past ten. Sayid, he tells me, always left his shoulder bag down beside the sofa. When Mitch checked, it wasn't there. This was seriously worrying. Sayid has very few friends. He likes his own company. Apart from running, he rarely goes out.

'I was tempted to call the police,' he says. 'But they get funny around people like Sayid.'

'So you waited?'

'Yes.'

'And?'

'The phone went. It must have been around eleven. Caller ID said Sayid. I can't tell you how relieved I felt. Then I realized it wasn't him. It was another voice. Nothing like Sayid. He said to go to the end of the road. It's a cul-de-sac. There's waste land at the foot of the railway embankment. The kids are always up there. Drinkers, too. Loads of empty cans.'

'What sort of voice?'

'Male. Rough.'

'Accent?'

'Might have been London. Might not. Hard to tell.'

Mitch left the house. He's not built for running but he says he was at the end of the road in seconds and I believe him.

'The street lighting's shit up there. I should have brought a torch but I didn't. I had a look round, couldn't see anything, and then a train came by. He was less than a metre away. I don't know how I missed him.'

'Who?'

'Sayid.' His head goes down. He covers his face with his hands. There are few sights harder to cope with than a big man sobbing.

I do my best. I tell him once again to take his time. Finally he reaches for a tea towel and mops his face. The tea towel is a souvenir from Weston-super-Mare. Odd how these details settle in your brain.

'At first I thought he was dead. They'd beaten him up, smashed his face. There was nothing left, nothing I could recognize. Then I realized he was still breathing. Jesus . . .'

He phoned 999. An ambulance, happily, had been on standby in Lewisham and arrived in minutes.

'The guys were amazing. Two of them. They worked on him for a minute or two and then got him into the ambulance. It was obvious he was in a bad way. They'd found a pulse but he wasn't responding. King's is close that time of night. We were there in less than ten minutes.'

I nodded. I know King's College Hospital. It's next door to the Maudsley. I used to pass it when I was visiting a friend with breast cancer.

'So what happened?'

'They fast-tracked him. A&E. He's still alive but only just.' His eyes settle on his mobile. 'They said they'd phone me if there were any developments.'

'You didn't stay?'

'The police arrived. Took a statement. They wanted to know everything about Sayid. Thank God he's legal.'

'So what are they going to do?'

'They'll obviously search the embankment. Beyond that they might do house-to-house down the road, CCTV, put out some kind of appeal, whatever. The bloke who did most of the talking told me not to hold my breath. He didn't actually say it but that's what he meant.' At last he looks me in the face. His eyes are red. His cheeks are shiny with tears. 'Foreign-looking bloke? Alone on the street at night? The story writes itself. You know that.'

'But it doesn't. You took a call. That's not some accident.'

'I know. That's exactly what I told them. Then they said they couldn't find a phone, not on Sayid. That's when they really started in. People he might have upset. Enemies. Some grudge or other. Maybe drug debts. *Drug debts? Sayid?* Are you kidding?'

It's a mess. But then, according to Mitch, it gets a whole lot worse.

'They wanted to know about our relationship. Whether we were close. Whether we'd had some monster row. Whether I had any previous for violence.'

'You're telling me you're a suspect?'

'I'm telling you exactly what happened. This morning we're going to go through it all again. Brixton police station. Half nine.' He nods at the card on the work surface. 'DC Brett Daltrey. My new buddy.'

'You want me to come?'

'No.' He shakes his head. 'Thanks, but no. You being here now is enough. I can't tell you. A time like this, you think you're going mad. It's like a nightmare. It's like a truly crap movie. Except you end up playing the key role and having bugger all to say. You know something? I used to think I'd been around a bit, seen a few things. I used to think nothing could ever take me by surprise, touch me, hurt me, whatever. And you know something else? I was wrong. Such a handsome guy. So gentle. So fucking *luminous*.'

I hold his gaze for a long moment. He looks wrecked. There's a big fat question mark hanging over this conversation and both of us know it. This isn't about Sayid at all. He's not the kind of person to upset anyone. It's about Mitch.

'So who made the call?' I ask him. 'And why?'

He won't answer the question and in a way I don't blame him. All he wants is to be held, to be comforted, to have someone reassure him that there might – one day – come an end to this madness. That Sayid will recover. That the damage will be repaired. That their relationship will take them somewhere safe and perhaps a little brighter.

I do the motherly things. I make tea. I offer toast. I lend a listening ear. Mitch has started to ramble a little. He wants to tell me about the good times he and Sayid have had, about the day they met at a seminar at the LSE, how they took a trip down the Thames on a tourist boat that same afternoon, surrounded by foreigners, how the relationship slowly deepened until Sayid was sharing Mitch's kitchen, and finally his bed. This is all depressingly past tense, as if Mitch's lover has died already, and I keep reminding him of the miraculous reach of modern medicine, how gifted strangers can reach deep into your brain and make you better.

'You'd know, of course. I'm sorry. That's crass.'

'No problem. I don't blame you. All I'm saying is that he's still alive. Otherwise they'd have phoned you.' I reach for his hand, give it a squeeze. 'No news? Good news? Isn't that the way it goes?'

I leave him around half past seven. He wants a shower and a change of clothes. He'll call me if anything happens. He hopes I don't mind. I tell him again that I'm there whenever he needs me. As I close the front gate behind me I can see him in the front room, a ghost of a man. He lifts a hand, waving goodbye, and then he's gone.

Instead of looking for a bus stop or summoning an Uber, I take a walk to the end of the cul-de-sac. Access to the railway embankment is barred by an unmarked white van and a length of Police Do Not Cross tape that dances in the wind. Three men in forensic jump suits are combing through the grass where the embankment meets the scabby tarmac. I watch one recover a can of what looks like cider and sniff it before putting it carefully to one side. An assortment of other cans and bottles await sorting and bagging. This little tableau is beyond bleak. No one deserves to have been fighting for his life in a setting like this.

Thoroughly depressed, I finally phone for an Uber. By nine o'clock, after the slow crawl north, I'm back in Holland Park. At this point I truly don't know what to do. Unless Mitch has upset someone else, there has to be a connection between the *Guardian* article and what happened last night.

Beating Sayid nearly to death is a very neat way of getting to Mitch. H and his mates may be much subtler than I've ever suspected but that puts us all in a place I never want to go. Maybe Spencer Willoughby was behind last night's attack. Maybe he's opened his cheque book and hired a bunch of thugs and made himself feel a great deal better in the process. Or maybe the Tories took umbrage at Tuesday's article and did something similar. Whatever happened, I realize I will go to very great lengths not to believe any of this has anything to do with Malo's new dad.

I've been at home for a couple of hours when the phone rings. Thinking it might be Mitch I don't bother to check caller ID. Silly mistake. Easily done. It's H.

'I'm just walking up your road,' he says. 'Posh-looking block? Four floors? Have I got that right?'

He's at the front door within the minute. I buzz him in and wait for the lift to arrive. He's whistling a tune I don't recognize. This is something new. I've never known H whistle before.

'Nice.'

I've invited him in. He's left his briefcase on the table and he's having a look round, exactly the way you might if you were planning to put in an offer. He seems very cheerful. The limp has gone and when I tell him to take a seat there's none of the caution that's attended every movement since he left hospital.

'Well?' He's beaming up at me. 'Everything shipshape?'

This is a stupid question and I suspect we both know it. Since the moment I heard his voice on the phone I've realized what a bizarre conversation this is going to be. Deep inside, no matter how hard I fight the conclusion, I know that H must have had something to do with Sayid. It carries his fingerprints. In TV drama cop-speak, it's very much his MO: clever, effective, ruthless. Yet the last thing I want to do is start any kind of conversation about Mitch Culligan because I've no idea where that might lead. Malo is still Malo. Malo is my son. And just now he seems to belong to H.

'Shipshape?' I query.

'Here.' H has opened his briefcase. Out comes a sheaf of artwork. He spreads it on the table. What happens next stretches my acting skills to their absolute limit. The artwork is themed around a single word: Remembrance. Sepia shots of *Persephone* under full sail bleed artfully into black and white photos from the D-Day landings. Five thousand pounds will buy you the once-in-a-lifetime chance to pay your respects to the hundreds of thousands of Allied soldiers who were killed in the Battle of Normandy, and to the countless millions who died in both World Wars. The voyage has the blessing of a number of regimental associations and all proceeds will go to Front Line. There's a promise of a four-course gourmet lunch afterwards in a top Normandy restaurant plus on-board silver service on both legs of the crossing. Better still, every guest will have the chance to get to know *Persephone*'s guest of honour. Me.

I tell H it looks great. I tell him Malo's done a fabulous job. I've absolutely no doubt that the voyage will be over-subscribed within weeks and that Front Line will be mega-grateful. I'm trying very hard to make this burst of enthusiasm sound both spontaneous and genuine, and I think I've done OK, but all I can really think of is that poor bloody man fighting for his life. Is now the time to step out of role and have a proper conversation? Or do we carry on the masquerade?

H is looking pleased. He wants to know what I think of the photo Malo has chosen. H has his finger on it.

'You mean of me?'

'Yeah.'

'It's fine. I was ten years younger then. Can these people handle disappointment?'

H thinks that's funny. He also thinks I'm wrong.

'Beautiful women always age well,' he says. 'And you're the

living fucking proof. It's not the boat that will get them on board. It's you.'

I'm not at all sure I want to be the come-on for a bunch of ageing war-junkies with time on their hands and far too much money, but that's a different conversation. For now, all I want is to get H out of my apartment. I suspect he senses this. His feral instincts never let him down.

'Busy, are we?'

I glance at my watch. Nearly half past eleven. I can invent a lunch date. I can say that Evelyn needs some groceries. I can bring this little exchange to an end in any way I like. Except I know it's not going to happen.

H has produced a bottle from his briefcase. It's very early for Krug.

'Little celebration,' he announces. 'A toast for the boy's efforts.'

I fetch glasses from the kitchen. This is a bit like surfing, I tell myself. You try and spot the right wave and then see what happens.

'Malo . . .' H has his glass raised.

I murmur my son's name and do the same.

'I'm surprised he's not with you,' I say.

'He is. I dropped him at a courier place. Woman called Clemenza? Ring any bells?'

'Of course. I've met her. He's crazy about her and I'm not surprised. She's lovely.'

'So why didn't you mention her before?'

'She's Malo's business, not mine. Or yours for that matter.'

This sounds harsher than I intended. I try and soften the impact with a smile but H seems oblivious.

'Anything else you haven't told me?'

'I don't know what you mean. In what respect? You want to know about my girlfriends? My marriage? My all-time favourite movie? Only this could take a while.'

'Don't play games, gal. Not with me.'

'Gal' is new. 'Gal' comes loaded with menace. This is a new H, a different H, and he is, to be frank, scary. I'm trying very hard not to think about that embankment. Or Wesley Kane. Or expanding foam.

I ask H how long he's been up in town.

'We came up last night,' he says. 'M and me. Took him to a casino. Stayed at a nice hotel. Made a fuss of the boy. He loved it.'

I nod, reach for my glass. No longer Malo but 'M'. My son has been rebadged. M.

'You should have phoned,' I tell him. 'You could have stayed here.'

'Didn't want to disturb you. Didn't know what you might be up to.'

He lets the sentence hang in the air.

'I could ask the same,' I say.

'But I just told you. Casino, something to eat, something to drink, another couple of twirls on the roulette table, then bye-byes.' He counts each of these little adventures on his thick fingers and then looks up. 'You want to check? See the receipts?'

'Why would I want to do that?'

'Fuck knows. Just wondering.' He gets up and starts to prowl around the living room, pausing to stare at the odd photo, inspecting my collection of paperbacks, my DVDs, my music. I know he's playing with me, building up a certain tension. It's a trick Berndt used to use all the time in his darker scripts. The victim was always a woman. The camera would linger on her face until the moment came when she broke. This would usually be the cue for some kind of confession. This isn't going to happen to me, I tell myself. Never.

H has ended up at the window. He has his back to me. Something in the street below seems to have caught his attention.

'I sent Jessie to Salisbury the other day,' he says. 'To look for some stuff she wants for that room upstairs.'

Salisbury. I briefly close my eyes. *Watch out for yourself*. Shit.

'And?' I say brightly. 'Any luck?'

'Yeah. She got it sorted quicker than she thought. Had a bit of time on her hands.' He turns round at last. 'So where do you think she went?'

'I've no idea. Should I?'

'Place called the White Hart. Place she dropped you the other week. And you know what? She popped in, had a word with the bloke behind the bar. Good girl, Jessie. Been on the team for a while.'

'What team?'

'My team. The top team. The A team. Is there any other?'

I ignore the dig. I know exactly what's coming next. Jessie warned me in the pub at the weekend. I should have sussed it. I should have thought it through. In a way she was trying to help me, trying to head off this scene. *Watch out for yourself.* I should have been bolder, cannier. We should have talked.

'Did she go for lunch?'

'Of course she fucking didn't.'

'Just a drink, then? One for the road? Something soft?'

H isn't amused. He leaves the window and steps towards me. He's very close now. I can smell mints on his breath.

'Your problem, gal, is that people remember you. Bloke behind the bar turns out to be a fan. That's nice unless you're dicking someone else around, especially if that someone happens to be me. Are we on the same page here or do I have to spell it out?'

I say nothing, telling myself there's too much at stake. H is back beside the window, checking the street again. Then he spots his son.

'M is back early,' he says. 'That makes you one lucky girl.'

'Is that a threat?'

'Not at all. It's over. Done. Sorted. We'll never mention it again. Deal?'

I don't know what to say. No script of mine ever ended this way. H is beyond clever. He holds all the aces, including my son. No wonder they went to the casino.

'You mean Malo is back early,' I say.

H is laughing. It's not a pleasant sound. At last he turns back from the window.

'No,' he says. 'I mean M.'

TWENTY-ONE

Malo joins us in the apartment. He missed Clemenza by minutes at the courier agency, which didn't please him, but he's made contact by phone and he's fixed to see her this afternoon before he and his dad go back home.

This news floors me completely. Home. Flixcombe fucking Manor. Malo has spotted the artwork on the table and wants to talk me through it. I notice that H doesn't offer him any champagne but is keen to top up my own glass. Our troubles, he seems to be saying again, are over. We've had a bit of a ruck, he's said his piece, we know where we stand, and that's the end of it. This begs one or two of the larger questions but for the time being I have no choice but to listen to my son.

He's explaining why this element works and this one doesn't. He's pointing out why some images are a lot stronger than others. And he's agreeing with his father that the lads in Yeovil have got to do a whole lot better. This is a new Malo, a Malo I've never seen in action before. He must have picked up a lot of this stuff in barely days and yet his grasp of the language is genuinely impressive. But what's also obvious is that he shares H's steeliness in negotiations, in business. There's no question of compromise, of accepting second best. The presentation has to be tone-perfect. It has to be exactly *right*.

I wander through to the kitchen to put the kettle on. The Krug has gone to my head and it's a moment or two before I realize that H has followed me.

'No hard feelings?'

I shoot him a look. At least he's not calling me 'gal'.

'This isn't the time or the place,' I tell him. 'And you know it.'

'For what?'

'For . . .' I shrug. I feel utterly hopeless. There's nothing I can say, nowhere I can go. H has ordered a punishment beating to mark Mitch's card, exactly the way that gangsters do, and Sayid may yet die. Most of this appears to be my fault and that makes me feel a whole lot worse. In the meantime, artfully, he seems to be holding my son hostage. Not only that but Malo is very visibly thrilled by the prospect. What the fuck have I done, I wonder. First Berndt, and now someone altogether more terrifying.

I pour Malo's tea and step past H to get back to the living room. When I suggest to my son that he might like to spend a night or two in his own bed he shakes his head. Wrong time. Too busy. Too fraught. He has to get the *Persephone* project up and running.

Barely weeks ago this would have been music to my ears. My son engaged. My son motivated. My son eyeballing a learning curve steep enough to delight any parent. Instead, though, I can think of nothing but getting him back. That and what's left of Sayid's face once H's thugs had filled their boots.

H and Malo leave within the hour. We have a civilized chat about where next for the *Persephone* project and I'm absolutely certain that Malo has no idea of what's happening between his mum and dad. Much of this, to be fair, is to H's credit. I've been around actors all my working life but I've never met anyone with H's gift for crashing the emotional gearbox without stripping a cog or two. Sunshine and rain, I think. The blackest of moods one moment, Mr Affable the next. Probably bipolar.

H and Malo gone, I collapse on the sofa. When I call Mitch, he's at the hospital. I ask about Sayid. 'How is he?'

'Still out.'

'Unconscious?'

'Yes.'

'So what are they telling you?'

'It's early days. He has hairline fractures to his skull and his brain has swollen but it's not bad enough to operate. I think that means they won't take the risk. Not at his best doesn't really cover it.'

This tiny spark of humour I take to be a good sign. When things get truly horrible, as I know only too well, there's nothing left but the darker jokes.

'What about his face?'

'They say they can rebuild most of it. Teeth. Nose. Cheekbones. Jaw. That's if he survives.'

'And the police?'

'Hang on . . .' I picture Mitch leaving Sayid's bedside. The rest of our conversation needs to be away from listening ears. 'You're still there?'

'Yes.'

The police, Mitch says, have started behaving like human beings. The case has escalated up the food chain and is now in the hands of a Detective Inspector. Mercifully, she's a woman. She appears to be ready to accept that Mitch may not have kicked his partner half to death, which comes as a relief, and she's also pursuing a number of lines of enquiry with some vigour.

House-to-house calls, alas, have yielded no sighting of an unfamiliar vehicle going up the cul-de-sac and a proper CCTV trawl is pointless without clues on a likely target. Officers are in the process of contacting train drivers in case they might have noticed a parked car or van at the foot of the embankment and the enquiry team may commission posters for display on stations along the line asking the travelling public to come forward with any information. The forensic team are still trying to tease some kind of timeline out of the scene while analysis on Sayid's phone hasn't thrown up any usage since the call to my number.

'They think the phone got binned after that. They can track these things if they're switched on but there's been no signal.' He pauses. I can hear a lift door clanging shut in the background. 'You know something really sick?'

'Tell me.'

'I'm starting to wonder if Sayid would get even better service if he'd died. The DI's virtually admitted it. We're talking GBH, maybe even attempted murder, but these attacks are ten a penny in London. Homicide would have moved things along. That's her phrase, not mine.'

I tell him I'm confused. The work the DI's already commissioned seems pretty impressive to me.

'That's because she knows I'm a journo. The Met aren't shy when it comes to getting a good press.'

Mitch wants to know whether I could bear to pay Sayid a visit.

'What would be the point?'

'He liked you. He liked you a lot. Apparently a voice at the bedside can make all the difference.'

'You mean he might come round? Wake up?'

'It's possible. It's unlikely but he might. Would you mind?'

'Of course not.'

We fix a time. He says early evening would be good. Sayid is in Critical Care. The very phrase makes me shudder. We agree to meet downstairs in the main reception area at half past six. He'll walk me up. I ask whether he'll be with me at Sayid's bedside.

'No,' he says. 'One visitor at a time. House rules.'

I get down to Brixton with half an hour to spare. Hospitals seem to have become part of my life recently. First me. Then H. Now Sayid. I step in through the main entrance. The waiting area is crowded, visitors everywhere, but I find myself a chair by the water cooler. I'm deep in a book by Stefan Zweig just now and in less than a paragraph or two I'm back in the dying days of the Austro-Hungarian Empire.

'*Beware of Pity*?'

It's Mitch. I look up, startled. He knows the book well, loves it, along with most of everything else Zweig wrote.

'You know his story? What happened at the end?'

To my shame I don't. Mitch squats clumsily beside me. He seems calmer, even rested. Zweig and his wife fled the Nazis, he says, and made a new life in Brazil.

'They were Jews?'

'Of course.'

'Home safe, then.'

'Hardly. They had a suicide pact. Died of barbiturate poisoning. Took their own lives. 1942.'

He delivers this news without any explanation, any hint of why they'd chosen such a death after surviving the attentions of the Nazis, and I'm still trying to work it out minutes later when we emerge from the lift and make our way to the Critical Care unit.

Mitch pauses at the entrance and points inside.

'Second to last bed on the left,' he says. 'They get a bit edgy if you stay too long.'

I nod. Even from here the atmosphere in the unit is sepulchral, a heavy silence punctuated by the whirring of what seems like a million machines. I've no plans to stay a moment longer than I have to. A couple of minutes of one-way conversation, of trying to penetrate the darkness of Sayid's swollen brain, and then an honourable retreat.

A nurse steps across from the nest of desks in the middle of the unit and asks my name. Mitch has already warned her to expect me and she leads me down the line of beds until we're looking at Sayid.

'How is he?' It's an instinctive question, almost a reflex.

'Not too bright, I'm afraid. He's ticking along. I don't think we're going to lose him. It's really a question of what's left if and when he comes round.'

What's left? I swallow hard. We're standing at the bedside. There's nothing of the handsome, playful Sayid I recognize. His face, the colour of ripe watermelon, has ballooned. Blood has crusted around his mouth and nose. God knows what they did to the rest of him but everything else is hidden beneath some kind of cage that has tented the single sheet. He's breathing with the aid of a tube inserted into his throat. Another feeds him nutrients. A catheter drips urine into a bottle beside the bed.

What's left? Yuck. This lovely man escaped the nightmare that was his home city and fled to what he judged to be the safety of what we still call civilization. Now this.

'Mitch wants me to have a word or two,' I tell the nurse. 'He thinks it might be useful.'

'By all means. Help yourself.'

I wait for her to go back to the nursing station before I draw up a chair and position myself beside the wreckage of Sayid's face. I've done countless acting exercises in my time, little games we play to explore this dramatic possibility or that, but I've never tried anything like this. Just what do I say? Just how might I trigger a memory or two?

'Aleppo,' I say very softly, then break the word down. '*Al–epp–oh*.'

Nothing. I try again. Same city. Same birthright. Same nest of memories. Intently, I watch what I can see of his face. Not a flicker. His eyelids are closed. His thin chest rises and falls beneath the fold of sheet.

'Syria.'

Nothing.

'Daeesh.'

Nothing.

'Islamic State.'

Nada.

Then I remember the Greek island where he and members of his extended family paused in their journey towards mainland Europe.

'Lesvos,' I say. Still nothing.

I'm concentrating very hard now, alert for the faintest sign of recognition and I haven't heard the nurse returning to Sayid's bedside.

'Try using your name,' she says. 'Pretend you're introducing yourself.'

Good advice.

'It's Enora,' I whisper. 'Enora Andressen. We met in the Italian restaurant. You had an orange juice. We went back to Mitch's place.' I pause. I'm wondering quite how far to take this. Then I have a brainwave.

'Enora. Enora Andressen,' I repeat. 'Cassini.'

The nurse is laughing softly. Like everyone else in the world she's seen the pictures on the TV. Saturn. Those amazing rings. The tiny twinkling dot that is planet Earth. And little Cassini.

'He made us a cake,' I tell her. 'It was beautiful. A work of art. Tasted even better.'

'Again.'

'What?'

'Again.' She's looking hard at Sayid.

'OK.' I lean forward, my mouth to Sayid's bloodied ear. 'Cassini,' I say. 'Cassini.'

'There.' It's the nurse.' She's definitely seen something. 'Say it again.'

'Cassini.'

'And again.'

'Cassini.'

Now I see it too. The merest flicker of an eyelid, something so brief, so fleeting, that you'd have to work in a place like this to recognize it.

'You think he heard that?'

'I do. Try again.'

I lift my head a moment and look down the ward. Mitch is still at the entrance, still watching our every move.

'Cassini.' I'm back with Sayid. For a second, maybe less, one eye flicks open. There's no sign of recognition, or understanding, and try as I might I can't trigger any sustained reaction. 'Cassini,' I whisper. Cassini, Cassini. But nothing happens.

It doesn't matter. The nurse is already on the phone to one of the doctors. Mr Abdulrahman is showing signs of consciousness.

I linger a moment at Sayid's bedside and then bend quickly and plant a kiss on what used to be his face. His flesh feels warmer than I'd expected. I only know one word of Arabic but it feels all too fitting.

'*Inshallah*,' I murmur. And leave.

TWENTY-TWO

Mitch and I find ourselves in a cafe-bar in the middle of Brixton, a five-minute walk from the hospital. It's small, and quietly intimate, and on a Tuesday evening we are, by far, the oldest people in the room. I'd shared the good news about Sayid the moment we left the ward but so far Mitch hasn't said much. He seems neither pleased nor relieved, a reaction I find, to say the least, troubling. Only when we've split the first carafe of red wine is he ready to talk.

'All this is my doing,' he says. 'My fault.'

'What is?'

'Sayid. What happened.'

I nod. This is hard to deny and I've no intention of letting him off the hook, not least because I'm equally to blame. No H, no snooping around Flixcombe Manor, no covert meetings in a Salisbury pub, and Sayid would still be pulling on his Nikes, still heading out into the half-darkness of Blackheath. Life is rarely fair. I've known that for a while. But why the wrath of H should descend on someone as blameless and lovely as Sayid is beyond me. Except that he matters a great deal to Mitch and that, in turn, has put this lover of his in the firing line.

I'm toying with my wine glass. Overpriced Rioja is doing absolutely nothing for me.

'H came round to my place this morning,' I tell Mitch. 'He's been in town since yesterday.'

'Surprise, surprise.'

'You think he did it?'

'Not personally, no. That's never been his style. He gets on the phone, calls in a favour or two. He'll have known my name from the Finisterra website. After that the script writes itself. All he needs is my address and someone keeping an eye on the house to join up all the dots.'

'Have you seen anyone?'

'No, but then I've been away a lot lately. And in any case I haven't been looking.'

'You never anticipated something like this?'

'No, not this extreme, not this clever. No.'

'*Clever?*'

'Getting to me through Sayid. People like Prentice know which buttons to press. That hurts. Believe me.'

This is a deeply strange thing to say, all the more bizarre because Mitch seems oblivious to its implications. Sayid, his lover, his best friend, has nearly died. Yet the real damage, Mitch is suggesting, lies closer to home.

'So was it worth it?'

'Of course not.'

'So will you stop now? Call a halt?'

'Stop what?'

'The book. Maybe another article. Bothering the world with all this Brexit shit. You told me recently the people in the shadows have the real power . . . and guess what?'

'I'm right.'

'Exactly. So might now be the time to admit it? To close the file? To move on?'

'Why on earth would I ever do that?' He looks genuinely shocked.

'Because it nearly got Sayid killed and even now he may end up a vegetable. Life support for the rest of his days? A bed in some godforsaken nursing home? Faces he'll never recognize at his bedside? Every meal spoon-fed? Is that really worth saving democracy or whatever else you've got planned?'

Mitch is staring at me. This is the last conversation he needs.

'It's what he would have wanted,' he says.

'How do you know?'

'Because we used to discuss everything together. I never came out of an interview or a session on the internet without sharing it with him. We were in lockstep. Total agreement.'

Mitch's use of the past tense is beginning to upset me. He clearly regards Sayid as dead already.

'What if he starts to get better? What if he needs lots of support, lots of looking after, lots of making things right again?'

'Then I'll be there for him.'

'But you don't think it's likely?'

'I don't know. I'm not a brain surgeon.'

'But do you really *want* him to get better?'

My question is beyond hurtful but I'm getting really angry now. So far there hasn't been the slightest hint that Sayid's recovery might be more important than the fucking cause. Mitch is giving obsession a bad name.

I press him again but he won't respond, won't answer my question. His glass is empty. He signals the woman behind the bar for another carafe. Then he's back with me.

'Finding Sayid on the embankment was the worst moment of my life,' he says slowly. 'You'd only understand that if you were me, if you had a relationship that close. The moment I saw him – saw him properly – I knew what I'd done. Putting him in harm's way. Making him a target like that. Do I regret it? Of course I do. Does it make me feel like shit inside? Yeah, just a bit. But is there any part of me that's tempted to chuck it all in? No. And you know why? Because those evil bastards will have won.' He pauses. Looks away. Shakes his head. Then he's back in my face. 'Edmund Burke? Ever heard of him?'

I shake my head. Never.

'*The only thing necessary for the triumph of evil is for good men to do nothing.* That's me, my creed. That's what I believe in. That's what made me phone you in the first place. You have to do something. Backing down isn't an option. And you know something else? If Sayid was here he'd be the first to agree.'

The woman behind the bar has arrived with a new carafe. Mitch spares her a glance and then pushes his chair back and gets to his feet.

'Second thoughts.' He shakes his head. 'I've had enough.'

The moment I get home I hit Google. The wealth of information on Edmund Burke doesn't help me at all. Rebellion in the American colonies. The impeachment of Warren Hastings. The iniquities of the French Revolution. The sanctity of property rights. But there, towards the end, is Mitch's quote again. *The only thing necessary for the triumph of evil is for good men to do nothing.*

This is heavy stuff. I've had a glimpse or two of what might be H's default setting, and it chilled me to the bone, but evil is a very big word and has some uncomfortable implications. If H belongs in this box, does that make Malo – H's precious M – the spawn of evil? And if it does, what should I be doing about it?

The short answer is that I've no idea. I toy with phoning H, with asking for a little honesty. *Did you really order the attack on Sayid? Were you there when it happened? Are you aware of the state of the man as we speak? And if so do you really think that Mitch's wretched article could ever justify that kind of savagery?*

A conversation like this is probably overdue, but however hard I try I can't for a moment imagine H bothering me with the truth. He's already made it clear that this is strictly business. It might, regrettably, have been over the top but it should never intrude into what he's now delighted to call family life. That's me. And Malo. And his gangster dad.

Around eight, I'm starting to drive myself crazy. I seem to be bound hand and foot in a situation way beyond my control. I feel totally helpless. Compared to this, another brain tumour might come as a merciful relief. I chase the questions around and around – about Mitch, about Sayid, about H, about M – until there's nowhere left to go. I need a wiser soul to tell me what to do.

Evelyn is watching television when I tap at her door. I apologize for the intrusion. Some other time, I say. Maybe tomorrow.

She takes one look at my face, reaches for the remote, and invites me in. The television goes off. I take a seat and say yes to a glass of Sauvignon blanc. Evelyn offers a cautious toast to good times, whatever that might mean, and asks me what's wrong. I tell her everything. If she's shocked it doesn't show. Just getting the right facts in the right order makes me feel slightly better. I'm probably kidding myself but I sense the return of some kind of control.

Evelyn wants to know how she can help. It's a reasonable question.

'Advice, maybe?' I suggest.

She nods. I have options, she says. I can do my best to support Mitch. I can share visits to the hospital, try and relieve the pressure he must be under. At the same time, I can have it out with H, demand an explanation for the attack on Sayid.

I tell her I know this won't work. But even if it did, and he admitted it, then what?

'You go to the police.'

'And Malo?'

'Malo will do whatever Malo does. If his dad gets arrested I'm assuming the lad will come back here.'

'I doubt that. The two of them are really close. Malo's the age when you're not thinking straight. It won't matter what his dad's done, Malo will stick by him.'

'Sweet.'

'But that's not the end of it. Malo will know why his dad's been arrested because his dad will tell him. I can hear it now. *That mum of yours grassed me up*. That will matter to Malo. I'll be lucky to see him again.'

Evelyn nods, sips her wine. I may have a point, she seems to think. In this sorry mess, motherhood trumps everything.

She asks again about Sayid. Exactly how bad is he?

I've already described the injuries. True, there's been just a flicker of consciousness, a brief flaring of the candle that was once Sayid. But the prognosis would still seem to be grim.

'You think he might die?'

'Yes.'

She nods again, glances at my glass, reaches for the bottle.

'So you think H must have had something to do with it and, on the evidence, you might be right. I'm assuming he has ways and means of ordering the beating. He doesn't have to do it himself. Let's imagine these people had killed Sayid in the first place. That's murder. What would you have done then?'

I'm staring at her. Oddly enough, in the chaos of the last couple of days, this is one scene I haven't played in my head. Sayid lying dead on the railway embankment. What then?

'The police,' I say slowly. 'In the end they'd probably come for me anyway so I wouldn't have an option.'

Evelyn is pouring more wine. I sense this part of the conversation has run its course. Then comes another toast.

'Here's to better times,' she says. 'For now I suggest you wait and see.'

* * *

Good advice. Sayid, mercifully, doesn't die. On the contrary, he begins to recover. I go to the hospital daily, always contacting Mitch to make sure our visits don't coincide. Mitch is pleasant enough on the phone, and readily accedes to the arrangement, but there's an unspoken acceptance on both our parts that what's happened to Sayid has come between us.

Try as I might, I can't forgive Mitch for not putting all his baggage aside. Crusaders, I want to tell him, belong on a horse. They wear fancy armour and head for Jerusalem and look no further than the tips of their bloody lances. Human beings, on the other hand, put friends and lovers first. This is a conversation we don't have, not yet, and every time I arrive at the hospital, I'm truly glad that Mitch won't be there.

Sayid leaves Critical Care after the third day. His injuries have turned out to be less serious than doctors at first feared. He's fully conscious now, and his various tubes have been withdrawn. His jaw is wired up, and he'll need the attentions of a dental surgeon before too long, but the hideous swelling that deformed his lovely face is fast subsiding. We're able to communicate by a series of grunts and gargles, nods and shakes of his head, and tiny squeezes of my hand. To my delight there seems to be no evidence of permanent brain damage and as the days pass, the light returns to his eyes. When his consultant judges that the time is right, two detectives spend an hour at his bedside but they leave the hospital no wiser about the thugs who put him there. Of the beating that left him on the embankment, he has no recollection.

Malo, meanwhile, appears to be going from strength to strength. He phones me on a near-daily basis, largely to keep me abreast of the latest developments in the *Persephone* project. His father is giving him a great deal of latitude, which gladdens me no end, and I start to anticipate his calls with something I can only describe as glee. He reminds me of a faithful spaniel, in from the weather, determined to share the spoils of the hunt. As our conversations get longer and longer, I can picture him down in Dorset, his eyes bright, his tail wagging, his coat ever more lustrous as the days spool by.

To begin with I suspected H's hand in all this. After what happened to Sayid he wants to make amends, to put the darkness back in its box, to glue our little family together again. No bad thing, I tell myself. Especially if it's turning my wayward son into someone I barely recognize. Those long-ago days of slagging off his teachers, of not bothering with his homework, of drifting round bits of West London instead of attending school, have gone. This is emphatically a new Malo.

On the day when Sayid takes his first hesitant steps down the ward to the lavatories at the end, I return home to find a message on the answerphone. Malo sounds breathless, excited.

'Cracked the Pompey end, Mum,' he reports. 'Give me a ring.'

Cracked the Pompey end? I barely have time to get my coat off before the phone rings. It's Malo. He's given up waiting for me to call.

'Where have you been all day?' he says.

'Out.'

'But you're always out.'

'I know. It's an actress thing. Never turn a job down.'

'Anything interesting?'

I invent a small lie about another BBC job. Malo will never listen to radio drama so I don't think I'll be caught out.

'Is there a date on this? I need to be sure you're still on for the cruise.'

'Of course I'm on for the cruise. It's in the diary. Rosa knows the dates are sacrosanct. Not only that but I'm looking forward to it.'

This is another lie. I'm not one of the world's natural sailors and the more I find out about the kind of weather we can expect in mid-November in the English Channel, the more queasy I feel. Googling 'Brixham trawlers' was another mistake. *Persephone* might have survived intact for over a century but these fishing boats were obviously built for men with a high tolerance for primitive living conditions, constant soakings, and the lurking possibility of the rogue wave that might end it all. Sooner rather than later we're going to have a serious conversation about safety drills and on-board comforts, but in the meantime Malo wants to tell me about Southwick House.

'Southwick what?'

'It's where they planned for D-Day. It's not in Portsmouth, not in the city. You drive north. It's about twenty minutes. General

Eisenhower was there. It's private now. It's some kind of college. But like I told you, I've cracked it.'

It turns out Malo has been worried about value for money. Five thousand pounds a berth, he says, is a lot, even though most of it is going to charity. He wants people to think they've had a proper time of it and so he's been thinking of ways to make more of the experience. He's started using phrases like 'the offer', and 'value-added'. This amuses me at first but I quickly realize he means it. Not only that, but he's very definitely got a point. H, in his blunt way, probably thinks it's enough to bang six strangers up in a quaint little bit of seafaring history and let them get on with it. Malo, on the other hand, is determined to do a whole lot better than that.

H has always referred to punters. Malo now calls them guests. They'll all meet, he says, in a place called the Camber Dock in Old Portsmouth. This will be the day before departure for Normandy. There's a D-Day museum on Southsea seafront and he's fixed a late-morning visit once everyone has got to know each other. After lunch at a seafront hotel a minibus will take the guests up to Southwick House on the mainland.

'For what?'

'I'm working on a presentation. There's a guy I've found at the university. He's a historian. He's really good. He knows all about D-Day, especially what happened at Southwick House. They've got a special room there. It was all about the weather. D-Day was supposed to be the fifth of June. They had to hang on for a day because it would have been so rough. Did you know that?'

I didn't. But what's far more interesting for little me is where all these ideas came from. Who dreamed up the visit to the museum? Who got the permission to use the special room at Southwick House? And who knocked on the university's door and recruited this star academic? The answer, in every case, is Malo. When I tell him how impressed I am he just laughs.

'It's easy,' he says. 'Dad says you just turn up and be nice to people. Having Front Line on board is good. People love all that. And me being so young is a massive help as well. It's like as soon as they see you, they want to make it happen.'

'And your guests? Is anyone confirmed yet?'

'Three, so far. There's a couple from London, Alex and Cassie. He's just retired. Dad said he was some kind of top civil servant. His wife is really nice. I've talked to her on the phone and guess what? She's a really big fan.'

'Of what?'

'You. That's why they're coming. It's her birthday. Like a present thing.'

'So what is she expecting? Apart from yours truly?'

'That's something else Dad says we need to talk about. There's a brand-new place in the Camber Dock, a huge building, really cool. I had a tour with Dad the other day. They built it for the America's Cup, the sailing thing, and I've managed to blag a little presentation suite they've got. So the night before we leave we have drinks and eats and stuff in a pub down there and then watch one of your movies. Just to get in the mood. We thought you might have an idea which one. Dad's favourite is *The Hour of Our Passing.*'

This doesn't surprise me. *The Hour of Our Passing* is a period piece set in Nantes during the wartime occupation. A couple of Resistance fighters gun down the top Nazi in the city. Berlin demands retribution and a hundred hostages are taken on pain of death unless the killers are delivered to justice. I play a Resistance heroine trying to keep the peace between the head-bangers – all communists – and the gentler souls. There's an important love interest and the script called for a scene towards the end when I and the male lead are together for the last time. It was my first taste of sustained on-screen nudity and according to Malo it made a big impact on his dad. Would I mind if he got hold of a copy to show the guests on the eve of departure?

A moment's thought prompts a yes on my part. I made the movie nearly fifteen years ago and I remember parking Malo with my mum in Brittany while I was doing the location shoots down in Nantes. I had a decent body in those days, firm in all the right places, and I'm genuinely curious to see how I made out in the scene. My recent tussle with the Grim Reaper has stirred many memories and I'm more than happy to make room for one more.

'And the other guest?'

'Sadie Devine. She's a journalist on the *Daily Telegraph.* She wants to write a piece about what people will do for Front Line and Dad thinks that's cool. She kicked up about the money but Dad wouldn't budge. He's happy to have her along but only if she pays up. Dad says the *Telegraph* is rolling in it. No freebies.'

I'm smiling. Sometimes my son can be so transparent, so guileless. Just now I can hear H in every phrase.

'Three down,' I say. 'Three to go.'

'No problem. Easy. We're spoiled for choice.' He says the first mailing produced over a hundred expressions of interest, of which he reckoned at least half were ready with a deposit.

'So what do you do? Audition? Score them out of ten?'

'Sort of. Dad and I met this Sadie woman when we were up in London. Dad has met the MoD man, too. He thinks it's all going to be down to the chemistry on board. All it takes is one bad apple, he says. Do you think he's right?'

I do, but this emphasis on careful vetting comes as another surprise. H's interest in inter-personal chemistry normally boils down to people doing what they're told. I saw him in action over the party weekend down at Flixcombe. This is not a man well-versed in the nuances of human behaviour. What H says, goes. End of.

'So your dad's happy with the way things are coming together?'

'Yeah, I think so.'

'And you're getting on OK?'

'Massively. We've put in for my driving test. If I pass, and this whole thing goes OK, he'll get me a car.'

'That's good to know,' I say lightly. 'You can be my chauffeur, drive me around.'

Malo laughs but says nothing. Then he mentions another trip he's planning to London.

'When's this?'

'Next week. There's a guy Dad wants me to go and see. He's an Indian businessmen. You know the *Hot* chain? That's him.'

I nod. *Hot* is a mega-successful bid to scoop up the younger generation and give them a life-long passion for Kashmiri cooking. *Hot* as in bird's-eye chillies. *Hot* as in eye-watering interior decor schemes, designer Mumbai lagers, and a soaring share price. The week before my tumour was diagnosed, Rosa took me to the branch in Covent Garden. *Pas mal.*

I wonder aloud whether Malo might find time in his busy schedule for a coffee and a proper chat.

'Yeah, Mum, and that's the point. Dad's bunged me a wad of money. He wants me to take you somewhere nice. This Indian guy has an office off the Tottenham Court Road.' He gives me an address and hangs on while I find something to write with. Then I'm back on the phone.

'Half twelve OK? Next Friday?' he says. And then he's gone.

TWENTY-THREE

Over the days to come I spend most of the morning at the hospital. Sayid has an aversion to television. He doesn't much like what's on screen and despite my offer to pay the exorbitant fees to give him access to any channel he fancies, he much prefers to read a book. The physical act of reading, he says, gives him headaches just now. This appears to be a consequence of what he's been through but his consultant says the headaches will resolve themselves as the brain settles down.

This is good news but has still left Sayid without a book. I offer to sort out some audio recordings but he shakes his head. What he really wants is for me to read to him. Might I do that? This is clever on his part because a longish book will bring me back to his bedside again and again, but by now I know I'm as dependent on our time together as he is and so I say yes. Watching his face recover day after day does wonderful things to my self-esteem. It was me, after all, who helped get him into this mess and the knowledge that I might be playing some small part in getting him better is a real tonic.

I've left the choice of book to Sayid. He settles on Stefan Zweig's *Beware of Pity*, partly because I've been raving about it, and partly because he likes the title. He's been through a great deal in his life, especially over the past few years, and he's had to learn the difference between practical charity – money, visas, clean water – and the hand-wringing ministrations of what Berndt once called the grief-porn lobby. The latter, according to Sayid, want to share your misery, to feel your pain, but not too much, and certainly not too often.

And so I've started the book all over again, an interesting exercise in its own right because knowing it well already has given me an insight into the more important characters, something I can carry into the way I read it aloud. The story itself is very simple. A young soldier in the Austro-Hungarian army, just months before the outbreak of the First World War, finds himself posted to a small provincial town. There he finds himself invited to the local castle for dinner. Afterwards there is to be dancing. The owner has two daughters. The voluptuous Ilona and her crippled sister, Edith. A little tipsy, and ignoring all the clues, our hero asks Edith whether she would care for a dance, an invitation which sparks a reaction of some violence. In one sense, the entire book is pre-figured in this single incident. The soldier blames himself for the crassness of his mistake and resolves to do whatever he can to make amends. This commitment, over the hundreds of pages to come, proves catastrophic on a number of levels. Hence the title.

This narrative device, of course, has parallels to my own presence at Sayid's bedside, though mercifully only I'm aware of it, and as the days go by, and the comforting beat of Zweig's prose takes us deeper into the soldier's quandary, a smallish knot of other patients enquire whether they, too, might draw up a chair and listen.

By now, Sayid is occupying a bed in a general ward and for his sake I'm delighted that Stefan Zweig has brought him new friends. One of them, slowly recovering from complications after an operation on her lower gut, takes a shine to Sayid. Formerly, in another of life's little ironies, she's been dead set against anyone foreign but the more glimpses she gets of Sayid's inner grace, the more she begins to change her views on immigration.

'If we had that referendum again,' she confides to me one afternoon, 'I'd be a bleeding Remainer.'

I carry this thought home and wonder about sharing it with Mitch. I know he's been going to the hospital most evenings because one of the nurses has told me so, but when I try and imagine the reality of even a phone conversation I know it wouldn't be fun for either of us. It's not that I dislike the man. On the contrary, I have some very fond memories of how generous and how kind he was. It's just that his single-mindedness, once glimpsed, has left a very deep impression. Beware of Causes, I tell myself. No matter how worthy.

Friday arrives. I limit myself to ten pages at Sayid's bedside and take a bus to Trafalgar Square. The Indian entrepreneur, whose name is Amit Iyengar, has a walk-up suite of offices in a narrow street off the Tottenham Court Road. There's no plaque on the door downstairs, no sign of the distinctive *Hot* logo, just an entry phone that connects to a woman with a heavy cold.

I give my name and wait for her to open the door. In keeping with his restaurants I was rather expecting something a little edgier from Mr Iyengar, certainly a lot grander, and while I wait for my son in the dingy cubbyhole that serves as a reception area I wonder why he hides himself away like this.

Malo, as ever, has the answer. This is a guy, he tells me over lunch, whose working day rarely starts without a death threat. For whatever reason, people out there hate success, especially when it comes in the shape of a great deal of money, and most especially if the lucky recipient happens to be foreign. 'Out there' is an interesting phrase, one I've not heard Malo use before, and I wonder if he's acquiring the same sense of embattlement I sometimes associate with his father. Either way, it certainly explains Mr Iyengar's interest in keeping his head down.

'So have you signed him up?'

'Big time. He thinks it's a great idea.' Malo pats the front of his leather jacket. 'He wrote us a cheque on the spot. When I looked at it properly I told him he'd got it wrong. It's five thousand, not ten, but he said it didn't matter, we could keep the change. He also offered me a job if I ever fancied it.'

'Deliveries? Like Clemenza?'

'Business development. He wants to open big in France.'

'But the French hate spicy food.'

'That's what I told him. When he realized I could speak the language he said I was wasting my time doing all this charity crap. If I was smart I'd work for him.'

'And?'

'I took the cheque and said thank you.'

I'm looking hard at the leather jacket. It's obviously brand new because I can tell from the smell but it also has that deceptive slightly rumpled pre-aged quality that comes with top-end leather. Hundreds of pounds, I think. Another little present from his doting father. Along with the off-white Nikes and the rather nice CR7 skinny jeans. Is this gangster chic? Or a look that Malo has conjured for himself?

I ask him about the rest of the guests. He tells me, with a

hint of pride, that every berth is now taken and paid for. Alex and Cassie, the recently retired couple from London, and Sadie, the journalist, I already know about. Next to make the cut is Ruth, a young lawyer with the Crown Prosecution Service. She works down in Bristol and she wants to make the trip in homage to her grandfather who died on one of the invasion beaches back in 1944.

'She wants to visit his grave, too, so that's something else we have to fit in.'

'You know where it is?'

'She does. She's been there before. She gave me the details.'

'Is she nice?'

'She's awesome. Quite pretty. But you wouldn't fuck with her. She even managed to keep Dad quiet. Impressive or what?'

'You both went to see her?'

'Yeah. That was Dad's idea. Two pairs of eyes. Two minds.'

'So why isn't he here today? Scoping out your new Indian friend?'

'I think they've talked on the phone a lot. Dad sussed he might be up for paying over the odds. I'm just here to collect the cheque.'

We're in The Ivy, one of my favourite restaurants, and I'm picking at my champagne and wild mushroom risotto. Malo, in keeping with his new jacket, is wolfing his Ivy hamburger. I've been keeping score as far as the passenger list is concerned and we're still short of a body.

'Rhys,' Malo says. 'He's Welsh and he's been around a bit. This time last year he was working on a shrimp boat in the Gulf of Mexico. He knows all the really cool places to go in New Orleans. Top bloke. Dad loves him.'

'I can imagine.'

'Seriously. We met him in this pub in Plymouth. He told us to come down by train and I never realized why until they started drinking. I had trouble getting Dad into the taxi afterwards. He'd have stayed there all night.'

I nod, watching Malo tidy the remains of his *pommes allumettes* and dill relish into a neat pile at the side of his plate. Not just a tour organizer. Not just a well-turned-out youth who knows how to sweet-talk the big money. Not just an instant expert on D-Day and the ins and outs of Brixham trawlers. But a mate and carer when his Dad hits the rocks.

'Impressive,' I say quietly.

By now we've finished our main course. Malo insists we call for the dessert menu.

'Dad says to fill your boots,' he says. 'And he's paying.'

I order a salted caramel fondant but only manage half of it while Malo tucks into Bramley apple crumble pie. There's an actress at a nearby table I haven't seen since we were both in a Tennessee Williams in Newcastle and on the way out I pause by her table and have a brief chat. She's much taken by Malo, whom I introduce as my toy boy. She looks him up and down with frank approval and hopes he knows how to make an older woman very happy. All three of us know it's a joke, and even Malo laughs.

We say goodbye on the street outside. This is the first time for a while that I've had my son to myself and I've loved every moment. Seeing him like this from time to time, dipping into his busy, busy life, I'm aware of him maturing in front of my eyes.

He gives me a hug and turns to leave but then stops. His hand goes to his jacket pocket.

'I nearly forgot. You have to fill this in, especially the medical bits. It's for the insurance.'

He gives me a form and I skim-read it quickly. All the usual questions – name, gender, DoB, contact details, then some more pointed stuff. *Do I know how to swim? Am I currently under medical supervision? Is there any other health reason why I shouldn't embark on the voyage?*

'You had this drawn up specially?'

'Yeah. Dad said not to take the medical bits too seriously. We don't want you to be turned down for the insurance.'

'You're asking me to lie?'

'Not me. Dad.'

He gives me a peck on the cheek and leaves me with the form. I watch him loping away down Litchfield Street, a moment – on my part – of real pride. When H took me to the pub that first time, and talked of the need to enrol our son in some project or other, I never dreamed it would be something as complex and challenging as this. At Malo's age I would have had neither the raw nerve nor confidence to put myself in front of all these people, to confect knowledge I probably didn't have, to make sophisticated judgements about their personalities, about their preparedness to take a risk or two, about how well they'd adapt to each other in what might turn out to be testing conditions.

Then, as he disappears round the corner at the end of the street, I tell myself I'm wrong. At seventeen I'd begun to think seriously about being an actress, a performer, downstage in front of hundreds of people, or – even more daunting – in front of a camera and a director and a technical crew. That took nerve, too. Plus the guile to play a part, to inhabit someone else's skin, someone else's persona, and all in the cause of making people laugh or cry or maybe just think a little bit harder.

That, in a sense, is what my boy is doing. The bravura, the *chutzpah*, the sheer cheek, he gets from H. But some of the rest surely comes from me. This is a thought I can live with for a while, a conclusion that makes me very, very happy, and as I head for the bus stop, still clutching the insurance form, it takes me a moment or two to realize that I haven't once thought about Sayid, or Mitch, or fucking Brexit for at least two hours.

Happy days, as H would say. From the top floor of the 94 bus, I send him a text. I thank him for The Ivy, trail five kisses across the screen, and then add a postscript. *We should both be very proud of Malo*, I text. *Just look what we've done.*

TWENTY–FOUR

Sayid is discharged from hospital two days before we're due to set sail. *Beware of Pity* is a long book – 454 pages – and once we know when Sayid is to go home I speed up the readings a little and spend longer at his bedside. We get to the final page on the eve of his departure and a relative of one of the group of fellow listeners turns up with a home-baked cake, suitably iced and decorated, to mark the occasion. A single toy soldier guards the lone candle. Sweet gesture.

By now, Sayid is on his feet, fully mobile. He won't be running for a while but he's learned to put a sentence together through his broken teeth and still-wired jaw. Mitch, he says, will be coming in the morning to take him back to Hither Green. I hold him for a long moment. We've shared a lot these past few weeks and I know he's grateful for me coming to the hospital so often, but as ever he's thinking not of himself but of me.

'Have fun on that boat of yours,' he says.

I've told him all about the voyage, and our complement of paying guests, and I've shared my misgivings about crossing the Channel. I've had a peek at the Met Office advance forecast for the next few days and although I'm no expert it doesn't look great. The surface pressure charts show a whorl of isobars in mid-Atlantic, a sign of strong winds, maybe even gale force. This depression – an interesting word – is coming our way and there appears to be another one brewing up behind it. On the phone, Malo has been dismissive when I played the feeble mum. Boats like *Persephone*, he told me, can live with any kind of weather. This may be true, and I didn't say anything on the phone, but it isn't *Persephone* I'm worried about.

Sayid accompanies me to the door at the end of the ward. He wishes me God speed and tells me to take care. Then he touches his own head.

'You're OK in there?'

'I'm fine.'

'And in here?' His hand slips down to his chest, covering his heart.

'In there is fine, too.'

I mean it. After the *Guardian* shitstorm and everything that followed, my daily visits to Sayid's bedside have achieved what a true penance should. I feel rested, calmer. I feel comfortable with myself. On an especially good day I can even see myself crossing the Channel without spending most of the voyage throwing up.

I leave the hospital and take the bus home. After consulting with *Persephone*'s skipper, Malo has circulated an information pack for all our guests plus myself and H. Among the items I need to take on board are a heavy-duty wind-and-waterproof jacket, trousers made of the same Gore-Tex, plenty of insulating layers for underneath, slip-proof footwear, and heavy woollen socks. A lifejacket will be waiting for me in my cabin.

We'll all be meeting on the morning of 9 November at the Camber Dock. Onboard safety and house-keeping brief. Pre-lunch drinks in the saloon with a lightish meal to follow. Visits to the D-Day museum and Southwick House in the afternoon followed by dinner in a local pub and a very special screening of *The Hour of Our Passing*. *Moi* to introduce the movie.

Persephone, Malo has written, loves to put herself in the hands of first-time matelots. She can be a challenge to sail – lots of sweat and effort – and once we've left Portsmouth on the following day there'll be plenty of opportunities to teach us novices exactly what to do. The skipper plans to slip from the Camber Dock in the late morning to make the most of a hefty spring ebb tide. Nautical twilight is expected around 17.01. Dawn, at 08.02 French time, should find us within thirty miles of landfall. With a fair wind and a measure of luck we should be berthing in Ouistreham in time for lunch. *Bon appetit!*

The information pack goes on to detail the schedule once we've arrived in France. Courtesy meetings with various civic dignitaries. *Un vin d'honneur* with members of the Caen/Portsmouth Twinning Association. A promising succession of meals. Our second day, 11 November, is to feature a wreath-laying on a beachside memorial at Arromanches, followed by a visit to a war graves cemetery for the benefit of Ruth. In the evening we shall be dining in a top local restaurant, a must for seafood buffs. At first light on the twelfth, weather permitting, we'll be slipping our moorings in Ouistreham and putting to sea for the return leg to Portsmouth. Home safe in the Camber Dock in time for breakfast the next day.

This was a lot to digest in one sitting and I remember being impressed yet again by Malo's attention to detail. The sheer precision of the schedule. His apparent mastery of time differences and tide tables. Phrases like *spring ebb* and *nautical twilight.* Just where did all this stuff come from? And how come my clever, clever son is such a quick study?

Back in Holland Park I complete my packing and re-check the train times for tomorrow. An early departure from Waterloo should put me beside Portsmouth Harbour by nine o'clock. Plenty of time to make my way to *Persephone*, meet my new shipmates, and join the trip to the D-Day museum and Southwick House. I book an Uber for 5.45 next morning and retire early. Malo texts me just after nine. He's laid hands on some new seasickness pills the skipper swears by. *Dors bien. A demain.*

Sleep well. See you tomorrow. His message makes me smile. My precious boy thinks of everything.

Next morning another text awaits me. It's Sayid this time, wishing me *bon voyage.* I take this as the best of omens and struggle downstairs in the darkness to throw my bags in the back of the waiting Uber. By half past nine, in a thin drizzle, I'm paying off another taxi driver down in Portsmouth on the very edge of the Camber Dock. *Persephone* rides easily below me, nudging the harbour wall. I appear to be the first to arrive.

I'm struck at once by *Persephone*'s good looks. Sturdy? Yes. Workmanlike? Definitely. But graceful, too, a lyrical poem in carefully varnished timber. Gifted shipwrights, I tell myself, must have evolved the shape of boats like these. Chosen the wood, laid down the keel, made her ready for the ocean and the decades of heavy-duty fishing that will follow. Malo has already blitzed me with dozens of sepia shots of trawlers like *Persephone* in their prime, wallowing down-Channel, hauling huge nets, the trapped fish attracting clouds of quarrelsome seagulls. In those days, he told me, the three-man crew had nothing but sails to see them through. No engine. No navigation aids. Just raw muscle and the kind of inbred knowledge that passes from generation to generation. These days, looking at the array of aerials on top of the mast, I suspect it must be a whole lot easier and for that I'm duly grateful.

I'm still debating how to get out of the rain when a slight, diminutive figure appears below. It turns out to be the skipper. Her name is Suranne. She's barefoot on the wetness of the deck. She's wearing patched jeans and puffer gilet over a black T-shirt. She has an elfin haircut, a strong jaw line, and is sucking on a thin roll-up. I throw her my bags and then, under her direction, clamber down the iron ladder.

'Welcome.' She's laughing. 'Haven't I seen your face before?'

She takes me through the doghouse where she says all the important stuff happens and then down another set of steps to the saloon. The key word here is cosy. There's a tiny galley and a biggish dining table lapped by leather-covered banquettes. There are two cabins at the back of the boat, another midships, and four berths in the bow. The crew, all four of them, live somewhere else.

The next hour or so, as crew and guests gather, is a blizzard of names I dimly recognize from conversations with Malo. Malo himself is in the very middle of it all, working the crowd, pressing the flesh, carefully deferring to the skipper. At first glance, Suranne might be in her late twenties. However, given the responsibility she carries, she must be a good deal older. She's direct, and likes to share a joke or two, but she's learned the knack of establishing the pecking order without a hint of pulling rank. Clever. And impressive.

Persephone's first mate is taller, and fuller, also a woman. Fresh-faced and super confident, Georgie divides us into smaller groups and walks us through a list of dos and don'ts. The key location are the two loos. She calls them heads. There's a pump arrangement for disposing of the effluent and she squats on the toilet seat in this sentry box of a lavatory and demonstrates exactly what we're expected to do. Numbers are evidently the key to a happy boat. Twenty pumps to fill the bowl. The flick of a switch. Twenty more to empty it. Alex, the ex-civil servant, happens to be in my group. Like all of us, this is his first brush with onboard authority.

'Not a motion more, not a motion less,' he murmurs.

I think it's funny. I'm not sure Georgie even heard the comment but one thing is already uncomfortably clear. In any kind of weather, or 'blow' as Georgie calls it, spaces like these will quickly become somewhere you'd probably best avoid.

By midday, after a full safety brief up on deck, we return to the saloon. The cook, a cheerful teenager called Esther, is making what she calls a winter soup. The fourth member of the crew is Jack. He's a strapping gap-year youth with hands the size of plates and a winning grin, and his pre-lunch job is to make sure we all have drinks. The soup is delicious and the rolls that come with it are hot from the oven.

By now, we're starting to get to know each other. Sadie the *Telegraph* scribe has yet to turn up while Alex and Cassie tend to hang back a little. Ruth, petite, pretty, has an icy smile and clearly dislikes direct questions. Whether this comes with being a CPS lawyer, I've no idea, but I make a mental note to avoid anything too playful or light-hearted. Amit Iyengar, the man the media have dubbed Mr Hot, does his best to hog every conversation and will clearly become a pain. Meanwhile Rhys, the Welsh roustabout, earns my undying affection for having exactly the same pre-voyage reservations as me.

We find ourselves side by side at the table, dunking rolls in the soup. He's taking a long look round, sizing up this cluttered space we're about to call home. Then he beckons me closer.

'Not a lot of room to throw up,' he says in his Welsh lilt. 'Best to kip on deck.'

We spend the afternoon being grown-ups at the seafront museum and later at Southwick House. We learn about the five invasion beaches and everything else that followed on behind. Floating harbours. Oil pipelines. And a huge July storm that did almost as much damage as the marauding Germans. The latter, at Southwick House, offers our university friend a chance to analyse the 1944 weather in detail. His PowerPoint presentation features surface pressure charts that shaped the forecast on 5 June. To me it could be a duplicate for what we might expect tomorrow. Except that tomorrow is November. And then was sunny June.

By now, Rhys and I are bosom buddies. He, too, has seen the latest pressure charts. Back in the day, Eisenhower called a twenty-four-hour delay. Why can't we do something similar?

'Fuck knows.' He looks resigned. 'That was a war. This is for charity. You tell me the difference.'

That evening we all eat fish and chips together at a harbour-side pub. Sadie, the *Telegraph* scribe, has at last arrived and it's obvious at once that Amit is besotted. He games the seating arrangements so they're sitting side by side. Judging by their conversation, they're old mates. When it comes to any excuse – the ketchup, the salt, the vinegar – he can't keep his hands off her.

Later, when I say a word or two ahead of the screening of H's favourite movie, I'm aware of Amit trying to canoodle in the half-darkness and I'm wondering when Sadie's going to do something violent. Then the lights go down and the movie begins and I watch H bend quickly and have a word in his ear. It takes less time than the brief exchange of dialogue in the opening scene, and after that Amit doesn't lift a finger.

That night I share a cabin with Sadie. She must be ten years older than me but she's managed to preserve her looks. I'm wondering how much of that is down to money and careful exposure to cosmetic surgery when she confesses that this assignment might be her last chance to live the dream.

'Dream?'

'Telly. Channel Five are looking for a presenter. It's a travel show. Your son's offered to shoot a trial piece to camera. Lovely boy that he is.'

A trial piece to camera? My son? I shake my head in wonderment. To my knowledge Malo has never done anything like this in his life but I'm sure that wouldn't stop him for a moment.

Later, once we've put the lights out, I mention Amit. She's obviously met him before. Nice guy.

'Wrong.' I hear her rolling over in the bunk below. 'Amit is an arsehole.'

We set sail next day earlier than planned. Our skipper has spotted what she calls a window in the onrushing weather and wants to be in Ouistreham at first light tomorrow. Accordingly we sail at high water, missing the ebb tide that Malo has promised. We slip out through the harbour narrows with the thump-thump of the engine beneath our feet. A gauzy sun is visible through the blanket of high cloud and the sea is the colour of pewter. There's a whisper of wind coming up the Solent from the west but Suranne decides to wait until we're off the Isle of Wight before hoisting the sails.

I sense we're all aware that this will be our first real test. Suranne eases the engine into neutral and we all gather beside the main mast for instructions. This is a totally new world, the realm of jackstays and Samson posts, of lizards and cleats, of scuppers and deadeyes and something – improbably – called 'baggywrinkles'. I've come across some challenging scripts in my life but this is the most bizarre.

We split into two-man teams. Georgie shows us how to hoist the sail up the mast by hauling on the main halyard, a thickish rope, one big tug at a time. This isn't as simple as it looks. Most of us are game enough but it's heavy work. Even H tries to lend a hand but his shoulder is still far from perfect and it's Malo who gently tells him to stand aside. I, thank God, am teamed with Rhys. He's done all this before, understands exactly where to put his weight, and our sail races up the mast.

By mid-afternoon, with the light beginning to die, we've cleared the Isle of Wight. There's a biggish swell from the south-west and *Persephone* wallows along at a stately five knots. It's cold in the freshening wind but most of us have preferred to stay on deck rather than go below. We huddle in little groups, white-faced, slowly getting to know each other.

I spend a couple of hours with Alex and Cassie, gladly sharing some of my experiences on various movies. Cassie is very astute when it comes to particular films of mine. Her judgements are more than sound and she's keen to understand what led to this interpretation or that. Like H she rates *The Hour of Our Passing* very highly, though probably for different reasons, and she's especially keen to know whether I had any input into the original script. When I say no she seems disappointed but all three of us agree to meet for drinks before we eat.

Dinner is at seven o'clock. It's been dark for a couple of hours. Down below, the roll and dip of the boat is harder to cope with and some of my shipmates are beginning to look queasy. The smell of roasting lamb from the galley doesn't help and I notice that Rhys is one of the few takers when Jack clatters down from the doghouse to enquire about drinks. Rhys is a Guinness fan. Most of us are wedged in various corners of the saloon, mostly around the big table, but he drinks on his feet, his long body effortlessly in tune with the motion of the boat. I envy him this and tell him so.

'Practice,' he says. 'I've been at this game a while.'

I remember Malo telling me about Rhys shrimping on the Gulf of Mexico. He says it's true. He did it for three years and his last trip clipped the hurricane season. The first big storm was called Anthea.

'We knew she was coming and we made a run for New Orleans. Worst night of my life. Just made it. Another couple of hours and we'd have been breakfast for the 'gators.'

The rest of us exchange glances. The wind is definitely picking up. We can hear it every time someone steps into the doghouse from the deck above us. It's a low keening noise in the rigging, punctuated by a wild flapping of sail every time the boat wallows. I catch Cassie's eye. Like me she's probably wondering whether there are alligators off the Normandy coast.

Supper comes and goes. Nobody, except Rhys and Malo, eats very much. Even H seems to have lost his appetite. Already both loos are occupied and when Sadie emerges, pale, haunted, I suspect all is far from well. She did her test piece to camera in the last of the daylight and it was crap so maybe she wasn't feeling too bright even then.

Malo's anti-seasickness tablets seem to be doing the trick. I risk a couple of mouthfuls of lamb and a baked potato washed down with red wine and feel fine. Malo, bless him, is doing his best to keep our spirits up and I'm happy to perform when he asks me for a couple of location stories. My new friends around the table are nice enough to listen and respond with a question or two and it's nearly midnight when my turn comes to stand watch.

This time I'm teamed with Amit and Ruth, the CPS lawyer. Ruth, I've noticed, has tucked into a decent plate of food but Amit hasn't touched a thing. For once, thank God, he has nothing to say.

We put on our life jackets and make our way up to the doghouse. Suranne is bent over a chart of the central portion of the English Channel, logging our progress south. Inconceivably we aren't even halfway across but she's anticipating a break in the weather and thinks we should still make landfall by breakfast time.

Mention of breakfast does nothing for Amit. I think I hear a groan when we make our way out into the heaving darkness. We're all wearing safety harnesses and we clip into lines laid specially on the deck. I can see Georgie standing at the big ship's wheel, fighting the wind and the waves, her face shadowed by the red glow of the binnacle. She gestures me closer, asks whether I might relieve her for a while. All I have to do is keep my eye on the compass.

'Steer one hundred and sixty-seven degrees,' she says, 'more or less. She responds very slowly so don't overdo any input. I'll be in the doghouse. Shout if you need me.'

Shout if you need me?

This is beyond belief. I tally up the numbers. Suddenly, without any warning, I have sole responsibility for twelve souls, in the middle of the night, with wave after wave roaring out of the darkness. The biggest of these break over the boat in an explosion of white spume, soaking us all.

Georgie is still beside me. She makes sure I'm comfortable at the big wheel, feet apart, body weight nicely balanced, and tells me not to grip the wheel too hard.

'Treat it like a friend,' she says. 'Be firm but gentle. Tough love, *quoi*?'

This is the first time I realize she speaks French. Nice. I like this woman. I love her competence. And I'm in awe at the degree of trust she seems to have.

She lingers for a second or two longer, eyeing the oncoming waves. Then she picks her moment and disappears into the doghouse, leaving me at the wheel.

I look round for my watch-buddies. Amit has tucked himself into a corner of the crescent of bench at the stern. His head is in his hands and I suspect he's just thrown up. Ruth, on the other hand, can't wait to have her go at the wheel. She's watching my every movement. She's probably been doing this half her life, I think, just like Rhys. She's probably marking me out of ten.

I concentrate very hard on the soft red glow of the binnacle, trying to keep the central line locked on 167 degrees. This is a lot harder than it might seem but as the minutes tick by I begin to get a feel for the boat. Georgie's right. It's a living thing. It needs constant attention, endless small course corrections. Then from nowhere arrives a truly big wave, smashing the bow sideways until the sails begin to flap, and I have to haul on the wheel to get her back to where she belongs.

By now, I'm soaking. Cold water has got in around the hood of my anorak and through the folds of scarf around my neck and I can feel it trickling down my back. I'm concentrating so hard that none of this really matters and I'm aware with something close to euphoria that I'm managing to hack all this when Georgie returns. When she checks the course on the binnacle compass I catch an approving nod of the head. Then it's Ruth's turn.

Our watch ends at four in the morning. We make our way back down to the saloon. Amit gives the hot chocolate a miss. There are bodies everywhere, most of them in a bad way, fighting the boat's motion. By now I've sensed that the key to survival in a situation like this is to roll with the punches, to break down every little task into its component bits. If you really need to go to the loo, do everything slowly, deliberately, consciously. Unpeel your waterproofs. Drop your knickers. Don't rush things. Relax.

To my delight, it seems to work. Forty pumps later I'm back in the saloon contemplating an hour or two's sleep in my cabin. When Rhys suggests that going back on deck with himself and Malo would be a wiser option I ignore him. A zig-zag course across the saloon, thrown left and right by the motion of the boat, takes me to bed. Sadie is showing no signs of life. I struggle out of my clothes, towel myself dry, pull on a new T-shirt and trackie bottoms, and wriggle into my sleeping bag. Moments later, no kidding, I'm asleep.

Dawn finds us in the Bay of the Seine. France is a low grey line on the horizon and the English Channel appears to be a friend again. The wind has dropped and the remains of the swell sweeps us towards Ouistreham.

We dock shortly after eight o'clock. The sound of Malo confirming our mooring with the marina boss in fluent French puts a huge smile on my face. H, especially, is deeply impressed. We're standing on the deck together while Georgie and Jack make fast our lines.

'Top job, my son,' he says, as Malo hops back on board.

'Our son,' I tell him.

The smell of frying bacon has revived flagging spirits. We all gather down below to attack Esther's breakfast. Crossing the Channel, us ladies agree, is a bit like childbirth. Bloody painful and immediately forgotten. Rhys begs to differ.

'You were a pro,' he tells me. 'I'd take you to the Gulf of Mexico any day.'

After breakfast, in a burst of team spirit, we help with the washing-up while Sadie sweet-talks Malo into a second take on the piece to camera. I watch from the deck as she preens on the dockside, lifting her face to the thin winter sun while Malo sorts out his new iPhone. He's using the phone to shoot the video and H smiles when I ask him where Malo got it.

'Another present?' I suggest.

'Present, bollocks. Essential tool is the way I look at it. Great screen. Brilliant camera. Shoots videos. Water-resistant. The lot. The boy's a credit to himself. Just look at him.'

H is right. Sadie has to be more than twice his age. According to H she's a seasoned feature journalist, a tough old bird who knows how to mix it with the best of them. Yet here she is, putty in my son's hands as he tells her to relax, to be more normal, to treat the camera like a friend.

This line seems to startle H.

'Where the fuck did he learn that?' he asks.

'He got it from Berndt,' I tell him. 'And from me.'

We spend the day in a minibus, driving from invasion beach to invasion beach and meeting an assortment of local worthies. Malo sits at the front, chatting with the French driver, telling him what we're up to. At Pegasus Bridge, where the first British paratroops dropped out of nowhere, we pause for coffee and inspect a display of military bits and pieces before Malo describes our D-Day heroes taking the Germans by surprise and establishing a position to protect the flank of the Allied bridgehead. Once again, with little visible effort, he seems to have mastered his brief. Fields of fire. Mortar crews. Artillery support. And all this delivered with his now-trademark confidence.

Gradually, our little gang is sorting itself out. Alex and Cassie are locked in conversation with Ruth, the young lawyer, while Sadie and Amit appear to be friends again. Jack, the ship's mate, has been given leave of absence from *Persephone* for the day and is eager to see the invasion beaches. He and Rhys are the naughty boys at the back of the bus, where I, too, have a seat. We laugh a great deal and get peaceably drunk over a four-course lunch at a seaside resort called St Auban.

By nightfall, we're all exhausted. Rhys proposes a game of Spoof around the saloon table after yet another meal while Cassie and I chat more about the movie business. To my relief she wants to move on from my own films and I'm more than happy to share gossip about actors and actresses that have always interested her. One of them is Liam Neeson, whom she suspects of wasting a huge talent, and I do my best to defend him. Liam, I tell her, parlayed his gifts into some fine movies and a small fortune, something much rarer in La La Land than she might suspect.

The weather, to our surprise, perks up for Remembrance Day. Bright sunshine greets us when we de-bus at Arromanches. Malo and H have paid a visit to a *fleuriste* in Ouistreham and the resulting wreath is a clever interlacing of deep red hypericum berries with green leaves and twists of white lilies. Now, it's Ruth who carries our offering up the steps of the D-Day memorial at Arromanches. This is where the Allies established their floating harbour and Malo has managed to attract a decent scatter of local print reporters plus a film crew from a Caen-based TV station.

One of my proudest all-time memories will be Malo and Ruth being interviewed together while the rest of us gathered in the background. Towards the end of the interview, asked why someone of his age would be remotely interested in an event that had happened seventy-three years ago, Malo paused for a moment and then suggested that bad things always happened when the world was looking the other way. This was a phrase that came straight from Berndt but I didn't care. It sounded respectful, and wise, and would play wonderfully on French regional TV.

That afternoon, we accompany Ruth to her grandfather's grave and then return to the boat. Our final run ashore takes us to the fish restaurant in Ouistreham that Malo has been promoting since our arrival. In a much-appreciated gesture of thanks, H insists that *Persephone*'s crew join us. The meal is sensational, especially the *loup de mer en croûte,* but the mood around the big table is oddly flat. I can't make up my mind whether we're all exhausted or simply dreading the journey back. The weather forecast, once again, is far from promising but back on *Persephone* Suranne assures us that a good south-westerly blow will have us home in no time. It's this thought I take with me to bed and it's a moment or two before I realize that Sadie and her two bags have gone.

I find Malo sitting in the doghouse with Suranne. Our skipper has spent the day attending to maintenance jobs around the boat and she's now plotting tomorrow morning's course.

I ask Malo about Sadie. He says she's had to take the train to Paris for a meeting. From there she'll return to London by plane.

'She bottled it, more like.' Suranne doesn't look up from the chart. 'Bloody journalists are all the same. Show them a bit of serious weather and they can't take it.'

Malo and I agree our skipper is probably right. From my point of view I couldn't care less whether Sadie made the return leg or not. Having our cabin to myself will be a deep pleasure.

I give Malo and Suranne a hug and retire to bed. Once again, I'm asleep in seconds. Eight hours later, I'm awoken by a movement beside the twin bunks. I rub my eyes and reach for my torch. The battery is low and it's several seconds before I get a proper look at the face beside me.

Mitch Culligan.

TWENTY-FIVE

'What are you doing here?'

He doesn't answer me. He hasn't shaved for a while and the beard has altered the shape of his face. He's inspecting the lower bunk. He wants to know if this belonged to Sadie.

'You know her?'

'Yes.'

'She's left something? You've come to pick it up?'

'I'm sailing back with you.'

'You're *what*?'

This is the last thing I expected. Not just Mitch Culligan appearing from nowhere. But Mitch Culligan choosing to share this boat with someone determined to hurt him.

I'm up on one elbow now, staring down at him as he opens his rucksack and tugs out a bundle of wet-weather gear.

'But why are you here?' I say. 'What's the point?'

'Every story deserves a proper end. Especially this one.'

'I don't understand. Are you pissed or something? You're making no sense.'

He doesn't answer. From somewhere he's found a pair of yellow wellies and a red woollen hat with a bobble on the top. Absurd.

'How did you get on board?'

'Your son met me on the quayside. Sadie phoned him late last night, explained I'd be taking her place on the voyage home. Nice lad. Looks just like his dad.'

'And Malo was OK with that?'

'He was fine. I'm paying. That's all that seems to matter.'

'And H? Does he know you're here?'

'I've no idea. I gave your boy a different name. Prentice won't have a clue who I am. Neither will Malo.'

'H will have seen photos.'

'I doubt it. I never use a picture on bylines and I'm not on social media. I'm Mr Nobody. Cadging a lift home.'

Worse and worse. Whose responsibility is it to tell H about the cuckoo in our nest? Mine? Then comes another thought, even more pressing.

'But what about Sayid? Who's looking after him?'

'He's back in his care home. As a client, this time.'

'And you're happy with that?'

Mitch shoots me a look before heading for the door.

'He is,' he says. 'And that's all that matters.'

We gather for breakfast as dawn breaks. Mitch is calling himself Larry Elliott. He offers a hand across the table by way of introduction and apologizes for not being Sadie. This sparks a ripple of laughter. Rhys, who's just seen the latest weather forecast, hopes our new friend has a sense of humour, while Cassie offers to share her dwindling supply of seasickness tablets. The mood, unlike last night, is buoyant. The trip is nearly over. It's been testing in parts, occasionally worse than that, but we've learned a lot and now we're going home.

H is the last to arrive for breakfast. There are still curls of bacon left, and a couple of croissants, and Esther offers to scramble more eggs but H shakes his head. I can tell he's had a bad night. His shoulder seems to be hurting him and there are a couple of tiny cuts on his chin where his razor slipped.

Rhys pours him a cup of coffee and makes room on the sagging banquette. H stays on his feet. He's looking at our new shipmate and something in his face tells me that he's far from convinced.

'This is Larry, Dad. He's a mate of Sadie's.'

'Is that right?' His eyes haven't left Mitch's face. 'He's here for breakfast?'

'We're taking him home.'

'Why?'

Mitch stirs. He's sitting beside me. The sight of H in the flesh hasn't disturbed him in the least.

'Sadie got sidetracked,' Mitch explains. 'I happened to be in Paris. I'm not due back in the UK until next week. EasyJet or a Brixham trawler under full sail? Tough call.'

I'm aware of people exchanging covert glances around the table. After the rigours of our recent crossing I suspect most of them would kill for a seat on easyJet.

'You've sailed before?' H asks.

'Never.'

H nods, eyeballs him a moment longer. No handshake.

'Good luck my friend.' He grunts. 'You're gonna need it.'

By mid-morning, we've cleared the French coast. The distant smudge of Le Havre lies off to our right, or starboard as I'm learning to call it, while the Normandy beaches are disappearing astern. We've broken sweat hauling up the sails and Suranne has dispensed with the engine. The wind is blowing from the south-west, exactly as predicted, and *Persephone* is making the most of a gathering swell.

I'm standing with Mitch up in the bow, away from listening ears. He's brought a pair of binoculars and he's looking for dolphins.

'Malo tells me you're the come-on for the punters,' he murmurs.

'That's right. My pleasure.'

'And all this money really goes to good causes?'

'That's what I'm told.'

'And you believe it?'

'Every word.'

'Really?' He spares me a sideways glance.

I nod. For most of our brief relationship Mitch has let me pretty close. He's been supportive when it mattered, frank about his private life, passionate about his political beliefs. But this is a different Mitch. He's become solitary, withdrawn. He's retired behind his new beard. And when I press him about his real reason for being on board, for becoming Larry Elliott, he won't answer me. Most of the time he avoids my gaze but just occasionally, like now, I sense real anger in his eyes. *The only thing necessary for the triumph of evil,* I remember, *is for good men to do nothing.*

I ask Mitch whether he's planning some kind of confrontation. He's abandoned the hunt for the dolphins.

'Confrontation?' He seems quite taken by the idea.

'You and H.'

'Because of Sayid?'

'Of course.'

'And you think that might help?'

'I've no idea. I'm asking you.'

He shrugs, turns back to the view, leaning on the rail, absorbing the slow roll of the boat. Then I'm aware of another figure making his way towards us. It's H. He's carrying two mugs of coffee. He pauses beside us.

'Here, gal.'

I take the coffee. Mitch turns back from the rail, expecting the other mug. H lifts it, takes a sip. His eyes haven't left my face.

'Cheers.' He wipes his mouth with the back of his hand. 'Happy days, eh?'

For the next few hours we pick our way carefully through the busy shipping lanes off the French coast. I spend some of this time alongside Suranne in the doghouse, watching her monitoring the screen that charts the movement of passing vessels. When their course and speed indicate the possibility of a collision, she gets on the radio to alert them to our presence in case they haven't seen us. Earlier generations of fishermen, I think, would have killed for technology like this.

By nightfall, as the wind begins to rise, we're forty-three nautical miles north of the French coast. In a couple of hours, God willing, we'll be halfway across. Esther announces an early supper for those who want it while Mitch, hunched in a corner of the saloon, makes occasional entries in the *Guardian* crossword. I'm sitting beside Ruth. She's deep in a John Grisham thriller when something prompts her to check the time.

'Shit,' she says.

She produces her mobile and scrolls through her directory before trying to make a call. It doesn't happen. Mitch pauses beside us.

'We're out of signal range,' he says. 'Patience is all.'

This could be a line from an Agatha Christie movie, a handful of near-strangers banged up on an ancient trawler at the mercy of the elements, and of each other. I've spent most of the day keeping a watchful eye on both H and Mitch but so far they've behaved themselves. This I interpret as a blessing but the real test, I suspect, will come later. Darkness, especially on a heaving deck, can invite all kinds of retribution and I have a quiet word with Suranne, suggesting that she try and keep H down below for the crossing. When she asks why I blame his shoulder.

'He'll never admit it,' I tell her. 'But he's in agony.'

Supper is soup with freshly baked rolls and fish pie to follow. Out here in mid-Channel we're fully exposed to a wind that appears to be rising. Rhys has volunteered to take the wheel, to give the crew a chance to eat. The swell has given way to a succession of big rollers, each breaking wave thumping into the wooden hull. The old trawler is corkscrewing now, plunging into one trough, shaking herself dry, then disappearing into another. The motion is acutely uncomfortable and there are few takers for either the soup or the fish pie. People are nervous again. You can sense it, even smell it. They know exactly what to expect because we've been here before and they dread the hours to come.

It's at this point that I hear a commotion on deck above our heads. Raised voices. Heavy footsteps. Someone screaming. Malo is on his feet, following Suranne up the steps to the doghouse. I stay put, waiting on events. No point getting in the way, I tell myself. No point playing the hero.

Around the table, apprehension about the weather has given way to alarm. Something has obviously gone badly wrong and we're in that scary zone where things abruptly become unreal. Cassie has found her husband's hand. Amit's eyes are huge in his face. Ruth is frowning. What's happened? What the fuck's going on? Only H and Mitch appear unmoved.

Another scream. Then comes the thump of something heavy from the doghouse and a blast of freezing air from the open door to the deck. Mitch is sitting beside me, rolling with the motion of the boat. He has perfect line of sight to the steps leading down from the doghouse.

The first pair of legs to appear belong to Rhys. He's obviously surrendered the ship's wheel. He's still in his lifejacket and wet-weather gear and he's dripping water everywhere. He gives us a nod and then turns back to the ladder as another body descends. Battered Nikes. Torn jeans, soaking wet. A filthy once-white T-shirt. And finally a head. His mouth is open, gasping for air. His teeth are very white in the blackness of his young face.

The boy stares at us. He's shaking with cold. He seems to be in shock. For a moment there's nothing but the howl of the wind in the rigging and a groan from the hull as another wave bears down. Then I get to my feet. My cabin is closest. Sadie has left a black quilted anorak hanging on the hook behind the door. I grab it and then judge the moment to get back to the saloon.

Rhys has already yelled at Esther to get something hot to drink. He has his arms around our surprise guest. It's a clumsy embrace because of the life jacket but it's exactly the right thing to do. I take over from Rhys. Mitch has found a towel from somewhere and I wrap Sadie's anorak around the newcomer before mopping his face with the towel. I need a name. When I ask in English he doesn't seem to understand.

'*Votre nom? Comment vous appelez-vous?*'

'Mbaye.'

'*Vous venez d'où?*'

'*Un petit village. Près de Dakar.*'

Mbaye. From Senegal. A stowaway.

I pass the message for everyone else's benefit. We make space at the table, sit him down, make him drink the hot tea. First he warms his hands on the mug before taking a sip or two. Then he looks at me. Huge eyes. A deep scar down the left side of his face. Tight curls of black hair laid flat against his skull. Young. Maybe Malo's age.

'*Sucre?*' he asks.

There's sugar on the table. I give him the bowl. He tips most of it into the mug. He's still shaking but he manages a nod of thanks.

'*Merci,*' he says.

H has been watching the boy carefully. 'How did he get onboard?' he asks.

I point out that we were all at the restaurant last night and that the boat was locked up but unattended. H isn't interested in my opinions. He wants me to put the question to the boy.

I ask him exactly what happened. He mutters an explanation in rapid, heavily accented French. I'm not used to the Senegalese *patois* but I manage to pick up most of it.

'I was right,' I tell H. 'He'd been watching the boat for a while. When we all went off to the restaurant he took his chance and got on board. He says there's a pile of tarpaulins up towards the bow. He hid under there until he got so wet and so cold he couldn't bear it any longer.'

'I saw him on deck just now,' Rhys says. 'Just a movement

in the darkness. I called Jack. He took care of it. There was a bit of a struggle but Jack calmed him down. Bloke could have been overboard in a flash, just like that.'

H nods, his eyes returning to the boy. H seems to have appointed himself Chairman of the Board, which in a way I suppose he is.

'Ask him if he's hungry.' I put H's question in French. The boy nods. 'Then give him something to fucking eat.'

There's still fish pie in the oven. Esther dishes it out. The boy eyes it for a moment. Then a pair of thin arms emerge from the folds of Sadie's anorak and he wolfs the entire plate. He's uncomfortable in our company but the shaking has stopped.

Ruth wants to know what happens next. Do we turn round? Take him back? Rhys laughs. With a wind like this turning round wouldn't be for the faint-hearted and after that we'd be facing hours of grief from the onrushing waves.

'Home, then? We hand him over when we dock?'

I'm aware of everyone looking at H. In law I imagine this must be Suranne's call. It is, after all, her boat, her kingdom, but by the sheer force of his presence at the table this decision appears to fall to H. Mitch is watching him with some interest.

At length, after a huge wave takes the boy's empty plate skidding off the table, H yawns and asks Esther for a coffee.

'Early days,' he says. 'Let's see what happens.'

Early days? I'm not the only one who finds this declaration puzzling. The situation couldn't be plainer. We have a stowaway on board, a refugee, an illegal. He's seen the flag on our stern, listened to our chatter from the quayside, knows we're English. From my mum, I've gathered that France is unpopular among refugees, even French speakers like Mbaye. Work is hard to come by and there's a lot of racism. Across the Channel, on the other hand, Mbaye could find some kind of job, maybe plead his case, maybe avoid deportation, start a whole new life. The attraction is obvious.

H is looking at Alex. He knows he was recently at the Home Office, bossed an entire department, understands the rules.

'What's the score then, Alex? With the boy here?'

Alex takes his time. I know he loathes heavy weather because Cassie's told me so, but he seems to welcome this new development.

Something to think about. Something to chew over. I also know the depths of his despair at a job he once cherished. Another confidence from Cassie.

'The regulations are tricky,' he says. 'How old is the lad?'

I put the question to Mbaye. *Dix-huit ans.* Eighteen.

'Then forgive me but it's black and white. If he's eighteen, he goes home unless there's some pressing reason he might come to grief there.'

I ask Mbaye why he left Senegal. He says his life was boring. He was a fisherman with his father and two of his brothers but the fish have gone. In town there was no work, nothing to do. With a bit of money you could buy somewhere to squat in the back of a truck and head north across the desert to Libya. Everyone was doing it.

Alex shrugs. He doesn't need a translation.

'Economic migrant,' he says. 'We'd put him on the plane.'

H wants to know if he's carrying a passport. Mbaye says no. H shoots a glance at Alex. I know what's coming next.

'Then what if he lies about his age?' H growls. 'Pretends to be sixteen? Fifteen? Whatever?'

'Then he can stay until he's eighteen,' Alex says.

'And how does he get by?'

'We look after him.'

'And then?'

'He makes an application to stay, which he'll lose, and then he goes home.'

Ruth intervenes. I can tell there are elements in this conversation she finds difficult. She is, after all, a lawyer.

'But the boy's eighteen.' She's looking at H. 'He's just told us so. Alex is right. It's black and white. Eighteen years old. Economic migrant. *Adieu.*'

Amit is nodding with some vigour. He, too, would march Mbaye to the airport and put him on a plane home. Odd, I think, how a fellow immigrant might be hardest of all to win over in this strange debate. Last in. First out.

Rhys has at last shed his life jacket. He steadies himself against one of the pillars in the saloon and then looks at H.

'You're right, boyo,' he says softly. 'We take care of the lad. Make sure he comes to no more harm.'

No one calls H 'boyo', but Malo's dad, our onboard gangster, looks strangely untroubled. Suranne has arrived from the doghouse. She slips on to the banquette alongside Mitch. Mitch updates her with the boy's name and the rest of Mbaye's story. When Ruth enquires again about returning our stowaway to French jurisdiction, Suranne confirms what Rhys has already told us. The barometer is falling fast. The weather is guaranteed to get worse, the wind strengthening and veering westerly to bring the waves beam-on. There's no question of danger, or even of heaving to and riding out the next few hours, but neither does it make any sense to turn round and go back. We're in for a rough night.

Ruth nods. Suranne has redrawn the watch list as I requested and Ruth will be up on deck between two and four, along with Mitch, Malo and myself. Suranne offers us all the chance to say no but none of us will hear of it. I've just taken the last of the tablets and my helping of fish pie is staying where it should, but I've learned enough now to know that Rhys is right: weather like this is best endured where you can see the next big wave coming.

Ruth hasn't finished. She wants Suranne to confirm that she'll be handing Mbaye over to the authorities when we berth in Portsmouth. I'm sure that Suranne hears the question but she has one ear cocked for noises that might indicate trouble. Every storm, I suspect, has a soundtrack of its own and the best skippers, the ones who survive, are the ones alert to every hint of danger.

Ruth puts the question again. She's beginning to get under H's skin. I can sense it.

'We have to,' Ruth repeats. 'We have to stick by the rules.'

Suranne nods absently. Her body language speaks volumes. At the moment, she has more than enough on her plate. First things first.

She tells Esther to make sure the boy rehydrates properly. If he's still hungry give him something else to eat. Unthinkingly, I pat Mbaye on the arm. He's our charge now, our responsibility. The boy returns my smile and huddles closer. Every time the boat lurches sideways, falling off a wave, I feel the thinness of his body pressing against mine but unlike most of us, now he's warmer and fed, he has no problem with the plunge and roll of the ancient trawler.

I ask him about life at home in Senegal. He's already told me he's worked on the family fishing boat. He has no fear of the sea, he says. It's always been a friend, a source of food, a living. You have no money? You fish. And that way, if the gods are kind, you survive.

H is nearby, doing his best to monitor our conversation. I translate as best I can but H is greedy to know more. How many brothers? How many sisters? Does he miss them all? What else does he know how to do apart from fishing? And how's life been since he jumped on the back of the lorry and headed north?

I put all these questions to the boy and he's happy to fill in the details. He has four sisters and five brothers, one of whom died recently in a traffic accident in Dakar. Yes, he missed his family, and his friends too, especially when the trip north got rough. The desert, he says, goes on forever. The smugglers rip you off. And beware of soldiers with guns.

I ask him about Libya. How was it?

'*Affreux*.' His head goes down and he covers his eyes. '*Vraiment*.'

Even H doesn't press him any harder at this point but what's really interesting is Alex, the ex-civil servant. I'm not at all clear what role he played at the Home Office but he certainly knows a great deal about asylum seekers, appeal tribunals, and the mechanics of deportation. At the same time I'm getting the impression that this may be the first time he's seen a refugee on the run, the first opportunity he's had to translate all those Home Office statistics into flesh and blood.

H, too, is aware of this. He's never less than direct.

'So what do you think, Alex?'

'About?'

'Our friend here. And all those mates of his. Are we right to put them on the plane? Send them back? Or have the Germans got it right?'

Alex has been watching Mitch out of the corner of his eye. He seems to be as interested in the reply as H.

'You want the truth? We're civil servants. We're there at the bidding of our masters. Our masters are the politicians. And to be frank, as far as immigration is concerned, they haven't got a clue. That's point number one. Point number two is related in a way. Because those same politicians always want something for nothing. They chop us down all the time. They call it austerity. But when it comes to the police, or the Border Force, or the prison service, God help it, there's no flesh left on the bone. We have fewer bodies behind fewer desks and so we're making bad decisions all the time. About a hundred people a month get letters ordering them to leave the country when they're perfectly entitled to live here. That's shameful enough but I'm afraid it's not just the Home Office. Whitehall is a shambles and it's about to get a whole lot worse. So it doesn't matter what the politicians think they want. It won't happen.'

'That's not what they tell us.'

'Of course it isn't. Most of them have only got one thing in mind. They need to stay in power. You want the immigration numbers down? They'll promise you the earth. Do they ever tell you how complex this stuff is? Never. Will they ever get their way? No.'

This is beyond blunt. Mitch is nodding. I bet he's memorized every word. H is only interested in Mbaye.

'So say our friend here makes it.'

'Makes it?'

'Gets ashore. What then?'

'That depends. If he's on a watch list the police might go through the motions. It's unlikely but they might.'

'And the numpties in the Border Force?'

'If he's already made it they'll be looking the wrong way.'

'Home safe, then?'

'Almost certainly.'

H nods. Says nothing. Ruth is staring at Alex. Then the boat rears up and falls sideways off a huge wave and we can hear nothing but the gunshot flapping of the sails. Some of us have tumbled off the banquette. Mbaye, Mitch, H, and myself are in a heap on the floor. I can hear water dripping through cracks in the decking overhead and sluicing down the companionway from the doghouse. Then the boat rights itself and comes slowly on to the wind again.

H and Mbaye are nose to nose. H is holding his shoulder.

'Fucking madness, mate,' he says. 'Love it.'

Mbaye doesn't understand. He looks to me for a translation. I shrug. I think H is talking about the Home Office but in truth I can't be sure. I glance at Mitch. He's eyeballing H.

'I could swear you're starting to bond,' he says softly.

'With fucking who?'

'Our friend here. Mbaye.'

'And so what if I am?'

'Nothing,' Mitch shrugs. 'I'm just surprised, that's all. Stowaway black kids aren't to everyone's taste. Neither are other refugees. Like Syrians, for instance.'

'You're talking about people like me?'

'Yes.'

'So what are you trying to say?'

Mitch won't answer the question. I know he has a name on his lips and I'm praying it gets no further. Sayid. Sayid Abdulrahman. Is this why Mitch has stolen aboard? To exact a little revenge for what happened at the foot of the railway embankment in Hither Green?

H has had exactly the same thought. I can see it on his face. The way his eyes have narrowed. The set of his jaw beneath the three-day stubble.

'You want to sort it out now, son? Up there? On deck?' He nods towards the doghouse as the boat takes yet another lurch. The fiction that is Larry Elliott hasn't fooled him for a moment.

Mitch holds his stare, then permits himself the faintest smile.

'No, thanks,' he murmurs. 'Not yet.'

We're on watch from two o'clock in the morning. The waves are beam-on now, exactly as Suranne predicted, and walls of water loom out of the darkness. We huddle on a bench on the lee side of the boat, sheltered by the doghouse from the worst of the weather, not really knowing what we're supposed to do. Georgie and Jack share turns at the wheel, trying to keep the heavy old boat on course, hanging on for dear life as *Persephone* lurches sideways. Conversation has died. Ruth sits beside Mitch, her hands dug between her thighs, her eyes shut tight.

Between me and Mitch is H. Despite my carefully laid plans he's refused to stay below but so far, to my immense relief, both men seem to have forgotten their earlier conversation. Maybe the sheer force of the storm has blunted their appetite for a helping of extra violence. Maybe there's some prospect of getting them both home in working order. Either way I'm not sure I care that much any more. Wet through, freezing cold, all I'm counting are the minutes between now and the moment when I can struggle below and get warm again.

Relief comes earlier than we expect. Suranne, bless her, steps around the corner of the doghouse with a flask of hot coffee. She says she's worried about our welfare. She doesn't want any of us to end up with hypothermia. We nod, too numb to say anything. Ruth is the first to get carefully to her feet. She doesn't want the coffee. She's clipped on to the safety line and when yet another wave nearly tosses her towards the boiling ocean it's Mitch who grabs her flailing arm and hauls her back. A glimpse of her face tells me everything. She's terrified.

I watch Mitch hauling her towards the doghouse, reaching out for the sliding door, making sure she gets safely inside. Then he turns again and heads back towards us, fighting the pitch and heave of the boat, and for the first time I realize that he isn't clipped on to the safety line. Another huge wave rolls the old trawler sideways. The deck on our side is awash, practically underwater, and there's nothing Mitch can do about it. He's on his knees now, half submerged, and the wire guard rail saves him from going overboard but I know that one more wave will pitch him into the darkness.

H is watching him carefully, still motionless, but the moment I lunge towards Mitch, both arms outstretched to try and help him, H is on his feet.

'Sit down, gal,' he shouts. 'Fucking leave this to me.'

Leave what? Leave my son's gangster dad to finish the job? To wait a second or two for the next wave? To toss this upstart journalist into oblivion?

I turn on H. No way will I let this happen. I'm trying to push him. I'm trying to fight him. I'm trying to get him back to the safety of the bench we've all been sharing. H's face is inches from mine. I swear he's grinning at me. Not because I amuse him. But because he is, in some unfathomable way, impressed.

'Leave it out, gal,' he growls. 'I'm going to save his fucking life. Not that he deserves it.'

And that's exactly what happens.

Moments later, after I've helped H wrestle Mitch back across the deck, I leave them both on the bench. Mitch, his head in his hands, is throwing up. H even has his arm around him. I watch this little tableau, still trying to make sense of what's just happened, but then decide to leave them to it. One day I might understand men but not tonight, and certainly not here.

With great difficulty, one step at a time, I make my way back to the doghouse, fighting the chaos around me. Suranne, as ever, is concentrating on the GPS read-out on one of the screens. We're twenty-one miles south of the Isle of Wight. Thanks to the sheer force of the wind, progress has been faster than she'd anticipated. Soon we should pick up the lighthouse on St Catherine's Point.

St Catherine's Point, I think. England.

I know I should stay on deck. For whatever reason peace appears to have broken out, but I know I should try and make sure that neither H nor Mitch or even both of them end up in the sea, unmissed for minutes, maybe longer, and almost certainly lost. But I'm simply too cold to think straight.

Down below looks like a scene from a war movie. There are bodies everywhere, pale-faced, slack-mouthed, grimly hanging on to anything solid. Mbaye is the only exception. He's wedged in a corner of the banquette and he appears to be asleep. Still cocooned in Sadie's anorak, he has a smile on his face. Remarkable.

I stay as long as I can. I make it to the cabin and towel my face dry. I'm tempted to get out of my wet-weather gear but if anything happens I know it would take me forever to put it on again. And so I slip on to the bottom bunk and close my eyes but it's impossible to sleep. Every next wave throws me against the side of the hull. Water is pouring through cracks in the deck, soaking the top bunk and then dripping on to my face. The pillow is wet. The bedding is wet. My whole world has surrendered to an eternity of wetness.

By now, I've run out of tablets. After a while, I've no idea how long, I know with a terrible certainty that the fish pie was a huge mistake. Either I get back out on deck or this slum of a cabin will be awash with vomit. I make it to the door. It takes forever to lay a course towards the steps up to the doghouse. I climb them one by one, taking advantage of those tiny moments between the boat's recovery and the next roll. Suranne is where I last saw her: squinting up at the GPS screen, making notes on a pad on the desk.

'Two miles off Ventnor,' she murmurs. 'Past Shanklin it might ease off.'

Ventnor? Shanklin? Ease off? I've no idea what she's talking about and now is no time to find out. I throw myself towards the door and wrestle it open in time to vomit into the darkness beyond. Slowly, a figure at the wheel swims into focus. It looks like Jack but I can't be sure. Then I feel a comforting pressure on my arm. It's Georgie, the first mate. She's looking concerned.

'You have to clip on,' she says.

I bend and try but I'm useless. My fingers, already stiff with cold, won't work. She drops to her knees and does it for me. Then she takes me by the arm again and begins to lead me into the shelter of the lee side. Sweet, sweet girl.

My stomach half-empty, I'm feeling a whole lot better. I'm trying to thank her when I hear a shout from the wheel. Definitely Jack.

'Starboard forrard,' he yells. 'Big container.'

A second or two later the boat shudders from a huge impact. Over the shriek of the wind I hear a splintering of wood. Instinctively I reach for the grab rail around the doghouse and peer into the teeth of the wind. Our precious *Persephone* seems immobile a moment and then slowly rises to the next wave as the enormous metal box disappears into the darkness.

Georgie tells me to hang on. Moments later she's back in the doghouse. The windows are streaming with water but I can see her and Suranne heading down the steps into the saloon. Jack is still on the wheel but he seems to be having trouble holding course. I want, very badly, to do something to help. The other end of the safety line runs to the wheel. Judging my moment, I lunge towards him, just make it.

'Take the wheel,' he yells. 'Steer three hundred and twenty-two.'

I do my best, peering at the compass. Three hundred and twenty-two degrees. We need four pairs of hands, I tell myself.

But then Jack has gone, swerving nimbly towards the doghouse. The door open, I can see that Suranne's back from the saloon. She has the VHF radio in her hand. I've no idea what's happening but the very pit of my stomach tells me that this is a real disaster. Not just a stowaway. Something infinitely worse.

I fight the wheel and the weather for what seems an eternity. The course I've been given, 322, quickly becomes a joke. With every wave the compass swings wildly left and right, and with every slow recovery, the boat feels heavier. Every few seconds, off to port, I can see the long white finger of what I presume is a lighthouse. It stabs into the darkness, slowly revolving, bathing us in a brief moment of clarity before moving on. Then, lurching out of the doghouse, comes another figure.

It turns out to be Rhys. He's come to relieve me at the wheel. When I ask him what's happening he says there's water pouring into the forepeak. Georgie and Jack will be back on deck in a moment to try and rig a spare sail over the damage in the hull, a kind of Band-Aid solution that just might work. For now, though, they're sorting out a pump.

'Pump?'

'The bilges are filling.' Rhys nods down at his feet. 'Technically we call that sinking.'

Shit. I'm staring at the compass: 354, then 350. Then, very, very slowly, 346. Rhys is right. We must be carrying tons of water already.

I snatch a look at the doghouse. Suranne has stirred the engine into life. I can feel it throbbing beneath our feet. Now, she's talking on the radio. A life boat, I think. A real fishing boat. A passing ferry. Santa fucking Claus. Anything to get us out of this mess.

Rhys puts me out of my misery. Suranne, he says, has a plan. She's called a Mayday but these things take a while. There are no harbours on this bit of the Isle of Wight, nowhere to find shelter, and so she wants us to head for the nearest beach which happens to be Ventnor. There, *inshallah*, she'll drive the boat up on to the pebbles.

A plan? I nod. Then I have another thought.

'Malo?' I yell.

Rhys has the wheel now. The new course is 299. He thinks

Malo might be with his dad on the lee side. I nod, grateful. I have to be with him. I have to be with my son.

Still clipped on to the safety line, I make my way around the doghouse. All the water down below has helped stabilize the boat. Movement is easier. Crouched together on the bench I find Malo with his dad, which is a huge relief. To my shame I don't even ask about Mitch.

H wants to know what's happening. I tell him about the container, the bilges filling, and Suranne's bid for us to end up on the beach at Ventnor.

'Where?'

'Ventnor.'

In the darkness I faintly detect what might be a smile on H's face. Then, above the shriek of the wind in the rigging, I hear my name. It's Rhys. He wants me back on the wheel. I get to my feet, steady myself, check my safety harness, and wait for that fleeting moment of steadiness when I can move again. Rhys has managed to maintain the new course. The pump is working, and so now he has to go below to help organize a chain of buckets while Georgie and Jack do their best with getting the spare sail over the hole in the hull.

'Two hundred and ninety-nine.' Rhys nods at the compass, gives me the wheel.

By now, I've lost all track of time. Dimly I'm half-expecting dawn in the east but whenever I check in that direction I can see nothing but darkness. The boat is wallowing, the water almost close enough to touch, but the wind is filling the sails and with the help of the engine we're still making progress. The lights of what I assume must be Ventnor lie directly ahead. I can just make out strings of coloured bulbs dancing on the promenade and in the throw of the streetlights I can see curls of surf pounding the beach.

A huge wave throws the boat sideways. Ventnor now lies off to port. The wind bellies the sails again and no matter how hard I try and wrestle *Persephone* back on course, it doesn't happen. The wheel seems locked. There's nothing I can do to shift it. Fighting the tons and tons of water below is beyond my strength.

Then, from nowhere, comes a movement. It's Mbaye. He's

emerged from the doghouse, still wearing Sadie's anorak, and he seems to dance towards me. He's nimble, light as a feather. He has the experience to anticipate the boat's every movement. He can read this mad score, recognize the music, conduct the orchestra. Unlike the rest of us, he's totally in tune.

Beside me now, he takes the wheel, tests it, looks up at the set of the sails, then peers out at the blackness of the night.

'*Vers les lumières,*' I yell, pointing at the lights of Ventnor off the port bow.

'Les lumières?'

'Oui.'

He nods. Judging his moment, waiting for a particular wave, he eases the boat to port, holds tight, then manages to spin the wheel again as another wave lifts us high. His touch, his knowledge, is instinctive. He doesn't spare the compass a second glance. His gaze is fixed on the lights in the darkness ahead. He's even smiling. White teeth in the darkness, I think. Our salvation.

I shout at him. I want to know how I can help. He shakes his head. *Tout va bien.* Everything's going to be OK. Trust me.

Ventnor is bang on the nose again. In the darkness up near the bow, Georgie and Jack are bent over the rail, wrestling with the spare sail. In the meantime, another figure has appeared in the open door of the doghouse. It's Alex, our honcho from the Home Office, our expert in winnowing the genuine refugees from the chaff. He's carrying a brimming bucket which he empties towards the lee side. Then he catches sight of Mbaye at the wheel. He pauses a moment, puts the bucket down, and then begins to applaud. It's a gesture I'll take with me to my grave.

The next few minutes bring more buckets, all emptied by Alex. There must be a succession of willing hands, I think, stretching down into the saloon, my shipmates doing their best to save us from sinking. Seconds later, another figure appears.

It's Suranne. She's animated, full of life. She makes it across from the doghouse and gives Mbaye a hug. Then she has the wheel. She's looking up. She's cursing the set of the sails, how untidy everything is.

Untidy? Shit.

She's yelling at Alex and Georgie. No more buckets. Forget

the spare sail. We're minutes from beaching. She wants everyone up in the bow, ready to jump.

The boat is corkscrewing again but very slowly. If this was an animal, I think, we'd be talking death throes. I find Malo and his dad where I left them.

'You OK?' I shout.

H is studying his phone. Malo nods.

'And Mitch?'

Malo shrugs, gestures back at the doghouse. Maybe he's below. Maybe. Then I feel a tug on my arm. It's Rhys again. He wants us all in the bow.

We make our way forward and join the tight little knot of crew and passengers crouched against the wind and the curtains of spray. The beach is very close now, yellow pebbles, steeply shelving. It must be high tide, I think vaguely, which might or might not be good news. There's no sign of police or an ambulance but in the distance, shredded by the wind, I can make out the howl of a siren.

I look at the faces around me. Amit appears to be talking to himself. Ruth is tight-lipped, a coil ready to spring. Alex and Cassie are holding hands. Then Mitch, to my huge relief, joins us. We swap glances. I'm glad he's still alive.

A hundred metres to go, maybe fewer. Jack is standing in the bow, ready to jump. Georgie is beside him. She'll be handing us up on to the lip of the bow. Her instructions, as ever, are very clear. We'll jump on her command, not before, not after.

'Land and roll on the pebbles,' she says. 'Make like a paratrooper.'

Amit, I can tell, isn't convinced. Suranne is driving the boat as hard as she can, riding a huge roller. I can feel the deck lifting beneath my feet. Then, very suddenly, the beach is upon us and there comes a grinding noise from deep in the hull as the old trawler spears into the pebbles, pushed to safety by the breaking wave.

On board, there's a moment of silence. All we can hear is the thunder of the surf astern. Jack must have jumped because he's gone. Georgie is looking over the bow. She steadies Amit and then – when he refuses to jump – gives him a push. Next is Ruth. Then Cassie and Alex. Malo, Mitch, H and I are looking at each other. Rhys has just disappeared over the bow. A big wave lifts the stern and then the boat settles at an angle. Georgie isn't pleased. Time is precious.

'For fuck's sake.' She's looking at me. 'Just do it.'

I clamber up on to the bow. See the ring of white faces beneath me. Shut my eyes. Brace myself. Then, without waiting for her command, I jump.

I hit the pebbles hard. Coldness. Wetness. And the enveloping roar of the surf. Unseen hands pull me up the beach. Then come more bodies jumping out of the darkness until we're all on our feet, still swaying slightly, orphans from the storm.

Suranne is doing a head count. Georgie goes from face to face, asking, checking, making sure we're OK. Then Suranne tallies up the numbers a second time before calling us all to order. We stare at her, still numb, still in shock.

'Mbaye?' she asks.

TWENTY-SIX

Dawn has broken. It's nearly eight o'clock. Suranne has notified the police about our missing stowaway and while coastguards begin to search the rocky foreshore beyond the beach at Ventnor, a lifeboat is en route from neighbouring Shanklin. The assumption, I suspect, is that Mbaye jumped from the boat and probably drowned but everyone is hoping against hope that he made some kind of landfall. In no time at all, this young asylum seeker has acquired an almost mythical status. A handful of us saw him at the wheel – assured, fearless, immensely competent – and when Alex wonders whether he might have helped save our lives, I find myself nodding.

As the tide recedes down the beach, Suranne organizes the offload of our gear from the boat. Now, dry at last and nearly warm, we take it in turns to inspect the damage to the hull.

The wound is ugly. According to Rhys, a corner of the container must have ripped into the hull just below the waterline. The planking is stove in, the wood splintered around the edges of the gash. The impact of the beaching shifted the mattress that the crew had used to try and plug the hole and if you stand on tiptoes at a certain angle you can see into the saloon. Had the hole been even ten per cent bigger, says Rhys, we would have been history because neither a pump nor all those buckets would have coped. As it was, we dug ourselves out of an early grave and this knowledge has produced a real bond. We did it. We survived.

A journalist has arrived. Mitch tells me she's a freelance stringer. These people apparently spend their working lives monitoring the emergency channels and now she's here to ask that eternal media question: what was it *like*?

We shuffle around on the shingle and exchange awkward glances while she readies her little camera, but the truth is we can't wait to get it all off our chests. Amit admits he was scared to death. He prayed for his wife and his two kids. Ruth describes moments when the storm was at its height, water everywhere, the boat thrashing around, while Alex makes a point of telling the journo what a brilliant job the crew did. There was absolutely no panic, he says. Simply a succession of practical measures carried out in the teeth of impossible conditions. Just watching the crew in action made you believe in your own survival. Outstanding.

Rhys agrees. He puts his arm round young Jack and gives him a kiss. On camera. On the lips.

'Saved our lives, boyo. Put *that* on your CV.'

Suranne happens to join us at this point. She's been up on the promenade working her phone. The charity that owns *Persephone*, she tells H, will be despatching a couple of carpenters from Pompey and they should be with us by mid-morning. Mercifully, the spring tides will be receding from now on and, barring another storm, the old trawler should be safe for the next couple of days while the hull is made watertight. At that point, she'll be refloated and returned to the mainland for a proper inspection.

Interviewed on camera, Suranne fends off any suggestion that attempting the return leg was reckless or even unwise. The winds, she points out, were forecast Force 4, gusting 5, conditions that most trawlermen would consider a breeze. In the event, the depression turned out to be a little deeper but even so *Persephone* would still have made it safely across. Had the container not intervened, by now we'd all be having breakfast in the Camber Dock.

The journalist is happy with her interviews. When she asks for onboard smartphone footage, a number of hands go up. Mitch's isn't among them. Interestingly, no one has mentioned Mbaye. As far as our young stowaway is concerned, we're all complicit.

Suranne organizes breakfast at a big seafront hotel called The Imperial. At H's insistence, the food is served in a function room away from the main restaurant. I'm not the only one to find this puzzling but the moment the waitress leaves us alone he's on his feet. He thinks this may be the last time we're all together. Arrangements are in hand to return us to the mainland and so now is the moment to thank Suranne and her crew for, in H's phrase, digging us out of the shit. Speaking personally, he's been in some dodgy situations in his life but nothing to compare to the last couple of hours. When he first set eyes on our skipper he thought she must be a truant bunking off school. Now, thank Christ, he knows different.

He gestures Suranne to her feet, then the rest of the crew, while we stand and applaud. It's a graceful gesture on H's part, nicely done, and I detect a glint of tears in Cassie's eyes when we exchange glances.

'I'm going to miss all this,' she whispers. 'Can you believe that?'

After breakfast, I meet H at reception. He's booking a couple of rooms.

'Why?' I ask him.

'The boy,' he says briefly. 'We need to find him.'

'You mean Mbaye?'

'Yeah.'

'But he's probably dead.'

'You really believe that?'

'I do, yes. You were there. Those waves? The state of the sea? Did I imagine all that?'

H shakes his head. He draws me aside. Not for the first time, he seems to know more than I'd assumed.

'He jumped at the last minute,' he says. 'I was with him. He went off the stern before we hit the beach.'

'And?'

'The lad's a survivor. He's been swimming all his fucking life.'

I nod. True, I think. But then I remember how thin he was, how vulnerable, skin and bone wrapped in Sadie's sodden anorak. It's November. It's freezing cold. He's wet through. Where would he go? How could he possibly survive?

I put these questions to H. We've found a sofa in a corner of the reception area beneath a potted palm. Outside, it's started to rain again.

'M gave him his mobile, told him to ring Fat Dave. We gave him money, too. Quite a lot of money.'

Fat Dave Munroe? The bent cop? I should have guessed. He lives here in Ventnor with his wheelchair and the new woman in his life. She runs a B&B. I try and imagine her coping with a surprise visitor in from the street. An open door, a hot bath and plenty of home comforts to follow.

'You've phoned him? You've been in touch?'

'Of course I fucking have. I belled him from the boat before we hit the beach, as soon as you told me about Ventnor. Fat bastard never answered so I had to send a text.'

'And now? You've phoned Dave again?'

'Three times.'

'And?'

'Dave's heard nothing. The boy's gone to ground. He's biding his time. Or maybe he's moved on already. Four hundred quid goes a long way.'

I gaze at him for a moment, then shake my head. Fantasy, I think. H isn't stupid. He can read my mind.

'You think it's wrong? Me doing all this? You think we should have just handed him over?'

'Not at all. I think what you've done is great. Let's hope he's alive. Let's hope he's on some ferry as we speak. Did you give him any other contacts?'

'Yes. Ours.'

'You mean Flixcombe?'

'Of course. He can stay forever. M told him that. Last night.'

Stay forever. I'm thinking of Sayid at the foot of the railway embankment. Sayid in hospital. Sayid trying to coax an entire sentence through his broken teeth. How, in H's teeming brain, does any of this add up? Then I notice Mitch coming in from the rain. He spots us in the corner and comes across. He's looking for Suranne.

I get to my feet. I happen to know where she is. H watches us as we cross the reception area and head for the manager's office. The two men haven't even looked at each other but I know H isn't pleased.

Suranne is sitting behind the manager's desk. She's on the phone when we step in. She gestures for us to take a seat before finishing the conversation.

'That was the Lifeboat secretary,' she says moments later. 'They're after search parameters.'

She's told them what little she knows. The last she saw of Mbaye was when she relieved him on the wheel. After that it was guesswork. He may have gone overboard a couple of hundred metres out. It may have been closer. He's certainly not still hiding aboard because she and Georgie have had a good look.

'He jumped when we were about to hit the beach,' I say quietly.

There's a moment of silence. Then Suranne asks me how I know.

'Someone told me.'

'Who?'

I shake my head. Another silence. Then I sense Mitch stirring beside me.

'Prentice,' he says. 'Has to be.'

'Is that true?' Suranne is still looking at me.

'Yes.'

'Why? Why did he let that happen? And why hasn't he told anyone?'

'Because he wants the boy to be safe.'

'*Safe?* In a sea like that?'

'He thought he'd make it. He liked him. He thought he had guts.'

'And he thinks he's still alive?'

'Yes.'

Suranne nods. She liked our stowaway, too. But she's a realist.

'I'm sure he had the best of intentions,' she says, 'but Hayden probably killed the lad.'

Mitch agrees. He's here to tell Suranne that he's been in contact with a diver he knows.

'His name's Joe Cassidy. He runs an outfit in Southampton. They have contracts in the oil business, mainly in the North Sea. He happens to be a friend of mine and he owes me a favour.'

Joe, he says, still dives himself. On the phone Mitch has explained about Mbaye and about the likelihood that the boy went overboard. Joe's getting some gear together and should be in Ventnor by mid-afternoon. If the coastguards and the lifeboat draw a blank he's happy to have a go himself.

'Go at what?'

'Finding him. Joe used to be in the Navy. Clearance diver. He knows all the tricks, believe me.'

'We're talking a body, right?'

'Yes.'

'And you want me to pass this on to the lifeboat people?'

'Yes, please.'

While Suranne reaches for the phone I ask Mitch whether this is the guy he wrote the book about. The one he gave me when we were at his agent's place. He nods.

'The man's a legend. If anyone can find the lad it'll be Joe.'

This seems to imply he'll be dead. I'm still trying to hang on to H's belief that Mbaye may somehow have survived but Mitch tells me I'm delusional. The conditions were evil. The least we owe Mbaye is a serious attempt to recover his body. Then we can pay our respects and say a decent farewell.

Suranne's off the phone. The lifeboat people, she says, are keen to stress that Mr Cassidy is welcome to conduct whatever search he might deem appropriate but will be working at entirely his own risk. Meanwhile, if there are any developments, Suranne will be the first to know.

Mitch and I leave the office but linger in the corridor outside.

'I gather you're staying the night,' he says.

'Who told you?'

'Malo. I'll be around, too. I've booked into a B and B. Maybe dinner together?' His smile is icy. 'All of us?'

Minutes later, back in reception, I find Malo and his dad on the sofa. Malo has shipped all our gear up from the beach and now needs to take it upstairs to the rooms H has booked. There are two of them, both doubles.

We take the lift and H unlocks the first of the bedroom doors. This is a sizeable suite, tastefully appointed. A Battle of Britain theme runs through the paintings on the wall and a notice beside the TV draws guests' attention to a collection of accompanying World War Two DVDs. While H inspects this little treasure trove, Malo and I drift across to the view.

The weather outside is still horrible but, even so, the long stretch of beach fills the big picture window. *Persephone*, our precious home for the past few days, sits forlornly on top of the pebbles. She has a list to port and local workmen have already shored her up with heavy baulks of timber. Both of us stare at the old trawler for a long moment. Then H tells Malo to leave our stuff by the bed.

'*Our* stuff?' I'm still looking at the view. No one says a word. Not to begin with. I glance back at H. He's still on his knees beside the library of DVDs. His face is a mask. Then Malo tells his dad they'll be sharing. There are two beds pushed together. He'll occupy one single, H the other. H shoots me a look.

'Is that what you want?' he asks me.

I nod. After *Persephone*, a bit of privacy is exactly what I want.

'No offence.' I bend to pick up my bags. Then, as an afterthought, I tell H about the diver we can expect. This is Mitch's idea. I'm still calling him Larry Elliott.

'When's he coming? This bloke?'

'This afternoon.'

'They're staying over? Him and your friend?'

'My friend?'

'Mitch fucking Culligan.'

Malo appears to have lost the plot. His gaze goes from each of our faces. He's never seen his Dad in a mood like this. So dark. So full of menace.

'Who's Culligan?' he says.

'Leave it out, son.' H is still staring up at me. 'I asked you a question, gal. Are these numpties here tonight?'

'Yes. As far as I know.'

'And they're staying in the hotel?'

'I doubt it.'

'OK.' H at last struggles to his feet. 'Then get them round here tonight. Half seven. I'm in the fucking chair.'

TWENTY-SEVEN

A sense of dread stays with me for the rest of the day. I rest up in my hotel room, lying full length on the bed while the last of the light drains from the sky outside. I have no idea what to expect from the hours to come but I'm certain none of it will be good. Mitch and H have been on a collision course since the *Guardian* article appeared. God knows what happens next.

We've planned to meet in one of the two hotel bars for drinks. The rest of Malo's guests are already off the island thanks to a hired minibus, while Suranne and her crew have found somewhere much cheaper to stay, waiting for the boat to be repaired. I've told Malo that Mitch Culligan is a journalist and a good man and he appears to be happy with that. Now he and I are reliving various moments of the storm when Mitch and his diver friend appear. H, according to Malo, is still upstairs watching *Reach for the Sky.*

I recognize Joe Cassidy from the photos in Mitch's book. He's a broad-set man, not an ounce of spare flesh. He has a bony face, receding hair, and gnarly old hands I last saw on a gardener H occasionally employs at Flixcombe. The jeans, boots and lumberjack shirt give him the air of someone who's wandered into this swank hotel by mistake, a role I suspect he enjoys. He walks with a certain stiffness, as if he'd overdone some exercise or other, but there's no indication that half his left leg is missing. He must be in his sixties at least but it doesn't show. This man, I tell myself, is full of mischief.

Malo fetches drinks from the bar, putting them on his dad's account. Mitch is drinking Perrier, not a good sign, while Joe sips at a brimming pint of Guinness. We're still discussing what might have happened to Mbaye when H appears. Both Malo and I can tell at once that he's not in a good mood. When Malo gets him a chair he shakes his head. He says he's asked for the function room again. No point hanging around.

Mitch refuses to be hurried. He's no longer pretending to be Larry Elliott and he's deep in conversation with Joe. His diver friend seems to know a great deal about the way the tides work around this bit of the island and he's still drawing lines on the tabletop with his huge forefinger when a waitress appears with a menu. Chef, she says, will be busy tonight with a coachload of Americans signed up for his tasting menu and would appreciate an early order. Heads bow around the table. H is first to make a choice. Fillet steak, rare.

'What are *pommes allumettes*?' he asks.

When Mitch shrugs, Malo steps in.

'String chips, Dad. You'll love them.'

The meal ordered, we follow H through to the function room. Use of this facility has been a recognition on the hotel's part of the ordeal we've all been through but I sense their patience is wearing thin. This, of course, is lost on H.

He sits us down. Shuts the door. In truth, I'm expecting the worst. This is the moment I've been dreading, the moment when the brittle civilities of the last few hours descend into something more visceral. H hates being taken for an idiot. The thin fiction of Larry Elliott was a serious mistake on Mitch's part.

'You're Culligan, aren't you? Just fucking admit it.'

'Yes.'

'The cunt that wrote the drivel in the *Guardian.*'

'Yes.'

'Then why fanny around? Why all this Larry crap?'

'Because I wanted to be on that boat when it left.'

'And you thought I'd throw you off?'

'That was a possibility, yes.'

'You're right, I might have done. But you know something else? I'd have been wrong. I had you down as a devious little bastard, the kind of twat who gets other people to front up and ask all those tricky questions you know I'd never answer direct. Spencer fucking Willoughby. UKIP. Stuff about the old days in Pompey. How I got all that money in the first place. What I do with it now. That was you talking to Enora, wasn't it? That was you helping yourself, taking fucking liberties, taking fucking *advantage.* The poor bloody woman's just been through brain fucking surgery and I bet you were there at her bedside. Flowers.

Chocolates. Whatever. Am I getting warm here, my friend? Isn't this the way it happened?'

I risk a glance round the table. Malo is spellbound, his glass untouched. Joe Cassidy is studying those huge hands of his. H, I suspect, should really be in Berndt's line of business. He has an instinctive nose for a compelling story well told. All the better if it happens to be true.

Mitch clears his throat. He wants to say something. Or maybe he feels the need to defend himself. But either way it doesn't much matter because H hasn't finished.

'So you made friends with her, right? You got her onside. And then you pointed her in my direction. What I need to know is why. That's what's been on my mind. That's why I invited you over tonight. That's what you're doing here. I need you to spell it out to thick old me. I need you to explain what made all that effort worthwhile. Because one of the things you don't seem to understand is that blokes like me go back awhile. We have history. We have mates. We've done stuff together. We're like shit on the shovel. We stick by each other. And so when one of these mates comes to me and says there's some numpty asking whether I've ever killed someone, not frightened them, not even hurt them, but fucking done a nut job on them, then I start to get seriously interested. Does that sound reasonable? Or am I the one taking liberties here?'

He lets the question hang in the air. I'm looking hard at the ceiling, trying to affect indifference. Tony Morse. The night of the party. The conversation in the barn. And the question I popped afterwards. Not on my account but Mitch's.

Fuck.

'So have you?' This from Mitch.

'Have I what?'

'Killed anyone?'

'No I haven't. And even if I had, you'd be the last person I'd ever fucking tell. You're doing some book, right? You want to make your name? You want to be famous? You want to be rich? I have no problem with any of that. But it's not fame, is it? And it's not even money. I know about you, Culligan. I know the kind of life you lead, the kind of company you keep. I know how to hurt you, too, but that's a different story. What this is about is why. Why go to all this trouble? Why lean on Malo's mum? Why put yourself in the fucking firing line? When all you're guaranteed is grief?'

Dimly, through the fog of obscenities, I begin to understand H's drift. Deep down, he's asking exactly the same question I put to Mitch after we left Sayid's bedside that first time at the hospital. What, when you peel everything else away, really matters?

'We're talking motivation, right?' Mitch is toying with his Perrier.

'Right.'

'Then it's simple. I do what I do because I believe in certain things. They include all the stuff about liberty and the rule of law we're lazy enough to take for granted. Just now the wreckers are at work. You're one of them, big-time. You're not at the sharpest end. That's for the politicians. But you help make it happen. Whether you realize the consequences of what you're doing, I dunno. In one way it probably doesn't matter because when the shit really hits the fan you and your rich mates will clean up. What's round the corner for the rest of us, millions and millions of us, is immaterial. We'll get poorer, just like the country. We'll lose interest in what really matters because we'll all wake up one morning and realize that we've been suckered, that we're helpless, that there's sod-all we can do, that we're in the hands of a bunch of gangsters who call themselves politicians but who couldn't care a toss. In posh-speak we're collateral damage. But in real life we're fucked.'

I resist the temptation to applaud. H, I suspect, is losing the plot.

'This is about UKIP?'

'Partly, yes.'

'Me and UKIP? Bunging them all that moolah?'

'Of course. Along with others.'

'But UKIP were always a joke.'

'Then why support them?'

'Because that way we could give all you posh twats a poke in the fucking eye.'

'Just for the fun of it.'

'Of course. What else is there?'

What else is there? Brilliant. I'm back at Flixcombe, back at the party, back watching a roomful of grown men weeping to Tina Turner. H, by accident or design, has got to the very heart of it. In a world growing more complex by the nanosecond, we're all helpless. All we can do, all we can manage, is to try and raise our voices, try and make a fuss, crack a joke, get briefly in the face of the people who make things happen. *What else is there?* Clever man. Clever, clever man.

So far, Joe Cassidy hasn't said a word. His glass is empty. He pushes it towards Malo who nods and leaves the room. Then he turns to Mitch.

'You know about me,' he says. 'And the proof's there in that book of yours. It was excellent. It was a terrific job. It even made me sound half-interesting. My wife, if she was still around, would have loved it. You listened to what I had to say about my Navy days, about all those Falklands dits, about clearance diving, about setting up in business afterwards, and all this is all the more remarkable to me because I wasn't altogether honest with you. There was stuff I never talked about. Stuff I withheld. Stuff that might make better sense on an evening like this.'

'Like what?' Mitch is watching him carefully.

'Like the Falklands War. You're a journalist. You guys love stories. We all know that. You also happen to have a bit of an agenda and that was obvious, too. So when you asked about certain incidents, I was happy to tell you what you wanted to know.'

'We're talking Sheffield?'

'Yes.'

There's a moment of silence. I haven't a clue what these two are talking about. Did the Argies bomb South Yorkshire?

H comes to my rescue. He's word-perfect on the details. He tells me that HMS *Sheffield* was the first Royal Navy ship to go down in the war. She was hit by an Exocet missile which set her on fire and killed twenty blokes. Everyone knew there were fuck-ups, because that's what happens in every war, but at the time the papers were full of our brave lads doing their best in the face of the enemy.

I ask Joe whether he was on the *Sheffield.* He says yes. He also says that fuck-ups were the order of the day. The ship wasn't prepared for a missile attack because no one could properly understand the huge intelligence assessment that had recently arrived onboard. The key warfare officer didn't believe the Argie aircraft had the range to launch any missiles. When a warning came from a neighbouring warship, this officer was having coffee in the wardroom while another was taking a leak. That left just two juniors on the bridge, neither of whom had a clue what to do.

According to Joe they just stared at the incoming missiles. No evasive action. No turning towards the threat to minimize the bulk of the ship. Just a huge bang in the galley and bodies and smoke everywhere.

'Shit.' This from H.

'Exactly. Mitch asked all the right questions. He knew we'd fucked up. But I never told him how bad things really were. Shiny *Sheff* was just one example. I could give you hundreds more.'

Mitch nods. He's beginning to look uncomfortable. H wants to know what any of this has to do with what we've been talking about.

'Everything.' Joe again. 'My point is simple. I've been around a bit. I watch. I listen. I remember. I pick things up. I join the dots. And what I learn tells me that government, big government, isn't too clever. They get far too many things wrong, far too often. After I left the Navy I went into business. Business teaches you everything. It teaches you to stand on your own two feet. It teaches you to get things *right.* And unless you do that you'll end up like *Sheff*, burned out, sunk, abandoned, royally fucked. So Mr Prentice here is more right than we might think. If this is about Brexit then you're looking at someone who voted yes. Why? Because it puts us back where we belong. On our own two feet.'

Another speech. H is looking at Joe with something close to wonderment.

'Shit,' he says quietly. 'Why aren't there any politicians like you?'

The door opens. In comes Malo with a pint of Guinness. Joe contemplates it for a moment and then asks what everyone else is drinking. I know from Mitch's face that he's itching to get stuck in again, to argue his case, to paint H as the bad guy, to make all the important points about the imperfections of democracy and the threat of the Deep State and the power of money and all the rest of it, but H has another idea. He leans across and puts his thick fingers on Mitch's forearm.

'There's one thing I never mentioned,' he growls. 'It took guts to get on that boat in France, no matter what fucking name you used. Believe it or not, I admired that and I still do. All the rest of what you're about is total bollocks and one day we'll meet in the street or some fucking place and you'll admit it, but for now let's get some serious drinking in.' He glances across at Malo. 'Son?'

Malo is well trained but I think he needs some help. I follow him out of the function room and back to the bar. It takes no time at all to organize four bottles of Moët. By the time we're making our way back towards the function room I can hear laughter through the half-open door.

Malo pops the first bottle and I'm on hand with the glasses. Joe is sticking to Guinness.

'A toast?' I suggest.

H nods. 'Here's to our little black fucker.' He raises his glass. 'Wherever he might be.'

TWENTY-EIGHT

The following morning, I have an early breakfast with Mitch at a cafe in Ventnor. The invitation has come from him and I've been very happy to accept. Joe, he says, has gone to talk to a local fisherman about the recent storm and about a reef along the coast. Later he'll take a look at the foreshore, hire a dive boat, and decide where to start.

'You think he'll find the boy?'

'Yes.'

'You *know* that?'

'Yes. Nothing that man has ever told me has ever been wrong.'

'Except by omission.'

'Of course. But even then, I can see his point. He knew what I was after at the time and he gave me enough to make the book work but there were other issues in play.'

'Like?'

'Loyalty.'

'To?'

'The Navy, I guess. It wasn't perfect but these were blokes he loved.'

This, from Mitch, is a major concession. Flesh and blood trumping Mitch's headlong gallop towards the truth.

'And last night? Brexit? All that?'

'He has his point of view. He's wrong, of course, but I'd never question his right to believe it.'

'And H?'

'H is different.'

I'm staring at him. This is the first time Mitch has called Hayden Prentice by his nickname. H. Progress indeed.

'Different how?'

'Different because I'm sure he did kill someone. His name was Darren Atkins. I can't prove it and there's no way H will ever discuss it, but the circumstantial evidence is pretty convincing. The guy disappeared, just went off the radar. He'd stitched up a guy called Mackenzie on a big consignment of cocaine. He'd brought it in from Aruba and about two hundred grands' worth had gone missing. He blamed it on the Colombians, who'd sold it in the first place, but Mackenzie knew that was bollocks. It turned out Mackenzie was right. He had H and a couple of other guys turn Atkins' place over. They found the stash in a lock-up in the north of the city after they'd had a little talk. No one ever saw him again. Apart from the remains of a left foot washed up on Southsea beach. Atkins used to be a decent footballer in his day. Always scored with his left foot.'

Mackenzie. Aruba. A little talk. And the signature warning on the wet pebbles for any apprentice drug runner with similar ambitions.

'Bazza,' I say softly. 'And a guy called Wesley Kane.'

'You know about this?' Mitch has stopped eating.

'Of course I do. I'm part of it now, whether I like it or not.'

'And do you?'

'What?'

'Like it?'

I give the question some thought. As ever I prefer the subjunctive mood to anything truly decisive.

'I might,' I say. 'It depends.'

It's mid-morning before Joe has the information he needs about the reef. Malo has acquired a boat from a local crab fisherman and Joe drives his van down to the promenade to offload all his gear. I help him and Malo carry the heavy equipment down to the waiting crabber while H looks on. According to Malo, his dad awoke with a thumping headache but a pint or two of water and a Percocet from yours truly seem to have done the trick.

Aside from the boat's owner, there's only room for three aboard. H and I find a seat on the prom while Joe strips to his swimming trunks. He wears a false leg below his left knee and we watch him perched on the side of the boat while Malo helps him with the dive suit. Next comes the heavy cylinder of compressed air, a weight belt, and a single fin. Joe's helmet has a built in LCD light with an attached camera. According to the crabber's skipper, the set of the offshore current at the time we speared into the beach would have taken Mbaye away to the east, towards the reef. After his walk along the foreshore, Joe has already decided on his first dive site. Now he pivots on the thwart and makes himself comfortable while Malo gives the boat a shove seawards and hops in. Moments later, Malo gives us a wave and then they're off.

H and I watch them in silence until the boat disappears. The carpenters from Pompey are at work on the trawler and from time to time the sound of their hammering echoes around the shallow curl of the bay. It's grey and miserable and there's a dampness in the air. H is wrapped in the big sea-going jacket he wore for the crossing. I think he's cold. I think he wants to go back to the hotel.

'Tell me about Sayid Abdulrahman,' I say. 'Did you do that yourself? Or did you leave it to someone else?'

H doesn't answer. He's staring out to sea. I put the question again. And I tell him he owes me an answer.

'Why?'

'Because I know that man. And what you did should never have happened. It was horrible. It was worse than horrible. It was evil.'

H nods. He appears to agree but I can sense there's no way he's going to take this conversation any further. For a few minutes we sit in total silence, watching the men at work on the trawler. Finally I try another name.

'Darren Atkins,' I say softly. 'Tell me about him.'

For a moment H doesn't react. He's still hunched on the bench, hands dug deep in the pockets of the jacket.

'You've been talking to Culligan,' he says at last. Statement, not question.

'You're right.'

'How much did he tell you?'

'Enough.'

'Enough for what?'

'Enough to make me wonder.'

'About me?'

'Of course.'

'And? Supposing whatever he said is true?'

I shut my eyes. This is typical. I'm on the back foot. Again.

'He thinks you killed him,' I say carefully. 'All I want is a yes or a no.'

'And if it's a yes? Will you go running off and tell Culligan?'

'Absolutely not.'

'So what difference will it make? To you and me?'

'I don't know.'

This isn't the answer he wants. He seems pained by it as if I've somehow let him down. He's about to ask me something else, to maybe take the conversation in a new direction, but then he has second thoughts.

He stirs on the bench and then struggles to his feet. I'm looking up at him.

'So is that a yes?' I ask. 'You're telling me you killed a man?'

He gazes down at me for a long moment and I realize how much the last few days have taken out of him. He looks haunted, gaunt, somehow diminished. The last thing he needs is me on his case.

'I'm going back to the hotel,' he grunts. 'Bell me if they find the kid.'

Joe and Malo return a couple of hours later. I've walked the length of the promenade twice and had a brief conversation with an elderly woman in a motorized buggy who very much liked a movie I did a couple of years ago. I've also made contact with Rosa in case she's picked up news of our adventures on *Persephone*. It turns out she hasn't but insists on a full report. I tell her most of it but skip the bit about Mbaye. My description of taking the wheel at the height of the storm has her in fits of laughter. I gave her a lift once when I was driving a little 2CV and she got out after less than a mile. We agree to talk again as soon as word comes from Montréal on the Canadian movie and before she hangs up she gives me a transmission date for the radio play.

'The male lead's been in touch,' she adds as an afterthought. 'He wants your contact details.'

'Tell him nothing,' I say.

Now, I hurry down the beach to meet the boat. Malo hops off and hands me a rope. Joe needs something from his van, parked up on the promenade. The first dive was fruitless and he had to abort the second because his regulator was malfunctioning. Waiting for Malo, Joe and I have a brief conversation. He says that visibility is better on the reef than he anticipated after the heavy weather. He has three more dive sites in mind, with a day in hand, after which the search might get problematic. I'm still wondering quite what this might mean when Malo returns with a replacement regulator. Seconds later, after a hefty push from yours truly, they're off again.

By now it's early afternoon. I make contact with Mitch, who is en route back to an important meeting in London, and confirm that so far we haven't found any trace of Mbaye. Then I find a seafront cafe with a view of the beach and settle down in front of a bowl of lentil soup.

My last conversation with H has disturbed me. I know I should never forgive him for what happened to Sayid, but even so I seem to have succumbed to something I find difficult to describe. Charm is too feeble a word. H himself would hate it. A kind of animal magnetism comes closer. This is a man with total self-belief, a man for whom life holds no fears. That makes him brave, as well as reckless, and it also makes him very rare. Like it or not, thanks to Malo, Hayden Prentice has become part of me. Never, I realize with a sudden jolt of surprise, have I regretted that evening in Antibes.

Leaving the cafe, I walk to the hotel to check that he's OK. I can't find him anywhere and when I enquire at reception the girl tells me he's retired for a nap. Strictly no calls, she says. I nod and return to the beach. This is worrying. H never naps.

By the time the boat returns for the second time, I've nearly given up waiting. A low duvet of cloud has hung over us all day and by half past four it's getting dark. I hear the boat before I see it, a putter-putter growing louder by the second. I walk down to the waterline, aware of the presence of *Persephone* on the shingle. The carpenters are doing a fine job on the gash but are now packing up. Queen of the Underworld, I think. As sorry-looking and beached as H.

The moment I see the expression on Malo's face, I know the news isn't good. He's crouched in the bow, ready with the line, but there's a tightness in his mouth that tells me everything I need to know.

'You've found him?'

He nods, throws me the rope. Joe has the underwater footage. There's a laptop in his van. We all need to watch.

I accompany Joe and Malo back up the beach, helping with the equipment. Joe has strapped on his false leg and manages the pebbles better than me. At the van I wait on the pavement while Joe stows the gear. I'm wondering whether to try and put a call through to H but before I can make any decision Joe emerges with his laptop. He's transferred the footage.

All three of us settle on a nearby bench. Joe opens the laptop and finds the footage. The camera is built into his dive helmet and acts like a second pair of eyes. The image on the screen is pin-sharp and I watch, fascinated, as he takes a final look at the crabber and then submerges. Light drains from the screen as he swims deeper until he triggers his LCD and suddenly we're above the reef in a world of yellows and dark greens. Sediment hangs in the water. Tiny creatures drift past. Then comes a glimpse of a pocket of sand as he touches the bottom. Ahead is a wall of darkness. Slowly, stroke by stroke, I can make out the shape of individual rocks, heavily barnacled. I see a starfish. Something brown with big eyes. A torn fragment of fishing net, a bright synthetic orange. Joe is moving slowly along the outer wall of the reef. His passage stirs tiny plumes of sand from the bottom. Then, very dimly, I sense a different shape, softer, slowly coming into focus. This has to be him, I think. And I'm right.

I sit back a moment, overwhelmed. Then I force myself to watch again. Mbaye has been trapped by one foot in a gap between the rocks. The rest of his thin body hangs in the yellow light, gently moving in the current. His head is down, his chin on his chest, while his arms float free. Joe is circling the body. The boy's eyes are closed and his mouth is slack. Those teeth, I keep thinking. The startling white of those teeth.

Joe has frozen the image. He wants to be sure. Malo confirms it's Mbaye. I nod, saying nothing. Joe closes the laptop. He's buoyed the site and tomorrow he'll inform the police.

'So what happens?' I ask. 'To Mbaye?'

'They'll recover him at low tide. The weather should be OK. There'll be a post-mortem, then an inquest. The body will belong to the Coroner. Once the legals are over, I imagine he'll release it.'

Release it. Not him. It.

I get up and walk to the edge of the promenade where it drops away towards the beach. I want very badly to be alone. It's dark now but in the throw of the seafront lights I can make out the shape of the old trawler on the shingle. I shut my eyes a moment. It takes nothing, no effort on my part, to be back on that heaving deck, with the waves roaring out of the darkness and nothing but fear in the pit of my stomach. I can feel the icy bite of the wind, hear the distant thunder of the surf, flinch at the crack of the sails, and then comes another image, altogether more comforting. That thin little figure at the wheel. How he took over. How clever he was. And how brave.

TWENTY-NINE

The same fisherman, with the aid of Joe and the police, recovers Mbaye the following day. Malo, H, and I are on hand when the grim little party carries the grey body bag up the beach towards the waiting undertaker's van. H has spent an anguished evening at the hotel trying to decide what to do next. When I tell him that an inquest could be months away he doesn't seem to be listening. He needs to put the lad to rest.

It's Malo who suggests a local funeral and then maybe cremation. We could scatter the ashes at sea. We could buy an urn, something nice, and find somewhere suitable to keep it. I like this idea but when I suggest Flixcombe, H shakes his head.

'We have to take him home,' he says. 'We have to take him back to his own people.'

This is easier said than done. All we have is a Christian name. No passport. No home address. No next of kin. Just Mbaye's word that he came from a big family in a fishing village near Dakar. All too sadly, Mbaye is just one in a squillion young refugees jumping on smugglers' trucks and disappearing north. If H really wants to find his family, just where do we start?

To this question H has no answer. We're in the hotel bar, having a night cap. H is staring glumly at his glass of malt when he suddenly brightens.

'Son?'

Malo is on his smart phone, scrolling through his emails. He doesn't look up.

'Dad?'

'It's gonna be down to you. Your project. Your call. I'll fund it. I'll pick up the costs. You find the boy's folks. Deal?'

Malo nods. He's replying to one of the emails.

'Deal,' he says absently.

I return to London the next day while H and Malo drive west to Flixcombe. All three of us have left our contact details with the police who will forward them to the Coroner. In due course, explains the desk sergeant, the Coroner's Officer will be in touch to get statements from each of us. We may also be summoned to attend the inquest. When H pushes for some kind of date, the sergeant shrugs. Could be a while, he says. Sad business, all these young lads.

In London, I pick up the threads of my former life. The events of the past week or so have made a far bigger impact than I ever anticipated and the more time slips by, the harder it is to rid my mind of some of the more vivid images. We could all have gone down with that boat, every single one of us, H, Malo and me, a little gene pool wiped out forever, and the more I think about this the gladder I am that we all managed to pull through. None of us are perfect, least of all H, but I like belonging to these people. They've brought colour and a whiff of something dangerous into my life, and that matters a great deal. H, very much on the mend now, also makes me laugh. Which I suspect is just as important.

Towards the end of November, Mitch gets back in touch. At his suggestion we meet at an Indian restaurant in Clerkenwell. He arrives late but I'm still pleased to see him. He tells me that Sayid has been back home for a while now and is almost completely recovered. Last week, for the first time, he put on his Nikes and managed a gentle mile and a half around the block. Next week, God willing, he'll try and make it to Blackheath. Only towards the end of the meal do I raise the issue of the book. Mitch has already told me that he's been working all hours to meet a deadline on a major piece on Universal Credit. After that he fancies something ecological. Scientists have reported a seventy-five per cent wipe-out of insects in nature reserves across Germany. This appears to be down to the widespread use of pesticides. If something similar is happening here the longer-term consequences to life on earth could be catastrophic.

This is the old Mitch, the crusading doomster who stepped into my life with the offer of breakfast and changed pretty much everything.

'And the book?' I ask.

'Book?'

'Brexit? UKIP? All of us fucked?'

Mitch shakes his head. This is a question he's probably anticipated but doesn't much like. He mumbles something about the cowardice of publishers in general and his in particular. Not his fault. Theirs.

'So what's happened?'

'They pulled out.'

'Why?'

'They said it was legally imprudent to carry on.'

'*Imprudent?* How does that work?'

'Prentice threatened to sue. They also had calls.'

'From who?'

'They never knew but that wasn't the point. These people were promising to set them on fire and they believed them.'

'You mean arson? Torching the premises?'

'Murder. Torching them.'

'Shit. Didn't you try any other publishers?'

'Of course I did. Same story. Publishing's a small world.'

'You mean they're frightened?'

'Terrified. Game, set and match to your H. Will all that stuff ever see the light of day? I doubt it. Not between hard covers. Does that matter? Of course it does. But there comes a time when you realize that no one cares, not out there where they should. Same old story. Shit wages. Shit telly. Shit lives. What else is there? Fuck all. Because nothing matters.'

Because nothing matters. This is resignation on stilts which must be tough for Mitch.

'So is that it? All that research? All that work? Having sources like me passing out on you over breakfast?'

Mitch offers me a tight smile. He's not a good loser and it shows.

'There's a satirical rag based in Paris I've been talking to,' he says. '*Le Canard enchaîné*. You may even have heard of it. They're braver than our lot and they might stretch to some kind of serialization. I think they want to stick it to *les rosbifs* and who can blame them. A French publication telling the truth about the Brexit lot?' He shakes his head. 'That's beyond irony.'

I nod. *Le Canard enchaîné* is venerated by certain sections of the French intelligentsia but its readership is tiny. If Mitch has been dreaming of mass-market sales figures, he's about to be disappointed. I want to sympathize but when I reach for his hand he flinches. He has no idea what's coming next and he finds that alarming.

'Don't be frightened.' I'm smiling at him. 'I just want to say thank you.'

'For what?'

'For H.' I give his hand a squeeze. 'And for Malo.'

In early December, bad news arrives from Montréal. Everything is on schedule for location filming to begin shortly after Christmas, but the principal backer, for reasons that no one can fathom, has suddenly pulled out. The shoot has therefore been postponed indefinitely until other funding is in place.

'Are we dead in the water?' I ask Rosa on the phone. 'Be honest.'

'We are, my precious,' she says. 'But something very interesting has just popped up in Belgrade.'

She doesn't volunteer any further information and I don't ask because I've long trusted this woman with looking after my best interests. Either a script will arrive in due course or it won't. Whatever happens, I have more important things on my mind.

My three-month check is due this same week. I attend first for an MRI scan. I've been through this procedure before and there's a real feeling of coming full circle as I make myself comfortable on the scanner table and allow the assistants to slide me into the machine. The next forty-five minutes are painless but incredibly noisy. It's important to keep your head still and I try very hard to remember what a yoga teacher once told me about trying to imagine every inch of my body in fierce detail. I've got as far as my left knee when the machine is switched off and I'm free to go.

I return to the hospital three days later. My neurosurgeon has had a cancellation on an earlier patient and he's waiting for me at his office door when I arrive. He invites me in and we sit down. My MRI scan is already hanging on his white board. He asks me how I've been recently and seems surprised when I say fine. He scribbles himself a note and then directs my attention to a darkish area around the site of the first operation. The news, he says, is not good. These are still early days but there appears to be an indication that the tumour has returned.

'Really?' I'm still looking at the scan. I don't feel the least bit shocked, or even disappointed. Just curious.

I ask him why this might have happened. He says he can't be sure. Maybe he didn't excise enough of the diseased brain tissue. Maybe we're dealing with a brand-new bunch of rogue cells. Either way there are decisions to be made.

'You want to operate again?'

'Not really. Not unless I absolutely have to. We'll try radiotherapy first. And a couple of cycles of chemo.'

I have a fear of chemo. Too many friends of mine have emerged from treatment with horror stories about hair loss, throwing up, and industrial strength exhaustion. My neurosurgeon is nice enough to assure me that the drugs are getting better every week. So far, to my relief, he hasn't mentioned the Grim Reaper.

'So when do you want all this to start?'

He consults his calendar. Christmas is fast approaching. Best, he says, to leave it until the New Year.

'Is that OK with you?' He's on his feet already, his hand outstretched.

'That's fine,' I say. 'Something to look forward to.'

I walk the three miles home, partly to test my newfound resilience in the face of bad medical news, and partly because I don't want any kind of company. I've learned a great deal about myself over the past few months and one of the conclusions I've drawn is the need to be ruthless about time. Time, in the end, is all we've got. And when someone in the know suggests it might be limited, it's better to spend it wisely.

I get home mid-afternoon. I know that Malo and H have been out in Senegal these last couple of weeks, trying to locate Mbaye's family. Malo has taken a couple of screen grabs from his smart phone, shots of Mbaye he took on the boat, and armed with these he and his dad have been driving along the coast from village to village, trying to find someone who might recognize the boy.

Malo, bless him, has been phoning me every other day with what began as progress reports but as the calls got shorter it became obvious that neither of them were falling in love with

West Africa. On the one occasion I found myself talking to H, he left little to the imagination. The people were OK, he said, but there were far too many of them. Dakar was a shit heap and some of the outer townships were worse. Malo had had a run-in with a rogue prawn and spewed his guts out for forty-eight hours. The weather was too hot, even in December, and they were both sweating like a bastard by midday.

'It's costing me a fortune,' he said before handing me back to Malo. 'And that's just for a new fucking shirt every morning.'

Now, I phone Malo. They've been back at Flixcombe since last night, plenty of time to recover. The phone rings and rings until my son finally picks up.

'Mum?' he says. 'Is that you?'

It's lovely to hear his voice. We haven't talked for nearly a week. I ask him how he is.

'Great. Knackered but OK.'

'Stomach all right?'

'Yeah.'

He badly wants to tell me about the back end of the trip, what they got up to those last few days in Senegal, but I ask him to save it.

'For what? For when?'

'For tomorrow,' I say, 'if you can make it. Maybe the day after if you can't.'

'Make what? What is this?'

I tell him I have a little expedition in mind. I want him, need him, to carve out a couple of days in his busy, busy life.

'What for? Where?'

'The Isle of Wight.'

'Is this the inquest? Have they brought it forward?'

'Not at all. It's got nothing to do with the inquest. They're talking the New Year now. It could still be months away.'

'Then what's up? Why the island?'

I'm not prepared to say. All I want him to do is promise to meet me at Portsmouth either tomorrow or the next day. The weather forecast is glorious. We'll take the ferry to the island. We'll travel by bus or cab. We'll be away for one night. Then he'll be back home again.

Something in my voice must have told him to say yes because

the barrage of questions abruptly comes to a halt. He agrees to meet me at Portsmouth Harbour station around noon tomorrow.

'Are you sure that's OK?' I ask him.

'That's fine, Mum. Twelve o'clock. Skinny guy. Quite suntanned.'

Next morning he's waiting for me on the platform at the harbour station. Expecting us to take the catamaran across the Solent to Ryde, he steers me towards the main exit. Outside I pause to admire the view. Last time I was here it was pouring with rain and I was about to embark on *Persephone*. Now, under a cloudless sky, sunshine dances on the big, blue expanse of the harbour. A tug fusses around a huge warship. A pair of yachts motor slowly towards the harbour mouth. Flags flutter and snap in the wind.

Malo is tugging me away. Time, it seems, is short.

'Where are we going?'

'The car ferry.'

We've come to a halt. I'm looking down at a smart Audi convertible, dark blue. It looks brand new.

'This is yours?'

'Yeah. Present from Dad. I think he wanted to make up for all the hassle on the boat.'

We get in. Firm leather seats. Fancy audio set-up. Even a camera at the back for getting out of tight spots, doubtless invaluable if you happen to be a son of H.

'Like it?' Malo is beaming.

'Love it. What time does the ferry go?'

We drive to the car ferry terminus. As the ferry slips her moorings and rumbles towards the harbour mouth, we're both on the upper deck, enjoying the sunshine. From here, we have a perfect view of the Camber Dock.

Malo spots her first. 'There, Mum.'

I follow his pointing finger. *Persephone* is tied up at her usual spot beside the harbour wall. Nothing I can see from the ferry would give me the slightest clue to the trauma she's been through. A line of washing from the doghouse aft indicates that someone must still be aboard. I look a little harder. Skimpy black knickers. A child-size T-shirt. Definitely Suranne. The ferry is making a turn now and as the old trawler begins to disappear I linger as long as I can on the big wooden steering wheel aft. Mbaye, I think. My last ever glimpse of him alive.

An hour later, we're bumping off the ferry boat ramp in Fishbourne. Malo gets rid of the roof and I settle down for the ride. The seats have electric heating. I feel glad, and cosseted, and very much alive. I'm also within touching distance of the most important man in my life, a bonus I've only ever dreamed about.

I've asked Malo to take us to Ventnor. We drive down through the town and park on the promenade. It's low tide. The beach is empty apart from a couple walking a dog and I fancy I can see an indentation in the pebbles where *Persephone* came to rest. Malo laughs and tells me I'm making it up. He's shading his eyes against the slant of winter sunshine. I suspect he thinks we're going to The Imperial. He's wrong.

'Freshwater Bay,' I say.

'Freshwater what?'

'Bay.'

'Anywhere in particular?'

'The Farringford.'

He punches the word into his sat nav, grunting with satisfaction when the directions appear. Then he pauses.

'I've heard that word before,' he says. 'Recently.'

'You have.'

'You made a film there? Am I right? When you were playing the photographer's model?'

'Yes.'

I'm delighted. Half my work is done. I want to draw my precious son to the very heart of me, the very middle of me, and this is the perfect start.

The sun is beginning to set as we head west. The road follows the line of chalk cliffs that stretch way out towards The Needles. At any time of year this is a spectacular journey but in mid-winter, in weather like this, it feels ultra special. We have the road to ourselves. At every next bend, with no sign of traffic, I'm starting to wonder whether we have the entire island to ourselves. Away to the right, on rising ground, the green smudge of a forest. On our left, still blue, still lightly flecked with waves, that same English Channel that nearly killed us. Ahead, softened by the haze on the far horizon, the dying orb of the sun.

Accommodation at the Farringford comes in the shape of self-catering cottages in the grounds. I've booked one of the two-bedroom versions. Access to the house itself is restricted at this time of year, which is a shame, but that's not why we're here.

Malo has parked the car in front of our cottage. There's still a blush of orange in the west but dusk is falling fast. Malo is looking back at the house. This once belonged to Alfred Lord Tennyson. To me it has perfect proportions, just two floors in soft grey stone. One end of the property is covered in ivy and a magnificent cedar tree frames the view towards Tennyson Down.

'This is where you stayed, right?'

'Yes. I had the bedroom at the top there, third window from the left. I got up early every morning to catch the sunrise. I'll never forget it.'

We eat that night at a pub in the village. At H's insistence, I'm spending Christmas down at Flixcombe Manor, but I want this evening to be our own special celebration, just Malo and myself. My son sorts out the drinks and takes our food order to the bar. There's a comfortable buzz of conversation from the surrounding tables and locals attend to the wood-burning stove when it needs a new log or two. Perfect.

Over loin of pork and roast potatoes, Malo fills me in about his two weeks in West Africa. Only towards the end of the trip, he says, did his dad really get any kind of handle on the place. Until then, he'd been certain that they could find Mbaye's family, pay their respects, bung them a quid or two, and offer to organize repatriation of the body. The latter gesture, I suspect, wouldn't be cheap and when I ask Malo how much he says nearly £3,000. The body must be embalmed. It must also be free from infection and transported in a zinc-lined coffin to provide a hermetic seal. Lots of documentation includes a copy of the death certificate and the findings of the Coroner.

'And you'll be organizing all this?'

'Yes. But that's not the point. We looked high and low for the right village. There are loads of them, boats drawn up on the beach, everyone sitting on their arse, kids everywhere. In the end we gave up on finding the right one and just chose the prettiest.'

'The *prettiest*?'

'Dad's idea. We tracked backwards. There was one beside an estuary, a bit nicer than the rest. Dad got me to go and find the priest. He turned out to be a really good guy, quite young. I explained about Mbaye, and not finding anyone who seemed to recognize him, and then I asked him whether it would be OK for him to do the honours.'

Honours? Once again I tell him I don't understand. Malo grins.

'Take care of him. Find a plot. Have a grave dug. Put a headstone in, something decent, something nice with just the name. Mbaye. Plus the date he left us.'

The date he left us. Monday 13 November. Written on my heart.

'And?'

'He was a bit doubtful at first but then Dad arrived and took him aside. You know the way he does that?'

I nodded. Oh, yes.

'Dad said he'd pay for it all, the grave and everything, plus he'd give some money to the village for whatever they needed, maybe like a medical centre or something, I dunno.'

'A medical centre? How much money are we talking here?'

'That's what I asked Dad. He said half a million.'

'Half a *million*?'

Malo nods. Half a million pounds. Instead of UKIP, I think, a medical facility in a remote West African village. Instead of mischief-making in the UK, something truly worthwhile. And all in the memory of a boy we'd known for less than a day.

I resist the temptation to phone Mitch. He'd hate news like this.

'So what happened?'

'The priest said yes. Only too happy. That night he introduced us to the elders in the village. Weird. I thought we might be in for some kind of ceremony, some tribal thing, but all they did was sit round and talk about how poor they were. Dad was really disappointed.'

I'm smiling. I can imagine this scene, H feeling short-changed by a bunch of West African whingers. Then another thought occurs to me.

'So we have no family for Mbaye?'

'That's right. None.'

'Then how are you going to lay hands on the body? The Coroner won't just give you the keys to the mortuary. It doesn't work that way.'

'I know. That's what I told Dad. I took advice. Mbaye's family have to come up with proof that he was their son.'

'Is H aware of this?'

'Of course. I told him. I got advice. I spelled it out before we went.'

'And?'

'He said it wouldn't be a problem. We'd find the family and sort it.'

'And now?'

'I dunno. He never listens, Dad. Maybe he'll just end up giving the priest the half million. Call it quits.'

'Conscience money? For getting the boy killed?'

'Maybe.'

We finish our main course and study the desserts board. Malo settles for treacle tart while I order a sorbet. When Malo returns from the bar he finds something waiting for him.

'What's this?'

'Open it.'

'But what is it, Mum?'

'Call it an early Christmas present. Call it what you like. Just open it.'

He nods, reaches for a knife, folds back the wrapping paper. Inside is a box. He lifts it out and studies it. It's American. I found it in a gift shop I especially like off the Cromwell Road.

Malo opens the box. Inside is a container the size of a dinner plate. It has two halves that fit together, both concave, both silvered on the inside. The top half has a circular hole cut in the top.

Malo is staring at it from every angle. He hasn't a clue what he's looking at. I delve in my bag for another present, much smaller this time.

'This goes with it,' I say. 'I'll show you how.'

Malo does some more unwrapping. This time it's a soapstone figure, jet black, beautifully carved, a bird-like shape with a long curved neck.

'It's a loon,' I tell him. 'Made by Inuits.'

'Is it precious?'

'Very.'

'So what do I do?' He gestures at the first present. I position it in the middle of the table and lift the top lid. The carving sits snugly inside. I give the top lid to Malo and tell him to replace it.

He does what he's told. The lighting, thank God, is perfect.

'Now sit back. Any angle. It doesn't matter.'

I hear the scrape of the chair. He's looking hard at the container, at the hole on the top, at the loon magically hovering above. He reaches forward for it, tries to trap it in his fingers, but it's made of nothing.

'It's an illusion,' I tell him. 'In French we'd call it a *trompe-l'oeil.* What you see isn't what you get.'

He's still staring at the loon, daring it to move, to disappear, but it's still there, a beautiful black nothing. I love the expression on his face. He could be seven again.

'Happy Christmas,' I say.

Next morning I get up before sunrise, find something warm to wear, and tiptoe out of the cottage. My feet are bare and the dew is icy on the grass. I've been canny enough to pack a heavy sweater as well as my jeans and I wait beneath the cedar for the clouds to part. The forecast doesn't let me down. The sky to the east is already on fire and I tilt my face to the sun as shadows slowly appear all round me. When I was on location here, so many years ago, it was high summer but the feeling of aloneness, of privilege, of somehow being the only human being on the face of the planet, is no less powerful. I linger for a while, enjoying the first hints of warmth, before returning to the cottage. My son, bless him, is still asleep and to my deep satisfaction he has the loon tucked beneath his pillow.

We set off for Tennyson Down an hour or so later. I lead Malo around the edge of two fields, exactly the path I took every morning all those years ago, and then I find the gap in the hedge that leads up on to the Down. Someone's installed a new-looking stile but apart from that I've stepped into a time warp.

The springy turf seems to go on forever. The sun is behind us now and our long shadows sometimes seem to coalesce into one. Gulls are soaring and wheeling overhead, turning to ride the wind that blasts up the cliff face. There are rabbits everywhere, scuttling away for cover, and twice Malo swears he's seen a stoat. These are magical moments, moments of pure bliss. Ahead of us lies the Tennyson Monument which marks the highest point of the Down.

We pause beside it. The briskness of the walk has brought colour to Malo's cheeks. He looks alive. He looks so well.

'So what do you think?' I nod at the view west. The Needles are invisible from here but the bony whiteness of the sheer cliffs are promise enough.

'Brilliant,' he says. 'No wonder you loved it so much.'

I nod. He's understood. That's all I wanted. And now he's shared that feeling, that wild exaltation, and that's even better.

'Come with me.'

I take his hand. It feels very natural. We're walking towards the cliff edge. He trusts me completely. I know he does. The wind is blowing in from the sea, hitting the cliff face, funnelling upwards. From maybe ten metres away, you can feel its force. Look up, and those same gulls are stationary above us, weightless, riding this invisible column of air.

I stop at the cliff edge and peer over. My son does the same. We can see the creamy swirl of breaking waves among the rocks at the bottom of the cliff. They look so small, so distant, so tempting. One step. Just one. Then oblivion. Cassini, I think. My batteries running out. Our final plunge. Then nothing but silence in the vastness of deep space.

Malo is shivering. I can feel it. He wants to know what happens next.

'Good question.'

We step back and then head on west towards The Needles. I want to keep this matter-of-fact. No drama. No tears. No concessions to anyone else's script. Just one promise from my lovely boy.

I motion for him to stop. The sun is on his face. I have some news, I tell him. I've been to see the consultant. I've had another scan and the results aren't great. I feel fine in myself, no real symptoms, not like the last time, but there appears to be a chance the Grim Reaper might come calling.

'The Grim what?'

'Reaper. It's code for death. If it does happen, if I die, I want my ashes thrown off that cliff edge.'

Malo follows my pointing finger. He frowns, and then shakes his head and puts his arms around me.

'I love you, Mum,' he says. 'We'll sort this out.'